Transport of Information-Carriers in Semiconductors and Nanodevices

Muhammad El-Saba
Ain-Shams University, Egypt

IGI GLOBAL
DISSEMINATOR OF KNOWLEDGE
www.igi-global.com

A volume in the Advances in Computer and Electrical Engineering (ACEE) Book Series

Published in the United States of America by
IGI Global
Engineering Science Reference (an imprint of IGI Global)
701 E. Chocolate Avenue
Hershey PA, USA 17033
Tel: 717-533-8845
Fax: 717-533-8661
E-mail: cust@igi-global.com
Web site: http://www.igi-global.com

Library of Congress Cataloging-in-Publication Data

Names: El-Saba, Muhammad, 1960- author.
Title: Transport of information-carriers in semiconductors & nanodevices / by Muhammad El-Saba.
Description: Hershey, PA : Engineering SCience Reference, 2017. | Includes bibliographical references.
Identifiers: LCCN 2016057816| ISBN 9781522523123 (hardcover) | ISBN 9781522523130 (ebook)
Subjects: LCSH: Electron transport. | Photon transport theory. | Semiconductors--Transport properties.
Classification: LCC QC611.6.E45 E423 2017 | DDC 621.3815/2--dc23 LC record available at https://lccn.loc.gov/2016057816

This book is published in the IGI Global book series Advances in Computer and Electrical Engineering (ACEE) (ISSN: 2327-039X; eISSN: 2327-0403)

British Cataloguing in Publication Data
A Cataloguing in Publication record for this book is available from the British Library.

For electronic access to this publication, please contact: eresources@igi-global.com.

Advances in Computer and Electrical Engineering (ACEE) Book Series

Srikanta Patnaik
SOA University, India

ISSN:2327-039X
EISSN:2327-0403

Mission

The fields of computer engineering and electrical engineering encompass a broad range of interdisciplinary topics allowing for expansive research developments across multiple fields. Research in these areas continues to develop and become increasingly important as computer and electrical systems have become an integral part of everyday life.

The **Advances in Computer and Electrical Engineering (ACEE) Book Series** aims to publish research on diverse topics pertaining to computer engineering and electrical engineering. **ACEE** encourages scholarly discourse on the latest applications, tools, and methodologies being implemented in the field for the design and development of computer and electrical systems.

Coverage

- Applied Electromagnetics
- Algorithms
- VLSI Design
- Analog Electronics
- VLSI Fabrication
- Qualitative Methods
- Electrical Power Conversion
- Computer Architecture
- Programming
- Chip Design

IGI Global is currently accepting manuscripts for publication within this series. To submit a proposal for a volume in this series, please contact our Acquisition Editors at Acquisitions@igi-global.com or visit: http://www.igi-global.com/publish/.

The Advances in Computer and Electrical Engineering (ACEE) Book Series (ISSN 2327-039X) is published by IGI Global, 701 E. Chocolate Avenue, Hershey, PA 17033-1240, USA, www.igi-global.com. This series is composed of titles available for purchase individually; each title is edited to be contextually exclusive from any other title within the series. For pricing and ordering information please visit http://www.igi-global.com/book-series/advances-computer-electrical-engineering/73675. Postmaster: Send all address changes to above address.

Titles in this Series

For a list of additional titles in this series, please visit: www.igi-global.com/book-series

Handbook of Research on Nanoelectronic Sensor Modeling and Applications
Mohammad Taghi Ahmadi (Urmia University, Iran) Razali Ismail (Universiti Teknologi Malaysia, Malaysia) and Sohail Anwar (Penn State University, USA)
Engineering Science Reference • copyright 2017 • 579pp • H/C (ISBN: 9781522507369) • US $245.00 (our price)

Field-Programmable Gate Array (FPGA) Technologies for High Performance Instrumentation
Julio Daniel Dondo Gazzano (University of Castilla-La Mancha, Spain) Maria Liz Crespo (International Centre for Theoretical Physics, Italy) Andres Cicuttin (International Centre for Theoretical Physics, Italy) and Fernando Rincon Calle (University of Castilla-La Mancha, Spain)
Engineering Science Reference • copyright 2016 • 306pp • H/C (ISBN: 9781522502999) • US $185.00 (our price)

Design and Modeling of Low Power VLSI Systems
Manoj Sharma (BVC, India) Ruchi Gautam (MyResearch Labs, Gr Noida, India) and Mohammad Ayoub Khan (Sharda University, India)
Engineering Science Reference • copyright 2016 • 386pp • H/C (ISBN: 9781522501909) • US $205.00 (our price)

Reliability in Power Electronics and Electrical Machines Industrial Applications and Performance Models
Shahriyar Kaboli (Sharif University of Technology, Iran) and Hashem Oraee (Sharif University of Technology, Iran)
Engineering Science Reference • copyright 2016 • 481pp • H/C (ISBN: 9781466694293) • US $255.00 (our price)

Handbook of Research on Emerging Technologies for Electrical Power Planning, Analysis, and Optimization
Smita Shandilya (Sagar Institute of Research Technology & Science, India) Shishir Shandilya (Bansal Institute of Research & Technology, India) Tripta Thakur (Maulana Azad National Institute of Technology, India) and Atulya K. Nagar (Liverpool Hope University, UK)
Engineering Science Reference • copyright 2016 • 410pp • H/C (ISBN: 9781466699113) • US $310.00 (our price)

Sustaining Power Resources through Energy Optimization and Engineering
Pandian Vasant (Universiti Teknologi PETRONAS, Malaysia) and Nikolai Voropai (Energy Systems Institute SB RAS, Russia)
Engineering Science Reference • copyright 2016 • 494pp • H/C (ISBN: 9781466697553) • US $215.00 (our price)

Environmental Impacts on Underground Power Distribution
Osama El-Sayed Gouda (Cairo University, Egypt)
Engineering Science Reference • copyright 2016 • 405pp • H/C (ISBN: 9781466665095) • US $225.00 (our price)

IGI GLOBAL
DISSEMINATOR OF KNOWLEDGE

www.igi-global.com

701 East Chocolate Avenue, Hershey, PA 17033, USA
Tel: 717-533-8845 x100 • Fax: 717-533-8661
E-Mail: cust@igi-global.com • www.igi-global.com

Table of Contents

Preface

During the last decade, rapid development of electronics has produced new high-speed devices at nanoscale dimensions. These nanodevices have tremendous applications in modern communication systems and computers.

This book, *Transport of Information-Carriers in Semiconductors and Nanodevices*, is intended to be the first in a series of 3 volumes titled *Semiconductor Nanodevices: Physics, Modeling, and Simulation Techniques.*

The main purpose of this course is to develop an appreciation and a deep understanding for the conceptual foundations underlying the operation of emerging nanoelectronic devices.

I've decided to dedicate the first volume to talk about transport modelling, which can serve both academicians and professionals. The next book will cover the *Modeling and Simulation Techniques*, and will be rather dedicated for professionals and postgraduate students in device simulation. The third book is about *Physics and Operation of Modern Nanodevices.* However, for the matter of completeness in each book, I squeeze other volumes in a single chapter or as illustrative case studies.

In this book, we study the transport models of information carriers (e.g., electrons and photons) in semiconductors and nanodevices. It contains a comprehensive discussion about carrier transport phenomena and includes some topics not previously assembled, altogether, in a single book.

I mean by information carriers, the particles or particle characteristics that carry and transport signals in semiconductor materials and solid-state devices. For instance, the electronic charge in conventional semiconductor devices, the electronic spin in spintronic devices and photons in optoelectronic devices. In fact, the characteristic of any particle may be utilized for information transport. For example, a quantum bit (or qubit) of information can be manipulated and encoded in any of several degrees of freedom, notably the photon polarization. In addition, other quasi particles, such as phonons (lattice vibration waves) may be considered as information carriers, because they are capable of transporting thermal energy from point to another in solid-state devices. In fact, some or all of these information carriers may interact in the same device. Indeed, electrons and phonons interact in all semiconductors devices. They also intervene, together with photons in photonic devices, like laser diodes. In the so-called spin light emitting diode (spin LED), the electron spin plays a basic role with all the aforementioned types of information carriers.

The main subject of this book is, therefore, focused around the transport equations, which govern the transport of information carriers. These transport equations form the physical device models of all semiconductor devices, including the emerging nanodevices. The TCAD (Technology Computer-Aided Design) tools make use of these transport models to simulate the behavior of solid-state devices and circuits, in terms of the device structure and external boundary conditions of bias voltage or current.

The utilization of TCAD tools is essential because they accelerate the R&D cycle and nowadays, they become essential more than ever. In fact, the device simulation has three main purposes; to *understand* the underlying physics of a device, to *depict* the device characteristics and to *predict* the behavior of new devices. Actually, the advent of new nanodevices has been an everyday occurrence. For example, some versions of the 6th generation of Intel Core processors, is manufactured using a 14nm process. Projecting the advance of semiconductor industry for the next few years, we expect to see nanodevices approaching the size of a few atoms (1nm). The devices at such nanoscale display special quantum properties which are completely different from the case of bulk systems. Therefore, the availability of powerful transport models, which account for the underlying quantum effects, is very important for the simulation of such nanodevices. Everybody working in the field of modeling and simulation of state-of-the-art devices feels that current TCAD tools should be pushed beyond their present limits.

Almost all scientists in the field of semiconductors, agrees that a rigorous study of carrier transport in nanodevices needs a many-body quantum description. Such a description requires the solution of a huge number of equations describing each carrier of the system. Actually, the description of transport in a real device should include the real number of carriers in both the device and its contacts to the external world, and this is beyond the ability of typical computing platforms. Therefore, many levels of approximation that sacrifice some vital information about the physics of transport process are necessary. The figure below illustrates the hierarchy of main transport approaches, which are used in describing carrier transport in semiconductors and nanodevices.

Many authors distinguish between three classes of transport models, namely;

- Quantum models,
- Kinetic models, and
- Macroscopic (fluid dynamics) models.

The quantum approach lies at the top level of transport theories, for many-body problems. To treat quantum problems, a mean-field (e.g., the Hartee-Fock potential) approximation is usually adopted to

Figure 1.

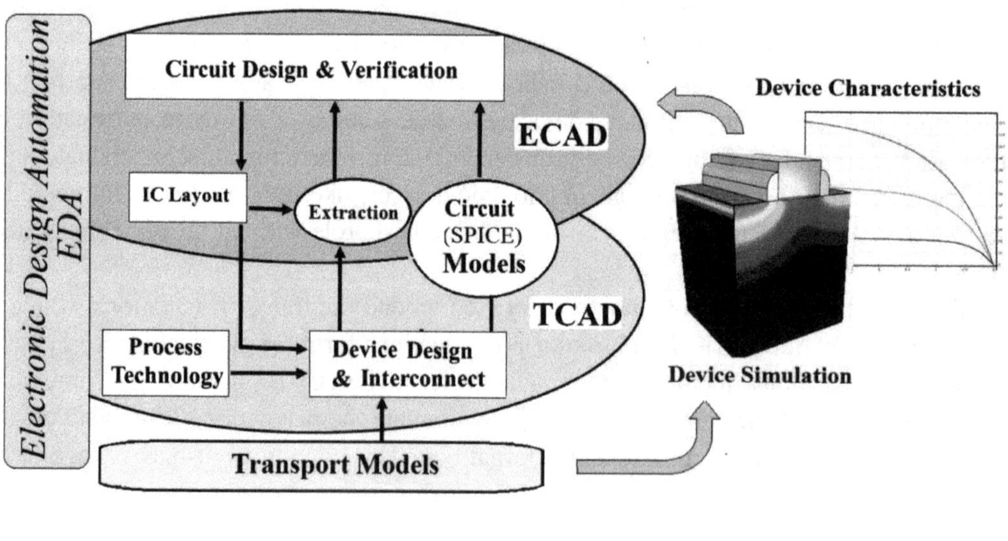

transform the many-body system into one-electron problem. The Non-equilibrium Green function (NEGF) method is very popular as a quantum approaches. Above this are quantum kinetic approaches such as the *Liouville*-von Neumann equation of motion for the density matrix, or Wigner distribution that contain quantum correlations but retain the form of semiclassical approaches. When we move from quantum to classical description of carrier transport, information concerning the phase of the electron and its non-local behavior are lost, and electronic transport is treated in terms of a localized particle framework.

The semiclassical transport theory is based on the Boltzmann transport equation (BTE), which represents a kinetic equation describing the time evolution of the distribution function of particle. The BTE has been the primary framework for describing transport in semiconductors and semiconductor devices with micro-scale dimensions. There are then approximations to the BTE, given by moment expansions of the BTE which lead to the hydrodynamic, the drift-diffusion, and relaxation time approximation approaches to transport. Finally, the so-called *compact models* come at an empirical level as circuit models for circuit simulation.

This book consists of 11 chapters, which are organized as follows.

Chapter 1: Introduction to Information-Carriers and Transport Models
Chapter 2: Semiclassical Transport Theory of Charge Carriers (Part I: Microscopic Approaches)
Chapter 3: Semiclassical Transport Theory of Charge Carriers (Part II: Macroscopic Approaches)
Chapter 4: Quantum Transport Theory of Charge Carriers
Chapter 5: Carrier Transport in LDS and Nanostructures
Chapter 6: Carrier Transport in Nanotubes and Nanowires
Chapter 7: Phonon Transport and Heat Flow
Chapter 8: Photon Transport
Chapter 9: Spin Transport and Spintronic Devices
Chapter 10: Polarons, Plasamons, and Polaritons Transport
Chapter 11: Carrier Transport in Organic Semiconductors and Insulators

Figure 2.

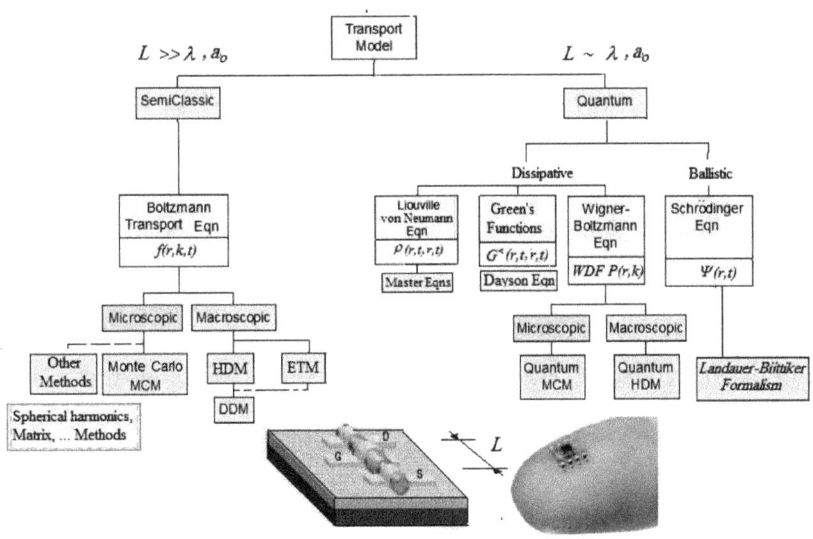

I start with the classical approaches and end with the quantum description for composite quasi particles, such as polarons, plasmons, and polaritons.

Each chapter starts with a recap of concerned concepts and provides the state of the art advances in the field as well as some case studies and overview of the literature. Some physical and mathematical notes are inserted (without interrupting the main context) to clarify the jargons, that are unavoidably utilized in such a specialized book.

In Chapter 1, I review the fundamental properties of semiconductors, and explain the transport phenomena within the framework of the classical Drudé model. The Drudé classical model is frequently introduced to describe the electrical conductivity in solids. This model is still very relevant because free particle picture can still be used as far as we can assume parabolic energy bands with a suitable effective mass, near equilibrium. In fact, the Drudé model succeeded to explain (to some extent) the electrical conductivity, the thermal conductivity, the Hall Effect, as well as the dielectric function and the optical response of solids. Everything we explain in this chapter about semiconductor properties and carrier transport is correct to the zero order approximation. In order to get into the details of carrier transport in semiconductor devices, we proceed in the following chapters, and search for a master transport equation, in two vertices, namely: the semiclassical and quantum transport theories.

In Chapter 2, I cover the essential aspects of charge carrier transport through solid materials, within the semiclassical transport theory. We start with a review of the semiclassical approaches that leads to the concepts of drift velocity, drift mobility, electrical conductivity and thermal conductivity of charge carriers in metals and semiconductors. The semiclassical transport theory is based on the Boltzmann transport equation (BTE). The Boltzmann transport equation can be derived from the Lowville equation, which describes the evolution of the distribution function changes in phase space and time. I discuss the various approximations and phenomenological approaches which make the equation useful and solvable for semiconductor devices. For instance, I present the spherical harmonic expansion (SHE) and the Monte Carlo (MC) stochastic Methods as well as the microscopic relaxation-time approximation (RTA), which leads to the conventional drift-diffusion model (DDM).

In Chapter 3, I discuss the hydrodynamic model (HDM) for semiconductor devices, which plays an important role in simulating the behavior of the charge carrier in nano devices. This model consists of a set of nonlinear conservation laws for the particle density, current density, and energy density. The hydrodynamic model for semiconductors is an inexpensive alternative tool for two- and three-dimensional device simulation. The set of hydrodynamic equations (HDEs), which is derived from the first few moments of the semiclassical BTE, is indeed more accurate than the conventional DDM and less complex than the direct solution of the BTE (by e.g., the SHE and Monte Carlo Methods).

In Chapter 4, I present the quantum transport approaches, which are necessary to simulate nanodevices including tunneling and other quantization effects. The quantum transport theory originates from several directions, including the quantum Liouville (von Neumann) equation, the Feynman path integral as well as the Wigner-Boltzmann transport equation (WBTE). The quantum Liouville equation describes the temporal evolution of the density operator. The density *matrix* operator is the favorite mathematical instrument in quantum statistical physics. The so-called *Pauli Master* equation (PME) is derived from the quantum *Liouville* equation. The PME is frequently used to describe irreversible processes in quantum systems. The kinetic equation for the Wigner distribution function including scattering effects is called the Wigner-Boltzmann transport function (WBTE). After solving the WBTE, and calculating the Wigner distribution function (WDF), we can calculate the spatial density of carriers and current, as well as the average value of any microscopic physical parameter.

Based on the WBTE, the quantum corrected Boltzmann equation, the quantum hydrodynamic model (QHDM), and the density gradient (DG) approximation can be obtained. Also, the WDF may be defined as the energy integral of the Green's function. The Green's function approach can be used to give the response of a system to a constant perturbation in the Schrödinger equation. The so-called non-equilibrium Green's function (NEGF) formalism is a very powerful technique to evaluate the transport properties of quantum systems in both thermodynamic equilibrium and non-equilibrium conditions. At the end of this chapter, I present the multi-band transport models and the major band structure calculation methods. This includes the ab initio models, such as the density functional theory (DFT), and the approximate methods, such as, the tight binding (TB) model, pseudopotential methods, as well as the GW approximation.

In Chapter 5, I demonstrate the carrier transport phenomena in low-dimensional semiconductors (LDS), where, free electrons are only permitted to move in one or two dimensions. I describe some LDS structures, such as quantum wells, quantum wires and quantum dots and the transport models of charge carriers across them. I discuss the conductance of LDS systems, using the Landauer formalism (for 2-terminal devices) or the generalized Landauer-Büttiker formalism (for multi-terminal devices). I also describe some quantum effects that take place in such nanostructures, such as quantum Coulomb blockade, Aharonov–Bohm, Shubnikov-De Haas oscillations and Kondo effects.

In Chapter 6, I handle transport across Carbon nanotubes (CNT's), which are one of the most interesting materials in nanotechnolog. Nanotubes and nanowires with dimensions on the nanometer length scale cannot be treated as classical conductors because their diameters are as small as the mean free path length (between collisions), but their length is large for the full quantum treatment. Therefore, such mesoscopic structures need a special framework of transport models, which we discuss in this Chapter.

In Chapter 7, I investigate phonon transport and thermal conductivity in semiconductor structures and nanodevices. Microscopic approaches such as the Peierls-Boltzmann transport equation (phonon BTE) and phonon Monte-Carlo simulation can capture quasi-ballistic phonon transport. These models are valid only when heat transport is diffusive and the characteristic length scales are much larger than the phonon mean free path. When phase coherence effects cannot be ignored, these semiclassical approaches fail and result in erroneous results. Therefore, I handle the topic of ballistic (non-diffusive) phonon transport for nanoscale structures and nanodevices.

Chapter 8 is dedicated for photonic transport models. Accurate modeling of photonic devices is essential for the development of optical components in fields like communications, sensing, biomedical instrumentation, consumer electronics and defense. The specific challenge of optoelectronic device simulation lies in the combination of photonics and electronics, including the sophisticated interaction of photons and electrons. Mathematical models for photon transport include the Monte Carlo simulation method, numerical solution of the semiclassical and quantum transport equations, as well as phenomenological models. Macroscopic photonic transport analysis requires the consideration of seven independent variables: three spatial directions, two angular directions, frequency and time.

The so-called radiative transport equation (RTE) is an integro-differential equation that describe the diffusion and scattering of photons inside matter. The diffusion approximation alleviates the solution of this equation. Also, the optical Bloch equations, which are sometimes called the Maxwell-Bloch equations (MBEs), describe the dynamics of two-state quantum systems interacting with the electromagnetic modes of an optical system. Within a semiclassical approach, where the light field is treated as a classical electromagnetic field and the material excitations are described quantum mechanically, all quantum effects can be treated microscopically on the basis of the semiconductor Bloch equations (SBEs). The quantum approaches are based on some sort of dynamic wave equations (Schrodinger-like or Heisenberg-like) in the microscopic level or the SBEs in the macroscopic level. The so-called dynamics-controlled truncation (DCT) formalism is another successful microscopic approach that describes coherent correlations in optically excited semiconductors. On the other hand, the most successful approach to study incoherent effects and correlations in highly excited semiconductors is the nonequilibrium Green's functions (NEGF) approach. We discuss these models with illustrative examples, to show the features and weakness of each model.

In Chapter 9, I present a full quantum and a semiclassical description of spin transport, which explains how the motion of carriers gives rise to a spin current. The so-called two-component spin-drift-diffusion model (SDDM) is a simple semiclassical and straight-forward method for spin transport modeling. The semiclassical models can be useful for investigation of a broad class of transport problems in semiconductors, but they do not include effects of a spin phase memory. The quantum approach of spin density matrix with spin polarization vector of a spin state is more appropriate for this case. The classical Bloch equations for spin transport are the analogue of the classical BTE for particle transport. They can be extended to time-dependent non-equilibrium quantum transport equations, using a suitable non-equilibrium quantum distribution function, like the spin magnetization quantum distribution function (SMQDF). The so-called spinor-BTE resembles the Boltzmann kinetic equation with spin-orbit coupling in a magnetic field together with spin-dependent scattering terms. By taking the macroscopic moments of the spinor-BTE, we can get a density-matrix based version of the SDDM. The last section of this chapter covers the latest proven spintronic devices, such as spin-FET, MRAM and spin LED.

In Chapter 10, I present the semiclassical and full quantum models of composite quasi particles, such as polarons, plasmons and polaritons. These information carriers play a significant role in the emerging nanoelectronic and nanophotonic devices and systems.

In Chapter 11, I focus the attention on the carrier transport in organic semiconductors and insulator materials. Organic semiconductors are hydrocarbon molecular crystals or polymers. In order to understand charge transport in organic solids, I review the transport and tunneling mechanisms in disordered materials. Thereupon, I discuss the recent transport models, such as the semiclassical and quantum formalisms of Marcus theory.

As the reader can see from the above description of this book, I tried to give a balanced amount of theory for almost all known transport models of charge carriers, phonons, photons and spin in semiconductors and nanodevices. However, I did my best to avoid drowning the reader into the minor mathematical details. I consider this a critical point, because even the specialized reader may get bored from the arcane mathematical proofs. In fact, the basic interest of the readers who have an engineering background, is to discover and then to know *when* and *how* to apply the different transport models. For these readers I wrote this book. However, when, some mathematical details are important I mention them, in brief, as a note, so that the reader can bypass them without interrupting the main subject.

Although this book is primarily theoretical in approach, I frequently refer to experimental results, which show the variation of transport parameters as well as their measurement methods. I also supplement each chapter, with one or two case studies of real devices that aid understanding of the treated theory in this chapter. The book has many illustrations and diagrams to clarify the presented transport models, and comprehensively referenced for further study.

This one-stop book (for almost all semiconductor transport models) is dedicated for engineers and researchers in solid-state physics and nanodevices, as well as students in nanoelectronics and nanotechnology.

Muhammad El-Saba
Ain-Shams University, Egypt

Chapter 1
Introduction to Information-Carriers and Transport Models

1. OVERVIEW AND CHAPTER OBJECTIVES

During the last decade, the rapid development of electronics technology has produced several new devices at nanoscale dimensions (nanodevices). Nanodevices, are the tiny devices whose dimensions are in the order of nanometers (or less than 100nm however). The information carriers in these devices are the particles or quasi particles that can carry and transport information objects or signals. The most famous example of an information carrier is the electron charge in conventional semiconductor devices. Also, photons in photonic and optoelectronic devices and the electron spin in spintronic devices can be considered as information carriers. In addition, other quasi particles, such as phonons (quasi particles associated with lattice vibration waves) may be considered as information carriers, because they are capable of transporting energy from point to another in solid-state devices. The recent research in nanodevices is focused around the control of such information carriers and to exploit their features to build new devices with superior characteristics in terms if speed and integration density. Naturally, great efforts have been dedicated to understanding the transport mechanisms of such information carriers in semiconductors and nanostructures.

The transport theory of information carriers forms the basis of any physical device model. The transport models are used in Technology/Computer-Aided Design (TCAD) tools to simulate the device behavior, in terms of its structure and geometry as well as external boundary conditions of voltage and current. In fact, the transport of information carriers is a non-equilibrium phenomenon, where the role of external forces plays a crucial role. External forces which drive the device out of equilibrium may be electromagnetic in origin, such as the electric fields associated with an applied bias, or the excitations of electrons by optical sources. Alternately, thermal gradients and electrochemical potentials may also provoke the transport of charge carriers and therefore create external currents and voltages drops, across the device. The Figure1 depicts the role of carrier transport models in TCAD simulation tools and how they are used to calculate the current-voltage (I-V) and capacitance-voltage (C-V) characteristics of a

DOI: 10.4018/978-1-5225-2312-3.ch001

Figure 1. Electronic design automation (EDA) lifecycle and carrier transport modeling in TCAD Tools; in the EDA section, DFT=Design for Testability, LVS=Layout versus Schematic, DRC=Design Rule Checking and GDSII is a design file format.

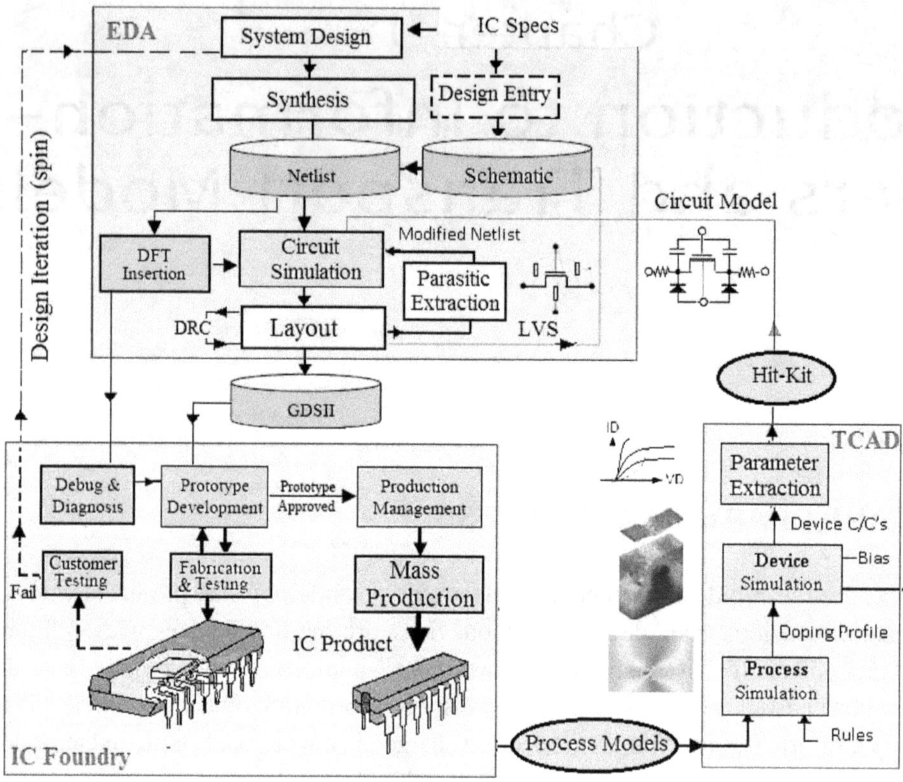

certain device and interact with other electronic design automation (EDA) tools. Also, Figure 2 depicts the different levels of transport models, in device simulation. As shown, the TCAD tools are based on semiclassical and quantum transport models. These models range from *ab-initio* physical models, which describe the transport of information carriers from first principles down to compact models that describe the outer behavior (usually the I-V and C-V) of devices and circuits. The success of nanotechnology to produce well-functioning nanodevices and systems is mortgaged by the availability of suitable and efficient transport models that meet the challenges at the nanoscale.

As shown in Figure 2, the transport models cover a wide scale, from classical to quantum transport, according to their accuracy and the required computational costs. Actually, a single description in the hierarchy of transport models may not be suitable to provide the correct behavior of all devices.

Depending on the device length scale, the carrier transport may be semiclassical or purely quantum. Nowadays, the most famous semiclassical approaches for the simulation of charge-carrier transport in semiconductor devices are the drift-diffusion model (DDM), the hydrodynamic model (HDM), the Spherical harmonic expansion (SHE) as well as the Monte Carlo method (MCM). DDM and HDM descriptions of particle transport are macroscopic in nature and enable a quick computation of device characteristics (in terms of macroscopic quantities like the carrier density). Depending on the particular application, the macroscopic transport models are applicable to devices with characteristic lengths in

Figure 2. Complexity (accuracy) of transport models versus computational time (cost)

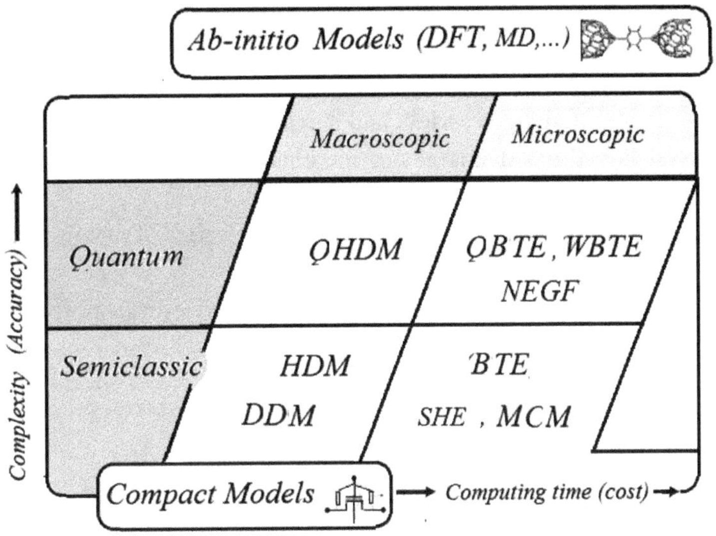

the range of micrometers or some hundred nanometers, where microscopic-size and quantum effects are not dominant. For even smaller devices, it is necessary to resort to microscopic approaches, which are based on the semiclassical Boltzmann transport equation (BTE) or its quantum counterparts, e.g., the quantum Liouville equation (QBTE) or the Wigner BTE (WBTE).

The solution of the BTE by MCM or SHE approaches may yield accurate results for the transport characteristics in many small devices. However, the semiclassical approaches (both microscopic and macroscopic) fail as soon as quantum mechanical effects dominate and a description of the information carriers as localized particles becomes invalid. Indeed, the description of carrier transport in modern nanodevices requires sophisticated many–body quantum approaches. Clearly, the full quantum description including the actual number of carriers in a device is beyond the ability of any computational platform nowadays[1]. Therefore, approximations are necessary to simulate and predict the behavior of such devices. In order to construct a successful approximation (model), we need to understand the phenomena behind the real problem, and under which physical limits, the approximation can be assumed.

Hence, successive levels of approximation, that sacrifice some information about the exact nature of transport, are sometimes utilized in any nanodevices. As shown in Figure 2, the quantum models range from *ab-initio* models, such as density-functional theory (DFT), and the tight-binding (TB) models that predict the band structure, to the quantum Liouville equation (QBTE) and its variant master equations as well as the non-equilibrium green functions (NEGF) to predict the device characteristics.

When the appropriate transport model is selected and utilized by a suitable device simulator, we can get the device input/output characteristics and understand the device behavior. Finally, the so-called compact models are non-linear circuit models that capture the device behavior, and are suitable for circuit simulation.

Although we assume a basic knowledge of solid-state physics in this Book, we start with the theoretical fundamentals of semiconductors. This Chapter is a general review of the fundamentals physics of charge carrier transport in semiconductors, with emphasis on the classic transport models.

Upon completion of this chapter, students will

- Understand the concept of transport modeling and information carriers in semiconductors and nanodevices.
- Be familiar with the different models of information carrier transport.
- Review the fundamentals of semiconductor physics, such as energy band structure, density of states, drift and diffusion of charge carriers and carrier scattering mechanisms.
- Explain the advantages and disadvantages of the classical transport theory of charge carriers in metals and semiconductors.
- Describe the electrical, thermal, magnetic and optical properties of metals and semiconductors, on the basis of the simple Drudé model.
- Decide what evidence can be used to support or refuse a carrier transport model.

2. CLASSIFICATION OF INFORMATION CARRIERS

The term Information Carriers has its origin in computer science and information technology and has been applied in many different ways. In computer science, an information carrier is a means to keep (store) information. However, I mean by information carriers in electronic devices, the particles or particle characteristics that can carry, transport or store signals within a device. For instance, the electron charge in conventional semiconductor devices and the spin of electrons in spintronic devices as well as photons in photonic devices are all examples of information carriers. In addition, other quasi particles, such as phonons (quasi particles associated with lattice vibration waves) may be considered as information carriers, because they are capable of transporting energy from point to another in solid-state devices. Other examples of information carriers are shown in Figure 3.

A charge carrier is a moving particle, which carries an electric charge. Examples are moving electrons, ions and holes. In a conducting medium, an electric field can exert work (force) on the free particles, causing a net motion of their charge through the medium; this is what is referred to as electric current. In metals, the charge carriers are electrons. Free (or more precisely quasi free) electrons in good conductors are able to move about freely within the material. Free electrons can also be generated in vacuum and act as charge carriers. As well as charge, an electron has another intrinsic property, called spin. A spinning charge carrier produces a magnetic field similar to that of a tiny bar magnet.

Figure 3. Examples of information-carriers in electronic, spintronic, optoelectronic, thermoelectric and quantum devices

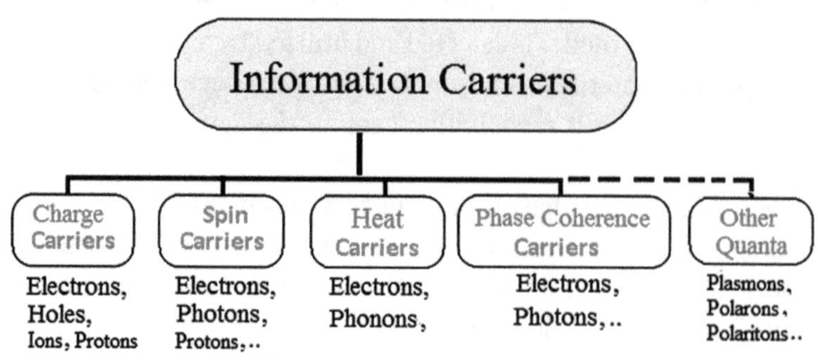

In melted ionic solids or electrolytes, such as salt water, the charge carriers are ions, atoms or molecules that have gained or lost electrons so they are electrically charged. Atoms that have gained electrons and become negatively charged are called anions, while atoms that have lost electrons become positively charged and called cations.

In semiconductors, electrons and holes (moving vacancies in the valence band) are the charge carriers. In fact, holes are considered as mobile positive carriers in semiconductors. In semiconductor devices, most of the electrical, thermal and electrical properties of interest have their origins from electrons (in the conduction band) and holes (in the valence band).

Of course electrons and holes carry electrical charges as well energy. Other important energy carriers are phonons (lattice vibrations). Actually, the thermal energy transport in crystals occurs primarily due to the vibration of atoms about their equilibrium positions. In semiconductors, the heat conduction process takes place, primarily, through lattice vibrations (phonons).

The property of coherence was originally connected with light propagation in optics but now it is defined in all types of waves. In quantum mechanics coherence is due to the nature of the wave functions, which are associated with moving particles. Coherence means that the phase difference between wave functions is kept constant for coherent particles. The delay over which the phase or amplitude wanders by a significant amount is defined as the coherence time (usually termed τ_c), as shown in Figure 5. The coherence length λ_c is defined as the distance the wave travels in time τ_c. The spatial coherence of a wave is defined as the cross-correlation between two points in the wave for all times. The most popular experimental technique which provides direct information about charge carrier coherence in semiconductors is four-wave-mixing (FWM) spectroscopy.

Figure 4. Information-carriers in electronics and spintronics

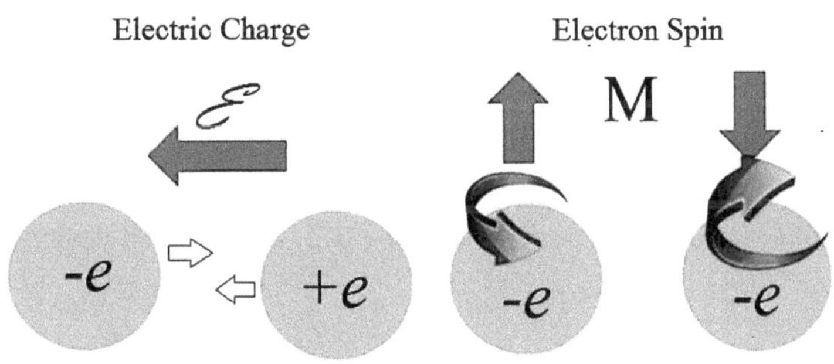

Figure 5. Illustration of the concept of phase coherence of the wave functions

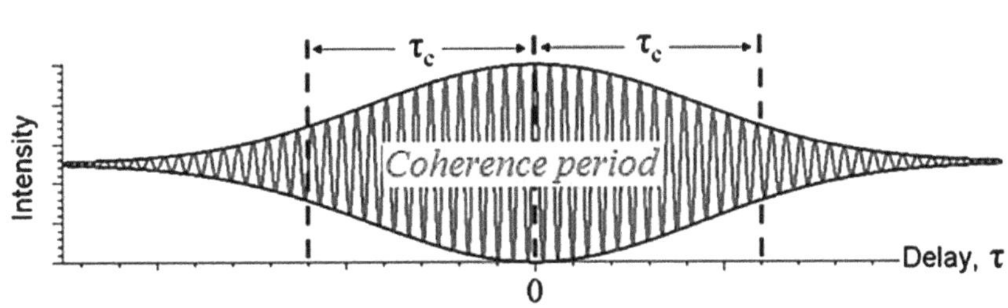

3. CLASSIFICATION OF TRANSPORT MODELS

The nature of transport in a semiconductor device depends on the characteristic length of the device active region. The carrier motion can be described with classical laws, when the length of the device active region is much larger than the corresponding carrier wavelength. When the device dimensions (or one of them) are comparable to the carrier wavelength, the carriers can no longer be treated as classical point-like particles, and the effects originating from the quantum- nature of propagation begin to determine transport.

The appearance of quantum effects can be determined by comparing the device size[2], L, to the electron mean-free path (λ_n), or the dephasing length (λ_ϕ) or the de Broglie wavelength ($\lambda_{dB} = h/p$, where h is Planck's constant and p is the electron momentum). The dephasing length (or phase coherence length), λ_ϕ, is a physical quantity which describes the quantum interference and may be defined as follows:

$$\text{Dephasing length} \rightarrow \lambda_\phi = \sqrt{(D_n \tau_\phi)} \tag{1}$$

where D_n is the electron diffusion constant and τ_ϕ is the dephasing (or phase-breaking) time. One way to obtain the dephasing time (τ_ϕ,) is to measure the magneto-resistance of the material (Pierret, 2003).

The quantum interference and strong coherence phenomena can be observed in nanostructures, when

$$\lambda_{dB} \approx L, \lambda_n << \lambda_\phi \tag{2}$$

As the temperature is raised, the quantum interference smears out and the coherent states start to appear and participate in conduction. The so-called "weak localization" is expected at the following conditions:

$$L > \lambda_T, \lambda_{dB} \approx \lambda_n < \lambda_\phi , \tag{3}$$

where λ_T is called the temperature length (or thermal correlation length) which is defined as follows:

$$\text{Temperature length} \rightarrow \lambda_T = < \hbar D_n / k_B T_L \tag{4}$$

where $\hbar = h/2\pi$ and k_B is Boltzmann's constant.

On the other hand, the semiclassical approach can be still used in small devices as long as:

$$\lambda_{dB} << \lambda_{n,,} , \lambda_\phi << L \tag{5}$$

Figure 6 depicts the hierarchy of transport models, which are currently known and utilized to describe electronic transport in semiconductors and nanodevices. At the top level we find the Schrodinger equation[3] for many-body problems, which are only tractable for tiny structures with a few numbers of electrons. In order to treat the many-body quantum problem, some sort of mean- approximation is necessary to transform the problem into an effective one-electron problem. This is done in the so-called Hartee-Fock (H-F) equation and other variant methods, such as the Kohn-Sham (K-S) functional approach.

Following to this level, we find the quantum kinetic approaches in terms of the Liouville equation of motion for the density matrix (QBTE), or Wigner transport equation (WBTE) that contains quantum correlations but have the form of semiclassical approaches.

Figure 6. Hierarchy of information-carrier transport models

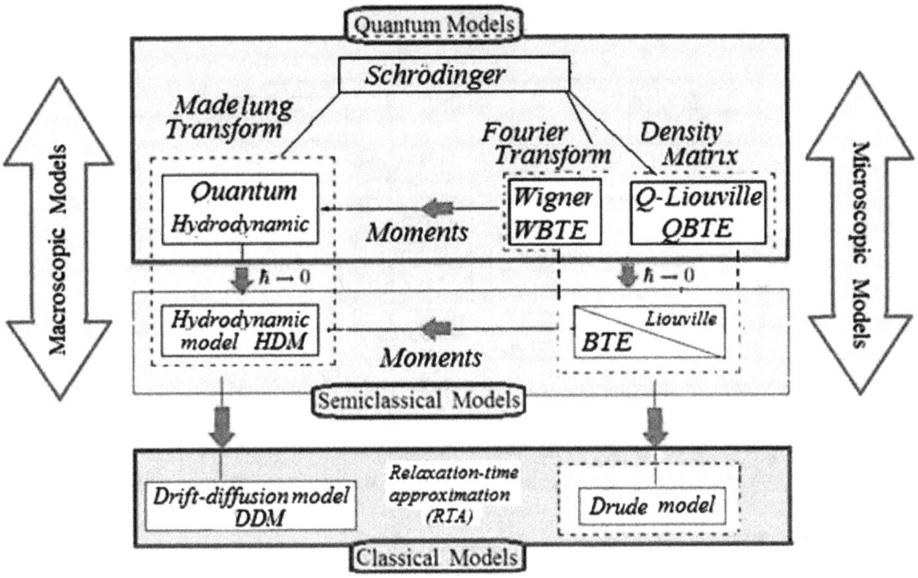

In semiclassical approaches, we assume that the spatial extent (Δx) of the wave packet, which is associated with the motion of an information carrier, is much smaller than the mean free path between collisions ($\Delta x << \lambda_n$) in the device area. This means that we can talk about the motion of localized (or point-like) quasi-particles. Therefore, the main feature size of the device (L) should be much greater than the mean free length between collisions ($L >> \lambda_n$). The motion should also be localized in the k-space so that we can talk about a mean wavenumber $\mathbf{k} >> \Delta k$ (satisfying the Heisenberg uncertainty principle $\Delta x . \Delta k \approx 1$). This is the level of the Boltzmann transport equation (BTE), which is a kinetic equation describing the time evolution of the distribution function of particles. The BTE has been the primary framework for describing transport in semiconductors devices down to submicron scale. There are then approximations to the BTE, given by hydrodynamic moments of the BTE which lead to the hydrodynamic model (HDM), the drift-diffusion model (DDM), and relaxation time approximation approaches (RTA).

Figure 8 illustrates the details of transport models of different sophistication levels to describe the transport of charge carriers in semiconductor devices.

Figure 7. Schematic of the extension of a particle wave packet and its Fourier transform in the k-space

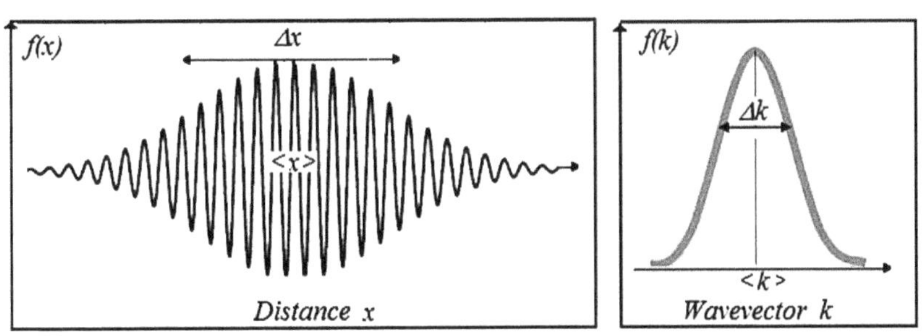

Figure 8. Detailed illustration of information-carrier transport models

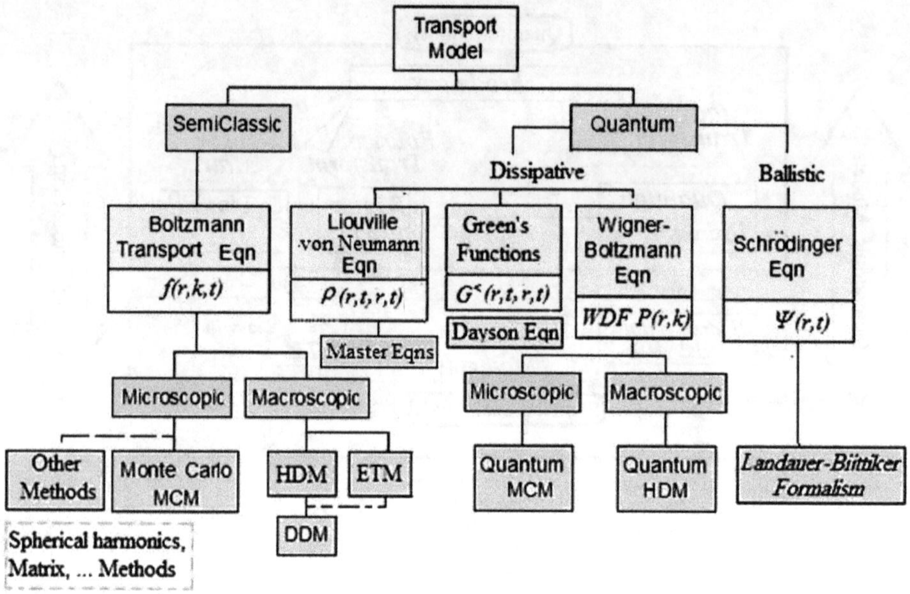

4. CHARGE-CARRIER TRANSPORT MODELS IN SEMICONDUCTORS

We know that any thermodynamic system is in *thermal equilibrium* forever unless it is acted upon by external forces, i.e. when no exchange of energy is done with the exterior. We may consider a semiconductor in state of thermal equilibrium, as long as it is not acted upon by any external force field (e.g., electric field, magnetic field, electromagnetic field or light). However, the individual atoms and electrons in a solid still exchange energy between themselves, even when no external force is applied. Therefore, the equilibrium state is called "*dynamic thermal equilibrium*".

4.1 Semiconductor Conductivity Model

A semiconductor is neither a true conductor nor an insulator, but half way between. The discovery of semiconductor properties, dated back to Michael Faraday (1839) who noticed that the conductivity of some materials decreases as temperature increases, inverse to the behavior of known metals. A variety of substances, such as germanium (*Ge*), silicon (*Si*) and gallium arsenide (*GaAs*), exhibit semiconducting properties. In this section we first review the model of conduction in semiconductors using the silicon as an example. In fact silicon was established as a good semiconductor material about 80 years ago (the 1930s). At this time, Alan Wilson applied Felix Bloch energy band theory to study the energy band structure of silicon. Actually, the *Si* atom has 14 electrons distributed over energy levels of different orbitals ($1s^2$, $2s^2$, $2p^6$, $3s^2$, $3p^2$). The incomplete outer shell of silicon atom contains 4 electrons ($3s^2$, $3p^2$). The silicon lattice has a diamond lattice and its atoms have tetrahedral covalent bonds as shown in Figure 9.

In pure silicon lattice all electrons are bound, in the valence band, and there are no free charge carriers (no free electrons!) at zero absolute temperature (0K). Therefore, behaves like an insulator and the application of an electric field does not result in electric current. In order to produce an electrical

Figure 9. Diamond lattice and covalent bonds in pure elemental semiconductors, like silicon

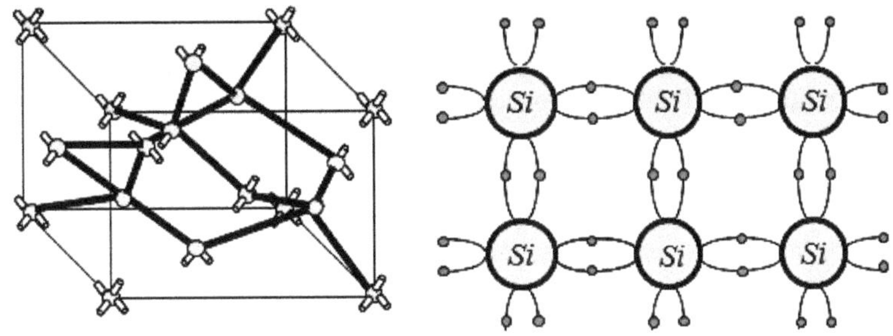

current in a semi-conductor, some valence electrons must be freed from their bonds. This can be done by supplying the crystal by external energy, usually in the form of heat or light. The minimum energy that is required to free an electron in a pure semiconductor is equal to the height of its energy gap E_g. In Si, the energy gap is about 1.2eV at 300K. Each free electron in a pure semiconductor leaves a broken bond (or a hole) as shown in Figure 10. Such a free electron roams everywhere in the crystal with equal probability in all directions. A free electron can also recombine with a vacant bond (a hole) to produce a bond, while transmitting its excess energy in the form of light quanta (photons) or lattice vibrations (phonons)

$$\bar{e} + hole \underset{\text{Generation}}{\overset{\text{Recombination}}{\rightleftarrows}} bond \tag{6}$$

If an electric field ζ is applied to a crystal, the free electrons will be acted upon by a force $F = -e.\zeta$ and they begin to drift against the field direction. If the concentration of free electrons in the conduction band is n electrons per unit volume (electrons/cm^3) and their average drift velocity is v_n, then the electron current density J_n (A/cm^2) is given by:

$$J_n = -e\, n\, v_n = \sigma_n .\zeta \tag{7a}$$

where $\sigma_n = -e\, n\, (v_n / \zeta)$ is called the electrical conductivity of electrons. Unlike metals the conductivity of semiconductors depends actually on many ambient parameters such as temperature, illumination, etc.

Figure 10. Generation of an electron-hole pair by breaking a covalent bond

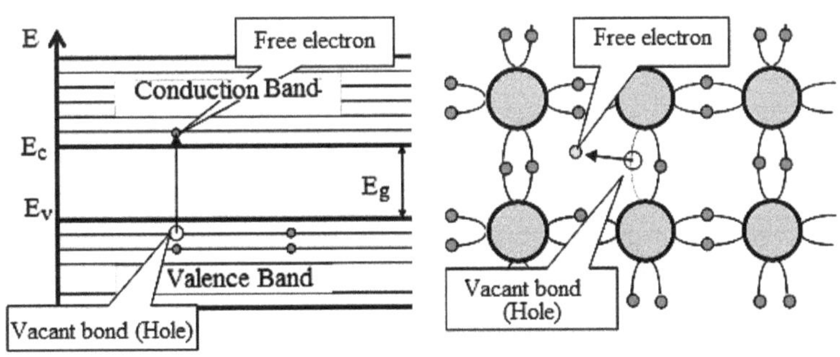

Regarding the valence band, it is more convenient to consider the motion of holes instead of the motion of valence electrons, as shown in Figure 11. This is because the number of holes is usually much less than the number of valence electrons[4]. If there are p holes (vacant bonds in the crystal lattice) per unit volume in the valence band, then the current produced by the motion of valence electrons to fill in these holes, against field direction, is equal to the current produced by the motion of holes, along the field direction. Therefore, the hole current density J_p is given by:

$$J_p = e\, p\, v_p = \sigma_p \zeta \tag{7b}$$

where $\sigma_p = e\, p\, (v_p / \zeta)$ is called the electrical conductivity of holes and v_p is their average velocity .

Therefore, there exist two types of charge carriers in semiconductors, Electrons in the conduction band, and Holes in the valence band.

The conduction electrons (and valence holes) in pure semiconductors can be produced by thermal or optical excitations. Extra conduction electrons (or valence holes) can also be obtained in semiconductors by doping them with impurity atoms. Accordingly, semiconductors are called intrinsic (pure) semiconductors or extrinsic (impure) semiconductors.

4.2 Concentration of Electrons and Holes

In intrinsic semiconductors, the charge carriers (electrons and holes) are mainly generated by thermal excitation of the valence electrons. When the supplied thermal energy is high enough, some covalent bonds are broken and electron-hole pairs are produced. Therefore, the number of broken bonds and hence the concentration of generated electron-hole pairs is proportional the ambient temperature. Consequently, the concentration of electrons (n) must equal to the concentration of holes (p) in intrinsic semiconductors:

$$n = p = n_i \tag{8}$$

where n_i is called the intrinsic carrier concentration. The intrinsic carrier concentration is temperature dependent and is given by:

$$n_i = A\, T^{3/2} \exp\, (-E_g / 2k_B T) \tag{9a}$$

Figure 11. Motion of electrons and holes in the valence band of a semiconductor

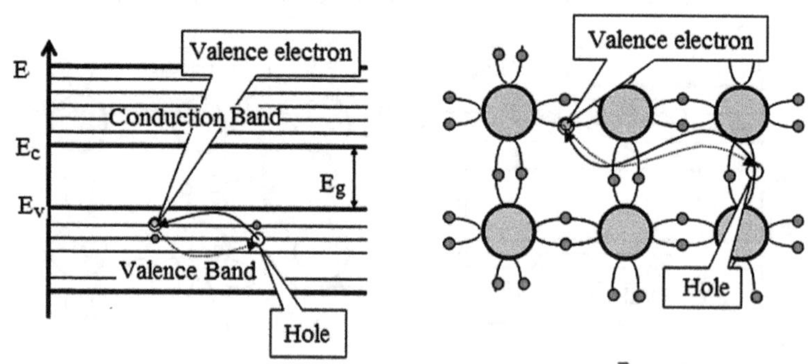

where A is a constant. Hence, the value of n_i is strongly dependent on temperature and the type of semiconductor material. For the matter of comparison, we can express n_i by the following relations for Si and 4H-SiC:

$$n_i\,(Si) = 3.67x10^{16}\,T^{3/2}\,exp\,(-7020\,/\,T) \tag{9b}$$

$$n_i\,(SiC) = 1.\,7x10^{16}\,T^{3/2}\,exp\,(-20800\,/\,T) \tag{9c}$$

In silicon, n_i is almost $1.38\ x10^{10}$ (electron/cm^3) at $T=\ 300\ K$. On the other hand, n_i is as small as $6.74\ x10^{-11}$ (electron/cm^3) or practically zero in SiC at $300\ K$. For this reason, SiC is more suitable for high temperature devices.

At thermal equilibrium, the process of electron-hole pair thermal generation is compensated by an opposite electron-hole recombination process, such that the rate of thermal generation g_{th} is equal to the rate of recombination R. Therefore, the net rate of change of electron-hole pair concentration $\partial n/\partial t$ $=(g_{th} - R)$ is null at thermal equilibrium and the intrinsic carrier concentration remains fixed.

In order to increase the number of free charge carriers (electrons or holes) in a semiconductor, and hence to increase its conductivity, semiconductors are usually doped with impurity atoms. In this case the semiconductor is called an extrinsic semiconductor. In extrinsic semiconductors extra conduction electrons are typically produced by doping the semiconductor with impurity atoms of the group V of the periodic table of elements, like phosphorous (P). This type of impurities is called *donors*. A semiconductor which is doped with donors is said to be of n-type.

Similarly, extra valence holes can be produced by doping the semiconductor with impurity atoms of the group III, like Boron (B). This type of impurities is called acceptors. A semiconductor which is doped with acceptors is said to be of p-type. The more abundant charge carriers in a piece of semiconductor are called majority carriers, which are primarily responsible for current transport. In n-type semiconductors majority carriers are electrons, while in p-type semiconductors they are holes. The less abundant charge carriers are called minority carriers. Minority carriers in n-type semiconductors are holes, while in p-type semiconductors they are electrons.

The density of electrons in the conduction band is equal to the density of occupied states[5]. Also the density of occupied states is equal to the density of states in the conduction band $g_c(E)$ multiplied by the probability of occupation which is given by the Fermi-Dirac energy distribution function for electrons $f_n(E)$. Therefore, the density of electrons is given by the following integration:

Figure 12. Basic bond pictures of n-type (a) and p-type (b) semiconductors

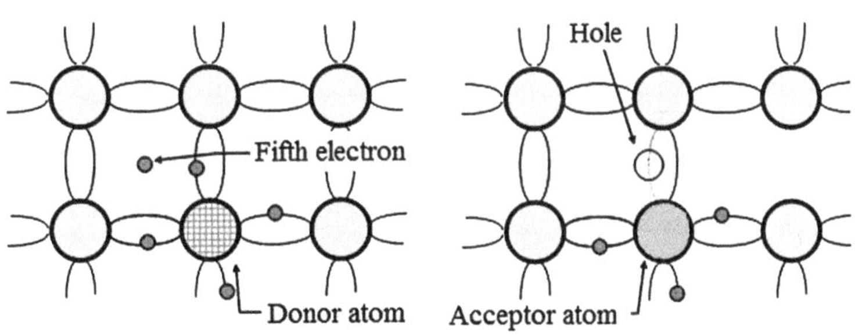

Figure 13. Taxonomy of semiconductors, according to doping type

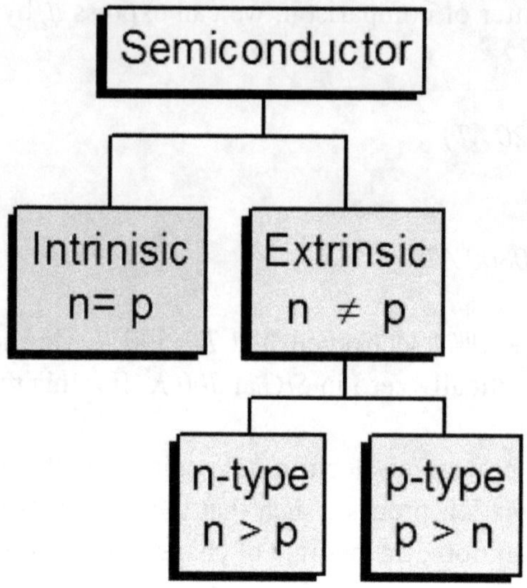

$$n = \int_{C.Band} g_c\left(E\right) f_n\left(E\right) dE \tag{10}$$

with

$$f_n(E) = \frac{1}{1+\exp\left[\dfrac{E-E_F}{k_B T}\right]} \tag{11}$$

The Fermi energy, E_F, refers to the energy of the highest occupied energy level at absolute zero temperature (0K). Also, the density of holes in the valence band is equal to the density of vacant states. The density of vacant states is equal to the density of states in the valence band $g_v(E)$ multiplied by the probability of *non-occupation* by electrons, which is given by the Fermi-Dirac distribution for holes $f_p(E)=1-f_n(E)$,

$$p = \int_{V.Band} g_v\left(E\right) f_p\left(E\right) dE \tag{12}$$

$$f_p\left(E\right) = 1 - f_n\left(E\right) = \frac{1}{1+\exp\left[\dfrac{E_F-E}{k_B T}\right]} \tag{13}$$

The density of states $g(E)$ in a certain band can be deduced from the E-k relation of the material, using the relation.

$$g\left(E\right)=\frac{1}{2\pi^3}\iint_{\text{Const.E.Surface}}\frac{\text{ds}}{\left|\nabla_k E\left(k\right)\right|} \tag{14}$$

Here, the surface integral is taken over a constant energy surface (CES), where $E(k)$ = constant. Figure 14 shows the density of occupied state, which is the product of the density of sates by the Fermi-Dirac distribution function of the concerned carriers. The Figure 15 illustrates this for electrons and holes.

Usually free charge carriers (electrons or holes) reside at the bottom of conduction bands or the top of valence bands. Therefore, we assume that the $E(\mathbf{k})$ relation of the semiconductor is almost quadratic close to extreme points, which concave up conduction bands or concave down valence bands. The approximated $E(k)$ relation is then similar to the dispersion relation of free electrons in free space ($E=p^2/2m_o=\hbar^2k^2/2m_o$). However, in order to account for the internal lattice field, the free electron mass (m_o) should be replaced with the carrier effective mass (m^*), which depends on the curvature of the semiconductor $E(k)$ relation. When the semiconductor is anisotropic, the carrier effective mass is a 2nd order tensor, whose components are given by:

Figure 14. Schematic illustration of the carrier occupation and carrier density

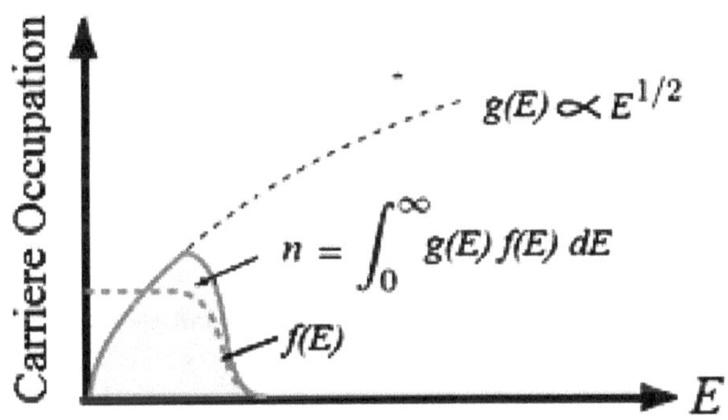

Figure 15. Product of the density of states and the Fermi-Dirac distribution function for electrons and holes in n-type (left graph) and p-type (right graph) semiconductors.

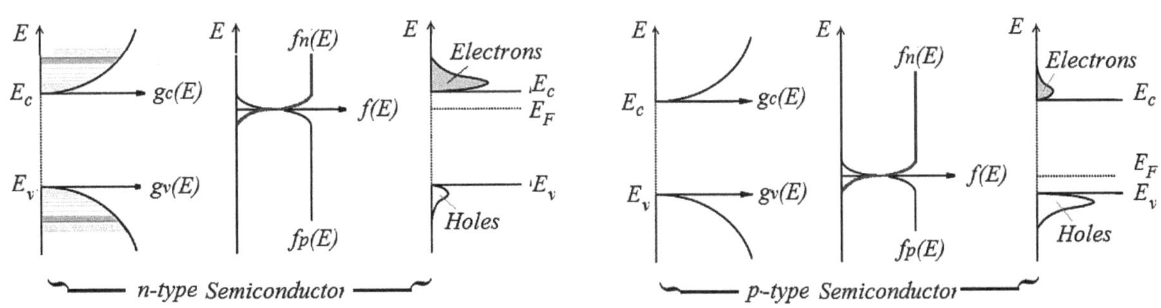

$$m_{ij}^{*-1} = \frac{1}{\hbar^2}\frac{\partial^2 E}{\partial k_i \partial k_j} \text{ with } (i, j = x, y, z) \tag{15}$$

When the semiconductor is isotropic then the effective mass tensor reduces to a scalar quantity (zero-order tensor), such that: $m^* = m_{xx} = m_{yy} = m_{zz}$ and other coefficients are null. Therefore, the inverse effective mass $m^{*-1} = (1/\hbar^2).\partial^2 E/\partial k^2$.

It comes from the above discussion that the effective mass and so many other characteristics of charge carriers, depend on the band structure *E(k)*, or more precisely, the shape of constant energy surfaces of the material. In cubic semiconductors, like *Si* and *GaAs*, we can distinguish three types of constant energy surfaces: spherical, ellipsoidal and warped energy bands. Figure 17 shows a general band structure model of cubic semiconductors (near main extreme points). Note that the energy gap may be direct or indirect. The Figure 18 is a schematic of the real band structure, *E(k)* of *Si* and *GaAs*, in certain directions of the k-space. Also, Figure 19 shows the shape of constant energy surfaces of main conduction and valence bands of such semiconductors. As we'll see in Chapter 7, the application of strain on a semiconductor shifts the energy levels of the conduction and valence bands and can remove the band degeneracy.

Case 1: Spherical Constant Energy Surfaces

In certain direct-gap semiconductors, like *GaAs*, the constant energy surfaces of the E-K relation are almost spherical and isotropic, near extreme points. The *E-k* relation in this case may be approximated as follows:

$$E = E_o \pm \frac{\hbar^2}{2m^*}\left(k - k_o\right)^2 \tag{16a}$$

Figure 16. Schematic representation of the conduction and valence bands around their extreme points in two-dimensional k-space

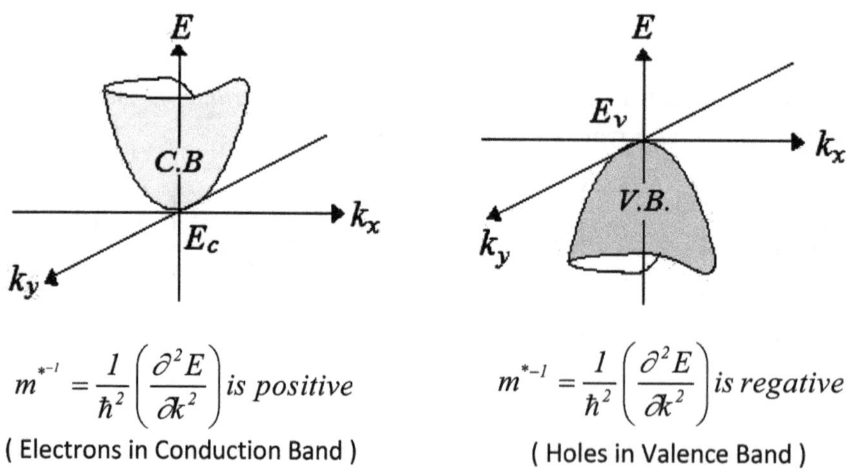

Figure 17. Schematic representation of the band structure of a cubic semiconductor

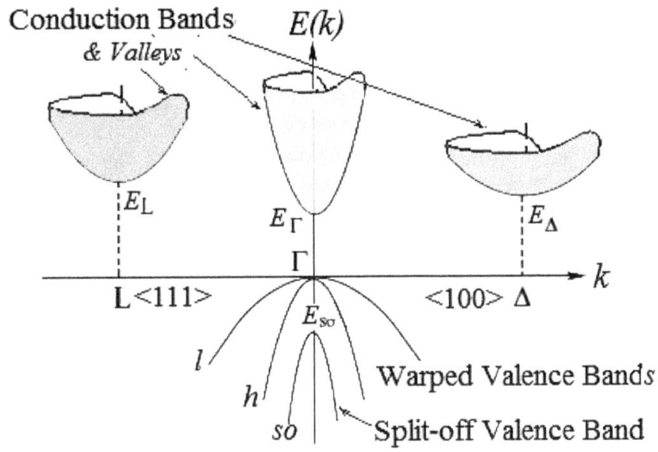

Figure 18. Energy band structure of Si and GaAs

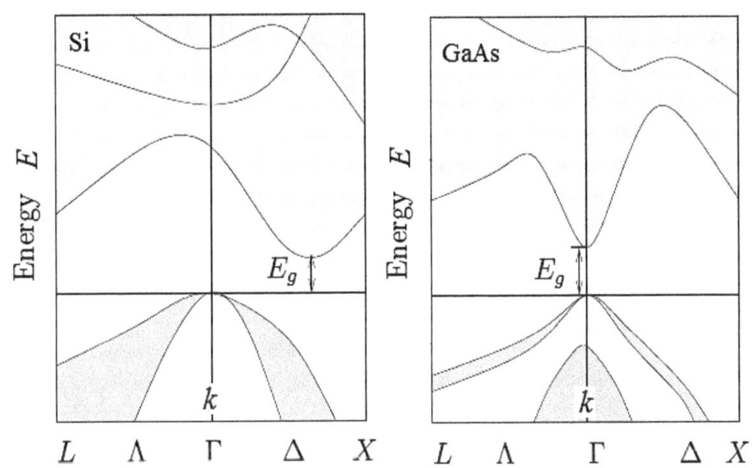

Figure 19. Shapes of the constant energy surfaces in cubic semiconductors

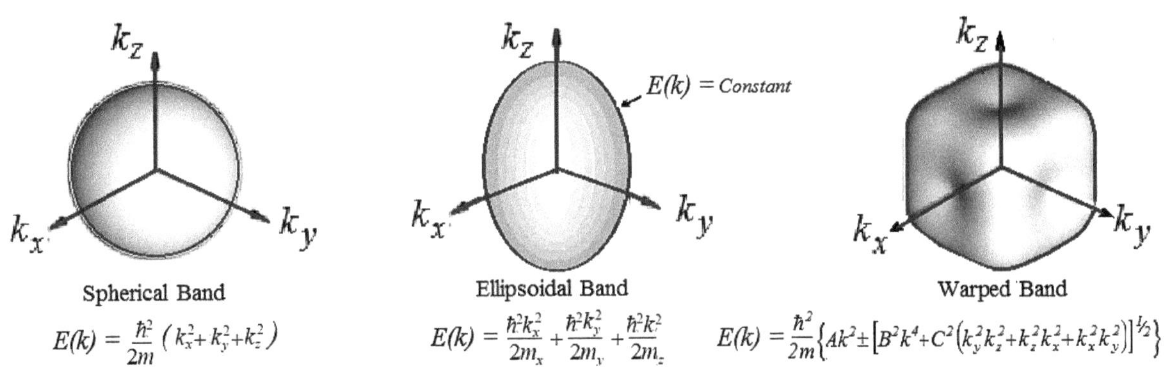

where E_o is a constant and the ± sign denotes either the conduction or the valence bands. If the number of equivalent minima (or valleys) in the conduction band is denoted by M_c, and the effective mass of electrons is denoted by m_n*, then the density of states in the conduction band is given by:

$$g_c(E) = 4\pi M_c \left(\frac{2m_n^*}{h^2}\right)^{3/2} \left(E - E_c\right)^{1/2} \text{ for } E > E_c \tag{16b}$$

Similarly, the density of states in the valence band is given by:

$$g_v\left(E\right) = 4\pi M_v \left(\frac{2m_p^*}{h^2}\right)^{3/2} \left(E_v - E\right)^{1/2} \text{ for } E < E_v \tag{16c}$$

where M_v is the number of maxima in the valence band and m_p* is the effective mass of holes there. Note that for *Si*, $M_c=6$ (six valleys in the main conduction band) and $M_v=1$ (if we only considered heavy and light hole valence bands).

Case 2: Ellipsoidal Constant Energy Surfaces

In indirect gap semiconductors, like *Si*, the constant energy surfaces are ellipsoidal (or approximated so). The *E-k* relation is hence a more complicated than the spherical isotropic case. The effective mass is no longer a scalar quantity but depends on the direction (a tensor). If the directions are chosen such that m* is a diagonalized tensor with diagonal elements m_{xx}*, m_{yy}* and m_{zz}*, then, the *E-k* relation is approximated as follows:

$$E = E_o \pm \frac{\hbar^2}{2}\left[\frac{\left(k_x - k_{xo}\right)^2}{m_{xx}^*} + \frac{\left(k_y - k_{yo}\right)^2}{m_{yy}^*} + \frac{\left(k_z - k_{zo}\right)^2}{m_{zz}^*}\right] \tag{17a}$$

where $E_o = E(k_{xo}, k_{yo}, k_{zo})$ is a constant. For instance, the main conduction band of some indirect-gap semiconductors (like Si) has $k_{xo}=k_{yo}=0$ and $k_{zo}=k_{min}$ such that

$$E = E_o + \frac{\hbar^2}{2}\left[\frac{\left(k_z - k_{\min}\right)^2}{m_l^*} + \frac{\left(k_x + k_y\right)^2}{m_t^*}\right] \tag{17b}$$

where we put $m_{xx}^*=m_{yy}^*=m_t^*$ (transverse effective mass) and $m_{zz}^*=m_l^*$ (longitudinal effective mass). The density of states in the conduction band is then:

$$g_c(E) = 4\pi\left(\frac{2m_{nd}^*}{h^2}\right)^{3/2}\left(E - E_c\right)^{1/2} \tag{18}$$

where m_{nd}^* is called the density-of-states effective-mass in the conduction band.

$$m_{nd}^* = M_c^{2/3}\left(m_{xx}^* m_{yy}^* m_{zz}^*\right)^{1/3}. \tag{19}$$

In Si, m_{xx}*= m_{yy} *= $m_t^* = 0.19m_o$ and m_{zz}*= $m_l^* = 0.916m_o$, so that m_{nd}* = 1.08 m_o, according to Singh (1993).

Case 3: Warped Constant Energy Surfaces

The valence bands of cubic semiconductors (like Si) are approximately quadratic. The constant energy surfaces of the two upper warped bands (for heavy and light holes) are fluted spheres. The *E(k)* relation of such semiconductors may be described by the following relations near *k= 0,*

$$E(k) = E_v - \frac{\hbar^2}{2m_o}\left\{Ak^2 \pm \left[B^2k^4 + C^2\left(k_y^2k_z^2 + k_z^2k_x^2 + k_x^2k_y^2\right)\right]^{1/2}\right\} \tag{20}$$

where *A*, *B* and *C* are constants and the ± signs correspond to light holes and heavy holes bands, respectively. For the valence band of light holes (with plus sign and designated by the letter *l*), the effective mass is usually denoted m_{lh}*. At the band edge, the light holes mass m_{lh}* (for *Si*) [6] is given by:

$$m_{lh}^* = m_o \Big/ \left(A + \sqrt{B^2 + 1/6\,C^2}\right) = 0.153\ m_o \tag{21a}$$

Figure 20. Constant energy surfaces of the principal conduction band of silicon. The symmetric points in the first Brillouin zone of the k-space are shown in the left figure.

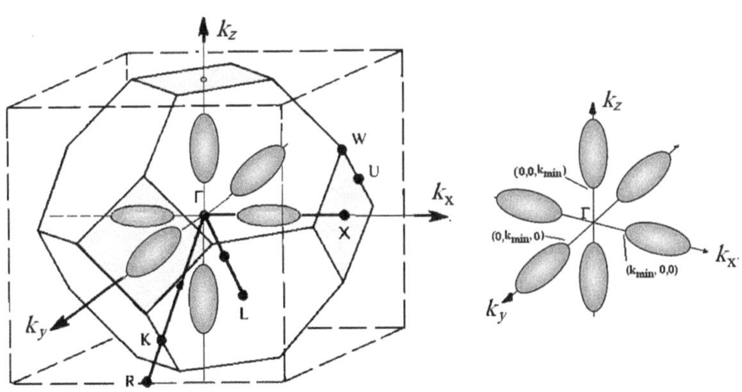

Also, for the valence band of heavy holes (with minus sign and designated by the letter *h*), the effective mass is usually denoted m_{hh}*. At the band edge, the heavy hole mass m_{hh}* (for *Si*) is given by:

$$m^*_{hh} = m_o \Big/ \left(A - \sqrt{B^2 + 1/6\ C^2} \right) = 0.537\ m_o \tag{21b}$$

Equation (20), which describes the *E(k)* relation of the two upper valence bands (the *l* and *h* valence bands) in cubic semiconductors, is sometimes written using polar *k*-coordinates as follows:

$$E(k,\theta,\varphi) = E_v - \frac{\hbar^2 k^2}{2m_v^*} g(\theta,\phi) \tag{22}$$

where m_v* is the isotropic hole effective mass, while *g(θ, ϕ)* contains the *l* and *h* valence bands anisotropy information. As shown in the following figure, the constant-energy surfaces of *l* and *h* hole bands are warped, like a cube with rounded corners and dented-in faces. This is more pronounced in heavy holes.

The third valence band in cubic semiconductors is called the split-off band (s-band). This band is only populated at higher hole energies, and its *E(k)* relation may be described by the following simple *parabolic* relation:

$$E(k) = E_v - \Delta_s - \frac{\hbar^2 k^2}{2m_o} A \tag{23}$$

where Δ_s is the shift between the top of the *s*-band and the top of the *l* and *h* valence bands. In Si, Δ_s=0.044 eV below the *l* and *h* bands. The effective mass of split-off holes in Si is given by $m_{h,so}* = 0.234\ m_o$.

The density of states in the valence band is generally given by:

Figure 21. Constant energy surfaces of light- and heavy-hole valence bands of silicon

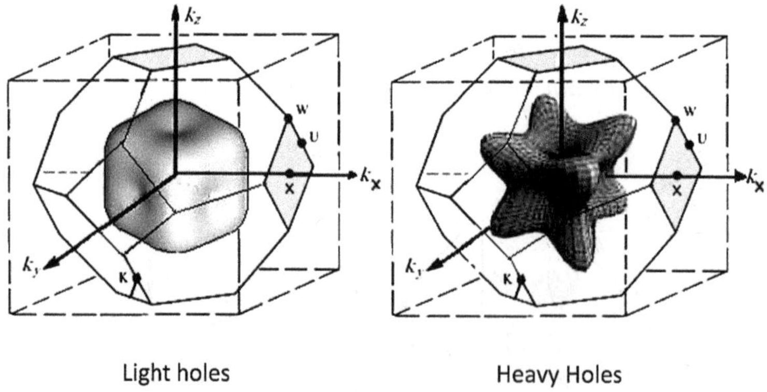

$$g_v\left(E\right)=4\pi\left(\frac{2m_{pd}^{*}}{h^2}\right)^{3/2}\left(E_v-E\right)^{1/2} \tag{24}$$

where m_{pd} is the density-of-states effective-mass in the valence band:

$$m_{pd}^{*3}=M_v^2\left(m_{lh}^{*\,3/2}+m_{hh}^{*3/2}\right)^2. \tag{25}$$

where m_{lh}* is the effective mass of light holes and m_{hh}* is the effective mass of heavy holes. For Si, m_{lh}* = *0.16* m_o and m_{hh}*=*0.49* m_o, so that m_{pd}*=0.81m_o.

Substituting $g_c(E)$ from Equation (18) into (10) yields the following expression for the density of electrons in the conduction band:

$$n=\int_{E_c}^{\infty}4\pi\left(\frac{m_{nd}^{*}}{h^2}\right)^{3/2}\frac{\left(E-E_C\right)^{1/2}dE}{1+\exp\left[\left(E-E_F\right)/k_B\ T\right]}=N_C\cdot F_{1/2}\left(\frac{E_F-E_C}{k_B\ T}\right) \tag{26a}$$

where N_C is called the effective density of states in the conduction band:

$$N_C=2\left(\frac{2\pi m_{nd}^{*}k_BT}{h^2}\right)^{3/2} \tag{26b}$$

Also $F_{1/2}(\zeta_n)$ is the Fermi-Dirac integral of order ½. The Fermi-Dirac integral is defined as follows[7]:

$$F_j\left(\zeta_n\right)=\frac{1}{\Gamma(j+1)}\int_o^{\infty}\frac{x^j dx}{1+\exp\ \left(x-\zeta_n\right)} \tag{27a}$$

with

$$x=E/k_BT \text{ and } \zeta_n=\left(E_F-E_c\right)/k_BT \tag{27b}$$

In non-degenerate semiconductors, where $(E-E_F) >> k_BT$, the number of electrons is much smaller than the effective density of states in conduction band. Then, we have $n << N_c$ (diluted gas of electrons) and the Fermi-Dirac distribution may be approximated by the Boltzmann (exponential) distribution. As shown in Figure 21, the Fermi integral may be then approximated as:

$$F_{1/2}\left(\zeta_n\right) \cong \frac{2}{\sqrt{\pi}} \int_o^{\infty} x^{1/2} \exp\left(\zeta_n - x\right) dx = \exp\left(\zeta_n\right) \tag{27c}$$

Therefore, the density of electrons may be written as follows:

$$n = N_c \exp\left[-\left(E_c - E_F\right)/k_B T\right] \tag{28a}$$

Similarly, the density of holes in the valence band, in non-degenerates semiconductors, is given by:

$$p = N_V \exp\left[-\left(E_F - E_v\right)/k_B T\right] \tag{28b}$$

where N_V is effective density of states in the valence band and given by:

$$N_V = 2\left(\frac{2\pi m_{pd}^* k_B T}{h^2}\right)^{3/2} \tag{28c}$$

Table 1 depicts the values of N_c and N_v as well as the density of states effective mass in the bulk of some semiconductors at *300K*.

In compound semiconductors and alloys, like *GaAs*, which have upper and lower energy valleys with different band edges (e.g., E_{c1} and E_{c2}), the density of states in the conduction band maybe expressed as follows:

$$N_c = 2\left(\frac{2\pi\ m_{nd}^*\ k_B\ T}{h^2}\right)^{3/2} \left[M_{c1} + M_{c2}.\exp(\frac{E_{c1} - E_{c2}}{k_B T})\right] \tag{29}$$

where M_{c1} and M_{c2} are the number of upper and lower equivalent valleys. For instance, $M_{c1} = M_{c2} = 1$ in *GaAs* and $M_{c1} = 4$, $M_{c1} = 2$ in $Si_{1-x}Ge_x$. In the latter case, the shift in energy levels (E_{c1} - E_{c2}) is expressed in terms of the mole fraction x, such that N_c becomes a function of x.

It worth notice that equations (28a) and (28b) are not ready for the calculation of the electron and hole concentrations (n, p) because we don't know yet the Fermi level position (E_c - E_F or E_F -E_v). In the

Table 1. Effective density of states in conduction and valence bands of some semiconductors at 300K

Semiconductor	E_g (eV)	m_{nd}/m_o	m_{pd}/m_o	N_c (cm^{-3})	N_v (cm^{-3})
Si	1.12	1.08	0.56	2.8 x10^{19}	1.04 x10^{19}
Ge	0.67	0.55	0.37	1.04 x10^{19}	6.0 x10^{18}
GaAs	1.42	0.067	0.48	4.7 x10^{17}	7.0 x10^{18}

According to Yu and Cardona (1996).

following sections we show how to determine the electron and hole concentrations in equilibrium, by an alternative method, from the mass action law and the neutrality condition.

4.3 Mass Action Law in Semiconductors

It follows from the above discussion that the concentration of electrons and holes depends on the location of Fermi level E_f. In thermal equilibrium, the *np* product is independent of the Fermi-level position and given by:

$$n.p(Thermal\ equilibrium) = n_o\ p_o = N_c N_v \exp\left(-\frac{E_g}{k_B T}\right) \tag{30}$$

Here, the subscript '*o*' denotes the values of *n*, *p* at the thermal-equilibrium state.

The above equation is called the mass-action law in semiconductors. According to this law, the *n.p* product is equal to a constant independent of time and of the type of added impurities. In intrinsic semiconductors we have $n_o = p_o = n_i$, then the mass-action law can be written as follows:

$$n_o p_o = n_i^2 \tag{31}$$

Figure 22. Normalized electron concentration (n/N_c) versus normalized Fermi energy $\zeta_F = (E_F - E_c).k_BT$, with different approximations at low- and high-doping concentrations

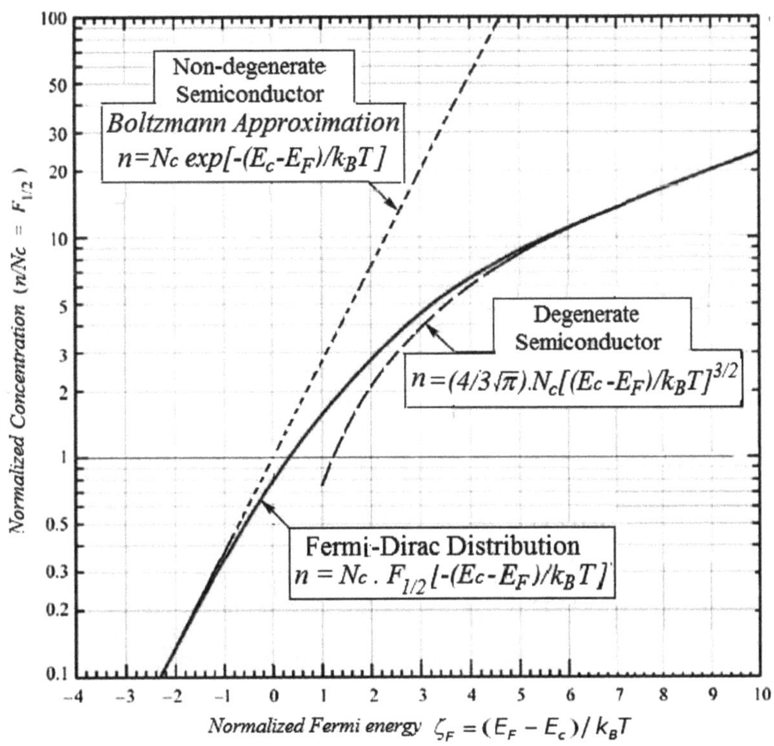

This means that the *n.p* product in semiconductor, at thermal equilibrium, is constant equal to the square of the intrinsic carrier concentration.

4.4 Neutrality Equation in Semiconductors

The electron and hole concentrations as well as the location of the Fermi-energy level in a semiconductor can be calculated by the aid of the so-called "*neutrality condition*". According to the neutrality condition, the total charge in a semiconductor at thermal equilibrium is zero. If the charge density (per unit volume) is labeled by ρ then:

$$\rho = \left[e(p_o + N_d^+) - e(n_o + N_a^-)\right] = 0 \tag{32a}$$

Therefore,

$$n_o + N_a^- = p_o + N_d^+ \tag{32b}$$

where N_d^+ and N_a^- are the densities of ionized donors (bound positive charges) and ionized acceptors (bound negative charges), respectively. For the calculation of the Fermi-level, each term in the above equation must be expressed in terms of E_F. For non-degenerate semiconductors, if the impurity atoms are localized and singly ionized, then the neutrality equation may be written as follows:

$$\frac{2N_c}{\sqrt{\pi}} F_{n1/2} + \frac{N_d}{1 + 2\exp\left(-\dfrac{E_d - E_F}{k_B T}\right)} = \frac{2N_v}{\sqrt{\pi}} F_{p1/2} + \frac{N_a}{1 + 4\exp\left(\dfrac{E_a - E_F}{k_B T}\right)} \tag{33a}$$

where we substituted the ionization ratios for donors and acceptors (N_d^+/N_d and N_a^-/N_a,, respectively) using Gib's law[8], for singly-ionized impurities (Sze, 1969):

$$N_d^+ = \frac{N_d}{1 + \gamma_d \exp\left(\left(E_F - E_d\right) / k_B T\right)} \tag{33b}$$

$$N_a^- = \frac{N_d}{1 + \gamma_a \exp\left(\left(E_a - E_F\right) / k_B T\right)} \tag{33c}$$

Here, E_d is the donor energy level and γ_d is the donor degeneracy factor (2 for *P* in *Si, Ge* and most semiconductors). Like most of V-group impurities E_d is close to the bottom of conduction band (44 meV for *P* in *Si*). Such impurities are called *shallow-level impurities.* Also, E_a is the acceptors energy level and γ_a is the acceptors degeneracy factor (4 for *B* in *Si*, due to spin and the folded valence band, for light and heavy holes). For most III-group impurities, E_a is close to the top of valence band (about 10meV), so that $N_a^- \approx N_a$ at room temperature.

The Figure 23 depicts shallow acceptor and donor levels in p-type and n-type semiconductors. Shallow impurities are of great interest in semiconductors, since they define the conductivity and the type of semiconductor.

The neutrality Equation (33a) can be solved graphically, to find out the Fermi level E_F. In special cases, the analytical solution of this equation may be simple.

4.5 Carrier Density and Fermi Level in Intrinsic Semiconductors

In intrinsic semiconductors, the number of electrons is equal to the number of holes (vacant places or broken bonds). Thus, in thermal equilibrium we have:

$$n_o = p_o = n_i \tag{34}$$

where n_i is the intrinsic-carrier concentration in the semiconductor

$$n_i = \left(np\right)^{1/2} = \left(N_c N_v\right)^{1/2} \exp\left[-\left(\frac{E_c - E_v}{2k_B T}\right)\right] \tag{35a}$$

Substituting both N_v and N_c expressions into (35) yields the following expression for n_i

$$n_i = 4.82 \times 10^{15} \left(\frac{m_{nd}^* m_{pd}^*}{m^2}\right)^{3/4} T^{3/2} \exp\left[-\frac{E_g}{2k_B T}\right] \tag{35b}$$

The Figure 24 shows the intrinsic carrier concentration of *Si*, *Ge*, and *GaAs* as a function of temperature. Naturally, the Fermi level in intrinsic materials is almost midway between the conduction and valence band ($E_F \approx E_i = E_V + ½ E_g$). More precisely, we have:

Figure 23. Energy levels of acceptor and donor impurities in p-type and n-type semiconductors

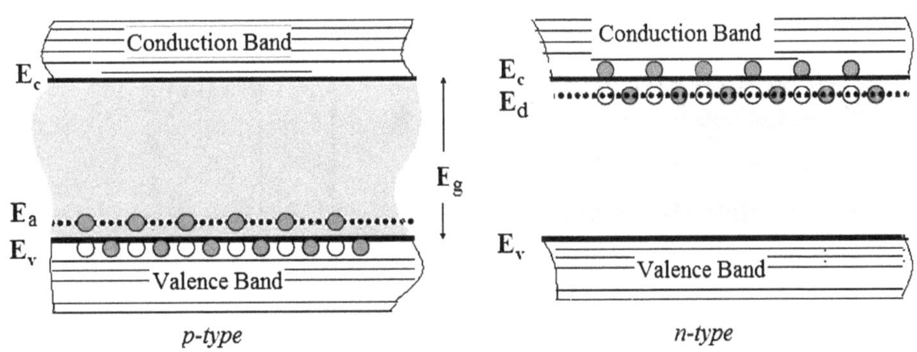

Figure 24. Intrinsic carrier concentrations of Si, Ge and G_aAs vs. temperature
Source: Semiconductors (Smith, 1979).

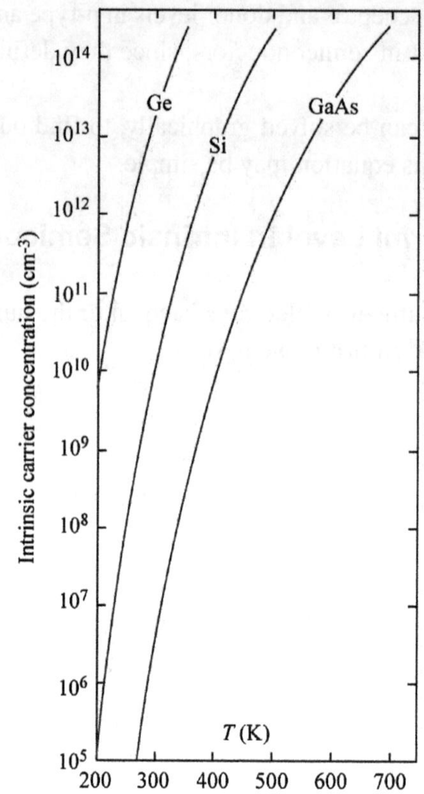

Figure 25. The product of the density of states and the Fermi-Dirac distribution function for electrons and holes in intrinsic semiconductors

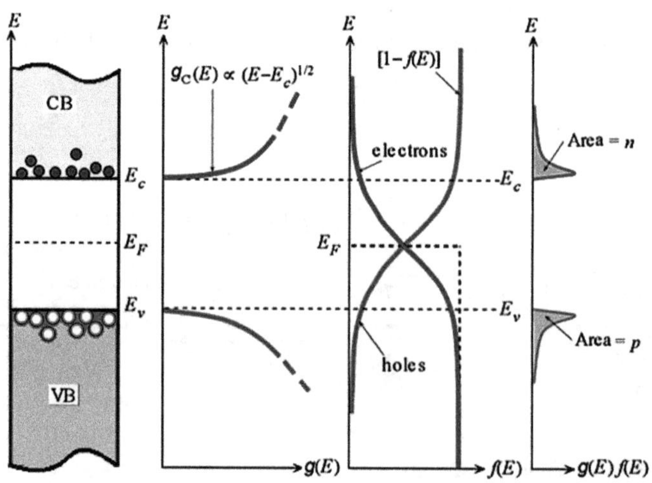

$$E_F \cong E_i + k_B T \ln\left(N_v / N_c\right) = E_i + \tfrac{3}{4} k_B T \ln\left(m^*_{pd} / m^*_{nd}\right). \tag{36}$$

Note 1: Meaning of the Fermi Level and Chemical Potential

The Fermi level is the term used to describe the top of the collection of electron energy levels in a solid at absolute zero temperature. At absolute zero temperature (T=0K), electrons pack into the lowest available energy states and build up a Fermi gas, just like a sea of energy states. The Fermi level is the surface of that sea at 0K where no electrons will have enough energy to rise above the surface.

The concept of the Fermi energy is a crucially important concept for understanding the electrical and thermal properties of solids. The Figure 26 shows the Fermi-Dirac energy distribution, *f(E),* at different temperatures over the energy band diagram of an intrinsic semiconductor.

According to statistical thermodynamics, the term (μ) that actually appears in the Fermi-Dirac distribution ($f(E)=1/[1+exp-(E-\mu)/k_BT]$), is called the chemical potential of the gas of electrons. However, the Fermi energy of a free electron gas is related to the chemical potential by the following equation (Kireev, 1979):

$$\mu = E_F \left[1 - \frac{\pi^2}{12}\left(\frac{k_B T}{E_F}\right)^2 + \frac{\pi^4}{80}\left(\frac{k_B T}{E_F}\right)^4 - \ldots\right]$$

Hence, the chemical potential is approximately equal to the Fermi energy at temperatures much less than the characteristic Fermi temperature $T_F = E_F/k_B$. At room temperature, the Fermi energy and chemical potential are essentially equivalent.

4.6 Fermi-Level and Carrier Density in Extrinsic Semiconductors

The density of charge carriers (electrons and holes) in an extrinsic semiconductor at thermal equilibrium can be calculated by solving two basic equations, namely, the mass-action law:

$$n_o p_o = n_i^2 \tag{37}$$

Figure 26. Fermi-Dirac distribution

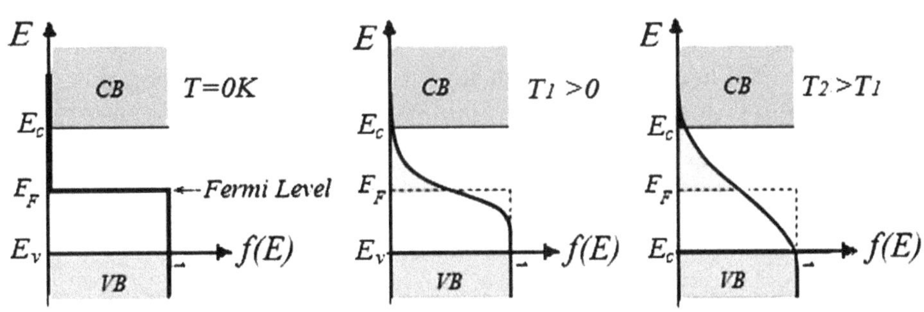

and the neutrality equation:

$$n_o + N_a^- = p_o + N_d^+ \tag{38}$$

Combining these two equations yields:

$$n_o + N_a^- = \frac{n_i^2}{n_o} + N_d^+ \text{ or } n_o^2 + n_o(N_a^- - N_d^+) - n_i^2 = 0 \tag{39}$$

By solving the above two algebraic equation, we get the equilibrium concentrations n_o and $p_o = n_i^2/n_o$

$$n_o = \frac{\left(N_d^+ - N_a^-\right)}{2}\left[1 \pm \sqrt{1 + \left(\frac{2n_i}{N_d^+ - N_a^-}\right)^2}\right] \tag{40a}$$

$$p_o = \frac{\left(N_a^- - N_d^+\right)}{2}\left[1 \pm \sqrt{1 + \left(\frac{2n_i}{N_a^- - N_d^+}\right)^2}\right] \tag{40b}$$

where the ± sign (inside the square brackets) stands for the type of majority carries. That is the + sign is taken when we calculate n_o from (40a) in n-type materials or p_o from (40b) in p-type materials. The minority carriers can be then calculated, simply from the relation $p_o n_o = n_i^2$.

Case 1: Fermi Level in n-Type Semiconductors

For *n*-type semiconductors, the electrons are majorities. At moderate temperatures, where all impurities are ionized ($N_d^+ = N_d$, $N_a^- = N_a$) we can consider $n_o = N_d - N_a$. Substituting n_o into (26) yields:

$$E_F = Ec - k_B T \ln\left(\frac{N_c}{N_d - N_a}\right) \tag{41}$$

Therefore, the Fermi level decreases linearly as the temperature is raised. However, at very low temperature, the Fermi level in n-type semiconductors rises initially with temperatures and reaches a maximum and then begins to go down towards the intrinsic level at high temperature. The initial rise of E_F with T is due to the fact that, the donor atoms are not all yet ionized at low temperature. Then, the Fermi level at low temperature may be described by the following relation:

$$n = N_c \exp\left(\frac{E_C - E_F}{k_B T}\right) = \frac{N_d}{1 + \gamma_d^{-1} \exp\left(\left(E_d - E_F\right) / k_B T\right)} \tag{42b}$$

This equation reduces to the following relation at very low temperature, where $k_B T << (E_c - E_d)$,

$$E_F(low\ temperature) \cong \frac{E_c + E_d}{2} + \frac{k_B T}{2} \ln\left(\frac{\gamma_d N_d}{N_c}\right) \tag{43}$$

Hence, the Fermi level lies at the mid-point between the conduction band edge E_c and the donor level E_d at $0K$, and raising the temperature will result in an increase in E_F. Substituting this E_F into Equation (26), gives us an expression of electron concentration, n, at very low temperatures.

$$n = N_C \exp\left[-\frac{\left(E_C - E_F\right)}{k_B T}\right] \cong \sqrt{\gamma_d^{-1} N_C N_d} \exp\left(\frac{E_d - E_c}{2 k_B T}\right) \tag{44}$$

Case 2: Fermi Level in p-Type Semiconductors

Similarly, for p-type semiconductors, holes are majorities. At moderate temperatures, where all acceptors are ionized ($N_a^- = N_a N_d^+ = N_d$) we can consider $p_o = N_a - N_d^-$. Substituting p_o into (27) yields:

$$E_F = E_v + k_B T \ln\left(\frac{N_v}{N_a - N_d}\right) \tag{45}$$

Thus, the Fermi level increases linearly as T is raised and E_F tends to E_i as T tends to high values.

At very low temperature, the fraction of ionized impurities is given by the *Gibbs* law and the Fermi level is given by:

$$p = N_v \exp\left(\frac{E_v - E_F}{k_B T}\right) = \frac{N_a}{1 + \gamma_a \exp\left(\frac{E_a - E_F}{k_B T}\right)} \tag{46}$$

This may be approximated as follows:

$$E_F(low\ temperature) \cong \frac{E_v + E_a}{2} + \frac{k_B T}{2} \ln\left(\frac{\gamma_a \cdot N_v}{N_a}\right) \tag{47}$$

Therefore, at $T=0$, the Fermi level in p-type materials lies midway between the valence band edge E_v and the acceptor level E_a. Substituting this E_F into Equation (27), gives us an expression of hole concentration, p, at very low temperatures.

$$p = N_V \exp\left[-\frac{\left(E_F - E_V\right)}{k_B T}\right] \cong \sqrt{\gamma_a \cdot N_V N_A} \exp\left(\frac{E_A - E_V}{2k_B T}\right) \tag{48}$$

The Figure 27 depicts the variation of Fermi level in the above three cases with temperature. Also, Figure 28 depicts the variation of electron and hole densities with temperature in n-type and p-type semiconductors.

4.7 Scattering of Charge Carriers

The motion of free charge carriers in a solid (e.g., electrons in a metal or electrons and holes in a semiconductor) is different from their motion in free space because of collisions with the vibrating nuclei of the solid, as shown in Figure 29. Carriers may also collide with impurity atoms or themselves as well as other crystal defects. These collisions cause the free electrons to scatter in different directions and the resulting motion of electrons is equally probable in all directions (random!) so that there is no net displacement.

Figure 27. Variation of the Fermi level position in semiconductors with temperature

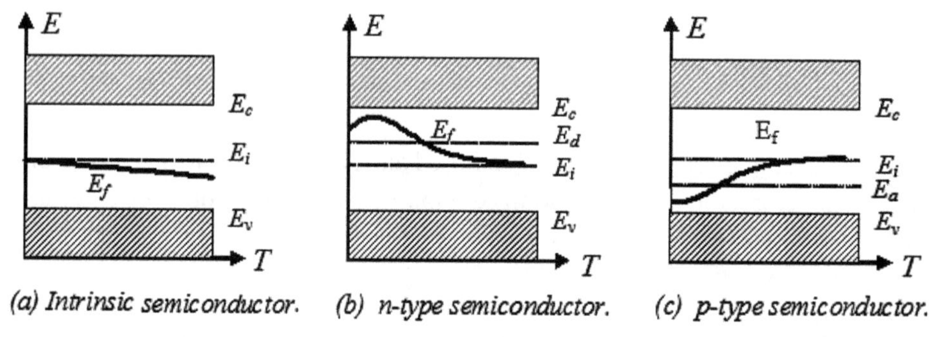

Figure 28. Variation of electron and hole concentration in semiconductors with temperature

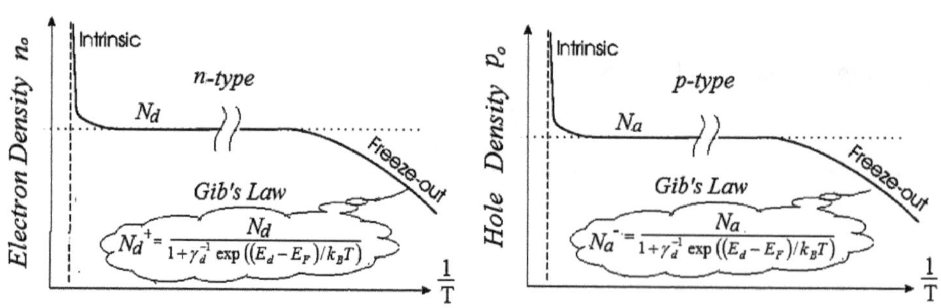

Figure 29. Schematic of the electrons scattering and the effect of electric field in a piece of semiconductor

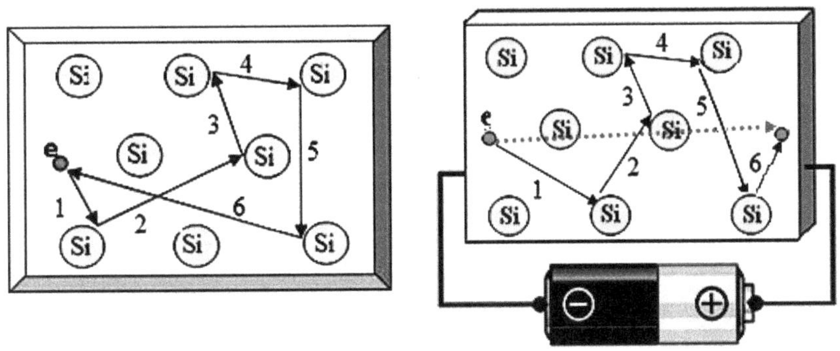

No net drift in thermal equilibrium. *Drift under non-equilibrium*

Thus, under no external bias the net current is zero. The thermal speed of charge carriers changes with temperature. For instance, the thermal speed of electrons and holes in a non-degenerate semiconductor is given by (Sze, 1981):

$$v_{th}(electrons) = \sqrt{\frac{3k_BT_L}{m_n^*}}, \quad v_{th}(holes) = \sqrt{\frac{3k_BT_L}{m_p^*}} \tag{49}$$

where T_L is the crystal lattice temperature, m_n* and m_p* are the effective masses of electrons and holes, respectively. At room temperature, the mean thermal velocity of electrons in a *Si* is about 10^7 cm/s.

The average distance, an electron travels between collisions, is called the *mean free path* (λ) and its order of magnitude is about *100*Å at room temperature. The average time between collisions (mean free timeτ) is related to the mean free path by the simple relation: $\lambda = v_{th}\,\tau$.

5. CLASSICAL TRANSPORT THEORIES (DRUDÉ MODEL)

The Drudé transport model is a classical transport model, which was initially introduced by Paul Drudé in 1900, to explain[9] the transport properties of electrons in metals (Drudé, 1900). This model assumes a free electron gas inside metals. The Drudé model, assumes independent and free particle approximation.

This model results from the application of the classical kinetic theory to electrons in a solid material. It assumes that the solid material contains immobile positive ions and an *electron gas* of classical non-interacting electrons, whose motion is damped by a frictional force, due to collisions with the ions.

If the electron density is n and their collisions with ions are characterized by a mean relaxation time τ then, under an applied electric field ζ, the conduction current density J is given by:

$$J = e^2\, n\, (\tau / m^*)\, \zeta = \sigma\zeta \tag{50}$$

where e and m^* are the electronic charge and effective mass inside the solid. The electrical conductivity σ is then given by:

Figure 30. Guided electron motion under the influence of electric field in a conductor

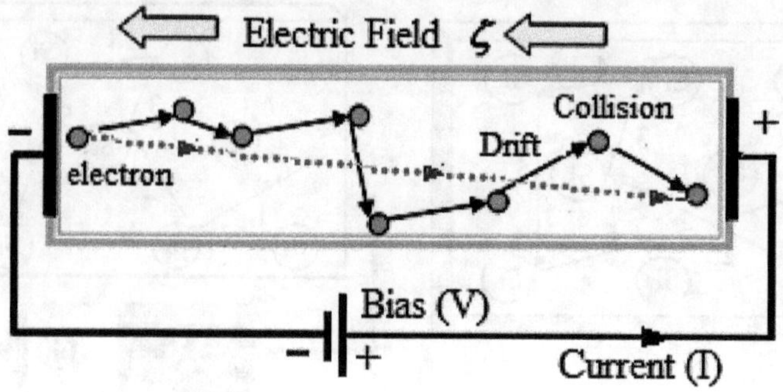

$$\sigma = e^2 n (\tau / m^*) \qquad (51)$$

This simple classical model leads to surprisingly good results which explain the electrical conductivity and some other physical phenomena, like the Hall Effect and the thermal conductivity in metals.

The Drudé model is based on the following classical equation of motion of free charge carriers (e.g., electrons in metals), under the effect of an external force $\boldsymbol{F}$ (electric or magnetic fields or both):

$$m^*(dv/dt) + m^*v/\tau = \boldsymbol{F} \qquad (52)$$

where $\boldsymbol{F}$ is the external force: $\boldsymbol{F} = \pm e (\boldsymbol{\zeta} + v_n \boldsymbol{x} \boldsymbol{B})$, and the ± sign stands for the type of charge carrier (- for electrons and + for holes). Also, τ is the mean free time between collisions and the medium term (m^*v/τ) represents the internal friction force, due to collisions. In steady state (where $dv/dt = 0$), the solution of this equation gives the average carrier velocity $\boldsymbol{v}$:

$$\boldsymbol{v} = \tau\, F / m^* = \pm (e\tau / m^*).(\boldsymbol{\zeta} + \boldsymbol{v}_n x\boldsymbol{B}) \qquad (53)$$

The charge carrier current density (flux) is given, by definition, as follows:

$$\boldsymbol{J} = \pm\, e\, n\boldsymbol{v} = e^2 n (\tau / m^*).(\boldsymbol{\zeta} + \boldsymbol{v}_n x\boldsymbol{B}) \qquad (54a)$$

or

$$\boldsymbol{J} = \sigma\boldsymbol{\zeta} = \sigma_o (\boldsymbol{\zeta} + \boldsymbol{v}_n x\boldsymbol{B}) \qquad (54b)$$

with the electrical conductivity: $\sigma_o = e^2 n (\tau / m^*)$.

For sinusoidal time-varying electric fields where $\zeta = \zeta_{o.} exp(jt)$, the electrical conductivity becomes complex and given by:

$$\sigma = \sigma_o / (1 - j\omega\tau). \qquad (55)$$

The Drudé theory has an unknown parameter (τ), which is the mean free relaxation time between collisions (of electrons with crystal lattice atoms and other collision centers). The problem is how to calculate this parameter? Actually, we don't know. But we can calculate it from the electrical conductivity measurements of different conductors, as listed in the following Table. Note in this table that the mean free time decreases with temperature in metals, where the electron collisions with crystal lattice vibrations (phonons) are dominant.

6. DRIFT MOBILITY OF CHARGE-CARRIERS IN SEMICONDUCTORS

It should be noted that the Drudé model was originally developed for metals. However, we may consider semiconductors as an extreme case, with both electron and holes as charge carriers, while considering the composite effect of different scattering mechanisms.

From the Drudé model, we can define the charge carrier drift velocity under the effect of electric field, as follows:

$$v_{drift} = \pm\,(e\tau / m^*).\zeta \tag{56}$$

Therefore, the steady-state carrier drift velocity is proportional to the electric field. The constant of proportionality between the carrier drift velocity v_{drift} and the electric field ζ is called the *carrier drift mobility* and termed by μ

$$v_{drift} = \pm\,\mu\zeta \text{ with } \mu = e\,\tau / m^* \tag{57}$$

As the effective mass of electrons and holes in a semiconductor are different, the mobility of electrons is not equal to the mobility of holes. The mobility of electrons is usually denoted by μ_n and the mobility of holes is denoted by μ_p. Therefore, we can write:

$$v_{n\,(drift)} = -\,\mu_n\zeta \text{ for electrons} \tag{58a}$$

Table 2. Mean free (relaxation) time from conductivity measurements (units of 10^{-14}s)

Element	τ at 77K	τ at 273K (0C)	τ at 373K (100∘C)
Cu	21	2.7	1.9
Ag	20	4	2.8
Au	12	3	2.1
Mg	6.7	1.1	0.74
Nb	2.1	0.42	0.33
Fe	3.2	0.24	0.14
Zn	2.4	0.49	0.34
Al	6.5	0.8	0.55
Pb	0.57	0.14	0.099

Source: Vapail, 1970.

$v_{p\,(drift)} = +\mu_p \zeta$ for holes (58b)

Figure 31 depicts the drift velocity of electrons and holes in Si at 300K versus electric field. As shown in figure, the drift velocity of electrons and holes is initially proportional to the applied field and then saturates at high electric fields to a saturation velocity v^{sat}. Therefore, the charge carrier drift mobility is constant at low fields ($\zeta < 10^3$ V/cm) and decreases at high fields. When the applied field is further increased, the carrier-drift mobility becomes field dependent and the drift velocity is approximately given by:

$$v_n = \frac{\mu_{no}\zeta}{1+\mu_{no}\zeta / v_n^{sat}}, \quad v_p = \frac{\mu_{po}\zeta}{1+\mu_{po}\zeta / v_p^{sat}} \tag{59}$$

where μ_{no} and μ_{po} are called the low-field mobility of electrons and holes. Also, v_n^{sat} and v_p^{sat} are the electron and hole saturation velocities. At high fields, the carrier drift velocity saturates to a maximum value.

It should be noted that the carrier saturation velocity is generally different from the carrier thermal velocity. Because of collisions with the semiconductor crystal lattice, charge carriers may loss part of their energy (phonon emission process) or gain additional energy (phonon absorption process). The collision with optical phonons becomes dominant at the onset of saturation regime and the carrier saturation velocity can be given by:

$$v_n^{sat} = \sqrt{\frac{2E_{op}}{m_n^*}}, \quad v_p^{sat} = \sqrt{\frac{2E_{op}}{m_p^*}} \tag{60}$$

Figure 31. Drift velocity of electrons and holes in Si at 300K versus electric field
After Arora, Hauser and Roulston (1982).

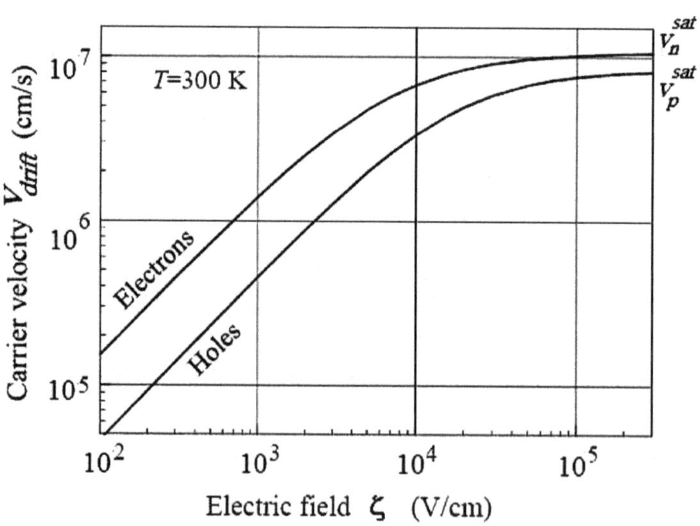

where E_{op} is the optical phonon energy (about *60* meV in *Si*). Generally speaking, the saturation velocity of charge carriers in semiconductors depends on the lattice temperature T_L and the semiconductor band structure. For instance, in *Si* at *300K* we've the saturation velocity of electrons $v_n^{sat} = 10^7 cm/s$ and that of holes v_p^{sat} is approximately $8x10^6$ cm/s.

At higher fields (in the order of 100 kV/cm), the energy gained from the field becomes greater than the optical phonon energy and the carrier drift velocity is already saturated. Hence, the drift energy (½ m^*v^2) reaches to its maximum value and any increase in carrier energy will no longer increase the carrier velocity, but rather increases the carrier temperature. In other words, the increase in carrier energy, at very high field, is not transformed into a drift energy ($1/2.m^*v^2$), but rather to a thermal energy ($3/2.k_BT$). Therefore, the charge carriers will be *hotter* than the lattice and therefore called *hot carriers*.

The behavior of electrons in some compound semiconductors, like *GaAs,* at high fields is different from *Si*. Specifically, the velocity-field curve of *GaAs* crystals reaches a peak value before saturation at a lower value of velocity. Therefore, there exists a region where the differential mobility (the slope of the velocity-field curve) is negative. According to Thornber (1980), the field dependent electron mobility in such materials can be modeled by the following camel-like form.

$$v_n(\zeta) = \frac{\mu_{no}\zeta + v_n^{sat}(\zeta/\zeta_C)^4}{1+(\zeta/\zeta_C)^2} \quad (61)$$

where ζ_c is the critical electric field. In *GaAs,* we may take $\zeta_c = 4\ kV/cm$. The carrier drift mobility is also dependent on both the impurity doping concentration (due to carrier scattering over impurity atoms) and the ambient temperature (due to carrier scattering over phonons) as shown in Figure 32.

As we have seen so far, the carrier current density is related to the carrier density and average velocity by the convection current relation (2). If the semiconductor is acted upon by an electric field ζ, then the carrier drift currents are given by the following relations:

$$J_{n(drift)} = -e\, nv_{n(drift)} \text{ for electrons} \quad (62a)$$

Figure 32. Schematic of the carrier drift mobility as function of temperature and doping

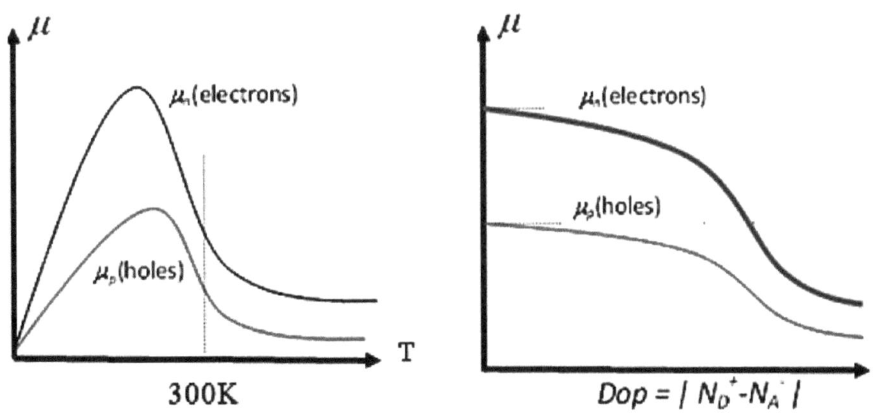

Table 3. Intrinsic drift mobility and saturation velocity of electrons & holes of some semiconductors (in their purest form) at 300K

Semiconductor	μ_{no} [cm²/Vs]	μ_{po} [cm²/Vs]	v_n^{sat} (Electrons) [cm/s]	v_p^{sat} (Holes) [cm/s]
Si	1500	450	10^7	$8.0\text{x}10^6$
Ge	3900	1900	$6.0\text{x}10^6$	$6.0\text{x}10^6$
GaP	110	7.5	-	-
3C-SiC	380	12.5	$2.5\text{x}10^7$	
C(diamond)	1800	1200	$2.7\text{x}10^7$	
GaAs	8500	400	$2\text{x}10^7$	
In Sb	80000	1250		

Source: Fundamentals of Semiconductors, Dargys & Kundrotas, 1994.

$$J_{p(drift)} = + e\, p v_{p(drift)} \text{ for holes} \tag{62b}$$

Substituting v_n *and* v_p from Equations (2) yields:

$$J_{n(drift)} = e\, n\, \mu_n \zeta = \sigma_n \zeta \tag{63a}$$

$$J_{p(drift)} = e\, p\, \mu_p \zeta = \sigma_p \zeta \tag{63b}$$

where $\sigma_n = en\mu_n$ and $\sigma_p = ep\mu_p$ are the electron and hole electrical conductivity, respectively.

The total drift current density is given by the sum of electron and hole drift current densities:

$$J_{drift} = J_{n(drift)} + J_{p(drift)} = \sigma\zeta \tag{64}$$

where $\sigma = \sigma_n + \sigma_p$ is the total conductivity of electrons and holes.

As shown in Figure 33, the current-field relation is linear at low fields as long as the mobility is constant, and saturates at high fields. Thus, the famous Ohm's law is satisfied in semiconductors at low electric fields. In fact the above relation is equivalent to the well-known Ohm's formula: $I = V/R$ where R is the sample resistance, V is the applied voltage and I is the resultant flowing current. The resistance of a semiconductor sample is related to the sample geometry and the resistivity $(\rho = 1/\sigma)$. Taking L and A as the length and cross section of a semiconductor rod, then its resistance R, at low fields, is simply given by: $R = \rho\, L/A$.

In a multi-valley semiconductor, the electrical conductivity is equal to the conductivity of individual valleys

$$\sigma = e\, \Sigma_i\, n_i\, \mu_i \tag{65}$$

Figure 33. Schematic of the drift current density - versus electric field in a semiconductor

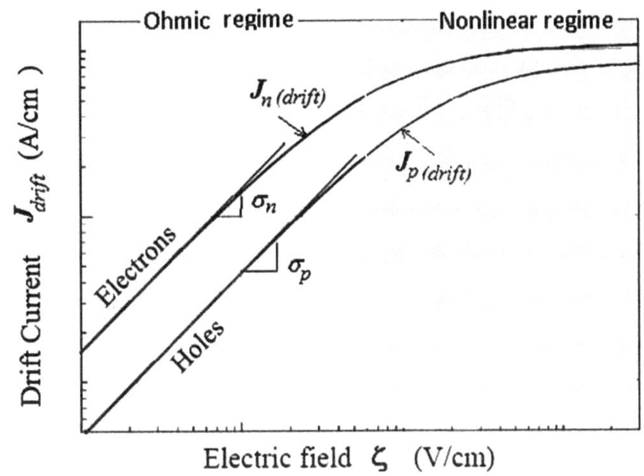

where n_k and μ_k are the charge carrier concentration and drift mobility of the kth valley of the semiconductor. Evidently, the total concentration of carriers is equal to the sum of carrier concentrations in each valley ($n=\Sigma\, n_k$).

Physical Note 2: Mobility and Conductivity Tensors

Generally speaking, the drift mobility is a 2nd order tensor (a 3x3 matrix), whose components depend on the direction of carrier drift.

$$\mu = \begin{bmatrix} \mu_{xx} & \mu_{xy} & \mu_{xz} \\ \mu_{yx} & \mu_{yy} & \mu_{yz} \\ \mu_{zx} & \mu_{zy} & \mu_{zz} \end{bmatrix}$$

The mobility tensor may be diagonalized (such that $\mu_{ij} = 0$ *for* $i\neq j$) by the appropriate choice of the *x, y, z* coordinates. On the other hand, the carrier conductivity ($\sigma_n = e\, n\, \mu_n$) is also a tensor such that:

$$\sigma = \begin{bmatrix} \sigma_{xx} & \sigma_{xy} & \sigma_{xz} \\ \sigma_{yx} & \sigma_{yy} & \sigma_{yz} \\ \sigma_{zx} & \sigma_{zy} & \sigma_{zz} \end{bmatrix}$$

Therefore, the electron drift current in the z-direction is given by:

$$J_{nz} = (\sigma_n \zeta)_z = en\mu_{zx}\zeta_x + en\mu_{zy}\zeta_y + en\mu_{zz}\zeta_z$$

It worth noting that the product of the conductivity tensor σ_n and the electric field $\boldsymbol{\zeta}$ is a vector. Thus, the electron current density $\boldsymbol{J}_n$ is a vector. The tensorial properties of the mobility and conductivity follow directly from the dependence of the conductivity on the reciprocal effective mass tensor.

It should be also noted that the effective mass, which is used in the calculation of the average electrical conductivity is called the conductivity effective mass.

The conductivity effective mass of electrons in a multi-valley semiconductor is given by:

$$m^*_{nc} = \frac{3}{\frac{1}{m_{xx}} + \frac{1}{m_{yy}} + \frac{1}{m_{zz}}} \tag{66a}$$

For instance, in *Si*, where $m_{xx} = m_{yy} = m_t = 0.19m_o$ and $m_{zz} = m_l = 0.98m_o$, the electron conductivity effective mass of electrons is hence given by:

$$m^*_{nc} = \frac{3}{\frac{2}{m_t} + \frac{1}{m_l}} = 0.26\, m_o \tag{66b}$$

However, the recent optical measurements show that $m_{nc} \approx 0.275m_o$ at T=300K (Rife, 2002). Also, the hole conductivity effective mass in *Si* is given by:

$$m^*_{pc} = \frac{m^{*\,3/2}_{lh} + m^{*\,3/2}_{hh}}{m^{*\,1/2}_{lh} + m^{*\,1/2}_{lh}} = 0.38\, m_o \tag{66c}$$

In thermal equilibrium, the average energy of electrons in the conduction band (or holes in the valence band) is given by:

$$w_o = \left\langle \tfrac{1}{2} m^* v^2(t) \right\rangle = \left\langle \tfrac{1}{2} m^* v^2_{th} \right\rangle = \tfrac{3}{2} k_B T_L \tag{67a}$$

where T_L is the lattice temperature. Therefore, the average carrier energy in a semiconductor in equilibrium is proportional to the semiconductor lattice temperature. This physical fact coincides with our definition of the lattice temperature (in Chapter 3), and its relation to the average energy of the gas of electrons in the heat bath of phonons, filling the lattice space.

When an electric field is applied on a semiconductor, the carrier drift velocity component has to be considered. Therefore, the average energy of electrons in the conduction band near equilibrium is given by:

$$w_n = \left\langle \tfrac{1}{2} m^*_n v^2(t) \right\rangle = \left\langle \tfrac{1}{2} m^*_n v^2_{th} \right\rangle + \tfrac{1}{2} m^*_n v^2_d = \tfrac{3}{2} k_B T_L + \tfrac{1}{2} m^*_n v^2_n \tag{67b}$$

Also, the average energy of holes in the valence band near equilibrium is:

$$w_p = \left\langle \tfrac{1}{2} m_p^* v^2(t) \right\rangle = \left\langle \tfrac{1}{2} m_p^* v_{th}^2 \right\rangle + \tfrac{1}{2} m_p^* v_d^2 = \tfrac{3}{2} k_B T_L + \tfrac{1}{2} m_p^* v_p^2 \tag{67c}$$

In the above equations, we assume that the gas of electrons is perfect and that its temperature is equal to the lattice temperature T_L. This assumption may be acceptable near equilibrium. However, at high electric fields, when the drift velocity reaches to its saturation value, and the electric field is further increased, the carrier temperature will be hotter than the crystal lattice.

7. DIFFUSION OF CHARGE CARRIERS

When there exists a spatial variation of carrier concentration (electrons or holes), the mobile charge carriers will move from the location of higher concentration to the location of lower concentration. This thermodynamic process is called the diffusion charge carriers.

7.1 Diffusion Current of Charge Carriers

The current resulting from the diffusion of electrons (or holes) is proportional to their concentration gradient (Elliot & Gibson, 1982) such that:

$$\boldsymbol{J}_{n,\,diffusion} = e\, D_n . \nabla n \tag{68a}$$

$$\boldsymbol{J}_{p,\,diffusion} = -e\, D_p . \nabla p \tag{68b}$$

where the constants of proportionality D_n and D_p are called the diffusion coefficients of electrons and holes, respectively. Also ∇ *is* the spatial gradient operator. Equation (68) is called the first Fick's law. It's a phenomenological description of the charge carrier diffusion process in semiconductors.

Note 3: Fick's Law

The diffusion coefficient of charge carriers is generally a second order tensor (3x3 matrix) whose components depend on the direction of diffusion.

$$D = \begin{bmatrix} D_{xx} & D_{xy} & D_{xz} \\ D_{yx} & D_{yy} & D_{yz} \\ D_{zx} & D_{zy} & D_{zz} \end{bmatrix}$$

Therefore, the electron diffusion current in the z-direction is given by first Fick's law:

Figure 34. Diffusion of electrons due to electron concentration gradient

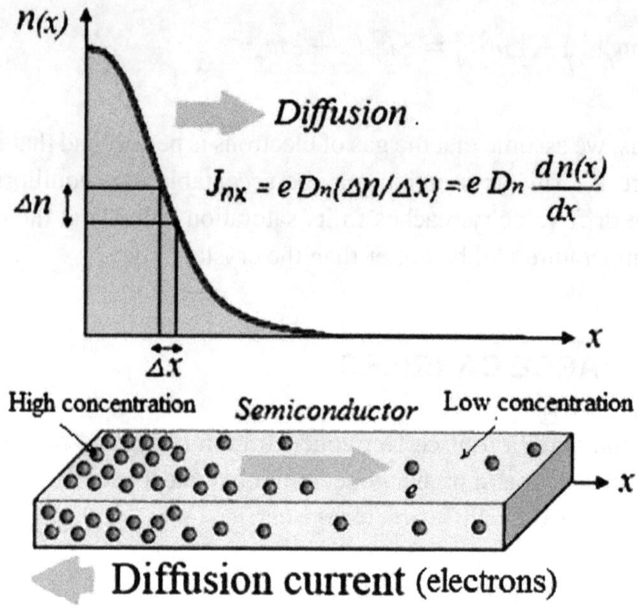

$$J_{nz} = eD_{zx}\frac{\partial n}{\partial x} + eD_{zy}\frac{\partial n}{\partial y} + eD_{zz}\frac{\partial n}{\partial z}$$

As we stated above, the Einstein relation is not valid at high fields (Nonlinear transport regime). A more precise definition of the diffusion coefficient is given by the following relation (in xy-plane):

$$D_{xy} = ½\, d/dt < (\mathbf{x} - <x>)(\mathbf{y} - <y>) >$$

where the angular brackets denote the statistical average. Also, the longitudinal diffusion coefficient (along the z-direction) is given by:

$$D_{zz} = ½\, d/dt < (\mathbf{z} - <z>)^2 >$$

According to Jacoboni & Regiani (1983), this may be further simplified as: $D_{zz} = ½t <(z - <z>)^2>$. We dully note that the diffusion coefficient tensor may be diagonalized (such that $D_{ij} = 0$ for $i \neq j$) by the appropriate choice of the x, y, z coordinates. The above relation is used in Monte Carlo, which is a statistical method for studying the carrier transport phenomena simulation (Fawcet, 1973; Cercignani, & Gabetta, 2007).

Note that the carrier diffusion coefficient is roughly equal to the product of the carrier mean-free path $\lambda = v_{th}\tau$ and the carrier thermal velocity v_{th}, such that $D = \lambda . v_{th}$. This is based on the fact that the carrier diffusion process is driven by the thermal energy. The diffusion constant in a semiconductor can be determined by the Haynes–Shockley experiment (Shockley, 1950). Alternatively, if the carrier mobility is known, the diffusion coefficient can be determined from the Einstein relation.

7.2 Einstein Relation

The Einstein relation expresses the ratio of drift mobility to diffusion coefficient, near equilibrium. In this condition, the carrier current density is almost null and the drift and diffusion currents are equal. Therefore, the carrier diffusion coefficient is proportional to the carrier drift mobility as follows (Ashcroft & Mermin, 1976):

$$D_n/\mu_n = e\,n\,/\,[dn\,/\,dE_F\,] \tag{69a}$$

Substituting the density of carriers at equilibrium ($n \approx n_o$) by the appropriate relation, results in the generalized Einstein relations for carriers. For instance, in the bulk of a 3D semiconductor, we can substitute (22) into (69a) to get:

$$D_n/\mu_n = V_T\,[2F_{1/2}(\eta_n)/F_{-1/2}(\eta_n)], \tag{69b}$$

$$D_p/\mu_p = V_T\,[2F_{1/2}(\eta_p)/F_{-1/2}(\eta_p)] \tag{69c}$$

where $V_T = (k_B\,T_L/e)$ is the thermal voltage, $F_{1/2}$ and $F_{-1/2}$ are the Fermi integrals and the arguments $\eta_n = (E_F-E_c)/k_BT_L$ and $\eta_p = (E_V - E_F)/k_BT_L$. The Fermi integral, $F_{1/2}(x)$, has been already defined by Equation (23). The definition of the Fermi integral, $F_{-1/2}(x)$ is similar to that of $F_{1/2}$, with $x^{½}$ is replaced with $x^{-½}$. The above relations are valid near the state of equilibrium, i.e., for small external perturbations (electric fields). Therefore, the Einstein relation is useful when the total current is very small compared with either drift or diffusion components.

For non-degenerate semiconductors, these relations can be reduced to the well-known Einstein relations (Ashcroft & Mermin, 1976):

$$D_n\,/\,\mu_n = D_p\,/\,\mu_p = V_T \tag{70}$$

Some authors claim that this relation is also valid for degenerate gasses in metals and other materials (Palenskis, 2013). However, the above relation may be simply derived from the phenomenological relations $D = \lambda v_{th}$ and $\mu_p = e\tau_p/m$, such that $D/\mu = mv_{th}^2/e = k_BT_L/e$, where we substituted $½\,mv_{th}^2 = \langle E\rangle = ½\,k_BT_L$.

7.3 Drift-Diffusion Model of Semiconductors

When both carrier-concentration gradient and electric field are present in a semiconductor, the resultant convection current is the sum of carrier-drift and carrier-diffusion currents.

$$\boldsymbol{J}_n = \boldsymbol{J}_{n,\,drift} + \boldsymbol{J}_{n,\,diffusion} = en\,\mu_n\zeta + e\,D_n\nabla\,n \tag{71a}$$

$$\boldsymbol{J}_p = \boldsymbol{J}_{p,\,drift} + \boldsymbol{J}_{p,\,diffusion} = e\,p\,\mu_p\zeta - e\,D_p\nabla p \tag{71b}$$

In fact, the currents in semiconductors are composed of two components, namely, the convection currents (J_n, J_p) due to moving carriers, and the displacement current due to time-varying fields. Hence, the total current density is equal to the sum of electron and hole currents, J_n and J_p, in addition to the displacement current J_d when the electric field is time variant.

$$J = J_n + J_{p,} + J_{d,} \tag{72a}$$

$$J_d = \varepsilon . \partial\zeta/\partial t \tag{72b}$$

where ε is the electrical permittivity of the semiconductor. The above current equations are the basis of the so-called drift-diffusion model (DDM) of semiconductors. This model is based on the assumption that all carriers have the same temperature as the crystal lattice, which contains them. Therefore, the drift-diffusion is sometimes called the isothermal transport model. The drift-diffusion model is only valid for small perturbations (for electric fields up to 10^4 V/cm in silicon). At high fields, carriers can acquire high energies and their temperature may be different from the lattice temperature.

8. THERMAL DIFFUSION AND THERMAL CONDUCTIVITY OF CHARGE CARRIERS

In our previous discussion, we assumed that the temperature is constant overall the semiconductor crystal. In the presence of a spatial temperature gradient, ∇T_L, across the semiconductor lattice, the charge carriers diffuse from hot to cold lattice points. This thermal diffusion mechanism results in an additional current component; called the thermal diffusion current. For electrons, the thermal diffusion current density is given by (Elliot & Gibson, 1982):

$$J_{n\,(thermal\ diffusion)} = e\, D_n^{th}\, \nabla T_L \tag{73a}$$

where D_n^{th} is the thermal diffusion coefficient of electrons. Similarly for holes, the thermal diffusion current is given by:

$$J_{p\,(thermal\ diffusion)} = -\, e\, D_p^{th}\, \nabla T_L \tag{73b}$$

where D_p^{th} is the thermal diffusion coefficient of holes. Therefore, in the presence of temperature gradients, the total electron current density becomes:

$$J_n = e\, n\, \mu_n \zeta + e\, D_n\, \nabla n + e\, D_n^{th}\, \nabla T_L \tag{74a}$$

Similarly, the total hole current density becomes:

$$J_p = e\, p\, \mu_n \zeta - e\, D_p\, \nabla p - e\, D_p^{th}\, \nabla T_L \tag{74b}$$

The thermal diffusion coefficient of charge carriers in a semiconductor are related to the thermal conductivity of the material (k^{th}). The thermal conductivity of the semiconductor is also related to the thermal conductivity due to charge carriers (k_n^{th}, k_p^{th}). In fact, the carrier thermal conductivity is a very small fraction of the material thermal conductivity k^{th}, in semiconductors. This is because the thermal energy is predominantly transported by lattice vibrations (phonons) in semiconductors. However, in metals the electron thermal conductivity is dominant. Generally, the material thermal conductivity may be expressed as follows:

$$k^{th} = k_1^{th}\,(phonons) + k_{n,p}^{th}\,(carriers) + k_3^{th}\,(\text{mixed conduction}) \tag{75}$$

8.1 Thermal Transport

The following figure depicts the shift of the Fermi surface (constant energy surface) caused by electrical conduction (with current density ***J***) and thermal conduction (with heat flux ***Q***). In electrical conduction, the Fermi surface is shifted by a small amount corresponding to the drift velocity due to the electric field. Therefore, an excess of electrons travel towards the right. In thermal conduction, hot electrons head right and cool electrons head left. The long arrows in the figure indicate phonon scattering events, which hinders the electrical and thermal currents. The short arrows represent the phonon scattering events at low temperatures. Note that the phonon scattering reduces the thermal current by warming up cold electrons and cooling hot electrons

8.2 Wiedemann-Franz Law

In 1853, two German physicists, Gustav Wiedemann and Rudolf Franz, found that the ratio of thermal and electrical conductivity for all METALS is constant ($k^{th}/\sigma = L$) at a given temperature (at room tem-

Figure 35. Representation of the alteration of the Fermi surface caused by electrical conduction (a) Current density J and thermal conduction (b) Heat flux Q; black circles represent filled states and white circles empty states.
After Vapaille (1970).

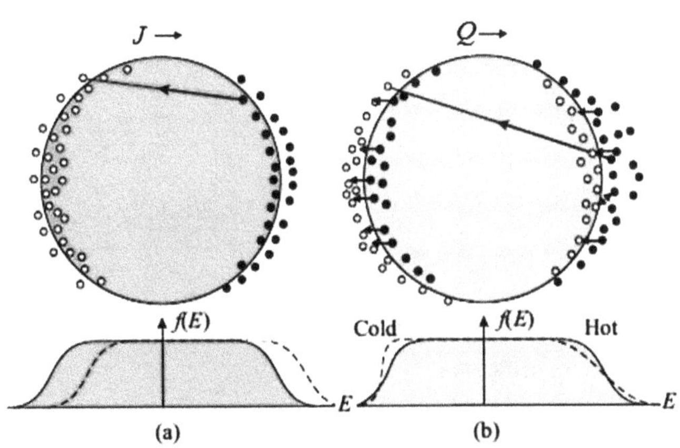

perature and above). Later it was found by Hendrik Lorenz that this constant is proportional to the temperature ($k^{th}/\sigma = L.T$). The origin of this empirical observation did not become clear until the discovery of the electron and the advent of quantum theory. A moving electron must carry both heat and charge: that is why the ratio does not vary from metal to metal. Therefore, thermal and electrical conductivity of charge carriers may be also related by the Wiedemann-Franz law, as follows:

$$k_n^{th} = \gamma_n (k_B/e)^2 \sigma_n T_n,\ k_p^{th} = \gamma_p (k_B/e)^2 \sigma_p T_p \tag{76}$$

where γ_n and γ_p are the Lorentz coefficients for electrons and holes. Also, T_n and T_p are called the electron and hole temperatures, which are equal to the lattice temperature near equilibrium. In metals, where the thermal conductivity is mainly due to electrons ($k^{th} \approx k_{n,}^{th}$) and the electron temperature can be replaced by the lattice temperature ($T_n = T_L$), the Lorentz coefficient is constant and given by the Somerfield value:

$$\gamma_n\,(metals) = (\pi^2/3) \tag{77}$$

In this case, the Wiedemann-Franz law takes the conventional form:

$$k_n^{th} / \sigma_n = (\pi^2/3).(k_B/e)^2 \,.\, T_L = L.\,T_L \tag{78}$$

where $L \approx 2.44x10^{-8}$ Watt Ω K^{-2} is the Lorentz number (for metals). For metals, the Wiedemann-Franz law is in good agreement with experiment, regardless of the energy band structure, as shown in the Figure 36.

Figure 36. (a) The Wiedemann-Franz law in metals and alloys; (b) variation of the Lorentz number with Fermi level
After Rowe (1995).

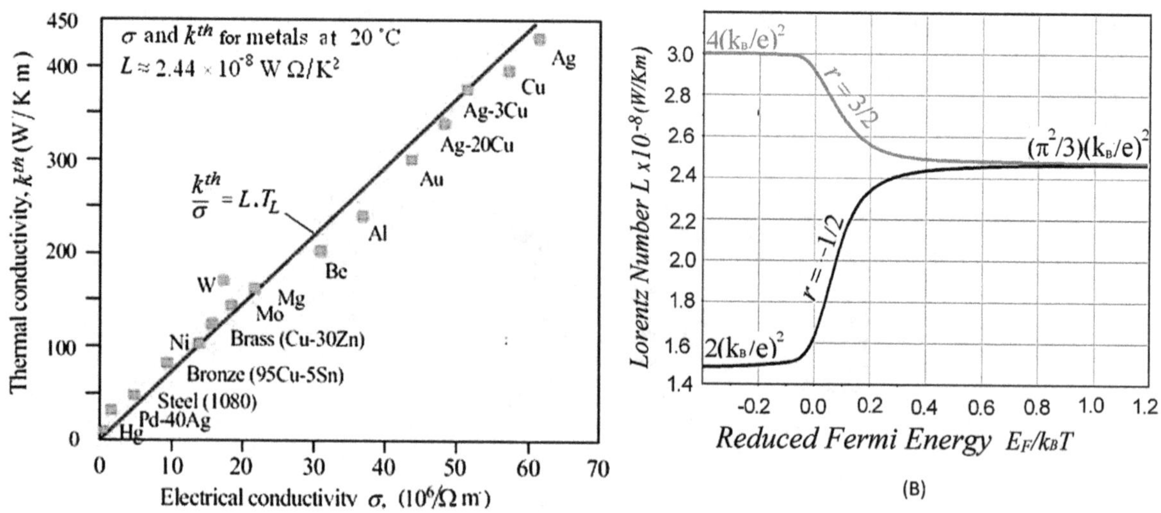

(A)

Nevertheless, for semiconductors, the Wiedemann-Franz coefficient is strongly correlated to the dominant collision mechanism and its dependence on the carrier energy. Therefore, the Lorentz coefficient may be expressed as follows:

γ_n *(semiconductor)* $= 5/2 + r_n$,

where r_n is called the scattering parameter. Figure 36b illustrates the variation of Lorentz number with Fermi level, as derived from the Fermi-Dirac statistics for two electron scattering mechanisms (ionized impurities scattering parameter $r_n = 3/2$ and acoustic phonons r_n=-1/2).

8.3 Heat Flow Equation

Semiconductors generate heat during use. This heat must be removed in order to prevent further heating and consequent destruction of the semiconductor. In this section we briefly describe the nature of heat transfer through semiconductors. Heat transfer refers to the movement of heat from one point to the other due to temperature difference. Generally speaking, the heat energy is transported in matter by either of three methods, namely:

- Conduction,
- Convection,
- Radiation.

Conduction is a method of heat transfer in which heat is exchanged between two elements that are situated close to one another or in still fluids or gaseous materials. Convection refers to heat transfer in a liquid or gas by the circulation or flow of the liquid or gas from one region to another. Radiation refers to heat exchange through electromagnetic heat exchange in the form of waves and rays. Radiation can also occur in a vacuum. Heat transfer through radiation increases as the fourth power of temperature and is only relevant for high temperatures (Steffan-Boltzmann law). In solids, the heat is mainly transported by conduction. The lattice heat flux $\boldsymbol{Q}_L$ is defined as heat energy per unit area per unit time (W/cm^2). It can be expressed by the classical Fourier relation:

$$\boldsymbol{Q}_L = - k^{th} \nabla T_L \tag{79}$$

where k^{th} is the total thermal conductivity of the material and T_L is the lattice temperature. Neglecting recombination losses and the Joule heating effects due to external-fields, the heat continuity equation is as follows:

$$\rho_L c_v \partial T_L/\partial t + \nabla.\boldsymbol{Q}_L = 0 \tag{80a}$$

where c_v is the lattice specific heat and ρ_L is the lattice density. Substituting the lattice heat flux $\boldsymbol{Q}_L$ = $-k^{th}\nabla T_L$, yields:

$$\rho_L c_v \partial T/\partial t - \nabla .(k^{th}\nabla T_L) = 0 \tag{80b}$$

When the Joule heating effects due to external electric fields $\boldsymbol{J}.\zeta$ and cooling due to net recombination-generation mechanisms $(R\text{-}G)$ are present, the heat continuity equation becomes:

$$\rho_L c_v \,\partial T/\partial t - \nabla .(k^{th} \nabla T_L) = H_s \tag{81}$$

The heat flux source term, H_s, may be modeled as follows:

$$H_s = \boldsymbol{J}.\,\zeta - e\, E_g\,(R - G) \tag{82}$$

Note that the energy lost in the recombination process should be considered. This energy may be converted to heat or light. In addition to the Joule heating term and the recombination cooling effect, one can also consider the heat radiation from the lattice and the relaxation of the lattice temperature.

9. THERMOELECTRIC EFFECTS

There exist a further number of relations between the thermal transport and electrical transport in conductors and semiconductors. We present in the following subsections a phenomenological summary of the so-called thermoelectric effects. We assume here that the temperature of carriers is equal to the temperature of the lattice. This approximation is acceptable at small electric fields near equilibrium. At such small fields, the transport is linear and one can make use of the relaxation time approximation. The Figure 37 depicts the most famous thermoelectric effects, namely the Seebeck effect and the Peltier effect.

9.1 Seebeck Effect

The Seebeck effect (Seebeck, 1822) refers to the built-in electrical potential which is induced as a result of applying a temperature gradient across a junction of two different conductors, while no current passing between them. The device that works on this principle is known as the thermocouple. When no current is passing ($J_n=0$) an electric field should appear across the junction to counter balance the effect of the temperature gradient. This induced field is proportional to the temperature gradient, such that:

Figure 37. Schematic illustration of the Seebeck and the Peltier thermoelectric effects metals

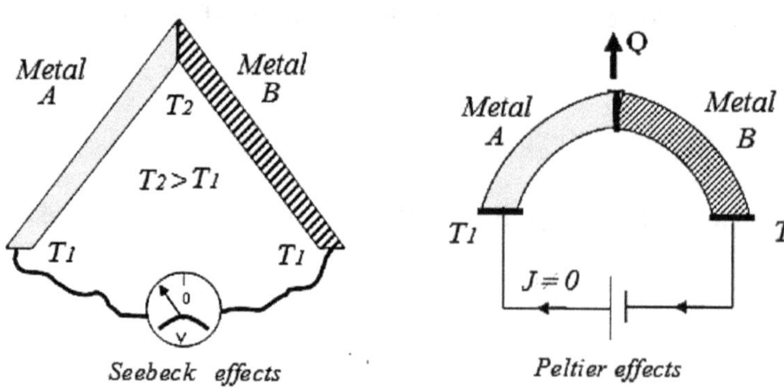

$$\zeta = - P^{th} \nabla T \tag{83}$$

where the proportionality factor P^{th} is called thermoelectric power or the Seebeck coefficient (in V/°C). The thermal voltage ΔV which is induced between two different conductors, due to a temperature difference is given by:

$$\Delta V = V_A - V_B = - \int \zeta \, dx = - \int P^{th} \, dT \tag{84}$$

The thermal voltage may be also produced across a sample of semiconductor whose two ends are kept at different temperatures, as shown in Figure 38. It should be noted that the thermoelectric power is negative for metals and n-type semiconductors and positive for p-type semiconductor. In fact, the silicon emerged as a promising material for thermoelectric devices due to its high thermoelectric power. Besides its larger Seebeck coefficient than any metal, its thermoelectric properties can be adjusted by altering its impurity concentration.

The Seebeck coefficient represents the entropy per charge carrier in the solid state and can be exploited to distinguish different materials. The thermoelectric *figure of merit* of a material is defined as follows (Rowe, 1995):

$$ZT = (P^{th})^2 \sigma T / (k_l^{th} + k_n^{th}) \tag{85}$$

Figure 38. Schematic illustration of the Seebeck thermoelectric in a semiconductor

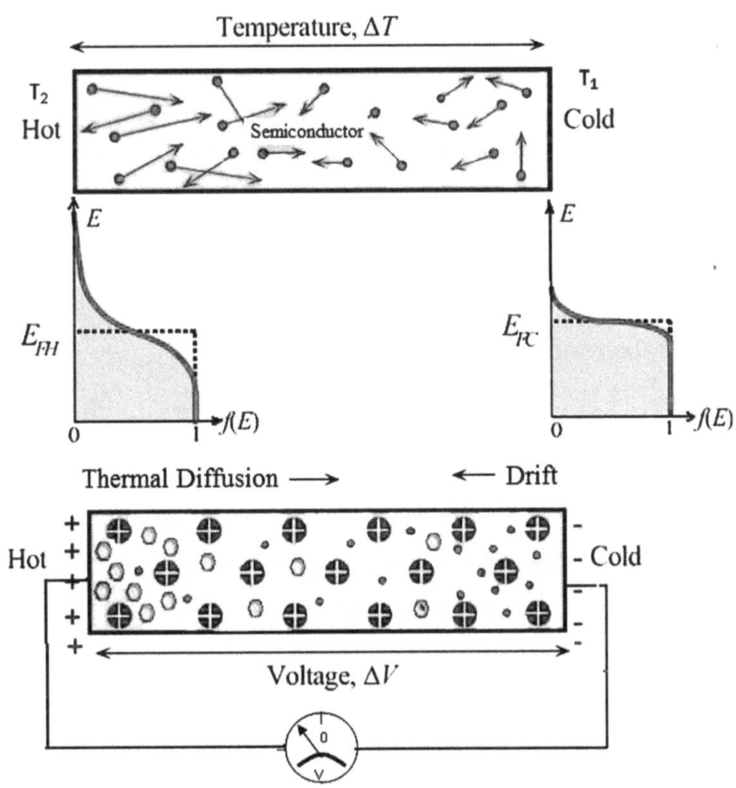

where P^{th} is the Seebeck coefficient, σ is the electrical conductivity, T is the absolute temperature, and k_l^{th} and k_n^{th} are the lattice (phonons) and electronic components of thermal conductivity, respectively. The ZT of a material has to be as high as possible in order to be used efficiently in thermoelectric applications. The Figure 39 depicts thermoelectric figure of merit (ZT) of different semiconductors.

9.2 Peltier Effect

The French physicist Peltier (Peltier, 1883) is a thermoelectric effect opposite to the Seebeck effect. The Peltier effect refers to the absorption of heat at the junction between two different materials, when a current passes between them while the temperature is maintained constant. The Peltier coefficient of a junction between two elements (say A and B) is defined as the ratio of absorbed heat flux to the passing current:

$$\pi_{AB} = Q/(e\,J) = T.\,P^{th} \qquad (86)$$

where $P^{th} = P^{th}_A - P^{th}_B$ is the thermoelectric power of the junction and T is the junction temperature. Note that the Peltier coefficient has the dimension of Volts, as shown in Figure 40. The above relation is sometimes called the Kelvin first law (Kittle & Kroemer, 1980). The so-called Peltier heat pumps are p-n junction cooling cells, which are implemented using different semiconductor materials (e.g., *PbTe, InSb, PbSn or Au-doped Ge or Si*). Figure 40 depicts the Peltier heat pump. Such a cell can be used for cooling down from –100°C to –200°C, depending on the type of used material and geometry. For obtaining more cooling power, several cells can be stacked together as shown in figure.

9.3 Thomson Effect

In 1857, Thomson discovered that heat is either absorbed or released in a conductor when a current passes from a material at one temperature to the same material at different temperature. The Thomson effect also refers to the absorption of heat at the junction of two different materials, when a current passes while a temperature gradient exists across them. Thus, the Thomson effect takes place in a thermoelectric circuit whether the current in the circuit is applied externally or developed internally due to

Figure 39. Thermoelectric figure of merit of different semiconductors
After Rowe (1995).

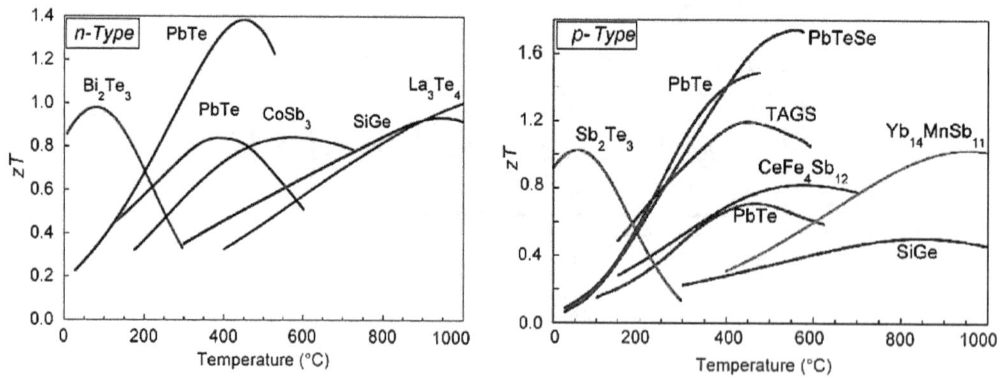

Figure 40. (a). The Peltier effect as a function of temperature, in both n-type and p-type silicon. The doping concentration is $10^{14}cm^{-3}$. (b) Peltier element and Peltier heat pump
After Vapaille (1970).

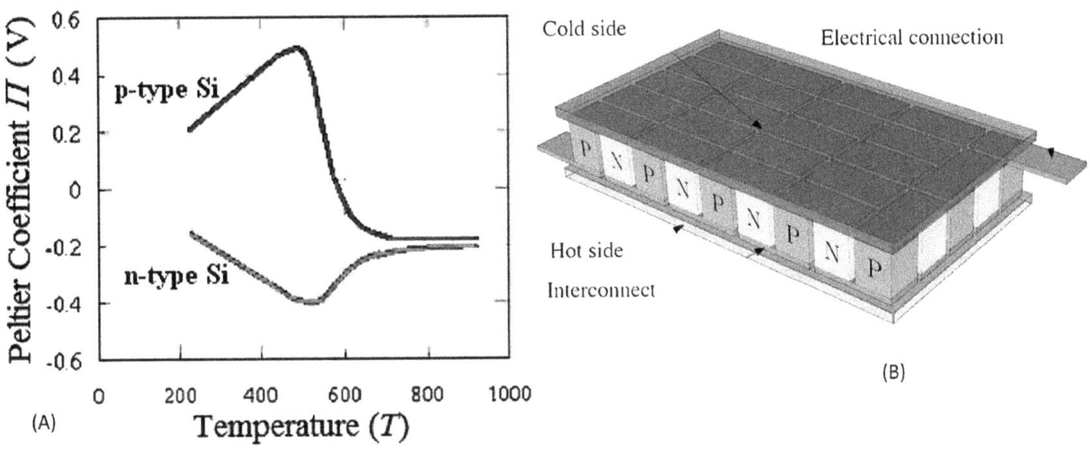

the Seebeck effect. The Thomson coefficient τ_{AB} is defined as the absorbed heat flux per unit current per unit temperature gradient:

$$\tau_{AB} = Q_n /(e\, J_n \nabla T) = T\, .\, (d\, P^{th} /dT) \tag{87a}$$

The above relation can be also put in the following form:

$$\tau_{AB} = (P^{th}_{A} - P^{th}_{B}) \,/\, ln\, (T_1/T_2) \tag{87b}$$

The Thomson coefficient is also referred as the specific heat of electricity because of its analogy with the thermodynamic specific heat. The thermo-dynamic specific heat represents the amount of heat transfer per unit temperature difference per unit mass; and the Thomson coefficient represents amount of heat absorbed per unit temperature difference per unit current.

Figure 41. Schematic illustration of the Thomson thermoelectric effect

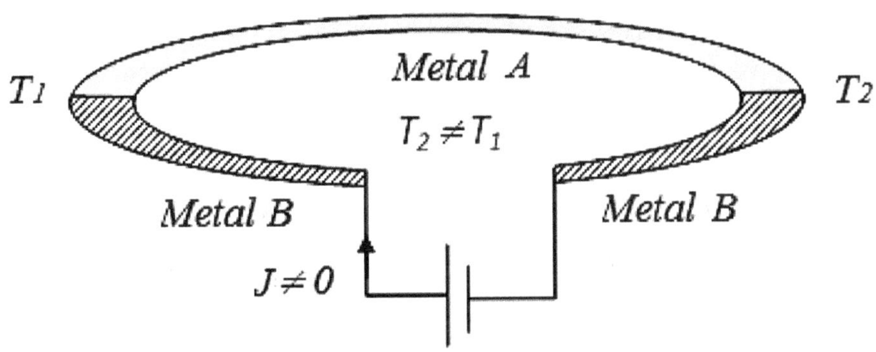

9.4 Joule Effect

The Joule effect is an irreversible thermoelectric effect that causes internal heating of a current carrying conductor. Unlike the Thomson effect, heat due to the Joule effect is always absorbed in a conductor irrespective of the direction of the current and the presence of temperature gradient. The magnitude of Joule heating depends on the current and the resistivity of the conductor. The rate of the internal heat generation in a conductor due to the Joule effect is given as: $Q_j = I^2R$ where I is the current passing through the conductor and R is the resistance of the conductor

10. EFFECT OF MAGNETIC FIELDS ON CHARGE CARRIER TRANSPORT

When electrons are drifted in a magnetic field B, they are subjected to a Lorenz force, which is given by:

$$\boldsymbol{F}_n = - e v_n \boldsymbol{x} \boldsymbol{B} \tag{88}$$

When a charge carrier has a component of velocity perpendicular to a strong magnetic field $\boldsymbol{B}$, it circles the field at the cyclotron frequency

$$\omega_c = e\, B/m^* \tag{89}$$

Figure 42 shows the circular (cyclotron) motion of electrons, under the effect of magnetic field. For semiconductors with laboratory magnetic fields, the cyclotron frequency is typically in the terahertz (THz) range. For example, in n-type *GaAs*, the cyclotron frequency varies about 0.42 THz per Tesla.

In the Drudé classical description of the cyclotron resonance, one can show that, if the semiconductor material is cubic and a static magnetic field $\boldsymbol{B}$ is directed along the material z-azis, then the conductivity tensor is given by:

Figure 42. Schematic illustration of the circular (cyclotron) motion of electrons, under the effect of strong magnetic field

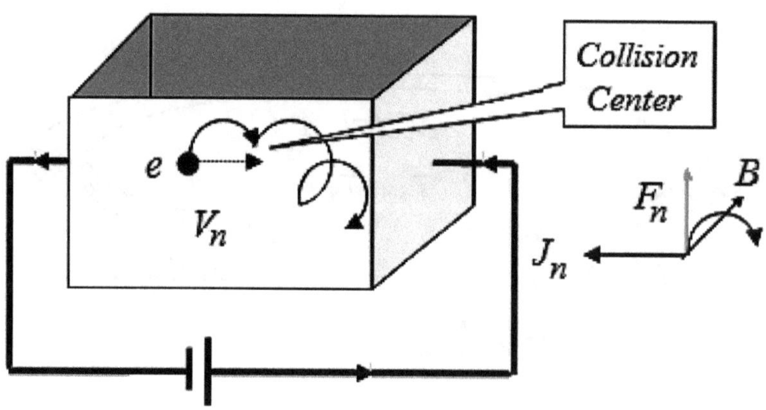

$$\sigma = \begin{bmatrix} \sigma_{xx} & \sigma_{xy} & 0 \\ \sigma_{yx} & \sigma_{yy} & 0 \\ 0 & 0 & \sigma_{zz} \end{bmatrix} \tag{90}$$

where

$$\sigma_{xx} = \sigma_{yy} = \sigma_o . \frac{j\omega\tau + 1}{(j\omega\tau + 1)^2 + \omega_c^2 \tau^2} \tag{91a}$$

$$\sigma_{xy} = -\sigma_{yx} = \sigma_o . \frac{\omega_c \tau}{(j\omega\tau + 1)^2 + \omega_c^2 \tau^2} \tag{91b}$$

$$\sigma_{zz} = \sigma_o . \frac{1}{(j\omega\tau + 1)} \tag{91c}$$

$$\sigma_o = en\mu_n = \frac{e^2 n\tau}{m_n^*} \tag{91d}$$

Here, $\omega_c = e\ B/m^*$ is the cyclotron resonance angular frequency and the applied electric field is a varying electric field $\zeta(t) = \zeta_o \exp(-j\omega\ t)$.

10.1 Hall Effect

The Hall Effect was discovered by Edwin Hall in 1879, during the preparation of his doctoral thesis in Physics (Ioffe, 1957). The phenomena can be summarized as follows. It is well known that a charged particle moving in a magnetic field feels a Lorentz force perpendicular to its direction of current (***J***) and the magnetic field (***B***). As a direct consequence of this force, charge carriers will accumulate to one side of the material in a perpendicular direction to both electric current and magnetic field. This effect is shown in Figure 43; it is called the Hall Effect. The Hall coefficient is given by the following expression (at low magnetic fields):

$$\underline{R}_H = \frac{\varsigma_H}{J.B} = \frac{p\mu_p^2 - n\mu_n^2}{e(p\mu_p + n\mu_n)^2} \tag{92a}$$

Figure 43. Schematic illustration of the Hall Effect in a semiconductor

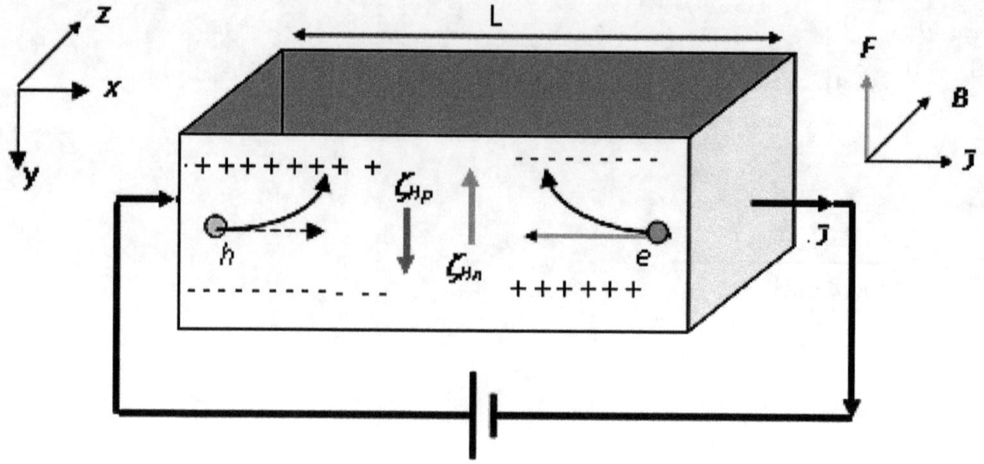

where the total Hall field ζ_H is the sum of individual field components due to deflection of electrons and holes ($\zeta_H = \zeta_{Hp} - \zeta_{Hn}$). Evidently, the above relation reduces to $\underline{R}_H = -1/en$ for n-type materials when $n>>p$ and reduces to $\underline{R}_H = 1/ep$ for p-type materials when $p>>n$. The voltage drop at right angles to the current is called the Hall voltage (V_H). Also, the current divided by Hall voltage is called the Hall conductance.

In the above calculations of $\underline{R}_H$, the electron collisions are not considered. A more elaborate formula for the Hall coefficient can be obtained by solving the equation of motion of charge carriers, while considering an energy-dependent carrier relaxation time (which is closely related to the mean free time of the Drudé model). This results in a similar expression of $\underline{R}_H$,

$$\underline{R}_H = \underline{r}.\frac{p\mu_p^2 - n\mu_n^2}{e(p\mu_p + n\mu_n)^2} \tag{92b}$$

where $\underline{r}$ is called the Hall scattering parameter. It can be obtained by averaging the relaxation time between dominant collisions ($\underline{r} = <\tau^2>/<\tau>^2$). In Si at 300K, $\underline{r}=1.18$ for electron-phonon scattering and $\underline{r}=1.93$ for electron-impurity scattering (at low and moderate magnetic fields).

10.2 Magnetoresistance

Magnetoresistance is a phenomenon where the application of a static magnetic field changes the resistance of a material. The phenomenon has been known for many years in ordinary metals, and is due to the conduction electrons being forced to move in helical trajectories about an applied magnetic field. The effect becomes evident only when the magnetic field is strong enough to curve the electron trajectory within a length equal to its mean free path. The magnetoresistance coefficient (M) is defined as the relative change of resistivity in presence of a strong magnetic field.

$$M = (\rho - \rho_o) / \rho_o = \Delta\rho / \rho_o \tag{93}$$

where ρ_o is semiconductor resistivity at zero magnetic field. M is dependent on the energy band structure and crystal orientation of the semiconductor material.

For n-type isotropic materials ($n>>p$), if the magnetic field $\boldsymbol{B}$ is directed normal to current flow (along z-axis), we have:

$$M\,(n\text{-}type) = \Delta\rho / \rho_o \approx \mu_n^2 B_z^2 \tag{94a}$$

Similarly, for p-type isotropic materials (p >> n), we have:

$$M\,(p\text{-}type) \approx \mu_p^2 B_z^2. \tag{94b}$$

The magnetoresistance effect occurs in metals only at very high magnetic fields and low temperatures. For example, in pure copper at 4K a field of 10 T produces a factor of 10x change in the resistance.

10.3 Other Magnetic Field Effects

In addition to the Hall Effect and magnetoresistance, there exist so many other magnetic field effects in semiconductors and nanostructures. For instance, the giant magnetoresistance (GMR) is a large change in resistance that is observed in ferromagnetic/ paramagnetic multilayer structures. Also, the quantum Hall effect (QHE) is a manifestation of collective behavior of electrons in 2-dimensions under the effect of magnetic field. In this effect, the Hall resistivity exhibits plateaus for integer values of h/e^2 (Ezawa, 2008). The Figure 44 depicts the so-called integer quantum Hall effect. This effect can be understood as follows. After applying a magnetic field, the energy states of electrons in the solid contract into separate levels, called Landau levels. As shown in Figure 44, the Hall resistivity is quantized at certain integer filling factors, which are equal to the number of filled Landau levels. This phenomenon is only observ-

Figure 44. Illustration of the quantum Hall Effect. The ratio $R_k = h/e^2 = 25.812807\ k\Omega$.

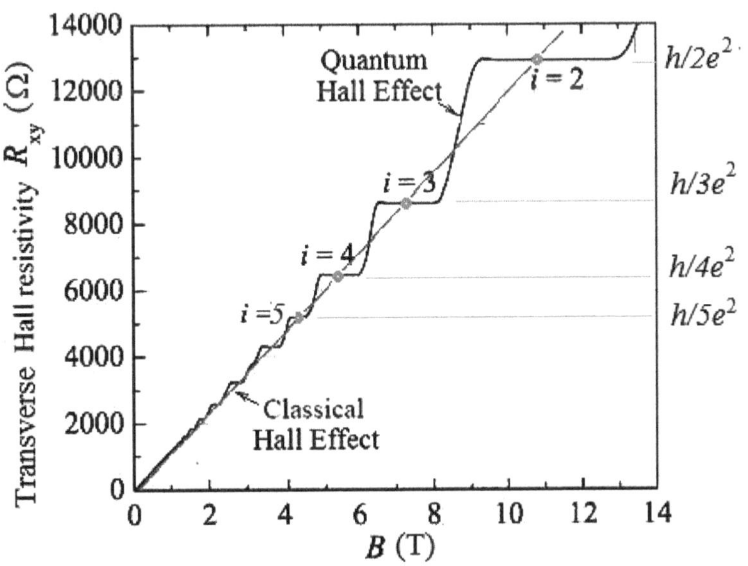

able on two-dimensional electron gas (2DEG) systems like the hetero-junctions or the inversion layer of MOSFET devices (Yoshioka, 2002).

Of course there exist other magnetic effects in semiconductors and nanostructures, which may be combined with thermal or optical excitations. For instance, the Nernst Effect, Ettinghausen Effect and Righi-Leduc Effect are well-known thermomagnetic effects, which happen when the semiconductor is subjected to temperature gradients and magnetic fields (Koviskii, 1962). Also, the Faraday orientation, Voigt orientation, Helicon waves and magetoplasma resonance are all magneto-optical effects (Elliot & Gibson, 1982). Some of these effects may be described, within the framework of semiclassical theory and others need a quantum mechanical explanation. Therefore, we review such effects, within the corresponding transport theories, in the next chapters. However, for the matter of completeness, we mention here the effect of magnetic field on the energy band structure, due to which many quantum effects have their origin. In a certain direction of the k-space, say k_z, the E-k relation becomes as follows:

$$E = E_n = \mathrm{E_c} + \frac{\hbar^2 k_z^2}{2m_z^*} + \hbar\omega_c(\mathrm{n} + \tfrac{1}{2}) \tag{95}$$

where $\omega_c = eB_z/m_z^*$ is the cyclotron frequency, B_z is the component of magnetic field along the z-axis of the lattice, m_z^* is the component of effective mass in the z-direction and E_c is a constant (energy reference). We note here an additional quantized term, in comparison with the simple parabolic E-k relation $E=E_c + \hbar^2k^2/2m^*$, with no applied magnetic field. We note also that the additional term $(\tfrac{1}{2}\hbar\omega_c)$ is only significant for strong magnetic fields, where ω_c is in range of microwave frequencies or higher. Therefore, the simple parabolic E-k relation of the quasi-free electrons is transformed into a family of quantized parabolas (sub-bands), in the presence of a strong magnetic field. The quantized energy levels described by (95) are called Landau levels.

Figure 45. Illustration of the Landau levels and effect of magnetic field on density of states

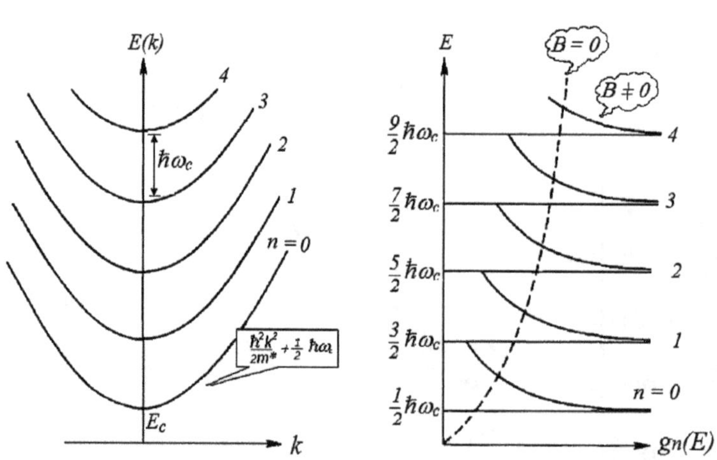

11. ONSAGER RELATIONS

So far we reviewed the effects of electric and magnetic fields on carrier transport. Now, we summarize the semiconductor parameters, under the simultaneous presence of electric and magnetic fields as well as temperature gradients. We assume the case where the semiconductor is homogeneous but anisotropic so that its transport coefficients are tensors. In this case the charge carrier current density $\boldsymbol{J}$ and energy flux $\boldsymbol{S}$ are given by the following relation (in the ith direction):

$$J_i = \sigma_{ij}\zeta_j + \mathcal{M}_{ij}\partial T_i / \partial x_j \tag{96a}$$

$$S_i = \mathcal{N}_j\zeta_j + \mathcal{L}_{ij}\partial T_i / \partial x_j \tag{96b}$$

where the indices $i,j = [x,y,z]$ in the Cartesian coordinates and the coefficients σ, $\mathcal{M}$, $\mathcal{N}$ and $\mathcal{L}$ are generally tensors. The above relations may be also written in the following form (Vapail, 1970):

$$\zeta_j = \rho_{ji} J_i + P_{ji}\, \partial T_j/\partial x_i \tag{96c}$$

$$S_j = \pi_{ji} J_i + \kappa_{ji}\, \partial T_k/\partial x_i \tag{96d}$$

where $\boldsymbol{\rho}$ is the semiconductor resistivity, $\boldsymbol{\pi}$ is the Peltier coefficient, $\boldsymbol{P}^{th}$ is the Seebeck coefficient and $\boldsymbol{k}^{th}$ is its thermal conductivity. The thermodynamic irreversibility, implies the following conditions upon these coefficients:

$$\rho_{ji}(B) = \rho_{ji}(-B),\ \sigma_{ji}(B) = \sigma_{ji}(-B) \tag{97}$$

$$T_i.P_{ji}(B) = \pi_{ji}(B), \tag{98}$$

$$\kappa_{ji}(B) = \kappa_{ji}(-B), \tag{99}$$

The above relations (96) are called the Onsager relations (Onsager,1931). They tell us that the transport coefficients are symmetric tensors in absence of magnetic fields. When the magnetic field is not null, the diagonal elements of these tensorial coefficients are even functions of $\boldsymbol{B}$.

12. OPTICAL EFFECTS

Semiconductor materials exhibit interesting characteristics that can be exploited for optoelectronic applications, such as luminescence and light absorption. The study of the optical properties of matter is a huge field and we will only be able to summarize the most basic effects, such as luminescence and absorption. More details about optical properties of matter are discussed in Chapter 8 of this book.

12.1 Photo-Luminescence

The *luminescence* property of some materials means their ability to emit optical radiation (ultraviolet, visible light or infrared) in response to an electronic excitation. In other terms, luminescence is the ability to convert electron energy to light, evidently in an efficient manner.

Luminescence is different from incandescence, which is purely due to heating (temperature) of the material. The efficiency of luminescence may be measured by the ratio of radiative to non-radiative transition (or recombination) rates. Therefore, direct-gap semiconductors, in which radiative recombination occurs with a high probability, are expected to be more efficient than indirect-gap semiconductors, in terms of luminescence.

The wavelength of emitted light from a semiconductor, due to direct transitions from conduction band to valence band, is related to the energy gap of the semiconductor by the relation:

$$\lambda(\mu m) \leq \frac{1.24}{E_g(eV)} \tag{100}$$

The human eye is sensitive to wavelengths between $0.7\mu m$ (red light) and $0.45\mu m$ (violet light), and has a maximum sensitivity for wavelengths around $0.55\mu m$ (green light). Thus, semiconductors having $E_g > 1.8$ eV, like *GaP, CdS, AlGaAs, InGaAsP, β-SiC* and *GaN* are typical candidates for luminescence devices, such as light emitting diodes (LEDs). Figure 46 illustrates the photoluminescence phenomenon

Figure 46. Illustration of the photoluminescence phenomenon in a semiconductor

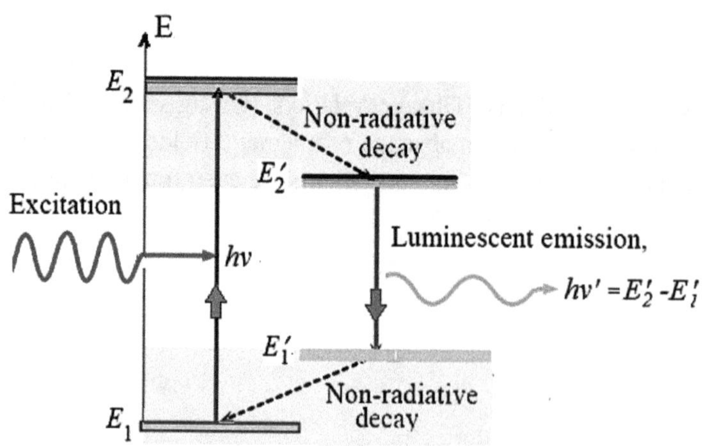

Figure 47. Spectrum of visible light and corresponding energy gap heights. Some semiconductors of corresponding energy gap heights are shown in the insert.

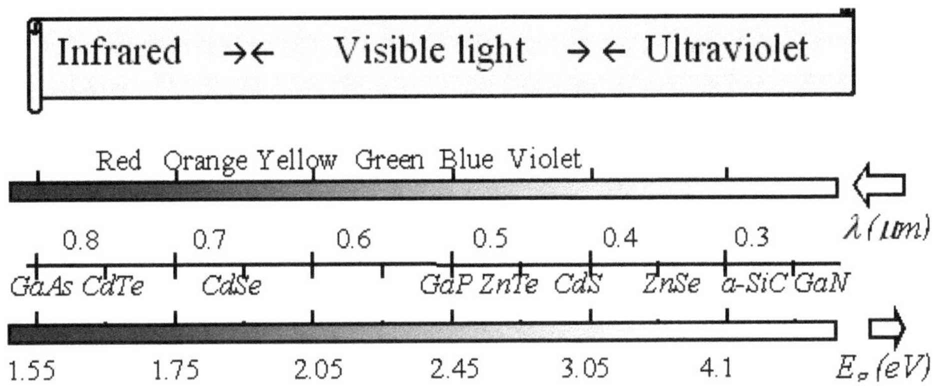

in a semiconductor and Figure 47 depicts the Spectrum of visible light and the corresponding energy gap heights.

12.2 Photo Absorption

If a semiconductor is excited by a flux of light quanta (photons) whose energy E_{ph} is greater than the energy gap E_g of the semiconductor, then the energy of incident photons is absorbed by valence electrons and helps them to break the valence bond and make transitions to the conduction band. In this case we have $E_{ph} = hc/\lambda > E_g$, where λ is the light wavelength, such that $\lambda(\mu m) < 1.24/E_g(eV)$. This phenomenon is known as the photo-generation or photo-absorption process. It is exploited in optical devices, like visible light and infrared detectors. When a semiconductor is subjected to a light flux, the light intensity I_{ph} (the number of incident photons per cm^2 per second) decays with distance x, according to the Beer-Lambert law (Beer, 1852), as follows:

$$I_{ph}(x) = I_o \exp(-\alpha_{op} x) \tag{101}$$

where α_{op} is the light absorption coefficient (cm^{-1}) in the semiconductor and I_o is the net intensity of light at the edge of semiconductor. Note that the ratio I_{ph}/I_o is equal to $1/e$ at $x = 1/\alpha_{op}$, so this distance is called the penetration depth. The penetration depths ($1/\alpha_{op}$) for water: 32 cm, glass: 29 cm, and graphite: 0.6 μm. Generally speaking, the light absorption coefficient α_{op} is a function of the incident light wavelength λ or the energy of incident photons. Figure 48 depicts the dependence of light absorption coefficient of some semiconductors on the wavelength of incident light.

Note 4: Direct and Indirect Gap Semiconductors

One of the most important observations on the energy-band diagram $E(k)$ of semiconductors is that the valence band maximum may be found at a different value of k than the principal conduction band. For instance, the silicon material has its top of valence bands at $k = 0$, while the bottom of principle conduction band lies at $k = 0.85(2\pi/a)$ in directions [100]. Such semiconductors are called indirect-gap

Figure 48. Light absorption in a semiconductor

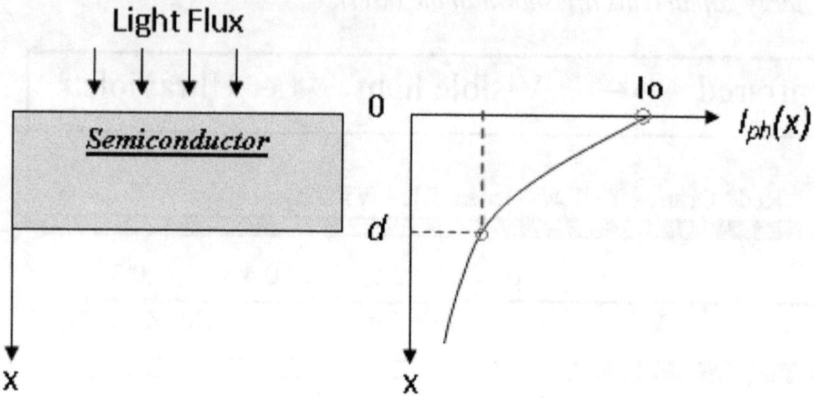

semiconductors. The electron transitions from the minimum point in the conduction band to the maximum point of the valence band in indirect-gap semiconductors require some change in k. Therefore, the conservation of crystal momentum (or wave vector) requires a phonon emission (or absorption) during the band-to-band transitions in indirect-gap semiconductors.

In direct-gap semiconductors, the excited electron momentum is conserved during the photo-excitation process. Then, the change in electron momentum $\Delta k = k_{ei} - k_{ef} = k_{ph} = 0$, where k_{ph} is the incident photon wave vector. Also, the energy conservation implies: $E(k_{ei}) - E(k_{ef}) = E_{ph}$. Hence, the absorption coefficient is given by:

$$\alpha_{op} = A_1 \sqrt{E_{ph} - E_g} \tag{102}$$

For *GaAs*, the constant $A_1 = 11700$ cm^{-1}eV$^{-1/2}$. In indirect-gap semiconductors, like *Si*, the conservation of momentum requires a phonon emission during the photo-excitation process. In silicon, both direct and indirect processes can take place simultaneously and the absorption coefficient is given by:

Figure 49. Direct and indirect gap semiconductors

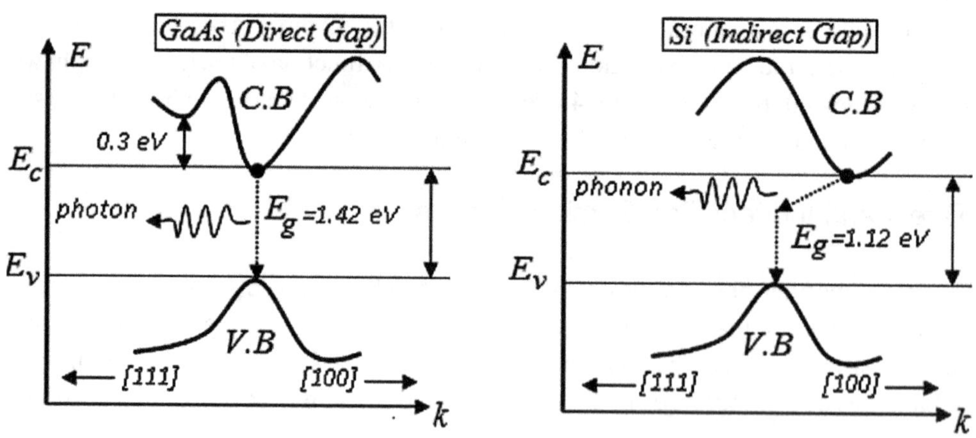

$$\alpha_{op}(Si) = 1.29x10^{6}\sqrt{(E_{ph} - 3.2)} + 6x10^{3}(E_{ph} - 1.1)^{2} + 8x10^{4}(E_{ph} - 2.5)^{2} \quad (103)$$

The wavelength at which the absorption coefficient is zero is called the cutoff wavelength and termed α_c. As shown in Figure 50, $\alpha_c = 1.12\ \mu m$ for *Si* and $0.86\ \mu m$ for *GaAs*.

13. GENERATION-RECOMBINATION MECHANISMS IN SEMICONDUCTORS

When the thermal equilibrium state is disturbed in a semiconductor, by the influence of external force (e.g. via an optical excitation), the charge carrier densities (n, p) change from equilibrium values (n_o, p_o) so that

$$n = n_o + \Delta n,\ p = p_o + \Delta p \quad (104)$$

where the quantities Δn and Δp are called the excess carrier concentrations. Hence at non-equilibrium:

$$n.p \neq n_i^2 \quad (105)$$

Both Δn and Δp may be positive (due carrier injection into the semiconductor) or negative (due to carrier extraction from the semiconductor). When there is an excess carrier concentration, the semiconductor will try to restore its equilibrium state by endeavoring carrier generation or recombination processes. At the thermal equilibrium state, the thermal generation rate (of electron-hole pairs), g_{th}, is

Figure 50. Absorption coefficient of some semiconductors vs. light wavelength
Source: Elliot & Gibson (1982).

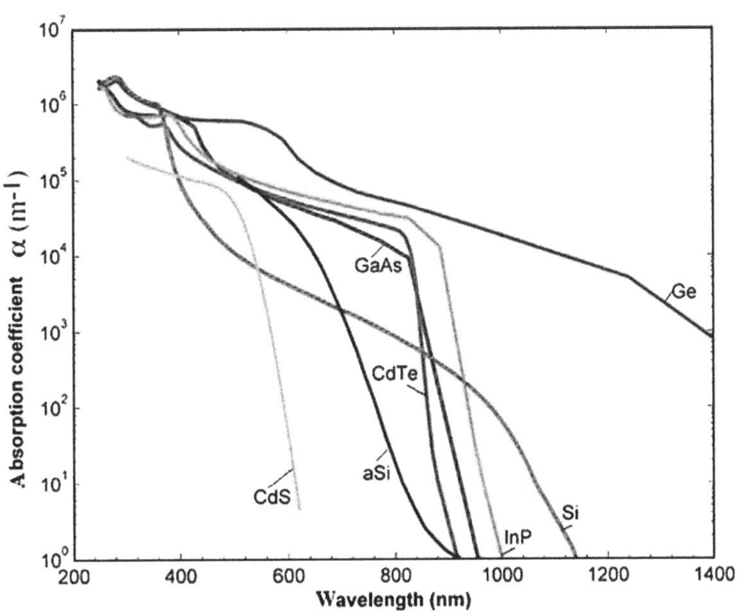

compensated by a default recombination rate (of electron-hole pairs), R_o, such that the net recombination rate $U = R_o - g_{th}$ is null (zero) and there exists no excess carriers.

When Δn and Δp are negative, meaning that $np < n_i^2$, then the thermal generation mechanism g_{th} (which takes place regardless of the presence of excess carriers) will dominate the default recombination R_o to restore the equilibrium state. On the other hand, when Δn and Δp are positive, meaning that $np > n_i^2$, then the recombination mechanism will dominate the thermal generation. Therefore, the net recombination rate of electrons and holes is given by:

$$U_n = R_n - g_{th} = \frac{\Delta n}{\tau_n}, U_p = R_p - g_{th} = \frac{\Delta p}{\tau_p} \tag{106}$$

where τ_n and τ_p are called the electron and hole lifetimes, respectively.

As shown in Figure 51, the recombination of charge carriers, may be direct (radiative) or indirect (non-radiative), via a trapping center. In the former case, the recombination is associated with emission of a photon with equivalent energy of the energy gap (E_g). In the latter case, the recombination, releases smaller energy which is transmitted to the crystal lattice vibrations. The quanta of the resultant lattice vibrations are called phonons.

13.1 Carrier Recombination Mechanisms in Semiconductors

The charge carrier recombination may happen in the bulk or on the surface of a semiconductor. Also, the mechanism of recombination may be direct band-to-band or via-trap levels, as shown in Figure 51. When a semiconductor is doped with impurities that produce profound energy levels in the energy gap (like gold, *Au*), these levels will act as trapping centers for carries. Therefore, carriers will prefer to transit for a while at these trapping levels during their transitions between conduction and valence bands. This indirect recombination process is sometimes called trap-assisted recombination or Shockley-Read-Hall (SRH) recombination.

Figure 51. Generation and recombination mechanisms in a semiconductor

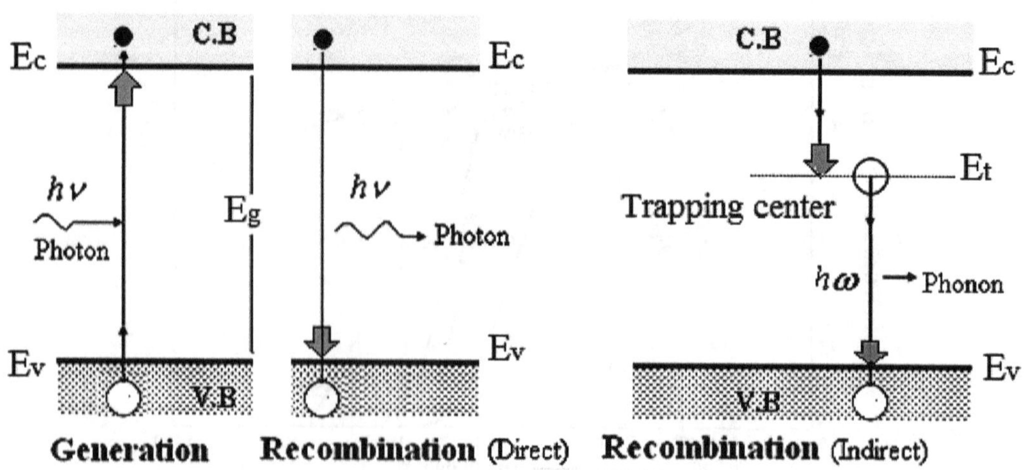

Band-to-Band Recombination

Direct band-to-band recombination (also called radiative recombination) is important in direct-gap semiconductors like *GaAs*. During *radiative recombination*, a form of spontaneous emission, a photon is emitted with the wavelength corresponding to the energy released. This effect is the basis of light emitting diodes (LEDs).

The rate of band-to-band recombination is proportional to the electron and hole concentrations and given by:

$$U\,(radiative) = R - g_{th} = \alpha_r\,(n\,p - n_i^2) \tag{107a}$$

where α_r is called the bimolecular recombination rate. For *GaAs*, we have $\alpha_r = 1.5x\ 10^{-10}$ cm^{-3}/s. In n-type semiconductors at low-level injection, where $n >> \Delta n$ and $n \cong n_o$, we have:

$$U\,(radiative) = \Delta p\,/\tau_{pr} = (p - p_o)\,/\tau_{pr} \tag{107b}$$

where τ_{pr} is the radiative carriers (holes) lifetime, which is given by:

$$\tau_{pr} = 1/(\alpha_r\, n_o) \tag{107c}$$

Shockley-Read-Hall (SRH) Recombination

The steady-state rate of indirect recombination in the bulk of a non-degenerate semiconductor is given by the Shockley-Read-Hall (1952) formula:

$$U_{SRH} = \frac{np - n_i^2}{\tau_{po}(p + p_1) + \tau_{no}(n + n_1)} \tag{108}$$

where τ_{no} and τ_{po} are constants (minority carrier lifetimes in a heavily-doped semiconductor).

The minority carrier lifetimes can be defined in terms of the density of traps N_t [cm^{-3}] and the capture rates of electrons and holes C_n, C_p [cm^3/s] as follows:

$$\tau_{no} = \frac{1}{N_t C_{no}}, \quad \tau_{po} = \frac{1}{N_t C_{po}} \tag{109}$$

The electrons and holes capture rates are related to the trap capture cross sections σ_{cn}, σ_{cp} [cm^2] and the thermal velocity v_{th} of carriers as follows:

$$C_n = \sigma_{cn}\, v_{th}, C_p = \sigma_{cn}\, v_{th} \tag{110}$$

Also n_1 and p_1 are defined in terms of the traps energy level E_t[10]:

$$n_1 = n_i Exp\left[-\frac{E_i - E_t}{k_B T}\right], p_1 = n_i Exp\left[-\frac{E_t - E_i}{k_B T}\right] \tag{111}$$

The parameters n_1 and p_1 are called *the trap equivalent concentration*, in conduction band and valence band, respectively. When the trap energy level E_t is midway between the conduction band and the valance band edges, such that $E_t = E_i$, then $n_1 = p_1 = n_i$.

Auger Recombination

In addition to the above basic processes, there exist other recombination mechanisms, such as the Auger recombination mechanism. Excess carriers decay by these recombination processes, and the energy of carriers is dissipated during this process by several ways, such as emission of a photon, heat generation in the lattice in the form of phonons, or the transmission of energy to a third particle. At the bulk of a semiconductor the following relation depicts the Auger recombination rate:

$$U_{Auger} = \Gamma_n \cdot n\,(n\,p - n_i^2) + \Gamma_p \cdot p\,(n\,p - n_i^2) \tag{112}$$

where Γ_n and Γ_p are Auger recombination coefficients for electrons and holes, respectively. For Si, $\Gamma_n = 2.8x10^{-31}$ cm^6/s and $\Gamma_p = 9.9\ x10^{-32}$ cm^6/s.

The Auger lifetime is defined as follows:

$$\tau_{p(AU)} = \Delta p / U_{Auger},\ \tau_{n(AU)} = \Delta n / U_{Auger} \tag{113}$$

Under low-level injection conditions (when $\Delta p << n_o$ in n-type or $\Delta n << p_o$ in p-type), the Auger lifetime is given by:

$$\tau_{p(AU)} \approx 1/(\Gamma_p p_o^2) = 1/(\Gamma_p N_a^2),\ \tau_{n(AU)} \approx 1/(\Gamma_n n_o^2) = 1/(\Gamma_p N_d^2) \tag{114}$$

At high injection levels (where $\Delta p \approx n_o$ in n-type or $\Delta n \approx p_o$ in p-type), the Auger lifetime is given by:

$$\tau_{p(AU)} \approx 1/[(\Gamma_p + \Gamma_n).\Delta p^2],\ \tau_{n(AU)} \approx 1/[(\Gamma_p + \Gamma_n).\Delta n^2] \tag{115}$$

In fact, the measurement of the Auger lifetime, in heavily excited Si by laser pulses (where $\Delta n = \Delta p$), indicates that: $\tau_{n,p(AU)} \approx 1/[3.4x10^{-31}).\Delta p^2]$.

Surface Recombination

The rupture of crystal periodicity at the surface of a semiconductor creates some trapping centers (or surface states) inside the energy gap. The surface recombination rate of electrons and holes in steady state is given by the Shockley-Read-Hall formula:

$$S_n = S_p = \frac{s_n\, s_p\left(n_s\, p_s - n_i^2\right)}{s_p(p_s + p_{1s}) + s_n(n_s + n_{1s})} \tag{116}$$

where n_s and p_s are the concentrations of electrons and holes at the semi-conductor surface, respectively. Also s_n and s_p are the electron and hole surface recombination velocities [cm/s], respectively. They are given by:

$$s_n = \sigma_n v_{th} N_{ts},\ s_p = \sigma_p v_{th} N_{ts} \tag{117}$$

where σ_n and σ_p [cm^2] are the trapping cross sections of electrons and holes, respectively, and N_{ts} [cm^{-2}] is the density of surface traps or surface states[11]. Bare *Si* surface has surface states density of about 10^{12} cm^{-2} while SiO_2 surface has 10^7–10^{10} cm^{-2}. In n-type semiconductors, where $n_s >> p_s$, the rate of surface recombination of holes reduces to the following simple form:

$$S_p = s_p\,(p_s - p_{so}) \tag{118a}$$

Similarly the rate of surface recombination of electrons in p-type semi-conductors, where $p >> n$, can be reduced to the following simple form:

$$S_n = s_n\,(n_s - n_{so}) \tag{118b}$$

where n_{so} and p_{so} are the surface concentrations of electrons and holes in thermal equilibrium (such that $p_{so}.n_{so} = n_i^2$ at the surface).

Carrier Lifetime

The carrier lifetime (recombination lifetime) is defined as the average time it takes an excess minority carrier to recombine. The overall lifetime of carriers due to: Shockley-Read-Hall, Auger, band-to-band and surface recombination mechanisms may be expressed as:

$$\frac{1}{\tau_{pe}} = \frac{1}{\tau_p} + \frac{1}{\tau_{p(AU)}} + \frac{1}{\tau_{pr}} + \frac{1}{\tau_{ps}} \tag{119}$$

where τ_{ps} is the surface lifetime, which is given by: $\tau_{ps} = 1/\beta^2 D_p$, with D_p is the holes diffusion constant at the surface, and β can be obtained by solving the equation: $tan(½\beta\, t) = s_p/\beta\, D_p$, where t is the semi-conductor thickness. When the surface recombination velocity s_p is very small, then $\tau_{ps} \rightarrow (t/2s_p)$. Also, when s_p is very high, then $\tau_{ps} \rightarrow (t^2/\pi^2 D_p)$. Figure 52 depicts the reduction of effective lifetime of electrons in p-type silicon, as a function of the doping concentration, according to (Schroder, 1990).

Figure 52. Minority carrier lifetime (holes in n-type) in Si, at 300K
After Schroder (1980).

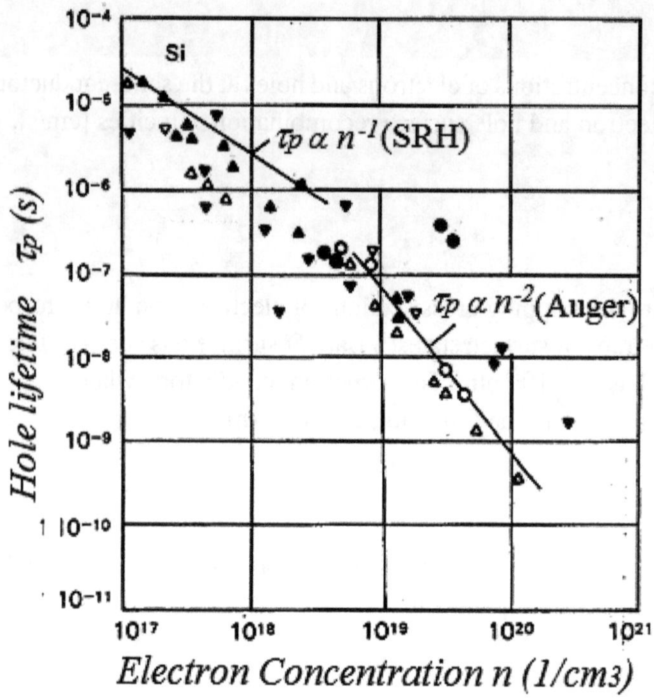

13.2 Carrier Generation Mechanisms in Semiconductors

The generation of charge carriers (electrons and holes) in a semiconductor is accomplished by the aid of different mechanisms, e.g., by the thermal and optical excitation as well as the impact ionization process.

Thermal Generation

The thermal generation is the default generation mechanism in semi-conductors. It is well known that the recombination rate is proportional to the product of electrons and holes densities (*n.p*). The thermal

Figure 53. Types of carrier generation mechanisms in semiconductors

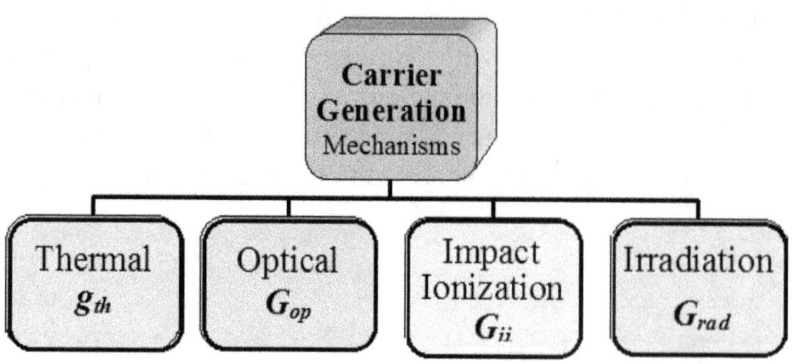

generation rate at thermal equilibrium is also proportional to the np product ($n_o p_o = n_i^2$) such that: $g_{th} = a_r n_i^2$ where a_r is constant.

Optical Generation

When a semiconductor is subjected to a flux of light quanta (photons), the energy of incident photons can excite some valence electrons. When the photon energy is greater than the energy gap of the semiconductor ($E_{ph} = h\nu = hc/\lambda > E_g$), a number of valence bonds are broken and electron-hole pairs are produced. The optical generation rate G_{op} (electron-hole pair/cm^3 sec) is proportional to the gradient of the light flux I_{ph} (photon/cm^2 sec) such that:

$$G_{op} = \alpha_{op} I_{ph} \tag{119}$$

where α_{op} is the light absorption coefficient (cm^{-1}) in the semiconductor.

Actually, the light absorption coefficient α_{op} is a function of the incident light wavelength λ (or the energy) of incident photons E_{ph}.

Generation by Impact Ionization

When a semiconductor is subjected to a high electric field such that the energy acquired by carrier from field *goes* beyond a threshold value (about 3/2 E_g, the energy gap width), then electron-hole pairs are produced by impact ionization. The electron-hole pair generation rate by impact ionization G_{ii} is given by:

$$G_{ii} = \alpha_n n v_n + \alpha_p p v_p = \alpha_n |J_n/e| + \alpha_p |J_p/e| \tag{120}$$

where α_n, α_p are the impact ionization (or Townsend's) coefficients of electrons and holes. These coefficients express the number of generated carriers by impact ionization per unit length (cm^{-1}) and are naturally functions of the electron and hole energies. However, the impact ionization coefficients usually expressed as functions of the local electric field intensity, as follows:

$$\alpha_n = \alpha_{no} exp(-\zeta_{cn} / |\zeta|),\ \alpha_p = \alpha_{po} exp(-\zeta_{cp} / |\zeta|) \tag{121}$$

where α_{no} and α_{po} are the impact ionization saturation values (at very high electric fields). Also ζ_{cn} and ζ_{cp} are the critical fields for electron and hole ionization. The impact ionization parameters change with temperature and depend on the crystal orientation.

14. CONTINUITY EQUATIONS IN SEMICONDUCTORS

When the thermal equilibrium state in a semiconductor is disturbed, the net recombination rate of carriers is non-zero. Therefore, the density of carriers will continue to change with time because of the generation-recombination processes as well as the divergence of carrier flux (current density). The continuity equations for electrons and holes in a semiconductor read:

Figure 54. Impact ionization coefficients of electrons and holes (α_n and α_p) in Si at 300K, measured as a function of electric field ζ. The straight lines depicts the exponential dependence of $\alpha_n(\zeta)$ and $\alpha_p(\zeta)$. According to van Overstraeten and de Man (1970) and Monte Carlo simulation (Karlowatz et al., 2006).

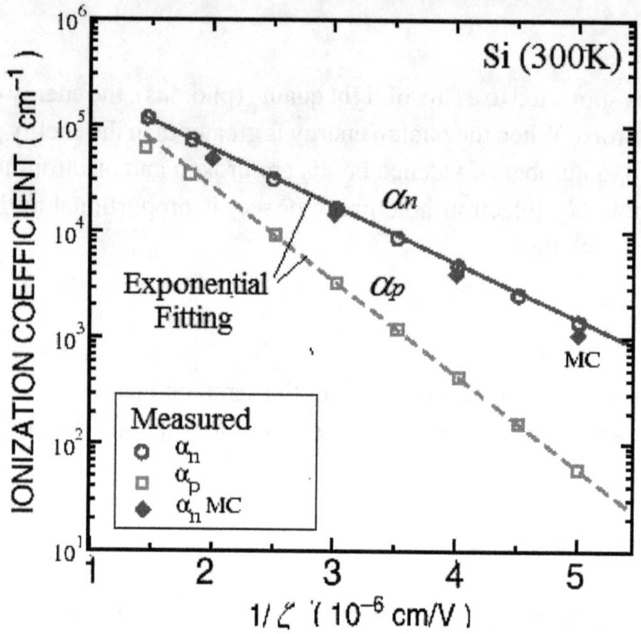

$$\frac{\partial n}{\partial t} = \left(G_n - R_n\right) + \frac{1}{e}\nabla \cdot J_n \tag{122a}$$

$$\frac{\partial p}{\partial t} = \left(G_p - R_p\right) + \frac{1}{e}\nabla \cdot J_p \tag{122b}$$

where, J_n and J_p, are the electron and hole current densities, respectively.

The continuity equations are fundamental tenet about electromagnetism and can be derived from the Maxwell's equations. Generally speaking, the continuity equations are statement of conservation of physical quantities such as the number or charge of carriers, carrier momentum or carrier energy.

The electron and hole current densities, (J_n and J_p), in the continuity equations may be expressed by the drift-diffusion equations (71a) and (71b), or any other suitable model. The resultant system of partial differential equations (PDEs) was first presented by Van Roosbroeck (1950) as the first full model to describe the transport of electrons and holes in semiconductors. This model has been known as the drift-diffusion model (DDM).

Figure 55. Illustration of the continuity equation in semiconductors

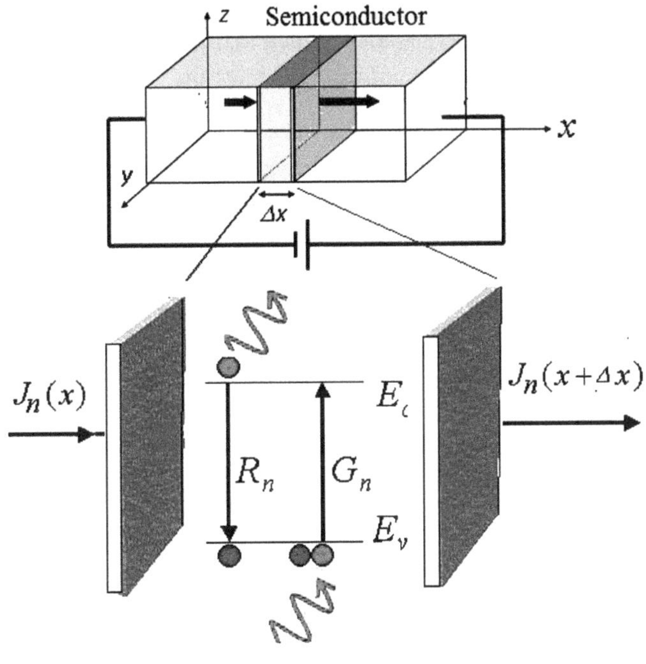

15. SUMMARY

The information carriers in nanodevices are the particles (or particle characteristics) that carry and transport information objects or signals within a device. The most famous example of an information carrier is the mobile electronic charges in conventional semiconductor devices (electrons and holes). The electric ccurrent is the rate at which charge carriers flow. The process by which these charged particles move is called transport.

As semiconductor devices have been continuously downscaled, the size of new devices became in the order of a few nanometers. Solid understanding of carrier transport mechanisms in such nanoscale devices is critical to accelerate device design and approach new horizons of speed and scales of integration.

Modern TCAD tools should have physically-based transport models to address challenges of the emerging nanodevices and systems. Such models must be prepared to adequately describe the quantum mechanical phenomena which determine transport in such devices.

In this chapter we reviewed the fundamental properties of semiconductors, and explained the transport phenomena within the framework of the classical Drudé model. The Drudé classical model is frequently introduced to describe the electrical conductivity in solids. This model is still very relevant because free particle picture can still be used as far as we can assume parabolic energy bands, near equilibrium, with suitable effective mass. In fact, the Drudé model succeeded to explain (to some extent) the electrical conductivity, the thermal conductivity, the Hall Effect, as well as the dielectric function and the optical response of solids. Therefore, we can describe the flow of electrons and holes in a semiconductor, under the effect of external forces, such as electric and magnetic fields. The two basic transport mechanisms in a semiconductor crystal are:

- **Drift:** The Movement of the charge due to electric fields.
- **Diffusion:** The flow of charges due to density gradients.

Everything we explain so far in this chapter about semiconductor properties and carrier transport is correct to the zero order approximation. In order to get into detailed transport in semiconductor systems we need to solve a master transport equation. Actually, there exist two theoretical approaches, for studying the charge carrier transport in semiconductors, according to the physical dimension of the problem in hand (i.e. the semiconductor device dimensions):

- The semiclassical transport theory, for macroscopic devices
- The quantum transport theory, for the microscopic devices.

REFERENCES

Adler, R., Smith, A., & Longini, R. (1964). *Introduction to Semiconductor Physics*. New York: Wiley & Sons.

Arora, N. D., Hauser, J. R., & Roulston, D. J. (1982). Electron and Hole Mobilities in Silicon as a Function of Concentration and Temperature. *IEEE Transactions on Electron Devices*, *29*(2), 292–295. doi:10.1109/T-ED.1982.20698

Ashcroft, N. W., & Mermin, N. D. (1976). Solid State Physics. New York: Holt, Rinehart and Winston.

Beer. (1852). Bestimmung der Absorption des rothen Lichts in farbigen Flüssigkeiten [Determination of the absorption of red light in colored liquids]. *Annalen der Physik und Chemie*, *86*, 78–88.

Ben Abdallah, N., & Degond, P. (1996). On a hierarchy of macroscopic models for semiconductors. *Journal of Mathematical Physics*, *37*(7), 3306–3333. doi:10.1063/1.531567

Blakemore, J. S. (1962). *Semiconductor Statistics*. New York: Pergamon Press.

Blatt, F. J. (1968). *Physics of Electronic Conduction in Solids*. McGraw-Hill.

Bloch, F. (1929). Uber die Quantenmechanik der Elektronen in Kristallgittern. *Zeitschrift fur Physik*, *52*(7-8), 555–600. doi:10.1007/BF01339455

Cercignani, C., & Gabetta, E. (Eds.). (2007). *Transport Phenomena and Kinetic Theory*. Boston: Birkhauser. doi:10.1007/978-0-8176-4554-0

Colclaser, R. A., & Diehl-Nagle, S. (1985). *Materials and Devices for Electrical Engineering and Physics*. Singapore: McGraw-Hill Co.

Dargys, A., & Kundrotas, J. (1994). *Handbook on Physical Properties of Ge, Si, GaAs and InP*. Vilnius: Science and Encyclopedia Publishers.

Debye, P., & Deut, V. (1913). Collected papers of Peter J.W. Debye. Interscience.

Deinega, A., & John, S. (2012). Effective optical response of silicon to sunlight. *Optics Letters*, *37*(1), 112. doi:10.1364/OL.37.000112 PMID:22212808

Dexter, R. N., & Lax, B. (1954). Effective Masses of Holes in Silicon. *Physical Review*, *96*(1), 223–224. doi:10.1103/PhysRev.96.223

Drabble, J. R., & Goldsmid, H. J. (1961). *Thermal Conduction in Semiconductors*. Oxford, UK: Pergamon Press.

Dresselhaus, G., Kip, A. F., & Kittel, C. (1955). Cyclotron Resonance of Electrons and Holes in Silicon and Germanium Crystals. *Physical Review*, *98*(2), 368–384. doi:10.1103/PhysRev.98.368

Dresselhaus, M., Dresselhaus, G., & Avouris, Ph. (Eds.). (2001). *Carbon Nanotubes: Synthesis, Structure Properties and Applications*. Berlin: Springer-Verlag. doi:10.1007/3-540-39947-X

Drudé, P. (1900). Zur Elektronentheorie der metalle. *Annalen der Physik*, *1*(3), 566–613. doi:10.1002/andp.19003060312

Durkop, T., Getty, S., Cobas, E., & Fuhrer, M. (2004). Extraordinary Mobility in Semiconducting Carbon Nanotubes. *Nano Letters*, *4*(1), 35–39. doi:10.1021/nl034841q

Elliot, R. J., & Gibson, A. F. (1982). *An Introduction to Solid State Physics and its Applications*. London: Macmillan Press Ltd.

Ezawa, Z. F. (2008). *Quantum Hall Effects - Field Theoretical Approach and Related Topics*. Singapore: World Scientific. doi:10.1142/6242

Faraday, M. (1839). Experimental Researches in Electricity (vol. 1). London: Richard and John Edward Taylor.

Ferry, D., & Goodnick, S. M. (1997). *Transport in Nanostructures*. Cambridge, UK: Cambridge University Press. doi:10.1017/CBO9780511626128

Fu, Y. (1999). *Magnus Willander. In Physical Models of Semiconductor Quantum Devices (Electronic Materials Series)*. Berlin: Springer. doi:10.1007/978-1-4615-5141-6

Gibbs, J. W. (1902). *Elementary Principles in Statistical Mechanics*. Oxford, UK: Bow Press.

Goldsmid, H. J. (1986). *Electronic Refrigeration*. London, UK: Pion.

Greenaway, D. L., Harbeke, G., & Pamplin, B. R. (1968). Optical Properties and Band Structure of Semiconductors. In *International Series in the Science of the Solid State*. Berlin: Elsevier.

Hall, E. H. (1879). On a New Action of the Magnet on Electric Currents. *American Journal of Mathematics, 2*(3), 287–292. doi:10.2307/2369245

Hall, R. N. (1952). Electron-hole Recombination in Germanium,'. *Physical Review, 87*(2), 387. doi:10.1103/PhysRev.87.387

Ioffe, A. F. (1957). *Semiconductor thermoelements and thermoelectric cooling*. London: Infosearch.

Jakoby, B. (2009). The relation between relaxation time, mean free path, collision time and drift velocity. *European Journal of Physics, 30*(1), 1–12. doi:10.1088/0143-0807/30/1/001

Jüngel, A. (2009). *Transport Equations for Semiconductors*. Springer Science & Business Media. doi:10.1007/978-3-540-89526-8

Karlowatz, G., Ungersboeck, E., Wessner, W., & Kosina, H. (2006). Full-band Monte Carlo analysis of electron transport in arbitrarily strained silicon. *Proceedings of Semiconductor Processes and Devices,* 63-66.

Kasap, S. O. (2005). *Principles of Electronic Materials and Devices* (3rd ed.). New York: McGraw-Hill.

Kireev, P. (1979). *Semiconductor Physics*. Moscow: Mir Publisher.

Kittel, C., & Kroemer, H. (1980). *Thermal Physics* (2nd ed.). Wiley.

Kovskii, T. (1962). *Thermomagnetic effects in Semiconductors*. Infosearch.

Landau, L., & Lifshits, E. (1959). *Statistical Physics*. London: Pergamon Press.

Lewis, R. A. (2007). Physical Phenomena in Electronic Materials in the Terahertz Region. *Proceedings of the IEEE, 95*(8), 1641–1645. doi:10.1109/JPROC.2007.898902

Li, Z. Q., & Li, Z. M. (2010). Comprehensive modeling of super-luminescent light-emitting diodes. *IEEE Journal of Quantum Electronics, 46*(4), 454–461. doi:10.1109/JQE.2009.2032426

Lundstrom, M. (1990). Fundamentals of Carrier Transport, vol. 10 of Modular Series on Solid State Devices. New York: Addison-Wesley.

Lundstrom, M. (2012). *Near-equilibrium Transport: Fundamentals and Applications (NET)*. World Scientific.

Morin, F. G., & Miata, J.-P. (1954). Electrical Properties of Silicon Containing Arsenic and Boron. *Physical Review, 96*(1), 28–35. doi:10.1103/PhysRev.96.28

Myers, H. P. (1997). *Introductory Solid State Physics* (2nd ed.). Taylor & Francis. doi:10.4324/9780203212554

Nag, B. R. (1972). *Theory of Electrical Transport in Semicondnuctors*. Oxford, UK: Pergamon.

Nakajima, S., Toyozawa, Y., & Abe, R. (1980). *The Physics of Elementary Excitations. Springer Series in Solid-State Sciences*. Berlin: Springer.

Onsager, L. (1931). Reciprocal Relations in Irreversible Processes. I. *Physical Review*, *37*(4), 405–426. doi:10.1103/PhysRev.37.405

Palenskis, V. (2013). Drift Mobility, Diffusion Coefficient of Randomly Moving Charge Carriers in Metals and Other Materials with Degenerated Electron Gas. *World Journal of Condensed Matter Physics.*, *1*(3), 73–81. doi:10.4236/wjcmp.2013.31013

Pankove, J. (1971). *Optical Processes in Semiconductors*. Prentice-Hall.

Phillips, J.C. (1968). Covalent bonds in crystals. I. Elements of a structural theory. *Phys. Rev., 166*, 832–838; II.

Pierret, R. F. (2003). *Advanced Semiconductor Fundamentals* (2nd ed.). Prentice Hall.

Riffe, D. M. (2002). Temperature dependence of silicon carrier effective masses with application to femtosecond reflectivity measurements. *JOSA B.*, *19*(5), 1092–1100. doi:10.1364/JOSAB.19.001092

Roosbroek, V. (1950). Theory of Flow of Electrons and Holes in *Ge* and other Semiconductors. *The Bell System Technical Journal*, *29*(4), 560–607. doi:10.1002/j.1538-7305.1950.tb03653.x

Rowe,D.M.(1995).*HandbookofThermoelectrics*.BocaRaton,FL:CRCPress.doi:10.1201/9781420049718

M. Saraniti, & U. Ravaioli (Eds.). (2006). Nonequilibrium Carrier Dynamics in Semiconductors. *Proceedings of the 14th International Conference*. Springer-Verlag.

Schafer, W., & Wegener, M. (2002). *Semiconductor Optics and Transport Phenomena, Advanced Texts in Physicsm*. Berlin: Springer. doi:10.1007/978-3-662-04663-0

Seeger, K. (2004). *Occupation probabilities of impurity levels. Semiconductor physics: An introduction* (9th ed.). Springer.

Shockley, W. (1950). *Electrons and Holes in Semiconductors*. Princeton, NJ: Van Nostrand.

Shockley, W., Read, W. T., & Hall, R. N. (1952). Statistics of the recombination of holes and electrons. *Physical Review*, *87*(5), 823–842. doi:10.1103/PhysRev.87.835

Smith, R. A. (1979). *Semiconductors* (2nd ed.). London: Cambridge Univ. Press.

Stratton, R. (1972). Semiconductor current-flow equations (diffusion and degeneracy). *IEEE Transactions on Electron Devices*, *10*(12), 1288–1292. doi:10.1109/T-ED.1972.17592

Sze, S. M. (1969). *Physics of Semiconductor Devices*. Wiley-Interscience.

Sze, S. M. (1981). *Semiconductor Devices* (2nd ed.). New York: John Wiley& Sons.

Tavernier, J., & Calcki, D. (1970). *Introduction aux Phenomenes de Transport Lineaires dans les Semiconducteurs*. Paris: Masson & Cie.

Thornber, K. (1980). Relation of Drift Velocity to Low-Field Mobility and High-Field Saturation Velocity. *Journal of Applied Physics*, *51*(4), 2121–2136. doi:10.1063/1.327885 van Overstraeten, R., & de Man, H. (1970). Measurement of the ionization rates in diffused silicon p-n junctions. *Solid-State Electronics, 13*(5), 583–608. doi:10.1016/0038-1101(70)90139-5

Vapaille, A. (1970). *Physique des Dispositifs a Semiconducteur*. Paris: Masson&Cie.

Vasileska, D., Goodnick, S., & Klimeck, G. (2010). *Computational Electronics: Semi-Classical and Quantum Device Modeling and Simulation*. CRC Press. doi:10.1201/b13776

Wilson, H. (1931). The Theory of Electronic Semiconductors. *Proceedings of the Royal Society of London. Series A, Containing Papers of a Mathematical and Physical Character, 134*(823), 277–287. doi:10.1098/rspa.1931.0196

Yoshioka, D. (2002). *The Quantum Hall Effect*. Berlin: Springer. doi:10.1007/978-3-662-05016-3

Yu, P. Y., & Cardona, M. (1996). *Fundamentals of Semiconductors*. Berlin: Springer. doi:10.1007/978-3-662-03313-5

Zeghbroek, B. V. (2011). *Principles of Semiconductor Devices*. Available at: http://ecee.colorado.edu/~bart/book/movie/movie.htm

Ziman, J. M. (1960). *Electrons and Phonons*. Oxford, UK: Clarendon Press.

Ziman, J. M. (1972). *Principles of the theory of solids*. Cambridge, UK: Cambridge University Press. doi:10.1017/CBO9781139644075

Žutić, I., Fabian, J., & Sarma, S. D. (2004). Spintronics: Fundamentals and applications. *Reviews of Modern Physics, 76*(2), 323–410. doi:10.1103/RevModPhys.76.323

ENDNOTES

1. Note that the density of mobile carriers (electrons and holes) in the bulk of pure Si, for instance, is about 10^{10} cm^{-3} at 300K. The density of mobile carriers in doped semiconductors is generally much more than this. Indeed, it is futile to solve the equation of motion for all carriers in a real device.
2. We mean by device size the smallest length of the device active region, such as the channel length of a MOSFET or the active layer thickness of a laser diode.
3. It should be kept in mind that the basic Schrödinger equation is not a fundamental theory but rather, a successful tool that results in good solutions, when applied to atoms and particles
4. Exactly like we observe the motion of gas bubbles in gaseous water, instead of the actual motion of liquid molecules. Exceptional things are more easily distinguished and counted.
5. The density of occupied states is equal to the density of occupied energy levels multiplied by 2 since each energy level can accept up to 2 electrons of different spin.
6. According to Dexter and Lax (1954), we may take $A=4.0 \pm 0.1$, $B=1.1\pm 0.4$, $C=4.1\pm0.4$.
7. This is the modern definition of the Fermi-Dirac integral. The original Sommerfeld definition omitted the pre-term $1/\Gamma(j+1)$, which is equal to $2/\sqrt{\pi}$ for $j=½$. The modern definition accepts negative values of j

8 This law can be derived from the mass action law and the Fermi-Dirac statistics.

9 The Drude model was actually introduced to explain the empirical Wiedermann-Franz law (1853). This law states that at a given temperature the ratio of the thermal to the electrical conductivity is the same for all metals

10 The trap density is frequently assumed to decay exponentially from the band edge into the band gap. We consider here indirect recombination via a single trap level. Multi trapping is also possible but rare in conventional (inorganic) semiconductors.

11 The surface states may differ in nature and number according to the state of the semiconductor surface.

Chapter 2
Semiclassical Transport Theory of Charge Carriers, Part I:
Microscopic Approaches

1. OVERVIEW AND CHAPTER OBJECTIVES

The semiclassical approach is suitable for semiconductor devices whose feature dimensions are much greater than the mean free path between collisions ($L >> \lambda_e$). In silicon devices, where the electron mean free path λ_e ~100Å, this approach can be safely used for devices having feature size $L>0.1$ µm. It helps us to calculate the main transport parameters (such as the charge carrier mobility, conductivity, etc.) and so to determine the electrical, thermal, magnetic and optical characteristics of a given semiconductor device.

In classical statistical mechanics, the state of a particle is completely defined at a certain time t as a point in a space of *six* coordinates, called the phase-space (sometimes called the μ-space). The *six* coordinates of the phase-space are the position and momentum coordinates (x, y, z, p_x, p_y, p_z) as shown in Figure 2. Physically, it is not possible to do so in large systems, where the number of particles is in the order of 10^{22} particle per cm. Then one has to use the concept of probability or the distribution function of the system under certain conditions, like the condition of *thermal equilibrium*[1]. Therefore, we usually speak about the so-called distribution function, which expresses the average number of particles at a certain position in the phase-space. The distribution function is often written in either of the following forms:

$f(\boldsymbol{x},\boldsymbol{p}, t)$ or $f(\boldsymbol{x},\boldsymbol{v}, t)$ or $f(\boldsymbol{x},\boldsymbol{k}, t)$

where:

$$\boldsymbol{p} = \left[p_x, p_y, p_z\right]^T \tag{1a}$$

DOI: 10.4018/978-1-5225-2312-3.ch002

Figure 1. Semiclassical approaches for charge carrier transport

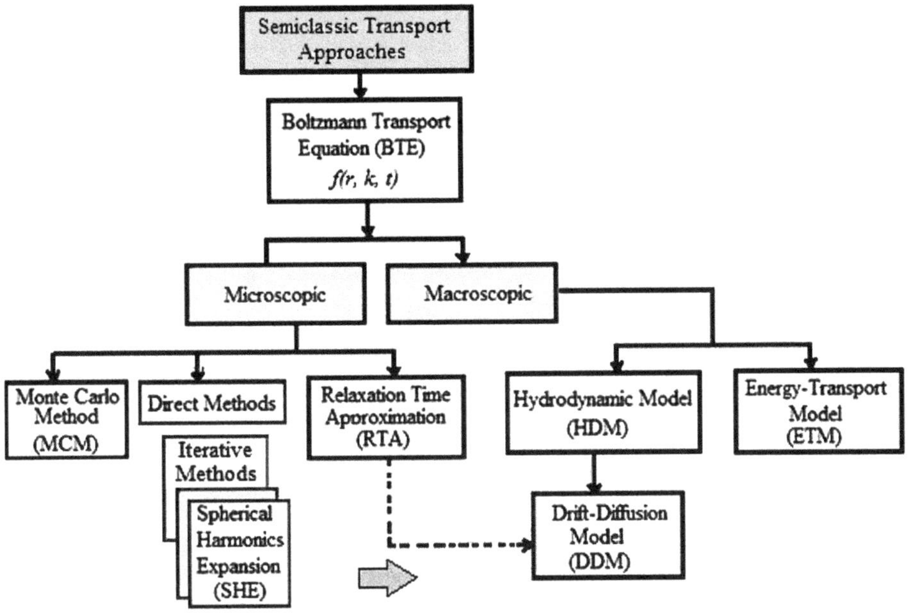

$$\boldsymbol{v} = \left[v_x, v_y, v_z\right]^T \tag{1b}$$

$$\boldsymbol{k} = \left[k_x, k_y, k_z\right]^T \tag{1c}$$

Evidently the momentum $\boldsymbol{p}$ and velocity $\boldsymbol{v}$ of a classical particle are interrelated by: $\boldsymbol{p} = m\,\boldsymbol{v}$. Also, the momentum of a semiclassical particle is related to the wave vector $\boldsymbol{k}$ by the De Broglie relation: $\boldsymbol{p} = h/\lambda = \hbar\boldsymbol{k}$. Note that, we may indifferently use $\boldsymbol{v}$, $\boldsymbol{k}$ or $\boldsymbol{p}$, as second argument of the carrier distribution function, in this book.

The semiclassical transport theory is based on the ensemble concept and the Boltzmann transport equation (BTE), which describes the evolution of the distribution function $f(x, k, t)$ of a gas of particles under nonequilibrium conditions. The ensemble concept shows that a macroscopic observation is consistent with a very large number of microscopic configurations (particles), which are represented by points in the phase space. The ensemble therefore is basically represented by a normalized density distribution function $f(x, k, t)$ in the phase-space.

The density distribution function $f(x, k, t)$ is employed to characterize the evolutions of a thermodynamic system of particles (e.g., electrons in a semiconductor material) in the phase space. If f is high at some point of space, this means there are many particles around this point, flying with the same momentum (or wave vector k). The physical meaning is that $f(x,k,t).\Delta x.\Delta k$ is the number of particles at a certain point x with wavevector k in the phase-space incremental volume $\Delta x.\Delta k$. When integrated over all wave vectors in the k-space, we get the particle density in physical space, and when further integrated over the x-space, we get the total number of particles.

Figure 2. Schematic illustration of a fiction distribution functions in the phase-space (**x** *=x,y,z, and* **p** *=p_x,p_y,p_z) at a certain moment $t = t_o$*

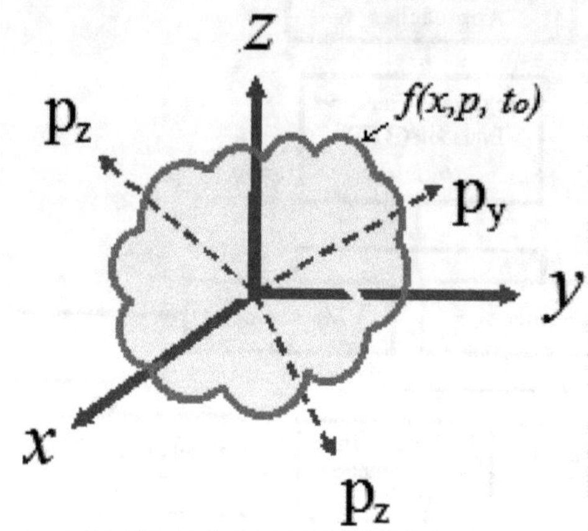

The term *ensemble* was coined by Gibbs (1902). In the ensemble concept, it is not important to know what each particle does, but rather try to know what the ensemble of particles do on the average, and thus a statistical treatment is needed[2]. In order to describe the evolution of a system, one considers the evolution of this density function, rather than the evolution of a single point (particle). The most important tool to investigate the evolution of such a density function is the Liouville theorem (1838).

Upon completion of this Chapter, the reader will be able to

- Explain the advantages and disadvantages of the semiclassical transport theory of charge carriers in semiconductors.
- Understand the scattering mechanisms in semiconductors and how to calculate their probabilities, starting from the Fermi golden rule.
- Understand the Boltzmann transport equation (BTE) and how to solve it under the effect of external electric, magnetic or electromagnetic fields.
- Describe the electrical, thermal, magnetic and optical properties of semiconductors, on the basis of the BTE.
- Understand the linear transport theory and the microscopic relaxation time approximation of the BTE
- Understand the treatment of carrier transport with Monte Carlo method.

2. BOLTZMANN TRANSPORT EQUATION (BTE)

The Boltzmann transport equation (BTE) depicts the statistical evolution of a thermodynamic system under non-equilibrium conditions (Boltzmann, 1896, 1898). The BTE can be derived from Liouville's theorem (1838), which is a generalized form of the continuity equation[3]. The Liouville theorem states that the total change (total derivative) of the particle distribution function in the phase space is null, so that the volume of any region in phase-space is invariant under Hamiltonian evolution.

$$d f(\boldsymbol{x},\boldsymbol{k}, t) / dt = 0 \tag{2a}$$

where $\boldsymbol{x}$ is the particle location: $\boldsymbol{x} = [x, y, z]^T$ and $\boldsymbol{k}$ is the particle wave vector: $\boldsymbol{k} = [kx, ky, kz]^T$. Note that in the semiclassical treatment, we utilize the wavevector, $\boldsymbol{k}$, rather than the particle momentum $\boldsymbol{p}$ and make use of the de Broglie relation ($p=h/\lambda=\hbar k$). By developing the total derivative of the particle distribution function in the phase space ($\boldsymbol{x}$ and $\boldsymbol{k}$), one can obtain:

$$d f / dt = \partial f/\partial t + (d\boldsymbol{x}/dt).\nabla f + (d\boldsymbol{k}/dt).\nabla_k f =0 \tag{2b}$$

or

$$\partial f/\partial t +\boldsymbol{u}_g.\nabla f + (1/\hbar)\,(d\boldsymbol{p}/dt).\nabla_k f = 0 \tag{2c}$$

where $\boldsymbol{u}_g = d\boldsymbol{x}/dt$ is the particle group velocity. We also have the rate of momentum change $\boldsymbol{dp/dt}$ is equal to the total applied force, such that:

$$\boldsymbol{dp}/dt =\boldsymbol{F}_{total} = \boldsymbol{F}(external) +\boldsymbol{F}_c(collisions) \tag{3a}$$

Unfortunately, it is difficult to express analytically the friction force $\boldsymbol{F}_c$, which arises from collisions between gas particles and the surrounding boundaries. However, one can substitute:

$$(1/\hbar)\boldsymbol{F}_c.\nabla_k f \rightarrow - [\partial f/\partial t]_{col} \tag{3b}$$

where $[\partial f/\partial t]_{col}$ represents the rate of change of the distribution function, due to collisions. Hence, we have:

$$\partial f/\partial t +\boldsymbol{u}_g.\nabla f + (\boldsymbol{F}/\hbar).\nabla_k f = [\partial f/\partial t]_{col.} \tag{4a}$$

The above equation is called the semiclassical BTE. In one dimensional space, the BTE reads:

$$\partial f/\partial t +\boldsymbol{u}_{gx}\, \partial f/\partial x + (\boldsymbol{F}_x/\hbar)\, \partial f/\partial k_x = [\partial f/\partial t]_{col.} \tag{4b}$$

2.1 Original Classical BTE

The original formulation of the Boltzmann equation (Boltzmann, 1872), for a gas of particles, gives a deterministic description for the distribution function of gas particles both in *real* and *velocity* space. It is obtained by equating the total time differential for the density distribution function $f(\boldsymbol{x},\boldsymbol{v},t)$ with the local production and loss rates, i.e.

$$\partial(f(\boldsymbol{x},\boldsymbol{v},t))/\partial t +\boldsymbol{v}.\nabla_r(f(\boldsymbol{x},\boldsymbol{v},t))+\boldsymbol{F}/M.\nabla_v(f(\boldsymbol{x},\boldsymbol{v},t))=q^{Ion}(\boldsymbol{x},\boldsymbol{v},t) - l^{Rec}(\boldsymbol{x},\boldsymbol{v},t)+C(\boldsymbol{x},\boldsymbol{v},t) \tag{5}$$

where $\boldsymbol{F}$ contains all external forces on the particle with mass M, i.e. electric, magnetic and gravitational forces. Also, q^{Ion}, l^{Rec} and C are the local production and loss rates due to ionization, recombination and collisions. When $(q^{Ion} - l^{Rec} + C) = 0$, the above equation is called the **Vaslov** equation.

Figure 3. Schematic of the particle velocity space

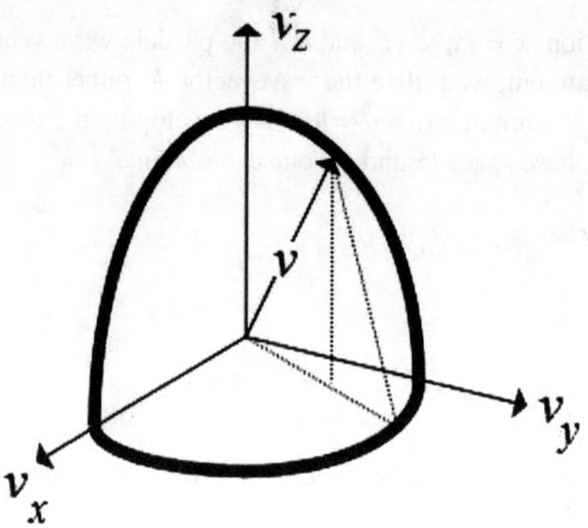

2.2 The H- Theorem

Boltzmann showed in his lectures on gas theory (1896, 1898) that no matter how the velocities of the particles in a gaseous system are initially distributed, the velocity distribution would evolve via successive collisions to the stationary, Maxwellian distribution. Therefore, when the BTE is valid, the following ***H***-function could only decrease in time.

$$H(t) = \int f.log(f)d^3v\, d^3x \tag{6}$$

This way Boltzmann established a kind of macroscopic irreversibility for a gas of particles. In fact the function ***H*** is mathematically equivalent to the formulations of the entropy of a gas of particles[4].

2.3 Boltzmann Transport Equation in Semiconductors

In the bulk of a semiconductor, the semiclassical Boltzmann transport equation (BTE) for a specific gas of charge carriers (4) may be written in the following form:

$$\frac{\partial f_v}{\partial t} + u_v \cdot \nabla f_v + \frac{F}{\hbar} \cdot \nabla_k f_v = \left[\frac{\partial f_v}{\partial t}\right]_{col.} \tag{7}$$

where $f_v = f_v\,(\boldsymbol{k}_v, \boldsymbol{x}, t)$ is the charge carrier distribution function in the phase space (v stands for n, for electrons or stands for p, for holes) and $\boldsymbol{k}_v$ is the charge carrier wave vector. Also $\boldsymbol{u}_v$ is the group velocity of charge carriers and $\boldsymbol{F}$ is the applied external force.

In presence of external electric field intensity ζ and magnetic field intensity $\boldsymbol{B}$, we have $\boldsymbol{F} = -e(\zeta + \boldsymbol{u}_v\, x\, \boldsymbol{B})$ for electrons and $\boldsymbol{F} = +e(\zeta + \boldsymbol{u}_v\, x\, \boldsymbol{B})$ for holes, with e is the elementary charge of electrons. The subscript '*col.*' means the rate of change of the distribution function due to collisions of charge carriers with the crystal lattice and other crystal defects. As we'll explain later on, the collision term can be evaluated using statistical analysis, within a quantum framework.

Once, the BTE is solved and the carrier distribution function $f_v(\mathbf{k}_v, \mathbf{x}, t)$ is known, the density of carriers can be determined.

$$n(x,t) = \int_{k-space} g(k).f_n(k,x,t)d^3k \tag{8}$$

where the integration is taken over the entire k-space and $g(k) = 1/(2\pi)^3$ is the density of states in the k-space (should be multiplied by 2, the number of carriers that each energy level can allocate[5]). The mean value of any observable physical quantity $Q(\mathbf{k}, \mathbf{x}, t)$, like the electron energy, can be obtained as follows:

$$Q(x,t) = < Q(x,k,t) > = \frac{\int_{k-space} g(k).f_v(k,x,t)Q(k,x,t)d^3k}{\int_{k-space} g(k).f_v(k,x,t)d^3k} \tag{9a}$$

Note that the de-nominator of the above equation is equal to the carrier density. For instance, the electron mean velocity is given by:

$$v_n(x,t) = < u_n(x,k,t) > = \frac{\int_{k-space} g(k).f_v(k,x,t)u_n(k,x,t)d^3k}{n} \tag{9b}$$

Evidently, the electron current density is then given by:

$$J_n(x,t) = -env_n = -e\int_{k-space} g(k)\cdot f_v(k,x,t)u_n(k,x,t)d^3k \tag{9c}$$

2.4 Collision Term in the BTE

The collision term in the BTE can be expressed as the sum of all contributions of the acting scattering mechanisms. It can be expressed using the statistical laws as follows:

$$\left[\frac{\partial f_v}{\partial t}\right]_{col} = -\int_{k-space} \left[S(k,k').f_v(k).(1-f_v(k')) - S(k',k).f_v(k').(1-f_v(k))\right].G(k).d^3k' \tag{10a}$$

where $S(\mathbf{k}, \mathbf{k}')$ is the total collision rate (probability density of collision per unit time) from initial carrier state $\mathbf{k}$ to final state $\mathbf{k}'$ in the k-space and $G(k)=V/(2\pi)^3$ is the density of states in the k-space and

V is the volume of crystal lattice (Note that $g(k)=G(k)/V$). The Pauli Exclusion Principle is taken into account via the $(1-f_v)$ terms. In non-degenerate semiconductors, we can neglect the $(1-f_v)$ terms so that:

$$[\partial f_v/\partial t]_{col.} \approx - V/(2\pi)^3 \int [S(\boldsymbol{k}, \boldsymbol{k}') f_v(\boldsymbol{k}) - S(\boldsymbol{k}',\boldsymbol{k}) f_v(\boldsymbol{k}')]\, d^3\boldsymbol{k}'$$

$$= - \lambda(\boldsymbol{k}) f_v(\boldsymbol{k}) - V/(2\pi)^3 \int S(\boldsymbol{k}',\boldsymbol{k}) f_v(\boldsymbol{k}')\, d^3\boldsymbol{k}' \qquad (10b)$$

where $\lambda(\boldsymbol{k}) = V/(2\pi)^3 . \int S(\boldsymbol{k}, \boldsymbol{k}') d^3\boldsymbol{k}'$ is the out collision rate (of carriers at the k-state). The second term in the above equation represents the in-collision rate. According to Ferry (2000), the above definition of $\lambda(k)$ is incomplete, since part or all of it may be absorbed in the second term of (10b).

The total collision rate operator $S(\boldsymbol{k}, \boldsymbol{k}')$, may be also given by the sum of individual rates $S_j(\boldsymbol{k},\boldsymbol{k}')$:

$$S(\boldsymbol{k}, \boldsymbol{k}') = \sum S_j(\boldsymbol{k}, \boldsymbol{k}') \qquad (10c)$$

Evidently, the calculation of the collision integral needs some information about each collision mechanism, and its dependence on the carrier wave vector $\boldsymbol{k}$ and/or energy $E(\boldsymbol{k})$. However, when the carrier initial state $\boldsymbol{k}$ and final state $\boldsymbol{k}'$ are not exactly situated on the symmetry points in the k-space (like X, L), where the $E(k)$ relation is well known, the determination of the collision rate $S_j(\boldsymbol{k}',\boldsymbol{k})$ of a collision mechanism becomes difficult. Also, the coupling potential, which is needed to calculate the carrier-phonon collision rate, is not well known at high electric fields. Therefore, one has to adopt some simplifying assumptions. For instance, collisions are assumed to be point-like and instantaneous in the conventional semiclassical picture. Therefore, we make use of the so-called first **Born** approximation of diluted gases, which is also known as the **Fermi** golden rule, to calculate the collision rates.

Physical Note 1: Elastic and Inelastic Collisions

An elastic collision is a collision where total momentum and total kinetic energy are both conserved. An inelastic collision is a collision in which total momentum is conserved but total kinetic energy is not conserved; the kinetic energy is transformed into other kinds of energy. Therefore, the total momentum of the colliding particles before inelastic collisions is the same as after the collision. But the total kinetic energy before and after the inelastic collision is different.

When an electron is scattered by lattice vibrations (phonons), the following relations depicts energy and momentum, when the collision is elastic.

Energy conservation dictates $E(k') = E(k) \pm \hbar\omega(q)$

Figure 4. Phonon absorption and emission

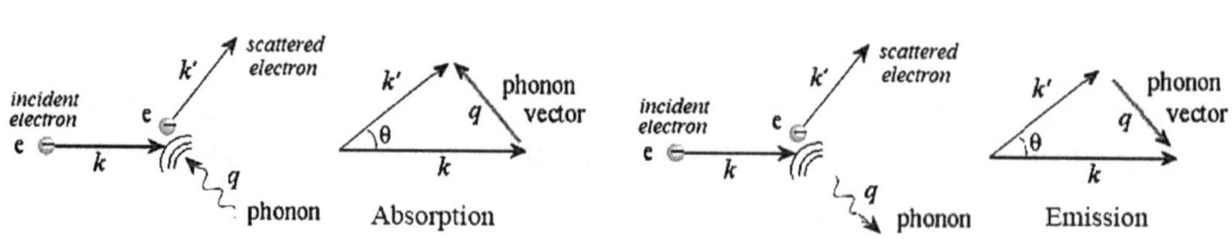

Momentum conservation dictates $k' = k \pm q \rightarrow k'^2 = k^2+q^2 \pm 2kq.\cos\theta$ where θ is the angle between k and k'. In the above relations, we assume a Normal scattering process, where electrons are scattered in the same Brillouin zone (both k, k' $\in$ BZ).

If the electron is scattered outside the first Brillouin zone (in the so-called Umklap process), then the momentum is not conserved and the above relation should be corrected by the reciprocal lattice vector ($k= k' \pm q+G$).

In equilibrium the distribution function is known and the left hand side of the Boltzmann equation is equal to zero. Therefore,

$$f_0(k')(1-f_0(k))S(k',k) = f_0(k)(1-f_0(k'))S(k,k') \tag{11a}$$

Using the explicit form of the Fermi-Dirac distribution function, we get:

$$S(k',k)\exp\left[\frac{E(k)}{k_B T}\right] = S(k,k')\exp\left[\frac{E(k')}{k_B T}\right] \tag{11b}$$

This equation relates the forward and backward probabilities of collisions and known as the principle of detailed balance. For elastic collisions we have $E(k) = E(k')$, and hence:

$$S(k',k) = S(k,k') \tag{11c}$$

2.5 Collision Matrix

We know that the dynamic perturbation theory in quantum mechanics is concerned with determining the energy eigenvalues of a particle in perturbed non-equilibrium state (Bardeen & Shockley, 1950). By applying this theory, we can obtain the energy eigenvalues of a perturbed crystal by solving the Schrödinger equation with the perturbed Hamiltonian $\hat{H}$:

$$\hat{H}(r) = H_o + e\boldsymbol{\zeta}.\boldsymbol{x} + H_{def} \tag{12}$$

where H_o is the unperturbed Hamiltonian of the crystal system (electrons and ions), $\boldsymbol{\zeta}$ is the external field and H_{def} is the Hamiltonian perturbation component due to system defects (e.g., lattice vibrations or impurities). We also know that according to the Born approximation (Born & Oppenheimer, 1927), the charge-carrier collisions may be considered as instantaneous, for weak interactions where collisional broadening is negligible. In this case, the probability of transition per unit time of charge carriers (from state k to state k'), due to a scattering mechanism may be approximated by the Fermi golden rule (Shockley, 1950):

$$S_j(\boldsymbol{k}, \boldsymbol{k'}) = (2\pi/\hbar) \, . \, |A(\boldsymbol{k}, \boldsymbol{k'})|^2 \, \delta(E - E') \tag{13}$$

where $E=E(k,q)$ and $E'=E(k',q')$ represent the initial (unperturbed) and final (perturbed) energy states, respectively. The Kronecker delta function in the above expression describes the conservation of crystal energy[6]. Also, $A(\mathbf{k},\mathbf{k'})$ is called the transition matrix element of the Hamiltonian perturbation operator H_{def}

$$A(\mathbf{k}, \mathbf{k'}) = \langle \mathbf{k},q \mid H_{def} \mid \mathbf{k'}, q' \rangle \tag{14a}$$

The symbols q and q' represent here the crystal lattice wave vectors before and after perturbation[7]. Taking the adiabatic approximation into consideration, the transition matrix element can be further reduced as:

$$A(\mathbf{k}, \mathbf{k'}) = \langle q \mid H_{def} \mid q' \rangle .\xi. I(k, k') \tag{14b}$$

where $I(k, k')$ is the overlap factor, and ξ is the polarization tensor. In summary, the scattering rate per unit time from state $E=E(k,q)$ to state $E'=E(k',q')$ for a given scattering mechanism, is given by:

$$S_j(\mathbf{k},\mathbf{k'}) = (2\pi/\hbar) .|\langle q| H_{def} | q'\rangle|^2.\xi .I(k,k').\delta (E - E') \tag{15}$$

The overlap factor, $I(k,k')$, depends on the band structure of the semi-conductor, and the type of involved electron transition:

$$I(k, k') = | N. \int u^*_{k'} (r)\, u_k(r) \exp (j\mathbf{G}.\mathbf{r})\, d^3r |^2 \tag{16}$$

where N is the number of unit cells in the crystal, $u_k(r)$ and $u^*_{k'}(r)$ are the electron Bloch functions, before and after scattering and G is the reciprocal lattice (k-space) translational vector. When the scattering process is normal (N-process), then $G=0$, and when the scattering process is umklapp (U-processes) then $G\neq0$. In normal processes and when the involved electrons have exact plane waves, the overlapping integral is usually set to unity. However, for some transitions between warped bands $I(k,k')\rightarrow I(\theta)$, where θ is the angle between k and k', and $I(\theta)$ may be different from unity.

Considering the electron collisions with lattice vibrations (phonons), the initial and final energy states may be expressed as follows:

$$E = E(k) + N_q \hbar\omega,\ E'= E(k') + (N_q\pm1)\hbar\omega \tag{17}$$

Here $\hbar\omega$ is the amount of exchanged energy between electron and the lattice vibrations (phonons), N_q is the initial number of phonons and $N_q\pm1$ is their final number after a collision. The ± sign depends if the collision involve emission or absorption of a phonon.

It should be noted that the scattering probability $S(\mathbf{k},\mathbf{k'})$, as defined by the Fermi golden rule, is not a measurable quantity since the measured probability should be a sum over all k' states. To find a measurable probability, we need to sum the probability over all final states. The integrated scattering probability of a certain collision mechanism of charge carriers per unit time $\Gamma_j(\mathbf{k})$ is defined as sum over all possible values of $\mathbf{k'}$ in the k-space:

$$\Gamma_j(\mathbf{k}) = \int S_j(\mathbf{k}, \mathbf{k'})\, G_v(\mathbf{k'})\, d\mathbf{k'} \tag{18}$$

Again, $G_v(\mathbf{k})=V/(2\pi)^3$ is the density of states in the k-space. Note the similarity between the definitions of $\lambda(k)$ and $\Gamma_j(k)$. However, the subscript '$_j$' means a particular collision mechanism.

The mean (scattering) time between collisions, $\tau_j(\mathbf{k})$, of a certain collision mechanism of charge carriers, is defined as the inverse of the integrated scattering probability per unit time $\Gamma_j(\mathbf{k})$. Therefore:

$$\tau_j(\mathbf{k}) = 1 / \Gamma_j(\mathbf{k}) \tag{19}$$

In summary, to calculate the scattering rate of a certain collision mechanism and the mean free time between collisions, we proceed as follows:

1. Consider the system Hamiltonian as: $H = H_o+e\zeta+H_1$, where H_o is the unperturbed Hamiltonian, ζ is the applied field and H_1 is the perturbation due to a crystal defect. H_o may be given by the effective mass Hamiltonian, which lumps the interatomic forces in the effective mass of carriers.
2. Find out the Bloch functions of the system (by solving the Schrödinger equation, with Hamiltonian H_o). In bulk semiconductors, H_o is simply the crystal potential, which upon solution yields the band structure $E(k)$ of the material. You can then calculate the density of states.
3. Now we revert our attention to the crystal defect and its perturbation potential H_1. Use a suitable model to express this potential. On the basis of the quantum perturbation theory, use the Fermi golden rule to express the scattering probability (or transition rate) from k to k', using (16).
4. Calculate the integrated scattering probability Γ_j for the scattering mechanism by summing over all available final state, using (18)
5. Calculate the mean free time between collisions (momentum relaxation time) for each collision mechanism $\tau_j = 1/\Gamma_j$
6. If you want to calculate the drift mobility (due to this scattering type), average the momentum scattering rate (over the carrier distribution function) $\tau_m = <\tau_j>$. This needs the knowledge of the carrier distribution function, which may be obtained by solving the BTE (or assuming a priori any suitable distribution function!). Then use the relation $\mu = e\,\tau_m/m^*$

3. SCATTERING MECHANISMS IN SEMICONDUCTORS

The lattice atom vibrations (phonons) hinder the motion of electrons in the crystal lattice and collide with them. As electrons travel through the material, they bounce off these vibrating atoms, giving rise to electrical resistance of the material. There exist many collision mechanisms, which may occur between the charge carriers (electrons or holes) and phonons as well as other quasi-particles and among the charge carriers themselves. Among the important collision mechanisms in solids, one can cite:

1. Carrier - Phonons (quanta of lattice atom vibrations) collisions,
2. Carrier - Impurity atoms collisions,
3. Carrier - Polarons (lattice-distortion polarization field) collisions,
4. Carrier - Magnons (quanta of spin coupling) collisions,
5. Carrier - Carrier (electron-electron) collisions,
6. Carrier - Plasmons (quanta of many electrons oscillations) collisions,
7. Carrier - Excitons (excited electron-hole pair) collisions and

8. Carrier - Covalent bond (impact ionization) collisions.

The electron–phonon collisions are the most common collision mechanism in semiconductors. Phonons may be acoustic or optical, according to the amount of exchanged energy during collision.

A polaron is a quasiparticle used to understand the interactions between a charge carrier (electron or hole) with a distortion of the surrounding charges of atoms in a dielectric or a polar semiconductor.

An exciton is a bound state of an electron and a hole which are attracted to each other by the electrostatic Coulomb force. Excitons are often the lowest energy excitation created by light absorption in semiconductors. The so-called biexcitons and trions are complex forms of multiple excitons.

Also, a magnon is a collective excitation of the electron spin in a crystal lattice. This is analog to phonons, which are collective excitation of the lattice vibrations.

3.1 Intra-Valley and Inter-Valley Phonon Collisions

There exist two basic types of carrier-phonon collisions, namely:

1. Intra-valley collisions (within the same energy valley),
2. Inter-valley collisions (between different energy valleys).

The intra-valley collisions involve transitions within the same energy valley in the E-k diagram. This type of collision is often elastic and involves the emission or absorption of acoustic phonons of small energy.

The inter-valley collisions involve transitions between different energy valleys (and sometimes different energy bands) in the E-k diagram. This type of collision is often inelastic and involves the emission or absorption of optical phonons. Figure 6 depicts the carrier-phonon mechanisms of collision. As shown in figure, the phonon scattering may be polar or involve some specific transitions, according to the semiconductor material.

Figure 6 depicts the basic mechanisms of collision in Si, according to the type of transition. The so-called *f*-type collisions involve transitions of electrons between two neighboring perpendicular valleys in the k-space, like the (100) and the (010) valleys of the *Si* main conduction band. Also the *g*-type collisions involve transition of electrons between two valleys in opposite directions in the k-space. For instance, transitions between the (100) and the (-100) valleys of the *Si* conduction band.

Figure 5. Schematic illustration of lattice vibrations (phonons), polarons and excitons in a crystalline semiconductor

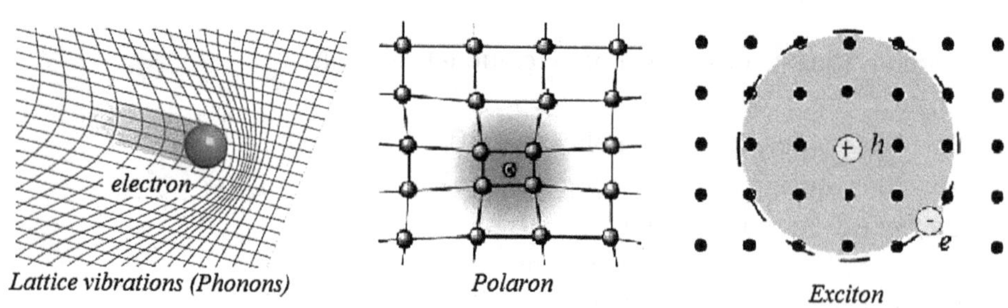

Figure 6. Carrier-phonon scattering mechanisms

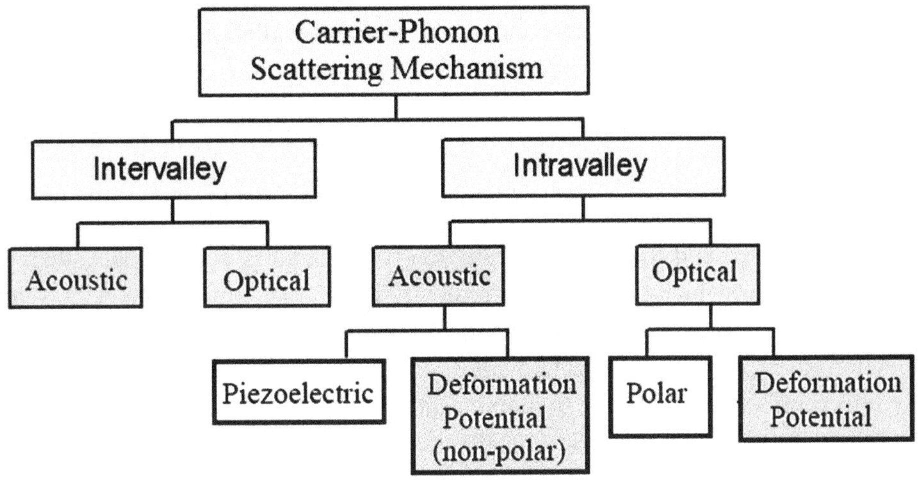

Figure 7. Fundamental scattering mechanisms in silicon

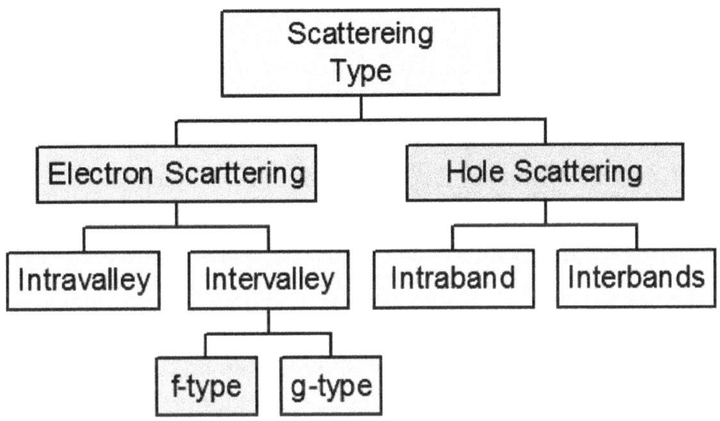

Figure 8. Schematic illustration of intravalley and intervalley transitions in silicon

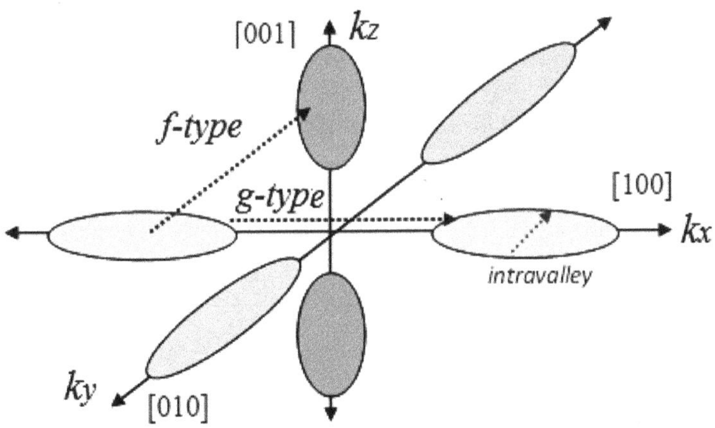

There exist three *f*-type mechanisms (denoted f_1, f_2 and f_3) and three *g*-type mechanism (denoted g_1, g_2 and g_3).). Similarly, holes may scatter within the same valence band or between different bands. When holes scatter within the same valence band, the scattering is called intraband. When holes scatter between different bands, the scattering is interband.

3.2 Selection Rules

The intravalley and intervalley transitions are not always permitted in any direction in the *k*-space. Some transitions are not permitted and the corresponding matrix element is zero. In fact, there exist some rules, called *selection rules*, which define the permitted types of transitions. In physics, selection rules define the transition probability from a quantum state to another. Selection rules may be divided into electronic selection rules, vibrational selection rules, and rotational selection rules. We are interested here about the selection rules of electronic transitions, during the carrier-phonon scattering process. For instance, we know that the main conduction band in *Si*, has 6 equivalent valleys along the directions <100>, which is usually termed as Δ. For a g-type intervalley transition in *silicon*, from the valley $(k_o, 0, 0)$ to the valley $(-k_o, 0, 0)$, where $k_o = 0.85(\pi/a_o)$, the transition rules dictates:

$$\Delta_1(k_o) ¤ \Delta_1(-k_o) = \Delta_1(2k_o) \tag{20}$$

where Δ_1 is the required symmetry of the associated electron wave function and ¤ represents the group convolution operator. Thus, in silicon, the required phonon must have a wave vector $q=2k_o = 1.7x(\pi/a_o)$ along the <100> direction (the Δ direction). However, such a vector is located outside the first Brillouin zone, beyond the X point, and the symmetry Δ_1 passes over into the symmetry $\Delta_{2'}$ in the second Brillouin zone. On the other hand, it's well known from the phonon spectrum in *Si*, that there are only LO phonons of such symmetry. Such a g-type intervalley transition is only permitted with LO phonons. For phonon-assisted transitions in *Si*, the selection rules allow acoustic intravalley scattering, *f*-type scattering with LA (longitudinal acoustic) and TO (transverse optical) phonons, and *g*-type scattering with LO phonons.

Table 1 depicts the selection rules for optical-phonons assisted transitions in both *Si*, and *GaAs*. For other III-V semiconductor compounds, same rules apply, except for the LO mode is changed to LA mode if the mass of the III-group atom is greater than that of the V-group atom. As shown in the table, the selection rules forbid intravalley transitions with optical phonons in *Si*. However, when the final state of a transition is not exactly situated on certain symmetric points in the *k*-space (like X and L), where the $E(\mathbf{k})$ relation is well known, the selection rules are less restrictive.

Table 1. Selection rules in Si and GaAs for optical mode transitions

Material	*Si*	*GaAs*	
Intravalley	Forbidden	Γ (LO)	
Intervalley	f-type: Σ_1 (LA, TO)	Γ → L: L (LO)	
	g-type: $\Delta_{2'}$ (LO)	Γ → X: X (LO)	
		L → L: X (LO)	

From Semiconductors, Ferry, 2000.

It worth notice that there are several experimental indications showing other intervalley scattering mechanisms in silicon. The contradiction of such mechanisms with the selection rules may be due to the fact that these rules are evaluated for electron transitions between points which are exactly situated on the symmetric points and along symmetric directions of the first Brillouin zone (Ferry, 2000). Actually, electrons may undergo transitions between points, which are far from these symmetry axes.

3.3 Collision Rates in Cubic Semiconductors

The detailed studies of the charge transport and collision rates in *Si* and other semiconductors have been started about 60 years ago. However, many scattering mechanisms in *Si* were not completely understood until recently (Cartier Fischetti, Ecklund, & McFeely, 1993). Figure 7 depicts the intravalley acoustic phonon and intervalley scattering rates in silicon, as a function of the electron energy. Also, Figure 8 depicts some reported results for the scattering rates in undoped (pure) *Si,* as a function of the electron energy.

As shown in Figure 9, so many research groups disagree on the magnitude of the electron-phonon scattering rates, particularly at high energies. Also, the published impact ionization rates vary over 3 orders of magnitude. In the following subsections, we present the expressions of the transition probability of the well-known collision mechanisms in silicon and other semiconductors, according to Jacoboni and Reggiani (1983), Fischetti and Lux (1993), and Seeger (1997). More recent treatment of scattering mechanisms in quantum structures and nanodevices, such as 2DEG systems, can be found in Davies' (1998) book (*Physics of low-dimensional semiconductors*). However, we present here the surface acoustic phonons and roughness scattering mechanisms, which are 2-dimensional scattering types. Other scattering mechanisms in low-dimensional semiconductors will be presented the in chapters 5, 6 of this book.

Intravalley Scattering (Case of Acoustic Phonons)

The intravalley acoustic phonon scattering is dominant at low energy and its calculation is carried out by the deformation potential. In the deformation potential interaction, the scattering between electrons

Figure 9. The electron collision rates in Si at 300K
After Jacoboni and Reggiani (1983).

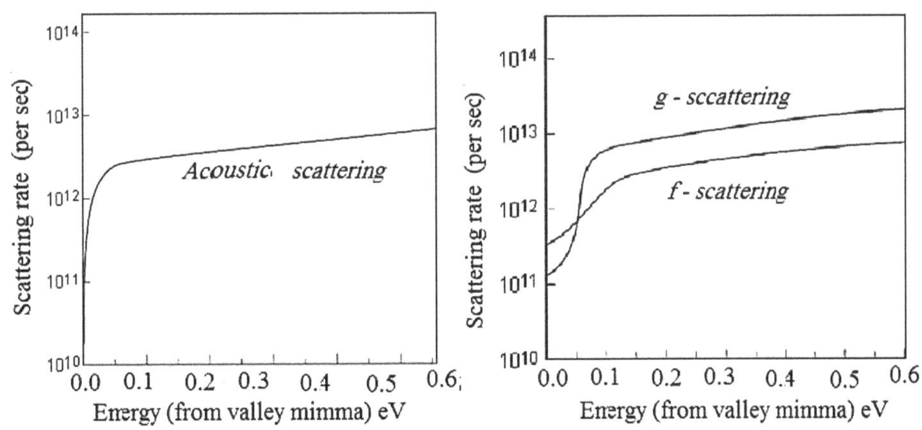

Figure 10. (a) Electron-phonon collision rates in Si at 300K, according to several research groups; (b) electron impact ionization rate. in Si at 300K, according to several research groups
After Fischetti, (1998).

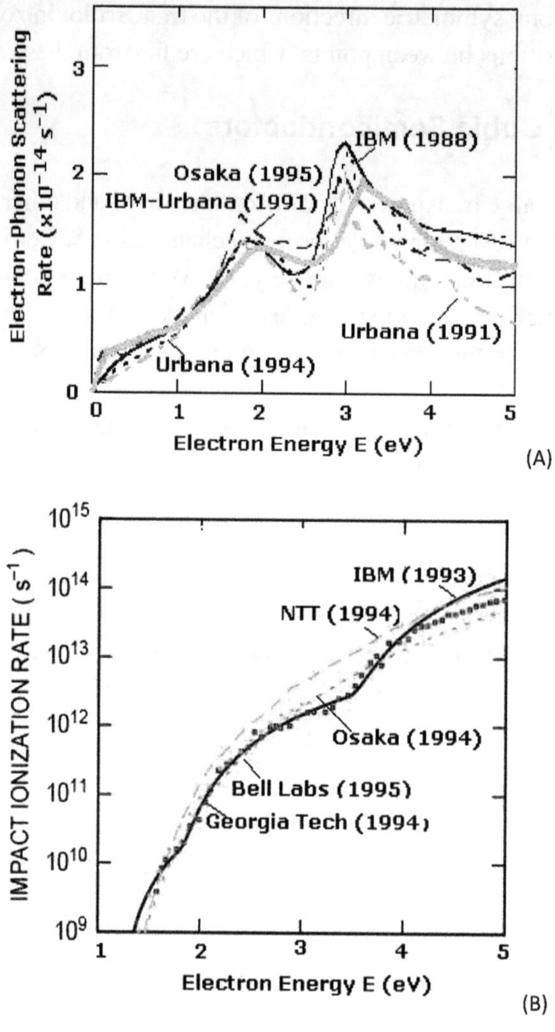

and phonons occurs when the electrons scatter off phonons (lattice vibrations). Such elastic waves are of the type $u(\mathbf{r},t)=u_o.exp(j\omega t\text{-}\mathbf{q}.\mathbf{r})$, where $u(\mathbf{r},t)$ is the atomic displacement from the equilibrium position at a given point $\mathbf{r}$ of the lattice, ω and $\mathbf{q}$ are the frequency and wavevector of the lattice waves. According to Bardeen and Shockley (1950), the interaction potential H_{def} due to small energy lattice vibrations (acoustic phonons) in the crystal is given by:

$$H_{def}(\mathbf{r}) = E_I \nabla u(\mathbf{r}) \tag{21}$$

where the atomic displacement $u(\mathbf{r})$ is generally a sum on all possible lattice vibration modes. Also E_I is the deformation potential constant (eV). The deformation potential is generally a tensor, i.e. a quantity which depends on the direction. However, when collisions are isotropic, E_l may be considered as a scalar quantity.

In intravalley transitions, the exchanged energy of phonons is so small compared to the carrier energy ($\hbar\omega \approx \hbar q v_s$, where v_s is the velocity of sound and $q=k-k'$ is the phonon wave-vector). Therefore, the intravalley collisions with acoustic phonons are often considered elastic and isotropic. Also, for spherical energy surfaces, we can assume that the deformation potential a scalar constant. This allows calculating the matrix element of the intravalley acoustic scattering mechanism, as follows:

$$|A(\mathbf{k}, \mathbf{k}')|^2 = ½\, E_l.k_B T /(V.\rho\, v_s^2) \tag{22a}$$

wher ρ and V are the density and volume of the crystal lattice, v_s is the velocity of sound. Using the Fermi golden rule, as indicated by (16), we can calculate the probability of collision of such a scattering mechanism. Therefore, the probability of collision of electrons with elastic phonons per unit time in intravalley transitions is given by the following relation:

$$S_j(\mathbf{k}, \mathbf{k}') = (\pi\, q\, E_l^2 /V \rho\, v_s).N_q .\delta(E(\mathbf{k}) - E(\mathbf{k}') - \hbar\, \omega) \tag{22b}$$

for the absorption of phonons, and

$$S_j(\mathbf{k}, \mathbf{k}') = (\pi\, q\, E_l^2 /V \rho\, v_s).(N_q+1).\delta(E(\mathbf{k}) - E(\mathbf{k}') +\hbar\, \omega) \tag{22c}$$

for the emission of phonons. Here, N_q is the phonon distribution function, which is assumed in equilibrium (Bose-Einstein distribution) and given by:

$$N_q = 1 / [\; exp(E(q)/k_B T -1\;] \tag{23a}$$

Here, $E(q)= \hbar\omega \approx \hbar\, q.v_s$ is the acoustic phonon energy and N_q may be approximated by a truncated Laurent expansion as follows:

$$N_q = 1/x - 1/12\; x - 1/720\; x^3 - ½ \text{ for } (\; x< 3.5),\; N_q = 0 \text{ elsewhere} \tag{23b}$$

with $x = \hbar q.v_s/k_B T$. As x is normally much smaller than unity (except at very low temperatures), N_q is sometimes approximated as $(k_B T/\hbar q\; v_s$ - ½). We can finally write the acoustic phonons scattering rate as follows:

$$S_j(k,k') = \frac{\pi q E_1^2}{V\rho v_s}\begin{bmatrix} N_q+1 \\ N_q \end{bmatrix} \cdot \delta(E(k') - E(k) \pm \hbar q v_s) \tag{24}$$

The ± sign depends if the collision involve emission (upper term) or absorption (lower term) of a phonon. The integrated scattering probability of acoustic phonons per unit time can be approximated by the following relation for both absorption and emission (Jacoboni & Reggiani, 1983):

$$\Gamma_j(k) = \frac{\sqrt{2}(m_n^*)^{3/2} E_1^2 k_B T}{\pi \hbar^4 \rho v_s^2} \left(E(k)\right)^{1/2} \tag{25a}$$

where m_n* is the electron effective mass and T is the lattice temperature.

In the above relation, we assumed a parabolic semiconductor, with spherical energy surfaces and constant effective mass. For non-spherical energy surfaces (where E-k is non-parabolic), we can take into account the non-parabolicity of the *E-k* relation by choosing a suitable non-parabolicity factor[8] (α) and we replace m_n* by the density-of-states effective mass m_{nd}* ($m_{nd}^* = (m_t^2\, m^l)^{1/3}$) to obtain:

$$\Gamma_j(k) = \frac{\sqrt{2}(m_{nd}^*)^{3/2} E_1^2 k_B T}{\pi \hbar \rho v_s^2} (1+\alpha E)^{1/2} \cdot (1+2\alpha E) \cdot E^{1/2} \tag{25b}$$

We dully note that the deformation potential is generally a tensor. Here, we neglect anisotropy and E_l is the average of the longitudinal (dilatation) and transverse (shear) components which are usually denoted by E_d and E_u. In addition, the sound velocity, v_s, is averaged over longitudinal and transverse valleys such that $v_s = ⅓(v_l+2v_t)$. Also, the total intravalley scattering rate should be calculated separately for the *LA* and *TA* phonons, as a function of the initial electron energy. In addition, the overlap factor is typically set to unity (*I(k,k')=1*) in intravalley scattering. Note also that the above expression is derived for bulk semiconductors. Similar expressions can be derived for the case of 2-D and 1-D semiconductor structures.

Note 2: Intravalley Deformation Potential

The intravalley deformation potentials have a general angular dependence, which can be written as follows (Herring & Vogt, 1956):

$\mathbf{E}_{LA}(\theta) = E_d + E_u \cos^2\theta$,

$\mathbf{E}_{TA}(\theta) = E_u \sin\theta \cos\theta$,

where θ is the angle between the phonon wave vector and the longitudinal axis of the conduction-band valley. Detailed calculations show that the influence of the angular dependence on the electron transport is relatively small. Hence, the intravalley deformation potentials can be averaged over the angle θ. The isotropic averaged deformation potentials are then:

$E_{LA} = \sqrt{(\pi/2)(E_d^2 + E_d E_u + ⅜E_u^2)}$,

$E_{TA} = ¼ \sqrt{\pi}\, E_u$

Therefore, E_l in (25a) and (25b) should be replaced by the appropriate deformation potential (E_{LA} or E_{TA}) for intravalley LA and TA phonons.

Intravalley Scattering (Case of Optical Phonons)

According Harrison (1956) and Lawaetz (1969) the crystal deformation potential H_{def} due to high energy phonons (optical phonons) is proportional to the atoms displacement ($\delta u = u_{n+1} - u_n$), not to its divergence.

$$H_{def} = D_t\, \delta u \tag{26}$$

where D_t is the optical phonons coupling constant [eV/cm]. The final formula of acoustic scattering rate may be then used, with replacing the deformation potential by the phonon coupling constant D_t, such that:

(Acoustic phonons Expression)→ (Optical phonons Expression)

$$E_u q \rightarrow D_t \tag{27a}$$

$$S_j(k,k') = \frac{\pi D_t^2}{V\rho\omega_{op}} \begin{bmatrix} N_{op}+1 \\ N_{op} \end{bmatrix} \cdot \delta(E(k') - E(k) \pm \hbar\omega_{op}) \tag{27b}$$

Here ω_{op} is the optical phonon angular frequency and N_{op} is the number of optical phonons (which is q-independent) and maybe approximated by the Boson-Einstein equilibrium statistics:

$$N_{op} = 1/[\, exp(\hbar\omega_{op}/k_B T) - 1\,] \tag{27c}$$

Also, the integrated probability of transitions due to scattering with such deformation potential optical phonons is given by:

$$\Gamma_j(k) = \frac{(m_n^*)^{3/2} D_t^2}{\sqrt{2}\pi\hbar^3 \rho\omega_{op}} \begin{bmatrix} N_{op}+1 \\ N_{op} \end{bmatrix} \cdot \left(E(k) \pm \hbar\omega_{op}\right) \tag{28a}$$

where the upper positive sign (with $N_{op}+1$) stand for absorption, while the lower negative sign (with N_{op}) stand for the emission rate. Again, the above relation assumes a parabolic semiconductor, with spherical energy surfaces and constant effective mass. In nonspherical energy surfaces, we can take into account the non-parabolicity of the E-k relation via a non-parabolicity factor (α). Thus, we replace the electron effective mass m_n* by the density-of-states effective mass m_{nd}* such that:

$$\Gamma_j(k) = \frac{(m_{nd}^*)^{3/2} D_t^2}{\sqrt{2}\pi\hbar^3 \rho\omega_{op}} \begin{bmatrix} N_{op}+1 \\ N_{op} \end{bmatrix} .(1+\alpha E')^{1/2}.(1+2\alpha E').E'^{1/2} \tag{28b}$$

where $E'=E(\mathbf{k}) \pm \hbar\omega_{op}$.

Intervalley Scattering

In intervalley transitions, the integrated probability of transition of carriers per unit time is given by:

$$\Gamma_j(k) = N_m \cdot \frac{(m^*_{nd})^{3/2} D_m^2}{\sqrt{2}\pi\hbar^3 \rho\omega_m} \begin{bmatrix} N_p + 1 \\ N_p \end{bmatrix} \cdot \left(E(k) \pm \Delta_m \pm \hbar\omega_m\right)^{1/2} \tag{29a}$$

where D_m is the coupling coefficient between the two involved valleys, Δ_m is the energy shift between them (Δ_m=0 for equivalent valleys) and $\hbar\omega_m$ is the associated phonon energy. The ± sign refers to the case of phonon emission or absorption. Note that N_p is the number of emitted phonons and N_p+1 is the number of absorbed phonons. Also N_m is the number of possible destination valleys, where transition can end. In *Si* we have $N_m = 4$ for *f*-type transitions and *1* for *g*-type transitions. We dully note that the above relation assumes a parabolic semiconductor, with spherical energy surfaces and constant effective mass. In nonspherical energy surfaces, we can take the non-parabolicity of the *E-k* relation into account via a non-parabolicity factor (α), such that:

$$\Gamma_j(k) = N_m \cdot \frac{(m^*_{nd})^{3/2} D_m^2}{\sqrt{2}\pi\hbar^3 \rho\omega_m} \begin{bmatrix} N_p + 1 \\ N_p \end{bmatrix} \cdot E'^{1/2} . \left(1 + \alpha E'\right)^{1/2} . (1 + 2\alpha E') \tag{29b}$$

where $E' = E(k) \pm \Delta_m \pm \hbar\omega_m$. Intervalley scattering can also include an overlap factor, but its value is typically merged into Δ_m. The coupling coefficient between different valleys (D_m) of a given semiconductor can be evaluated from the phonon spectrum[9]. For instance, the coupling coefficient between the *X* and *L* valleys of *Si* conduction bands amounts to $3x10^8$ *eV/cm.* Table 2 illustrates the deformation potential, and the coupling coefficients of some known collision mechanisms in *Si* and *GaAs*.

Collisions with Ionized Impurities

Impurity scattering is very important for low energy carriers in highly-doped regions, like the source and drain regions in MOSFET devices. The carrier collision with ionized impurities (donors and acceptors) is often intravalley elastic Coulomb scattering. The most popular models for ionized impurity scattering are Conwell and Weisskopf (CW) model (1950) and the Brooks and Herring (BH) model (1951). Figure 11 illustrates the impurity scattering process according to these models. Conwell and Weiskopf suggested truncating the Coulombic potential, $V(r)= Z(e^2/4\pi\, \varepsilon_s\, r)$ where *Z* denotes the number of charges of the impurity, at a value which corresponds to half the distance between two impurities. This approach is adopted because the deflection from a straight path is governed by the closer impurities. When the doping is homogeneous, the average distance between two impurities is just the cube root of the impurity density ($N^{1/3}$). Therefore, the integrated transition probability, due to impurities, may be given by the following relation (for parabolic bands):

$$\Gamma_j(\mathbf{k}) = \pi\, (2 / m^*)^{1/2}\, N^{1/3}\, E^{1/2} \tag{30}$$

Figure 11. Illustration of the scattering of a point charge in a Coulomb potential. Here, Ze is the impurity ion charge, b is the screening length, and θ is the scattering angle.

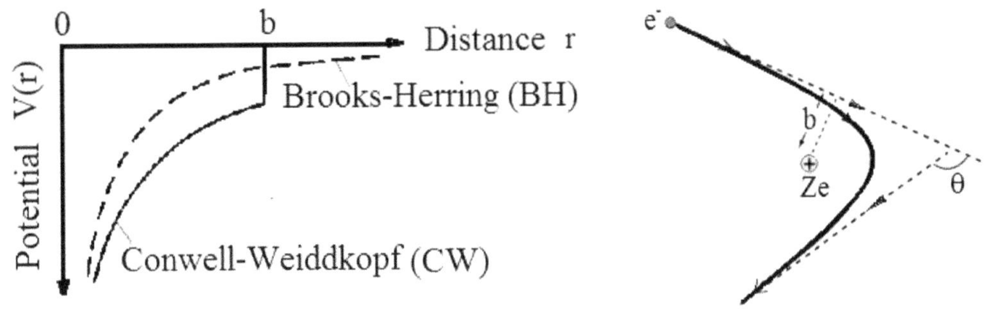

where N is the impurity concentration and E and m^* are the carrier energy and effective mass, respectively.

The other model, which is based on the screening potential, was developed by Brooks and Herring. According to the screened potential theory, the perturbation (deformation) potential due to ionized impurities is given by:

$$H_{def}(\mathbf{r}) = Z(e^2/4\pi\,\varepsilon_s\, r).exp(-r/\lambda_D) \tag{31a}$$

Here λ_D is the screening length (or the Debye length), due to the Coulomb scattering:

$$\lambda_D=\sqrt{(\varepsilon_s k_B T_L/e^2 n_o)} \text{ for } T_L>>0 \text{ and } \lambda_D=\sqrt{(2\varepsilon_s E_F/3e^2 n_o)} \text{ for } T_L\rightarrow 0 \tag{31b}$$

where n_o is the equilibrium electron density, E_F is the Fermi energy level and ε_s is the dielectric constant of the semiconductor. Therefore, we can express the impurity scattering rate as follows:

$$S_j(k,k') = \frac{z^2e^4N}{V\hbar^3\varepsilon_s^2}\cdot\frac{1}{(q^2+q_D^2)}\delta(E(k)-E(k')\pm\hbar\omega) \tag{31c}$$

where $q_D = 1/\lambda_D = \sqrt{(e^2 n_o/\varepsilon\, k_B T)}$ is the inverse Debye length, V is the crystal volume and $q=|\mathbf{k}\text{-}\mathbf{k}'|$. In this case, the integrated transition probability due to impurity scattering may be given by the Brooks-Herring relation:

$$\Gamma_{BH}(k) = \frac{2\pi.z^2e^4N}{\hbar\varepsilon_s^2}\cdot\frac{g_c(E)}{q_D^2(4k^2+q_D^2)} \tag{31d}$$

where $g_c(E)$ is the corresponding density of states (e.g., in the conduction band for donor impurities). However, Mahan (2000) showed that the electron wavefunction at large impurity densities is spread almost uniformly and without *piling up* around donor impurities (or away from acceptors). In this case the screening length ($1/q_D$) becomes comparable to the distance between impurities (about 1-3 nm in *Si* at 10^{18} to 10^{19} cm^{-3}). Hence, it is better to go to the Conwell-Weisskopf model, which assumed that the

Coulomb potential is unscreened. The so-called Ridley model has attempted to find a model reconciling the BH and CW approaches, as follows:

$$\Gamma_R = (v_g/d)\,[1\text{-}exp\,(d\Gamma_{BH}/v_g)] \tag{32}$$

where $d = 1/(2\pi N_d)^3$ is average distance between impurity atoms and v_g is electron group velocity.

A more accurate model, which is also based on the screening potential theory, expresses the integrated transition probability due to impurity scattering as follows:

$$\Gamma_j(\mathbf{k}) = (e^4\, z^2\, N/4\pi\, \hbar^4 \varepsilon_s{}^*)\, E^{1/2}.\ I(k,\theta) \tag{33a}$$

where θ is the angle between $\mathbf{k}$ and $\mathbf{k}'$ such that $q^2 = 2k^2(1\text{-}cos\theta)$ and $I(k,\theta)$ is given by the following integral over all possible values of θ:

$$I(k,\theta) = \int_{-\pi}^{\pi} \frac{\sin\theta.d\theta}{q^2 + q_D^2 \cdot F(\xi,\mu))} \tag{33b}$$

with $\xi = \hbar q/(8m^*k_B\,T_L)^{1/2}$, $\mu = E_F/(k_B T_L)$ and $F(\xi,\mu)$ is given by:

$$F(\xi,\mu) = \int_0^{\infty} \ln\left(\frac{x-\xi}{x+\xi}\right)\left(\frac{x.dx}{1+\exp(x-\mu)}\right) \tag{33c}$$

where $x=\hbar k/(2m^*k_B\,T_L)^{1/2}$. For nonparabolic bands the term $E^{1/2}$ has to be replaced by $(1+\alpha E)^{1/2}(1+2\alpha E)$ $E^{1/2}$, where α is the non-parabolicty factor. Also we replace m^* by the density-of-states effective mass.

Collisions with Neutral Impurities

The collision with neutral impurities (as oxygen) has a small contribution to the carrier mobility, only at very low temperatures and high concentration.

Intraband Hole Scattering

The intraband holes interaction with acoustic-phonons may be considered similar to the intravalley electron interaction with acoustic-phonons (16). However, due to warping and degeneracy of valence bands in covalent semiconductors, an overlap factor appears in the scattering rate. Thus, the intraband scattering rate of holes with acoustic phonons is given by:

$$S_j(k,k') = \frac{\pi q E_2^2}{V\rho v_s}\begin{bmatrix} N_q + 1 \\ N_q \end{bmatrix} \cdot I(\theta)\cdot \delta(E(k') - E(k) \pm \hbar q v_s) \tag{34a}$$

where $I(\theta)$ is the overlapping factor, with θ is the angle between k and k' and E_2 is the average deformation potential constant of the valence band. For Si, $E_2 = 5.0$ eV. Thus, $I(\theta)$ depends on the nature of valence band structure, and hence on the semiconductor material. For instance, for parabolic warped valence bands, one can use the following expression for the overlap factor of light and heavy holes (Wiley, 1971):

$$I(\theta) = \frac{1}{4}(1+3\cos\theta) \tag{34b}$$

Similarly, the intraband scattering rate of holes with optical phonons is given by:

$$S_j(k,k') = \frac{3\pi D_o^2}{2V\rho\omega_{op}a_o^2}\begin{bmatrix} N_{op}+1 \\ N_{op} \end{bmatrix} \cdot \delta(E(k') - E(k) \pm \hbar\omega_{op}) \tag{35}$$

where D_o is the optical deformation potential and a_o is the lattice constant. For Si, we have $D_o = 26.6$ eV and $a_o = 5.43$ Å.

Interband Hole Scattering

The interband scattering is only important when we use a two-band model for the valence band (for heavy and light holes). The expression of interband collision rate is quite complex, and may be found in the specialized literatures, for instance in Wiley (1971).

Impact Ionization Collisions

Impact ionization or pair production is a high field scattering mechanism. It consists of an energetic conduction electron hitting an electron in the valence band, exciting this electron across the band-gap, thus generating an electron-hole pair and returning to a lower energy state. The impact ionization phenomena influence so many important characteristics of all semiconductor devices, like the breakdown voltage and the gate leakage current in MOSFET devices.

The impact ionization collisions may be considered as a sort of interband transitions. The microscopic rate of this type of collisions may be given by the Keldysh formula (1965):

$$\Gamma_{ii}(\mathbf{k}) = (P/\tau_{ii})\,[\,(E(\mathbf{k})\,/\,E_v^{th}) - 1]^n \tag{36a}$$

where E_v^{th} is the carrier threshold energy for impact ionization, P is a coupling constant (usually $P>100$), n is a constant (2 for cubic semiconductors with parabolic bands) and $\tau_{ii}(\mathbf{k})=1/\Gamma_{ii}(\mathbf{k})$ is the total phonon collision rate at the threshold energy (at onset of ionization). In Si, E_v^{th} amounts to $3/2E_g$ for both electrons and holes (about 1.8eV in unstrained Si) and τ_{ii} is in the order of 10^{-14} sec. The Keldysh model has been improved by Thomas et al (1991), and reformulated as follows, for Si:

$$\Gamma_{ii}(\mathbf{k}) = 8.75x10^{12}\,(E(\mathbf{k}) - 1.128)^3 \text{ for } 1.128\ eV < E < 1.75\ eV \tag{36b}$$

$$= 6.64610^{13}\,(E(\mathbf{k}) - 1.128)^3 \text{ for } E > 1.75\ eV$$

Another Keldysh-like formula was derived by Cartier et al. (1993) by fitting Monte Carlo results to experimental data. The impact ionization rate for silicon is then given by:

$$\Gamma_{ii}(k) = \sum_{j=1}^{3} P_j[(E(k) / E_j^{th}) - 1]^2 . U(E(k) - E_j^{th} \qquad (36c)$$

where E_j^{th} =1.2, 1.8 and 3.45eV, P_j =6.25x10^{10}, 3 x10^{12} and 6.8x10^{14} s^{-1}, for j=1, 2, and 3, respectively, and U is the Heaviside unit step function. This multi-component Keldysh-type model is further generalized by (2005), as follows:

$$\Gamma_{ii}(k) = \sum_{j=1}^{n} P_j[(E(k) / E_j^{th}) - 1]^{aj} . U(E(k) - E_j^{th} \qquad (36d)$$

The fitting parameters are obtained from MC simulation. For $n=3$, the fitting parameters for electron ionization rate are E_j^{th} =1.13, 1.6 and 2.6eV, P_j =2x10^{12}, 2.3x10^{14}, and 1.8 x10^{16} 6.8x10^{14} s^{-1}, a_j =2.981, 2.978 and 2.49, for $j=1,2,3$, respectively. For the case of hole impact ionization rate in Si, we may take $n=1$, with E_j^{th} =1.33, P_j =6.58x10^{13} and a_j =4.172.

Once the microscopic ionization rate $\Gamma_{ii}(\mathbf{k})$ and the carriers distribution function are known, the macroscopic impact ionization rate can be evaluated. Therefore, the macroscopic impact ionization rate for electron is given by:

$$a_n(\mathbf{x}, t) = (1/n\, v_n) \int g(k) f_n(\mathbf{x}, \mathbf{k}, t)\, \Gamma_{ii}(\mathbf{k})\, d\mathbf{k} \qquad (36e)$$

where n and v_n are the electron density and average drift velocity at a given point of the physical space, respectively .

Figure 12. Impact ionization rates (of electrons and holes). in Si at 300K, according to several research groups
After Seeger (1997).

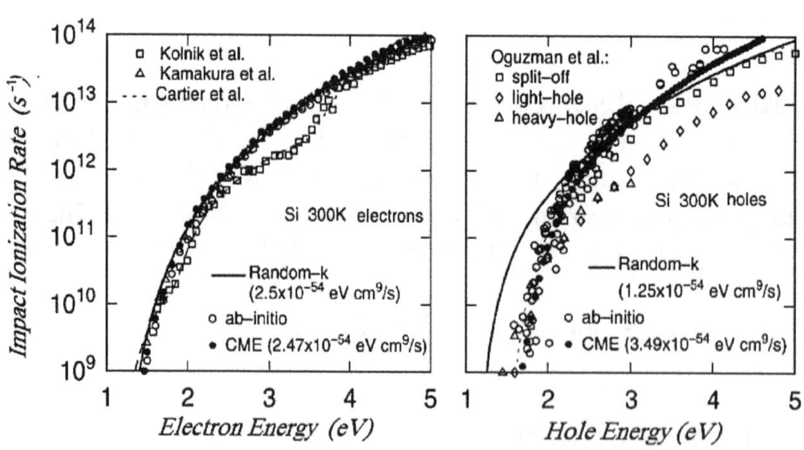

3.4 Additional Collision Rates in Polar Semiconductors

Beside the charge carrier scattering via deformation potential and impurity atoms, the scattering by polar optical phonons may be dominant in compound semiconductors (like *GaAs*). This scattering mechanism is not present in elemental semiconductors (like *Si* or *Ge*), since a sub-lattice structure is necessary for optical modes of vibrations. The following figure shows an example of a polar semiconductor material, with sub-lattice structure. In fact, phonons scatter charge carriers by two mechanisms in polar semiconductor materials:

1. Polar mode scattering mechanism, which is due to the polarization of atoms within the unit cell. This may be polar (optical phonon) scattering or piezoelectric (acoustic phonons) scattering.
2. Deformation potential scattering mechanisms. The scattering rate of this mechanism is similar to that of acoustic phonons. Deformation potential scattering mechanism. Here, the deformation of the lattice by phonons perturbs the dipole moment between atoms, and this results in an electric field that scatters charge carriers. The scattering rate of this mechanism is similar to that of acoustic phonons.

Polar Optical Phonon Scattering

The scattering rate of polar optical phonons is related to the dielectric constant of the material as follows:

$$S_j(k,k') = \left(\frac{2e^2}{q^2}\right)\left(\tfrac{1}{4}\hbar\omega_{op}\right)\cdot\left(\frac{1}{\varepsilon_\infty}-\frac{1}{\varepsilon_s}\right)\cdot\begin{bmatrix} N_{op}+1 \\ N_{op}\end{bmatrix}\cdot\delta(E(k')-E(k)\pm\hbar\omega_{op}) \tag{37a}$$

where phonon wave-vector ***q=k-k'***, ω_{op} is the angular frequency of (emitted or absorbed) optical phonon, Also, ε_s and ε_∞ are the dielectric constant of the polar material at low and optical frequencies, respectively. Note that the polar coupling constant is given by the Frohlich formula

Figure 13. Zincblende lattice of GaAs and its sub-lattice structure

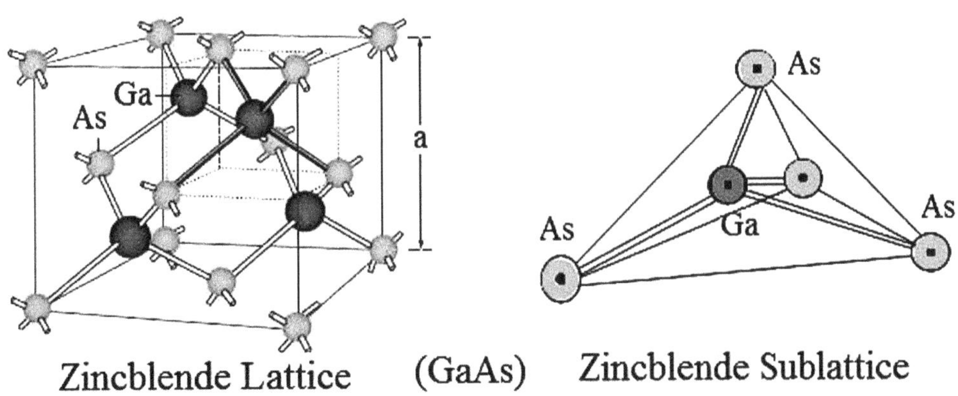

$D_p = ¼\hbar\omega_{op} (1/\varepsilon_\infty + 1/\varepsilon_s)$ (37b)

Piezoelectric Acoustic Phonon Scattering

Some polar semiconductors, (like *GaAs*, *ZnO*, *GaN*, *InN*, and *ZnS*) have non-central symmetry and exhibit the piezoelectric effect (charge accumulation under strain). In these materials, the polar scattering due to acoustic phonons is termed as piezoelectric scattering. In such piezoelectric crystals, the polarization vector is given by:

$$P_i(r) = \sum_{jk} e_{ijk} \frac{\partial u_j(r)}{\partial r_k} \tag{38a}$$

where e_{ijk} is the piezoelectric tensor and $u(\mathbf{r})$ is the displacement of atoms associated with lattice vibrations (phonons). The Zincblende semi-conductors, like *GaAs,* are cubic and there is only one independent piezoelectric constant $e_{123}=e_{132}=e_{213}=e_{231}=e_{312}=e_{321} = e_z$. This polarization corresponds to a piezoelectric field $\zeta=-\boldsymbol{P}/\varepsilon_0$. After calculating the interaction potential we can reach to the scattering probability for piezoelectric phonons for both absorption and emission, as follows:

$$S_j(k,k') = \frac{2\pi e^4 e_z^2 k_B T}{V\hbar\rho v_s^2 \varepsilon_s^2} \cdot \left[\frac{q^2}{(q^2+q_D^2)}\right]^2 \delta(E(k)-E(k')\pm\hbar\omega) \tag{38b}$$

where $q_D = 1/l_D$ is the inverse screening (Debye) length. The integrated scattering probability is obtained by Integration over the possible final states to get:

$$\Gamma_j(k) = \frac{m^{*1/2} e_z^2 k_B T}{2\sqrt{2}\pi\rho v_s^2 \varepsilon_s^2} \cdot \left[\ln\left(1+\frac{8m^* E}{\hbar^2 q_D^2}\right) - \frac{8m^* E}{8m^* E+\hbar^2 q_D^2}\right]^2 E^{1/2} \tag{38c}$$

3.5 Other Collision Mechanisms in Semiconductors

Beside the above mentioned scattering mechanisms, there exist other mechanisms, which may be dominant in a given semiconductor under certain conditions of doping, or near the surface of the semiconductor.

Electron-Electron Binary Collisions

This type of collisions is only important at high doping concentrations. Electron-electron scattering is the most complicated mechanism to treat, because it makes the BTE nonlinear. In the short range, the electron gas may be considered as a collection of particles that interact weakly by means of screened Coulomb force. Hence, electron-electron scattering rate may be calculated by the Brooks-Herring for-

Figure 14. Electron scattering rats in GaAs as function of electron energy, at 300K
After Jacoboni and Reggiani (1983).

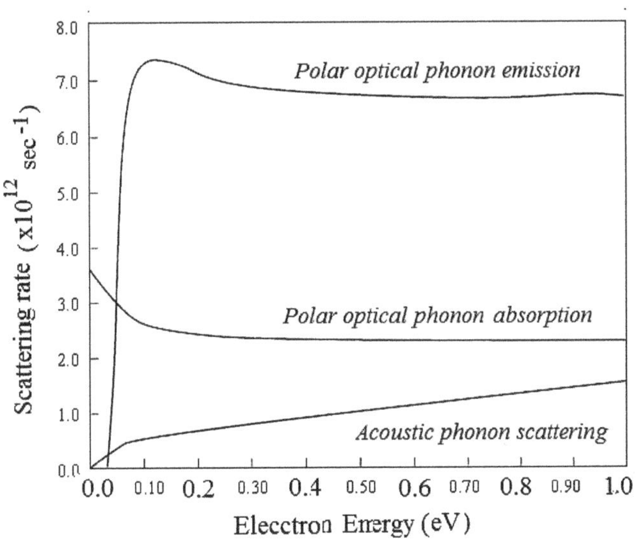

mula (24b) for ionized impurity scattering, with $z=1$ and replacing N with the local electron density (n). Therefore, the collision rate of short-range electron-electron (binary) scattering is given by (Fischetti & Higman, 1991):

$$\Gamma_j(k) = \sum_{k'} f(k') \frac{m^* e^4 n}{4\pi\hbar^3\varepsilon_s^2} \cdot \frac{|k-k'|}{q_D^2\left(|k-k'|^2 + q_D^2\right)} \tag{39}$$

where $q_D = 1/\lambda_D = \sqrt{(e^2 n/\varepsilon k_B T)}$ is the inverse Debye length, k' is the momentum if the second electron and $f(k')$ is the electron distribution function. There are several approaches to treat with the distribution function term, $f(k')$, that appear in this formula, due to Lugli and Ferry (1983), the rejection algorithm (Lawson, 1988) and real-space molecular dynamics approach (Jacoboni, 2010). Note that the e-e scattering is an elastic process and both momentum and energy are conserved. Therefore, momenta of electrons are interchanged and homogenized triggering the distribution function to a drifted/heated Maxwellian form (Jacoboni & Reggiani, 1983).

Electron-Plasmon Collisions

In semiconductors with high electron density an additional type of electron-plasmon scattering is possible for large distances (larger than the Debye length). The source of this scattering is the electron plasma oscillations. A plasmon is a quantum of plasma oscillation, just as phonons are quanta of lattice vibrations. Plasmon scattering of electrons represents the collective long-range part of the electron-electron interaction. For a non-parabolic band structure, the electron-plasmon scattering rate is given by the following expression (Jacoboni and Lugli, 2012):

$$\Gamma(k) = \frac{m^* e^2 \omega_{pl}}{4\pi\hbar^2 k} \begin{Bmatrix} N_{op}+1 \\ N_{op} \end{Bmatrix} n \left(\frac{q_c}{q_{\min}} \right) \tag{40}$$

The final electron energy $E_f = E_i \pm \hbar\omega_{pl}$, and the plasma frequency is given by

$$\omega_{pl} = \sqrt{\frac{e^2}{\varepsilon} \sum_v \frac{n_v}{m^*_{dv}}} \tag{41}$$

Here the sum is assumed over all possible valleys, and n_v is the v- valley contribution to the electron density. For instance, the plasmon energy $(\hbar\omega_{pl})$ in *GaAs* with concentration of 5×10^{17} cm^{-3} is about 30 meV. The cut-off wave vector q_c is defined as follows:

$$q_c = min\,(q_{max}, \lambda_D^{-1}) \tag{42}$$

Also, q_{min} and q_{max} stand for the limits of the momentum transfer:

$$q_{min} = |k - k_f|\,,\; q_{max} = |k + k_f| \tag{43}$$

Note that k_f is the final wave vector and $N(\omega_{pl})$ is the average number of the plasmon excitations defined by the equilibrium Bose-Einstein statistics:

$$N_{op}\,(\omega_{pl}) = 1/[\, exp(\hbar\omega_{pl}/k_B T) - 1\,] \tag{43}$$

Collisions with Surface Acoustic Phonons

We have presented so far the rate of collision of electrons with acoustic phonons in the bulk of silicon and other cubic semiconductors (18). However, near the surface of a semiconductor, this rate is affected by surface scattering and quantum confinement of electrons near the surface. The scattering rate of surface acoustic phonons may be written as follows:

$$\Gamma_j(k) = \frac{\sqrt{2}(m^*_{nd})^{3/2} E^2_{Is} k_B T}{\pi\hbar^4 \rho_s v_s^2} (1+\alpha E)^{1/2} \cdot (1+2\alpha E) \cdot E^{1/2} \tag{44a}$$

where E_{Is} is the deformation potential of surface acoustic phonon and the charge density ρ is replaced with the surface charges density ρ_s (per unit surface). Integrating this matrix element over one subband (of the surface potential well) yields the following relaxation time:

$$\tau_j^{-1}(k) = \frac{m^*_{nd} E^2_{Is} k_B T}{\pi\hbar^4 x_{av} \rho v_s^2} \tag{44b}$$

where we assumed $\rho_s = x_{av}\,\rho$ and x_{av} is the average thickness of the surface inversion layer. According to Lombardi and Schwartz (1999), who considered the average thickness of inversion layer is inversely proportional to $\zeta_\perp$, the relaxation time due to surface acoustic phonons is given by:

$$\tau_j^{-1}(k) = \frac{em_{nd}^{*}E_{Is}^{2}k_BT_L}{\hbar^3\rho v_s^2}\left[\frac{3k_BT_L}{2\zeta_\perp} + \delta\cdot\left(\frac{e^2\hbar^2}{m_{nd}^{*}\zeta_\perp}\right)^{1/3}\right]^{-1} \tag{45}$$

where *d* is a constant (about 2.9 in Si). The second term in the above equation reflects the quantum effect of triangular potential well at the surface of semiconductor.

Surface Roughness Collisions

Surface roughness is due to surface irregularities at the interface of a semiconductor. The charge-carrier collision with surface roughness is an important mechanism in field effect devices, like MOSFET (Metal-Oxide-Semiconductor Field-Effect Transistor). It is also important in low-dimensional semiconductor structures and nanodevices, where the ratio of device surface to volume is important. The surface roughness collision rate may be described by the following relation (Esseni, 2004):

$$\Gamma_j(k) = \frac{\pi m^{*4}\,L^2\Delta^2e^4}{2\varepsilon_s^4\hbar^3}\cdot\left(n_s + N_{ds}\right)^2.\exp\left[-\frac{k^2L^2}{2}\right]\cdot I_o\left(\frac{k^2L^2}{2}\right) \tag{46a}$$

where n_s is the surface carrier concentration, N_{ds} is the charge per unit area (usually beneath the surface area), Δ is the roughness average displacement (about 30Å in *Si*), *L* is the roughness correlation length (about 20Å in *Si*) and I_o is the Bessel function of zero order. The surface charge $(n_s + N_{ds})$ may be replaced by $(\varepsilon_s\zeta_\perp/e)$ where $\zeta_\perp$ is the effective normal electric field:

$$\Gamma_j(k) = \frac{\pi m^{*4}\,L^2\Delta^2e^2}{2\varepsilon_s^2\hbar^3}\exp\left[-\frac{k^2L^2}{2}\right]\cdot I_o\left(\frac{k^2L^2}{2}\right)\cdot\zeta_\perp^2 \tag{46b}$$

This expression is very useful in the case of MOSFET devices. We note that surface roughness scattering has a much stronger dependence on the transverse effective field compared to surface phonon scattering. For the case of nanodevices involving 2-dimensional electron gas (2DEG), interface roughness scattering becomes severe at high carrier densities.

Alloy Scattering in Compound Semiconductors

The alloy scattering is an important mechanism in compound semi-conductors (e.g., $Si_{1-x}Ge_x$, $Ga_{1-x}Al_xAs$) and heterojunction devices. In such alloy semiconductors, one of the constituent atoms (e.g., *As*) occupies the sites of one of the two interpenetrating FCC lattices of the zincblende crystal. The other constituent atoms of the alloy (e.g., *Ga, Al*) are distributed on the sites of the second FCC lattice. This leads to a certain disorder (perturbation) of the crystal potential, which induces an additional scattering of charge carriers. The alloy scattering rate may be described by the following relation, for parabolic band structure (Harrison, 1976):

$$\Gamma_j(k) = \frac{3\pi\sqrt{2}\,(m^*)^{3/2}}{64\hbar^4}\left[x(1-x)\right]\cdot a_o^2 \cdot U_{alloy}^2 \cdot \sqrt{E} \qquad (47)$$

where x is mole fraction of the semiconductor alloy, a_o is the lattice constant and U_{alloy} is the alloy potential fluctuation, due to alloy disorder. U_{alloy} may be taken as the conduction band offset. For instance, U_{alloy} = 0.8V for electrons in *SiGe* alloys. For nonparabolic bands, the square root of E is replaced with the square root of $E(1+\alpha E)$ multiplied by $(1+2\alpha E)$, where α is the nonparabolicity factor. Figure 15 shows the electron and hole collision rates of some semiconductors, as function of the carrier energy at 300K.

4. BTE IN MULTIVALLEY AND COMPOUND SEMICONDUCTORS

At high fields, charge carriers can possess high energy and may scatter between different valleys of the semiconductor. Therefore, it is often very important to consider the real band structure of the semiconductor, including all the valleys which are close to the main conduction band. For, instances, *GaAs* has two valleys in the conduction band (called the upper satellite valley and the lower central valley), which are separated by 0.3eV. The *Si* itself has 4 conductions bands, with several valleys, which are not far from the reach of the main conduction band electrons at high fields.

In multi-valley semiconductors, with N_v valleys, the semiclassical BTE can be written as follows in the ith valley:

Figure 15. The electron-phonon and hole-phonon collision rates in different semiconductors as function of electron kinetic energy, at 300K
After Fischetti (1993).

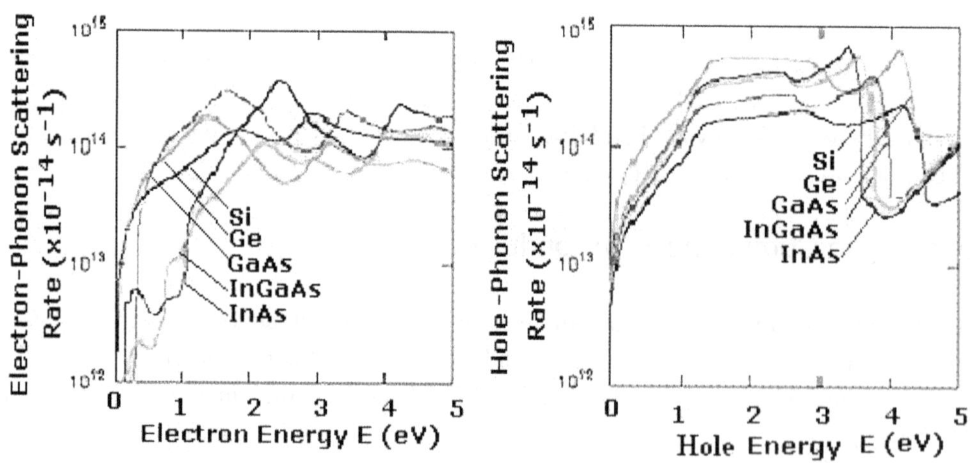

$$\frac{\partial f_i}{\partial t} + u_i \cdot \nabla f_i + \frac{F}{\hbar} \cdot \nabla_k f_i = \left[\frac{\partial f_i}{\partial t}\right]_{intra} + \sum_{j=1.j\mp i}^{Nv} \left(-\left[\frac{\partial f_i}{\partial t}\right]_{j,inter} + \left[\frac{\partial f_j}{\partial t}\right]_{i,inter} \right) \tag{48}$$

where i and j stand for the ith and the jth valleys, respectively. The intravalley collision term of the ith valley is given by:

$$[\partial f_i/\partial t]_{intra} = V/(2\pi)^3 . \int [S_{ii}(\mathbf{k}_i,\mathbf{k}_i')f_i(\mathbf{k}_i) - S_{ii}(\mathbf{k}_i',\mathbf{k}_i)f_i(\mathbf{k'}_i)]\, d^3k_i' \tag{49}$$

where we dropped the Pauli-exclusion terms (1-f), for the sake of simplicity. This approximation is acceptable at high energy, where f is much smaller than unity.

The intervalley collision terms of the ith valley is given by:

$$[\partial f_i/\partial t]_{i,\, inter} = \int S_{ji}(\mathbf{k}_j,\mathbf{k}_i')f_j(\mathbf{k}_j)\, d^3k_j' \tag{50a}$$

$$[\partial f_i/\partial t]_{j,\, inter} = \int S_{ij}(\mathbf{k}_i,\mathbf{k}_j')f_i(\mathbf{k}_i)\, d^3k_i' \tag{50b}$$

The following table depicts the deformation potential and coupling coefficients in silicon and other materials. The other parameters of *Si*, which appear in the scattering rate relations, are given in Table 2(c).The data are taken from the manual of Damocles simulator, from IBM Corp.

5. BTE SOLUTION METHODS

The analytical solution of the Boltzmann transport equation is only possible under very restrictive approximations. On the other hand, the direct numerical methods for device simulation have been limited by the complexity of the BTE. Thus, the fundamental methods which have been used for solving the BTE and finding the carrier distribution function can be summarized into the following basic categories:

Figure 16. Intervalley scattering between central and satellite valleys in GaAs

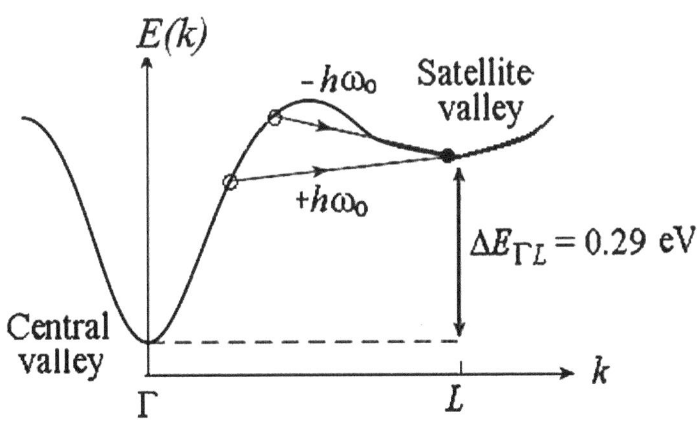

Table 2a. Conduction band deformation potentials in eV (acoustic) and 10^8eV/cm (optical) in Si, Ge and other semiconductors

Mechanism	*Si*	*Ge*	*AlAs*	*AlP*	*GaAs*	*GaP*	*GaSb*	*InAs*	*InP*	*InSb*	$In_{0.53}Ga_{0.47}As$
Acoustic intra-band1	1.2	1.5	7.0	7.0	5.0	5.0	5.0	3.4	5.0	5.0	5.4
Acoustic interband	1.5	1.0	7.0	7.0	3.5	5.0	5.0	2.4	5.0	5.0	6.0
Optical intra-band	1.75	2.0	2.0	3.0	2.1	1.0	3.0	1.1	2.0	2.0	2.0
Optical interband	1.9	1.5	2.0	3.0	1.5	1.0	3.0	1.1	2.0	2.0	2.5

Table 2b. Valence band deformation Potentials (acoustic and optical) in Si, Ge and other semiconductors

Mechanism	*Si*	*Ge*	*GaAs*	*InAs*	$In_{0.53}Ga_{0.47}As$
Acoustic (eV)	4.6	4.6	6.3	6.3	6.3
Optical (10^8 eV/cm)	6.6	9.0	11.3	11.3	11.3

Table 2c. Coupling coefficients in Si

Mechanism	Coupling Coefficient D_m *[eV/cm]*	Phonon Temperature T_p [K] = $\hbar\omega / k_B$	
Intervalley Phonons			
Intervalley f1	1.5 x10^7	210	*TA*
Intervalley f2	3.4 x10^8	500	*LA*
Intervalley f3	4 x10^8	600	*TO*
Intervalley g1	5 x10^7	140	*TA*
Intervalley g2	8 x10^7	210	*LA*
Intervalley g3	3.0 x10^8	700	*LO*

According to Jacoboni and Reggiani (1983).

Table 2d. Other parameters of silicon

Parameter	Symbol	Value	Reference
Lattice constant	a_o	*5.431 À*	Agrain & Balkanski, 1961
Density	ρ	2.329 g/cm^3	Jacoboni & Reggiani, 1979
Sound velocity	v_s		
Longitudinal:	v_l	9.0 x10^5 cm/s	Jacoboni & Reggiani, 1979
Transverse:	v_t	5.4 x10^5 cm/s	Jacoboni & Reggiani, 1979
Dielectric constant	ε_s	11.7	Jacoboni & Reggiani, 1979

- Assuming *a priori* analytical form of the distribution function,
- Approximate linearized methods or drift-diffusion model (DDM),
- Moment or hydrodynamic methods (HDM),
- Direct Methods
- Stochastic solution methods (Monte Carlo Methods)

The direct solution of the BTE can be performed by means of several approaches, such as expansion into spherical harmonics (SHE), and the iterative methods. The so-called weighted essentially non-oscillatory (WENO) Boltzmann schemes are also a sort of direct solution methods of the BTE. Direct methods consist in discretizing the BTE in phase space and solving the resulting difference equations, over a mesh of discrete points, by appropriate numerical methods (Banoo & Lundstrom, 2000). Some semiconductor solvers are also based on the stochastic solution methods, which involve the simulation of particle trajectories rather than the direct solution of differential equations (Jacoboni & Reggiani, 2000). These statistical methods are referred to as particle simulation by Monte Carlo method (MCM).

5.1 A Priori Solutions of the BTE

There are many approximations, which are sometimes used to guess the non-equilibrium distribution functions of charge carriers. Among them, one can cite:

- The Fermi-Dirac distribution function
- The heated Maxwellian distribution,
- The displaced Maxwellian distribution,

Evidently, the Fermi-Dirac distribution is only valid at thermal equilibrium. Therefore it may be used to calculate transport near to the equilibrium state.

Heated Maxwellian Distribution

The heated Maxwellian distribution (at electronic temperature) is given by:

$$f_M(E_n) = C.[n/(k_B T_n)^{3/2}].exp\,(-E_n/k_B T_n) \tag{51}$$

where n and T_n are the electron average density and temperature. Also, $E_n = ½m_n{}^*u_n{}^2$ is the electron energy, u_n is the electron group velocity, $m_n{}^*$ is the electron effective mass and C is a constant. This distribution is only valid if the collisions are totally randomized such that the average electron velocity is negligible. However, this assumption is not acceptable in practical devices.

Displaced Maxwellian Distribution

According to Stratton (1962), the electronic temperature has no physical meaning, unless the distribution function is a displaced Maxwellian. The displaced Maxwellian distribution at the electronic temperature is given by:

$$f_{DM}(k) = C'.[n/(k_BT_n)^{3/2}].exp\,(-\tfrac{1}{2}m_n^*(u_n(k)-v_n)^2/k_BT_n) \quad (52)$$

where $v_n=<u_n(k)>$ is the electron drift velocity and C' is a constant.

The drifted (displaced) Maxwellian distribution can be expanded into symmetric and asymmetric parts, as shown in the following figure. However, the exact numerical solution of the BTE has shown that the energy distribution is not Maxwellian at high electron energy. Figure 17 shows the drifted Maxwellian velocity distribution. Figure 18 depicts the electron energy distribution inside an n-MOSFET as obtained by MC simulation.

The distribution is calculated at 10nm of the drain junction by Higman et al. (1989). The displaced Maxwellian, which best fits the MC simulation (at 500K) is also illustrated. As shown in figure, the tail of energy distribution function cannot be approximated by a Maxwellian distribution. This result

Figure 17. The heated and drifted Maxwellian velocity distribution functions

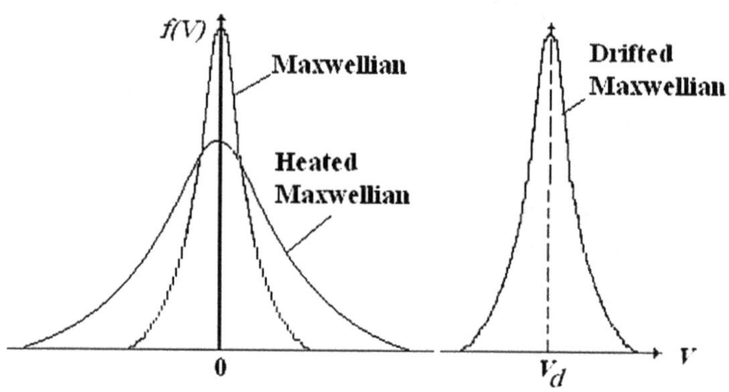

Figure 18. The drifted Maxwellian velocity distribution function, its symmetric and asymmetric components

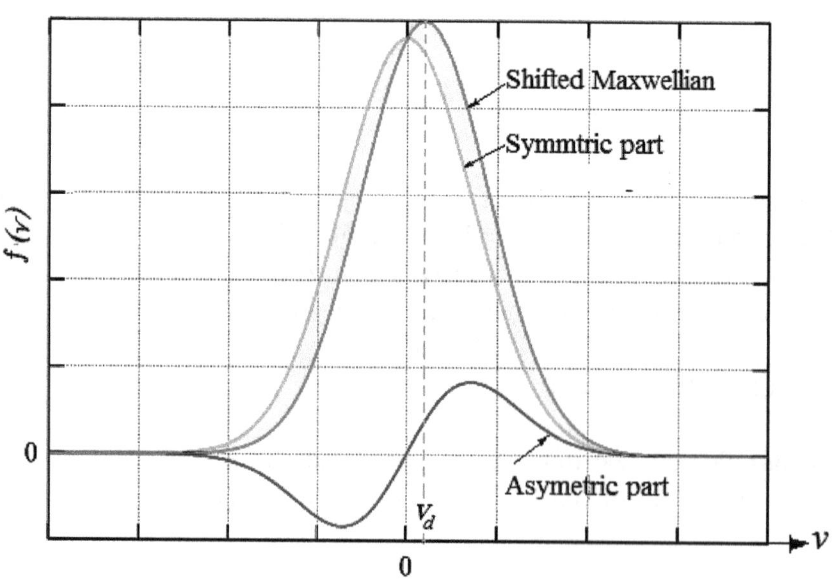

Figure 19. Electron energy distribution function jn a MOSFET by Monte Carlo method
After Higman et al. (1989).

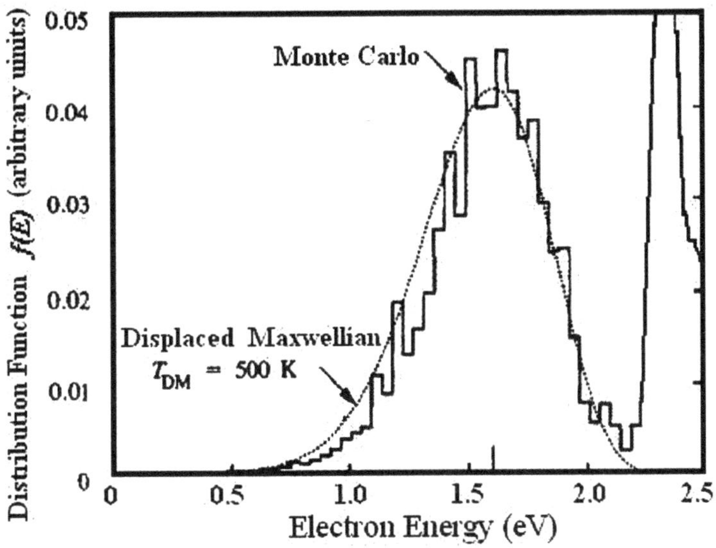

has been later confirmed, by experiment, using the so-called photon energy distribution (PED) method (Lanzoni et al., 1991).

Non-Maxwellian Distributions

The distribution function $f_n(E)$ is sometimes assumed with a non-Maxwellian factor γ, such that:

$$f_n(E_n) = (1+\gamma.E_n/k_B T_n).f_m(E_n) \tag{53}$$

Here, $f_m(E)$ is the heated Maxwellian distribution function at an elevated temperature T_n. This non-Maxwellian distribution function is frequently used in the so-called energy transport model (ETM),

Other a priori distribution functions, which assimilate the tail of the real distribution function at high carrier energy, have been proposed in the literature. These distribution functions will be presented in the context of impact ionization models, in Chapter3. For the matter of completeness, we mention here the two-component distribution, which was proposed to account for both cold and hot carriers populations in nanoscale devices:

$$f_n(E_n) = C_1 \{ exp[-(E_n/E_{ref})^b] + C_2 exp[-(E_n/k_B T_2)] \} \tag{54}$$

Here, C_1, C_2, E_{ref} and b are adjustable parameters, for tuning the hot carrier component, and the cold carrier (Maxwellian) component. Of course all these analytical distributions are not physical at all. However, they are sometimes used for fitting the simulation results of high-field phenomena in semiconductors solution methods (e.g., MCM) with real measurements.

5.2 Linear Transport Theory (Relaxation-Time Approximation)

The solution of the BTE can be greatly simplified by expressing the collision term in terms of a microscopic relaxation time as follows:

$$[\partial f_v/\partial t]_{col.} = -(f_v - f_o)/\tau_{Dv} \tag{55}$$

Here, τ_{Dv} is called the carrier microscopic relaxation time and f_o is the carrier distribution function at thermal equilibrium. Therefore, the BTE becomes linear in f. The carrier distribution function at thermal equilibrium may be expressed by the Fermi-Dirac distribution function or the Maxwell-Boltzmann distribution (at lattice temperature). Thus, for electrons we have:

$$f_o = n\,(m_n^*/2\pi\, k_B T_L)^{3/2}.\, exp[-E_n/k_B\, T_L] \tag{56}$$

where E_n is the microscopic electron energy (measured from the conduction band edge E_c), T_L is lattice temperature. This approximation is valid at low fields where the carrier velocity and current density are linear with field. Generally speaking, τ_{Dv} lumps all the acting collision mechanisms of charge carriers. It may be given by the following expression (for electrons):

$$\tau_{Dn}(E_n) = \tau_o\,(k_B\, T_L/E_n)^r \tag{57}$$

where k_B is the Boltzmann constant. The two scattering parameters r and τ_o are dependent on the lattice temperature T_L as well as the dominant collision mechanism.

The above approximation[10] involves a linearization of the distribution function in two terms, such that $f_n = f_o + f_1$, where f_o is symmetric and f_1 is non-symmetric in the k-space. The linearized BTE can be written as follows (for electrons, under the effect of external electric field $\boldsymbol{\zeta}$):

$$\partial f_n/\partial t + \boldsymbol{u}_n.\nabla f_n - (e\boldsymbol{\zeta}/\hbar).\nabla_k f_n = -(f_n - f_o)/\tau_{Dn} \tag{58}$$

where $\boldsymbol{u}_n(k)=(1/\hbar)\nabla E(k)$ is the electron group velocity. If we only consider the time scales which are longer than the collision time τ_{Dn}, (in the order of 0.1ps in *Si*), therefore, we can neglect the $\partial f_n/\partial t$ term, with respect to f_1/τ_{Dn}. Then, the solution of the linearized BTE can be expressed as follows:

$$f_n = f_o - \tau_{Dn}\boldsymbol{u}_n.\nabla f_n + e\,\tau_{Dn}\,(\boldsymbol{\zeta}/\hbar).\nabla_k f_n \tag{59a}$$

Putting $\nabla_k f_n = (\partial f_n/\partial E_n).(dE_n/dk)$, $\partial f_n/\partial E_n \approx \partial f_o/\partial E_n$, and then substituting $\boldsymbol{u}_n = (1/\hbar)\, dE_n/dk$ yields:

$$f_n \approx f_o - \tau_{Dn}\boldsymbol{u}_n.\nabla f_n + e\,\tau_{Dn}\boldsymbol{u}_n.\boldsymbol{\zeta}(\partial f_o/\partial E_n) \tag{59b}$$

For homogeneously-doped semiconductors $(\nabla f_n=0)$, we can write:

$$f_n = f_o\,(1 - e\,\tau_{Dn}\boldsymbol{u}_n\boldsymbol{\zeta}/k_B\, T_L) \tag{60}$$

Here, we substituted $(\partial f_o/\partial E_n) = -(f_o/k_B T_L)$. This solution is only valid for small perturbations (for elastic scattering events, near equilibrium) where the electric field satisfies the condition:

$$e\,\tau_{Dn}\boldsymbol{u}_n \cdot \boldsymbol{\zeta} << k_B T_L \tag{61}$$

Thus, if we admit that the mean free path of electrons $(\lambda_n = \tau_{Dn} u_n)$ to be in the order of $200\acute{A}$, then the above condition is only verified when the electric field $\boldsymbol{\zeta}$ is much smaller than 10^4 V/cm.

Physical Note 2: What's the Relation between the Microscopic Relaxation Time and the Collision Rate?

We have seen that the calculation of the BTE collision term we sum equation (10) over all possible states k'. For small fields witch change the distribution function f only little from its equilibrium, we have $f=f_0+f_1$. We have also $S(k',k)f_0(k')=S(k,k')f_0(k)$, from the detailed balance principle. Therefore we can write,

$$\left(\frac{\partial f}{\partial t}\right)_{coll} \approx -f_1(k)\int S(k,k')\left[1-\frac{f_1(k')}{f_1(k)}\right]d^3k' = -\frac{f_1(k)}{\tau_D(k)}$$

where $\tau_D(k)$ is microscopic relaxation time; this is NOT the same as the mean free time between collisions (used in the definition of the drift mobility $\mu = e\tau/m^*$). If the scattering process is assumed elastic, it does not change the energy but only the direction of the particles. Therefore, the derivative of f_0 with respect to energy is the same for k and k'; consequently it cancels out of the ratio. In addition we may use a parabolic relationship between energy and wavevector near equilibrium $(E=\hbar^2k^2/2m^*)$. Next, we choose a coordinate system where the initial wavevector (k) is along the z direction and the electric field (ζ) is contained in the yz plane.

Reporting all these assumptioins into the above integral, leads to $f_1(k')/f_1(k) = \zeta.k/\zeta.k'=cos(\theta)$, where θ is the scattering angle between k and k'. Then, the square brackets in the above integral reduce to [1-cos θ]. This is detailed in (Tritt, 2005, pp.47-48).Theb, the relation between microscopic relaxation time τ_D and the microscopic scattering rate $1/\tau(k) = \Gamma(k)=\int S(k,k')d^3k'$ is the average of this bracket $<1-cos(\theta)>$ in the k-space.

Current Density and Energy Flux

For a semiconductor with arbitrary doping profile, the current equations can be deduced from the linearized BTE (40) as follows:

$$\boldsymbol{J}_n = -e\int \boldsymbol{u}_n f_n(E_n)\, g_c(E_n)\, dE_n \tag{62}$$

where $g_c(E_n) = C.E_n^{1/2}$ is the density of states for electrons in the conduction band and C is a normalization constant. The normalization constant can be calculated using the relation:

$$n = \int f_n(E_n)\, g_c(E_n)\, dE_n \tag{63}$$

Figure 20. Scattering coordinate system

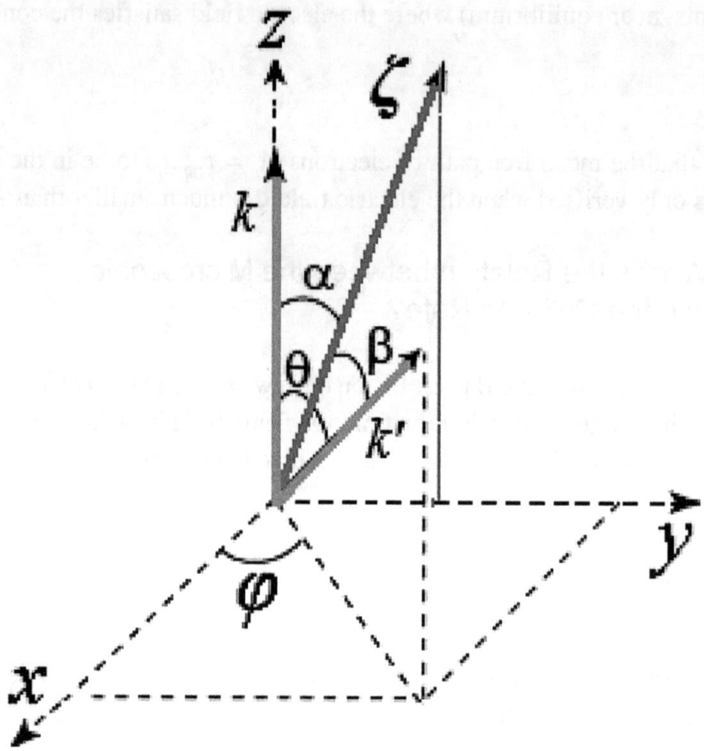

Substituting $f_n(E_n) \approx f_o(E_n)$, we can easily find that $C = 2\pi(2/m_n)^{3/2}$. Note that E_n is measured from the conduction band edge and hence the integration is carried out for $E_n = 0 \rightarrow \infty$. Also, equation (40b), which describes the linearized distribution function, may be further simplified as:

$$f_n \approx f_o - \lambda_n (e\,\zeta f_o / k_B T_L + \nabla f_o) \tag{64}$$

where we substituted $\lambda_n = \tau_{Dn} u_n$. Now, substituting f_n from (40a) and $g_c(E_n) = 2\pi(2/m_n)^{3/2}.E_n^{1/2}$ into (41a) and integrating yields:

$$J_n = -e\int_0^\infty \sqrt{\frac{2E_n}{m_n}} \cdot \left[f_o - \lambda_n \left(f_0 \frac{e\zeta}{k_B T} + \nabla f_0 \right) \right] \cdot 2\pi \left[\frac{2}{m_n} \right]^{\frac{3}{2}} \sqrt{E_n} \cdot dE_n \tag{65a}$$

Note that we also substituted $\boldsymbol{u}_n = \surd(2E_n/m_n^*)$. Now, assuming that both the effective mass m_n^* and the mean free path λ_n are energy-independent (or slow functions of energy) and substituting f_o from (35), we get:

$$J_n = 2en\lambda_n \left(\frac{2k_B T_L}{\pi m_n^*} \right)^{1/2} \cdot \left(\frac{e\zeta}{k_B T_L} + \frac{\nabla n}{n} + \frac{\nabla (k_B T_L)^{1/2}}{(k_B T_L)^{1/2}} \right) \tag{65b}$$

Note also that the integration over the first symmetric term, f_o, in (64) is null. Similarly, we can calculate the electrons energy flux $\mathbf{S}_n$ as follows:

$$\mathbf{S}_n = \int \mathbf{u}_n E_n f_n(E_n)\, g_c(E_n)\, dE_n$$

$$= \int_0^{\infty} \sqrt{\frac{2E_n}{m_n}} \cdot E_n \left(f_0 - f_0 \lambda_n \frac{e\zeta}{k_B T} - \lambda_n \nabla f_0 \right) \cdot 2\pi \left[\frac{2}{m_n} \right]^{\frac{3}{2}} \sqrt{E_n} \cdot dE_n$$

$$= -8n\lambda_n \frac{\left(k_B T_L\right)^{3/2}}{\left(2\pi m_n^*\right)^{1/2}} \left(\frac{e\zeta}{k_B T_L} + \frac{\nabla n}{n} + \frac{\nabla (k_B T_L)^{3/2}}{(k_B T_L)^{3/2}} \right) \tag{66a}$$

Therefore, the energy flux can be written as the sum of convective and conduction parts, as follows:

$$S_n = -2\,(k_B T_L/e).\mathbf{J}_n - 8\, n\, \lambda_n\, (k_B T_L/2\pi\, m_n^*)^{1/2}.\nabla (k_B T_L)] \tag{66b}$$

When there is just one type of charge carriers, the above relation may be put in the form ($S_n = Q_n + E_F.J_n$), where E_F is the Fermi energy (chemical potential). Therefore, if energies are measured with respect to the Fermi energy level (chemical potential), energy current (S_n) becomes a conduction heat current (Q_n). In this case, the heat current maybe defined as follows:

$$\mathbf{Q}_n = \int \mathbf{u}_n\, (E_n - E_F).f_n(E_n)\, g_c(E_n)\, dE_n \tag{67}$$

Transport Parameters

We finally investigate the well-known drift-diffusion current relation for electrons and the transport parameters as follows. Consider first the carrier current density in homogenous semiconductor, when there is no carrier nor temperature gradients:

$$\mathbf{J}_n(\nabla n{=}0, \nabla T_L{=}0) = \sigma_n \zeta = 2\lambda_n (2k_B T_L/\pi\, m_n^*)^{1/2} (e^2 n\zeta/k_B T_L) \tag{68a}$$

Thus, the electron conductivity σ_n is given by:

$$\sigma_n = e\, n\, \mu_n = e^2 n\, [4\lambda_n/(2\pi\, m_n^* k_B T_L)^{1/2}\,]. \tag{68b}$$

Hence the electron drift mobility μ_n is given by:

$$\mu_n = 4e\lambda_n/(2\pi\, m_n^*\, k_B\, T_L)^{1/2} \tag{68c}$$

Also, the carrier diffusion coefficient D_n can be found as follows:

$$J_n(\zeta=0, \nabla T=0) = eD_n \nabla n = 2e\lambda_n (2k_B T_L/\pi m_n^*)^{1/2} \nabla n. \qquad (69a)$$

Consequently, the carrier diffusion coefficient D_n is given by:

$$D_n = 4\lambda_n [k_B T_L/(2\pi m_n^*)^{1/2}] = \mu_n (k_B T_L/e) \qquad (69b)$$

This equation satisfies the Einstein relation ($D_n/\mu_n = k_B T_L/e$). Also, the lattice thermal k_n^{th} (due to electrons) can be calculated at zero current density, as follows:

$$S_n(J_n=0) = -k_n^{th} \nabla T_L = -8\lambda_n (k_B T_L/\pi m_n^*)^{1/2} n \nabla (k_B T_L) \qquad (70a)$$

Therefore, k_n^{th} is given by:

$$k_n^{th} = 8\lambda_n (k_B T_L/2\pi m_n^*)^{1/2} n k_B = 2(k_B/e)^2 \sigma_n T_L \qquad (70b)$$

The above equation satisfies the well-known Weidman-Franz law, which been already introduced in Chapter 1, as follows:

$$k^{th}/\sigma = L.T_L = \gamma (k_B/e)^2 T_L \qquad (70c)$$

Here k^{th} is the thermal conductivity of the semiconductor material and $L = \gamma (k_B/e)^2$ is the Lorentz number and γ is the Lorentz coefficient. Note that the thermal conductivity due to charge carriers (k_n^{th} and k_p^{th}) is just a part of the total thermal conductivity k^{th} of a semiconductor ($k^{th} = k_n^{th} + k_p^{th} + k_{ph}^{th}$ + mixed terms). We implicitly assume here that the electron temperature is equal to the lattice temperature and the thermal conductivity of the lattice is mainly due to the electron contribution, which is the case in metals.

Actually, the Weidman-Franz law is experimentally verified in metals. However, in semiconductors, phonons play a significant role in the heat conduction process. Thus, the Weidman-Franz law may be considered as a phenomenological relation in the case of semiconductors.

Now, by substituting the transport coefficients (drift mobility and diffusion constants) in the current equation (38) we obtain:

$$J_n = e n \mu_n \zeta_+ e D_n \nabla n + e D_n^{th} \nabla T_L \qquad (71a)$$

where $D_n^{th} = n(D_n/2T_L)$ is the thermal diffusion coefficient.

In a more elaborate derivation, we consider the variations of the Fermi-level with position such that $\nabla E_F \neq 0$ in the expression of ∇f_o and we integrate equation (38) starting from the edge of the conduction band E_c. Then the current equation may be put in the following form:

$$J_n = e n \mu_n \zeta_+ e D_n \nabla n - n \mu_n P_n^{th} \nabla T_L \qquad (71b)$$

where $P_n^{th} = -(k_B/e)[5/2 - r + (E_c - E_F)/k_B T_L]$ is the thermoelectric power of electrons. This relation is one of the Onsager relations that we presented in Chapter 1, and maybe derived, from the first principles (Drude's model).

More Accurate Derivation of J_n and S_n

In the above derivations we assumed that both the effective mass and mean free path $\lambda_n = \tau_{Dn} u_n$ are independent of energy. This means that we assumed $r=½$ in the relaxation time expression $\tau_{Dn} = \tau_o(k_BT_L/E_n)^r$, because $u_n=\sqrt{(2E_n/m_n^*)}$. As the mean free path is not constant and $r \neq ½$ in general, we should better use the relaxation time expression and derive $\boldsymbol{J}_n$ and $\boldsymbol{S}_n$ as well as all other parameters, in terms of r. Starting with the linearized distributed function: $f_n \approx f_o - \tau_{Dn}\boldsymbol{u}_n(e\zeta f_o/k_BT_L + \nabla f_o)$, then substituting τ_{Dn} and integrating yields:

$$\boldsymbol{J}_n = -e \int \boldsymbol{u}_n f_n(E_n)\, g_c(E_n)\, dE_n = n\frac{\alpha\tau_o}{m_n^*}\left(k_BT_L\right).\left(\frac{e\zeta}{k_BT_L} + \frac{\nabla n}{n} + \frac{\nabla(k_BT_L)^{1-r}}{(k_BT_L)^{1-r}}\right) \tag{72}$$

$$\boldsymbol{S}_n = \int \boldsymbol{u}_n E_n f_n(E_n) g_c(E_n) dE_n = -(5/2-r).(k_BT_L/e)\boldsymbol{J}_n - (5/2-r)(k_B/e)^2.\sigma_n T_L \nabla T_L \tag{73}$$

where $\alpha = 4\Gamma(5/2-r)/\sqrt{\pi}$. This expression is similar to (38), except for the numerical coefficient. Here, the mobility $\mu_n = \alpha.(e\tau_o/m_n^*)$ instead of $4e\lambda_n/(2\pi m_n^* k_BT_L)^{1/2} = (4/\pi).(e\tau_o/m_n^*)$. Note also that the Lorentz number $\gamma = (5/2-r)$, which is a well-known result in the bulk of semiconductors.

Drift-Diffusion Model (DDM)

In isothermal conditions, where the semiconductor crystal is maintained at a constant temperature such that $\nabla T_L = 0$, then equation (38) reduces to the conventional drift-diffusion current equation (for electrons):

$$\boldsymbol{J}_n = e\, n\, \mu_n \boldsymbol{\zeta} + e\, D_n \nabla n \tag{74a}$$

A similar relation can be deduced for holes.

$$\boldsymbol{J}_p = e\, p\, \mu_p \boldsymbol{\zeta} - e\, D_p \nabla p \tag{74b}$$

These relations are exactly equivalent to equations (71) from Chapter 1 within the framework of the drift-diffusion model (DDM).

Semiconductor Equations

In order to make use of the drift diffusion model, we substitute the current equations (65) into the corresponding continuity equations of electrons and holes to get the semiconductor equations. The resultant differential equations should be coupled with the Poisson equation, and solved together to get the main transport variables (the electric potential ϕ, the electron density, n, and the hole density, p). We write this system, below, for the matter of completeness:

$$\frac{\partial n}{\partial t} = (G_n - R_n) + \nabla \cdot (n\mu_n \zeta + D_n \nabla n) \tag{75a}$$

$$\frac{\partial p}{\partial t} = (G_p - R_p) + \nabla \cdot .(p\mu_p \zeta - D_p \nabla p) \tag{75b}$$

$$\nabla^2 \varphi = -\nabla \zeta = -\frac{e}{\varepsilon}\left(p - n + N_d^+ - N_a^-\right) \tag{76}$$

These equations describe the transport of charge carriers inside homo-geneous semiconductors. This set can be modified to account for heterogeneous devices, which consist of different semiconductors. However, these equations are based on the DDM, which is only valid at low electric fields.

The model parameters, such as the carrier mobility and diffusion coefficient are generally dependent on the main variables (ϕ, n, p) or their derivatives. Therefore, the above system is actually a nonlinear 2nd order system of differential equations. Though, the device dimensions are currently in the nanometer regime and the nonlocal quantum effects are dominant, the DDM is still used (occasionally, with some quantum corrections) to estimate the performance of many semiconductor devices. The quantum corrections have been extensively discussed in the literature, e.g., in (Jerome, 2009) and will be presented in Chapter 4 of this book,

Modeling of Physical Parameters

The modeling approaches of physical parameters in semiconductor materials can be divided into three categories:

1. **Physically-Based Models:** These expressions capture the physical model into a closed form solution in which the parameter dependencies are obtained from fundamental calculations, with simplifying assumptions.
2. **Semi-Empirical Models:** This approach arises because in practice it is seldom the case that the physically-based models conform to the experiment. In order to reconcile the model with experimental data, the coefficients in the physically-based model are allowed to vary from their original values.
3. **Empirical Models:** The empirical models are those in which all parameters are allowed to vary. They generally tend to obscure the important physics behind the phenomena.

For instance, the drift mobility of charge carriers is defined as the ratio of the magnitude of carrier drift velocity over the magnitude of electric field ($v_n = \mu_n \zeta$). We have also seen in the development of the classical transport theory that the expression of the carrier drift mobility may be given by:

$$\mu_n = e\,\tau / m_n^* \qquad (77a)$$

where τ is the mean time between collisions, averaged over all collision mechanisms of electrons in the semiconductor material.

$$\tau^{-1} = \sum_i \left(\bar{\tau}_j\right)^{-1} \qquad (77b)$$

Here $\bar{\tau}_j$ is obtained by averaging the mean free time (or relaxation time[11]) due to a certain collision mechanism, weighted by the carrier energy (E), over the distribution function, as follows (Elliot & Gibson, 1980):

$$\bar{\tau}_j = \frac{\int_{CB} \tau_j(E) \cdot g_c(E) \cdot E \cdot f_n(E) \cdot dE}{\int_{CB} g_c(E) \cdot E \cdot f_n(E) \cdot dE} \qquad (77c)$$

As the distribution function is not generally known, it is usually replaced by the Maxwell-Boltzmann distribution function, in equilibrium $f_n \approx f_{o'}$. Also, the band structure is usually assumed parabolic. Therefore, the average scattering time is calculated near equilibrium (at low electric fields).

The calculations details for each scattering mechanism may be found in many textbooks, such as Vapaille (1970), Tavernie and Calcki (1970), Kittle (1980) and Blatt (1992). We remind the reader that $\tau_j(E)$ of a certain collision mechanism, may be also defined as the inverse of the integrated scattering probability per unit time $\Gamma_j(E)$, as indicated by (12d). In a previous section, we have illustrated the expression of $\Gamma_j(k)$ of almost all collision mechanisms of interest in semiconductors. Note that we assume that these mechanisms are statistically independent in (49b). Substituting (49b) into (49a) results in the Mathiessen rule:

$$\mu^{-1} = \Sigma \mu_j^{-1} \qquad (78)$$

Applying the above integration results in some analytical models of charge carrier mobility (Tavernie & Calcki, 1970), as shown in Table 3.

Note that, at low electric fields, the ionized impurity scattering and phonon scattering predominate in semiconductors and should be considered, within the DDM simulation framework. However, the overall mobility models, which are usually employed in device simulation, are generally semi-empirical, and involve many fitting parameters (Del Alamo, 1985). The following figure depicts the experimental and theoretical (semi-empirical) electron and hole mobility in Si at 300K, according to several authors.

5.3 Direct Solution Methods

Recently, some researchers have used powerful computational platforms to attack the iterative numerical solutions of the BTE, by several techniques, such as the spherical harmonic expansion of the distribution function.

Figure 21. Experimental and theoretical (semi-empirical) electron and hole mobility in Si at 300K, according to several authors
Data from (Saso et al., 2009).

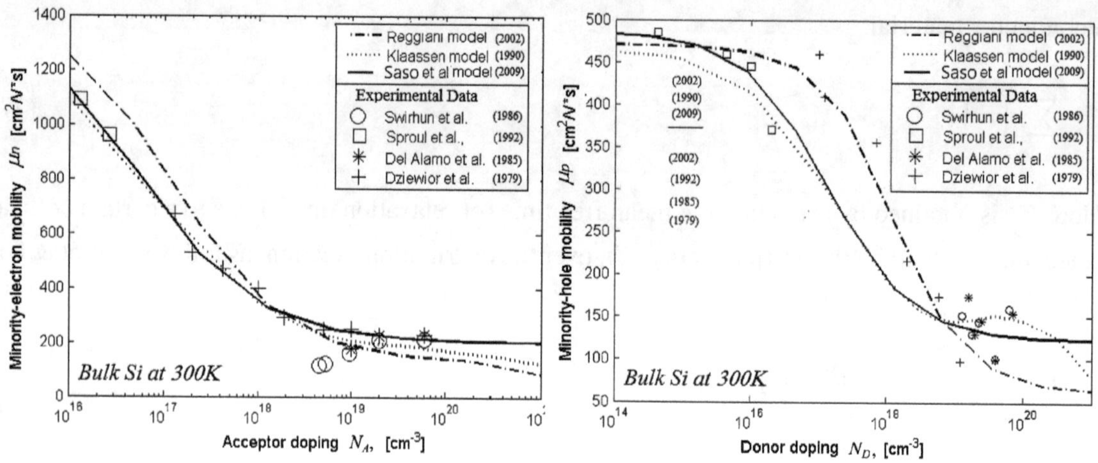

Table 3. Models of the charge carrier mobility, due different scattering mechanisms. Here μ_{vo} is a constant (different for each mechanism), m_{v*} is the carrier effective mass, m_o is the electron rest mass and T_L is the lattice temperature.

Scattering Mechanism	Mobility Model
Scattering over acoustic phonons	$\mu_v = \mu_{vo}\,(m_v{*}/m_o)^{-5/2}\,.(T/T_L)^{-3/2}$
Scattering over ionized impurities	$\mu_v = \mu_{vo}\,(m_v{*}/m_o)^{-1/2}\,.(T/T_L)^{3/2}$
Scattering over optical phonons	$\mu_v = \mu_{vo}\,(m_v{*}/m_o)^{-3/2}\,.(T/T_L)^{1/2}$

Direct Solution by Iterative Methods

The iterative methods for solving the BTE are based on the following integration formula for the distribution function (in a homogenous material·)

$$f(k,t)=\int_0^{\infty} dt'\exp\left\{-\int_0^{t'}\lambda\left[K(y)\right]dy\right\}f(k',t-t')S(K(t'),k') \qquad (79)$$

where $\lambda(k) = (V/2\pi)^3\int S(\mathbf{k},\mathbf{k}')d\mathbf{k}'$ is the out-collision rate, as defined by (10b). This integral form is sometimes called the Chamber path integral (Chambers, 1952) or the Chamber integral form of the BTE (Budd, 1966). The iterative method consists of substituting an initial distribution function $f_o(k,t)$ into the right-hand side of the above equation and evaluating $f(k,t)$, as demonstrated by Rees (1968, 1972). The new distribution function is again substituted into the right-hand side of this equation and this procedure is repeated until $f(k,t)$ converges to its solution with a given accuracy, A review of iterative methods for the solution of the BTE, can be found in Hammer (1977).

Direct Solution by Spherical Harmonic Expansion (SHE)

The spherical harmonic expansion (SHE) of the BTE has been widely used in the literature to calculate the electron distribution function, particularly at high energy. The SHE method is based on the development of the distribution function by the Legender polynomials, $P_{l,m}(\theta,\phi)$, in the k-space, as follows (in steady state):

$$f_n(r,\boldsymbol{k}) = f_n(r, k, \theta, \phi) = \sum_{l=0}^{\infty}\sum_{m=-l}^{l} f_m(r,E_n)\cdot P_{l,m}(\cos\theta,\sin\phi) \tag{80}$$

where $E_n=E_n(k)$ is the electron energy and k is the magnitude of the wave vector. The zero-order term $f_o(E_n)$, where $P_{0,0}=1$, is the spherically symmetric term of $f(r,\boldsymbol{k})$ in thermal equilibrium. The other expansion terms, which depend on polar angles θ and ϕ, represent the anisotropic components of the distribution function in presence of external fields. In one-dimensional physical space the angle θ represents the angle between the carrier wave vector k and the physical space axis $\boldsymbol{x}$. Here, we omit the spacial coordinate (r or x) for simplicity. After substituting the SHE into the BTE we get an infinite series of differential equations. For instance, the first-order term of the BTE expansion in steady state reads:

$$[u_n.\nabla_r - (e\zeta/\hbar).\nabla_k] f_o + 2/5\,[u_n.\nabla_r - (e\zeta/\hbar).\nabla_k] f_2 = (\partial f_f/\partial t)_{col} \tag{81}$$

where $(\partial f_f/\partial t)_{col}$ is the collision term which is associated with $P_{1,0}(cos\theta)$. After substituting the collision terms (using appropriate approximations), the solution of the set of differential equations (truncated to a certain order) results in the carrier distribution function $f_n(\boldsymbol{k})$.

Figure 22 shows the velocity distribution of an N-i-N structure, as calculated with MCM and SHE simulations. Considering just one *Legendre* polynomial (LP,1), the SHE result is different from the MC simulation, whereas for LP> 9, both simulations are in good agreement. Therefore, the SHE simulation with enough polynomials is a good alternative to the MCM. However, the biggest disadvantage of this method is the high memory requirement if a 2D or 3D real space is considered. It is interesting to note that taking just the first *Legendre* polynomial of the SHE, in a homogeneous system at low-field, is equivalent to the drift-diffusion model.

There exists a variety of device simulators, which are based on the solution of the BTE by the SHE. For instance, ViennaSHE is a multi-dimensional (1D, 2D and 3D) semiconductor device simulator on the basis of the SHE method. This simulator is an open source with its C++ code. The major challenge of the SHE method is the huge memory requirements when simulating electronic devices in two or three physical dimensions

Mathematical Note 3: Spherical Harmonics

The conventional spherical harmonics, $Y_{l,m}(\theta,\phi)$ are defined as follows:

$$Y_{lm}(\theta,\varphi) = (-1)^m\left[\frac{2l+1}{4\pi}\cdot\frac{(l-m)!}{(l+m)!}\right]^{1/2} P_{lm}(\cos\theta)\cdot\exp(jm\varphi)$$

Figure 22. Velocity profile of an N-I-N structure calculated with a device MC simulation and with a SHE simulator taking 1, 5, 9, and 15 Legendre polynomials (LP) into account
Hong, Pham & Jungemann, 2011.

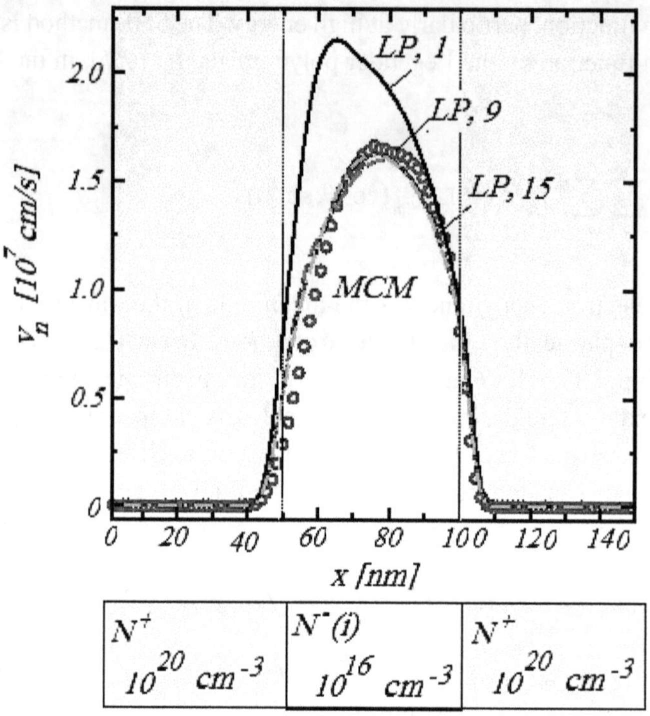

which is valid for $m \geq 0$, while the eigenfunctions corresponding to negative values of m are obtained from:

$$Y_{l,-m} = (-1)^m Y^*_{lm}$$

The functions $P_{l,m}$ are the associated Legendre polynomials, which are defined by:

$$P_{lm}(u) = (-1)^{l+m} \frac{(l+m)!}{(l-m)!} \cdot \frac{(l-u^2)^{-m/2}}{2!.l!} \frac{d^{i-m}}{du} \left(1-u^2\right)^l$$

where we put $u = cos\theta$ and $(1\text{-}u^2)^{1/2} = sin\theta$. The above relation is valid for $m \geq 0$, and the values for negative m are given by:

$$P_{l,-m}(u) = (-1)^m \frac{(l-m)!}{(l+m)!} P_{l,m}(u)$$

The carrier distribution function can be then expanded in the phase space as follows:

$$f(r,k)=\sum_{l=0}^{\infty}\sum_{m=-1}^{l} f_{lm}(r,k)Y_{lm}(\theta,\phi)$$

where $Y_{l,m}$ are the orthonormal, real-valued spherical harmonics on the unit sphere and the coefficients $f_{lm}(r, k)$ are given by:

$$f_{lm}(r,k)=\int_0^{\pi}\int_0^{2\pi} f(\theta,\phi)Y_{lm}^{*}(\theta,\phi)\,d\Omega$$

The average (moment) of a microscopic quantity (v) is then given by:

$$nV(r)=n\langle v\rangle=\sum_{l=0}^{\infty}\sum_{m=-1}^{l}\int_0^{\pi}\int_0^{2\pi} vf_{lm}(\theta,\phi)Y_{lm}(\theta,\phi)\,d\Omega$$

5.4 Monte Carlo Method

The Monte Carlo method (MCM) is a stochastic method, which can be used to solve integro-differential equations like the BTE and can offer great advantages over traditional approaches. The MC method was introduced in solid-state physics by several researchers and scientists. For instance, it was employed by Kurosawa (1966) to study nonlinear electron transport in semiconductors. The utilization of the MC method permits the calculation of the statistical mean values of all transport quantities, such as the mean electron velocity ($v_n = <u_g(k)>$) and the carrier mean energy ($\omega_n = <E(k)>$). The MC method can be used either for modeling of physical parameters (like drift mobility and impact ionization) or as simulation tool for whole semiconductor devices.

Since no *a priori* assumption is needed on the form of the carrier distribution (in real space and k-space), the Monte Carlo simulation is a reliable tool for the investigation of physical phenomena that critically depend on the shape of the distribution function, or on the details of its tail at high energies. Therefore, the Monte Carlo technique allows us to focus on certain physical mechanisms that might be of importance on the device performance, like the carrier generation by impact ionization or the hot carrier injection phenomena in MOSFET devices.

When the purpose of the analysis is the investigation of a steady-state, phenomenon in homogeneous bulk of a semiconductor (without electron–electron collision), it is sufficient to simulate the motion of one single carrier. The one-particle Monte Carlo (OPMC) simulation can be performed on a given fixed potential. However, when the transport is time or space dependent, which is generally the case in semiconductor devices, it is necessary to simulate a large number of carriers and follow their dynamical histories. The alternative technique is called Ensemble Monte Carlo or EMC (Lebwhol and Price, 1971). However, the stability requirements for EMC are more severe than the OPMC method (Kosina, Nedjalkov and Selberherr, 2000).

The Ensemble Monte Carlo (EMC) and the Full-band Monte Carlo (FBMC) methods have been widely used to study the proprieties of semiconductor devices. In FBMC, the band structure of the material is evaluated, usually with a pseudo-potential method, in a grid over the k-space grid, and stored as a look-up table. During MC simulation, the band structure, $E(k)$, and the density of states are called from the look-up tables when needed (e.g., for the calculation of the scattering rates). The Monte Carlo methods are usually coupled with the Poisson equation (in linearized form) via the carrier density. Once the basic physics involved in the transport of a device are known, the MC simulation results can be used to determine the device characteristics.

According to our knowledge, Hockney and coworkers applied the first self-consistent MC simulation to a whole semiconductor device (Hockney & Eastwood, 1988). For large class of devices, which are characterized by substantial areas of low electric field and retarding barriers, the direct simulation in such regions is time consuming. In contrast, the traditional simulators, based on the drift-diffusion model (DDM) can be applied to such situations. Therefore, a hybrid (MC-DDM) has been proposed, by relying on the fast DDM simulators for low field areas, and on the MC simulation where potential gradients heat-up carriers (Bandyopadhyay et al., 1987). Although excellent in principle, the hybrid MC-DDM technique requires accurate handling of boundary conditions at the interface regions (Higman, Hess & Dutton, 1989).

In the called *cellular automaton Monte Carlo* (CA/MCM), the entire Brillouin zone is discretized using a non-uniform mesh, and a transition table is generated between all initial and final states on the mesh, which greatly simplifies the final state selection of the conventional Monte Carlo method (Saraniti & Goodnick, 2000). However, this speed-up is obtained at the cost of large memory requirements to store the entire scattering tables.

Monte Carlo Algorithm for Device Simulation

The basic idea of the MC method is to simulate the motion of one or many carriers in the phase space, according to a specific equation of motion (like the BTE in semiclassical transport). The motion consists of a drift by an electric field, which is followed by a collision. The drift and the collision rate as well as the final states are controlled by the laws of probability. The basic steps of MC simulation can be summarized as follows:

1. **Initialization:** The physical space is divided into mesh of discrete points (grid). The choice of the grid (ΔX) size and time step (ΔT) depends on the Debye length (λ_D) and the plasma frequency (ω_p). The numerical discretization is usually carried out by the finite difference method (FDM).
2. **Charge Assignment:** In charge-carrier transport, the charge of each carrier is assigned to a particular mesh point. Since it is not possible to simulate all carriers in a real device, each simulated carrier represents a cloud of carriers in order to estimate current and field.
3. **Calculate Potential Distribution:** The Poisson equation is solved to determine the electrostatic potential at the mesh points. The solution can be obtained in several ways; the most efficient one is the Fourier analysis cyclic reduction (FACR) and the direct matrix inversion. The electrostatic field distribution and acting forces are then obtained from the potential.
4. **Simulate Carrier Flight and Scattering:** Each simulated carrier is treated as an individual particle, and undergoes a sequence of free flights and scattering, while it is subject to the local field. The MC sequence stops at fixed times, when the field is adjusted. The scattering probability $S(\boldsymbol{k}, \boldsymbol{k}')$ that an electron undergoes a transition from an initial state $\boldsymbol{k}$ to a final state $\boldsymbol{k}'$ is usually calculated within first order perturbation theory, using the Born approximation and the Fermi golden rule. Several scattering mechanisms can be included in the MC simulator, among which acoustic, optical inter-

valley and inter-band phonons, ionized and neutral impurities, surface roughness, alloy fluctuations and collision with other carriers. Also, impact ionization, Auger, and generation-recombination, processes can be taken into account. The time of flight (Δt) and the length of a free flight ($\Delta x = u_g \Delta t$) and the k-space trajectory of each particle can be calculated.

5. **Calculate Carrier Trajectory:** The carrier distribution function is calculated from the carrier trajectory in both the physical and k-space at different times.

During the charge-carrier free flight, it is accelerated by the electric field as a semi-classical particle such that:

$$\boldsymbol{F} = \Delta \boldsymbol{p} / \Delta t = -e\boldsymbol{\zeta}, \text{ (for electrons)}, \Delta \boldsymbol{p} = \hbar \, \Delta \boldsymbol{k} = \hbar(\boldsymbol{k} - \boldsymbol{k}') \tag{82}$$

where the electric field $\boldsymbol{\zeta}$, can be calculated from the charge carrier distribution, using Poisson's equation. Thus, from (82) we can write:

$$\boldsymbol{k} = \boldsymbol{k}' - e\,(\boldsymbol{\zeta}/\hbar)\,\Delta t \tag{83}$$

The flight time Δt is related to the probability of not-being-scattered r_s, which is selected randomly ($0 < r_s < 1$), according to the relation:

$$r_s = 1 - \exp(-\Delta t\, \Gamma_T) \tag{84}$$

where $\Gamma_T(\boldsymbol{k}) = \Sigma\, \Gamma_j(\boldsymbol{k}) = V/(2\pi)^3 \int \boldsymbol{S}(\boldsymbol{k},\boldsymbol{k}') d^3\boldsymbol{k}'$ is the integrated total scattering rate. Since $\boldsymbol{k} = \boldsymbol{k}(t)$, the above relation is usually written as follows:

$$r_s = 1 - \exp\left[-\int_0^{\Delta t} \Gamma_T(k(t'))dt'\right] \tag{85}$$

As r_s and $(1-r_s)$ have identical distributions, then, the flight time Δt may be calculated from the following integral

$$-\ln(r_s) = -\int_0^{\Delta t} \Gamma_T(k(t'))dt' \tag{86}$$

The above integral is not trivial and cannot be performed analytically, unless Γ_T is constant. In addition, we must use the above formula after every collision and for any particle. As the choice of random flights is performed so many times in MC simulation, several methods have been introduced to facilitate the solution of (86). Among these methods one can cite the self-scattering method (Rees, 1968), where a virtual scattering process is added such that the total Γ_T is constant. When the self-scattering mechanism is selected, the energy and wave vector of the electron are not changed. Therefore, we can choose the free flight duration Δt according to the rule

$$\Delta t = -\frac{1}{\Gamma_T}\ln\left(r_s\right) \tag{87}$$

Thus, every electron has a proper time Δt which is generated after every collision using the random number r_s. Therefore, the particle trajectory in the k-space ($\boldsymbol{k} = [k_x, k_y, k_z]^T$) can be calculated using (83). In the same manner, the particle trajectory in the physical space ($\boldsymbol{x} = [x,y,z]^T$) can be calculated using the following relation:

$$\boldsymbol{x} = \boldsymbol{x}' + \boldsymbol{u}_g(\boldsymbol{k}).\ \Delta t \tag{88}$$

where $\boldsymbol{u}_g(\boldsymbol{k}) = (1/\hbar)\ dE(\boldsymbol{k})/dk$ is the particle group velocity and $E(\boldsymbol{k})$ is energy dispersion, which may be approximated by a parabolic relation. However, it is possible to use a full band structure, which can be calculated, for instance, by the pseudo-potential method (Al-Said & Walter, 2008).

Calculate Averages (Mean Velocity and Energy): The mean carrier velocity can be written as

$$v_d = \langle u_g \rangle = (1/e.\zeta.T).\Sigma(E_f - E_i)$$

where $T = \Sigma\Delta T$ is the total simulation time and E_i and E_f are the initial and final carrier energy at the start and end of the electron flights. Also, the mean carrier energy can be calculated as

$$\omega = \langle E \rangle = ½\,(E_f - E_i).$$

A flowchart of a generic EMC device simulation is shown in Figure 23.

The Figure 24 depicts the electron energy distribution function in *GaAs*, at 300K, as obtained by the MC method, under different values of electric field, according to Fawcett, Boardman and Swain (1970).

Monte Carlo Simulation Programs

Together with the measurement techniques, the MC gives a microscopic description of several parameters in semiconductor devices. Today there is a variety of 1D, 2D and 3D Monte Carlo simulators like DAMOCLES from IBM, and MOCASIM from Silvaco and VMC from the Institute for Microlelectronics (Technische Universität Wien). Such Monte Carlo simulators can solve the BTE in bulk materials with full band structure.

Figure 25 depicts the electron energy distribution function at different electric fields in *Si*, as obtained by Damocles FBMC simulator (Fischetti, 2004). The Figure 26 depicts the electron mean velocity and mean energy, as calculated by Monte Carlo method.

Advantages and Disadvantages of MCM

The following figure shows a comparison between Monte Carlo and SHE simulation methods, in the bulk of Si, according to Bina and Rupp (2015). The figure shows the effect of electron-electron scat-

Figure 23. (a) Flow chart of Monte Carlo Method; (b) time discretization in EMC algorithm

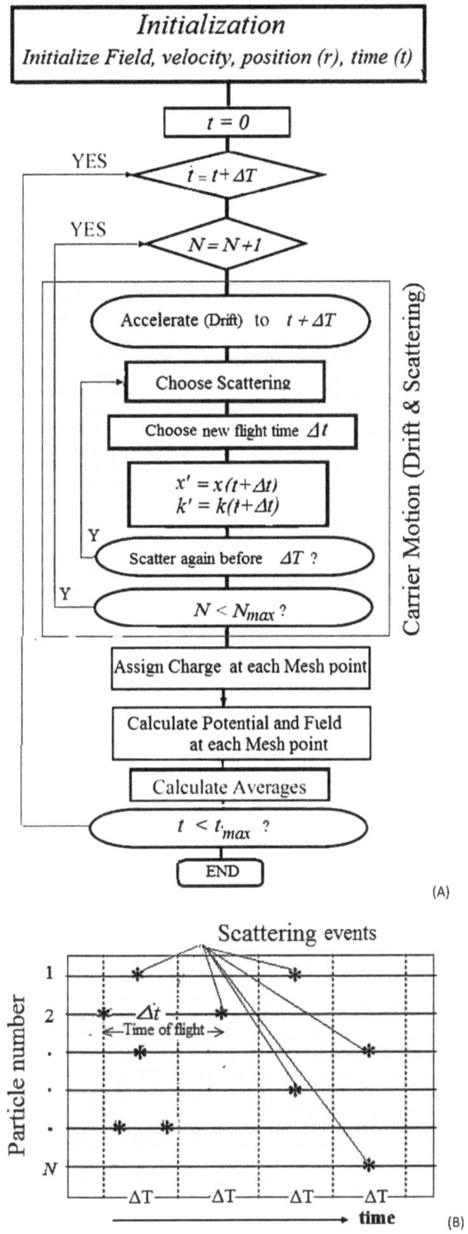

tering, which renders the BTE nonlinear and harder to solve by direct methods. In spite of the accuracy advantages of the MCM, it is expensive, in terms of computer resources, especially in three spatial dimensions. In addition, a closer look on numerical accuracy can easily identify some problems that are hard to solve with a MC algorithm. For instance, the calculation of the linear response near equilibrium for a MOSFET in weak or strong inversion is not trivial. This is an easy task for DDM-based simulators if the usual techniques are employed. However, for MC algorithm the situation is completely different. For a self-consistent multi particle solution of BTE and Poisson's Equation with the MC algorithm, the

Figure 24. Electron energy distribution function in GaAs at 300K, under different values of electric field, as obtained by the MC simulation; The Maxwellian distribution corresponds to the equilibrium state (at ζ=0).
Fawcett et al, 1970.

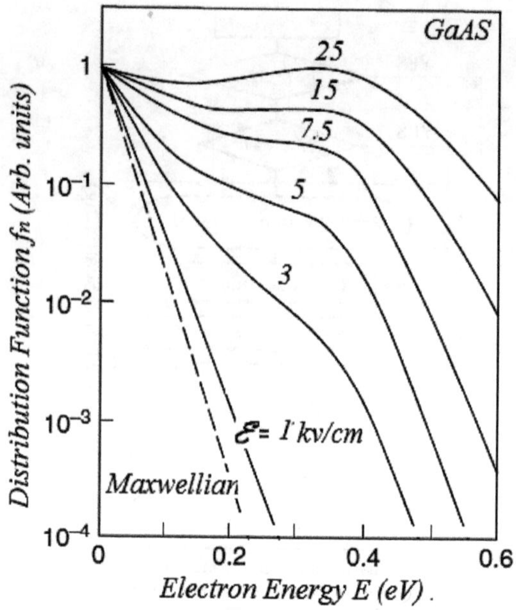

following equation holds for the CPU-time (TCPU) necessary to reach a relative error (r) for the DC drain current (I_D) with a probability of 95%

$$T_{CPU} = 8\alpha\,(V_{th}/V_D.I_D).(Q_t/r^2) \tag{89}$$

where V_{th} is the thermal voltage, Q_t is the total charge in the device, V_D is the drain voltage and α is the ratio of the CPU time and the simulated time divided by the number of simulated particles. For 1GHz CPU, $\alpha \approx 2.5\text{x } 10^{10}$.

This formula shows clearly that $r \sim 1/\sqrt{T_{CPU}}$, for all MC simulations. Moreover, it is clear that T_{CPU} can vary depending on the drain current and the required accuracy r which is not the case for DDM-based simulations.

6. LIMITS OF THE BTE

We have seen so far the Boltzmann transport equation is a powerful tool to study the transport phenomena in semiconductor devices. However, the BTE is only valid as long as the following assumptions are acceptable:

- Collisions are localized in space, so that the mean free path of charge carriers is much greater than the de-Broglie wavelength ($\lambda >> \lambda_{DB} = h/p$),

Figure 25. (a) Electron energy distribution function in Si, as obtained by the full-band MC method, under different values of electric field, by Damocles; (b) electron mean energy in bulk Si, by MCM according to several authors
Fischetti, 2004.

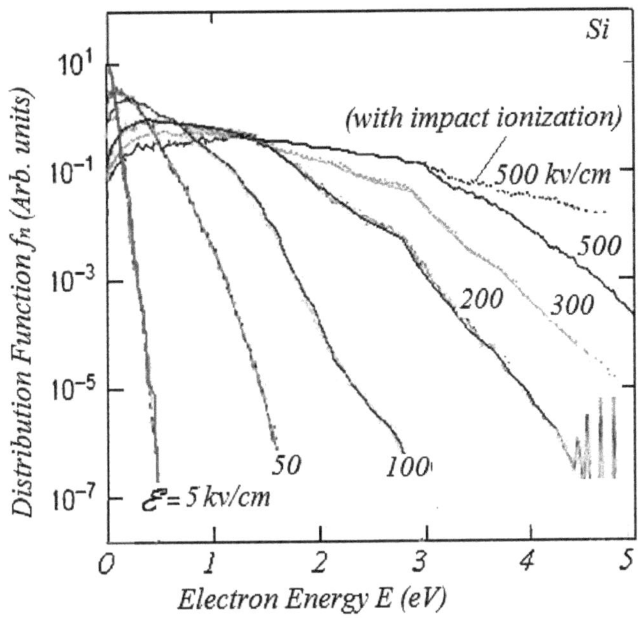

(A)

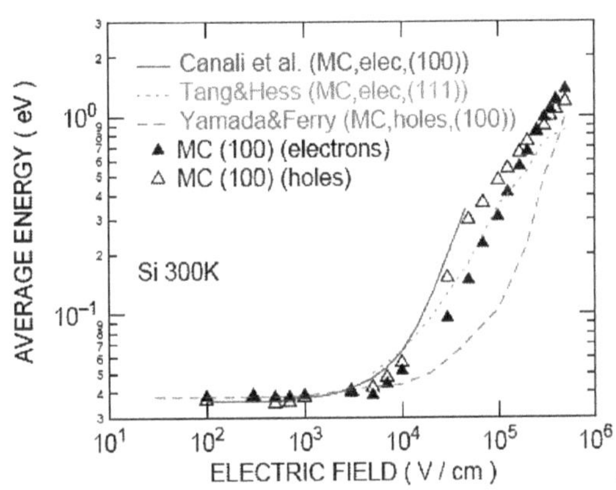

(B)

- Collisions are instantaneous in time such that the collision time τ_{cc} is much smaller than the time between subsequent collisions ($\tau_{cc} << \tau$),
- The external electric field force does not vary greatly over small distances in the order of wave packet length λ_{DB},
- The carriers are weakly correlated and the many-body effects are negligible. This is called the Bogoliubov assumption (Bogoliubov, 1946). For strongly ionized gasses, and degenerate semi-conductors, where carrier- carrier collisions are not negligible, the Boltzmann transport equation should be replaced by another more suitable model, like the so-called Fokker-Planck equation (Tome, 2006).

Figure 26. Electron drift velocity in Si, by MCM, according to several authors

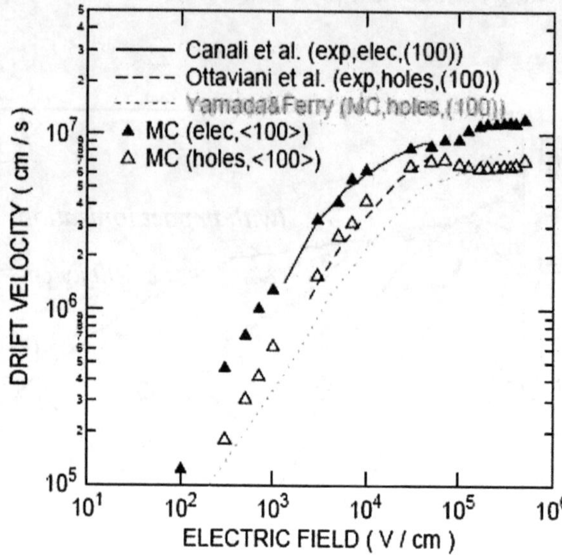

Figure 27. Comparison between Monte Carlo and SHE simulation methods
After Bina & Rupp (2015).

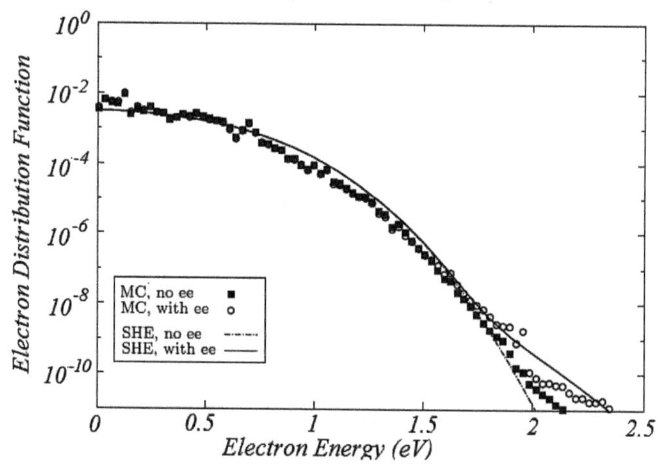

Mathematically speaking, the BTE is valid as long as the following inequality is realized (Ferry, Barker & Jacobini, 2012):

$$\left|\frac{e\varsigma\hbar q\tau}{2m^{*}}\right| \leq \frac{2\hbar}{\tau_{cc}} \tag{90}$$

where ζ is the applied electric field, τ_{cc} is collision time and τ is mean free time between collisions and $q = k–k'$ is the wave vector of involved phonons (emitted or absorbed) during the collision. The above condition is sometimes called the Barker equation. The right-hand side of this condition ($2\hbar/\tau_{cc}$) is sometimes called the collision bandwidth and is termed by Γ.

When collisions are not instantaneous ($\tau_{cc} \neq 0$) then the collision bandwidth $\Gamma=2\hbar/\tau_{cc}$ is finite, and the delta function in the Fermi golden rule (of scattering rate) should be replaced by a Lorentzian as follows:

$$S_j = \lambda_j \delta(E - E') \Rightarrow \lambda_j \frac{\Gamma^2}{\Gamma^2 + (E - E')^2} = \frac{\lambda_j}{1 + \frac{1}{4}\tau_{cc}^2(\omega - \omega')^2} \quad (91)$$

In this case, the quantum transport formulation should be used in order to overcome the above-indicated shortcomings and limitations of the BTE.

7. CASE STUDY 1: BTE SIMULATION IN A MOSFET (BY SHE)

In the early work on long channel MOSFET modeling, the analytical methods have been utilized to derive its I-V characteristics, assuming uniform doping profile. However, the MOSFET analytical models couldn't interpret the two-dimensional effects and the realistic MOS characteristics. To get rid of the problems of analytical models, the researchers have applied the numerical techniques. The numerical solution of semiconductor transport equations has been widely utilized with the rapid improvements of the computer performances in the last two decades. Using the numerical methods, an accurate solution can be achieved for the assumed transport model and physical parameters. Yet, there exist some challenges in the simulation of nanostructures and some bulk devices. For instance, the simulation of avalanche breakdown during the transient regime of power devices is not a trivial job. Our case study depicts the simulation of a lateral double-diffused MOSFET (LDMOSFET). This is the preferred control device

Figure 28. Illustration of the effect of collision bandwidth broadening

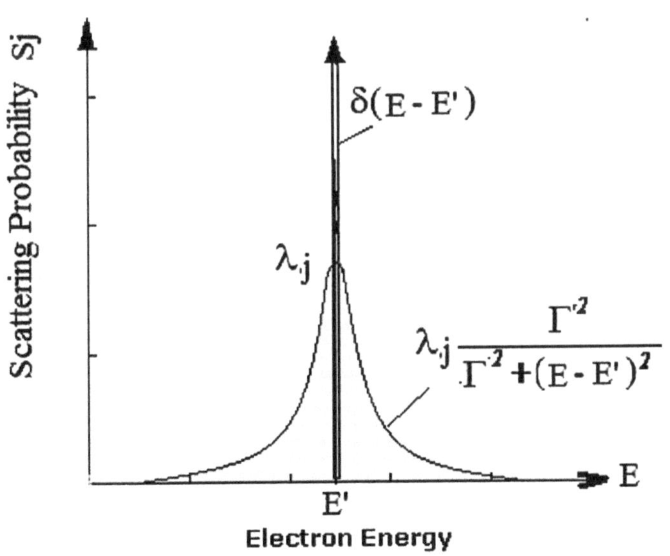

Figure 29. The structure of an LDMOSFET

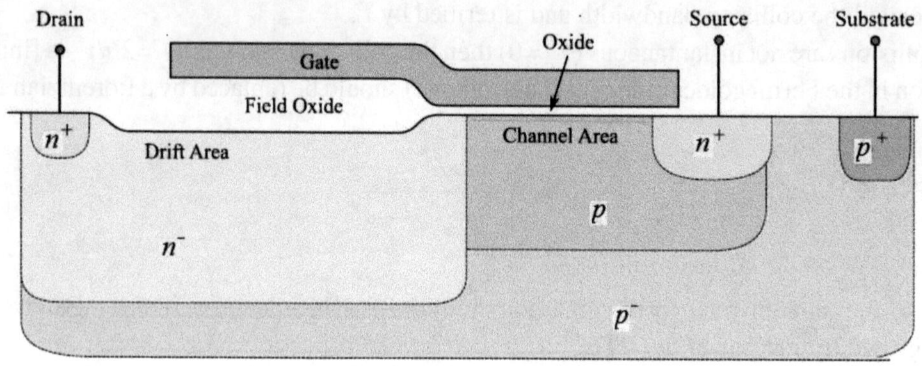

in smart power ICs. The results are taken from (Liang, Goldsman & Mayergoyz, 1998) and (Hong & Jungermann, 2009).

Simulation is performed by directly solving the BTE for electrons, and the Hole continuity equation together with the Poisson equation, self consistently. The method relies on the spherical harmonics expansion (SHE) of the distribution function *f(**x**, **k**, t)* in the form:

$$f(x,k,t) = \sum_{l=0}^{\infty} \sum_{m=-l}^{l} f_{l,m}(x,\varepsilon,t) Y^{l,m}(\theta,\varphi) \tag{92}$$

This expression describes the temporal evolution of electrons with wave vector $\boldsymbol{k}$ in the physical space $\boldsymbol{x}$. Note that φ is an angle, while ϕ dentoes the electrostatic potential, which is the solution of the Poisson equation. As for any system of partial differential equations, the BTE needs to be equipped with boundary conditions in order to be completely specified. In noncontact boundary regions, homogeneous Neumann boundary conditions with respect to the spatial coordinate are imposed. At the device contacts, Dirichlet boundary conditions are imposed by enforcing a Maxwell equilibrium distribution of carriers at room temperature. This leads to a sharp boundary layer of the distribution function near the contact (Schroeder et al, 1992). At the gate area, the boundary conditions yield a transverse electric field effect due to the gate bias. The two-dimensional boundary conditions for the region linking the oxide and semiconductor yield a Robin-type condition:

$$\zeta \cdot \hat{n} = \nabla \phi \cdot \hat{n} = \frac{1}{2\pi\delta} \frac{\varepsilon_{ox}}{\varepsilon_{si}} \left(\phi - V_g - \frac{\delta^2}{3} \right) \tag{93}$$

where ζ is the electric field, $\tilde{n}$ is the normal unit-vector to the interface between the oxide and the semiconductor layers. Also, V_g is the electrostatic potential at the gate contact; ε_{ox} and δ are the oxide permittivity and thickness, respectively. In the remainder of boundary regions, we impose elastic boundary refection, such that *f(x,k) = f(x, k')* with *k'=k-2k..ñ.*

The BTE solver reads the device mesh from an input file. The initial guess solution is obtained using the drift-diffusion model (DDM). As soon as the simulation is completed, results can be written to

Figure 30. Flowchart of the BTE-Poisson solver in a MOSFET

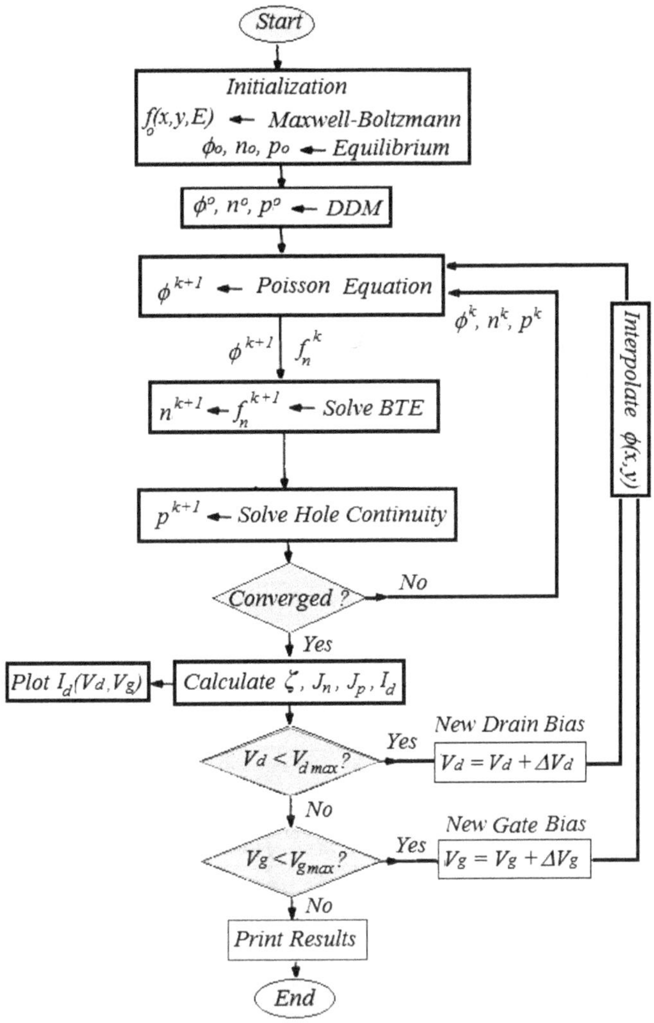

an output file. The solution provides the electron distribution function, electrostatic potential, and the concentrations of electrons and holes of the MOSFET in physical space. The maximum voltages for this device are $V_g = -25$V and $V_d = -50$V with the other terminals grounded. Figures 31 and 32 depict the distribution function of electrons, the electric field distribution and the electron concentration at equilibrium and at maximum bias conditions.

8. SUMMARY

In this chapter we cover the essential aspects of charge carrier transport through solid materials, within the semiclassical transport theory. We start with a review of the semiclassical approaches that leads to the concepts of drift velocity, drift mobility, electrical conductivity and thermal conductivity of charge carriers in metals and semiconductors.

Figure 31. The electron distribution function inside the LDMOSFET at maximum bias conditions (V_d=-50V, V_g=-25V), as obtained by the solution of the BTE, with Poisson's equation
Source: Liang, Goldsman & Mayergoys, 1998.

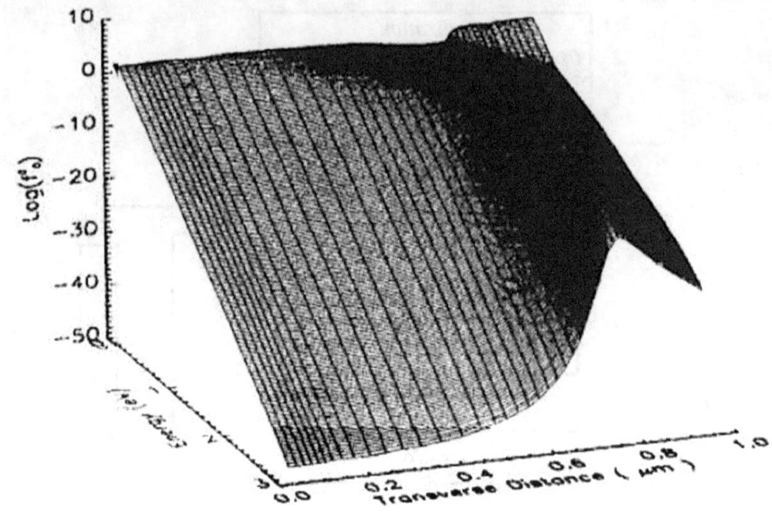

Figure 32. Electric field and electron concentration in the LDMOSFET
Source: Hong & Jungermann, 2009.

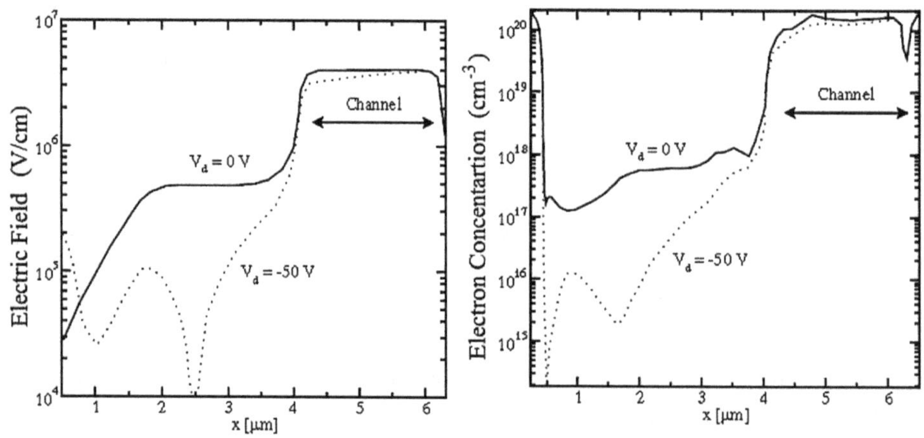

In Classical Mechanics, the complete state of a particle can be given by its physical coordinates (x,y,z) and momenta (p_x, p_y, p_z). The semiclassical transport theory is based on the Boltzmann transport equation (BTE). The Boltzmann transport equation can be derived from the Lowville equation, which describes the evolution of the distribution function changes in time. There are various approximations and phenomenological approaches which make the equation useful and solvable for semiconductor devices. The equation can describe macroscopic phenomena such as the electrical conductivity, Hall Effect, and diffusion process.

According to the BTE, the time evolution of the distribution function of localized particles is given by:

$$\partial f/\partial t + u_g.\nabla f + (F/\hbar).\nabla_k f = [\partial f/\partial t]_{col.}$$

The distribution function $f = f(x,k,t)$ describes the probability to find an electron in the infinitesimal space element d^3k around k. The collision term depends on the sum of the different scattering mechanisms. The value of $f(k)$ is increased by scattering of electrons from some state k' into k or decreased by scattering from k into some state k'.

$$\left(\frac{\partial f}{\partial t}\right)_{coll} = \int d^3k' \left\{S\left(k',k\right)f\left(k'\right)\left[1-f\left(k\right)\right] - S\left(k,k'\right)f\left(k\right)\left[1-f\left(k'\right)\right]\right\}$$

The function $S(k',k)$ describes the probability of a scattering event taking place in unit time. For the calculation of the collision term we must sum over all possible initial states k', weighted by its occupancy, and we must take into account the availability of the final state. According to the Fermi Golden Rule, the scattering rates $S(k',k)$ are given by:

$$S\left(k,k'\right) = \frac{2\pi}{\hbar}\left|\left\langle k\left|V\right|k'\right\rangle\right|^2 \delta\left(E_k - E_{k'}\right)$$

where V is the perturbation potential and k and k' refer to the particle wavevector states. The scattering mechanisms include acoustic intravalley scattering, intervalley scattering via three f and three g phonons, ionized impurity scattering, surface-roughness scattering at SiO_2/Si interface. For *SiGe* and other compound semiconductors, we should include the alloy scattering into account. The following figure depicts the basic types of carrier scattering mechanisms in semiconductors.

The BTE is valid under assumptions of semi-classical transport: effective mass approximation (which incorporates the quantum effects due to crystal periodicity); Born approximation for the collisions, in the limit of small perturbation for the electron-phonon interaction and instantaneous collisions; no memory effects. The phonons are typically treated as in equilibrium, although the condition of non-equilibrium phonons may be included through an additional transport equation.

The essential assumption of this theory is the Markovian behavior of the scattering process, that is, each scattering process is fully completed and independent of any other process. Mathematically speaking, the BTE is valid as long as the following inequality is realized:

$$\left|\frac{e\zeta\hbar q\tau_c}{2m^*}\right| \leq \frac{2\hbar}{\tau_{cc}}$$

where $\hbar q$ is the momentum exchanged in the scattering process ($q = k'-k$) and $1/\tau_{cc}$ is phonon collision rate.

Analytical solutions of the Boltzmann equation are possible only under very restrictive assumptions. Direct numerical methods for device simulation have been limited by the complexity of the equation. Most semiconductor simulations have been based on stochastic solution methods (like Monte Carlo method MCM), which involve the simulation of particle trajectories rather than the direct solution of partial differential equations. The MCM has been applied to the simulation of many semiconductor devices.

The BTE may be simplified by using the relaxation time approximation.

$$\left(\frac{\partial f}{\partial t}\right)_c = \frac{\Delta f}{\tau} = -\frac{f - f_0}{\tau}$$

which is based on the assumption that for small changes in f carriers return to equilibrium in a characteristic time τ, dependent on the dominant scattering mechanisms. Using the relaxation time approximation, one can obtain the diffusion drift transport equations:

REFERENCES

Abramo, A., Baudry, L., Brunetti, R., Castagne, R., Charef, M., Dessenne, F., & Yoshii, A. (1994). A comparison of numerical solution of the Boltzmann transport equation for high-energy electron transport silicon. *IEEE Transactions on Electron Devices, 41*(9), 1646–1654. doi:10.1109/16.310119

Al-Saidi, W. A., Walter, E. J., & Rappe, A. M. (2008). Optimized norm-conserving Hartree-Fock pseudopotentials for plane-wave calculations. *Physical Review B: Condensed Matter and Materials Physics, 77*(7), 075112. doi:10.1103/PhysRevB.77.075112

Arora, N. D., Hauser, J. R., & Roulston, D. J. (1982). Electron and hole mobility in Si as function of concentration and temperature. *IEEE Transactions on Electron Devices, 29*(2), 292–295. doi:10.1109/T-ED.1982.20698

Ashcroft, N. W., & Mermin, N. D. (1976). *Solid State Physics*. Philadelphia: Saunders College.

Aubert, J-P., Vassiere J-C. & Nougier, J-P. (1984). Matrix determination of the stationary solution of the Boltzmann Equation for Hot Carriers in Semiconductors. *J. Appl. Phys., 56*(4), 1128-1132.

Auer, C., & Schürrer, F. (2006). Efficient time integration of the Boltzmann-Poisson system applied to semiconductor device simulation. *Journal of Computational Electronics, 5*(1), 5–14. doi:10.1007/s10825-006-7914-6

Bandyopadhyay, S., Klausmeier-Brown, M. E., Maziar, C. M., Datta, S., & Lundstrom, M. S. (1987). A rigorous technique to couple Monte Carlo and drift-diffusion models for computationally efficient device simulation. *IEEE Transactions on Electron Devices, ED-34*(2), 392–399. doi:10.1109/T-ED.1987.22935

Banoo, K., & Lundstrom, M. (2000). *Direct Solution of the Boltzmann Transport Equation in Nanoscale Si devices*. IEEE. doi:10.1109/SISPAD.2000.871204

Bardeen, J., & Shockley, W. (1950). Deformation Potentials and Mobilities in Non-Polar Crystals. *Physical Review, 80*(1), 72–80. doi:10.1103/PhysRev.80.72

Bina, M., & Rupp, K. (2015). The Spherical Harmonics Expansion Method for Assessing Hot Carrier Degradation. In T. Grasser (Ed.), Hot Carrier Degradation in Semiconductor Devices. Springer International Publishing.

Bogoliubov, N. (1947). Problemi Dynamititi. *Journal Physics USSR*, *11*, 23.

Boltzmann, L. (1872). Weitere Studien über das Wärmegleichgewicht unter Gasmolekülen. *Sitzungsberichte Akademie der Wissenschaften.*, *66*, 275–370.

Boltzmann, L. (1896, 1898). Lectures on Gas Theory (S. G. Brush, Trans.). Berkeley, CA: University of California Press.

Born, M., & Huang, K. (1954). Dynamical Theory of Crystal Lattices. Oxford Univ. Press.

Born, M., & Oppenheimer, R. (1927). Zur Quantentheorie der Molekeln. *Ann of Physics (Leipzig)*, *84*(20), 457–484. doi:10.1002/andp.19273892002

Brooks, H., & Herring, C. (1951). Scattering by ionized impurities in semiconductors. *Physical Review*, *83*, 879.

Budd, H. F. (1966). Proc. Intern. Conf. on Phys. Semiconductors, Kyoto. *Journal of the Physical Society of Japan*, *21*(suppl.), 420.

Buet, C., Dellacherie, S. & Sentis, R. (2001). Numerical solution of an ionic Fokker-Planck equation with electronic temperature. *SIAM*, *39*(4), 1219-1253.

Canali, C., Maijni, G., Minder, R., & Ottaviani, G. (1975). Electron and Hole Drift Velocity Measurements in Silicon and their Empirical Relation to Electric Field and Temperature. *IEEE Transactions on Electron Devices*, *ED-22*(11), 1045–1047. doi:10.1109/T-ED.1975.18267

Cartier, E., Fischetti, M. V., Eklund, E. A., & McFeely, F. R. (1993). Impact ionization in silicon. *Applied Physics Letters*, *62*(25), 3339–3341. doi:10.1063/1.109064

Cassi, D., & Ricco, B. (1990). An analytical model of the energy distribution of hot electrons. *IEEE Transactions on Electron Devices*, *37*(6), 1514–1521. doi:10.1109/16.106247

Caughey, D. M., & Thomas, R. E. (1967). Carrier mobilities in silicon empirically related to doping and field. *Proceedings of the IEEE*, *55*(12), 2192–2193. doi:10.1109/PROC.1967.6123

Cerciganani, C. (1988). *The Boltzmann Equation and its Applications.* Vienna: Springer-Verlag. doi:10.1007/978-1-4612-1039-9

Chambers, R.G. (1952). Article. *Proc. Phys. Soc. London. A*, *65*, 458.

Conwell, E., & Weisskopf, V. (1950). Theory of impurity scattering in semiconductors. *Physical Review*, *77*(3), 388–390. doi:10.1103/PhysRev.77.388

Davies, J. (1998). *The Physics of Low-dimensional Semiconductors. An Introduction.* Cambridge, UK: Cambridge Press.

Del Alamo, J. A., Swirhun, S. E., & Swanson, R. M. (1985). Measurement and modeling of minority carrier transport in heavily-doped Si. *Solid-State Electronics*, *28*(1-2), 47–52. doi:10.1016/0038-1101(85)90209-6

Dorkel, J. M., & Leturcq, P. H. (1981). Carrier Mobilities in Si Semi-Empirically Related to Temperature, Doping and Injection Level. *Solid-State Electronics*, *24*(9), 821–825. doi:10.1016/0038-1101(81)90097-6

Esseni, D. (2004). On the Modeling of Surface Roughness Limited Mobility in SOI MOSFETs and its Correlation to the Transistor Effective Field. *IEEE Transactions on Electron Devices*, *51*(3), 394–401. doi:10.1109/TED.2003.822344

Fawcett, W., Boardman, A. D., & Swain, S. (1970). Monte Carlo Determination of Electron Transport Properties in GaAs. *Journal of Physics and Chemistry of Solids*, *31*(9), 1963–1990. doi:10.1016/0022-3697(70)90001-6

Fermi, E. (1927). Un metodo statistico per la determinazione di alcune proprietà dell'atome. *Rend. Accad. Naz. Lincei*, *6*, 602.

Ferry, D. (2000). *Semiconductor Transport*. CRC Press.

Ferry, D., Barker, J. R., & Jacobini, R. C. (2012). *Physics of Nonlinear Transport in Semiconductors*. Technology & Engineering. Springer Science.

Fischetti, M. (1998). Theory of Electron Transport in Small Semiconductor Devices using the Pauli Master Equation. *Journal of Applied Physics*, *83*(1), 270–291. doi:10.1063/1.367149

Fischetti, M. V., & Higman, J. M. (1991). Theory and calculation of the deformation potential electron-phonon scattering rates in semiconductors. In K. Hess (Ed.), *Monte Carlo Device Simulation: Full Band and Beyond*. Boston, MA: Kluwer. doi:10.1007/978-1-4615-4026-7_5

Fischetti, M. V., & Laux, S. E. (1988). Monte Carlo analysis of electron transport in small semiconductor devices including band-structure and spacecharge effects. *Physical Review B: Condensed Matter and Materials Physics*, *38*(14), 9721–9745. doi:10.1103/PhysRevB.38.9721 PMID:9945794

Fischetti, M. V., & Laux, S. E. (1993). Monte Carlo study of electron transport in silicon inversion layers. *Physical Review B: Condensed Matter and Materials Physics*, *48*(4), 2244–2274. doi:10.1103/PhysRevB.48.2244 PMID:10008617

Fischetti, M. V., & Laux, S. E. (1996). Band structure, deformation potentials, and carrier mobility in strained Si, Ge, and SiGe alloys. *Journal of Applied Physics*, *80*(4), 2234–2252. doi:10.1063/1.363052

Fokker, A. D. (1914). Die mittlere Energie rotierender elektrischer Dipole im Strahlungsfeld. *Annals of Physics*, *348*(4), 810–820. doi:10.1002/andp.19143480507

Garcias-Salva, P., Lopez-Gonzalez, J. & Part, L. (2001). A Comparison between Monte Carlo and Extended Drift-Diffusion Models for Abrupt InP/InGaAs HBTs. *IEEE Trans. Electron Devices, 48*(6).

Gibbs, J. (1902). *Elementary Principles in Statistical Mechanics*. Oxford Bow Press.

Harrison, J. W., & Hauser, J. R. (1976). Alloy scattering in ternary III-V compounds. *Physical Review B: Condensed Matter and Materials Physics*, *13*(12), 5347–5350. doi:10.1103/PhysRevB.13.5347

Harrison, W. A. (1956). Scattering of Electrons by Lattice Vibrations in Nonpolar Crystals. *Physical Review*, *104*(5), 1281–1290. doi:10.1103/PhysRev.104.1281

Herring, C., & Vogt, E. (1956). Transport and deformation-potential theory for many-valley semiconductors with anisotropic scattering. *Physical Review*, *101*(3), 944–961. doi:10.1103/PhysRev.101.944

Hess, K. (Ed.). (1991). *Monte Carlo Device Simulation: Full Band and Beyond*. Boston: Kluwer Academic Press. doi:10.1007/978-1-4615-4026-7

Hess, K. (2000). *Advanced Theory of Semiconductor Devices*. IEEE Press.

Higman, J. M., Hess, K., & Dutton, R. W. (1989). Coupled Monte Carlo-Drift Diffusion Analysis of Hot Electron Effects in MOSFETs. *IEEE Transactions on Electron Devices*, *36*(5), 745–751. doi:10.1109/16.299675

Hockney, R. W., & Eastwood, J. M. (1988). *Computer Simulation Using Particles*. Philadelphia: Adam Hilger. doi:10.1887/0852743920

Hong, S. M., & Jungemann, C. (2009). A fully coupled scheme for a Boltzmann-Poisson equation solver based on a spherical harmonics expansion. *Journal of Computational Electronics*, *8*(3-4), 225–241. doi:10.1007/s10825-009-0294-y

Hong, S.-M., Pham, A.-T., & Jungemann, C. (2011). *Deterministic Solvers for the Boltzmann Transport Equation*. Springer Science & Business Media. doi:10.1007/978-3-7091-0778-2

Jacoboni, C., & Lugli, P. (1989). *The Monte Carlo Method for Semiconductor Device Simulation*. Vienna: Springer-Verlag. doi:10.1007/978-3-7091-6963-6

Jacoboni, C., & Lugli, P. (2012). *The Monte Carlo Method for Semiconductor Device Simulation*. Springer Science & Business Media.

Jacoboni, C., & Reggiani, L. (1983). Bulk Hot-Electron Properties of Cubic Semi-conductors,''. *Advances in Physics*, *28*(4), 493–553. doi:10.1080/00018737900101405

Jacoboni, C., & Reggiani, L. (1983). The Monte Carlo method for the solution of charge transport in semiconductors with applications to covalent materials. *Reviews of Modern Physics*, *55*(3), 645–705. doi:10.1103/RevModPhys.55.645

Jerome, J. W. (2009). Mathematical Advances and Horizons for Classical and Quantum-Perturbed Drift-Diffusion Systems, Solid State Devices and Beyond. *Journal of Computational Electronics*, *8*(2), 132–141. doi:10.1007/s10825-009-0269-z

Jungemann, C., & Meinerzhagen, B. (2003). *Hierarchical Device Simulation: The Monte-Carlo Perspective*. Springer. doi:10.1007/978-3-7091-6086-2

Keldysh, L. (1965). Concerning the Theory of Impact Ionization in Semiconductors. *Soviet Physics, JETP*, *21*, 1135–1144.

Kittel, C. (1980). *Introduction to Solid State Physics, Univ. of California*. Berkeley, CA: Wiley.

Kosina, H., Nedjalkov, M., & Selberherr, S. (2000). Theory of the Monte Carlo Method for Semiconductor Device Simulation. *IEEE Transactions on Electron Devices*, *47*(10), 1898–1908. doi:10.1109/16.870569

Kurosawa, T. (1966). Monte Carlo calculation of hot electron problems. *Journal of the Physical Society of Japan*, *21*, 424–426.

Laux, S. E., Fischetti, M. V., & Frank, D. J. (1990). Monte Carlo analysis of semi-conductor devices: The DAMOCLES program. *IBM Journal of R & D*, *34*(4), 466–494. doi:10.1147/rd.344.0466

Lawaetz, P. (1969). Long-Wavelength Phonon Scattering in Nonpolar Semiconductors. *Physical Review*, *183*(3), 730–739. doi:10.1103/PhysRev.183.730

Lebwhol, P. A., & Price, P. J. (1971). Article. *Solid State Communications*, *9*, 1221. doi:10.1016/0038-1098(71)90015-9

Liang, W., Goldsman, N., & Mayergoyz, I. (1998). A New Self-Consistent 2D Device Simulator Based on Deterministic Solution of the Boltzmann, Poisson and Hole-Continuity Equations. *VLSI Design*, *6*(1-4), 251–256. doi:10.1155/1998/57195

Liouville, J. (1838). Article. *Journal de Mathématiques Pures et Appliquées*, *3*, 349.

Lombardi, C., Manzini, S., Saporito, A., & Vanzi, M. (1988). A physically-based mobility model for numerical simulation of nonplolar devices. *IEEE Transactions on Computer-Aided Design of Integrated Circuits and Systems*, *7*(11), 1164–1171. doi:10.1109/43.9186

Lundstorm, M. (1990). Fundamentals of Carrier Transport. In *Modular Series on Solid State Devices*. Reading, MA: Addison-Wesley.

Lundstrom, M. (2000). *Fundamentals of Carrier Transport*. Cambridge Univ. Press. doi:10.1017/CBO9780511618611

Mahan, G. D. (2000). Many-Particle Physics. New York: Academic Press.

Majorana, A., & Pidatella, R. (2001). A finite difference scheme solving the Boltzmann-Poisson system for semiconductor devices. *Journal of Computational Physics*, *174*(2), 649–668. doi:10.1006/jcph.2001.6929

Price, P. J. (1979). *Semiconductors and Semimetals*. New York: Academic Press.

Pruschke, T. (2014). Advanced Solid State Theory. In Transport theory. Morgan & Claypool Publishers. doi:10.1088/978-1-627-05328-0

Rees, H. D. (1968). Article. Solid-State Comunications, 26A, 416.

Reggiani, L. (Ed.). (1985). *Hot Electron Transport in Semiconductors*. Berlin: Springer. doi:10.1007/3-540-13321-6

Rupp, K., Jungeman, C., Hong, S.-M., Bina, M., Grasser, T. & Jünge, A. (2016). A review of recent advances in the spherical harmonics expansion method for semi-conductor device simulation. *J. Comput. Electron.*

Saraniti, M., & Goodnick, S. M. (2000). Hybrid Fullband Cellular Automaton/Monte Carlo Approach for Fast Simulation of Charge Transport in Semiconductors. *IEEE Transactions on Electron Devices*, *47*(10), 1909–1916. doi:10.1109/16.870571

Sasso, G., Rinaldi, N., Matz, G., & Jungemann, C. (2009). Accurate Mobility and Energy Relaxation Time Models for SiGe HBTs Numerical Simulation. *IEEE Transactions on Electron Devices*, *56*(1), 1–4.

Schenk, A. (2012). *Advanced Physical Models for Silicon Device Simulation- Technology & Engineering*. Academic Press.

Schroeder, D., Ventura, D., Gnudi, A., & Baccarani, G. (1992). Boundary conditions for spherical harmonics expansion of Boltzmann equation. *Electronics Letters*, *28*(11), 995–996. doi:10.1049/el:19920632

Seeger, K. (1997). *Semiconductor Physics* (6th ed.). Berlin: Springer-Verlag. doi:10.1007/978-3-662-03347-0

Shockley, W. (1950). *Electrons and Holes*. Princeton, NJ: D. Van Nostrand Company, Inc.

Stratton, R. (1962). Diffusion of hot and cold electrons in semiconductor barriers. *Physical Review*, *126*(6), 2002–2014. doi:10.1103/PhysRev.126.2002

Tang, J. Y., & Hess, K. (1983). Impact ionization of electrons in silicon (steady state). *Journal of Applied Physics*, *54*(9), 5139–5144. doi:10.1063/1.332737

Tang, T., & Gan, H. (2000). Two Formulations of Semiconductor Transport Equations based on Spherical Harmonic Expansion of the Boltzmann Transport Equation. *IEEE Transactions on Electron Devices*, *47*(9), 1726–1732. doi:10.1109/16.861583

Tavernier, J., & Calcki, D. (1970). *Introduction aux Phénomènes de Transport Linéaires dans les Semiconducteurs*. Paris: Mason.

Thoma, R., Peifer, H. J., Engl, W. L., Quade, W., Brunetti, R., & Jacoboni, C. (1991). An improved impact-ionization model for high-energy electron transport in Si with Monte Carlo simulation. *Journal of Applied Physics*, *69*(4), 2300–2311. doi:10.1063/1.348711

Tomé, T. (2006). Entropy production in nonequilibrium systems described by a Fokker-Planck equation. *Brazilian Journal of Physics*, *36*(4a), 1285–1289. doi:10.1590/S0103-97332006000700029

Tritt, T. M. (2005). *Thermal Conductivity*. Springer-Verlag.

Ungersböck, E., & Kosina, H. (2006). Monte Carlo Study of Electron Transport in Strained Silicon. *Journal of Computational Electronics*, *5*(2-3), 79–83. doi:10.1007/s10825-006-8823-4

Van Roosbroeck, W. (1950). Theory of flow of electron and holes in germanium and other semiconductors. *The Bell System Technical Journal*, *29*(4), 560–607. doi:10.1002/j.1538-7305.1950.tb03653.x

Vapaille, A. (1970). *Physique des Dispositifs a Semiconducteur* (Vol. 1). Paris: Masson & Cie.

Vlasov, A. (1938). On Vibration Properties of Electron Gas. *Journal of Experimental and Theoretical Physics*, *8*, 444–470.

Wiley, J. D. (1971). Polar Mobility of Holes in III-V Compounds. *Physical Review B: Condensed Matter and Materials Physics*, *4*(8), 2485–2493. doi:10.1103/PhysRevB.4.2485

Zawadzki, W. (1982). Mechanics of electron scattering in semiconductors. In T. Moss (Ed.), *Handbook of Semiconductors*. New York: North-Holland.

Ziman, J. M. (1972). *Principles of the Theory of Solids*. Cambridge, UK: Cambridge Press. doi:10.1017/CBO9781139644075

ENDNOTES

1 A closed thermodynamic system is in *thermal equilibrium* forever unless it is acted upon by external forces. The state of thermal equilibrium is the most probable state of a thermodynamic system

2 Actually, there exist three ensembles types, namely; Microcanonical, Canonical and Grand canonical ensembles. The most common ensemble, namely the Canonical ensemble, is used when the system of interest is put in contact with a thermal bath. The canonical ensemble is able to exchange any energy with the surrounding heat reservoir.

3 The Liouville theorem, named after the French mathematician Joseph Liouville, is a basic theorem in classical statistical and Hamiltonian mechanics

4 The entropy S of a thermodynamic system is a logarithmic measure of the number of accessible energy states (microstates) of a system in a small interval of energy. the entropy, S, may be thought of as a measure of disorder or lack of knowledge of a system.

5 Sometimes the factor of 2 is included in the definition of the density of states, such that $g(k) = 2/(2\pi)^3$

6 When, we have many possible final states (e.g., bands) the Kronecker delta is usually replaced with the final density of states in the expression of transition probability.

7 Many references utilize the symbol 'c' instead of 'q' to denote the lattice wavevector.

8 The non-parabolicity factor $\alpha = (1-m_{nd}/m_o)/E_g$ is 0.5eV^{-1} for Si and 0.7 eV^{-1} for $GaAs$.

9 The ω-q relation or the dispersion relation

10 This approximation is sometimes called the diffusion approximation because it results in a transport equation, which resembles the diffusion equation in mathematics. However, we'll note the presence of both diffusion and drift terms in this model.

11 The correlation between the relaxation time and the mean free time is illustrated in Note 1.

APPENDIX

Note About Reference Material

The Boltzmann papers on the statistical interpretation of thermo-dynamics, the H-theorem, transport theory, thermal equilibrium, the equation of state of gases, and other subjects, occupy about 2,000 pages in the proceedings of the Vienna Academy in 1896.

Boltzmann, L. (1896, 1898). Lectures on Gas Theory. Trans. S.G. Brush. Berkeley: University of California Press, 1964. Reprinted by Dover Publication, in 2011.

Chapter 3
Semiclassical Transport Theory of Charge Carriers, Part II:
Macroscopic Approaches

1. OVERVIEW AND CHAPTER OBJECTIVES

During the last two decades, a considerable interest has been devoted for the study of high-field transport because of the scaled-down dimensions of devices. The scaled-down dimensions imply the existence of high electric fields. With high electric field, the velocity-field characteristics become non-linear and the velocity eventually saturates at high applied fields. The nonlinear response of the carrier velocity to a high electric field has been extensively explored, both theoretically and experimentally. As we have pointed out earlier, the high-field transport of charge carriers can be described within the framework of Boltzmann transport equation (BTE). However, for large perturbations due to high fields, the collision term, in the BTE, cannot be linearized using a simple microscopic relaxation time. Also, the carrier effective mass concept does not hold valid as long as the inter-valley and interband transitions are concerned, which are common at high fields. Therefore, several other methods have been explored to solve the BTE, by the first order Chapmanp-Enskog (C-E) expansion or the spherical harmonic expansion (SHE) or the Monte Carlo (MC) method or the hydrodynamic (moment) method.

Upon completion of this Chapter, the readers and students will be able to:

- Be acquainted with the treatment of non-linear transport with hydrodynamic moments of the BTE,
- Understand the hydrodynamic model (HDM) of semiconductors and its ability to address the non-local transport effects
- Understand the energy-dependent transport parameters and their underlying physics and where and how they are used in TCAD simulation programs.

DOI: 10.4018/978-1-5225-2312-3.ch003

2. HYDRODYNAMIC MODEL (HDM)

The hydrodynamic moment method was proposed by Grad (1949) as an alternative approach to solve the Boltzmann transport equation (BTE) in the study of aero-dynamic flow. The hydrodynamic model (HDM), which is based on the moment method, was suggested by Bløtekjaer (1970) to study the hot electron transport in semiconductor devices. Unfortunately, Blotekjaer's formulation, upon which most hydrodynamic models are based, made use of the heated Maxwellian distribution to calculate the collision terms. This assumption is not valid at high electric fields (in hot carrier nonlinear regime). Since then, several contributions have been introduced to improve the hot carrier transport models in semiconductors. For instance, Shur and Eastman (1979) could describe the velocity overshoot of hot carriers by simple conservation equation.

The so-called energy-transport model (ETM), which is a reduced version of the HDM with no convection terms, was introduced by Cook and Frey (1982). The ETM is actually based on the early work of Stratton (1962,1972), about transport of hot and cold carriers. The Stratton model and variant ETM's usually make use of a priori distribution functions to calculate the collision terms and transport parameters.

A simple hydrodynamic model for studying the velocity overshoot in semiconductor structures was carried out by Baccarani and Wordmann (1985). The non-parabolicity of energy bands has been included for the first time by Thoma et al (1991). All these models were concentrating on the electron transport in field-effect transistors (FET). A generalized HDM for bipolar devices was introduced by El-Saba, Morel and Chante (1991) to investigate the impact ionization phenomenon in semiconductor devices. Also, Tang, Ramaswmy and Nam (1993) introduced an improved HDM. Another HDM, with six moments, was suggested by Grasser et al. (2001), to investigate the sub-100nm MOSFET devices. Unfortunately, almost all the added improvements to the HDM were associated with additional physical parameters, whose carrier energy dependence is generally unknown. Several attempts to derive simple models for the HDM parameters, mostly on the basis of Monte Carlo simulation in the bulk, have been carried out. In particular, several attempts have been carried out to derive an energy-dependent impact ionization model, e.g., Quade and Schol (1994). Also, Gonzaliz et al. (1999) proposed an HDM-based model for the impact ionization rate in semiconductor devices. Other formulation of a non-parabolic HDM was developed by Anile et al. (2000) and Romano (2001), on the basis of the maximum entropy principle. A survey about the HDM and the decisive role of different parameters was compiled by several authors, e.g., Jerome (2000) and Grasser et al. (2003). Yet other hydro-dynamic models, with a few correlated transport parameters were proposed to get rid of the pitfalls of previous models (e.g., El-Saba, 2012).

The hydrodynamic (HD) description of the gas of particles consists in the study of the evolution of certain macroscopic quantities (or moments), which have significant physical meaning and can be eventually measured. The hydrodynamic model (HDM) in semiconductors is constructed by multiplying the semiclassical BTE with weight functions (ψ_i) and integrating the whole equation over the k- space, so that a set of differential equations in physical space and time is obtained, as follows:

$$\frac{\partial}{\partial t}\int_{k-spce} f_v \psi_j d^3k + \int_{k-spce} \left(\boldsymbol{u}_v \cdot \nabla f_v\right)\psi_j d^3k + \int_{k-spce} \left(\frac{\boldsymbol{F}}{\hbar} \cdot \nabla_k f_v\right)\psi_j d^3k = \int_{k-spce} \left[\frac{\partial}{\partial t} f_v\right]_{col} \psi_j d^3k \tag{1a}$$

where $f_v = f_v(\boldsymbol{x},\boldsymbol{k}, t)$ is the carrier distribution function, $\boldsymbol{u}_v$ is the carrier group velocity and $\boldsymbol{F}$ is the applied force. The resulting set of macroscopic equations will look like the following equation (El-Saba et al, 1991):

$$\frac{\partial}{\partial t}n\langle\psi_j\rangle+\nabla\cdot n\langle\boldsymbol{u}\otimes\psi_j\rangle-n\left\langle\frac{\boldsymbol{F}}{\hbar}\cdot\nabla_k\otimes\psi_j\right\rangle=\left[\frac{\partial}{\partial t}n\langle\psi_j\rangle\right]_{col} \tag{1b}$$

where $<\psi_j>$ denotes the macroscopic average of the weight functions (ϕ_j) over the k-space, $n = \int z f_v.dk^3$ is the average carrier density and $z=2/(2\pi)^3$ is the density of states in the k-space (multiplied by 2 to account for Pauli exclusion principle).

$$n<\psi_j> = \int z.\psi_j f_v.dk^3). \tag{1c}$$

The weight functions (ψ_j) are usually chosen as powers of increasing order of the carrier vector $\boldsymbol{k}_v$, or carrier group velocity $\boldsymbol{u}_v$, or even the momentum $\boldsymbol{p}_v$. The powers of these microscopic parameters are usually formatted using some appropriate scaling factors to get physically meaningful quantities. For instance, if we adopt the powers of the carrier group velocity as weight functions (e.g., $\psi = 1$, $\boldsymbol{u}_v$ and u_v^2), then we may use the carrier energy $E_v=½m^*u_v^2$, instead of u_v^2 as the second-order multiplier.

The zero order moment ($\psi_0=1$) gives the carrier density balance equation (which is the same as carrier continuity equation), the first moment ($\psi_1=u_v$) gives the carrier momentum balance equation (almost similar to drift-diffusion current equation) and second order moment ($\psi_2=½m^*u_v^2$) gives the carrier energy balance equation. This method is an averaging process that leads to the loss of some microscopic information within the distribution function. However, in many practical cases the resultant average (or macroscopic) parameters retained by the equations in the physical space are sufficient to capture the essential features of the carrier transport phenomena.

Note 1: Proof of The General Moment Equation Form

Consider the distribution function $f(x,k,t)$, which obeys the BTE. Consider $A(k)$ a microscopic quantity, which is a function of the wave vector $\boldsymbol{k}$. $A(k)$ may be one of system physical quantities like electron group velocity $\boldsymbol{u}_g(k)$ or electron energy $E(k)$. Now, let's multiply the BTE by $A(k)$ and integrate the two sides of the equation over the entire k-space. Then one obtains the following result:

$$\int A\frac{\partial f}{\partial t}d^3k+\int Av_g\cdot\nabla_x f\cdot d^3k+\int\frac{A}{\hbar}\cdot\nabla_k f d^3k=\int A\left(\frac{\partial f}{\partial t}\right)_c d^3k$$

We have also, the mean value of $A(k)$, which we note by $\langle A\rangle$ is given by:

$$\langle A\rangle=\frac{\int A(k)f(k,r,t)d^3k}{\int f(k,r,t)d^3k}=\frac{\int A(k)f(k,r,t)d^3k}{n}$$

So, the 1st term of the L.H.S of equation (J-1a) becomes:

$$\int A\frac{\partial f}{\partial t}d^3k = \frac{\partial}{\partial t}\int Afd^3k = \frac{\partial(n\langle A\rangle)}{\partial t}$$

Also, the 2nd term of the L.H.S of equation (1) becomes:

$$\int Av_g \cdot \nabla_x fd^3k = \nabla_x \cdot \int Av_g fd^3k - \int \nabla_x A \cdot v_g fd^3k = \nabla_x \cdot \left(n\langle Av_g\rangle\right)$$

Also, the 3rd term of the L.H.S of equation (J-1) becomes:

$$\int \frac{AF}{\hbar} \cdot \nabla_k fd^3k = \sum_j \left|\frac{f\left(k,x,t\right)AF_j}{\hbar}\right|_{-\infty}^{\infty} - \int f\nabla_k \cdot \left(\frac{AF}{\hbar}\right)d^3k$$

where we effectuated an integration by parts on the k-variable. The index j of the force component F_j appears for any orthonormal basis (e.g. for Cartesian coordinates, $j= x,y, z$). The totally integrated term in equation disappears because $Lim_{k\to\pm\infty} f\left(k,x,t\right)A\left(k\right)F_j = 0$. This means that all the moments of the BTE are finite. On the other hand, the second term may be written as follows:

$$\left\langle \nabla_k \cdot \left(\frac{AF}{\hbar}\right)\right\rangle = \left\langle\left(\frac{F}{\hbar}\cdot\nabla_k\right)A\right\rangle + \left\langle \Delta\nabla_k \cdot \left(\frac{F}{\hbar}\right)\right\rangle = \left\langle\left(\frac{F}{\hbar}\cdot\nabla_k\right)A\right\rangle$$

where we put $\nabla_k \cdot F = 0$, which is true if $\boldsymbol{F}$ is independent of k or when $\boldsymbol{F}$ *is* normal to $\boldsymbol{k}$. Fortunately, this is the case for electric field force ($F = -e\zeta$) and Lorentz's force ($F = -ev_g xB$).

Now regrouping all the L.H.S components we get:

$$\frac{\partial}{\partial t}n\langle A\rangle + n\left\langle\left(\frac{F}{\hbar}\cdot\nabla_k\right)A\right\rangle + \nabla_x \cdot \left(n\left\langle v_g A\right\rangle\right) = \left[\frac{\partial}{\partial t}n\langle A\rangle\right]_c$$

By replacing $A(k)$ by 1, $v_g\left(k\right)$ and $E_n\left(k\right)$ we obtain the 3 moment equations of the BTE. Table 1 summarizes the first moments of the conservation law of a physical quantity $A(k)$, according to El-Saba (1993).

2.1 Generalized Hydrodynamic Model

We now proceed to develop a generalized set of conservation equations of carrier concentration, momentum[1] and energy. This set can be obtained by multiplying the BTE by some weighting functions (1, $\boldsymbol{u}_v$ and E_v), and then integrating the two sides of the resulting equations over the whole k-space. For the sake of

Table 1. Moments of the BTE

$A(k)$	$n\langle A\rangle$	$\frac{1}{\hbar}\langle(F\bullet\nabla_k)A\rangle$	$\nabla\bullet(n\langle\upsilon_g A\rangle)$
1	n	0	$\nabla\bullet(n v_n)$
$u_g(k)$	$n\langle v_g\rangle = nv_n$	$\frac{1}{\hbar^2}\langle\nabla_k^2 E\rangle F = \langle m_n^{*-1}\rangle F = m_n^{-1}F$	$\nabla\bullet(n\langle\upsilon_g\otimes\upsilon_g\rangle)$
$E_n(k)$	$n\langle E_n\rangle = n\omega_n$	$n\,v_n.F$	$\nabla\bullet(n\langle E_n\upsilon_g\rangle) = \nabla\bullet S_n$
$E_n(k)u_g$	S_n	$\langle n\langle m_n^{*-1}F.E + F\upsilon_g\otimes\upsilon_g\rangle$	$\nabla\bullet(n\langle E\upsilon_g\otimes\upsilon_g\rangle)$

simplicity and brevity we write here only the electron equations in a single-valley semiconductor. Similar equations for holes can also be deduced. The electron density-, momentum- and energy-conservation equations, in a semiconductor, can be put in the following form (El-Saba, 2012):

$$\partial n/\partial t + \nabla.(nv_n) = [\partial n/\partial t]_{col.} \tag{2a}$$

$$\partial(nv_n)\partial t + \nabla.(m_n^{-1}P_n)_+ \nabla.(nv_n \otimes v_n) = n\, m_n^{-1}.\boldsymbol{F} + [\partial(nv_n)\partial t]_{col} \tag{2b}$$

$$\partial(n\,\omega_n)/\partial t + \nabla.\boldsymbol{S}_n = nv_n\,.\,\boldsymbol{F} + [\partial\,(n\omega_n)/\,\partial t]_{col.} \tag{2c}$$

where n, v_n, P_n and ω_n are the electron concentration, mean velocity, pressure tensor and mean energy, respectively.

At the bottom of conduction band the average mass will be denoted by m_{no} and its value in Si, according to optical measurements, is about 0.275 m_o. The subscript '*col*' in the above equations means the rate of change of the quantity between brackets due to electron gas collisions with the semiconductor lattice vibrations and other defects. Also, $\boldsymbol{S}_n$ is the electron energy flow and m_n^{-1} is the mean inverse mass of electrons. Note the presence of the same mass term in both the LHS and the RHS of (2b), with the same definition. We dully note that we do not adopt the effective mass approximation.

The mean inverse mass of carriers is generally a tensor whose components are functions of the mean carrier energy:

$$m_{n\,(i,j)}^{-1} = (1/\hbar^2)\,\langle\partial^2 E_n/\partial k_i \partial k_j\rangle \tag{3}$$

When the energy band structure of the semiconductor material is assumed parabolic, m_n^{-1} is a constant and equal to the conventional electron inverse effective masse $(m_n^*)^{-1}$ at the bottom of conduction band. This assumption, though adopted in many hydrodynamic models in the literature, is only valid near equilibrium. In our analysis we consider m_n as a scalar quantity equal to the average conductivity mass (averaged over all the semiconductor valleys) such that $m_n^{-1} = ⅓\Sigma\, m_{n\,(i,j)}^{-1}$.

The sign ⊗ stands for the tensorial product of two terms such that:

$$\nabla.(nv_n \otimes v_n) = v_n \nabla.(nv_n) + (nv_n . \nabla)v_n. \tag{4a}$$

The electro-kinetic pressure tensor of the electron gas, P_n, is defined here as follows:

$$m_n^{-1} P_n = n <(u_n - v_n) \otimes (u_n - v_n)> \tag{4b}$$

The carrier energy flow (or energy flux) is given by the following expression, for electrons:

$$S_n = n<E_n u_n> = Q_n + (n\omega_n + P_n).v_n \tag{5}$$

where Q_n is called the electron heat flux. This expression can be easily derived by expanding the definition of $S_n = n<E_n u_n>$ and using the definition of the electron gas pressure (4b). The vector Q_n is defined as the third central moment of the electron distribution function f_n, as follows:

$$Q_n = n<\frac{1}{2} m_n^*(u_n - v_n).|u_n - v_n|^2)> = \int \frac{1}{2} m_n^* |u_n - v_n|^2 (u_n - v_n) f_n d^3k \tag{6}$$

Therefore, the heat flux component $Q_{n(i)}$ is the flux of random thermal energy across a surface with its normal oriented in the ith direction. As the electron distribution function is not known we don't make use of this equation in our model.

In absence of magnetic field, the external applied force on electrons is given by $F = -e.\zeta$, where the electric field ζ is equal to the negative gradient of the electrostatic potential ϕ. In heterogeneous devices, the applied force on electrons is given by $F = -\nabla E_c$, where $E_c = (E_{co} - e\phi)$ and E_{co} is the edge of the conduction band. Similarly, the applied force on holes is $F = \nabla E_v$, where $E_v = (E_{vo} + e\phi)$ and E_{vo} being the edge of the valence band.

The electrostatic potential distribution can be found by solving the Poisson equation:

$$\nabla.(-\varepsilon \nabla \phi + P) = e(p - n + Dop + N_t) \tag{7}$$

where p is the hole concentration, ε is the dielectric constant, P is the material polarization vector, N_t is the density of traps and $Dop = |N_d^+ - N_a^-|$ is the net ionized concentration in the semiconductor.

2.2 Closing the System of Hydrodynamic Equations

In order to close the system of conservation equations for charge carrier density, average momentum and average energy, we usually adopt some assumptions to approximate the terms which depend on the knowledge of the distribution function. The main assumptions are usually concerned with the following items:

- Definition of the carrier gas pressure and temperature,
- Approximation of collision terms,
- Modeling the heat flux term (3rd cental moment of f_v), and
- Modeling any higher moment terms, if moments aew greater than 3.

Definition of the Carrier Gas Pressure

The formal definition of the gas pressure tensor $\underline{P}_{n(ij)}$ is the flux of the ith component of the momentum across a surface element moving with the mean velocity oriented with its normal in the jth direction (Landau & Lifshitz, 1997):

$$\underline{P}_{n(i,j)} \equiv < \rho_{n(i,j)} (\boldsymbol{u}_{ni} - \boldsymbol{v}_{ni})(\boldsymbol{u}_{nj} - \boldsymbol{v}_{nj})> \tag{8}$$

where ρ_n is the gas mass density. In ideal gas theory, the gas mass density is a macroscopic quantity and ρ_n is taken outside the averaging brackets.

$$\underline{P}_n = \rho_n < (\boldsymbol{u}_n - \boldsymbol{v}_n) \otimes (\boldsymbol{u}_n - \boldsymbol{v}_n)> \tag{9a}$$

In classical transport models, the electron gas mass density is usually related to the electron effective mass by the relation $\rho_n = n.m_n^*$. This is only correct when the band structure is parabolic and m_n^* is energy independent. In order to rectify this, we define the gas mass density as follows:

$$\rho_n = n\, m_n, \tag{9b}$$

where m_n is the mean mass of electrons, as defined by (3). In this sense, our definition of the gas pressure, P_n, in (4b) is equivalent to the above formal definition. However, this encompasses some approximation, because our definition of the gas pressure, P_n, becomes different from the classic definition of electron gas pressure tensor, $\underline{P}_n$, with $\rho_n = n.m_n^*$

$$\underline{P}_n = n < m_n^*(\boldsymbol{u}_n - \boldsymbol{v}_n) \otimes (\boldsymbol{u}_n - \boldsymbol{v}_n)> \tag{10}$$

However, the electrokinetic pressure tensor P_n may be related to the classic definition of $\underline{P}_n$ via a correction factor, g_n, such that:

$$g_n = P_n \cdot \underline{P}_n^{-1} = [m_n <(\boldsymbol{u}_n - \boldsymbol{v}_n) \otimes (\boldsymbol{u}_n - \boldsymbol{v}_n)>] . [<m_n^* (\boldsymbol{u}_n - \boldsymbol{v}_n) \otimes (\boldsymbol{u}_n - \boldsymbol{v}_n)>]^{-1} \tag{11}$$

We call this the HDM **g**-factor. This factor can be calculated from MC simulation in a semiconductor with arbitrary band structure. Obviously, our formal definition of the carrier gas pressure is equivalent to the classic definition in the case of parabolic band structure, and then $\boldsymbol{g}_n = 1$

Definition of the Carrier Temperature

In this section we compare the different approaches for modeling the carrier temperature. Formally, temperature is a property which governs the transfer of thermal energy, or heat, between one system and another. The electronic temperature concept was initially introduced in 1947 by Fröhlich, and since then it has been exploited by several authors to model the high-field transport in semiconductors. In the isothermal drift-diffusion model (DDM), the carriers temperature is assumed equal to the lattice temperature T_L. However, the so-called static electron temperature T_n may be roughly estimated from the electric field distribution in several ways. For instance, the static temperature may be defined as follows:

$$T_n\,(static) = T_L\,(\mu_{no}/\mu_n(\zeta)) = T_L\,[\,1+(\mu_{no}\,\zeta^{//}/\,v_n^{sat})^2\,]^{1/2} \quad (12)$$

where μ_n is the electron drift mobility, v_n^{sat} is the electron saturation velocity and μ_{no} is the low-field drift mobility.

Also, $\zeta^{//}$ is the parallel component of the electric field to the electron current density such that $\zeta^{//} = \boldsymbol{\zeta}.\boldsymbol{J}_n/|\boldsymbol{J}_n|$. The above equation is based on the assumption that the parallel component of the diffusion coefficient is almost constant and independt of the electric field. Such static forms of the carrier temperature exclude the lag effect between the driving electric field and the carrier temperature response, and hence the velocity overshoot phenomenon will not be observed. However, the definition of the carrier temperature in the Monte Carlo simulation is given by:

$$3/2\,k_B\,T_n = \langle E_n(k)\rangle \quad (13)$$

Also, the carrier temperature in the conventional hydrodynamic models, where the effective mass approximation is assumed and the anisotropy of the distribution function is ignored, is defined by:

$$3/2\,k_B\,T_n = \tfrac{1}{2}\,m_n^{*}\langle|\boldsymbol{u}_n-\boldsymbol{v}_n|^2\rangle \quad (14a)$$

It is well known, that the above two definitions result in different temperatures. In fact, the above hydrodynamic definition (14a) is implicitly based on the displaced Maxwellian distribution at the electronic temperature.

$$f_m\,(k) = A.[n/(k_BT_n)^{3/2}].exp\,(-\tfrac{1}{2}\,m_n^{*}(\boldsymbol{u}_n(k)-\boldsymbol{v}_n)^2/k_BT_n) \quad (14b)$$

where A is a constant. According to Stratton (1970), the electronic temperature has no physical meaning, unless the distribution function is a displaced Maxwellian as described by equation (14). However, the Maxwellian distribution is only valid near equilibrium, and in highly doped regions, where carrier-carrier collisions are dominant. In fact, it has been confirmed by several authors that the energy distribution function is not Maxwellian at high electron energies (Higman & Hess, 1986).

In the so-called improved energy transport model (Chen et al., 1992), the nonparabolicity of the band structure is taken into account and the distribution function $f_n(E)$ is appended by a non-Maxwellian factor γ, such that:

$$f_n(E_n) = (1+\gamma.E_n/k_BT_n).\,f_m\,(E_n) \quad (15)$$

with $f_m(E)$ is the heated Maxellian distribution function at an elevated temperature T_n. Then, a modified (or equivalent) carrier temperature T_m is defined as follows:

$$T_m = T_n\,(1+\gamma) \quad (16)$$

However, referring to our system of HDE's, we make use of the formal definition for the electron temperature tensor, such that:

$$P_n = n\,k_B\,T_n = n\,m_n\,\langle\,(\boldsymbol{u}_n-\boldsymbol{v}_n)\otimes(\boldsymbol{u}_n-\boldsymbol{v}_n)\rangle \quad (17)$$

where we assumed a perfect gas law ($P_n = n\, k_B\, T_n$). When the carrier temperature tensor is reduced to a scalar quantity (such that $T_n = T_n \boldsymbol{I}$ where $\boldsymbol{I}$ is the unity tensor), then the above definition may be written as follows:

$$3/2\, k_B\, T_n = ½\, m_n < |\boldsymbol{u}_n\text{-}\boldsymbol{v}_n|^2> \approx ½\, g_n < m_n{}^* \, |\boldsymbol{u}_n\text{-}\boldsymbol{v}_n|^2 > \qquad (18)$$

In this case the g-factor may be calculated from the following relation:

$$g_n = P_n \,.\, \underline{P}_n^{-1} = [\, m_n <|u_n - v_n|^2>]/[<m_n{}^*|u_n - v_n|^2>] \qquad (19)$$

Figure 1 depicts the variation of T_n according to MC (with a non-parabolicity factor 0.5 eV^{-1}) and the classic hydrodynamic model (14a) and to the definition (18). Note that assuming $g_n \approx 1$ brings an error less than 3% in T_n, for electric fields up to 300 kV/cm. This error is less drastic than the error encountered when neglecting the convective part in the calculation of the electron energy.

In fact, most of the commercial simulators neglect totally the convection part of the average carrier energy. In this case the utilized formulation is called energy transport model (ETM). In conclusion, the scalar temperature definition in our HDM reads:

$$3/2\, k_B\, T_n = ½\, m_n <|\boldsymbol{u}_n\text{-}\boldsymbol{v}_n|^2> \qquad (20)$$

Figure 1. Electron temperature T_n as a function of electric field in homogeneous silicon, according to full-band Monte Carlo (FBMC), classic HDM with constant effective mass (HDM) and according to EL-SABA HDM model
El-Saba, 1993, 2012.

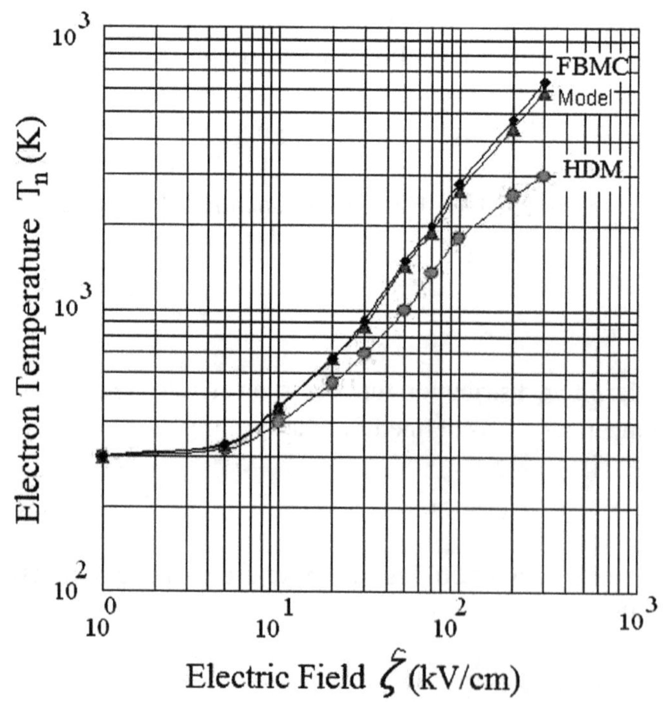

This definition means that we assume the kinetic pressure tensor $\underline{P}_n = nk_B\underline{T}_n$, and may be considered as a closure condition for the set of HDEs. If later full-band MC (FBMC) showed a significant deviation of the g-factor from unity at very high fields, it could be easily retained in the proposed HDM to get more accurate results.

Average Carrier Energy

By definition, we have the electron average energy $\omega_n = <E_n>$. Therefore, we can write (within the semi-classical approach):

$$\omega_n = <E_n> =<½\, m_n{}^*\, u_n{}^2> = <½\, m_n{}^*\mathbf{v}_n{}^2> + <½\, m_n{}^*(\mathbf{u}_n\text{-}\mathbf{v}_n)^2 > \tag{21}$$

Substituting the carrier gas pressure, we get:

$$< ½\, m_n{}^*(\mathbf{u}_n\text{-}\mathbf{v}_n)^2 > = n^{-1}\underline{P}_n = ½\, g_n\, n^{-1}P_n \tag{22}$$

Using the relations of the perfect gas law with pressure, we can easily prove that the mean electron energy ω_n is related to the scalar electron temperature T_n and scalar average mass m_n as follows:

$$\omega_n = <E_n> = ½\, m_n\, v_n{}^2 + 3/2\, g_n\, k_B\, T_n \tag{23}$$

where we substituted the average mass tensor $m_n = <m_n{}^{*-1}>$. Note that the factor of 3 comes from reducing the tensor temperature to a scalar quantity by the *trace* operator, while the tensorial average mass is reduced to an average scalar mass (e.g. the conductivity mass), which is already a sort of sum of the carrier mass components[2].

Note 2: Temperature Tensor

When the anisotropy of the semiconductor is not negligible, we may admit a temperature tensor, which is related to the pressure tensor by the perfect gas low

$P_n = nk_B\, T_n$.

By appropriate selection of coordinates, T_n may be expressed as a diagonal matrix.

$$T_n = \begin{bmatrix} T_{xx} & 0 & 0 \\ 0 & T_{yy} & 0 \\ 0 & 0 & T_{zz} \end{bmatrix}$$

Nevertheless, when the anisotropy of the material is negligible, the tensorial temperature T_n can be *reduced* to a scalar temperature T_n. This is done by taking the trace of the tensor temperature tensor, as follows:

T_n = trace (T_n) = 1/3 (T_{xx} + T_{yy} + T_{zz})

Macroscopic Relaxation-Time Approximation

As for the collision terms, we typically make use of the macroscopic or ensemble relaxation time approximation (E-RTA). Mathematically speaking, this approximation has not been verified yet (Brugger, Schenj & Fichtner, 2006). However, some studies showed that the E-RTA, with average energy dependence, is reasonably accurate for non-equilibrium transport in semiconductors (Hess, 2012, p.247). In this approximation, the collision term in the momentum conservation equation is expressed in terms of a macroscopic momentum relaxation time τ_{mn}, as follows:

$$[\partial n v_n / \partial t]_{col.} = - n v_n / \tau_{mn} + v_n [\partial n / \partial t]_{col.} \quad (24)$$

Also, the collision term in the energy conservation equation can be expressed in terms of a macroscopic energy relaxation time τ_{wn}, as follows:

$$[\partial n \omega_n / \partial t]_{col.} = -n (\omega_n - \omega_o) / \tau_{wn} + \omega_n [\partial n / \partial t]_{col} \quad (25)$$

where $\omega_o = 3/2\, k_B T_L$ is the electron energy at thermal equilibrium and T_L is the lattice temperature. Note that the collision term in the above two equations is decomposed into two terms (and two relaxation times, however[3]). The momentum and energy relaxation times (τ_{mn} and τ_ω), include the effect of collisions over phonons and other defects while the effect of impact ionization collisions is included in the $[\partial n/\partial t]_{col}$. This approach is exact if the relaxation times are considered as a function of all the BTE moments (n, v_n, ω_n, etc.). However, one can obtain these relaxation times, as functions of the mean carrier energy by Monte Carlo simulation in the bulk of semiconductor, as a second order approximation (El-Saba, 1993). A discussion about the validity of the macroscopic relaxation time can be also found in (Stettler, Alam & Lundstrom, 1993) as well as (Lundstrom, 2000). The collision term $[\partial n/\partial t]_{col.}$, which appears explicitly in the carrier density conservation equation, can be expressed as follows:

$$[\partial n / \partial t]_{col.} = - (n - n_o) / \tau_n = - U \quad (26)$$

where ($U=R-G$) is the net recombination-generation rate in the semiconductor. We dully note that the recombination rate, R, regroups the relevant recombination mechanisms in the semiconductor device (e.g., direct radiative, Shockley-Reed-Hall and Auger recombination mechanisms). Also, the generation rate, G, regroups the applicable generation mechanisms in the semiconductor device (e.g., thermal and optical generation as well as the generation by impact ionization mechanism).

Heat Flux Approximation

As shown in our previous discussion, the carrier energy flow (energy flux) can be put in the following form, for electrons and holes:

$$S_n = Q_n - (\omega_n + k_B T_n).(J_n / e) \quad (27a)$$

$$S_p = Q_p + (\omega_p + k_B T_p).(J_p / e) \tag{27b}$$

As indicated by (45b), the carrier heat flux terms $Q_{n,p}$ are defined as the third central moment of the carrier distribution function. As the carrier distribution function is unknown, the carrier-heat flux is usually modeled by the following Fourier relation:

$$Q_n = - k_n \nabla T_n \tag{28a}$$

where k_n is the electron thermal conductivity. The carrier thermal conductivity is usually modeled by the Wiedemann-Franz law at the electronic temperature:

$$k_n = \gamma_n (k_B / e)^2 T_n \sigma_n \tag{28b}$$

where σ_n is the electrical conductivity and γ_n is the Lorenz coefficient (for electrons). In the original Wiedemann-Franz formulation (using the microscopic relaxation time approximation) γ_n is given by:

$$\gamma_n = 5/2 + r \tag{28c}$$

where r is an exponent in the microscopic relaxation time τ_{Dn} expression in terms of the electron energy (E_n) and lattice temperature (T_L):

$$\tau_{Dn} = (E_n / k_B T_L)^r \tag{28d}$$

A more rigorous approach for modeling the carrier heat flux term (Q_n) consists in taking the third moment of the BTE into account. This procedure increases the number of differential equations to be solved and the number of parameters of the HDM. However, we show in section 5.2, how to derive a more accurate model of the carrier heat flux term and the Lorentz number, on the basis of the third moment of the BTE, without increasing the number of differential equations to be solved in the HDM.

3. SET OF HYDRODYNAMIC EQUATIONS (HDEs)

Using the relaxation-time approximation and our definition of the carrier temperature, the set of hydrodynamic equations (HDE's) can be formulated as follows for a single valley semiconductor:

$$e\, \partial n/\partial t - \nabla . J_n = e(G-R) \tag{29a}$$

$$J_n - n \mu_n m_n d[v_n]/dt = e\, n\, \mu_n . \zeta + k_B \mu_n m_n \nabla .(n\, m_n^{-1} . T_n) \tag{29b}$$

$$\partial(n\, \omega_n)/\partial t + \nabla . S_n = \zeta . J_n + \omega_n (G - R) - n(\omega_n - \omega_o) / \tau_{wn} \tag{29c}$$

The electron drift mobility μ_n is defined as follows:

$$\mu_n = e\tau_{mn}\, m_n^{-1} \tag{30}$$

Similar equations can be written for holes. We dully note that the average electron energy ω_n is measured *above* the conduction band edge *Ec*, while the average hole energy ω_p is measured *under* the valence band edge *Ev*.

In steady state (up to the terahertz frequency range in Si and several semiconductors), the carrier momentum conservation equations can be further simplified by suppressing the partial time-derivatives so that:

$$\boldsymbol{J}_n - n\,\mu_n.m_n\,(\boldsymbol{v}_n.\nabla)(\boldsymbol{v}_n) = e\,n\,\mu_n.\boldsymbol{\xi} + k_B\,\mu_n.m_n\nabla.(n\,m_n^{-1}.\,T_n) \tag{31a}$$

$$\boldsymbol{J}_p + p\,\mu_p.m_p\,(\boldsymbol{v}_p.\nabla)(\boldsymbol{v}_p) = e\,p\,\mu_p.\boldsymbol{\xi} - k_B\,\mu_p.m_p\nabla.(p\,m_p^{-1}.\,T_p) \tag{31b}$$

The above conservation equations are similar to the current equations of the isothermal DDM, except for the acceleration and thermal diffusion terms. Also the carrier energy flow (or energy flux) can be put in the following forms, for electrons and holes:

$$\boldsymbol{S}_n = \boldsymbol{Q}_n - (\omega_n + k_B\,T_n).(\boldsymbol{J}_n / e) \tag{32a}$$

$$\boldsymbol{S}_p = \boldsymbol{Q}_p + (\omega_p + k_B\,T_p).(\boldsymbol{J}_p / e) \tag{32b}$$

4. MOMENT EQUATIONS IN A MULTIVALLEY SEMICONDUCTORS

As we know, some semiconductors (like *GaAs*) have more than one valley that may significantly intervene in the transport process, at relatively high electron energies. The moment equations in multivalley semiconductors can be obtained in much the same manner we utilized so far for one-valley semiconductors. However, in addition to the intrinsic relaxation times in each valley (which substitute the intra-valley collision terms) we make use of additional relaxation times to model the inter-valley collision terms. Therefore, the moment equations in multivalley semiconductors may be described by the following relations (after substituting the collision terms with the appropriate relaxation times):

$$\partial n_i/\partial t + \nabla.(n_i\boldsymbol{v}_i) = (G_i - R_i) + \sum (n_j/\tau_{ji,} - n_i/\tau_{ij,}) \tag{33a}$$

$$\partial(n_i\boldsymbol{v}_i)/\partial t + \nabla.(m_i^{-1}\boldsymbol{P}_i) + \nabla.(n_i\boldsymbol{v}_i \otimes \boldsymbol{v}_i) = n_i\, m_i^{-1}\boldsymbol{F} - n_i\boldsymbol{v}_i/\tau_{mii} - \sum n_i\boldsymbol{v}_i/\tau_{mij} \tag{33b}$$

$$\partial(n_i\,\omega_i)/\partial t + \nabla.\boldsymbol{S}_i = n_i\boldsymbol{v}_i.\boldsymbol{F} - n_i(\omega_i - \omega_o)/\tau_{wij} + \sum(n_j\omega_{ij}/\tau_{wiji} - n_i\omega_i/\tau_{wiij}) \tag{33c}$$

where n_i, $\boldsymbol{v}_i$, and ω_i denotes respectively the average electron density, velocity and energy, in the ith valley. The $1/\tau_{ij}$ and $1/\tau_{ji}$ are the charge carrier (electron) transition rates from valley *i* to the valley *j* and vice versa. The τ_{mji} and τ_{mij} are the intervalley momentum relaxation times, due to carrier scattering-in and scattering-out the ith valley from the jth valley and vice versa. Also, $\tau_{\omega ij}$ and $\tau_{\omega ji}$ are the intervalley energy relaxation times, due to carrier scattering-in and scattering-out the ith valley from the jth valley and vice versa.

5. HIGHER MOMENT EQUATIONS

The set of hydrodynamic equations can be extended to any number of moments of the BTE. Although, the first three moments are widely used, because of their clear physical meaning (conservation of number, momentum and energy of charge carriers), additional moments are theoretically helpful[4]. The following set depicts the first six moments of the BTE for the case of nearly parabolic bands. The six moment equations (ϕ_0 through ϕ_5) represent the conservation equations of carrier density, momentum, energy, energy flux, kurtosis, and kurtosis flux, respectively. Note that the energy distribution function in equilibrium is symmetric. As the even weight functions (ϕ_0, ϕ_2, ϕ_4) are symmetric (in k and v) and the odd weight functions (ϕ_1, ϕ_3, ϕ_5) are anti-symmetric (in $\boldsymbol{k}$ and $\boldsymbol{v}$), therefore, the odd moments will vanish and only the even moments of the equilibrium distribution function will exist.

$$\phi_0 = 1, \frac{\partial n}{\partial t} + \nabla \cdot n\langle \boldsymbol{u} \rangle = -\frac{n - n_0}{\tau_n} = G - R \tag{34a}$$

$$\phi_2 = \boldsymbol{E}, \frac{n\langle \boldsymbol{E} \rangle}{\partial t} + \nabla \cdot n\langle \boldsymbol{uE} \rangle - q\varepsilon \cdot n\langle \boldsymbol{u} \rangle = -\frac{n\langle \boldsymbol{E} \rangle - n\langle \boldsymbol{E} \rangle_0}{\tau_e} \tag{34b}$$

$$\phi_4 = \boldsymbol{E}^2, \frac{\partial \langle \boldsymbol{E}^2 \rangle}{\partial t} + \nabla \cdot n\langle \boldsymbol{uE}^2 \rangle - 2q\varepsilon \cdot n\langle \boldsymbol{E} \rangle = -\frac{n\langle \boldsymbol{E}^2 \rangle - n\langle \boldsymbol{E}^2 \rangle_0}{\tau_{e2}} \tag{34c}$$

$$\phi_1 = \hbar \boldsymbol{k}, \frac{2}{3}\nabla n\langle \boldsymbol{E} \rangle + qn\varepsilon = -m\frac{n\langle \boldsymbol{u} \rangle}{\tau_m} \tag{34d}$$

$$\phi_3 = \boldsymbol{uE}, \frac{2}{3}\nabla n\langle \boldsymbol{E}^2 \rangle - \frac{5}{3}qn\varepsilon\langle \boldsymbol{E} \rangle = -m\frac{n\langle \boldsymbol{uE} \rangle}{\tau_S} \tag{34e}$$

$$\phi_5 = \boldsymbol{u}\boldsymbol{E}^2, \frac{2}{3}\nabla n\left\langle \boldsymbol{E}^3\right\rangle - \frac{7}{3}qn\varepsilon\left\langle \boldsymbol{E}^2\right\rangle = -m\frac{n\left\langle \boldsymbol{u}\boldsymbol{E}^2\right\rangle}{\tau_{S2}} \tag{34f}$$

where q is the carrier charge (q=-e for electrons and q=+e for holes) and m is the carrier effective mass, which is assumed constant (independent of carrier energy),

Note that we utilized the macroscopic relaxation time approximation in these moment equations. Here, τ_m, τ_e, τ_s, τ_{e2}, τ_{s2} are the relaxation times for momentum, energy, energy flux, kurtosis, and kurtosis flux, respectively. Also the time partial derivatives of the odd moments are dropped, assuming that their relaxation times are much smaller than the times of interest. These relaxation times may be obtained from Monte Carlo simulation or modeled using a priori distribution functions (e.g., displaced Maxwellian). However, the Maxwellian distribution is not acceptable at hot carrier nonlinear regime, except at special cases of high carrier concentrations where carrier-carrier scarpering dominates and homogenizes the carrier distribution.

Figures 2-4 show the variation of the relaxation times, as obtained by Monte Carlo simulation (Grasser et al, 2001). The relaxation times of the odd moments (τ_m, τ_s and τ_{s2}) are replaced with the corresponding

Figure 2. Electron energy and kurtosis relaxation times in Si, by MCM
Grasser et al, 2001.

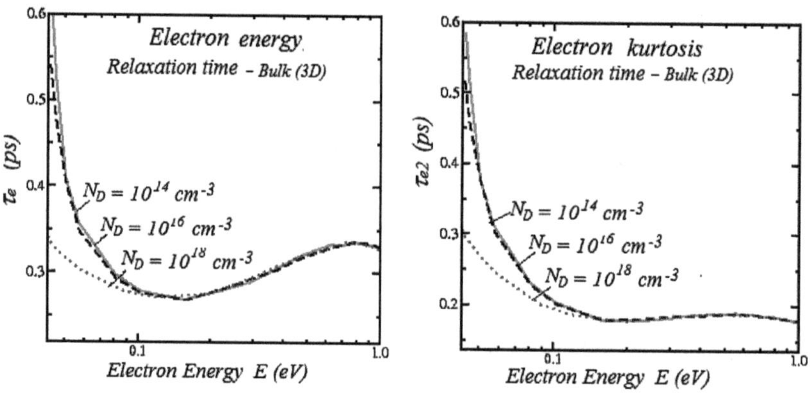

Figure 3. Electron mobility, and energy flux mobility, in Si, by MCM
Grasser et al, 2001.

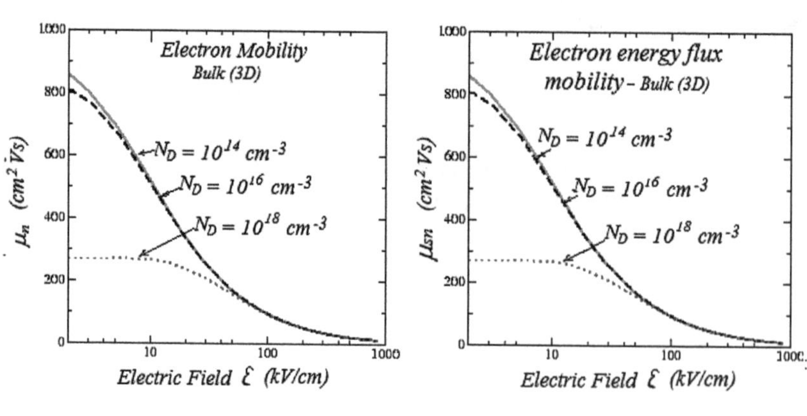

Figure 4. Electron kurtosis flux mobility in Si, by MCM
Grasser et al, 2001.

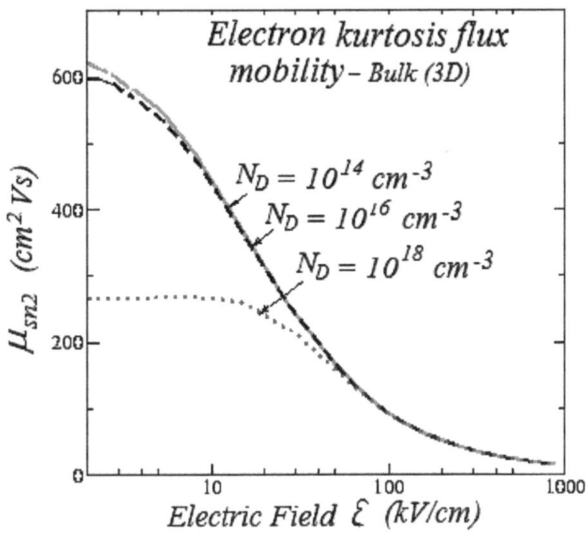

mobility ($\mu_n = e\tau_m/m$, $\mu_s = e\tau_s/m$ and $\mu_{s2} = e\tau_{s2}/m$). Of course, the above six-moments of hydrodynamic equations can be extended to include the effects of real band structure, but the resulting expression will be very complicated.

6. ENERGY-DEPENDENT PHYSICAL PARAMETERS

The transport properties of hot carriers under high-field conditions are of great interest for high-speed devices in general and nanodevices in particular. Therefore, the accurate modeling of transport parameters at high fields is vital in any transport model. In the DDM, the transport parameters are usually expressed as functions of the local electric field. Nevertheless, some parameters such as the high-field mobility and impact ionization rate involve energy exchange mechanisms that should better be expressed as functions of the carrier energy, rather than the local electric field. Fortunately, the solution of the HDM produces such information about the carrier energy and permits to express these parameters in a physical manner, along the semiconductor device.

6.1 Energy-Dependent Carrier Drift Mobility ($\mu_{n,p}$)

In the drift-diffusion model, the high-field drift mobility is usually expressed by the Caughey-Thomas relation (1967):

$$\mu_n(\zeta) = \mu_{no} / [1 + (\mu_{no}\, \zeta / v_n^{sat})] \tag{35a}$$

where μ_{no} is the low-field mobility of electrons and v_n^{sat} is their saturation velocity. This direct algebraic relation doesn't account for the nonlocal phenomena, such as the carrier velocity overshoot (El-Saba,

1998) in hot carrier nonlinear regime. However, in the hydrodynamic model where the carrier average energy can be calculated, the high-field mobility can be expressed in terms of the carrier mean energy $\omega_{n,p}$. The energy-dependent drift mobility relation may be written in the following form:

$$\mu_n^{-1} = \mu_{no}^{-1}\left[1 + \frac{\mu_{no}}{e\left[v_n^{sat}\right]^2}\left(\frac{\omega_n - \omega_o}{\tau_{\omega n}}\right)\right] \tag{35b}$$

This nonlocal model resembles the model of Hänsch and Mattausch (1986), but here the energy relaxation time is energy dependent and there is no a priori assumption about the distribution function (El-Saba, 1993). Figure 5 shows the variation of electron mobility as a function of electron energy, according to the above model and according to the model of Hänsch-Mattausch. In order to compare this model with the field-dependent phenomenological relations, one can substitute equation (20a) into the set of HDE's in the steady state homogeneous case (where the electric field ζ is constant) to obtain:

$$(\mu_{no}/\mu_n) = (\mu_n/\mu_{no})\,(\mu_{no}\,\zeta/v_n^{sat})^2 + 1 \tag{36}$$

The solution of the above quadratic equation (in μ_n) is given by the following relation:

$$\mu_n(\zeta) = \frac{2\mu_{no}}{1 + \sqrt{1 + \left[2\mu_{no}\zeta / v_n^{sat}\right]^2}} \tag{37}$$

Obviously, this result resembles the phenomenological relation of Caughey-Thomas. As shown in Figure 6, our model coincides with the conventional field-dependent mobility models of Caughey-Thomas (1967) as well as the measured drift mobility, according to Canali et al (1975). It should be noted that when the electric field is very high, the Einstein relation is not correct because we cannot assume quasi-equilibrium anymore. Based on Monte-Carlo (MC) simulations, the Einstein relation may be corrected as follows:

$$D_n = \frac{k_B T_n}{e}\mu_n\left(\frac{2}{5} + \frac{3}{5}\cdot\frac{\mu_n}{\mu_{no}}\right) \tag{38}$$

where $\mu_n = \mu_{no}(T_n)$ is carrier-temperature-dependent mobility.

Note 3: Derivation of the Energy-Dependent Drift Mobility Model

The above presented model of drift mobility is derived from the general set of hydrodynamic equations without making use of the carrier-momentum relaxation-time approximation (El-Saba, 1993). The electron momentum- and energy-conservation equations in homogeneous semiconductor read

Figure 5. Electron drift mobility in Si at 300K

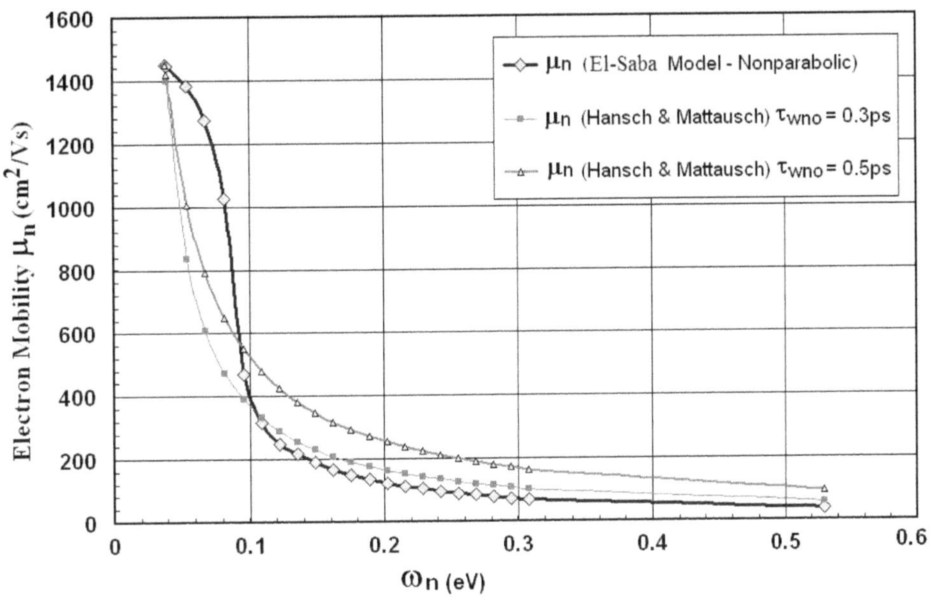

Figure 6. Electron drift mobility in bulk Si at 300K, according to several authors

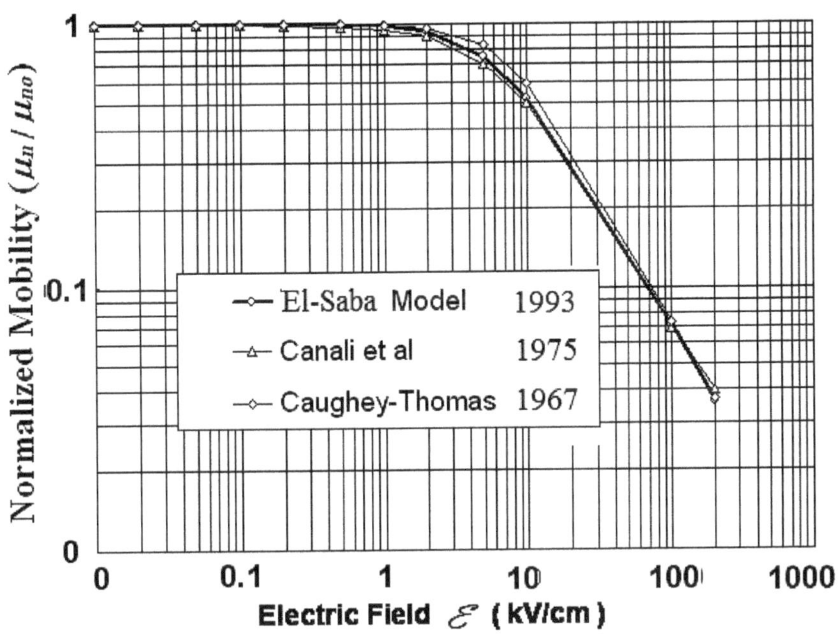

$\partial(nv_n)/\partial t = n\, m_n^{-1}\zeta + [\partial(nv_n)/\partial t]_{col}$

$\partial(n\,\omega_n)/\partial t = nv_n.\zeta + [\partial(n\,\omega_n)/\partial t]_{col}$

Multiplying the momentum conservation in steady state by $\boldsymbol{v}_n$ (dot product) and starting from the basic definition of the carrier drift mobility ($v_n = \mu_n\,\zeta$), we get:

$$\mu_n = -\frac{em_n^{-1}\boldsymbol{v}_n^2}{\boldsymbol{v}_n \cdot [\partial \boldsymbol{v}_n / \partial t]_{col}}$$

Obviously, this definition coincides with the classic definition $\mu_n = e\tau_{mn}\, m_n^{-1}$ when adopting the momentum relaxation time approximation $[\partial v_n/\partial t]_{col} = -\boldsymbol{v}_n/\tau_{mn}$ where τ_{mn} is the momentum relaxation time. Instead, we can derive an expression for the carrier mobility by searching a relation between the two collision terms in the carrier momentum- and energy-conservation equations. The momentum-collision term can be expressed in terms of the energy collision term, by eliminating the electric field ζ rom the above two conservation equations in steady state. Therefore, we get:

$m_n\, v_n.(\partial v_n/\partial t)]_{col.} = (\partial\omega_n/\partial t)]_{col.}$

The energy collision term can be expressed in terms of an energy relaxation time τ_ω such that $(\partial\omega_n/\partial t)_{col.} = -(\omega_n - \omega_o)/\tau_\omega$. This approach is exact if τ_ω, is considered as a function of all moments (n, $\boldsymbol{v}_n$, ω_n, etc.). Substituting, the last relation into the definition of the carrier mobility results in the following expression:

$$\mu_n^{-1} = \frac{1}{ev_n^2}\left(\frac{\omega_n - \omega_o}{\tau_{\omega n}}\right)$$

This simple analytic relation expresses the carrier drift mobility in terms of the carrier mean velocity and mean energy. However, the use of this analytic relation is not accurate at small fields. In fact, the relaxation time approximation, where τ_ω, is expressed as a function of only the carrier mean energy, is not accurate at small energies (Canali, 1975). Therefore, we postulate that the carrier drift mobility can be expressed as follows:

$1/\mu_n = 1/\mu_{no} + 1/\mu_{n\infty}$

where μ_{no} is the low-field mobility and $\mu_{n\infty}$ has the same expression given by but with replacing the carrier mean velocity by the saturation velocity v_n^{sat}.

$$\mu_{n\omega}^{-1} = \frac{1}{e\left[v_n^{sat}\right]^2}\left(\frac{\omega_n - \omega_o}{\tau_{\omega n}}\right)$$

Finally, the carrier drift mobility can be expressed as follows:

$$\mu_n^{-1} = \mu_{no}^{-1}\left[1 + \frac{1}{e\left[v_n^{sat}\right]^2}\left(\frac{\omega_n - \omega_o}{\tau_{\omega n}}\right)\right]$$

6.2 Energy-Dependent Impact Ionization Coefficient ($\alpha_{n,p}$)

As we pointed out earlier (in Chapter 1), the impact ionization phenomenon happens when colliding electrons have enough energy to ionize valence band electrons. The impact ionization mechanism was identified, about 60 years ago, by McKay (1954). In fact, the electron-hole pair generation by impact ionization, G_{ii}, influences so many characteristics of semiconductor devices such as the breakdown voltage and hot carrier injection currents. The simulation of such high field phenomena is still a challenge for the existing device simulators. This is not only attributed to the numerical difficulties, but more importantly, to the lack of accurate physically-based models of the impact ionization mechanism.

Indeed, the generation-recombination term in the carrier continuity equations should include the effect of generation by impact ionization.

The macroscopic impact ionization rate (per unit volume per unit time) for electrons is given by:

$$G_{ii} = \int g(k) f_n(\mathbf{k}, r, t)\, \Gamma_{ii}(\mathbf{k})\, d^3\mathbf{k} \tag{39a}$$

where $g(k) = 2/(2\pi)^3$ is the density of states in the k-space and Γ_{ii} is the microscopic rate of impact ionization, which may be expressed by the Keldish formula (1965)[5]. The macroscopic impact ionization rate G_{ii} is usually expressed in terms of the impact ionization coefficients (Townsend's coefficients) of electrons and holes, $\alpha_{n,p}$

$$G_{ii} = \alpha_n\, |J_n/e| + \alpha_p\, |J_p/e| \tag{39b}$$

Since the carrier distribution function is not generally known in the drift-diffusion and hydrodynamic models, we usually search for a suitable model for the impact ionization rates in terms of known variables such as the local electric field or the carrier average energy.

In the DDM, the impact ionization coefficients ($\alpha_{n,p}$) are usually expressed as functions of the local electric field (ζ). For instance, Wolff (1954) used the first two terms of the Legendre expansion of the distribution function, and derived the following electron impact ionization coefficients:

$$\alpha_n(\zeta) = \alpha_{no}\, exp\,[\, - C / \zeta^2\,] \tag{40}$$

where α_{no} and C are constants.

According to Baraff (1962), this procedure is only justifiable at very high electric fields where electron-phonon collisions are frequent and the distribution function is isotropic. Also, Keldysh (1965) solved analytically the BTE at high energy (assuming parabolic band structure) and calculated the impact ionization rate using the symmetric part of the distribution function. The Keldysh solution is interesting, because it tends to the Wolff relation ($exp[-C/\zeta^2]$) at high fields and tends to a simple exponential ($exp[-C'/|\zeta|]$) at moderate fields, which coincides with the experiment. In fact, the measurements of Chynoweth (1958), Lee et al (1964) and Van Overstraten and De Man (1970) showed that the impact ionization rate at moderate fields follow a simple exponential function. According to Grant's model (1973), the electron impact ionization coefficient follows the following exponential relation:

$$\alpha_n(\zeta) = \alpha_{no}\, exp\,[\,-\,\zeta_{cn}/|\zeta|\,] \tag{41a}$$

where the pre-exponential term is a constant and the critical field ζ_{cn} is related to the threshold energy for ionization of electrons, E_n^{th}, and the mean free path between successive collisions, λ_n, by the relation:

$$\zeta_{cn} = E_n^{th}/e.\lambda_n \tag{41b}$$

The numerical values for α_{no} and ζ_{cn} according to the accurate measurement of these parameters for electrons and holes in silicon and other semiconductors can be found in the literature e.g., (Maes et al., 1990).

Shockley (1961) tried to find out an explanation of the exponential relation of the ionization coefficients in P-N junctions. According to Shockley, the impact ionization mechanism is mainly due to "lucky electrons", which can escape from collisions and the ionization coefficient of such electrons can be expressed in the following exponential form:

$$\alpha_n(\zeta) = \alpha_{no}\, exp\,[\,-\,d/\lambda_n\,] \tag{42}$$

where $d = (E_n^{th}/e\zeta)$ is the minimum path an electron should travel to gain the threshold energy and λ_n is the mean free path between collisions with optical phonons.

In order to find out a more accurate solution of the BTE at high energy, and find out a suitable physical model of the impact ionization phenomenon which interprets the measurements, Baraff solved numerically the BTE, assuming parabolic energy bands (Baraf, 1964). Baraff plotted universal curves of α_n in the following form (1962):

$$\alpha_n = F(\zeta/\zeta_{cn},\, E_n^{th}/e\zeta\lambda_n,\, E_{ph}/E_n^{th}) \tag{43}$$

where E_{ph} is the average optical phonon energy. The above field-dependent exponential relations describing the impact ionization rates have been used in DDM-based simulation for long time.

The Lucky electron model (LEM) has been utilized to describe the breakdown phenomenon as well as the hot carrier injection and leakage currents in MOSFET devices in DDM-based simulation for long time. Unfortunately, the parameters of the lucky-electron exponential model are dispersed among the different measurements of different authors, as shown in Figure 7. For instance, we can find different values for λ_n from 50Å to 100Å in Si in the literature. In fact, the impact ionization phenomenon happens when colliding electrons have enough energy to ionize the valence electrons. Therefore, it is more physical to express impact ionization rates in terms of the carrier energy, rather than the local electric field value.

Schöll-Quade Impact Ionization Model

The model of Schöll and Quade (1987) for impact ionization rate is based on the heated Maxwellian approximation for the symmetric part of the distribution function. According to this model, the impact ionization coefficient of electrons is given by:

$$\alpha_n(\omega_n) = \frac{1}{v_n^{sat}\tau_o\sqrt{\pi}}\left[\sqrt{\varpi_n}\cdot\exp\left(-\frac{1}{\varpi_n}\right) - \sqrt{\pi}\cdot erfc\left(\frac{1}{\sqrt{\varpi_n}}\right)\right] \tag{44}$$

where $\varpi_n = k_B T_n / E_n^{th}$, E_n^{th} is the electron ionization threshold energy, τ_o is a constant ($\tau_o = \varepsilon_o^2\, \hbar^3\pi^3/3e^4 m_n^*$ $\approx 0.0126\, fs$) and $erfc(x)$ is the complementary error function. Note that this model considers a parabolic energy band (constant effective mass m_n^*). It was previously shown in literature that the analytical model of Scholl and Quad for the impact ionization rate under-estimates the generation rate at high fields. In order to get adequate results at moderate fields, both E_n^{th} and τ_o are usually considered as adjustable parameters, with values which are extremely far from their physical meaning. For example, Souissi et al. (1993) considered $E_n^{th} \approx 4$eV and $\tau_o \approx 0.14 fs$, in order to fit this model with experiment.

Figure 7. Impact ionization coefficients of electrons (α_n) in Si at 300K, measured as a function of electric field ζ, according to several authors

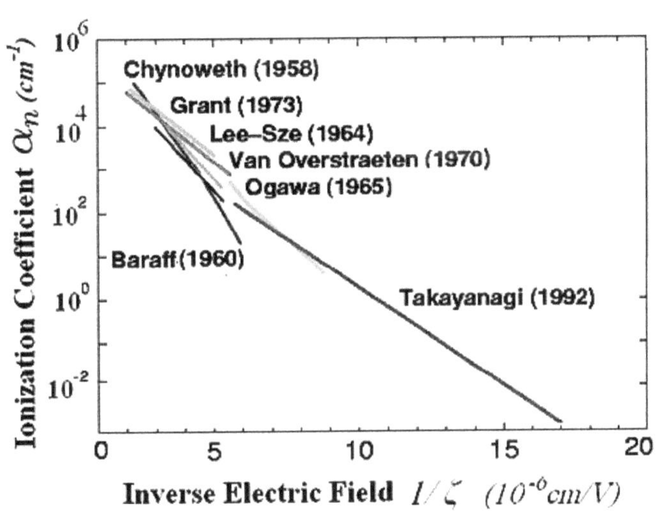

Baccarani and Stork Impact Ionization Model

The model of Baccarani and Stork (1990) belongs to the category of semi-empirical models, which are based on the Shockley lucky-electron model. According to Baccarani and Stork, the rate of impact ionization is given by:

$$\alpha_n(\omega_n) = c\lambda_w \cdot \exp\left(-\frac{3\lambda_w}{5\lambda_{op}}\frac{E_n^{th}}{(\omega_n - \omega_o)}\right) \tag{45}$$

where c, λ_ω, λ_{op} and E_n^{th} are constants. According to Baccarani and Stork, the ionization threshold energy E_n^{th} =1.1eV, the energy relaxation length $\lambda_\omega=(5/3)v_n^{sat}\tau_{\omega no}=500\mathring{A}$, the mean free path between optical-phonon collisions $\lambda_{op}=65\mathring{A}$ and the adjustable parameter c is equal to $2x10^{12}$ cm^{-2}. It was also reported that the semi-empirical model of Baccarani and Stork underestimates the impact ionization rate at high fields, when it is employed in a self-consistent manner in the hydrodynamic simulation (Crabbe et al, 1990). However, Souissi et al (1993) used the model of Baccarani and Stork in their investigation about impact ionization in bipolar Si transistors and found that the best fit with measured values implies $E_n^{th}=5.077$eV, which is excessively high (typically 1.8eV). It has been shown in the latter work that both the Schöll-Quade and Baccarani-Stork models may be used to predict the multiplication factor ($M=\int\alpha_n\,dx$) across the base-collector zone with reasonable error, only over a limited range of applied bias.

Non-Maxwellian Distribution–Based Impact Ionization Models

In order to get rid of the disadvantages of the above mentioned models, some authors tried to calculate the impact ionization rate as a function of carrier energy, starting from its microscopic definition, on the basis of non-Maxwellian distribution functions. For instance, Matsuzawa, Komahara and Wada (1991) suggested the following form for the hot carrier tail of the electrons distribution function:

$$f_n(E_n) = C_1\, exp\,[\,-\,½\,(E_n / k_BT_n)^{3/2}\,] \tag{46}$$

where C_1 is a constant. Also, Sonoda et al. (1996) and later Grasser et al (2001) assumed the following two-components distribution, to account for both cold and hot carriers populations in nanoscale MOS-FET devices:

$$f_n(E_n) = C_1\,\{\,exp\,[\,-(E_n / E_{ref})^b] + C_2\, exp[-(E_n/k_BT_2)]\,\} \tag{47}$$

Here, C_1, C_2, E_{ref} and b are adjustable parameters, which are used to tune the hot carrier distribution component, and the cold carrier (Maxwellian) component, in order to obtain reasonable results over a wide range of carrier energies. Therefore, when $C_1=0$, the distribution function becomes a heated Max-wellian and when $C_2=0$ it tends to the tail distribution function. Using this four-parameter distribution function and assuming parabolic energy bands, Morris, Pass and Abebe (2004) introduced a simple formulation for the impact ionization rate on the basis of Gauss-Laguerre approximation. It should be noted that several efforts have been devoted for the development a nonlocal impact ionization model on the basis of the hot-electron subpopulation. For instance, Tang and Nam (1998) have introduced a simplified form of such hot-carrier subpopulation models. The impact ionization coefficient is then fitted to homogeneous MC simulation in the bulk of silicon.

Advanced Impact Ionization Model

Unfortunately, the success of the above mentioned energy-dependent models of impact ionization is limited to specific cases of small and moderate fields. In addition, these models make use of a number of adjustable parameters, whose fitting values have sometimes irrelevant physical meaning. Indeed, one source of errors is their lake to a link with the realistic band structure of the semiconductor material. The following model (El-Saba, 1999) accounts for the energy band structure and can be used, with no need to fitting of its physical parameters.

$$G_{ii} = (n / \tau_{in}) + (p / \tau_{ip}) \tag{48}$$

where $1/\tau_{in}$ and $1/\tau_{ip}$ are the electron and hole average ionization rates (per unit time). Starting from the Chynoweth experimental law (1958) in its scaled form, according to the Thornber scaling theory (1981):

$$\alpha_n(\zeta)=(1/v_n\tau_{in})= (e\zeta^{//}/E_n^{th})\, exp[\ -\zeta_{cn}/\ \{\zeta^{//}\ (1+\zeta^{//}/\zeta_p)+\zeta_T\}\ ] \tag{49}$$

where $\zeta_{cn} = E_n^{th}/e\lambda_n$ is the critical field at the onset of impact ionization, ζ_T and ζ_p are the critical fields at which the electrons surmount acoustic and optical phonons scattering thresholds. Assuming that the ionization rate can be fully described by the kinetic moments (like, v_n and ω_n), one can write the electron energy conservation equation in the following form in static field conditions:

$$en\zeta.v_n = -(\partial n\omega_n/\partial t)_{col} \approx n(\omega_n - \omega_o)/\tau_{in} - \omega_n(n-n_o)/\tau_n + n(\omega_n-\omega_o)/\tau_{\omega n} \tag{50}$$

The first term in the above equation represents the generation by impact ionization, the second term represents the loss of electron energy due to recombination (τ_n being the electrons lifetime and n_o is the equilibrium density) and the third term represents the loss of electron energy due to collisions with phonons. The second term may be dropped from the above equation as $\tau_{\omega n} << \tau_n$. Also the first and third terms may be combined, with an effective energy relaxation time, which includes the effect of impact ionization at high energies. According to Schöll and Quade (1987), this effect is not significant. Substituting the product $e\boldsymbol{\zeta}.\boldsymbol{v}_n$ from (30) into (29b), we get:

$$\alpha_n(\omega_n) = \frac{1}{\tau_{in}v_n} = \frac{(\omega_n - \omega_o)}{\tau_{wn}v_n E_n^{th}} \cdot \exp\left(- \frac{\eta_n E_n^{th}}{(\omega_n - \omega_o)\left[1 + \frac{(\omega_n - \omega_o)}{\eta_n E_{ph}}\right] + \eta_n E_{ac}} \right) \tag{51}$$

where $\eta_n = (v_n \tau_{wn}/\lambda_n)$, E_{ph} and E_{ac} are the mean optical and acoustical phonon energy. Note that λ_n is the electron mean free path between collisions (of all scattering modes). Therefore, λ_n is related to the mean free path between ionizing collisions (λ_i) and the mean free path between collisions with acoustical and optical phonons, (λ_{ac}, λ_{op}) as follows:

$$\lambda_n^{-1} = \lambda_{op}^{-1} + \lambda_i^{-1} + \lambda_{ac}^{-1} \tag{52}$$

We do not presume values for these parameters, but rather we consider λ_n as a continuous function of the mean carrier energy in the bulk of semiconductor. Near the surface of semiconductor, additional reduction of λ_n, due to surface acoustic phonons and surface roughness scattering mechanisms, should be considered. Actually, the measurements of impact ionization in MOSFET's and CCD's showed that the impact ionization coefficient near the Si/SiO_2 interface is smaller than the bulk case. Slotboom et al (1987) attributed this reduction of the impact ionization to the effect of the shorter mean free path near the Si/SiO_2 interface. However, as we are interested here in the calculation of the impact ionization rate in the bulk of semiconductor, we may discard the surface effects on the mean free path.

In the region of very low energies, where $(\omega_n - \omega_o) \le \eta_n E_{ac}$, our impact ionization model reduces to:

$$\alpha_n(\omega_n) \approx \alpha_{n\infty} \cdot \exp\left[-\frac{E_n^I}{(\omega_n - \omega_o) + E_{ac}^I}\right] \tag{53a}$$

where $\alpha_{no} = (\omega_n - \omega_o)/(E_n^{th} \tau_{wn} v_n)$ and $E_{ac}^{\ I} = \eta_n E_{ac} = E_{ac}(\tau_{wn} v_n)/\lambda_n$. In fact, the study of low-voltage impact ionization has been recently a subject of debate. As shown in the above relation, the acoustic phonons energy, which *adds* to the carrier energy, may interpret this phenomenon. This means that at low carrier energies, carriers may acquire additional energy from the crystal lattice heat bath (by acoustic phonon absorption), in addition to the absorbed energy from electric field, and this may increase their energy to the threshold of impact ionization. This coincides with the theory of Su, Goto, Sugii and Hu (2002), which presumes that thermal activation of carriers may be the origin of the low voltage impact ionization in deep sub-micrometer MOSFETs. Actually, Impact ionization was observed in short-channel silicon SOI devices at drain-source bias as low as 0.8 V, where it was not expected. From the first sight, such very low applied bias is apparently not sufficient to supply electrons with necessary kinetic energy over the impact ionization threshold. However, impact ionization at such low voltages was observed in small SOI MOSFET devices, through the kink in drain-source current-voltage characteristics.

It should be also noted that the ionization rate at low energies is greatly influenced by both the energy relaxation time and the carrier velocity distribution. Thus, in dynamic operation, the velocity overshoot is expected to enhance the impact ionization rate in very small MOSFET devices.

At relatively moderate energies, where $\eta_n E_{ac} < (\omega_n - \omega_o) < \eta_n E_{ph}$, our model can be approximated to the following simple exponential form:

$$\alpha_n(\omega_n) \approx \alpha_{n\infty} \cdot \exp\left[-\frac{E_n^I}{\omega_n - \omega_o}\right] \tag{53b}$$

where $E_n^{\ I} = \eta_n E_n^{th} = E_n^{th}(\tau_{wn} v_n)/\lambda_n$. At high energies, where $\eta_n E_p << (\omega_n - \omega_o)$, this model can be approximated as follows:

$$\alpha_n(\omega_n) \approx \alpha_{n\infty} \cdot \exp\left[-\frac{E_n^I E_{ph}^I}{\left(\omega_n - \omega_o\right)^2}\right] \tag{53c}$$

where $\alpha_{no} \approx \omega_n/(E_n^{th}\tau_{wn}\, v_n^{sat})$ is the asymptotic value of impact ionization rate and $E_{ph}^{I} = \eta_n E_{ph}$. Note that we consider $\tau_{wn}=\tau_{wn}(\omega_n)$ and $\lambda_n=\lambda_n(\omega_n)$. However, even at low electric fields, the simple exponential realation: $exp[-E_n^{I}/(\omega_n-\omega_o)]$ does not imply a Maxwellian distribution, because E_n^{I} is energy dependent.

The energy dependence of the energy relaxation time, which implicitly includes the energy band structure effects, can be obtained from MC simulation or by measurement. Unlike classical models, the hard threshold energy (E_n^{th}) is replaced with a soft (energy-dependent) mean threshold energy (E_n^{I}) in our model. This coincides with the conclusions of Bude and Hess (1992) about the impact ionization soft threshold.

The calculation of the mean free path λ_n as a function of the electron energy may be carried out by MC simulation in the bulk, using the following relation:

$$\lambda_n(\omega_n) = \langle u_n(k) / \Gamma_n(k) \rangle \tag{54a}$$

where $\Gamma_n(k)$ is the total carrier scattering rate including impact ionization. The mean free path may be also estimated by fitting the so-called lucky-electron exponential model LE-EM, with experimental results. However, we can consider the following expression for λ_n, which includes the effect of acoustic and optical phonons as well as ionization collisions:

$$1/\lambda_n(\omega_n) = 1/\lambda_{ph} + 1/\lambda_i = \lambda_{ph} / [1+ (\omega_n -\omega_o)/ E_n^{I}] \tag{54b}$$

where $\lambda_{ph} = 1/(1/\lambda_{op} + 1/\lambda_{ac})$ depicts the mean free path due to both acoustical and optical phonons. We also make use of the following phenomenological relation to express of λ_{ph} as a function of carrier energy. This relation is based on Cromwell and Sze model (1966) at the electron temperature[6]:

$$\lambda_{ph}(T_n) = \lambda_o \tanh(E_{ph}/2k_B T_n) \tag{54c}$$

Here, λ_o is the mean free path between collisions at very low temperature (when $T_n \approx 0$). Figure 29 depicts the variation of λ_n and α_n in Si, according to the above relations. In our model, we take E_{ph}= 60meV, λ_o=4000Å and extrapolate the available MC data of $\tau_{wn}(\omega_n)$. For the matter of comparison, we plot α_n according to Baccarani-Stork and Scholl-Quade models, beside EL-Saba model (1999).

The Scholl-Quad model produces excessively high rates when the real physical value of τ_o is taken into account. So, we set τ_o equal to 12.6 fs, about 3 orders of magnitude greater than its physical value. Note that the value of α_n in EL-Saba scaled model is greater than all other models and continues to increase at high energies. In fact, the experimental values of α_n may exceed 10^6cm^{-1}. Also, the measurement of Kotani and Kawazu (1981) and Takayanagi et al. (1992) of $\alpha_n(\zeta)$ are actually fitted with $\alpha_{no} \geq 10^6$ cm^{-1}.

It worth noting that the impact ionization rate at high energies (in the saturation region) is greatly influenced by the roll-off of the energy relaxation time, while E_n^{th} and λ_n have a relative influence at low and medium energies. This explains the failure of the previous models, to properly estimate α_n at high energies, even with adjusting E_n^{th} and pre-exponential factors with non-physical values, while fixing both λ_n and τ_{wn}..

Figure 8. Electron mean free path between optical phonon collisions in Si at 300K, according to the LE-EM and the Crowel-Sze model at electronic temperature

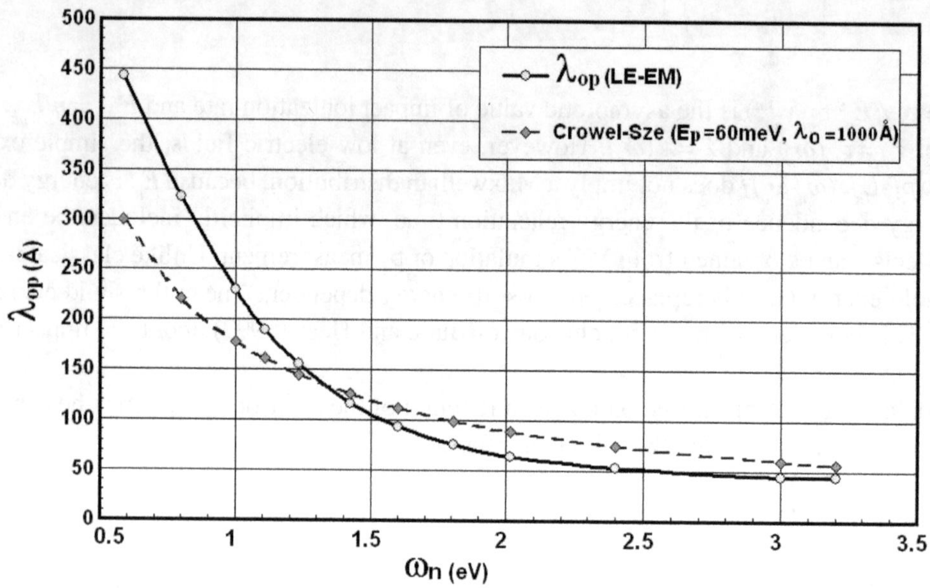

Figure 9. Electron impact ionization rate in Si at 300K, as a function of average electron energy, according to Baccarani-Stork model (λ_{ω}=500Å, λ_{op} = 65Å), Scholl-Quade model (τ_o= 14ps) and the model of the book author. In all models E_n^{th} =1.2 eV.

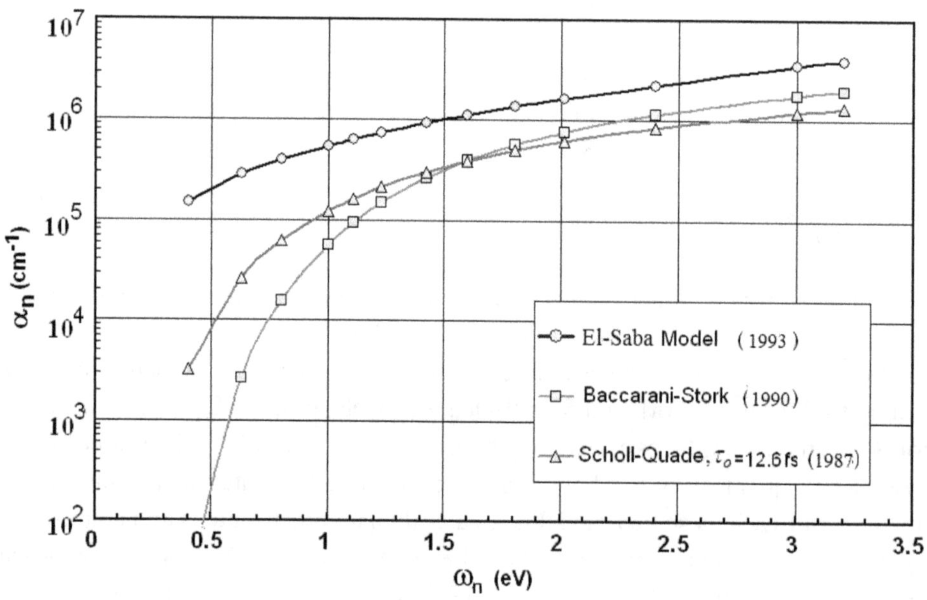

As we pointed out in the last section, the mean ionization energy E_n^I in El-Saba model of impact ionization has a soft distribution around E_n^{th}. In order to evaluate the broadening of mean ionization energy, this model defines impact ionization energy broadening ΔE_n^I as follows:

$$\alpha(\omega_n) = |E_n^I(\omega_n) - E_n^{th}| = E_n^{th} \, | 1 - v_n \tau_{wn} / \lambda_n | \tag{55}$$

Figure 10 depicts the ionization energy broadening ΔE_n^I as a function of mean electron energy. It is clear that the energy broadening is mainly due to the variation of the energy relaxation time and the mean free path at high energies. Curiously, one may wonder if there is a reason to *relate* this phenomenon to the broadening in energy levels due to finite lifetime of energy states between scattering events at high energy. In fact, it has been shown experimentally that there exist a line broadening transition, representing the transition from phonon scattering dominated to impact ionization dominated transport not only in silicon but also in any material with a bandgap (Bude, 1995).

In addition, the effect of collisional broadening on the impact ionization rate was studied in several articles. For instance, Madureira et al (2001) showed that the energy broadening leads to a lowering of the ionization threshold which is significant for low static fields. The theoretical studies of the impact ionization on the basis of quantum transport theory[7] showed that a fixed impact-ionization threshold does not exist, and the impact-ionization scattering rate is drastically enhanced around the semiclassical threshold by the intra-collisional field effect. Of course, the broadening factor may be better evaluated by precise MC simulation, with an improved impact ionization model. However, with our simple model of $\lambda_n(\omega_n)$ and the exponentially extrapolated $\tau_{wn}(\omega_n)$, we found that ΔE_n^I in Si is centralized around a

Figure 10. Average ionization energy in Si at 300K, with E_n^{th}=1.12 eV

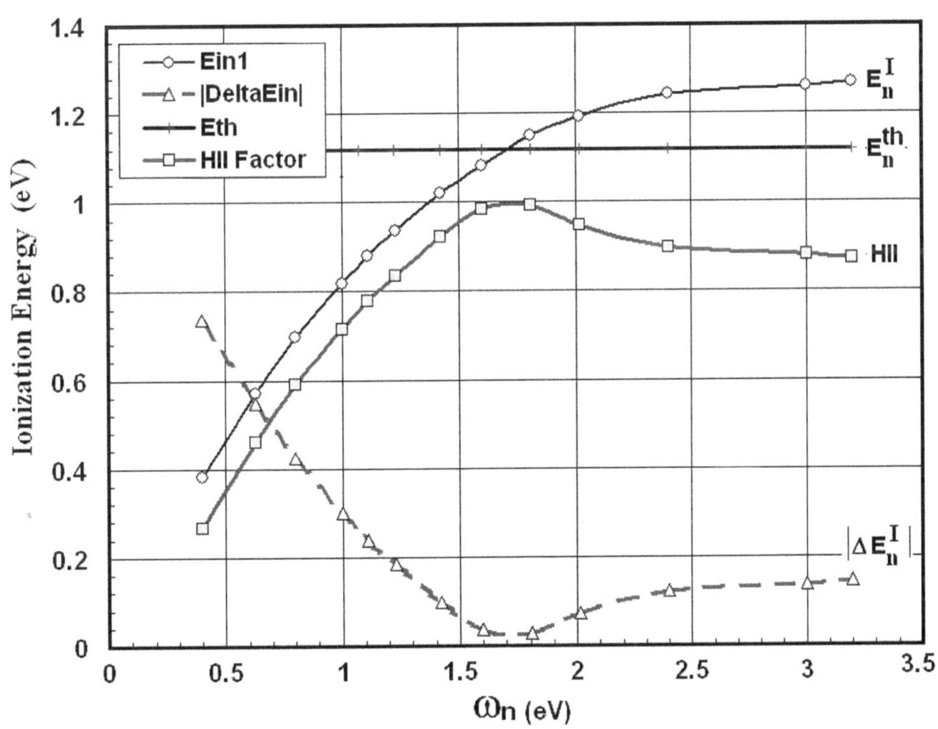

minimum (almost zero) at $\omega_n \approx 1.7$eV, which is interestingly equal to about 3/2 E_n^{th}. It should be noted here that we chose λ_o such that the distribution of $\lambda_{ph}(\omega_n)$ is as close as possible from the experimental LE-EM curve near the ionization threshold. As shown in figure 10, the two curves almost intersect at $\omega_n = E_n^{th}$ =1.12eV in Si.

Finally, we note that our model include the effect of phonon–assisted impact ionization rate, through the inclusion of energy-dependent mean free path $\lambda_n(\omega_n)$. The phonon–assisted impact ionization causes threshold shift and may have more significant effect on the hole impact ionization rate (Takayanagi et. al., 1992).

6.3 Carrier-Energy Relaxation Time ($\tau_{wn,,p}$)

The choice of the carrier-energy relaxation time is crucial in the HD simulation because it influences both the static and dynamic characteristics of the device to be simulated. According to the hydrodynamic theory, the exact modeling of the carrier energy relaxation time (and all other relaxation times) should be expressed as function of all the carrier moments (average quantities, like n, v_n and ω_n, etc). However, as we usually truncate the series of moments and satisfy ourselves with the few first moments, it is believed that much of the information about the energy relaxation time is tightly related to the carrier mean energy and density. Therefore, the carrier energy relaxation time is usually modeled as function of the carrier mean energy, as well as the lattice temperature and doping concentration.

According to Jacoboni and Reggiani (1999), the energy relaxation time should decrease in a phenomenological way representing the tendency of the isotropic part of the distribution function to decay towards its equilibrium value. Thus, they made use of the following formula, to model the carrier energy relaxation time for warm electrons:

$$\tau_{wn}(\omega_n) = (\omega_n - \omega_{no}) / e\, v_d\, \zeta \tag{56a}$$

This formula, which is sometimes called the Seeger formula (1973), is derived from the energy continuity at steady state:

$$\partial\omega_n/\partial t = e\, v_d\, \zeta + (\partial\omega_n/\partial t)_{col} = e\, v_d\, \zeta - (\omega_n - \omega_o)/\tau_{wn} = 0 \tag{56b}$$

According to the Seeger static formula, the energy relaxation time can be determined by substituting the static field relations $v_d(\zeta)$ and $\omega_n(\zeta)$ from MC simulation. This static relation has been used in the literature to evaluate the energy relaxation time from MC data or its fitting parameters. However, when the electron distribution function is warm enough and the electron velocity tends to saturate, the energy relaxation time can be considered as a constant. This is a good approximation when the band structure is parabolic.

For non-parabolic bands, the energy relaxation time continues to decrease when electron energy is further increased, as shown in Figure 11. This is due the fact that the intervalley phonons becomes more effective in dissipating the electron energy at high energies, which lead to decreasing the electron energy relaxation time, at higher energies.

For accurate modeling of the energy relaxation time, I proposed a phenomenological relation (El-Saba, 2012) between the average electron effective mass m_n and energy relaxation time τ_{wn}, for energies above the equilibrium value ($\omega_n > \omega_o$):

$$\tau_{wn}(\omega_n) = \tau_{wno} [m_{no} / m_n(\omega_n)] \tag{57}$$

where m_{no} is the density of states effective mass at the bottom of the conduction band. According to this relation, we only need the knowledge of the low-field value of the electron energy relaxation time and the average mass as a function of electron energy $m_n(\omega_n)$ to obtain $\tau_{wn}(\omega_n)$.

The low-field value of the average electron mass, m_{no}, is equal to the conductivity effective mass at the conduction band edge, and can be measured using the optically-detected cyclotron resonance (ODCR). The ODCR method is more accurate than the classical CR and has been recently used to measure the effective mass of several semiconductor compounds. According to the measurements of Riffe (2002) the conductivity effective mass at 300K $m_{no} = 0.275m_o$ and $m_{no} = 0.365m_o$ as shown in Figure 12.

The ultrafast laser based pump-probe techniques have been used to measure the energy relaxation as well as the momentum relaxation times in Si (Tsen, 2001). Using this technique, a time constant of *32 ± 5fs* associated with momentum relaxation and an electron-phonon energy relaxation time of *260 ± 30 fs* have been extracted from the coherent-transient variations in (001) Si at 300K.

In order to calculate the average electron mass as a function of the average electron energy, we use the MC simulation results and compare it with the measured ratio of (m_{no}/m_n). The measured values of m_{no}/m_n are taken according to Rife (2002). As shown in figure, the normalized value of the average inverse mass decreases as the electron energy increases. In a next step, we calculate the energy relaxation time $\tau_{wn}(\omega_n)$, using the above phenomenological relation (above ω_o, when electrons start to warm up).

As shown in figure 32, the discrepancy between the energy relaxation time, according to MC and the calculated one according to our model is within the measurement error (~10%) for a wide range of carrier temperatures (up to 2000K and may be linearly extrapolated to much more higher values). Alternatively, El-Saba model (2012) can be used to calculate the energy dependence of the average electron mass $m_n(\omega_n)$ from the energy dependence of the electron energy relaxation time $\tau_{wn}(\omega_n)$.

Figure 11. Electron average mass and energy relaxation time in Si at 300K

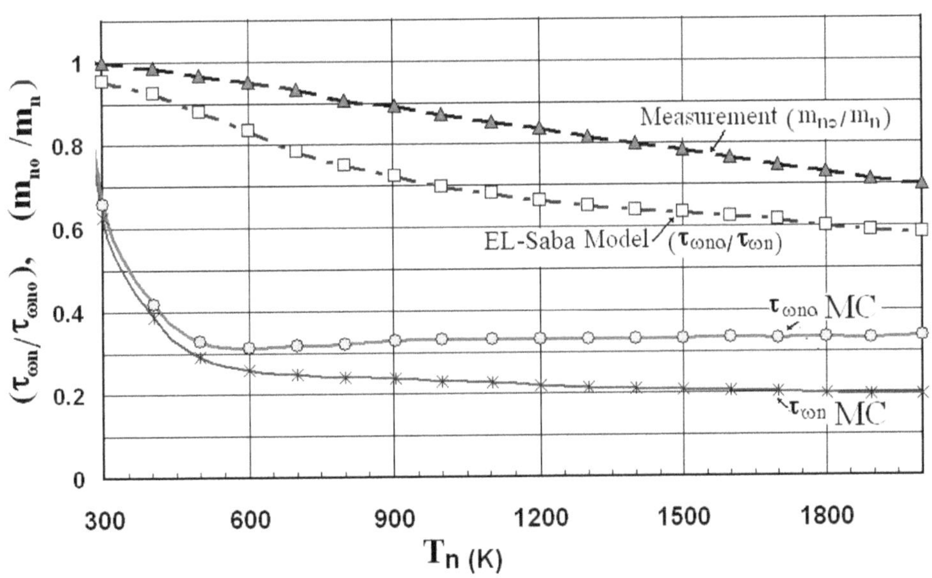

The energy dependence of τ_{wn} of several semiconductors (e.g., *AlGaAs/GaAs, AlGaN/AlN/GaN*) has been measured using the microwave-noise method. This method is based on the measurement of the noise temperature, which is equal to the electron temperature T_n in the semiconductor, according to the relation:

$$\tau_{wn} = 3/2\, k_B\, (T_n - T_L)\, /e\, v_d\, \zeta \tag{58}$$

Here we neglect the convection part of the average electron energy and consider $\omega_n \approx 3/2\, k_B\, T_n$. Because of the lake of sufficient experimental data about the energy relaxation time in semiconductors (as a function of carrier energy), the above model, which correlates $\tau_{wn}(\omega_n)$ with the measured $m_n(\omega_n)$ should be very useful.

Figure 13 depicts the experimental values of the energy relaxation time and optical phonons lifetime in *Si* at 293K, according to the recent optical measurement of Matulionis et al (2002). As shown in figure, the electron mass increases monotonically as its energy increases. On the other hand, the hole mass relaxes from an initial peak and then increases monotonically with its energy.

Figure 14 depicts the effect of the energy relaxation time on the hydrodynamic simulation results (the I-V characteristics of sub 0.1 µm MOSFET). Note that the solution, which is obtained using the HDM with lower $\tau_{\omega n}$ tends to the classical DDM solution. This means that for shorter energy relaxation times, the electrons are more efficient in dissipating their energy. For the limiting case when $\tau_{\omega n}$ tends to zero, the electrons will tend to dissipate their gained energy at once and their average energy is equal the lattice energy ω_o. This is exactly the case of DDM, where we consider the electron temperature equal to the lattice temperature $(T_n = T_L)$ at all fields.

$$\omega_n = \omega_o + e.\, \tau_{wn}\, v_d\, \zeta \tag{59}$$

Figure 12. Measured electron average mass in Si
After Rife (2002).

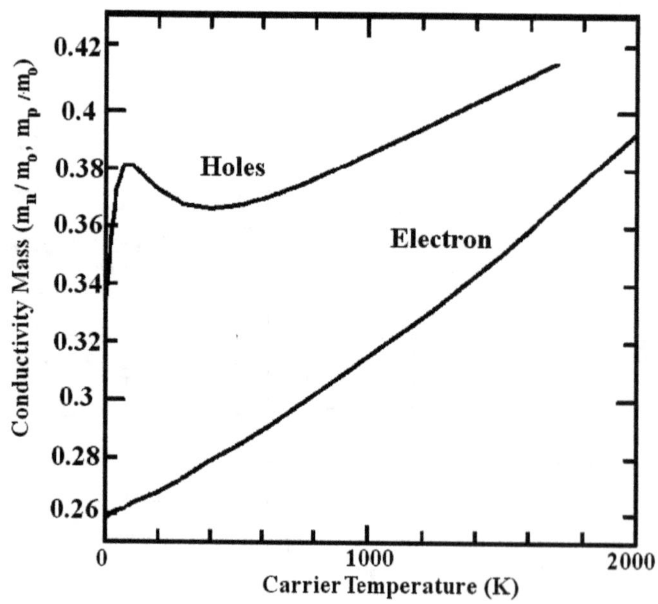

Figure 13. Electron energy relaxation time and optical phonons lifetime in Si at 293K, according to the optical measurement of Matulionis et al. (2002)

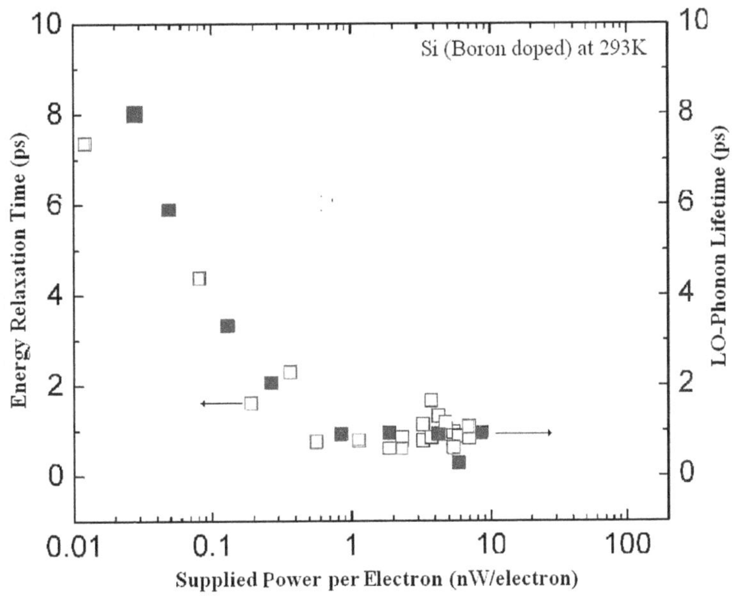

Figure 14. I-V characteristics of a 0.05μm MOSFET, at VGS=1.5V, showing the effect of energy relaxation time, as obtained by classic HDM
After Munteanu et al.

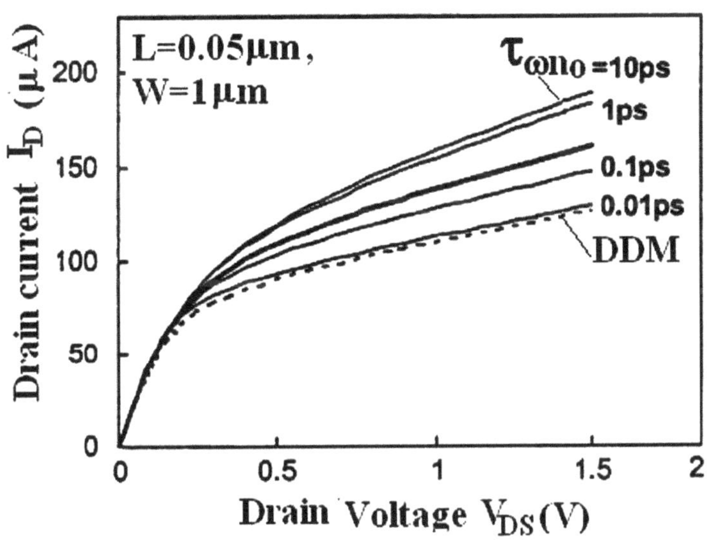

From the above, we may conclude that the DDM is a hypothetical case of the HDM where we assume $\tau_{\omega n} = 0$ and hence $\omega_n = \omega_o$, and this case means a virtual semiconductor whose collisions are so efficient that electrons relax to the equilibrium energy in almost no time.

$$DDM = Lim_{\tau_{wn} \to 0} HDM \tag{60}$$

Therefore, the electrons in such a virtual semiconductor (with zero energy relaxation time) will not be heated even at high fields and the mean electron energy will be always equal to the equilibrium value.

6.4 Advanced Heat Flux Model

The heat flux term in the majority of previous hydrodynamic models was either neglected or modeled by the Fourier relation. The energy transport model (ETM) was reformulated by Chen, Kan and Ravaioli (1992), who assumed non-Maxwellian and non-parabolicity correction factors, to solve the contradiction involved in evaluating a non-zero heat flux term for a Maxwellian distribution. However, it has been shown that the Fourier diffusion model is not accurate for modeling the heat flux across semiconductor interfaces in general and p-n junctions in particular (EL-Saba, 1996). This has been also discussed in the article of Kan et al (1995), where it is indicated that the Wiedmann-Franz law is suspected to be responsible for the spurious velocity overshoot.

In order to increase the accuracy of the HDM, I proposed in my article (EL-Saba, 1996) the following model, which is based on the third moment of the BTE in steady state:

$$nF \cdot \left\langle \left(m_n^*\right)^{-1} E_n + u_n \otimes u_n \right\rangle + \nabla_x \cdot \left(n \langle E_n \ u_n \otimes u_n \rangle\right) = \left[\frac{\partial S_n}{\partial t}\right]_c \approx -\frac{S_n}{\tau_{sn}} \tag{61a}$$

where I suppressed the time derivative $(\partial S_n/\partial t)$ and used the relaxation time approximation (τ_{sn} is the energy flux relaxation time). The above conservation equation (of energy flux), may be further simplified by approximating the average terms, as follows:

$$<(m_n^*)^{-1} E_n + \boldsymbol{u}_n \otimes \boldsymbol{u}_n> \approx 5/2\ \alpha_n <\boldsymbol{u}_n \otimes \boldsymbol{u}_n>, \tag{61b}$$

$$<E_n \boldsymbol{u}_n \otimes \boldsymbol{u}_n> \approx ½\, \beta_n\, m_n\, (<\boldsymbol{u}_n \otimes \boldsymbol{u}_n>)^2 \tag{61c}$$

Here, the correction factor α_n encompasses the anisotropy of the distribution function and the correction factors β_n expresses how much the distribution function is deviated from the Maxwellian distribution. These parameters can be estimated by MC simulation in the bulk of semiconductor. Then we can write the energy flux as the sum of a conduction part $\boldsymbol{Q}_n$ and a convection part (where $\boldsymbol{J}_n$ is a multiplier):

$$\boldsymbol{S}_n = \boldsymbol{Q}_n + 5/2\ (\mu_{sn}/\mu_n)\ (k_B\ T_n).\boldsymbol{J}_n \tag{62a}$$

Here $\mu_{sn} = (e\, m_n^{-1} \tau_{sn})$ and the electron heat flux Q_n is given by:

$$Q_n = - k_n' \, \nabla T_n - k_n'' \, [\nabla (nT_n) / n)] \tag{62b}$$

where

$$k_n' = 5/2\, (k_B^{\,2}/e)\, n\, \mu_{sn}\, \beta_n\, T_{n,} \tag{63a}$$

$$k_n'' = 5/2\, (k_B^{\,2}/e)\, n\, \mu_{sn}\, (\beta_n - \alpha_{n'})\, T_n \tag{63b}$$

Note that the convection energy flux is similar to the 3-moments formulation, except for the correction factor (μ_{sn}/μ_n) and that $(\omega_n + k_B T_n)$ is replaced with $5/2(k_B T_n)$. However, we just called the 3rd moment equation to derive a more accurate formula of the heat flux term, which includes the effect of both carrier density and carrier temperature gradients.

6.5 Modified Wiedman-Franz Relation

When the anisotropy of the distribution function is negligible, we may substitute $\alpha_{n'} \approx 1$. Then, the heat flux term can be put in the following form (El-Saba, 1996):

$$Q_n = - 5/2(k_B^{\,2}/e)\, \mu_{sn}\, (2\beta_n - 1) \nabla (nT_n) \tag{64a}$$

Figures 15 and 16 depict the heat flux beta factor and the relative energy flux mobility (μ_{sn}/μ_n) as a function of carrier energy for electrons. We can calculate these parameters from MC data in the bulk of Si at 300K, with non-parabolicity factor $0.5V^{-1}$. When the carrier-concentration gradient is negligible

Figure 15. Energy flow mobility in Si at 300K and various doping concentrations

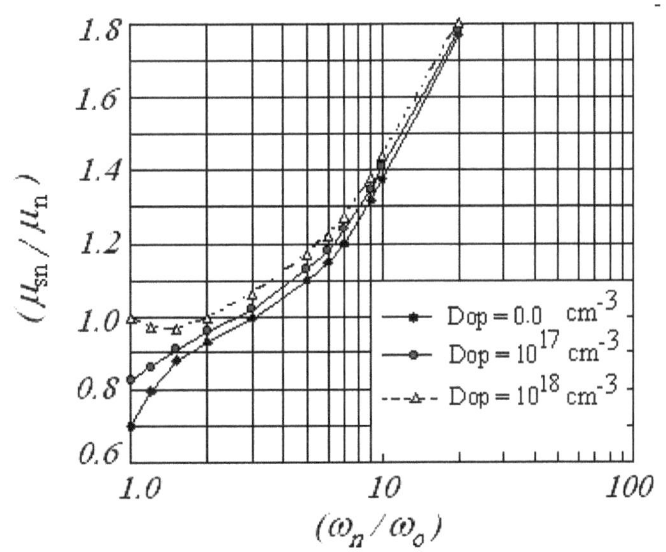

Figure 16. The heat flux beta factor in Si at 300K and various doping concentrations

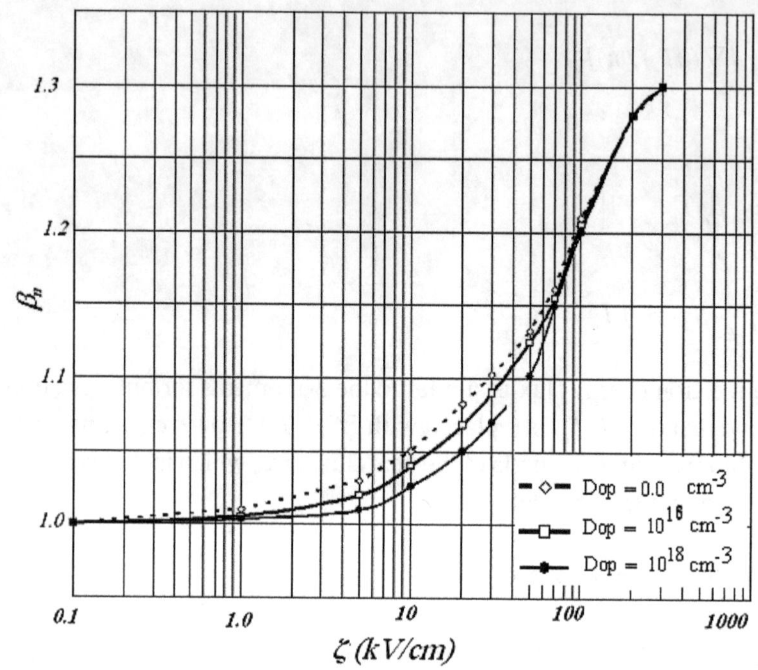

(with respect to the temperature gradient), the above formula may be further reduced to the conventional Fourier relation, as follows:

$$k_n = 5/2\ (k_B{}^2/e)\ n\ \mu_{sn}\ (2\beta_n\ \text{-}1)\ T_n \tag{64b}$$

The above equation is equivalent to the Wiedeman-Franz relation with the following '*modified*' energy-dependent Lorenz parameter:

$$\gamma_n = 5/2\ (\mu_{sn}/\mu_n)\ (2\beta_n\ \text{-}1) \tag{65}$$

In the case of very low doping and very low electric field, where $\mu_{sn}=\mu_n$ and $\beta_n=1$ then $\gamma_n=5/2$ which corresponds to $(5/2+r)$ in the classic model.

7. LATTICE HEAT CONSERVATION EQUATION

The lattice heat equation, which depicts the temporal and spatial distribution of lattice temperature along the semiconductor, has been already introduced in chapter 1. This equation is not only important for power devices but also for silicon-on-insulator (**SOI**) devices as well as nanodevices. In fact, the recent studies of SOI devices by standard HDM, showed certain anomalies due to lack of a heat evacuation mechanism at the insulator interface. In order to fix the heat evacuation problem, some authors suggested to consider a tensorial temperature and modifying the closure condition for energy flux term. Other authors suggested a more rigorous approach consists in solving the BTE for phonons (Pop, 2004).

In fact it is well known that the longitudinal optical (LO) phonons have small group velocities (~10^5 cm/s) and anharmonically decay into faster acoustic phonons which in turn transport the energy out off the semiconductor device. Therefore, we can add phonon BTE to the set of transport equations to solve this problem. The phonon BTE may be simplified by assuming a three-fluid phonon transport model involving a stationary optical reservoir mode and a propagating acoustic mode. The phonon BTE and its solution methods are presented in Chapter 7 of this book, in the context of phonon transport theory.

8. CASE STUDY 2: HYDRODYNAMIC SIMULATION OF P-I-N DIODE

Assume the following P-N junction geometry and doping profile, as shown in Figure 17. The p^+ region has a Gaussian profile with maximum doping of 10^{19} acceptor/cm^3 and the n-region is uniformly-doped with 10^{14} donor/cm^3. The buffer n-region, which follows the n-region is also uniformly doped with 10^{15} donor/cm^3. In order to obtain the electrostatic potential distribution in the device, the Poisson equation should be solved. In order to get the distribution of charge carriers (electrons, n, and holes, p) across the device and subsequently calculate the current density at its contacts, the continuity equations should be solved. In the hydrodynamic transport model, carrier temperatures are allowed to be different from the lattice temperature. The basic equations are augmented by energy balance equations which determine the carrier temperatures.

Figure 18 shows the distribution of electrons and holes obtained by both the DDM and HDM through the axial direction. As shown in Figure 18, there exists an undershoot in the HD simulation of the hole distribution along the p-i-n diode which means the presence of a velocity overshoot. However, such overshoots are not significant in our case because of the substantial length of the simulated device. Figure

Figure 17. The doping concentrations in the simulated p-i-n diode

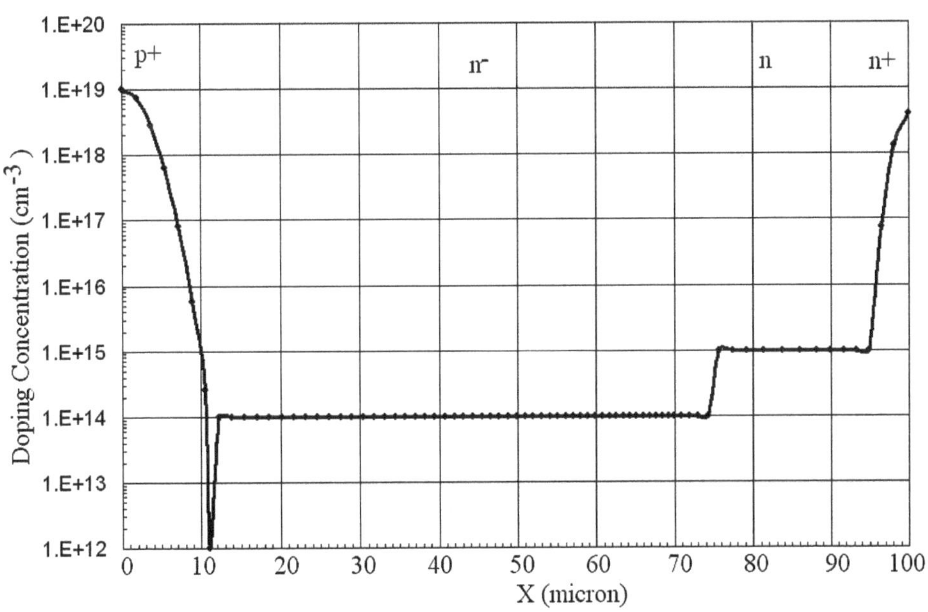

Figure 18. Electron and hole concentrations, across the p-i-n diode at -75V reverse bias

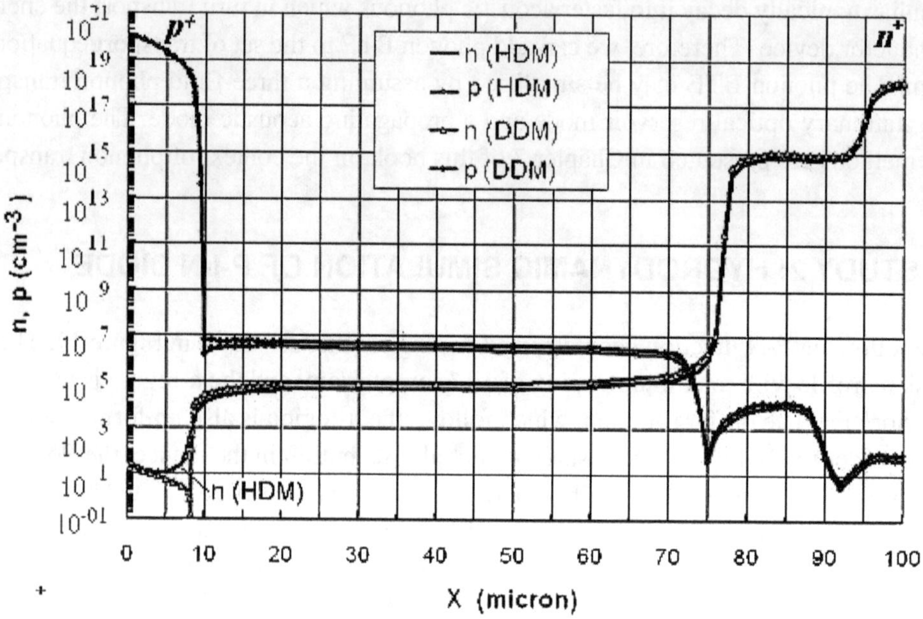

21 depicts the energy flux of electrons and holes across the P-I-N device under reverse bias. Note that the energy fluxes are not symmetric and have different directions.

The simulation results show that the density of electrons in the negative space charge region increases ($n^{HDM} > n^{DDM}$), and the density of holes in the positive space charge region decreases ($p^{HDM} < p^{DDM}$) because of the thermal diffusion effects. I attributed the discrepancy in the carrier density distribution in the HD simulation to the hot carrier backward diffusion or rebelling effect (EL-Saba, 2006). Figure 19 shows the normalized electron temperature and electric field distributions under reverse bias conditions.

9. CASE STUDY 3: HYDRODYNAMIC SIMULATION OF NanoMOS

The multi gate MOSFET structures have been receiving great attention because they help making Si MOSFETs even smaller (i.e., channel lengths under 20 nm). Such very short channel devices permit computers and portable appliances, to be operated at high frequency and powered from low-voltage supplies. On the other hand, the HDM has been recognized as more accurate: than the drift-diffusion model (DDM) for simulating hot-carrier effects in such small semiconductor devices. However, it has been reported in the literature that the HDM, has several drawbacks such as its exaggerated estimation of the electron velocity overshoot, when its parameters, such as the carrier energy-relaxation time in particular, are not properly modeled. Figure 22 depicts the structure of a double-gate (DG) MOSFET as well as its simulated potential distribution.

The simulated DG-MOS, described in figure above, has a gate length L of 24nm. The gate oxide t_{ox} is 1nm and the Si body thicknesses t_{Si} is 9nm. The doping density is $N_D = 5\times10^{19}$ cm^{-3} in the N^+ source–drain regions and $N_A = 2\times10^{15}$ cm^{-3} in the body.

Figure 19. Electron and hole temperatures, across the p-i-n diode, at -75V reverse bias

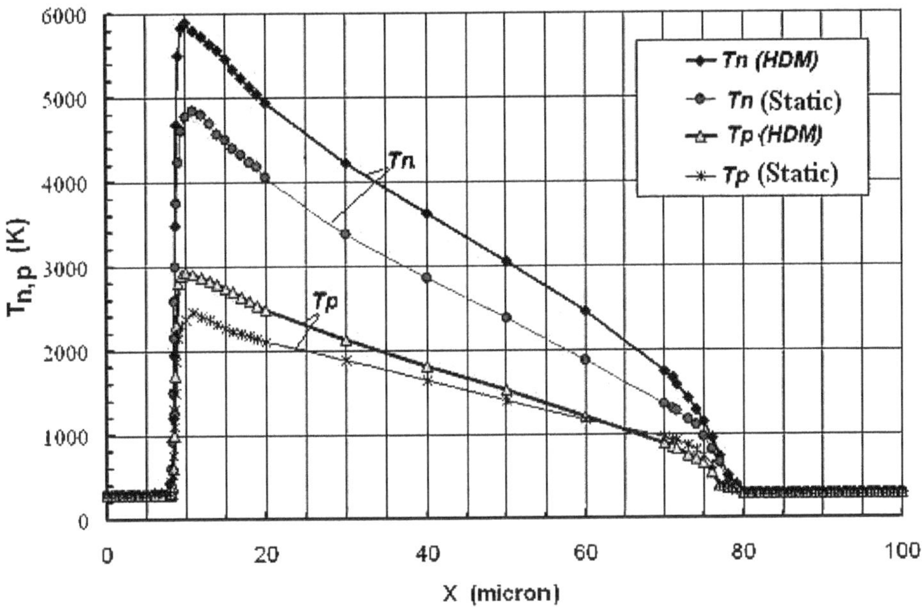

Figure 20. The electron and hole temperature distribution at reverse bias of -75V

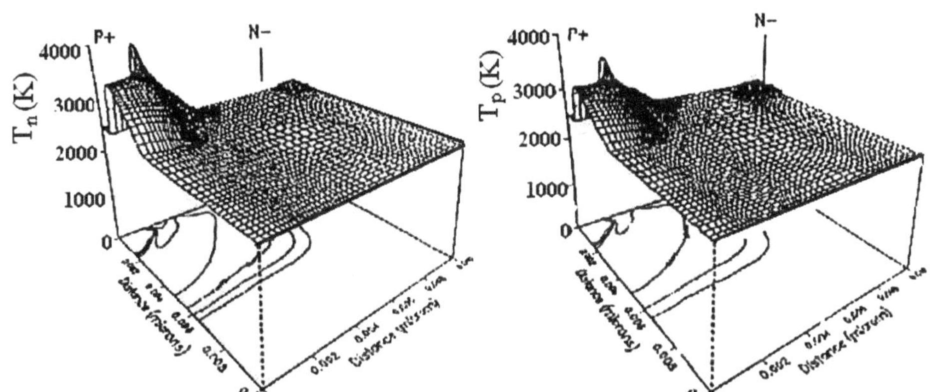

10. SUMMARY

The hydrodynamic description of the gas of charge carrier in semiconductors consists in the study of the evolution of certain macroscopic quantities, whose physical meaning is significant and whose quantity is eventually measurable. The hydrodynamic model (HDM) for semiconductor s plays an important role in simulating the behavior of the charge carrier in submicron semiconductor devices. The HDM consists of a set of conservation laws for the particle density, current density, and energy density.

The set of hydrodynamic equations (HDEs), which is derived from the first few moments of the semiclassical Boltzmann transport equation (BTE), is indeed more accurate than the conventional drift-

Figure 21. Energy flux of electrons and holes at reverse bias of -75V

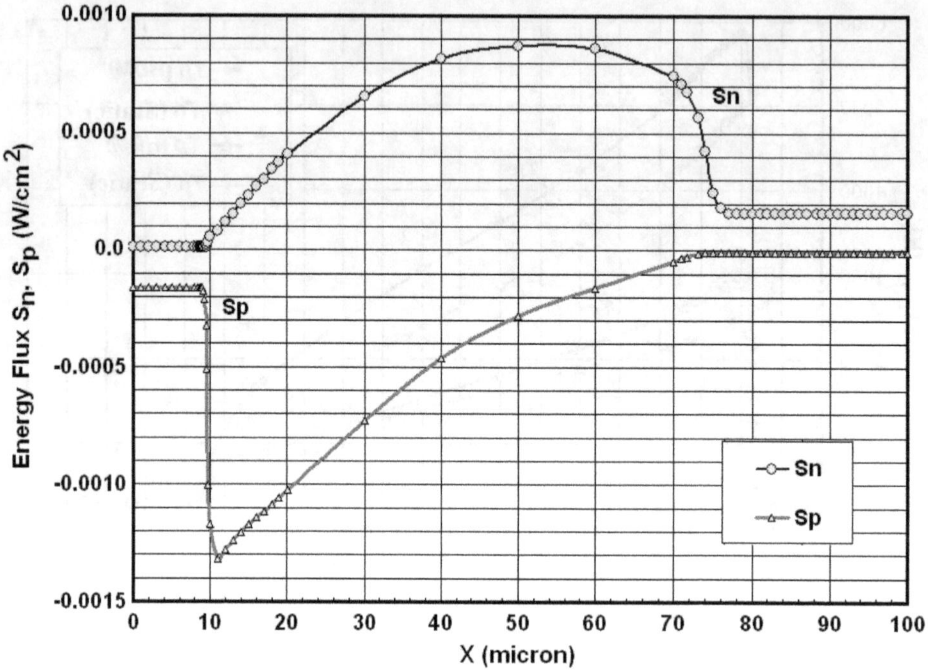

Figure 22. Simulated structure and electrostatic potential distribution of the DG-MOSFET

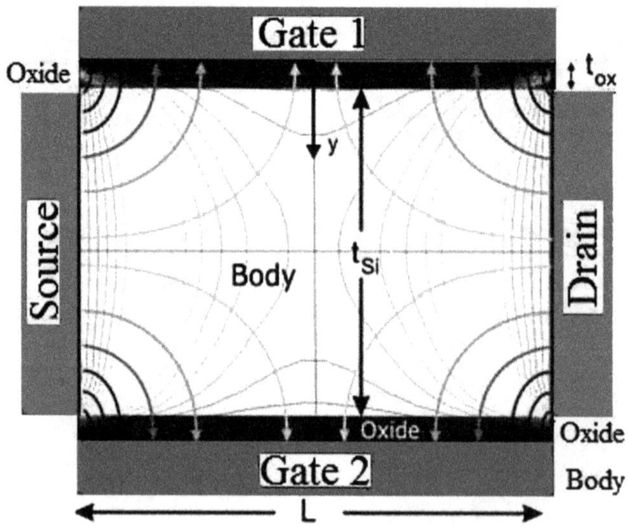

Figure 23. Three-dimensional distributions of electron density and electron temperature

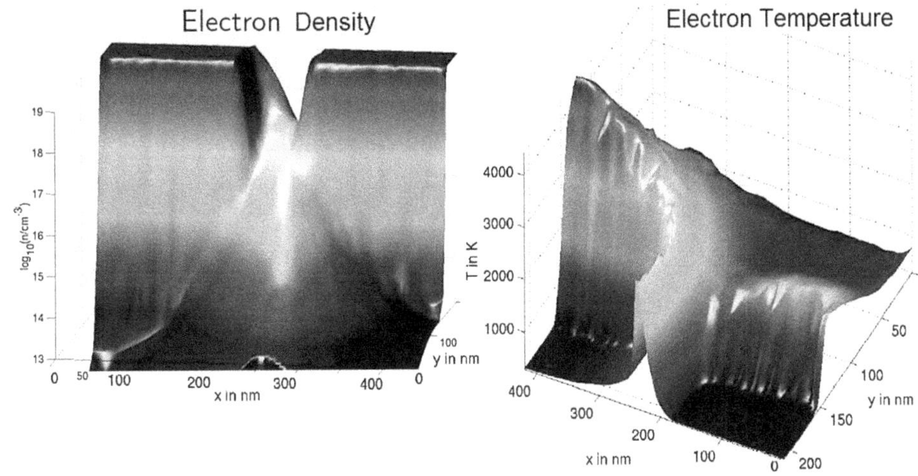

Figure 24. The Drain voltage Id vs. drain voltage Vd at Vg=0.7V, according to MCM, HDM and DDM

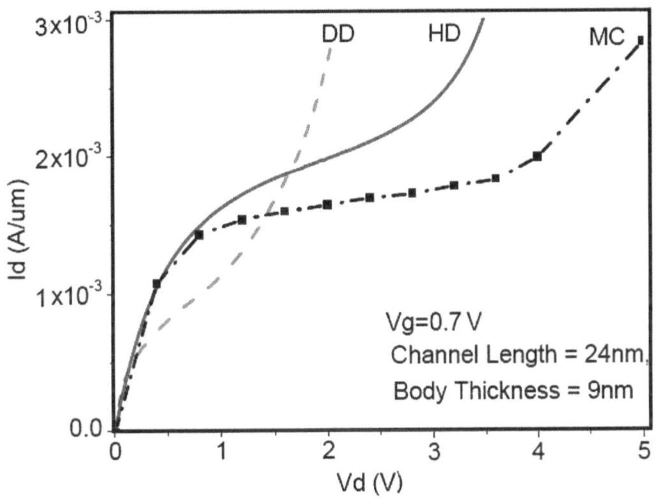

diffusion model (DDM) and less complex than the direct solution of the BTE, and Monte Carlo Methods. The following electron density, momentum- and energy-conservation equations in a semiconductor, can be deduced from the first three moments of the BTE:

$$\frac{\partial n}{\partial t} = \frac{1}{e}\nabla \cdot J_n + (R - G)$$

$$J_n = n\mu_n m_n \frac{dv_n}{dt} + e\mu_n n\zeta + k_B\mu_n \nabla(nT_n)$$

$$\frac{\partial(n\omega_n)}{\partial t}+\nabla\cdot S_n=\zeta\cdot J_n+\omega_n(R-G)-n\frac{(\omega_n-\omega_o)}{\tau_{\omega n}}$$

where n, T_n, v_n and ω_n are the electron density, temperature, mean velocity and mean energy, respectively. Also, $\tau_{\omega n}$ is the electron energy relaxation time, μ_n is the electron drift mobility, k_B is the Boltzmann constant and R and G are the recombination and generation rates. In addition, $J_n = -e\, n\, v_n$ is the electron current density, ζ is the electric field and S_n is the electron energy flow:

$$S_n=Q_n-\frac{J_n}{e}\left(\omega_n+k_BT_n\right)$$

Q_n is the electron heat flux and the electron mean energy ω_n is related to the electron temperature T_n by the following relation:

$$\omega_n=\tfrac{1}{2}m_nv_n^2+\tfrac{3}{2}k_BT_n$$

Also, the electron mean energy at thermal equilibrium ω_o is related to the lattice temperature T_L, such that $\omega_o = 3/2\, k_B\, T_L$.

Since the validity of the semiclassical transport approaches for simulating extremely scaled devices (sub 0.1 μm) and ultrafast devices (working up to the terahertz range) has been lately a subject of debate, we just remind here the main assumptions on which the BTE in semiconductors is based.

- Collisions are localized in space, so that the mean free path of charge carriers is much greater than the de-Broglie wavelength,
- Collisions are instantaneous in time such that the collision time is much smaller than the time between subsequent collisions,
- External electric field force does not vary greatly over small distances in the order of wave packet length,
- Particles are weakly correlated (Bogoliubov assumption) and the many body effect is negligible.

REFERENCES

Agarwal, P., Goossens, M. J., Zieren, V., Aksen, E., & Slotboom, J. W. (2004). Impact ionization in thin silicon diodes. *IEEE Electron Device Letters*, *25*(12), 807–809. doi:10.1109/LED.2004.838557

Aluru, N.R., Law, K.H., & Dutton, R.W. (1995). An Analysis of the Hydrodynamic Semiconductor Device Model-Boundary Conditions and Simulations. *COMPEL - The International Journal for Computation and Mathematics in Electrical and Electronic Engineering, 14*(2/3), 157 – 185.

Anantram, M. P., Lundstrom, M.S., & Nikonov, E. (2008). Modeling of Nanoscale Devices. *Proceedings of the IEEE Trans., 96*(9).

Anile, A. M., Romano, V., & Russo, G. (2000). Extended hydrodynamic model of carrier transport in semiconductors. *SIAM Journal on Applied Mathematics, 61*(1), 74–101. doi:10.1137/S003613999833294X

Ansgar, J., & Matthes, D. (2005). A derivation of the isothermal quantum hydrodynamic equations using entropy minimization. *Applied Numerical Mathematics*.

Baccarani, G. (Ed.). (1993). *Process and Device Modeling for Microelectronics*. Amsterdam: Elsevier.

Baccarani, G., & Wordmann, M. (1985). An Investigation of the Steady-State Velocity Overshoot in Silicon. *Solid-State Electronics, 28*(4), 407–416. doi:10.1016/0038-1101(85)90100-5

Ballinger, R. A., Major, K. G., & Mallinson, J. R. (1973). Impact ionization thresholds in semiconductors. *Journal of Physics. C. Solid State Physics, 6*(16), 2573–2585. doi:10.1088/0022-3719/6/16/011

Baraff, G. A. (1962). Distribution function and ionization rates for hot electrons in semiconductors. *Physical Review, 128*(6), 2507–2517. doi:10.1103/PhysRev.128.2507

Baraff, G. A. (1964). Maximum anisotropy approximation for calculating electron distributions; Application to high field transport in semiconductors. *Physical Review, 133*(1A), A26–A33. doi:10.1103/PhysRev.133.A26

Bløtekjaer, K. (1970). Transport equations for electrons in two-valley semiconductors. *IEEE Trans, ED-17*(1), 38–47. doi:10.1109/T-ED.1970.16921

Bonitz, M., Kremp, D., Scott, D. C., Binder, R., Kraeft, W. D., & Kohler, S. (1996). Numerical analysis of non-Markovian effects in charge-carrier scattering: One-time versus two-time kinetic equations. *Journal of Physics Condensed Matter, 8*(33), 6057–6071. doi:10.1088/0953-8984/8/33/012

Brugger, S. C., Schenk, A., & Fichtner, W. (2006). Moments of the Inverse Scattering Operator of the Boltzmann Equation: Theory and Applications. *SIAM Journal on Applied Mathematics, 66*(4), 1209–1226. doi:10.1137/050633275

Brunetti, R. A. (1989). *New Model of Si for Charge Transport at High Electron Energy. NASECODE VI* (J. Miller, Ed.). Duplin.

Bude, J., & Hess, K. (1992). Threshold of impact ionization in semiconductors. *Journal of Applied Physics, 72*(8), 3554–3561. doi:10.1063/1.351434

Bude, J. D. (1995). Monte Carlo simulation of impact ionization feedback in sub-micron MOSFET technologies. *Proceedings of SSDM*, 228–230 doi:10.7567/SSDM.1995.B-4-1

Bude, J. D. (1995). Gate current by impact ionization feedback in sub-micron MOSFET technologies. *Proceedings of Symposium VLSI Tech.*, 101-102. doi:10.1109/VLSIT.1995.520877

Canali, C., Maijni, G., Minder, R., & Ottaviani, G. (1975). Electron and Hole Drift Velocity Measurements in Silicon and their Empirical Relation to Electric Field and Temperature. *IEEE Transactions on Electron Devices, ED-22*(11), 1045–1047. doi:10.1109/T-ED.1975.18267

Cassi, D., & Riccò, B. (1990). An analytical model of the energy distribution of hot electrons. *IEEE Transactions on Electron Devices, 37*(6), 1514–1521. doi:10.1109/16.106247

Caughey, D. M., & Thomas, R. E. (1967). Carrier mobilities in silicon empirically related to doping and field. *Proceedings of the IEEE, 5*(12), 2192–2193. doi:10.1109/PROC.1967.6123

Cerciganani, C. (1988). *The Boltzmann Equation and its Applications*. Springer. doi:10.1007/978-1-4612-1039-9

Chapman, S. (1916). On the law of distribution of molecular velocities, and on the theory of viscosity and thermal conduction, in a non-uniform simple monatomic gas. *Philosophical Transactions of the Royal Society of London, 216*(538-548), 279–348. doi:10.1098/rsta.1916.0006

Chen, D., Kan, E. C., Ravaioli, U., Shu, W.-C., & Dutton, R. W. (1992). An improved energy transport model including nonparaboilicty and non-Maxwellian distribution Effects. *IEEE Trans. Electron Dev. Lett., 13*(1), 26–28. doi:10.1109/55.144940

Cheng, M., & Kunhardt, E. E. (1990). A Theory of Nonequilibrium Carrier Transport in Multivalley Semiconductors. *Journal of Applied Physics, 67*(4), 1907–1914. doi:10.1063/1.345620

Chynoweth, A. G. (1958). Ionization rates for electrons and holes in Si. *Physical Review, 109*(5), 1537–1540. doi:10.1103/PhysRev.109.1537

Conwell, E. M. (1967). High Field Transport in Semiconductors. *Solid-State Physics*, (s9).

Cook, R., & Frey, J. (1982). An efficient technique for 2-D simulation of velocity overshoot effects in Si & GaAs devices. *COMPEL, 1*, 65–87. doi:10.1108/eb009966

Costato, M., & Reggiani, L. (1973). Electron energy relaxation time in Si and Ge. *Journal of Physics and Chemistry of Solids, 34*(3), 547–564. doi:10.1016/0022-3697(73)90050-4

Crowell, C., & Sze, S. M. (1966). Temperature dependence of avalanche multiplication in semiconductors. *Applied Physics Letters, 9*(6), 242–244. doi:10.1063/1.1754731

Egley, J., Polsky, B., Min, B., Lyumkis, E., Penzin, O., & Foisy, M. (2000). SOI related simulation challenges with moment-based BTE solvers. Simulation of Semiconductor Processes and Devices, Seattle, WA. doi:10.1109/SISPAD.2000.871253

El-Saba, M.H., Morel, H., & Chante, J-P. (1991). Simulation of Bipolar Transport in Semiconductor p-n Junctions Using the Generalized Hydrodynamic Equations. *COMPEL, 10*(4), 289-299.

El-Saba, M.H., Morel, H., & Chante, J-P. (1992). A Novel Discretization Scheme for the Integrated HDM/DDM simulation of Semiconductor Devices. In Computing Methods in Applied Science and Engineering. New York: Nova Science.

El-Saba, M.H., Morel, H., & Chante, J-P. (1992). Simulation of Band-structure Dependent Transport and Impact Ionization in Semiconductor P-N Junctions Using an Improved Multivalley Hydrodynamic Model. *Proceeding NASCODE 8*, 107-108.

El-Saba, M.H. (1993). *Modélisation et Simulation Hydrodynamique des phénomènes de Transport de porteurs chauds et de L'ionisation par Impact dans les Dispositifs à Semiconducteur* (Ph.D.). INSA Lyon, France, Order No.93 ISAL 0072.

El-Saba, M. H. (1996). Accurate Non-Local Modeling of Hot-Carriers Drift Mobility in Semiconductor Devices. *Proc. 8th Int. Conf. on Microelectronics-ICM.*

El-Saba, M.H. (1996). *Problems Related to the Semiconductors Hydrodynamic Model: Modeling the Carrier-Heat Flux Term*. Academic Press.

El-Saba, M. H. (1998). Accurate Estimation of Electron Velocity Overshoots in Sub-0.1 micron Silicon Structures and MOSFET Devices. *Proceedings of the 15th Radio Science Conference.*

El-Saba, M. H. (1999). Accurate modeling of energy-dependent impact ionization rate for hydrodynamic simulators of semiconductor devices. *Proceedings of the IEEE Radio Science Conference.*

El-Saba, M. H. (2006). Investigation of Hot Carrier Repelling Effect in Semiconductor Devices, Using an Analytical Solution of the Hydrodynamic Model. *IEEE Transactions on Electron Devices*, *53*(7), 1615–1622. doi:10.1109/TED.2006.876040

El-Saba, M. H. (2012). Yet another Hydrodynamic Model for Silicon Devices with Correlated Parameters, Scientific & Academic Publishers. *Microelectronic Devices and Solid-State Electronics*, *1*(5), 118–147. doi:10.5923/j.msse.20120105.03

Enskog, D. (1917). *Kinetische Theorie der Vorgänge in mässig verdünnten Gasen.* Uppsala Univertitet.

Ferry, D. K., & Barker, J. R. (1980). Balance equation formulation and memory effects in retarded high-field semiconductor transport. *J. Appl. Phys. Chem. Solids*, *3*, 106–141.

Fourier, J. (1822). Théorie Analytique de la Chaleur. Gabay.

Fröhlich, H. (1947). On the Theory of Dielectric Breakdown in Solids. *Proceedings of the Royal Society of London*, *188-A*(1015), 521–532. doi:10.1098/rspa.1947.0023

Giovanni, M. (2015). A hydrodynamic model for silicon semiconductors including crystal heating. *European Journal of Applied Mathematics*, *26*(04), 477–496. doi:10.1017/S0956792515000157

Gnudi, A., Ventura, D., Baccarani, G., & Odeh, F. (1993). Two-Dimensional MOSFET Simulation by Means of a Multidimensional Spherical Harmonics Expansion of the Boltzmann Transport Equation. *Solid-State Electronics*, *36*(4), 575–581. doi:10.1016/0038-1101(93)90269-V

Goldsman, N. Henrickson, L. & Frey, J. (1990). Reconciliation of a Hot-Electron Distribution Function with Lucky Electron Exponential Model in Silicon. *IEEE Electron Devices Lett., EDL-11*(10).

Goldsman, N., Henrickson, L., & Frey, J. (1991). A physics-based analytical/ numerical solution to the Boltzmann transport equation for use in device simulation. *Solid-State Electronics*, *34*(4), 389–396. doi:10.1016/0038-1101(91)90169-Y

Gonzalez, B., Palankovsk, V., Kosina, H., Hernandez, A., & Selberherr, S. (1999). An analytical Energy relaxation time model for device simulation. *Solid-State Electronics*, *43*(9), 1791–1795. doi:10.1016/S0038-1101(99)00132-X

Grad, H. (1949). On the kinetic theory of rarefied gases. *Communications on Pure and Applied Mathematics*, *2*(4), 331–407. doi:10.1002/cpa.3160020403

Grant, W. (1973). *Electron and hole ionization rates in epitaxial silicon at high electric fields*. Academic Press.

Grasser, T., Kosik, R., Jungemann, C., Meinerzhagen, B., Kosina, H., & Selberherr, S. (2004). A Non-Parabolic Six Moments Model for the Simulation of Sub-100nm Semiconductor Devices. *Journal of Computational Electronics*, *3*(3), 183–187. doi:10.1007/s10825-004-7041-1

Grasser, T., Kosina, H., Heitzinger, C., & Selberherr, S. (2002). *An impact ionization model including an explicit cold carrier population*. Modelling and Simulation of Microsystems.

Grasser, T., Kosina, H., & Selberherr, S. (2001). Influence of the distribution function shape and the band structure on impact ionization modeling. *Journal of Applied Physics*, *90*(12), 6165–6171. doi:10.1063/1.1415366

Grasser, T., Tang, T.-W., Kosina, H., & Selberherr, S. (2003). A Review of Hydrodynamic and Energy-Transport Models for Semiconductor Device Simulation. *Proceedings of the IEEE*, *91*(2), 251–274. doi:10.1109/JPROC.2002.808150

Gritsch, M., Kosina, H., Grasser, T., & Selberherr, S. (2002). Revision of the Standard Hydrodynamic Transport Model for SOI Simulation. *IEEE Transactions on Electron Devices*, *ED-49*(10), 1814–1820. doi:10.1109/TED.2002.803645

Hansch, W., & Miura-Mattausch, M. (1986). The Hot-Electron Problem in Small Semiconductor Devices. *Journal of Applied Physics*, *60*(2), 650–656. doi:10.1063/1.337408

Hartnagel, H., Katilius, R., & Matulionis, A. (2001). *Microwave Noise in Semiconductor Devices*. John Wiley & Sons.

Hess, K. (1985). Comment on effect of collisional broadening on Monte Carlo simulation of high-field transport in semiconductor devices. *IEEE Electron Device Letters*, *EDL-2*, 297–298.

Hess, K. (1989). Comment on effect of collisional broadening on Monte Carlo simulation of high-field transportin in semiconductor devices. *IEEE Electron Device Letters*, *EDL-2*, 297–298.

Hess, K. (2000). *Advanced Theory of Semiconductor Devices*. Piscataway, NJ: IEEE Press.

Higman, J., & Hess, K. (1986). Comment on the use of the electron temperature concept for non-linear transport problems in semiconductor p-n junctions. *Solid-State Electronics*, *29*(9), 915–918. doi:10.1016/0038-1101(86)90013-4

Jacoboni, C., & Reggiani, L. (1983). The Monte Carlo method for the solution of charge transport in semiconductors with application to covalent materials. *Reviews of Modern Physics*, *55*(3), 645–705. doi:10.1103/RevModPhys.55.645

Jerome, J. W. (2000). *Analytical and Computational Advances for Hydrodynamic Models of Classical and Quantum Charge Transport*. Academic Press.

Jungemann, C., Grasser, T., Neinhilus, B. & Meeinerzhagen, B. (2005). Failure of Moments-Based Transport Models in Nanoscale Devices Near Equilibrium. *IEEE Trans. Electron Devices, ED 52*(11).

Kan, E. C., Yu, Z., Datong, R.W., Chen, D., & Ravaioli, U. (1995). Formulation of Macroscopic Transport Models for Numerical Simulation of Semiconductor Devices. VLSI Design, 3(2), 211-224.

Katayama, K., & Toyabe, T. (1989). A new hot carrier simulation method based on full 3D hydrodynamic equations. *Technical Digest International Electron Devices Meeting (IEDM)*, 135-138.

Keldish, L. V. (1965). Concerning the theory of impact ionization in semiconductors. *Soviet Physics, JETP*, *21*, 1135–1144.

Kotani, N., & Kawazu, S. (1981). A Numerical analysis of avalanche breakdown in short-channel MOSFETs. *Solid-State Electronics*, *24*(7), 681–687. doi:10.1016/0038-1101(81)90199-4

Landau, L. D., & Lifshitz, E. M. (1997). Fluid Mechanics (2nd Ed.). Butterworth Heinemann.

Lee, C. A., Logan, R. A., Latdorf, R. L., Klimack, J., & Wiegmann, W. (1964). Ionization rates of holes and electrons in Si p-n junctions. *Physical Review A.*, *134*(3A), 761–773. doi:10.1103/PhysRev.134.A761

Levinstein, M., Rumyantsev, S., & Shur, M. (Eds.). (1996). Handbook Series on Semiconductor Parameters (Vol. 1). Singapore: World Scientific.

Lundstrom, M. (2000). *Fundamentals of Carrier Transport*. Cambridge Uni. Press. doi:10.1017/CBO9780511618611

Madureira, J. R., Semkat, D., Bonitz, M., & Redmer, R. (2001). Impact ionization rates of semiconductors in an electric field: The effect of collisional broadening. *Journal of Applied Physics*, *90*(2), 829–836. doi:10.1063/1.1381554

Maes, W., de Meyer, K., & Van Overstraeten, R. (1990). Impact ionization in silicon: A review and update. *Solid-State Electronics*, *33*(6), 705–718. doi:10.1016/0038-1101(90)90183-F

Matsuzawa, K., Kamohara, I., & Wada, T. (1991). *Device simulation including energy transport with improved Physical Models*. In J. J. Miller (Ed.), *NASECODE VII Trans* (pp. 173–174). Academic Press.

Matulionis, A., Liberis, J., Ardaravicius, L., Ramonas, M., Matulioniene, I., & Smart, J. (2002). Hot-electron energy relaxation time in AlGaN/GaN. *Semiconductor Science and Technology*, *17*(3), L9–L14. doi:10.1088/0268-1242/17/3/101

McAndrew, C. C., Heasell, E. L., & Singhal, K. (1987). A Comprehensive Transport Model for Semiconductor Device Simulation. *Semiconductor Science and Technology*, *2*(10), 643–648. doi:10.1088/0268-1242/2/10/003

McKay, K. G. (1954). Avalanche breakdown in Si. *Physical Review, 94*(4), 877–884. doi:10.1103/PhysRev.94.877

Meinerzhagen, B., & Engl, W. (1988). The Influence of the Thermal Equilibrium Approximation on the Accuracy of Classical 2-D Numerical Modeling of Silicon Submicrometer MOS Transistors. *IEEE Transactions on Electron Devices, ED-35*(5), 689–697. doi:10.1109/16.2514

Meinerzhagen, B., & Engl, W. L. (1988). The influence of the thermal equilibrium approximation on the accuracy of classical two-dimensional numerical modeling of silicon submicrometer MOS transistors, IEEE Trans. *Computer Aided Design, 35*(5), 689–697.

Morris, H. C., De Pass, M. M., & Abebe, H. (2004). Analytic formulae for the impact ionization rate for use in compact models of ultra-short semiconductor devices. *Proceedings Nanotechnology Conference*, 2, 140-143.

Nag, B. (1980). Electron Transport in Compound Semiconductors. Springer.

Ning, T. H., Osburn, C. M., & Yu, H. N. (1977). Emission probability of hot electrons from silicon into silicon dioxide. *Journal of Applied Physics, 48*(1), 286–293. doi:10.1063/1.323374

Nougier, J. P., Vaissiere, J. C., Gasquet, D., Zimmermann, J., & Constant, E. (1981). Determination of the Transient Regime in Semiconductor Devices using Relaxation Time Approximation,'. *Journal of Applied Physics, 52*, 825–832. doi:10.1063/1.328423

Peifer, H. J., Meinerzhagen, B., Thoma, R., & Engl, W. L. (1991). *Evaluation of impact ionization modeling in the framework of hydrodynamic equations. IEDM91 Tech.* Washington, DC: Digest.

Pop, E. (2004). *Self-Heating and Scaling of Thin Body Transistors* (Doctoral Thesis). Stanford University.

Pop, E., Sinha, S., & Goodson, K. E. (2006). Heat Generation and Transport in Nanometer-Scale Transistors. *Proceedings of the IEEE, 94*(8), 1587–1601. doi:10.1109/JPROC.2006.879794

Potz, W., & Kocevar, P. (1992). *Hot Carriers in Semiconductor Nanostructures: Physics and Applications* (J. Shah, Ed.). Boston: Academic Press.

Quade, W., Scholl, E., Rossi, F., & Jacoboni, C. (1994). Quantum theory of impact ionization in coherent high-field semiconductor transport. *Physical Review B: Condensed Matter and Materials Physics, 50*(11), 7398–7412. doi:10.1103/PhysRevB.50.7398 PMID:9974719

Quade, W., Schöll, E., & Rudan, M. (1993). Impact ionization within the hydrodynamic approach to semiconductor transport. *Solid-State Electronics, 36*(10), 1493–1505. doi:10.1016/0038-1101(93)90059-Y

Rees, H. D. (1973). Computer simulation of semiconductor devices. *Journal of Physics. C. Solid State Physics, 6*(2), 266–273. doi:10.1088/0022-3719/6/2/008

Rife, D.M. (2002). Article. *J. Optical Soc. America, vol. B, 19*, 1092.

Romano, V. (2001). Non-parabolic band hydrodynamical model of silicon semiconductors and simulation of electron devices. *Mathematical Methods in the Applied Sciences, 24*(7), 439-471.

Sabbah, A. J., & Riffe, D. M. (2002). Femtosecond Pump-Probe Reflectivity Study of Silicon Carrier Dynamics. *Physical Review B: Condensed Matter and Materials Physics*, *66*(16), 165217–165228. doi:10.1103/PhysRevB.66.165217

Sano, N., Aoki, T., Tomizawa, M., & Yoshii, A. (1990). Electron transport and impact ionization in Si. *Physical Review B: Condensed Matter and Materials Physics*, *41*(17), 12122–12128. doi:10.1103/PhysRevB.41.12122 PMID:9993667

Sano, N., & Yoshii, A. (1992). Impact–ionization theory consistent with a realistic band structure of silicon. *Physical Review B: Condensed Matter and Materials Physics*, *45*(8), 4171–4180. doi:10.1103/PhysRevB.45.4171 PMID:10002029

Schöll, E., & Quade, W. (1987). Effect of Impact Ionisation on Hot-Carrier Energy and Momentum Relaxation in Semiconductors. *Journal of Physics. C. Solid State Physics*, *20*(31), L861–L867. doi:10.1088/0022-3719/20/31/002

Scrobhaci, P.G., & Tang, T.-W. (1994). Modeling of the hot electron sub-population and its application to impact ionization in submicron silicon devices- PART I: Transport equations. *IEEE Trans. Electron Devices, 41*(7), 1197.

Seeger, K. (1989). *Semiconductor Physics, Berlin*. New York: Springer-Verlag.

Sermuknis, E., Liberis, J., & Matulionis, A. (2007). Microwave Noise Technique for Measurement of Hot-Electron Energy Relaxation Time and Hot-Phonon Lifetime. *Lithuanian Journal of Physics*, *47*(4), 491–498. doi:10.3952/lithjphys.47423

Shockley, W. (1961). Problems related to p-n junctions in silicon. *Solid-State Electronics*, *2*(1), 35–67. doi:10.1016/0038-1101(61)90054-5

Shur, M. S., & Eastman, L. F. (1979). Ballistic transport in semiconductor at low temperatures for low-power high-speed logic. *IEEE Transactions on Electron Devices*, *ED-26*(11), 1677–1683. doi:10.1109/T-ED.1979.19671

Slotboom, W., Streutker, G., van Dort, M. J., Woerlee, P. H., Pruijmboom, A., & Gravesteijn, D. J. (1991). *Non-local impact ionization in silicon devices. IEDM91 Tech.* Washington, DC: Digest.

Sonoda, K., Dunham, S. T., Yamaji, M., Taniguchi, K., & Hamaguchi, C. (1996). Impact ionization model using average energy and average square energy distribution. *Japanese Journal of Applied Physics*, *35*(2B), 818–825. doi:10.1143/JJAP.35.818

Souissi, K., Odeh, F., Tang, H., Gnudi, A., & Lu, P.-F. (1993). Investigation of the impact ionization in the hydrodynamic model. *IEEE Transactions on Electron Devices*, *40*(8), 1501–1507. doi:10.1109/16.223711

Stettler, M., Alam, M., & Lundstrom, M. (1993). A Critical Examination of the Assumptions Underlying Macroscopic Transport Equations for Silicon Devices. *IEEE Transactions on Electron Devices*, *40*(4), 733–740. doi:10.1109/16.202785

Stratton, R. (1962). Diffusion of hot and cold electrons in semiconductor barriers. *Physical Review*, *126*(6), 2002–2014. doi:10.1103/PhysRev.126.2002

Stratton, R. (1972). Semiconductor current-flow equations (diffusion and degeneracy). *IEEE Transactions on Electron Devices, 10*(12), 1288–1292. doi:10.1109/T-ED.1972.17592

Su, P., Goto, K-I., Sugii, T., & Hu, C. (2002). A thermal activation view of low voltage impact ionization in MOSFETs. *IEEE Electron Devices Letters, 23*(9),550-552.

Takayanagi, I., Matsumoto, K., & Nakamura, J. (1992). Measurement of electron impact ionization coefficient in bulk silicon under a low-electric field. *Journal of Applied Physics, 72*(5), 1989–1992. doi:10.1063/1.351625

Tang, J., Shichijo, H., Hess, K., & Iafrate, G. J. (1981). Band-Structure Dependent impact ionization in Si and GaAs. *Proceedings of the 3rd International Conf. on Hot Carriers in Semiconductors.*

Tang, T.-W., & Nam, J. (1998). A simplified impact ionization model on the average energy of hot-electron subpopulation. IEEE Electron Devices Lett., 19(6).

Tang, T.-W., Ramaswamy, S., & Nam, J. (1993). An Improved Hydrodynamic Transport Model for Silicon. *IEEE Transactions on Electron Devices, 40*(8), 1469–1477. doi:10.1109/16.223707

Thoma, R., Emunds, A., Meinerzhagen, B., Peifer, H.-J., & Engl, W. L. (1991). Hydrodynamic Equations for Semiconductors with Nonparabolic Band Structure. *IEEE Transactions on Electron Devices, ED-38*(6), 1343–1352. doi:10.1109/16.81625

Thornber, K. K. (1981). Application of scaling to problems in high-field electronic transport. *Journal of Applied Physics, 52*(1), 279–290. doi:10.1063/1.328490

Van Overstaten, R., & De Man, H. (1970). Measurement of the ionization rates in diffused Si p-n junctions. *Solid-State Electronics, 13*(5), 583–608. doi:10.1016/0038-1101(70)90139-5

Vasicek, M. (2009). *Advanced Macroscopic Transport Models* (Dissertation). Technische Universität Wien.

Wolff, P. A. (1954). Theory of multiplication in silicon and germanium. *Physical Review, 95*(6), 1415–1420. doi:10.1103/PhysRev.95.1415

Wolodkin, G., & Frey, J. (1992). Overshoot effects in the relaxation time approximation. *Proceedings NASCODE 8*, 107-108.

Yamaguchi, K. (1979). Field-Dependent Mobility Model for Two-Dimensional Numerical Analysis of MOSFETs. *IEEE Transactions on Electron Devices, 26*(7), 1068–1074. doi:10.1109/T-ED.1979.19547

Yao, C. S., Ahn, J. G., Park, Y.-J., Min, H.-S., & Dutton, R. W. (1995). Formulation of a tail electron hydrodynamic model based on Monte Carlo results. *IEEE Electron Device Letters, 16*(1), 26–29. doi:10.1109/55.363210

Ziman, J. (2001). *Electrons and Phonons.* Oxford, UK: Oxford University Press. doi:10.1093/acprof:oso/9780198507796.001.0001

ENDNOTES

1 We may multiply by $m^*\boldsymbol{u}_v$ in the momentum equation. However, m* is energy-band structure dependent and will complicate the integration procedure. In addition, we express the momentum as $p = \hbar k$, rather than $p = m^*\boldsymbol{u}_v$ in our semiclassical approach.

2 For this reason, we use a single symbol m_n for both the scalar mass and the mass tensor. Being a scalar or a tensor is understood from the form of other terms in the equation.

3 This means that the perturbed distribution function will relax exponentially to the equilibrium function with two time constants when the perturbing field is removed.

4 Mathematically speaking, the BTE can be expanded into an infinite number of its moments.

5 It should be noted that the Keldysh formula was developed assuming parabolic bands at zero electric field. However, the band structure of the material can be parameterized in the constants of this model.

6 The original Cromwell and Sze relation, the mean free path is related to the lattice temperature, rather than the carrier temperature.

7 The consistent way is solving the Kadanoff–Baym equations (equation of motion of none-quilibrium Green's functions NEGF), which can be used to construct broadened non-Lorentzian spectral functions, instead of the typical chronicle delta function, in the Fermi golden rule. The replacement of sharp (delta) spectral functions by broadened Lorentzian shape, have been found to be inadequate, as they violate the energy conservation rules (Bonitz et al, 1996).

Chapter 4
Quantum Transport Theory of Charge Carriers

1. OVERVIEW AND CHAPTER OBJECTIVES

Although the Schrödinger formalism is the de-facto standard, other possible approaches to quantum mechanics are available. The quantum transport theory, which is based on the statistical quantum mechanics, is the most rigorous method describing the transport phenomena in solids. The basis of the quantum transport theory is as strong as that of the statistical mechanics of systems in equilibrium, so that the BTE itself can be deduced from it! Indeed, different but mathematically equivalent formulations have eventually been developed, with their respective advantages and disadvantages, among which we have the significant works of Wigner, Feynman and Keldysh. Interestingly enough, these approaches are not based on the concept of wave-functions, but on rather different mathematical objects such as quasi-distribution functions (WDF), path integrals (Feynman), nonequilibrium Green functions (NEGF), and still they provide the very same predictions as the Schrödinger equation. In fact, these approaches are capable of handling both quantum-coherent propagation and dissipative scattering effects. Actually, there exist several approaches in the quantum transport theory, among them one can cite:

- Density matrix formulation,
- Wigner distribution function,
- Madelung transform,
- Feynman path integral,
- Green's Functions (Kadanoff-Baym & Keldish) approaches.

These quantum approaches may be different in mathematical formalism, but a transform is always available to go from formalism to another and vice versa. The hierarchy of quantum transport models, for both ballistic and dissipative systems, is shown in Figure 1.

DOI: 10.4018/978-1-5225-2312-3.ch004

Figure 1. Hierarchy of quantum transport and band structure models

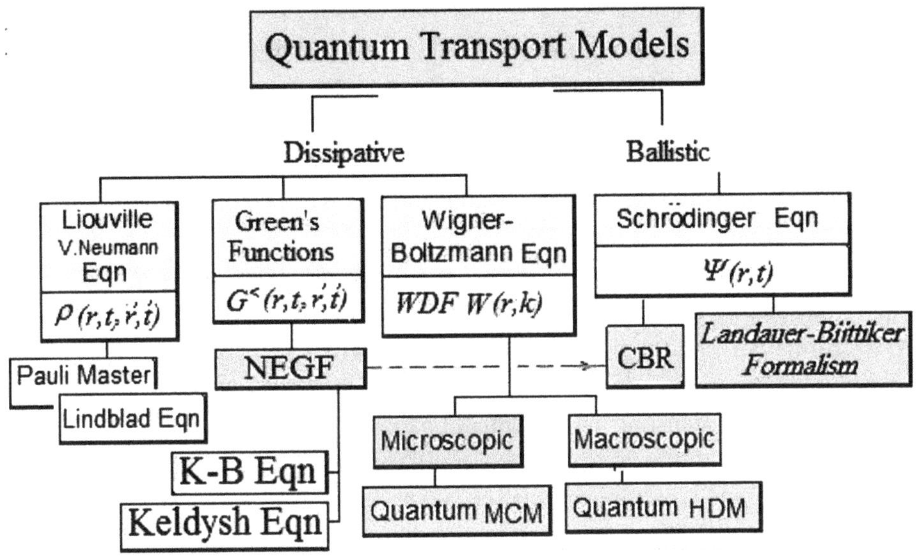

The Landauer-Büttiker approach describes the conductance of quantum system in the ballistic (collision-less) regime, where carriers are not scattered. The non-equilibrium Green's function (NEGF) formalism can handle open quantum systems as well as external contacts. The method can be used in both ballistic and a scattering dominated regime. Under ballistic conditions, the NEGF method is mathematically equivalent to solving the Schrödinger equation with open boundary conditions. There are two basic NEGF formulations, namely; the Kadanoff-Baym (K-B) approach and the Keldysh method. The K-B approach (1970) has been extended by Mahan and Hansch (1983). Nowadays this approach is sometimes referred to the quantum Boltzmann transport equation (QBTE). The so-called Contact-Block Reduction (CBR) method is an approximate method to solve the Schrodinger equation self-consistently with Poisson's equation in open systems (Mamaluy et al., 2005). The CBR may be considered as a variant of the NEGF for ballistic transport, using an equilibrium distribution function.

The Wigner Boltzmann transport equation (WBTE) can also be used in both ballistic and scattering dominated regimes. However, the WBTE is less or more similar to the NEGF method.

The quantum Liouville (or Liouville von Neumann) equation describes the quantum evolution of the density matrix and forms the fundamental equation of quantum transport. The so-called Pauli Master equation (PME) is derived from the quantum *Liouville* equation. The *Pauli* master equation is a frequently used to describe irreversible processes in quantum systems and can be also used in the ballistic and scattering-dominated regimes. The *Lindblad* master equation[1] is a generalized form of the *Liouville* equation. It describes the non-unitary evolution of the density matrix of a system.

In addition to the above microscopic quantum transport approaches, one can cite the macroscopic quantum approaches, such as the quantum hydrodynamic model (QHDM). All these transport models can be solved self-consistently with the Poisson equation, to predict the electrical (and sometimes, the thermal and optical) properties of semiconductor and nanodevices. In fact, these quantum approaches have many variants, and some of which are not only applicable for charge carriers, but also to phonon and photon transport. Finally, it worth notice that the quantum transport models can be solved using the stochastic Monte Carlo methods.

At the end of this Chapter, we also present the major band structure calculation models, such as the density functional theory (DFT), the tight binding (TB) model as well as the pseudopotential methods. The combination of the NEGF and the DFT in the so-called GW approximation (GWA) is also presented, because of its precision in predicting the energy gap of semiconductors.

Upon completion of this Chapter, the readers and students will be able to:

- Understand quantum nonequilibrium dynamics of distribution functions
- Understand the state-of-the-art quantum transport models, such as the Wigner distribution function (WBF) formalism and the nonequilibrium Green's function (NEGF), as well as the K-B and Keldysh formulations.
- Analyzing the coherent transport phenomena in nanodevices.
- Investigate the multiband transport models, on different size scales of semiconductor devices.
- Review the energy band structure calculation methods in bulk and molecular structures; both ab initio (such as H-F and DFT methods) and approximate models (such as TB, pseudopotential and GW methods).
- Know and decide when and how to utilize the solution techniques of the quantum transport models, using conventional numerical techniques and stochastic methods.

2. NEED FOR QUANTUM TRANSPORT

While the classical theory of transport is useful in understanding the motion of electrons and constructing Ohm's law, there are a number of inherent flaws in this theory. For instance, the resistivity of a metal, according to the classical theory, gives a value that is about seven times the measured value at 300K. There are also some flaws from the statistical perspective as well.

Applying the Boltzmann statistics and the Maxwell distribution of speeds, where electrons are viewed as particles will give an average kinetic energy of $3k_BT/2$. In addition, the molar heat capacity of a metal is expected to be constant ($C_v=3R$), according to Dulong-Petit law and doesn't change with temperature. Nevertheless, this is not observed experimentally, except for the range of ambient temperature. Also, the Wiedmann-Franz relation between the electrical and thermal conductivity is only verified in metals, because the role of phonons is neglected in the classical theory. Some of these problems were solved in the semiclassical transport approach, which is based on the Boltzmann-transport equation (BTE). In fact, solving the BTE for both electrons and phonons, and considering the crystal heat path of phonons could interpret some heat exchange problems.

However, there exist other problems, which are based on the wave-like nature of electrons, and the classical theory makes no mention of such properties. Among those problems, one can cite: the tunneling through thin energy barriers and the interaction phenomena between matter and electromagnetic waves, such as absorption and emission of light. In addition, there exist two significant transport phenomena in nanostructures, namely: ballistic transport and coherence phenomena.

Ballistic transport of electrons means collision-less motion, where the phase of the electron wavefunction is important. This enables us to observe the conduction interference effects. In fact, quantum interference influences so many characteristics of nanodevices. However, the phase of an electron wavefuction is destroyed by thermal fluctuations due to electron collisions. Recently, the study of electron current fluctuations and dephasing became a major topic of research. A great number of coherent phenomena have been studied, such as the nonequilibrium dynamics of optically created interband and intraband polarizations.

The dephasing time, τ_ϕ, of a given material characterizes how long a quasi-particle maintains its phase information. Loss of this phase information, or de-coherence, is caused by inelastic scattering or spin-flip events, which randomize the phase of the particle. The dephasing time is important because it determines the length on which the electron stays coherent and has the same phase. Actually, the most significant obstacle to build a quantum computer is de-coherence.

3. REVIEW ABOUT BASIC QUANTUM CONCEPTS

In order to explain black body radiation, Max Planck suggested (in 1900) that electromagnetic energy of radiation could only be emitted in quantized form, i.e. the energy could only be a multiple of elementary units (quanta) whose energy is proportional to the frequency of radiation

$$E = hf = \hbar\omega \tag{1}$$

where the proportionality factor h is the Planck constant and $\hbar = h/2\pi = 1.054571726\text{x}10^{-34}$ J.s According to this observation, Planck postulated that the emitted radiation is composed of bunches or packets of discrete quanta whose total energy is $E_n = n.hf = n.\hbar\omega_n$

The quantum idea was used by A. Einstein, in 1905, to interpret some experimental observation on the photoelectric effect. Historically, the important argument for the *necessity* of light quanta was given by Henri Poincare, in 1911 and 1912.

The idea that atoms are small solar systems was proposed in 1901, by Jean Perrin. In 1913, Niels Bohr used this idea, combining it with the Planck's quantization idea, to formulate his model for the hydrogen atom. Using some postulates, Bohr derived expressions for the discrete orbital radii of the hydrogen atom r_n and their energies E_n as well as the frequencies of absorbed or emitted light from an excited hydrogen atom. However, while the Bohr model was a good step towards understanding the quantum theory of atoms, it is not in fact a correct description of the nature of electron orbitals. In addition, it fails to provide any justification of why certain spectral lines are brighter than others. There is no mechanism for the calculation of transition probabilities in this model.

Until 1924, it seemed that matter has a particular nature. Louis de Broglie suggested in his doctoral dissertation in 1923, that all forms of matter have wave as well as particle properties, just like light. The wavelength, of a particle is related to its momentum, by the same relationship of photons:

$$\lambda = h / p \tag{2}$$

where p is the momentum of the particle and λ is the wavelength of the associated wave. It follows form this relation that the momentum $\boldsymbol{p}$ of a particle is related to the associated wave vector $k = 2\pi/\lambda$ by:

$$p = \hbar k \tag{3}$$

The de-Broglie hypothesis was confirmed experimentally three years later by Davisson and Germer in 1927 and later by G. Thomson in 1928.

In 1927, Werner Heisenberg introduced the uncertainty principle which states that: "It is impossible to specify precisely and simultaneously the value of both members of particular pairs of physical variables that describe the behavior of an atomic system". According to Heisenberg, we have:

$$\Delta p_x \cdot \Delta x \geq \hbar / 2, \Delta E \cdot \Delta t \geq \hbar / 2 \tag{4}$$

The Heisenberg uncertainly principal (Heisenberg, 1927) became an important issue in all disciplines, which involve extremely small dimensions.

Since particles have a wave nature, they can be described by wave functions. Our knowledge of a particle wave in a specific region of a system is defined by its eigenfunctions $\Psi(r,t)$. As we'll see soon, the wave function may be found by solving the Schrödinger equation. Once the boundaries of the system are known, then the probability of finding the particle somewhere (between r_1, r_2) is calculated as follows (Born, 1926):

$$P\left(r_2 < r < r_1\right) = \int_{r1}^{r2} \Psi\Psi^* d^3r \tag{5}$$

The square root of the probability density is called the probability amplitude:

$$A = \left|\Psi \cdot \Psi^*\right|^{1/2} = \left|\Psi^2\right|^{1/2} \tag{6}$$

The Copenhagen interpretation of quantum mechanics shows that the above modulus of square of the wave function gives the *probability density function (PDF)* of a quantum system.

Note 1: Quantum Operators

Quantum operators are mathematical instructions which replace the classical mechanical variables, such as position, momentum and time. For instance, the classical momentum component $p_x = m.v_x$ is replaced with the operator $j\hbar\, \partial/\partial x$. The momentum vector $\boldsymbol{p} = m.\boldsymbol{v}$ is replaced with $\hat{p} = j\hbar\nabla$. Also the total energy E is replaced with $\hat{E} = j\hbar\, \partial/\partial t$. The so-called Hamiltonian operator is the sum of kinetic and potential energy of a system: $\hat{H} = \hat{p}^2/2m + V(r) = -(\hbar^2/2m).\nabla^2 + V(r)$. Note the cap '^', which is often used to distinguish operators from their classical counterpart variables. Every linear operator $\hat{O}$ can be represented as a matrix O. The *Hermitian operators* in quantum mechanics are physically observable quantities such as $\hat{E}$ and $\dot{p}$. An *observable* quantity is a quality of a particle that can be observed or measured in the laboratory.

In quantum mechanics, the most famous differential equation that describes the position probability of a particle is known as the Schrödinger equation:

$$j_{\hbar} \frac{\partial \Psi}{\partial t} = \left[\frac{\hbar^2}{2m} \nabla^2 + V(r) \right] \Psi \tag{7}$$

where *V(r)* represents the acting potential energy in which the particle is moving. This is the time-dependent Schrödinger, which governs the time evolution of the wavefunction ψ on the configuration space. This relation can be rewritten in a simple form using the Hamiltonian operator $\hat{H}$ as follows:

$$j\hbar\, \partial\Psi/\partial t = \hat{H}\Psi \tag{8}$$

where the Hamiltonian differential operator $\hat{H}$ represents the total energy of the system. and given by:

$$\hat{H} = \left(\frac{p^2}{2m} + V(r) \right) = \left[\frac{-\hbar^2}{2m} \nabla^2 + V(r) \right] \tag{9}$$

The Schrodinger equation can be also written in the following form:

$$\hat{E}\,\Psi = \hat{H}\,\Psi \tag{10}$$

where $\hat{E}$ is the total energy operator ($\hat{E} \to j\hbar.\partial/\partial t$). If the wave function $\Psi(r, t)$ is monochromatic, it may be written as follows:

$$\psi(r,t) = \psi(r) \exp\left(-j \frac{Et}{\hbar} \right) \tag{11}$$

Therefore, the Schrödinger equation may be written in the following time- independent form:

$$\frac{\hbar^2}{2m} \nabla^2 \psi + \left[E - V(r) \right] \psi = 0 \tag{12}$$

Here, the potential energy *V(r)* is assumed to be only space-dependent.

It should be noted that the basic Schrödinger equation is based on certain assumptions, which are sometimes drastic. Among these assumptions one can cite:

- The phenomena of particles generation-recombination are neglected.
- The particles are assumed to move with velocities much smaller than the light velocity. The, particles are described in non-relativistic manner.

For time-independent Hamiltonians the solution to Schrödinger equation (SE) can be cast as an eigenvalue problem. This allows us to solve SE in some cases exactly. Therefore, the solution of the time-independent SE for a particular system gives the sets of eigenfunctions (ψ_n) of the Hamiltonian $\hat{H}$ and the corresponding eigenvalues E_n such that:

$$E_n \Psi_n \hat{H} \Psi_n \tag{13}$$

The starting point in the solution is to determine the Hamiltonian operator of the system for which we seek the wave functions. The potential energy *V(r)* of the system will specify the physical problem in hand. For an electron in an isolated hydrogen atom the potential energy is simple as $V(r) = -e^2/(4\pi\varepsilon_o r)$, wherever for a crystal it is a complicated periodic function.

As a matter of fact, only few problems in quantum mechanics can be solved exactly. Analytical solutions are only available in special cases, as for constant or linear potentials and potential steps. Figure 2 shows the approximate methods to solving the Schrödinger equation (SE). As shown, there exist many approximations to solving this equation, among which the adiabatic approximation is the most widely used.

Before going so far, we would like to describe a number of mathematical ideas that are used a lot in the literature of quantum & semiclassical physics. In Dirac notations, the eigenfunction ψ_j is considered as a vector and represented by the symbol *ket* $|j\rangle$ whereas its complex conjugate ψ_j* is called co-vector, and represented by the symbol *bra* $\langle j|$. The inner (scalar) product of two functions ψ_i andψ_j is produced by taking the product of a *bra* and a *ket* to produce a *bra(c)ket* $\langle i|j\rangle$.

$$\left(\psi_i, \psi_j\right) = \int_{position\ space} \psi_i^* \psi_j d^3x = \int_{momentum\ space} \psi_i^* \psi_j d^3p \rightarrow \langle i|j\rangle \tag{14a}$$

Figure 2. Different methods for solving the Schrödinger equation

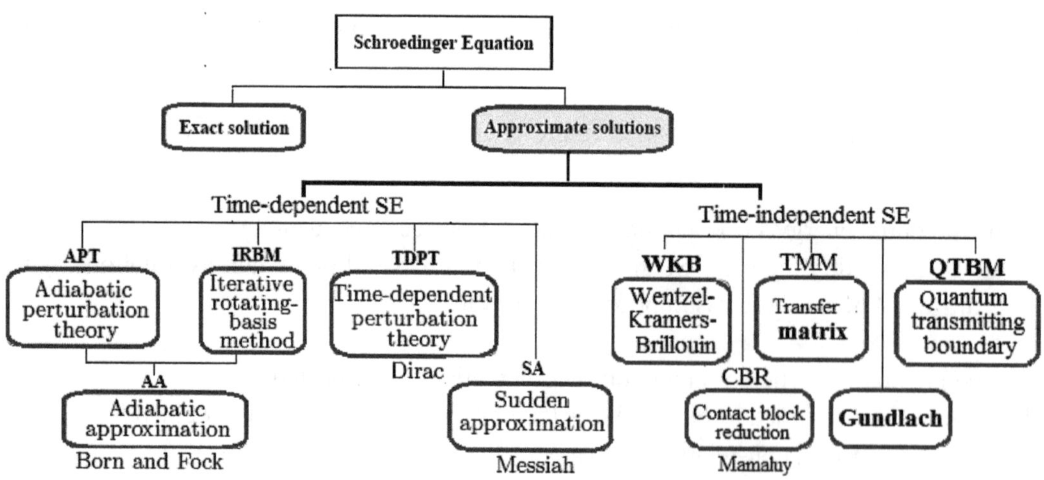

Therefore, the bracket $\langle i | j \rangle$ is equal to the inner (scalar) product of the vector ψ_i* and the co-vector ψ_j*. The bracket $\langle i | j \rangle$ is also a sign equivalent to: the probability amplitude of transition. The expression in the right of vertical line is always the starting condition (start state i) and the expression in the lift of the vertical line in the final condition (final state j). It should be noted that Dirac's notations are valid for both the position-dependent and the momentum-dependent wave functions. Therefore, we simply write ψ, rather than ψ(x,t) or ψ(p,t).

According to the Dirac notations, the average (expectation) value of an observable operator $\hat{A}$ is given by:

$$\langle \psi | \hat{A} | \psi \rangle = (\psi, \hat{A}\psi) = \langle \hat{A} \rangle \tag{14b}$$

Finally, it should be noted that the wave functions lives in a well-defined space, called the Hilbert space, which encompasses the sets of basis vectors. Mathematically speaking, a Hilbert space is a vector space of functions with a defined inner (scalar) product. Any function *f(x)* in the Hilbert space can be represented as an abstract vector $| f \rangle$ which may be expanded as a series of basis functions or components as follows:

$$| f \rangle = \sum_{i=0}^{\infty} c_i \phi_i \tag{15a}$$

The jth component of this expansion (*cj*) can be found by operating the bra $\langle j |$ as follows:

$$c_j = \langle j | f \rangle = \langle \phi_j | f \rangle \tag{15b}$$

4. QUANTUM REPRESENTATION PICTURES

In quantum mechanics there exist three basic representation methods or pictures, namely: Schrödinger picture, Heisenberg picture and the interaction (or Dirac) picture

The Heisenberg picture is a formulation of quantum mechanics in which the operators (observables and others) evolves in time, but the state vectors are time-independent. It stands in contrast to the Schrödinger

Table 1. Quantum pictures; the subscripts '$_S$' and '$_H$' mean Schrödinger and Heisenberg pictures, respectively

Schrödinger Picture	Heisenberg Picture	Interaction (Dirac) Picture
$\psi_S(t) = \psi_S(0).exp(-jHt)$	$\psi_H(t) = \psi_S(0)$	$\psi(t) = \psi_S(t).exp(-jHt)$
State $|\psi\rangle$ evolves in time	State $|\psi\rangle$ is fixed	State $|\psi\rangle$ evolves in time
Operators are fixed	Operators evolve in time	Operators evolve in time

picture in which the operators are fixed, and the states evolve in time. However, Schrödinger operators may come in two types; time-independent ones (like $\dot{x}$, $\dot{p}$) and time-dependent ones (like Hamiltonians with time-dependent potentials). For each of those types of operators we can associate a Heisenberg operator.

The Heisenberg picture is the formulation of matrix mechanics in an arbitrary basis, in which the Hamiltonian is not necessarily diagonal. It further serves to define a third, hybrid picture, formulated by Paul Dirac, called the Interaction picture. A detailed explanation of the three pictures is presented in text Books of Quantum Books, like Messiah (1963)

5. CRYSTAL SYSTEM HAMILTONIAN

Now, let's conclude our review by defining the Hamiltonian for the entire problem of solid state physics. It consists of the kinetic energy of all particles in the solid and of their interaction (potential) energies.

$$\hat{H} = H_{el} + H_{ion} + H_{e\text{-}i} + H_{ext} \tag{16}$$

where we omitted the operators cap sign for the sake of simplicity. For a non-perturbed system we have $H_{ext} = 0$. For electrons we have:

$$\mathrm{H_{el} = H_e\ (K.E.) + H_{e\text{-}e}\ (electron\text{-}electron\ interaction)} \tag{17a}$$

$$= \sum_k \frac{p_k^2}{2m} + \frac{1}{4\pi\varepsilon_o} \sum_{k,k`} \frac{e^2}{\left|r_k - r_{k`}\right|}$$

where r_k and $r_{k`}$ are the coordinates of the $k^{th}$ and $k`^{th}$ electrons, respectively. Similarly, for ions, we have:

$$\mathrm{H_{ion} = H_i\ (K.E.) + H_{i\text{-}i}(ion\text{-}ion\ interaction)} \tag{17b}$$

$$= \sum_i \frac{P_i^2}{2M_i} + \frac{1}{2}\sum_{i,j} V_i\left(R_i - R_j\right)$$

where R_i and R_j are the coordinates of the ith and j^{th} ions, respectively. Concerning the electron-ion interaction energy, it may be written as:

$$H_{e-i} = \sum_{k,i} V_{e-i}\left(r_k - R_i\right) \tag{18a}$$

The ion-ion interaction term $H_{i\text{-}i}$ may be subdivided into two terms; the first accounts to the interaction when the ions are in their equilibrium positions and the other to account for the vibrations of lattice ions (phonons).

$$H_{i-i} = H^{o}_{i-i} + H_{ph}, H_{e-i} = H^{o}_{e-i} + H_{e-ph} \tag{18b}$$

Unfortunately it is not always possible to rigorously solve the quantum mechanical problem. Approximations have to be made. In solid state theory, two simplifications are commonly made. In solving a given problem individual terms of the Hamiltonian may be neglected or only partially considered or handled subsequently as perturbations.

5.1 Adiabatic Approximation

The atoms in a crystalline solid are vibrating around their equilibrium positions, but their motion is so slow with respect to the speed of electrons. Therefore, atoms may be regarded as stationary, and decoupled from the study of electron transport. On this basis, one can divide the solid state problem into two components:

- Movement of electrons (electron gas) in a stationary lattice, and
- Movement of ions in a uniform space-charge of electrons.

This is the basis of the so-called Born-Oppenheimer *(adiabatic)* approximation, which is often used to provide a justification for system decoupling (Born & Oppenheimer, 1927). The Born-Oppenheimer approximation takes the wavefunction of an atom or molecule to be the product

$$\Psi = \psi_{electrons} \cdot \psi_{nuclei} \tag{19}$$

This product consists of independent wavefunctions for the electrons and the nuclei. The electron wavefunction $\psi_{electrons}$ obeys an approximate many-electron Schrödinger equation with Hamiltonian H_{el}, as given below.

5.2 Jellium Model

According to the adiabatic approximation, the many-electron Schrödinger equation Hamiltonian H_{el} can be written as follows:

$$H_{el} = \sum_{k} \frac{\hbar^2}{2m} \nabla_k^2 + \frac{1}{8\pi\varepsilon_o} \sum_{k,k`} \frac{e^2}{\left| r_k - r_{k`} \right|} + H^{+ve} \tag{20a}$$

where H^{+ve} includes the effect of the ions positive space charge and the interaction of the electrons with this space charge. The electron gas is considered here to be embedded in a positive charge background. On the other hand, the movement of ions can be treated in a similar way. This approach is usually called the Jellium model. Many of the properties of solids can be described by mean of this approximation. In this way, the Hamiltonian, H_{ion} can be written as:

$$H_{ion} = \sum_i \frac{\hbar^2}{2M_i} \nabla_i^2 + \frac{1}{2} \sum_{i,j} V_i \left(R_i - R_j\right) + H^{-ve} \tag{20b}$$

where H^{-ve} includes the effect of negative space charge of electrons and the interaction of this charge with ions. It should be noted that the two terms have H^{-ve} and H^{+ve} exactly cancel each other. As far as the motion of electrons is concerned, we can adopt a Schrödinger equation for electrons, like this:

$$\left(H_{el} + H_{e-i}\right)\psi = E_e \psi \tag{21}$$

Here H_{e-i} is considered with fixed coordinates of ions. The remaining coupling between electrons and ions vibration is the interaction H_{e-ph}. This coupling can be dealt with in most cases by the perturbation theory.

5.3 Hartree Equation (One-Electron Approximation)

In the absence of electron-electron interactions, the electron gas problem (Jellium model) could be de-coupled to one-electron problems, which describe the motion of individual electrons in a given potential field. In this case, the Hamiltonian term which accounts for the electron-electron interaction is neglected, and we can write the system Hamiltonian as follows:

$$\hat{H} = -\sum_k \frac{\hbar^2}{2m} \nabla_k^2 + \sum_k V\left(r_k\right) = \sum_k H_k \tag{22}$$

Then, the Schrödinger equation can be reduced, by substituting:

$$\sum_k H_k \psi = E\psi \tag{23a}$$

and

$$\psi\left(r_1..r_N\right) = \psi_i\left(r_1\right)\psi_2\left(r_2\right)...\psi_N\left(r_N\right) \tag{23b}$$

into the following one-electron equations:

$$H_k \psi_k\left(r_k\right) = E_k \psi_k\left(r_k\right) \tag{24}$$

Therefore, the many-body problem (the electron gas which is embedded in a homogenous, positively charged medium) could be reduced to one-electron problem, which includes an energy term to account for the electron interaction. Such a reduction is achieved by the Hartree (or Hartree-Fock) approximation. In this approximation, the Schrödinger equation for a single particle writes (Hartee, 1928; Fock, 1930):

$$\left[-\frac{\hbar^2}{2m}\nabla^2 + V_H(r)\right]\psi_k(r) = E_k\psi_k(r) \tag{25a}$$

where V_H is the Hartree potential, which is given by:

$$V_H(r) = V(r) + \frac{e^2}{4\pi\varepsilon_o}\sum_{j\neq k}\int\frac{\left|\psi_j(r')\right|^2}{\left|r-r'\right|}dV \tag{25b}$$

Here r and r` are the positions of the ith and k^{th} electrons, respectively. The above equation is called the Hartree equation. It describes the motion of the kth electron at the position r in the potential field $V(r)$ of the lattice ions, and in the Coulomb potential of an average distribution of all other electron: ($j=1, 2, .., N$, with $j\neq k$)

It should be noted that the Hartree equation does not account for the Pauli Exclusion Principle (Pauli, 1925). The spin of electrons is not considered in this equation. If two electrons are interchanged, the sign of the total wave function will remain unchanged or

$$\psi = \psi_1(r_1)....\psi_N(r_N) \tag{26}$$

This *Hartree product* representation may be suitable for Boson particles[2]. However, electrons belong to another type of particles, called Fermions, which have a spin and obey Pauli Exclusion Principle. This Principle states that no two electrons can have the same state and if they occupy the same energy level, they should have different spins. Therefore, the total wave function ψ must change its sign when two electrons are interchanged.

5.4 Hartree-Fock Equation

In order to account for the Pauli-exclusion principle we replace the Hartree product of the total electronic wave function ψ by the following Slater determinant (Slater, 1937), which satisfies the Pauli-exclusion principle:

$$\psi = \frac{1}{\sqrt{N!}}\begin{vmatrix} \psi_1(q_1) & .. & \psi_N(q_i) \\ . & & . \\ . & & . \\ \psi_i(q_N) & .. & \psi_N(q_N) \end{vmatrix} \tag{27}$$

Here, the coordinates q_k include the spatial coordinates r_k as well as the spin coordinates s_k of electrons along the quantization direction (typically chosen as the z-axis). The factor in front of the determinant is added for normalization purposes. If two electrons are interchanged, two columns of the determinate are interchanged, and ψ changes sign. Also if two electrons have the same coordinates, two columns will be identical, and ψ will vanish. Now the interaction term accounts for the Pauli exclusion term and the Schrödinger equation reads:

$$\left[-\frac{\hbar^2}{2m}\nabla^2+V(r)+\frac{e}{4\pi\varepsilon_o}\int\frac{\rho(r')-\rho^{-HF}(r,r')}{|r-r'|}dV'\right]\psi_k(r)=E_k\psi_k(r) \tag{28a}$$

where

$$\rho(r')=\sum_k\rho_k^H=\sum_k -e|\psi_k(r)|^2 \tag{28b}$$

and

$$\rho^{-HF}=-e\frac{\sum_{j,k}\psi_j^*(r`)\psi_k(r`)\psi_k^*(r)\psi_j(r)}{\sum_k\psi_k^*(r)\psi_k(r)} \tag{28c}$$

The above form of Schrödinger's equation is called the Hartree-Fock equation. Also, the following potential is called Hartree-Fock potential.

$$V'(r)=V(r)-\frac{e}{4\pi\varepsilon_o}\int\frac{\rho(r')-\rho^{-HF}(r,r')}{|r-r'|}dV' \tag{29}$$

The Hartree-Fock equation may be written in the following simple form:

$$\left[-\frac{\hbar^2}{2m}\nabla^2+V'(r)\right]\psi_k(r)=E_k\psi(r) \tag{30}$$

The set of Hartee-Fock equations forms a quite complicated nonlinear integro-differential system. It can be shown that *V`(r)* in a crystal lattice is a periodic function of lattice coordinates, *V(r)=V(r+**R**)*, where *R* is the lattice vector. This feature is exploited in the so-called Bloch functions that facilitate the manipulation of the problem. The post Hartee-Fock methods, like the tightly-bound (TB) model and the density functional theory (DFT), have been developed in such the way that they either neglect or approximate some of the involved integrals in the Hartree-Fock model.

6. DENSITY MATRIX FORMULATION

The quantum transport theory is based on the concept of the density operator (usually termed by $\hat{\rho}$). The density operator is an alternative way of representing quantum states. It is the quantum-mechanical analogue to the distribution function in classical statistical mechanics. The need for a statistical description via a density operator (or density matrix) arises because it is not possible to describe a quantum system that undergoes several measurements, using exclusively its eigenstates. If this system consists of a huge number of interacting atoms in a field, we should solve for a huge system of wavefunctions and probability amplitudes, which is beyond the capability of any computing platform.

Unlike the Schrödinger picture, the density matrix does not require a complete wavefunction for a system and we can treat a huge number of atoms interacting with a field via statistical means. This makes the density matrix appealing in such situations. The density operator for a quantum system having m vectors (or *kets* $|\psi_m>$) of independent pure states, is defined as follows:

$$\hat{\rho}(t) = \sum_m P_m \left|\psi_m\right\rangle\left\langle\psi_m\right| \tag{31a}$$

where P_m is the probability of finding the system in one of its pure m states. Note that a state described by a state vector $|\psi_m\rangle$ is called a *pure state*. In general, a system is said to be in a pure state, when the state is not reducible to a convex combination of other states. The density matrix is obtained from the density operator through the following relationship:

$$\rho_{ij} = \left\langle\varphi_i\left|\hat{\rho}\right|\varphi_j\right\rangle \tag{31b}$$

where ϕ_{ij} is a complete orthonormal basis set for the vector $|\psi_m>$ such that

$$\left|\psi_m\right\rangle = \sum_{\text{i}} c_{im}\left|\varphi_i\right\rangle \tag{31c}$$

Here the coefficients c_{im} are the probability amplitudes of the mth state. For the matter of demonstration, consider a system with a set of orthonormal orbitals $|\psi_m|$ and occupation numbers P_m (the probabilities to find the system in one of its eigenstates).The density-matrix in the coordinate representation is then given by:

$$\rho(r,r') = \left\langle r\left|\hat{\rho}\right|r'\right\rangle = \sum_m P_m\psi_m(r)\cdot\psi_m^*(r') \tag{32a}$$

The elements on the diagonal of the density matrix $\rho(r, r)$ are real numbers which correspond to the probability of being in one of the system eigen-states (pure states). That's to say the probability to find a particle at a certain point r, in the physical space of a single-particle system, is given by:

$$P(r) = \rho(r,r) = \langle r|\hat{\rho}|r\rangle = \sum_{m} P_m |\psi_m(r)|^2 \tag{32b}$$

The trace of the density matrix ($Tr\{\rho\}$ which is the sum of the diagonal elements divided by the number of particles, should be unity by the normalization criterion.

$$n = \Sigma_i \rho_{ii} \tag{33}$$

The off-diagonal elements, on the other hand, are complex numbers whose meaning is related to the phase of the wave-function. When external forces are removed, the density matrix is reduced to a diagonal matrix with diagonal elements equal the equilibrium distribution functions (e.g., the Fermi-Dirac distribution) at the system possible states. The thermal equilibrium density matrix in the quantum mechanical setting is given by:

$$\rho_{eq} = exp\,(-H / k_B T) \tag{34a}$$

For instance, the density matrix is given by the following form (for a system of n- pure states: *1,2,..n*)

$$\rho_{eq} = \begin{bmatrix} f(E_1) & & \\ & \vdots & \\ & & f(E_n) \end{bmatrix} \tag{34b}$$

It is worth noting that the mean value of any observable *A*, is given by:

$$\langle A \rangle = \sum_{i,j} c_{ij}^{*} A_{ij} c_{jk} = Te\{\rho A\} \tag{35}$$

Also, the entropy of a quantum system (Von-Neumann entropy), which is a measure of disorder or lack of knowledge of a system, is given by:

$$S = -k_B\, Tr\,\{\, \rho\, ln\, \rho \,\} \tag{36a}$$

In a diagonal representation of the density matrix, the entropy ***S*** reads:

$$S = -k_B\, \Sigma\, p_i\, ln\, p_i \tag{36b}$$

Therefore, the density matrix ρ is sufficient to describe the state of any quantum system.

7. QUANTUM LIOUVILLE'S EQUATION

The quantum Liouville equation (which is sometimes called the Liouville-von Neumann or mixed-state Schrödinger's equation) describes the temporal evolution of the density operator. It can be obtained by substituting the definition of the density operator, into the Schrödinger equation ($E\Psi = \hat{H}\Psi$) or:

$$-j\hbar\frac{\partial\left|\psi_k\right\rangle}{\partial t} = \widehat{H}\left|\psi_k\right\rangle \tag{37a}$$

Therefore, one can immediately obtain the equation of motion of the density matrix:

$$-j\hbar\frac{\partial\hat{\rho}}{\partial t} = \left[\widehat{H},\hat{\rho}\right] \tag{37b}$$

where $\hat{\rho} = \hat{\rho}(r,r',t)$ and the commutator is given by:

$$[\,\hat{H},\,\hat{\rho}\,] = \left[-\frac{\hbar^2}{2m}\left(\frac{\partial^2}{\partial r^2} - \frac{\partial^2}{\partial r'^2} + V(r) - V(r')\right)\right]\hat{\rho} \tag{38}$$

In order to simplify the equation, the limit of weak interaction is applied. Coarse-graining in time results in a Markovian system. The resulting quantum Liouville equation for reduced density matrix describes the quantum evolution of a system (or a subsystem) with losses (collisions).

$$-j\hbar\frac{\partial\hat{\rho}}{\partial t} = \left[\widehat{H},\hat{\rho}\right] + L\{\hat{\rho}\} \tag{39}$$

where $L\{\rho\}$ represents the losses (collision) operator. Since the density matrix is defined in position space, it is difficult to include scattering events in the above formalism, which is usually described in the momentum (***p*** or ***k***) space. It should be noted that the quantum Liouville equation (39) which describes the evolution of the density operator, together with (35) which describes the average value of any physical operator, represent the formal quantum transport theory in its compact form.

Under certain conditions the system which is initially in a non-equilibrium state $|\alpha\rangle$ will go over into an equilibrium state after some time, through a relaxation process, such that:

$$-j\hbar\frac{\partial\hat{\rho}}{\partial t} = \left[\widehat{H},\hat{\rho}\right] - j\hbar\left(\frac{\partial\hat{\rho}}{\partial t}\right)_{res} - j\hbar\left(\frac{\hat{\rho}-\hat{\rho}_{eq}}{\tau_s}\right) \tag{40}$$

where we added a collision term describing the interaction with the electrical contacts (reservoir) for open systems. The last relaxation term on the RHS of (40) expresses the weak interaction with the surroundings.

Note 2: Commutators

The commutator of two quantum operators $\hat{A}$ and $\dot{B}$ is defined as follows: $[\hat{A},\dot{B}] = \hat{A}.\dot{B} - \dot{B}.\hat{A}$. Therefore, $[\hat{A},\dot{B}] = -[\dot{B},\hat{A}]$.

When $[\hat{A},\dot{B}] = 0$, such that $\hat{A}.\dot{B} = \dot{B}.\hat{A}$, the two operators are said to be commuted (Born, 1925). If the two operators are canonically conjugated (one of them is the Fourier transform of the other, like $\dot{x}$, $\dot{p}$), then they are not commuted and their canonical commutator is given by: $[\hat{A},\dot{B}] = j\hbar$

The Poisson bracket is the classical version of the commutator in quantum mechanics. The Poisson bracket (or Lagrange) of two variables A, B is defined as:

$$\{A, B\} = \Sigma_i (\partial A/\partial x_i)(\partial B/\partial p_i) - (\partial A/\partial p_i)(\partial B/\partial x_i)$$

7.1 Resemblance with the BTE

Equation (39) resembles the semiclassical BTE for a system of N particles, by replacing the commutator [,] with the Poisson bracket or Lagrange **{, }**

$$\partial f/\partial t = \{ \mathrm{H}, f\} \tag{41}$$

where H is the classical Hamiltonian and f is the carrier distribution function. In fact, one can prove that $f(r,t)$ is nothing but the diagonal elements of the density matrix (i.e., $< r \mid \rho \mid r >$). Hence, the diagonal terms of the density matrix represent the carrier density variation throughout the system, whereas the off-diagonal parts represent the spatial correlation's that exist in the system. For the above reasons, we refer to the quantum Liouville (or Liouville-von Neumann) equation as the quantum BTE (QBTE).

7.2 Heterostructure Hamiltonian

The Hamiltonian of a homogeneous system of electrons, with constant effective mass m^* and subjected to an external electric field $\zeta(r)$, is given by:

$$\hat{H} = \left[-\frac{\hbar^2}{2m^*} \nabla^2 + V\left(r\right) \right] \tag{42a}$$

where the effective mass m^* lumps the internal lattice field and the potential energy $V(r)$ is expressed in terms of the electrostatic potential $\phi(\mathrm{r})$ as:

$$V(r) = e \int \zeta(r).dr = -e\phi(r) \tag{42b}$$

We dully note that the effective mass of carriers can be deduced from the energy band structure $E(k)$, using the relation ($m^* = \hbar^2/d^2E(k)/dk^2$). We also know, that the effective mass approximation reflects the expansion of the $E(k)$ relation near extreme points (top of valence band or bottom of conduction band) and hence near equilibrium.

For heterostructures, where the energy band structure, $E(k)$, may have discontinuities in space, $\Delta E(r)$, and the effective mass is space dependent, $m^{*}= m^{*}(r)$, the Hamiltonian may be written as follows:

$$\hat{H} = \left[-\frac{\hbar^2}{2} \nabla \left(\frac{1}{m^*(r)} \nabla \right) + e\varphi(r) + \Delta E_{co}(r) \right] \tag{42c}$$

where ΔE_{co} includes the variation of the conduction band edge, due to material inhomogeneity (heterostructure), at zero external bias.

7.3 Pauli Master Equation (PME)

The so-called Pauli master equation (PME) describes the evolution of the diagonal elements of the density matrix in nanodevices. The PME plays a fundamental role in statistical quantum transport theory. Consider a device system which is open and in contact with the surrounding environment (e.g. device contacts). When the electron dephasing length λ_ϕ in the contacts[3] is larger than the device length, the electrons are said to be larger than the device. In this case, the contacts inject only diagonal elements of the density matrix ρ. According to Van Hove's (1955), the time needed to build the off-diagonal elements of the density matrix, in such open system, is much longer than the carrier relaxation and transit times. Therefore, the off-diagonal terms of ρ are built up long time after the electron transport across the device. The so-called Pauli master equation (PME) describes the evolution of the diagonal elements of the density matrix in nanodevices.

Considering only the diagonal matrix elements over the eigenstates $|\beta\rangle$, of the Hamiltonian of the QBTE, and after some mathematical manipulations, we can get the PME (Fischetti, 1998).

$$\frac{d}{dt}\rho_{\beta\beta}(t) = \sum_{\gamma\neq\beta} \rho_{\gamma\gamma}(t)\cdot S_{\beta\gamma} - \rho_{\beta\beta}(t)\sum_{\gamma\neq\beta} S_{\gamma\beta} \tag{43}$$

where $S_{\beta\gamma}$ represents the transition probability per unit time, for the transitions from state $|\gamma\rangle$ to state $|\beta\rangle$, and can be expressed using the Fermi golden rule.

Therefore, we can consider the PME as a simplified form of the master equation (Von-Neumann Liouville equation or QBTE). In fact, the PME is easier in implementation and its complexity scales down to $O(N)$, where N is number of device grid points in numerical solution.

The Pauli Master equation (PME) plays a fundamental role in statistical quantum transport theory. Note the resemblance of the PME to the semiclassical Boltzmann transport equation (BTE). This resemblance facilitates the manipulation of the PME by the Monte Carlo methods to simulate carrier transport in nanodevices (Fischetti, 1999).

8. WIGNER-BOLTZMANN TRANSPORT EQUATION (WBTE)

The Wigner formulation of quantum transport is an attractive approach which allows the comprehension and prediction of quantum transport phenomena in terms of quasi-probability distribution functions.

The Wigner quasi-probability distribution was introduced by Wigner in 1932 to study quantum corrections to classical statistical mechanics. The goal was to replace the wave-function in the Schrödinger equation with a probability distribution in the phase space. The Wigner distribution function *W(x; p; t)*, is sometimes referred to as the Wigner function. The Wigner distribution function (WDF) depicts the probability distribution function[4] of a system of *N* particles, whose wave eigenfunctions (pure state functions) ψ has the coordinates $r = r_1 ... r_N$ and momentum operator $p = p_1 ... p_N$, such that:

$$W(\boldsymbol{r},\boldsymbol{p},t) = \frac{1}{\left(\pi\hbar\right)^{D.N}} \int_{-\infty}^{\infty} \psi(r - \frac{r'}{2}, t) \cdot \psi^* \left(r - \frac{r'}{2}, t\right) \cdot \exp\left[-\frac{jp.r'}{\hbar}\right] dr' \tag{44a}$$

This integration is *N*-times multiple such that $\boldsymbol{dr}=dr_1 ... dr_N$ and $\boldsymbol{p.r}=p_1 dr_1 + ... + p_N dr_N$. Also, *D* is the system dimension (1, 2, 3). Note that the above definition of the Wigner distribution resembles the Fourier transform of a certain combination of $\psi\psi^*$ (the density operator) to the momentum domain[5]. Therefore, this equation can be written in terms of the density operator function $\rho(r,r',t)$ as follows:

$$W(\boldsymbol{r},\boldsymbol{p},t) = \frac{1}{\left(\pi\hbar\right)^{DN}} \int_{-\infty}^{\infty} \rho(r,r',t) \cdot \exp\left[-\frac{jp \cdot r'}{\hbar}\right] dr' \tag{44b}$$

Although the Wigner function is a quasi-distribution function it can be used to calculate the average (macroscopic) value of any microscopic variable $A(\boldsymbol{r}, \boldsymbol{p}, t)$.

$$A(\boldsymbol{r},t) = \left\langle A(r,p,t) \right\rangle = \int_{-\infty}^{\infty} A(r,p,t) \cdot W(r,p,t) dp \tag{45a}$$

$$\int_{-\infty}^{\infty} W(r,p,t) dp = \psi \cdot \psi^* = \left|\psi(r,t)\right|^2 \tag{45b}$$

The dynamic behavior of the WDF can be obtained by applying the Wigner-Weyl transformation to the Liouville equation.

Note 3: Wigner–Weyl Transform

The Wigner–Weyl transform rule is a basic mathematical mapping tool. This transform is an isomorphism from the space of linear operators. For instance, the Weyl map of a given a quantum operator $\hat{A}(\hat{r},\hat{p})$, is expressed in the phase space as follows (Weyl, 1927):

$$A(\boldsymbol{r},\boldsymbol{p}) = M_W \left\{\hat{A}(\hat{r},\hat{p})\right\} = \frac{\hbar}{2\pi} \int d\xi \int Tr\left[\hat{A}(\hat{r},\hat{p}) \cdot \exp\left[j(\xi\hat{r} + \eta\hat{p})\right]\right] \cdot \exp\left[-j(\xi r + \eta p)\right] d\eta$$

Using this formalism, the Wigner distribution function is defined as the Weyl transform of the density matrix operator times a normalization factor,

$$W(\boldsymbol{r},\boldsymbol{p},t)=\frac{1}{2\pi\hbar}M_{W}\left\{\rho(t)\right\}$$

The evolution equation of the WDF can be then obtained as follows:

$$\frac{\partial W(r,p,t)}{\partial t}=\left[H,W(r,p,t)\right]_{M}$$

where the subscript '$_M$'of the commentator *([H,W] = HW-WH)*, represents the Moyal formalism (Moyal, 1949).

Alternatively, the dynamic behavior of the WDF can be evaluated using the Schrödinger equation as follows (supposing that the wave-functions $\psi(r, t)$ is a solution of the Schrödinger equation):

$$\frac{\partial W}{\partial t}+\frac{\boldsymbol{p}}{m}\cdot\nabla_{r}W=\Theta\cdot W(r,p) \tag{46a}$$

with

$$\Theta\cdot W(r,p)=\sum_{\lambda=1,3,5,..}^{\infty}\frac{(-1)^{\lambda}}{\lambda!}\left(\frac{\hbar}{2j}\right)^{\lambda-1}\nabla_{r}^{\lambda}V(r)\cdot\nabla_{p}^{\lambda}W(r,p) \tag{46b}$$

The operator Θ is a complicated operator, such that ∇_r operates on the total potential $V(r)$ and ∇_p operates on $W(r,p)$. The above expression is sometimes called the Moyal expansion. The operator Θ may be further simplified as follows:

$$\Theta.\Big|W\approx-\frac{2}{\hbar}\sin\{\frac{1}{2}\hbar\nabla_{r}\nabla_{p}\}V(r).W(r,p) \tag{46c}$$

Note that in the limit $\hbar\rightarrow 0$, the Wigner equation (46a) reduces to the classical Vlasov equation (collision-less BTE).

By separating the potential field V(r) in the above equation into a collision field (Vin) and an external field (Vex) such that V(r) = Vin+Vex, the right-hand side of (46a) may be further divided into two terms as follows:

$$\frac{\partial W}{\partial t}+\frac{\boldsymbol{p}}{m}.\nabla_{r}W+\theta.W=\left(\frac{\partial W}{\partial t}\right)_{C} \tag{47a}$$

where the operator θ resembles the operator Θ, with V(r) replaced with the external field potential Vex. Also, the internal interaction field Vin is included in the term (∂W/∂t)c to account for collisions, which are not considered in the one-particle Schrödinger equation. The first order approximation of this term may be expressed as follows:

$$\theta.W \approx -\nabla_r V_{ex}(r).\nabla_p W(r,p) \tag{47b}$$

This includes the effect of classical forces (F = -∇Vex), like the semiclassical BTE. Other high order terms retains the quantum nonlocal effects. The above equation (46) is called the Wigner-Boltzmann transport equation (WBTE).

Once the WBTE is solved, the electron density n and current density Jn can be determined from the Wigner distribution, as follows:

$$n(r, t) = \int W(r, p, t)\, dp \tag{48a}$$

$$J_n(r, t) = -e \int (p/m^*).W(r, p, t)\, dp \tag{48b}$$

The mean value of any other operator in the phase space, A(r, p, t), can be determined from the Wigner distribution function as indicated by (44).

In order to obtain a self-consistent solution of the WBTE, it should be coupled with the Poisson equation. The electrostatic potential (f) in Poisson's equation is related to the external potential energy (Vex), which is acting on carriers, and should include the spatial variation of energy band structure:

$$Vex(r) = ef(r) - e\,[\chi(r) - \chi(0)] \tag{48c}$$

where χ(r) is the electron affinity along the semiconductor device.

It comes from the above discussion that the WBTE is similar to the semiclassical Boltzmann equation, except for a nonlocal potential term. Several attempts for solving the WBTE for semiconductor devices have been reported e.g., by Frensley (1987), Buot & Jensenand (1990), Shifren & Ferry (2002), Gehring (2003) and Kosina, Nedjalkov & Selberherr (2003). The simplest method to include the scattering in the WBTE simulation is to consider the semiclassical collision operator (of the BTE), applied to the Wigner distribution function (Querlioz & Dollfus, 2010). Therefore, we can write the WBTE in the following form:

$$\frac{\partial W}{\partial t} + \frac{1}{\hbar}\nabla_k E(k)\nabla_r W = \int V_W(r, k-k', t) W(r, k-k', t)\, dk' \tag{49a}$$

where VW(r, k, t) is the Wigner potential, which is given by:

$$V_W(\boldsymbol{r},\ \boldsymbol{p},t) = \frac{j}{\left(2\pi\hbar\right)} \int_{-\infty}^{\infty} \left[V(r - \frac{r'}{2},t) - V^*(r + \frac{r'}{2},t) \right] \exp\left[-jk.r'\right]\ dr' \tag{49b}$$

The iteration procedure to solve the WBTE with Poisson's equation starts from an initial guess, and the iteration process continues until a specific convergence criterion is met. The carrier concentration is used to update the potential in the device by solving the Poisson equation. The stochastic methods, which are based on the association of trajectories of a single or an ensemble of particles, have been adapted to solve the BTE (Midjakov and Midov, 2010). In this work, the concept of Wigner paths is introduced with an extension of the Wigner MC method to the density functional theory (DFT), as a way to simulate many-body quantum problems. A review, which provides a detailed introduction to the Wigner-Boltzmann transport theory, along with a quantum Monte Carlo (QMC) simulation method for time-dependent quantum systems, is given by Sellier, Nedjalkov & Dimov (2015).

The Figure 3 depicts a flowchart of the QMC method, which is based on the WBTE6. In much the same manner as the semiclassical MC method, the two major steps are the calculation of scattering rates and the determination of the time of flight of the simulated particles.

The duration of the time of flight τ is determined by random numbers, and the time interval is typically chosen to be a few femtoseconds. The number of simulated electrons N and the simulation time t_s are chosen to give reasonable results. Taking N = 10000 and t_s = 10ps is usually sufficient to obtain good results.

The Wigner MC method gives accurate results for nanoscale semiconductor structures as well as purely quantum devices such as resonant tunneling diodes (Gehring, 2003; Kosina, Nedjalkov & Selberherr, 2003). The typical output of a *GaAs* resonant tunneling diode is shown in Figure 4.

Figure 3. Flowchart of quantum MC method, based on Wigner-Boltzmann transport equation
Adapted from Ellinghaus (2010)

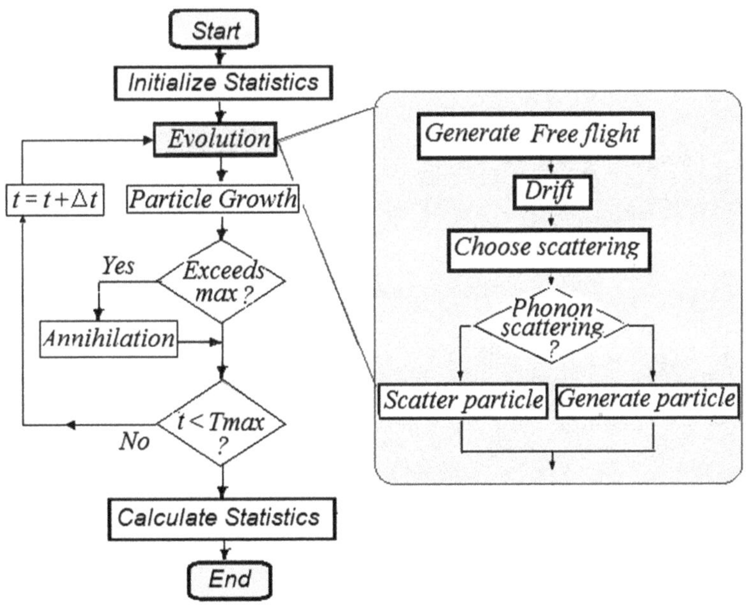

Figure 4. Typical IV curve of a resonant tunneling diode, calculated self-consistently (solid line), as obtained by WBTE-based Monte Carlo simulation
After Kosina, Nedjalkov & Selberherr (2003).

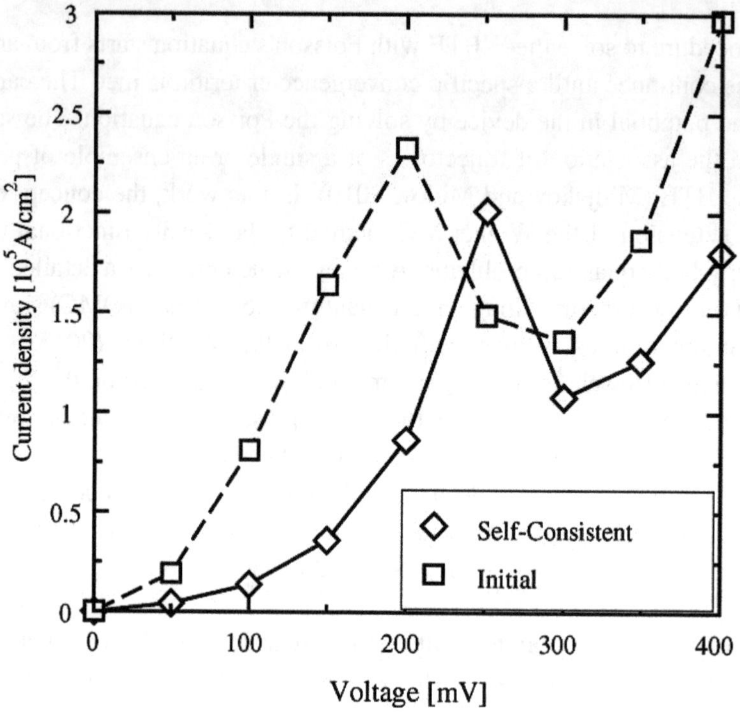

9. QUANTUM HYDRODYNAMIC MODEL (QHDM)

The quantum hydrodynamic model (QHDM) can be derived by a variety of methods, such as the first few moments of the WBTE or the Madelung transform or the principle of minimum entropy. We start here by the WBTE moments method. By multiplying the WBTE by *1*, ***p*** and ½ p^2/m_n* and integrating the result over the entire phase space, we get the following conservation equations (for electrons):

$$\frac{\partial n}{\partial t} + \nabla.(n m_n^{-1} \underline{p}_n) = \left(\frac{\partial n}{\partial t}\right)_{col.} \tag{50a}$$

$$\frac{\partial (n\underline{p}_n)}{\partial t} + \nabla.(n m_n^{-1} \underline{p}_n \otimes \underline{p}_n) = \left(\frac{\partial (n\underline{p}_n)}{\partial t}\right)_{col.} \tag{50b}$$

$$\frac{\partial (n\omega_n)}{\partial t} + \frac{1}{2}\nabla.(n\underline{p}_n m_n^{-1} \underline{p}_n^2) = n\underline{p}_n m_n^{-1} F + \left(\frac{\partial (n\omega_n)}{\partial t}\right)_{col.}$$

where $\underline{p}$n= $<p_n>$ and ω_n are the electron mean momentum and mean energy, respectively. Also m_n is the mean effective mass of electrons ($m_n = <m_n{*}>$). The electron mean energy ω_n is related to the scalar electron temperature T_n as follows:

$$\omega_n = \frac{1}{2}\frac{\left\langle p_n^{\ 2}\right\rangle}{m_n} = \frac{1}{2}\frac{p_n^{\ 2}}{m_n} + \frac{3}{2}k_B T_n + V_q \tag{51a}$$

where V_q is called the quantum potential, and given by:

$$V_q = -\frac{h^2}{8m_n}\nabla^2[\ln n] \tag{50b}$$

Equation (51a) is sometimes arranged in the following manner:

$$\omega_n = \frac{1}{2}m_n v_n^{\ 2} + \frac{3}{2}k_B T_{qn} \tag{51c}$$

where T_{qn} is called the quantum electron temperature. The quantum electron temperature is related to the semiclassical electron temperature by the following equation:

$$T_{qn} = T_n + \frac{2V_q}{3K_B} \tag{52}$$

10. QUANTUM CORRECTIONS

By comparing (51) with the corresponding semiclassical HDM expression, we note the quantum correction term V_q. This term is only significant when the electron concentration n varies rapidly over small distances in the order of de-Broglie wavelength (h/p). The quantum correction appears also in the current equation (which is a reduced form of the momentum conservation equation) in the form of an additional current J_q. The following electron current equation depicts this correction term:

$$J_n = enm_n\zeta + k_B m_n \nabla(nT_n) + J_q \tag{53}$$

$$J_q = n\mu_n\left(\frac{\hbar^2}{6m_n^*}\right)\cdot\nabla\left(\frac{\nabla^2\sqrt{n}}{\sqrt{n}}\right) \tag{54}$$

Note that we assumed a single parabolic energy band (with constant effective mass) and neglected the acceleration terms in the above current equation. Adding the above mentioned corrections to the semiclassical macroscopic models (like DDM) to account for quantum effects, is referred to as the *density gradient* (**DG**) correction method. Alternatively, one can replace the semiclassical temperature in the current relations (and energy conservation relations in HDM) by the quantum temperature, T_q, as defined by (52).

11. MADELUNG TRANSFORM

The quantum hydrodynamic equations can be also directly derived from the Schrödinger equation using the so-called Madelung transform. In the Madelung transform (Madelung, 1927), the electron wavefunction can be decomposed into an amplitude n and phase Φ as follows:

$$\psi = \sqrt{n}.\exp\left(j\frac{m^*}{\hbar}\Phi\right) \tag{55a}$$

The electron density $n = |\psi|^2$ and the electron current density $\boldsymbol{J}_n$ is given by:

$$J_n = -\frac{e\hbar}{m_n^*}\operatorname{Im}\left[\psi^*\nabla\psi\right] = -en\nabla\Phi \tag{55b}$$

By comparing the above expression of J_n with the conventional definition $J_n = -env_n$, we notice that v_n is replaced with $\nabla\Phi$. Thus, Φ may be considered as the velocity field. Note also that J_n is closely related to the probability current density, defined by.

$$J(r,t) \equiv -\frac{j\hbar}{2m}\left[\Psi^*\nabla\Psi - \left(\nabla\Psi^*\Psi\right)\right] \tag{56}$$

The following set of quantum hydrodynamic equations can be obtained by substituting ψ, as defined by (55), into the Schrödinger equation and separating the real and imaginary parts.

$$\frac{\partial n}{\partial t} - \frac{1}{e}\nabla \mathrm{J}_n = 0 \tag{57a}$$

$$\frac{\partial J_n}{\partial t} - \frac{e^2 n\zeta}{m_n^*} - e\nabla\left(n\cdot v_n \otimes v_n\right) + \frac{e\hbar^2}{2m_n^{*2}} n\cdot\nabla\left(\frac{\nabla^2\sqrt{n}}{\sqrt{n}}\right) = 0 \tag{57b}$$

where we substituted $v_n = \nabla \Phi$ such that $J_n = -env_n$. The above conservation equations resemble the semi-classical hydrodynamic equations, without the collision terms. This is due to the fact that the Schrödinger equation doesn't take collisions into account. In particular, the current equation (58b) looks like the zero temperature Euler equation in fluid dynamics (with a corrected potential term). However, this conservation equation *lakes* the electron pressure (or temperature) term. The combined potential term in this equation (due to electric field and quantum correction) is called the Bohm potential.

$$V_B = V - \frac{\hbar^2}{2m_n^{*2}} \cdot \frac{\nabla^2 \sqrt{n}}{\sqrt{n}} \tag{58}$$

Note that the potential energy $V(r)$ is related to the electric field by $-e\zeta = \nabla V$.

A more elaborate derivation of the above conservation equations consists in considering the case of a sequence of single states $\psi_k = \sqrt{n_k}\, exp(-jm_n{}^*\Phi_k/\hbar)$ with distributed occupation probability P_k such that $\Sigma P_k = 1$. This quantum system will have a total electron density $n = \Sigma n_k$ and a total current density $J_n = \Sigma J_{nk}$, where $J_{nk} = -e(\hbar/m_n{}^*).Im[\psi_k{}^* \nabla \psi_k]$. Again, substituting the eigen-functions ψ_k into the one-electron Schrödinger equation $(\hat{H}_k \psi_k = E_k \psi_k)$ and adding a collision term to each resulting conservation equation, we obtain:

$$\frac{\partial n}{\partial t} - \frac{1}{e}\nabla J_n = \left(\frac{\partial n}{\partial t}\right)_c \tag{59a}$$

$$\frac{\partial J_n}{\partial t} - \frac{e^2 n\zeta}{m_n^*} - e\nabla\left(n \cdot v_n \otimes v_n\right) - \nabla \cdot \left(\frac{P_n}{m_n^*}\right) + \frac{e\hbar^2}{2m_n^{*2}} n \cdot \nabla \left(\frac{\nabla^2 \sqrt{n}}{\sqrt{n}}\right) = \left(\frac{\partial J_n}{\partial t}\right)_c \tag{59b}$$

where the pressure tensor $P_n = nk_B T_n$ and T_n is the electron temperature. Note that the quantum correction term in the current equation (59b) is similar to the correction term in (54), when adopting the relaxation time approximation $(\partial J_n/\partial t)_c = -J_n/\tau_{mn}$ and substituting $\mu_n = e\tau_{mn}/m_n$.

12. QUANTUM THEORY OF PATH INTEGRAL

The theory of path integral of quantum mechanics has been formulated by Richard Feynman in 1948, as a description of quantum theory corresponding to the action principle of classical mechanics. It mimics the classical notion of single unique history of a system with a functional integral over an infinite number of possible histories to calculate the probability of a quantum process. In this theory, Feynman proposed the following postulates:

1. The probability for any fundamental event is given by the absolute square of complex amplitude.
2. The amplitude for some event is given by adding together all the histories which include that event.

3. The amplitude a certain history contributes is proportional to *exp[jI(H)/ħ]* where *I(H)* is the action of that history, or time integral of the Lagrangian.

In order to find the overall probability amplitude for a given process, then, one adds up, or integrates, the amplitude of postulate 3 over the space of *all* possible histories of the system between the initial and final states, including histories that are absurd by classical standards. In calculating the amplitude for a single particle to go from one place to another in a given time, it would be correct to include histories in which the particle describes elaborate curlicues, histories in which the particle shoots off into outer space and flies back again, and so forth. The path integral includes all these histories. It assigns all of them, no matter how bizarre, with amplitudes of *equal size.*

Note 4: The Least Action Principle

The fact that nature follows a '*law of least action*' has a long history. The optical laws of reflection and refraction can be derived from the principle that light follows a path that minimizes the transit time. In the 1700s, the law was reformulated to require the dynamics of mechanical systems to minimize the action defined as Energy-Time. In the 1800s, Hamilton stated the general form: A dynamical system will follow a path that minimizes the action defined as the time-integral of the Lagrangian (L=T-V), where *T* and *V* are the kinetic and potential energy of the system. A Legendre transformation of the Lagrangian produces the total energy of the system in the form of the Hamiltonian. The Lagrangian and Hamilton play central roles in quantum theory. The Schrödinger equation can be derived from the classical Hamiltonian by replacing the classical variables with operators. The Feynman path integral provides a beautiful formulation of the quantum principle by incorporating the integral over all possible paths of the action.

12.1 Feyneman Path Integral Formulation

Feynman showed that his formulation of quantum mechanics is equivalent to the canonical approach to quantum mechanics. Any amplitude computed according to Feynman's principles will also obey the Schrödinger equation for the Hamiltonian corresponding to the given action. For example, consider the case of the drift of a single particle. the path integral can be formally thought of as the small-step limit of an integral over zigzags: for instance, for one-dimensional motion of a particle from position x_0 at time 0 to x_{n-1} at time *t*, the time interval can be divided up into little segments of duration Δt and the path integral can be computed as follows:

$$A = \lim_{\Delta t -->0} \int_{-\infty}^{\infty} \exp\left(\frac{jI[H(x,t)]}{\hbar}\right) dx \tag{60}$$

where the integral is *n*-time multiple, $dx = dx_0.dx_1...dx_j...dx_{n-1}$ and *H* is the entire history in which the particle zigzags from its initial to its final position linearly between all the values of $x_j = x(j\Delta t)$.

12.2 Effective Potential Correction

In the contact of Feynman Path Integral Formulation of quantum transport, the quantum correction method was developed by Feynman and Hibbs. In this method the effective potential is derived by calculating the contribution to the path integral of the fluctuations a quantum particle around its classical path. The contribution can be calculated by a variational method and a trial potential to first order. The effective classical potential in the average point on each path is then given by:

$$V_{eff}(x) = \frac{1}{\sqrt{2\pi a}} \int_{-\infty}^{\infty} V(x') \cdot \exp\left[-\frac{(x' - x)^2}{2a^2} \right] dx' \quad (61a)$$

where

$$a^2 = \hbar^2 / 12 m^* k_B T \quad (61b)$$

The effective potential approach can be used in particle-based device simulation, like Monte Carlo method, to account for quantum phenomena (Vaseliska and Ahmad, 2004). However, according to Ferry et al (2002), the use of the effective potential to incorporate quantum effects, can only account for limited effects, such as charge set-back and threshold shifts in MOS devices, but tunnelling cannot be well handled by this approach, or by any other local potential approach. However, we can resort to the approximate methods for solving the quantum tunneling current (e.g., transfer matrix method, see Figure 28), in a post processing step.

Generally speaking, there are three important manifestations of quantum mechanical effects in nanoscale devices, namely:

- Tunneling
- Size quantization
- Quantum interference & coherence effects

In order to incorporate all these effects self consistently in the device simulation, a full quantum transport approach, such as the QBTE or WBTE or Green's functions, is necessary.

13. GREEN'S FUNCTIONS APPROACH

The Green's function method is a very efficient quantum approach to calculate the carrier densities and the terminal currents in nanoscale devices and molecular structures. The concept of Green's functions was first developed as a mathematical tool to solve differential equations, by the George Green in the 1830's. Actually, Green's functions are distributions, not necessarily proper functions.

13.1 Definition of the Green Functions

The aim of this section is to provide intuitive explanations of the Green function, and how it is related to the density matrix. We discuss first the basic definition of the Green functions that are not very familiar to most researchers in solid-state physics and electronic device engineers.

Mathematical Definition of Green's Functions

Mathematically speaking, the Green function is used to solve inhomogeneous differential equations subject to initial and boundary conditions. Actually, the Green function is the integral kernel of a linear operator which inverts a differential operator ***L***. The Green function solves (weakly) a differential equation of the form:

$$L\{G(r, t)\} = \delta(r, t) \tag{62a}$$

where δ is the Dirac delta (impulse) function. Therefore, if we've a nonlinear non-homogeneous differential equation in the form:

$$\boldsymbol{L}\{ y(x) \} = f(x) \tag{62b}$$

where ***L*** is any linear differential operator, for instance:

$$\boldsymbol{L}\{y(x)\} = a_o\, y(x) + a_1\, dy(x)/dx + a_2\, d^2y(x)/dx^2 + .. \tag{62c}$$

Then the solution, using Green's functions is carried out as follows:

1. Find the Green's Function, $G(x)$, such that

$$\boldsymbol{L}\,\{\boldsymbol{G}(x)\} = \delta(x) \tag{63a}$$

2. Integrate $G(x)$ with $f(x)$ over the entire solution domain ($x_1<x<x_2$) to find $y(x)$:

$$y(x) = \int \boldsymbol{G}(z).f(z)dz \text{ for } x_1 < z < x_2 \tag{63b}$$

The concept of Green's function-based solution relies on the fact that any function $f(x)$ can be represented as an infinite sum of delta functions at a series of discrete points, whose separation tends to zero:

$$f(x) = \int .f(z)\, \delta(x-z)dz \text{ for } -\infty < z < \infty \tag{63c}$$

When a delta function, $\delta(x)$, is applied to a linear system, it will generate an impulse response, $h(x)$. Likewise, the linear operator can be modeled as a linear system such that when it operates on the Green's Function, it will result in a delta function. Hence, we can write:

$$y(x) = \int f(z)\, h(x-z)dz \text{ for } -\infty < z < \infty \tag{63d}$$

Therefore, the Green's function is the impulse response which describes how the system will react to a single point source. And the following integration (sum) of the Green's function will produce the required solution:

$$y(x) = \int f(z)\, G(x-z)dz \text{ for } -x_1 < z < x_2 \tag{63e}$$

However, the infinite sum is replaced by a bounded integration in (63e). For this reason, the Green's function method produces a weak solution for the given differential equation. For instance, the Green's function of the Helmholtz wave equation is:

$$G(r) = -(1/4\pi\, r).exp(-j\mathbf{k}.\mathbf{r}) \tag{64a}$$

Hence the solution of this equation ($[\nabla^2+k^2]\psi(r) = Q$) is given by:

$$\psi(r) = \int G(r-r_o).Q(r_o)d^3r_o \tag{64b}$$

where Q is the source term. As a particular case of Helmholtz equation, the Schrodinger equation may have the following solution:

$$\psi(r) = \psi_o(r) - \int (1/r_o).exp(-j\mathbf{k}.\mathbf{r}_o).[(2m/\hbar^2).V(r_o).\psi(r_o)]\, d^3r_o \tag{64c}$$

where we substituted $Q = (2m/\hbar^2).V.\psi$ and $k^2 = (2m/\hbar^2).E$.

Physical Definition of Green's Functions

The Green's functions are used in many applications in physics such as the quantum field theory and statistical field theory. In fact, the Green's function method has been used since 1950s to describe the many-body problems in quantum physics.

In coordinate representation the Green's function, $G(r_1,t_1;r_2,t_2)$, depends on two position arguments r_1, r_2 and two time arguments t_1, t_2, representing the non-locality in space and time. For instance, the Green function $G(r_1,t_1; r_2,t_2)$ may be defined as the probability amplitude that a particle created at time t_2 at place r_2 moves to place r_1 at time t_1. In the Heisenberg representation the Green's function for the particle is given by:

Figure 5. Mathematical representation of the Green's function

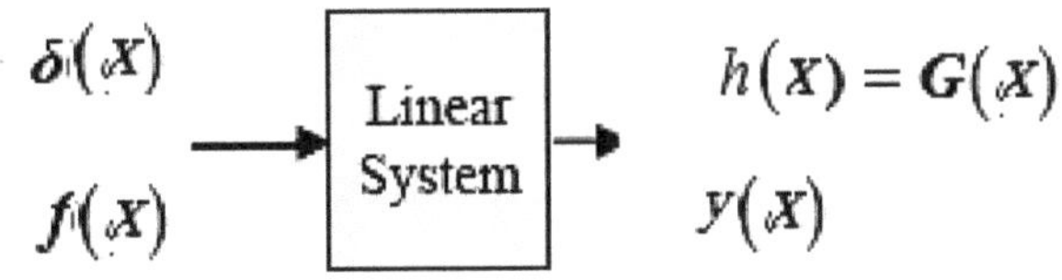

$$G(r_1,r_2,t_1,t_2) = -j\hbar^{-1}\left\langle T(\psi(r_1,t_1)\cdot\psi^{\dagger}(r_1,t_1))\right\rangle \tag{65}$$

where $\boldsymbol{T}$ is the time-ordering (or contour-ordering) operator and $\psi^{\dagger}$ and ψ are the particle creation and annihilation operators, respectively. Note that we use the second quantization language in this definition. The time ordering operator $\boldsymbol{T}$ orders operators from right to left in ascending time order and works as follows: $T(\psi(t_1).\psi(t_2))= \psi(t_1).\psi(t_2)$ when $t_1>t_2$ otherwise $T(\psi(t_1)\psi(t_2)) = -\psi(t_2).\psi(t_1)$ when $t_1<t_2$. At finite temperature, the quantum average <.> is taken over the equilibrium density matrix of the system. The above definition means that we consider the Green's functions as expectation values of the Heisenberg field operators.

Sometimes, it is preferable to change the independent variables (r_1,r_2,t_1,t_2) to the center of mass coordinates: $(r,t) \leftarrow [½(r_1+r_2),\ ½\,(t_1+t_2)]$ and the relative coordinates: $(u,\tau) \leftarrow [(r_1-r_2),(t_1-t_2)]$. Therefore, the Green's function with respect to the relative coordinates takes the form $G(u,\tau,r,t)$ and its Fourier transform $G(k,\omega,r,t)$ is defined as:

$$G(k,\omega,r,t)) = \int_{-\infty}^{\infty}\int_{-\infty}^{\infty} G(u,\tau,r,t)\exp[j(\omega\tau + k\cdot u)]du d\tau \tag{66}$$

The quantum BTE is in fact a kinetic differential equation for the Green's function $G(k,\omega,r,t)$. Under steady state conditions, the Green's functions depend only on time difference. Therefore, we can use the Fourier transform to change the time difference coordinate, $\tau=(t_1-t_2)$, to frequency coordinate (ω) or energy $(E = \hbar\omega)$ such that $G(r_1,r_2;E) = \int exp(jE\tau/\hbar).G(r_1,r_2;\ \tau).(d\tau/\hbar)$. In fact, it is more convenient to work with Green's functions in the frequency (or energy) domain. The Green's function may be then defined as the response of a system to a constant perturbation in the Schrödinger equation:

$$(E - H)\ G(r_1,r_2) = \delta(r_1,r_2) \tag{67a}$$

where the system Hamiltonian $\boldsymbol{H}$ is defined as:

$$H = H_o + H_{\text{int}} = -\frac{\hbar^2}{2}\nabla\cdot\left[\frac{1}{m(r)}\nabla\right] + V(r) \tag{67b}$$

Figure 6. Physical meaning and representation of the Green's function

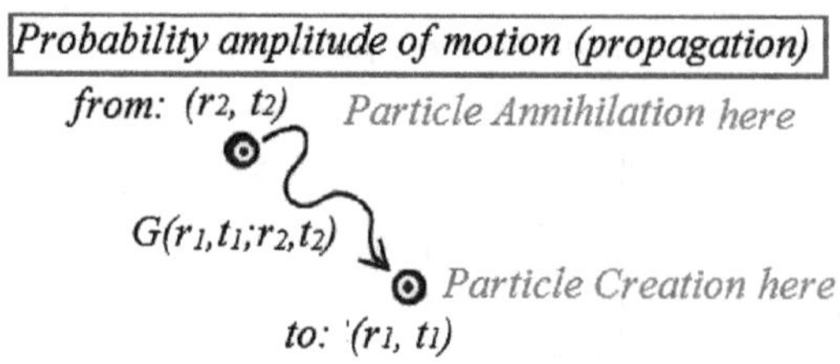

Here H_0 is the non-interacting Hamiltonian and H_{int} contains the interaction and external potential terms. The power of the Green's function formalism lies in its ability to approximate the effects of interaction H_{int}. The many body interactions are taken into account through the so-called self-energy, which are casted into the H_{int} term. There exist several methods to solve the above equation and find G in presence of perturbation. Among these methods, one can cite: Wick's theorem, and using Feynman diagrams

13.3 Wick's Theorem

The Wick theorem allows a perturbation expansion of Green's functions. This expansion allows us to evaluate the exact Green's functions G as a perturbation expansion involving expressions of the free (non-interacting) Green's functions G_o and the perturbation potential $V(r)$. This is illustrated in Figure 7.

13.4 Feynman Diagrams

Feynman introduced the idea of representing different contributions obtained from the Wick theorem by drawings, called Feynman diagrams. The Feynman diagrams provide an illustrative way to solve the many-particle problems and the perturbation expansion of the Green's functions. A diagram dictionary for Fermions (like electrons), and Bosons (like phonons), is shown in Table 2. Note that diagrams for electrons are in coordinate-time space, while phonon diagrams are in momentum-energy space.

The Green's function is interpreted as the creation of a particle at (r', t') in space-time, and the propagation of the corresponding perturbation to the point (r, t), where the particle is annihilated. Hence, the full Green's function G is represented by a double line joining these two points. The free Green's function G_o is characterized by a single line. Intermediate variables describe events taking place between the two space-time arguments of the Green's function. The factor M refers to the electron-phonon interaction matrix elements. The Coulomb potential is represented by a wavy line with two inputs and outputs which can be coupled to describe a self-interaction. The Coulomb interaction is assumed instantaneous. Each time a Fermion loop appears, the corresponding perturbation expression to its diagram must be multiplied by a factor^{-1}. For phonons it is more convenient to work on momentum-energy rather than in space-time

13.5 Dyson's Equation

By the aid of the Dyson equation, it is possible to obtain a perturbation expansion of the Green functions that allows us to evaluate them at times far from the initial equilibrium state (Dyson, 1978). This perturbation expansion may be formulated in terms *Feynman diagrams*. Starting from the Dyson equation, the quantum Boltzmann equation can be derived. The Dyson equation can be achieved by classifying the various contributions in arbitrary Feynman diagrams.

Using Dyson's equation we can summarize the Feynman-Dyson perturbation theory in a compact form. The exact Green's function can be written as the non-interacting Green's function G_o plus all the connected terms with a non-interacting Green's function at each end, as shown in the Figure 7.

By introducing the concept of self-energy Σ, the structure in Figure 7 takes the form shown in Figure 8.

The self-energy Σ consists of the energy level shift $(\Delta = Re\Sigma)$ and the energy level broadening or line width $(\Gamma = -2Im\Sigma)$, due to the interaction with the surrounding system (contacts). The corresponding expression of Green's function is given by:

Table 2. Feynman Diagrams for electrons (Fermions) and phonons (Bosons)

Expression	Description	Diagram
$G(r,r',t,t')$	Full Green's function	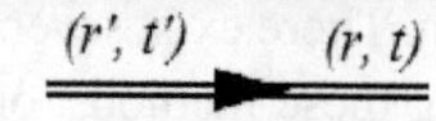
$j\hbar\, G_0(r,r',t,t')$	Free Green's function	
$V(r_1, r_2)$	Coulomb interaction	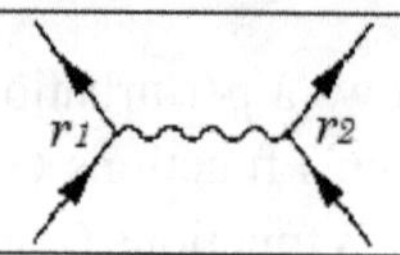
$\int r_1 \int t_1$	Intermediate variable(s)	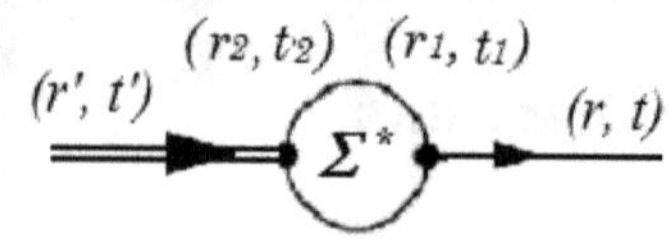
$Factor^{-1}$	Any Fermion loop	
$j\hbar\, D_0(q,\omega)$	Free phonons	
$-jM/\hbar$	Electron-phonon interactio	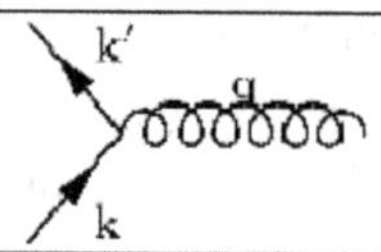

Figure 7. Green's functions expanded in terms of Feynman diagrams

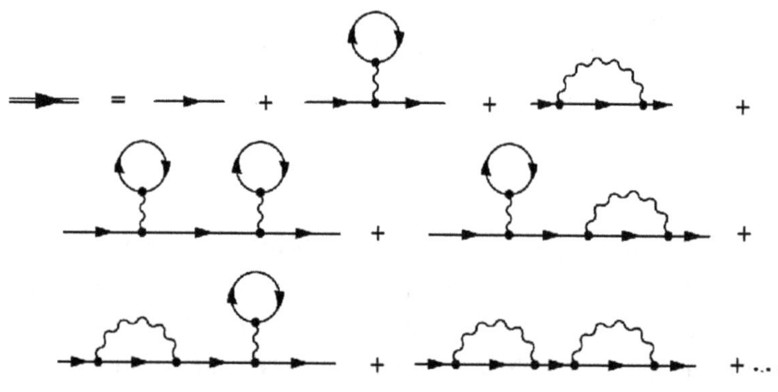

Figure 8. Illustration of the Dyson equation with Feynman diagrams

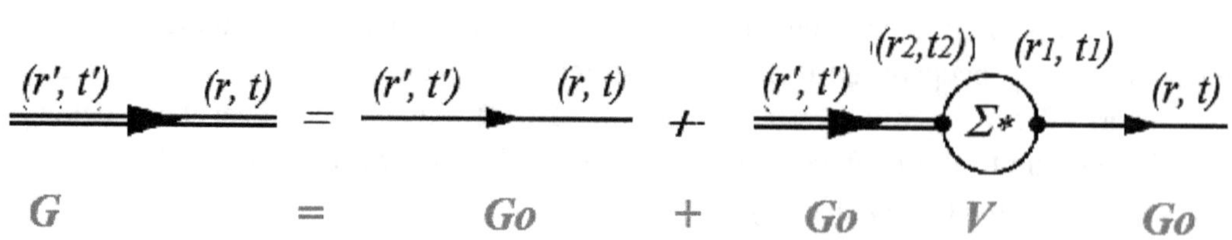

$$G(r,t,r',t') = G_o(r,t,r',t') + \int\int G_o(r,t;\ell_1)\,\Sigma^*(\ell_2)\,G_o(\ell_2;r',t')d_2.d_1 \tag{68}$$

Here, we use the abbreviation $\ell 1=(r_1,t1)$ and $\int d_1=\iint dr_1.dt1$ and so on. The Dyson equation determines the renormalized Green's function G, starting from its equilibrium (non-interacting) form G_o. The self-energy in this equation describes the renormalization of single-particle states due to the interaction with the surrounding many-particle system. Note that the quantum Boltzmann equation (QBTE) can be derived, starting from the Dyson equation

14. NON-EQUILIBRIUM GREEN'S FUNCTION (NEGF)

The Non equilibrium Green's function (NEGF) is a microscopic quantum method that allows us to study the time evolution of many-particle dissipative quantum systems. The origin of NEGF method was initiated by Schwinger to treat the Brownian motion in quantum oscillators (Schwinger, 1961). The NEGF theory has been extended by Kadanoff, and Baym and has been recently applied for modeling transport phenomena in semiconductor devices. Actually, the NEGF technique has been widely used to study quantum transport of electrons and holes as well as phonons and photons in a variety of materials and devices.

The initial motivation for the development of the NEGF formalism in semiconductor devices was the calculation of tunneling currents in metal-insulator-metal (MIM), hetero-structures, and resonant tunneling devices. In 1988, Kim and Arnold were the first to apply the NEGF formalism to such systems. The theory was adapted to address quantum optics, high-field transport, nanotube FETs and molecular devices.

There are six types of Green's functions that can be defined as the non-equilibrium statistical ensemble averages of the single carrier correlations. These are the greater, lesser, advanced, retarded, time-ordered (causal) and anti-time-ordered Green's functions. The first four types are more famous than others. They are defined as follows (Jacononi, 2010b):

$$G^>(r,t,r',t') = -j\hbar^{-1} <\psi(r,t)\,\psi^\dagger(r',t')> \tag{69a}$$

$$G^<(r,t,r',t') = +j\hbar^{-1} <\psi^\dagger(r',t')\psi(r,t)> \tag{69b}$$

$$G^r(r,t,r',t') = -j\hbar^{-1}\,\theta(t-t') <\{\psi(r,t),\psi^\dagger(r',t')\}> \tag{69c}$$

$$G^a(r,t,r',t') = +j\hbar^{-1}\theta(t'-t).<\{\psi(r,t),\psi^\dagger(r',t')\}> \tag{69d}$$

We dully note that $\psi^\dagger(r,t)$ and $\psi(r',t')$ are the carrier creation and destruction operators, which map a wavefunction at time t' to a wavefunction at time t and vice versa. The angular brackets mean the average over the equilibrium density matrix ρ_o. Also, $\theta(t)$ is the Heaviside unit step function, which is defined as $\theta(t)=1$ if $t\geq 0$ and 0 otherwise. The so-called time-ordered (causal) or anti-time-ordered Green's functions are defined as follows:

$$G^{t}(r,t,r',t') = \theta(t-t').G^{>}(r,t,r',t') + \theta(t'-t).G^{<}(r,t,r',t') \tag{69e}$$

$$G^{at}(r,t,r',t') = \theta(t'-t).G^{<}(r,t,r',t') + \theta(t-t').G^{>}(r,t,r',t') \tag{69f}$$

NEGF allows the calculation of scattering states (represented by the retarded Green's function G^r) and their non-equilibrium occupation (represented by the lesser Green's function $G^<$). Note that the Green's functions satisfy the relations: $G^r - G^a = G^> - G^<$ and $G^t + G^{at} = G^< + G^>$. In fact, $G^>$ and $G^<$ are related to observables and kinetic properties like particle density and current.

The NEGF formalism provides a means to handle open quantum systems. An open system is connected to surrounding reservoirs and has non-zero boundary conditions for the wave functions. The Hamiltonian of such open systems can be written as follows:

$$\begin{bmatrix} H & C \\ C^{+} & H_R \end{bmatrix} \tag{70a}$$

where H and H_R denote the Hamiltonian of the system (device) and the surrounding reservoirs (contacts) and C represents a coupling matrix. In real systems, the dimension of H_R is usually much larger than the dimension of H. Therefore, the corresponding single-particle Green's function reads:

$$\begin{bmatrix} G^{r} & G^{>} \\ G^{<} & G_R \end{bmatrix} = \begin{bmatrix} E-H & -C \\ -C^{+} & E+H_R \end{bmatrix} \tag{70b}$$

where $G^>$ and $G^<$ refer to the coupling of the system (device) to/from the reservoir (leads), and G_R describes the reservoir itself. Also, the retarded Green function, G^r, which represents the Green's function in the device, becomes:

$$G^r = (E - H - \Sigma)^{-1} \tag{71}$$

The spatially distributed self-energy Σ describes the scattering events inside the device (for example electrons–phonons and/or electrons–impurities scattering mechanisms). The reason for calculating the Green's function is that it is easier than solving the Schrödinger equation directly. Also, the Green's function of the device can be calculated separately without calculating the whole system Green's function. In the absence of external fields the NEGF reduces to the equilibrium Green function G_o

The so-called *spectral function* of Green's function is given by:

$$A = j\,(G - G^{+}) \tag{72}$$

The spectral function gives information about the density of states of the system as well as all the solutions of the Schrödinger equation. Note that we sometimes write G^r simply as G and G^a as its conjugate (G).

15. QUANTUM KINETIC EQUATIONS

Here, we introduce the equations of motion for the non-equilibrium Green's functions (NEGF). Actually, there are two different, but equivalent, NEGF formulations, namely:

- The Kadanoff-Baym approach and
- The Keldysh method.

In this section, we treat these two important formulations and derive the so-called quantum Boltzmann equation (QBTE). The QBTE has been derived by Kadanoff and Baym (1970) and extended by Mahan and Hansch (1983). This equation is valid for weak applied potentials, which are slowly varying in space and time. Therefore, the QBTE is particularly useful for linear transport in stationary systems.

15.1 Kadanoff-Baym (K-B) Approach

The Kadanoff-Baym (K-B) quantum theory of transport at non-equilibrium has been developed in the 1970's on the basis of the non-equilibrium Green's function (NEGF). The K-B equations represent the equations of motion that determine the time evolution of the Green's functions. These equations of motion can be derived from the differential form of the Dyson equation.

$$\left[+j\hbar\frac{\partial}{\partial t1} - \hat{H}_o(1)\right]G^{\lessgtr}(12) = \int \Sigma^r(13).G^{\lessgtr}(32).d3 + \int \Sigma^{\lessgtr}(13).G^a(32).d3 \quad (73a)$$

$$\left[-j\hbar\frac{\partial}{\partial t2} - \hat{H}_o(2)\right]G^{\lessgtr}(12) = \int \Sigma^{\lessgtr}(13).G^r(32).d3 + \int \Sigma^a(13).G^{\lessgtr}(32).d3 \quad (73b)$$

Here, H_o is the single carrier Hamiltonian (energy). The right hand sides of the K-B equations contain the collision integrals due to other particles. Collision integrals should describe all scattering and correlation effects of other interacting particles[7]. The K-B equations do not include any information about the system initial values. In order to have a closed set of equations, K-B equations should be supplemented with the Dyson equations for G^r and G^a. For instance, the equations satisfying G^r are:

$$\left[+j\hbar\frac{\partial}{\partial t_1} - \hat{H}_o(1)\right]G^r(12) - \int \Sigma^r(13).G^r(32).d3 = \delta_{1,2} \quad (74a)$$

$$\left[-j\hbar\frac{\partial}{\partial t_2} - \hat{H}_o(2)\right]G^r(12) - \int \Sigma^r(13).G^r(32).d3 = \delta_{1,2} \quad (74b)$$

Similar relations hold for the advanced Green's functions (G^a).

Under steady state conditions, the Green's functions are usually written as $G(r_1,r_2;E)$, or simply $G(E)$, and the quantum kinetic equations may be written as follows, in terms of the total energy E and total self energy Σ:

$$\left[E - \hat{H}_o(r_1)\right] G^r(r_1, r_2, E) - \int \Sigma^r(r_1 r_3; E).G^r(r_3 r_2; E).dr_3 = \delta_{r1,r2} \quad (75a)$$

$$G^{\gtrless}(r_1, r_2, E) = \int \left[\int G^r(r_1, r_2; E).\Sigma^{\gtrless}(r_3 r_4; E).G^a(r_4 r_2; E).dr_4 \right] dr_3 \quad (75b)$$

Then, the K-B technique consists in solving the transport equations in non-equilibrium Green's functions, which include the retarded and advanced Green's functions G^r, G^a as well as the correlation Green's functions $G^<$ and $G^>$. Therefore, the quantum transport problem may be described by the following two equations in steady state:

$$[E - H - \Sigma^r]G^r = I \quad (76a)$$

$$[E - H - \Sigma^r]G^< = \Sigma^< G^a \quad (76b)$$

where Σ is the self-energy matrix and $\boldsymbol{H}$ is the system Hamiltonian matrix. The Hamiltonian ($H = H_o + V$) contains the single electron Hamiltonian H_o and the effect of external potential energy ($V = - e\phi$). In open systems, H is not Hermitian, like closed systems, and it therefore admits complex energy eigenvalues[8]. Note also that the self-energy Σ is related to the scattering rate. Actually, Σ^r represents out scattering (to contacts and other channels) while $\Sigma^<$ represents in-scattering (from contacts and other channels). The scattering rate line width (Γ) is given by the imaginary part of the self-energy:

$$\Gamma^r = j\,(\Sigma - \Sigma^+) \quad (47)$$

After solving equations (45), all physical quantities of interest, such as the electron density, current density can be obtained from $G^<$. For instance, the electron density can be obtained from the lesser Green function $G^<(E_{n'}, r, r')$ by integration, as follows:

$$n(r) = -\frac{1}{2\pi}\int_{-\infty}^{\infty} G^<(E_n, r, r')\,dE_n \quad (78)$$

In equilibrium, the Green functions of charge carriers are related to the Fermi-Dirac distribution $f_{n'}$ by the following relation:

$$G^{<}(E_n,r,r') = 2f_n(E_n).G^r(E_n,r,r') \tag{79}$$

Therefore, the electron density near equilibrium is given by:

$$n(r) = -\frac{1}{\pi}\int_{-\infty}^{\infty} G^r(E_n,r,r').f_n(E_n)dE_n \tag{80}$$

Note the resemblance between Green's functions and the electron density of states (DOS). In non-equilibrium (with bias), Green's function consists of the electron flux across a semiconductor, according to the following relation:

$$\begin{aligned} G^{<}(E_n,r,r') = &-j.f_{nR}(E_n).\int_{\partial\Omega R} G^r(E_n,r,r_R).\Gamma_R.G^a(E_n,r',r_R').dr_R dr_R' \\ &-j.f_{nL}(E_n).\int_{\partial\Omega L} G^r(E_n,r,r_L)\Gamma_L.G^a(E_n,r',r_L').dr_L dr_L' \end{aligned} \tag{81}$$

where f_{nL} and f_{nR} are Fermi-Dirac distributions at the device left and right boundaries (in equilibrium). Also, the advanced Green function G^a and the scattering rates at left and right leads (Γ_L, Γ_R) include the scattering effect of leads into account. After obtaining the electron density, *n(r)*, by iterations, we can determine the electron current density through the device as follows:

$$J_n(r) = \frac{2e}{h}\int_{-\infty}^{\infty} T(E_n).\left[f_{nL}\left(E_n\right) - f_{nR}\left(E_n\right)\right]dE_n \tag{82a}$$

where the transmission probability $T(E_n)$ is given by:

$$T(E_n) = \int_{\partial\Omega L}\int_{\partial\Omega R} \Gamma_L.G^r\Gamma_R G^a dr_R dr_R' dr_L dr_L' \tag{82b}$$

Here, the integration is taken over all the system boundaries (left and right). Therefore, the above K-B formulation leads to a set of linear equations for the coefficients of the Green's functions. Figure 9 illustrates the utilization of Green functions for the simulation of transport across a 1-D nanodevice. As shown, the one-dimensional nanodevice is connected to two semi-infinite leads and divided to three separate areas. In equilibrium, the Fermi level is constant all over the three areas. In non-equilibrium the Fermi levels in the leads differ by the bias voltage. The leads electron density and potential can be then solved self-consistently in the central region.

When we want to take the total coupling between the lesser and the retarded Green's function in a more rigorous manner, we should solve four coupled partial differential equations. For the case of the electron-phonon interaction, under the effect of electric field, these equations are:

$$(E - H_0 - e\phi - \Sigma^r)G^r = 1, \tag{83a}$$

$$\Sigma^r = G^r D^r + G^r D^< + G^< D^r \tag{83b}$$

$$G^< = G^r \Sigma^< G^{r\dagger} \tag{83c}$$

$$\Sigma^< = G^< D^< \tag{83d}$$

where $\boldsymbol{H}_0$ is the single-electron Hamiltonian, G^r represents the retarded Green's function, Σ^r denotes the self-energy (due to out scattering), $\boldsymbol{D}$ denotes the phonon Green's function, and ϕ is the applied electrostatic potential.

Actually, it is not possible to include all interactions to infinite order and only first-order self-energies can be included. Using the Born approximation, the self-energies for the electron–phonon interactions may be expressed as follows (Pan, 2015):

$$\Sigma_{e-ph}^{\lessgtr}\left(r,r',E\right)=\frac{1}{\left(2\pi\right)^3}\sum_j\int D_{q,j}\exp\left(iq\left(r-r'\right)\right)\left(n_{q,j}+\frac{1}{2}\pm\frac{1}{2}\right)G^{\lessgtr}\left(r,r',E\pm\hbar\omega_{q,j}\right)dq \tag{84a}$$

Figure 9. Illustration of the utilization of the Green function method in device simulation

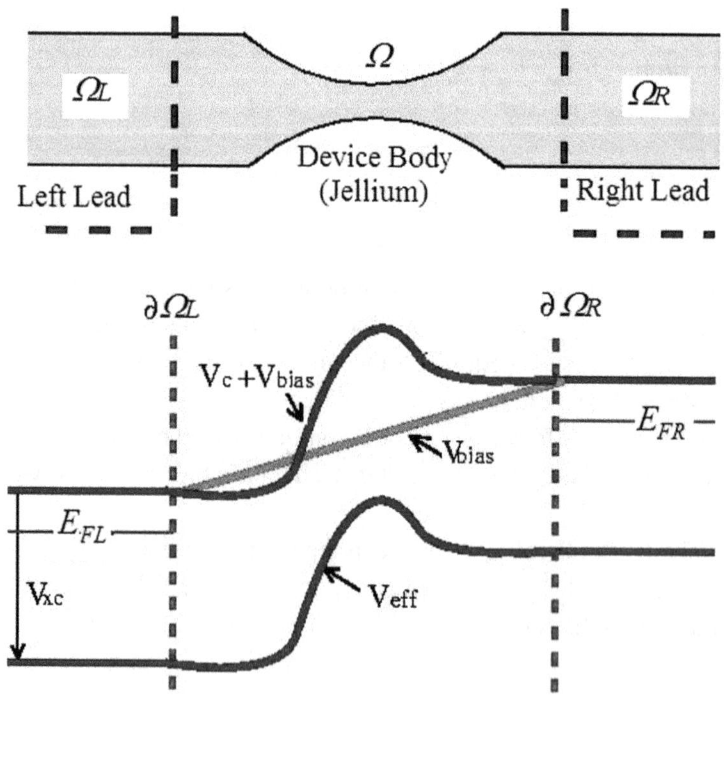

$$\Gamma_{\text{e-ph}} = j\,(\Sigma_{\text{e-ph}} - \Sigma_{\text{e-ph}}^{+}) \tag{84b}$$

$$\sum\nolimits_{e-ph}^{r}\left(r,r',E\right) = -\frac{i}{2}\Gamma_{e-ph}\left(r,r',E\right) + \mathrm{P}\int \frac{\Gamma_{e-ph}\left(r,r',E\right)}{2\pi\left(E-E'\right)}dE' \tag{84c}$$

where $\hbar\omega_{q,j}$ denotes the phonon energy of the jth branch j at the wavevector q, $n_{q,j}$ is the phonon occupation number (Bose-Einstein distribution, assuming equilibrium phonon population), $D_{q,j}$ is the electron–phonon interaction strength, and $\Gamma_{e\text{-}ph}$ is the scattering rate line width. Also, $\Sigma^{>}_{\text{e-ph}}$ and $\Sigma^{<}_{\text{e-ph}}$ are the energy broadening, and ***P*** represents the principal part of the integration (excluding singular point at $E=E'$). The ± sign denotes the phonon emission or absorption processes, respectively. The electron-phonon scattering processes may be elastic (energy conserving) or inelastic. The corresponding electron–phonon coupling strength may be expressed as follows:

$$D_{q,j}\,(Elastic) = (k_B T/2\, n\, m_c\, v_j).|M_j|^2, \tag{85a}$$

$$D_{q,j}\,(Inelastic) = (\hbar/2\, n\, m_c\, \omega_j).|M_j|^2, \tag{85b}$$

where M_j is the matrix element (for elastic or optical phonons), m_c is the crystal lattice atomic mass and v_j is the sound velocity (for elastic phonons). Similarly, the electron-impurity interactions can be expressed as follows:

$$\sum\nolimits_{e-i}^{\lessgtr}\left(r,r',E\right) = \rho\int\left|V(q)\right|^2\cdot\exp\left[jq\left(r-r'\right)\right]\cdot G^{\lessgtr}\left(r,r',E\right)dq \tag{86}$$

where, ρ is the density of impurities (N_i/V) and $V(q)$ is the Fourier transform of the Coulomb potential $V(r)$.

In order to illustrate the influence of coupling, Kubis and Vogl (2007) implemented a self-consistent NEGF simulation to study the quantum transport in several quantum structures. The Figure 10a depicts the simulation in the intrinsic region of an *N-I-N* structure. As shown in figure, the total length of the device is 50nm with 16nm intrinsic region between two *n*-type regions of 17nm length and doping of 10^{18}cm^{-3}. In the intrinsic region, there is a 12nm $In_{0.14}Ga_{0.86}As$ quantum well of 150meV height.

At zero bias, the full NEGF correctly yields the equilibrium Fermi-Dirac distribution function $f(E)=G^{<}(E)/\,[G^{r}(E)-G^{r\dagger}(E)]$. In contrast, the decoupled solution violates the Pauli principle at low energies. The state occupation exceeds the physical limit of 1 at energies below the Fermi level that (at 0eV). Figure 10b shows the I-V curve of a 40nm $GaAs/Al_{.3}Ga_{.7}As$ resonant tunneling diode with two 3nm wide $Al_{.3}Ga_{.7}As$ barriers and a 4nm quantum well in the center. Around the barriers, there is a 3nm intrinsic regions and a 12nm *n*-doped region with $n = 2\times10^{17}$cm^{-3},

15.2 Keldysh Formulation

Instead of working directly with the K-B equations, it may be useful to consider their integral forms. Historically, Leonid Keldysh derived his alternative form almost simultaneously with Kadanoff and Baym. However, the Keldysh and Kadanoff-Baym formalisms are equivalent.

According to Keldysh, the Green's functions should satisfy the following integral equation.

$$G^{\lessgtr}(1,2) = \int \left[\int G^{r}(1,3)\Sigma^{\lessgtr}(3,4).G^{a}(4,2).d4 \right] .d3 \tag{87a}$$

In matrix form, the Keldysh equation may be written as follows:

$$G^{<} = G^{r}\, \Sigma^{<}\, G^{a} \tag{87b}$$

where $\boldsymbol{G}^{<}$ is the lesser Green's function and $\Sigma^{<}$ is the lesser self-energy. This is equivalent to equation (49) but so much easier to apply. We dully note that the retarded and advanced Green's matrices (G^{r} and G^{a}) can be obtained from the inverted Hamiltonian matrix:

$$G^{r} = [E.O - H - \Sigma^{r}]^{-1} \tag{88}$$

where $\boldsymbol{H}$ is the Hamiltonian matrix of atoms in the scattering region ($H = H_{o} - e\phi$), $\boldsymbol{O}$ is the overlap matrix, and Σ^{r} is the retarded self-energies. The self–energies can be decoupled and related to the scattering rates line widths (Γ_{L}, Γ_{R}), which describes the coupling between the leads in the left and right sides with the device, as follows:

$$\Sigma^{<} = \Sigma^{<}_{L} + \Sigma^{<}_{R} = j.f_{L}.\Gamma_{L} + j.f_{R}.\Gamma_{R} \tag{89}$$

Here f_{L} and f_{R} are the equilibrium Fermi-Dirac energy distributions, in the left and right terminals, respectively. Once the Green's functions are determined by the Keldysh NEGF, one can construct the nonequilibrium carrier statistics, via the carrier density matrix:

$$\rho(E) = -\frac{j}{2\pi}\int_{-\infty}^{\infty} G^{<}(E)dE \tag{90}$$

Also, the Wigner distribution function can be obtained from the energy integral of the Green's function, $W(k,r,t) = \int G(k,\omega, r, t).d\omega$.

Figure 11 shows the flowchart of the self-consistent solution of NEGF-Poisson method using the Keldysh formulation.

Figure 10. (a) Results of simulation of an N-I-N structure by Green's functions (at zero bias); (b) I-V characteristics of a tunneling diode by NEGF
After Kubis & Vogl (2007)

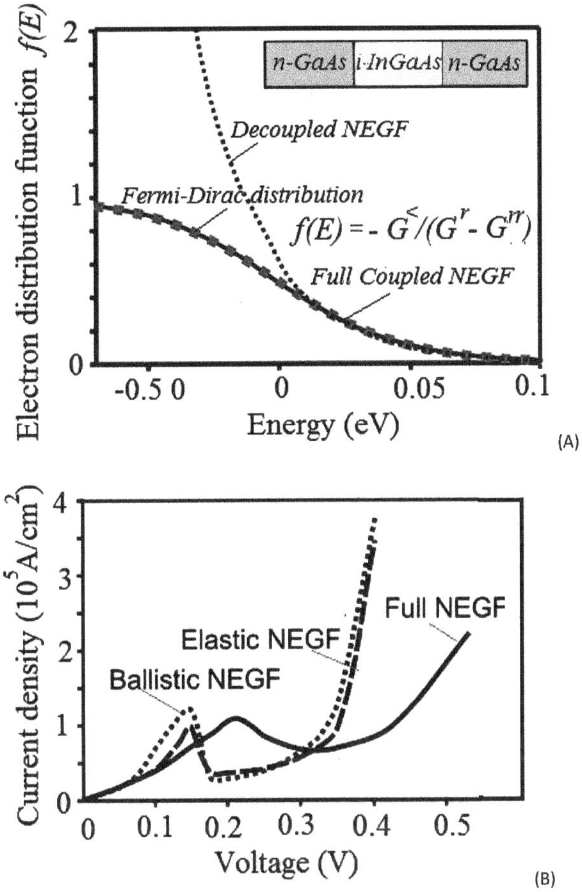

15.3 NEGF Calculation Procedure

The NEGF formalism can be applied to finite structures, to calculate the device characteristics as follows. The main calculation of the NEGF consists of a self-consistent computation of the Poisson equation and the quantum transport equation.As shown in figure, the domain is divided into left contact, right contact and the device. The calculations start with the NEGF transport equation (Kinetic equation). From the transport equation, the charge density (ρ) can be obtained and substituted into the Poisson equation. The Poisson equation computes the self-consistent potential (V_{sc}). It is important to note that since we have an infinite system, we obtain two types of solutions for the Green's functions, the retarded (G^r) and the advanced (G^a) solutions corresponding to outgoing and incoming waves in contacts. The retarded and the advanced Green's functions (G^r, G^a) describe the dynamics of carriers. Also, The greater and lesser Green's function, ($G^>$, $G^<$) deal with the statistics of carriers (e.g., carrier density) and can be derived from G^r, G^a, using the relations: $G^> = G^r\Sigma^> G^a$ and $G^< = G^r\Sigma^< G^a$. The reason for calculating the Green's function is that it is easier than solving the Schrödinger equation. Also, the Green's function of

Figure 11. Flowchart of the self-consistent NEGF-Poisson method by the Keldysh formulation

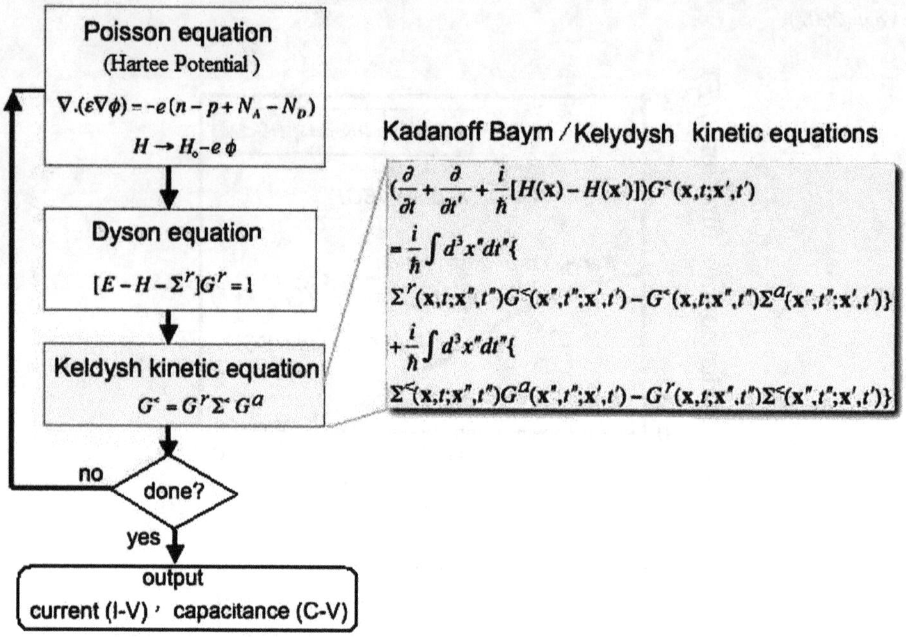

the central device region can be calculated separately without calculating the whole Green's function of the system (device + contacts).

The detailed procedure of the NEGF method to model nanoscale devices is as follows.

1. The first step is to identify a suitable set of basis functions and the Hamiltonian matrix (H) in the isolated device (central) region. This region may be the channel of a MOSFET or a carbon nanotube in a CNT-FET. The self-consistent potential (V_{sc}), is a part of the Hamiltonian matrix. The Hamiltonian matrix H describes the system at hand (e.g., MOSFET). It is usually a sparse matrix with connectivity only between neighboring mesh nodes, except for nodes at the boundary of the device which may have a non-local coupling (non-reflecting boundary condition).
2. The second step is to compute the self-energy matrices (Σ), which describe how the central device region couples to the left/right contacts and to the scattering process.
3. After identifying the Hamiltonian matrix (H) and the self-energies (Σ), the third step is to compute the retarded Green's function (G^r). Once the retarded Green's function is known, one can then calculate other Green's functions and determine the physical quantities (e.g., carrier density and current).

15.4 Boundary Conditions and Surface Green's Functions

The treatment of boundary conditions requires a special care in quantum transport problems in general, and the NEGF method in particular. The utilization of bulk Green's functions may be a reasonable approximation to capture the important characteristics of the contact electrodes. However, surface Green's functions (g_x) and the corresponding self-energies (Σ_X) are a more proper way to construct the contact electrode Green's functions. In fact, the contact electrodes constitute a semi-infinite system as compared to the infinite crystal assumed in the construction of bulk Green's function

A common problem in NEGF is to define Ohmic contacts of a given device (Frensley, 1990). In NEGF, we may define contacts as Ohmic if the current is controlled by the interior of the device (rather than by serial resistances or interface states). According to Kubis & Vogl (2009), this definition has the following implications:

- There must be a smooth transition in the density of states between the device and leads.
- The same scattering mechanisms must act within the device and leads.
- The carrier distribution within the leads must be a suitable Fermi-Dirac distribution to ensure the current continuity.

These conditions should be then applied to avoid quantum reflections and charge accumulation at the device interface regions.

15.5 Advantages and Disadvantages of the NEGF

The non-equilibrium Green's function (NEGF) approach is a state-of-the-art modeling tool in predicting performance and designing emerging nanoscale devices. The NEGF is a microscopic quantum theory for quantum transport that may include all particle interactions and scattering mechanisms. For instance, the electron-electron interaction can be considered through the Hartree-Fock approximation (correlation & exchange integrals). Also, the electron-phonon interaction can be considered through a self-consistent solution for the self-energy term in the NEGF formulation. Ballistic transport (coherence) of charge and energy carriers can be also considered in this method. However, scattering is typically incorporated in a phenomenological way using Büttiker probes. The connection between the self-energy and physical scattering mechanisms in this case is not straight forward. In comparison with the WBTE simulation methods, the NEGF simulation requires the consideration of eight dimensions (x1,y1,z1, x2,y2,z2, t1,t2) while the WBTE requires seven dimensions (x,y,z, px,py,pz, t). The higher dimensionality of the NEGF simulations makes it more demanding for computational resources, but allows the description of temporal correlations.

The NEGF method is able to deal with explicitly time-dependent as well as with stationary problems. The time-dependent calculations requires the knowledge of the four basic types of Green's functions (G^r, G^a, $G^<$, $G^>$), while stationary problems needs 2 types of Green's functions (G^r, $G^<$).

However, there is no rigorous self-consistent scheme to implement the NEGF, with Poisson's equation in real devices, without adopting a series of approximations. Also, there are no self-consistent forms of phonon self-energy without additional approximations (Pal et al, 2012). In comparison with semiclassical calculations, much more effort is required to properly take into account the physics of contacts. This is a consequence of the nonlocal nature of quantum mechanics and the nature of scattering in open quantum systems. In addition, the charge and current conservation are not satisfied once approximations are introduced.

Finally, it should be noted that the computational costs of the full NEGF simulations are prohibitive, in terms of CPU time and other computer resources. In fact the computational complexity of the NEGF scales as $O(N^3)$ where N is the number of elements (atomic orbitals or grid points)[9]. In addition, when applying NEGF method to transient analysis to calculate current as a function of time, the computational cost scales with $O(N^3)$ where N, in this case, is the number of time steps (Stokpro, 2008).

The most expensive calculation is computing some of the entries of the retarded Green's functions matrix G^r: $Gr = [EI - H - \Sigma]^{-1} = A^{-1}$ and the less-than matrix $G^<(E) = G^r S^<(G^r)^+$. By the aid of certain exotic techniques, the problem of computing density of states and transmission coefficient (and hence the carrier density and current) in the NEGF can be reduced to finding certain entries from the inverse of a large and sparse matrix, like A. The recently proposed methods, such as FIND algorithm (Li et al, 2008), together with massive parallel processing capabilities, have enabled the NEGF approach to attack real devices in 2D and 3D (Steiger et al., 2015).

16. CASE STUDY 4: QUANTUM SIMULATION OF AN N-I-N, USING NEGF

In this section we show some details of the NEGF quantum simulation of an N-I-N nano structure, similar to the one presented so far in Figure 10. This simple example shows the power of the NEGF quantum approach. The simulated N-I-N structure is 50nm long and has 12nm *InGaAs* quantum well as intrinsic zone.

The main calculation in the NEGF method consists in computing Green's function matrices. The NEGF method has been above described and more details can be found in the literature, e.g., Datta (2005). In brief, we divide the device into three regions (Left contact, Central device, and Right contact). The system Hamiltonian can be then written as:

$$\begin{bmatrix} H_L & \tau_L & 0 \\ \tau_L^+ & H_C & \tau_R^+ \\ & \tau_R & H_R \end{bmatrix} \tag{91a}$$

where H_L, H_C and H_R denote the Hamiltonian matrices of the left, central and right parts, respectively. Also, τ_L and τ_R are the matrices involving the interaction between the left/right contacts and the central region. We assume that there is no direct tunneling between leads. The retarded Green's function G^r corresponding to this Hamiltonian is obtained by inverting the matrix equation $(E^+.O - H)G^r = I$, which can be written as:

$$\begin{bmatrix} E^+.O_L - H_L & \tau_L & 0 \\ \tau_L^+ & E^+.O_C - H_C & \tau_R^+ \\ & \tau_R & E^+.O_R - H_R \end{bmatrix} \cdot \begin{bmatrix} G_L & G_{LC} & 0 \\ G_{CL} & G_C & G_{CR} \\ 0 & G_{RC} & G_R \end{bmatrix} = \begin{bmatrix} I & 0 & 0 \\ 0 & I & 0 \\ 0 & 0 & I \end{bmatrix} \tag{91b}$$

where O_L and O_R are the overlap matrices. If the center region is long enough, there will be no overlap elements in O or H between the left and the right electrodes. Also, $E^+=E+j\eta$ contains a small infinitesimal imaginary (broadening) energy, η, to facilitate the computations. Once Green's function matrices are computed, the electron density and the current at the device terminals can be obtained.

The Figure 13 shows the simulated N-I-N structure and the carrier density distribution, along the device, with different methods, when the applied voltage is 150 mV. Also, Figure 14 depicts the electron energy contours, along the device at the same bias. The calculations are carried out with the NEGF simulator NEMO, which is a famous NEGF simulator (Klimeck et al., 2016). Here, the calculation of charge-carrier density is performed with and without scattering, and compared with self-consistent solution by the semiclassical Boltzmann equation.

The ballistic NEGF calculation (without collisions) is equivalent to the solution of Schrödinger equation of an open system (without contacts). This reflects a coherent, energy conserving, ballistic transport. In this case, the carrier density within the device is fully determined by the overlap of the lead wave functions with the device. Since there are no lead carriers below the *GaAs* band edge, the quantum well states in the intrinsic region remain unoccupied. Thus, the electron density within the quantum well only stems from lead electrons, which explains the oscillatory density in the intrinsic region.

17. MULTIBAND QUANTUM MODELS

The quantum microscopic models, that we presented so far, are all concerned with the transport of conduction electrons and their associated current. These models may be divided into two basic categories, namely:

- Non-statistical (Pure-state) methods, which are Schrödinger equation-based and
- Statistical (Mixed-state) methods, such as density matrix-based and Wigner function-based methods

The majority of these quantum models were developed assuming a single parabolic band. The hole current has been only included in the macroscopic level (i.e., the QHDM). Therefore, the possible interband currents are not considered in such models. The so-called multiband quantum transport models (MBQTM) are concerned with interband transitions (e-g., conduction-to-valence and vice versa) and their associated currents. MBQTM models are particularly important for high-speed devices, such as resonant-tunneling diodes (RTD's) and quantum cascade laser (QCL).

Figure 12. The self-consistent procedure of the NEGF, with Poisson's equation

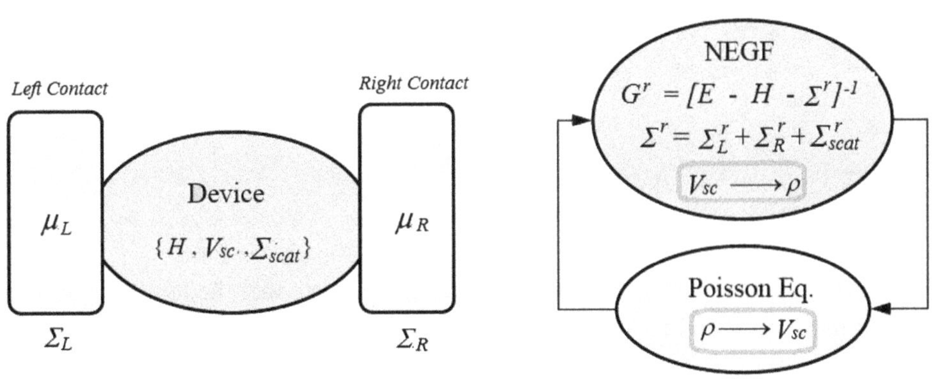

Figure 13. Electron density distribution
After Kubis et al (2009).

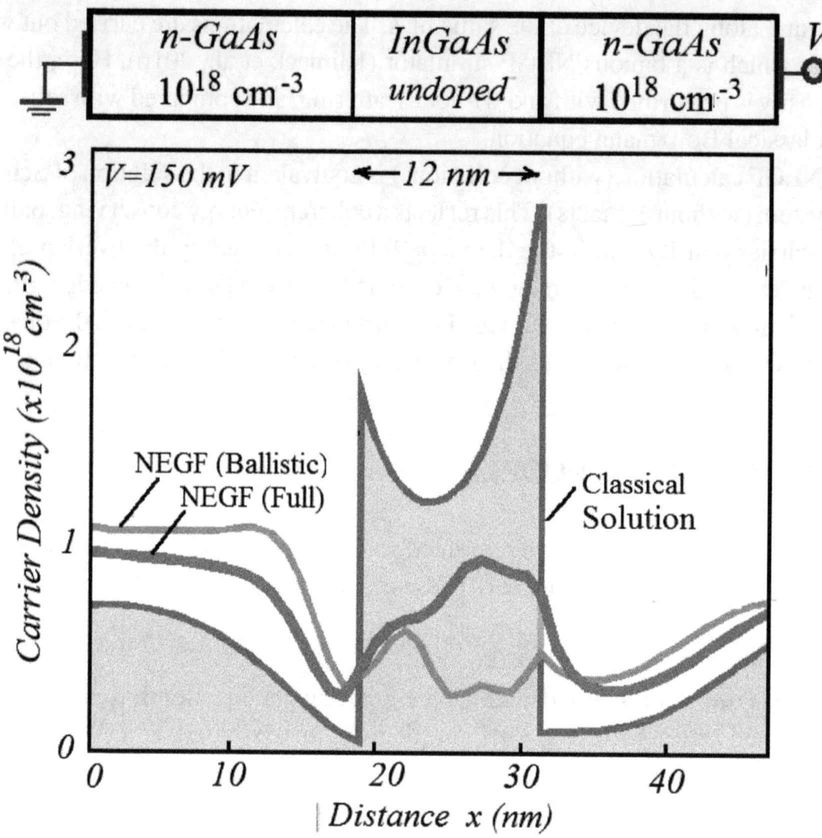

17.1 Multiband Models, Based on the Schrodinger Equation

Most of mathematical models of quantum transport in semiconductor devices make use of the effective-mass approximation (EMA). This consists in replacing the true Hamiltonian

$$\hat{H} = \left[-\frac{\hbar^2}{2m_o} \nabla^2 + V(r) \right] \tag{92a}$$

with

$$\hat{H} = \left[-\frac{\hbar^2}{2} \nabla \left(\frac{1}{m^*(r)} \nabla \right) + V_{ex}(r) \right] \tag{92b}$$

where m^* is the effective-mass tensor: $m^{*-1} = \nabla \otimes \nabla E(k) \,|_{k=0}$. Note that the total potential energy $V(r)$ includes both the internal crystal field as well as the external potential $V_{ex}(r)$ due to conduction band edge profile that includes band offsets, band shifts due to strain, and the electrostatic potential. In such approximation the charge carriers (electrons or holes) belong exclusively to the low energy levels of conduction or valence bands. However, at higher energies the coupling between bands becomes relevant, and the EMA breaks down. In this case, the inclusion of nearby bands of interest, by any band structure calculation method, is necessary.

Ideally the electronic structure can be self-consistently calculated using *ab initio* methods, such as the density functional theory (DFT), or even approximate methods, such as the tight-binding (TB), but such methods are time-consuming.

A common compromise is to use the so-called ***k.p*** theory (Luttinger & Kohn, 1955) to include band coupling within relatively small basis. The ***k.p*** method allows the extrapolation of the band structure over the entire Brillouin zone from the energy gaps and matrix elements at the zone center. One can improve the accuracy of the ***k.p*** method by including the strongly coupled bands and treating the influence of distant bands as perturbation.

There is a hierarchy of ***k.p*** models including 2-band, 4-band, 6-band and 8-band Hamiltonians. The simplest one is the Kane 2-bands model (Kane, 1956). It includes one conduction band and one valence band and consists of two coupled Schrödinger-like equations, with a coupling term. The Hamiltonian of the kane 2-band model consists of a 2×2 matrix as follows:

$$H = \begin{bmatrix} \frac{p^2}{2m_o} + \frac{1}{2}E_g & -j\frac{\hbar}{m_o}\boldsymbol{k}\cdot\boldsymbol{p} \\ j\frac{\hbar}{m_o}\boldsymbol{k}\cdot\boldsymbol{p} & \frac{p^2}{2m_o} - \frac{1}{2}E_g \end{bmatrix} = \begin{bmatrix} -\frac{\hbar^2}{2m_o}\Delta + \frac{1}{2}E_g & -\frac{\hbar^2}{m_o}\boldsymbol{k}\cdot\nabla \\ \frac{\hbar^2}{m_o}\boldsymbol{k}\cdot\nabla & -\frac{\hbar^2}{2m_o}\Delta - \frac{1}{2}E_g \end{bmatrix} \tag{93a}$$

Here, $\boldsymbol{k}$ is the matrix element of the gradient operator between the Bloch functions of the two bands (u_c, u_v), such that $\boldsymbol{k}=\int u_c(x)\nabla u_v(x)dx$, and the integration is taken over the lattice unit cell. Also, E_g is the band-gap between the two bands and m_o is the electron mass. The next stage in the *k.p* hierarchy is the 4-band Luttinger-Kohn-Hamiltonian, which describes the band-mixing between the heavy and light holes. This 4-band model neglects the spin-orbit coupling

$$H = \begin{bmatrix} E_{1,0} + \frac{\hbar^2 k^2}{2m_o} k\cdot p_{1,1} & \frac{\hbar}{m_o} k\cdot p_{1,2} & \frac{\hbar}{m_o} k\cdot p_{1,3} & \frac{\hbar}{m_o} k\cdot p_{1,4} \\ \frac{\hbar}{m_o} k\cdot p_{2,1} & E_{2,0} + \frac{\hbar^2 k^2}{2m_o} k\cdot p_{2,2} & \frac{\hbar}{m_o} k\cdot p_{2,3} & \cdots \\ \frac{\hbar}{m_o} k\cdot p_{3,1} & \frac{\hbar}{m_o} k\cdot p_{3,2} & E_{3,0} + \frac{\hbar^2 k^2}{2m_o} k\cdot p_{3,3} & \cdots \\ \frac{\hbar}{m_o} k\cdot p_{4,1} & \cdots & \cdots & \cdots \end{bmatrix} \tag{93b}$$

Alternatively, one can use the Kane 8-band Hamiltonian to include the spin-orbit coupling (Kane, 1966). In this case, we consider an additional term H_{so} that takes into account the relativistic effect of spin:

$$H_{so} = \frac{\hbar^2}{4m_o^2c^2}\nabla(V_{ex}xp)\cdot\sigma \tag{94a}$$

where σ is the vector of the Pauli matrices $\sigma = (\sigma_1; \sigma_2; \sigma_3)^T$. For instance, the zincblende semiconductor structures have the following spin-orbit interaction Hamiltonian (Hinckley & Singh, 1990):

$$H_{SO} = \frac{1}{3}\Delta_{SO}\begin{pmatrix} 0 & 0 & 0 & 0 & 0 & 0 & 0 & 0 \\ 0 & 0 & 0 & 0 & 0 & 0 & 0 & 0 \\ 0 & 0 & 0 & -i & 0 & 0 & 0 & 1 \\ 0 & 0 & i & 0 & 0 & 0 & 0 & -i \\ 0 & 0 & 0 & 0 & 0 & -1 & i & 0 \\ 0 & 0 & 0 & 0 & -1 & 0 & i & 0 \\ 0 & 0 & 0 & 0 & -i & -i & 0 & 0 \\ 0 & 0 & 1 & i & 0 & 0 & 0 & 0 \end{pmatrix} \tag{94b}$$

where Δ_{so} is the spin-orbit split-off energy.

Usually, the ***k.p*** method is utilized in combination with the so-called envelope function approximation. The envelope function ansatz (Ehrhardt & Koprucki 2014) is based on Bloch's theorem, according to which the electron wave function in a crystal with translational symmetry can be separated into an oscillating Bloch part over atomic distance and a smooth envelope function which varies on a larger mesoscopic scale.

In layered semiconductor heterostructures, the ***k.p***-Schrödinger equation approach for d bands, enables us to write the eigenfunctions $\psi(r)$ as

$$\Psi_{k//}(r) = \exp\left(jk_{//}\cdot r_{//}\right)\sum_{n=1}^{d}\varphi_n(z,k_{//})\cdot u_n(r) \tag{95a}$$

where z denotes the growth direction of the semiconductor layers and the index // indicates in-plane vectors, such that in-plane reduced wave vector $k_{//} = (k_x, k_y)$. Also, $u_n(r)$ are lattice-periodic Bloch functions and $\phi(z, k_{//})$ are the corresponding envelope functions describing the variation of wave-functions on a larger meso-scale. Using the perturbation theory, the rapidly oscillating Bloch functions can be eliminated from the Hamiltonian, and the resulting electron Hamiltonian only contains the envelope functions. Therefore, the envelope functions $\phi = (\phi_1, .. \phi_d)$ fulfill the ***k.p***-Schrödinger equation

$$j\frac{\partial\varphi}{\partial t} = H\left(-j\frac{\partial}{\partial z}, k_{//}\right)\varphi \tag{95b}$$

The following figure depicts the subband structure of 8.5nm thick *InGaAs* quantum well from TB, EMA and 4-band k.p assuming infinite potential boundary conditions. The material parameters and doping profile of this structure are based on (Pan, 2015).

Recently, other pure-state multiband models have been proposed (Barletti, Demeio & Frosali, 2015), on the basis of Luttinger-Kohn L-K (Luttinger & Kohn,1955) and Wannier-Slater W-S (Wannier, 1962) envelope functions. The book of Voon and Willatzen (2009) provides an overview on all types of the k.p Hamiltonians for bulk and nanostructured semiconductors._

17.2 Multiband Models, Based on the Wigner Distribution

The kinetic models of quantum transport that make use of the Wigner function are less complex than other quantum statistical approaches, such as the density-matrix approach and the Green's-function method. In addition, the inclusion of collisions to the WBTE is less complicated than adding them into other statistical models of quantum transport. However, all transport models for semiconductors, which are based on the Wigner function, assume a single parabolic conduction-band with EMA. In order to tackle with devices in which interband transitions are important, the EMA is not satisfactory and a well-defend Wigner function should include the populations of all bands involved in the transport process.

The statistical multiband model, which is based on the Wigner formalism, starts from an envelope-function model, such as the Kane model and then applies the Wigner/ Weyl transformation directly to the envelope functions (Borglie et al., 2003). For a 2-band model this yields a 2×2 matrix of Wigner functions (Wigner matrix). The basic idea of this model is to evolve the entire Wigner matrix rather than just the diagonal elements in the conventional case. The off-diagonal elements of the Wigner matrix can handle the multiband (non-adiabatic) transitions. For the case of 2-band model (with 2×2 Wigner matrix), this approach yields four coupled WBTE-like equations (Chai et al. 2014):

$$\frac{\partial W_{cc}}{\partial t}+\frac{p}{m_o}.\nabla_r W_{cc}+\frac{j}{\hbar}\theta_{cc}W_{cc}=-\frac{j\hbar k}{2m_o}.\nabla_r\left(W_{cv}-W_{vc}\right)-\frac{k.p}{m_o}\left(W_{cv}+W_{vc}\right) \tag{96a}$$

Figure 14. Electron energy contours in the N-I-N structure
After Kubis et al. (2009)

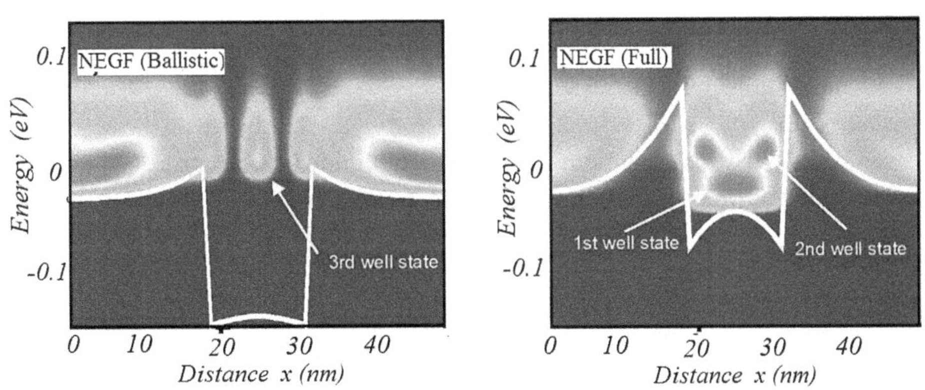

$$\frac{\partial W_{cv}}{\partial t}+\frac{p}{m_o}.\nabla_r W_{cv}+\frac{j}{\hbar}\theta_{cv} W_{cv}=+\frac{j\hbar k}{2m_o}.\nabla_r\left(W_{cc}+W_{vv}\right)+\frac{k.p}{m_o}\left(W_{cc}-W_{vv}\right) \tag{96b}$$

$$\frac{\partial W_{vc}}{\partial t}+\frac{p}{m_o}.\nabla_r W_{vc}+\frac{j}{\hbar}\theta_{vc} W_{vc}=-\frac{j\hbar k}{2m_o}.\nabla_r\left(W_{cc}+W_{vv}\right)+\frac{k.p}{m_o}\left(W_{cc}-W_{vv}\right) \tag{96c}$$

$$\frac{\partial W_{cc}}{\partial t}+\frac{p}{m_o}.\nabla_r W_{cc}+\frac{j}{\hbar}\theta_{cc} W_{cc}=-\frac{j\hbar k}{2m_o}.\nabla_r\left(W_{cv}-W_{vc}\right)+\frac{k.p}{m_o}\left(W_{cv}+W_{vc}\right) \tag{96d}$$

where the operator θ_{ij} is acting on Wigner matrix elements (W_{ij}) as follows:

$$\theta_{ij} W_{ij}(r,p)=\frac{1}{(2\pi\hbar)^3}\cdot\int V_{ij}(r,x).W_{ij}(r,p).\exp\left[-\frac{j(p-p').x}{\hbar}\right]dp'dx \tag{97}$$

with the index $i,j = \{v, c\}$ denoting the valence and conduction bands and the potential V_{ij} is defined as:

$$V_{ij}(r,x)=(E_i+V)\left[r+\frac{1}{2}x\right]-(E_j+V)\left[r-\frac{1}{2}x\right] \tag{98}$$

17.3 Multiband Hydrodynamic Models

The macroscopic hydrodynamic models (HDM or its quantum version QHDM) are less complex than the kinetic models (BTE or WBTE) and more accurate than the DDM. The majority of these models are either based on single parabolic conduction band or separate parabolic multivalleys or separate conduction-valence (bipolar) nonparabolic bands[10].

On the basis of the multiband kinetic models, many researchers have attempted to develop a multiband QHDM, e.g., Wilson, (1988), Gardner (1994), Barletti, Demeio & Frosali (2007), and Barletti, Brogioli & Frosali (2014). In fact, such models have faced many difficulties, such as the absence of the physical meaning of the concerned macroscopic quantities and the corresponding moment equations. Are they really conservation equation? Of What exactly?

One of the successful approaches to develop a multiband QHDM, is based on the Madelung approach, which consists in writing the wavefunctions in terms of amplitude (n) and phase (Φ) quantities[11]. In the case of a two-band model, we have the following envelop functions:

$$\psi_a=\sqrt{n_a}.\exp\left(j\frac{m}{\hbar}\Phi_a\right) \tag{99}$$

where $a \in \{c, v\}$, m is a scaling factor and the corresponding densities are:

$$n_c = |\psi_c|^2,\ n_v = |\psi_v|^2,\ n_{cv} = \psi_c{}^*\psi_v,\ n_{vc} = \psi_v{}^*\psi_c \tag{100}$$

The quantities n_c *and* n_v correspond to the position probability densities of the valence and conduction band electrons. The total electron density is then $n = n_c + n_v = |\psi_c|^2 + |\psi_v|^2$. The complex densities n_{cv} and n_{vc} may be written in the following form:

$$n_{cv} = \sqrt{n_c}.\sqrt{n_v}.exp\,[j\delta\,],n_{vc} = \sqrt{n_c}.\sqrt{n_v}.exp\,[-j\delta\,], \tag{101}$$

where $\delta = \Phi_v - \Phi_c$. The corresponding quantum current densities are:

$$J_c = -e(\hbar / \alpha)\,\mathrm{Im}\left[\psi_c^*\nabla\psi_c\right] = -en_c\nabla\Phi_c, \tag{102a}$$

$$J_v = -e(\hbar / \alpha)\,\mathrm{Im}\left[\psi_v^*\nabla\psi_v\right] = -en_v\nabla\Phi_v \tag{102b}$$

Also, the complex currents J_{cv} and J_{vc} may be written as follows:

$$J_{cv} = -e(\hbar / \alpha).\,\mathrm{Im}\left[\psi_c^*\nabla\psi_v\right] = -en_{cv}u_v, \tag{103a}$$

$$J_{vc} = -e(\hbar / \alpha)\,\mathrm{Im}\left[\psi_v^*\nabla\psi_c\right] = -en_{vc}u_c \tag{103b}$$

where u_v and u_c are some sort of complex velocities, which are given by:

$$u_c = (\hbar/\alpha).(\nabla\sqrt{n_c}/\sqrt{n_c}) + j\nabla\Phi c,\ u_v = (\hbar/\alpha).(\nabla\sqrt{n_v}/\sqrt{n_v}) + j\nabla\Phi v \tag{74}$$

For the case of 2-band Kane model, this approach yields four coupled continuity-like equations, for n_c, n_v, J_c and J_v (Chai et al. 2015):

$$-e\frac{\partial n_c}{\partial t} + \nabla\cdot J_c = 2eK\cdot\mathrm{Im}\left(n_{cv}n_v\right) \tag{105}$$

$$-e\frac{\partial n_v}{\partial t} + \nabla\cdot J_v = 2eK\cdot\mathrm{Im}\left(n_{cv}^*n_c\right) \tag{106}$$

where K is the interband coupling constant. Adding these equations yields:

$$-e\frac{\partial n}{\partial t}+\nabla\cdot J_n=2eK\cdot\mathrm{Im}\left(n_{cv}^{*}n_c+n_{cv}n_v\right) \tag{107}$$

Similarly, we can get the following coupled equations for J_c and J_v;

$$\frac{\partial J_c}{\partial t}-\nabla.\left(\frac{J_c\otimes J_c}{en_c}\right)+en_c\nabla\left(\frac{\hbar^2}{\alpha^2}.\frac{\Delta\sqrt{n_c}}{2\sqrt{n_c}}\right)-en_cV_c=-\left(\frac{e\hbar}{\alpha}\right).\mathrm{Re}\left(n_{cv}K.u_v\right)+2e\,\mathrm{Re}\left(n_{cv}K.u_vu_c^{*}\right) \tag{108a}$$

$$\frac{\partial J_v}{\partial t}-\nabla.\left(\frac{J_v\otimes J_v}{en_v}\right)+en_v\nabla\left(\frac{\hbar^2}{\alpha^2}.\frac{\Delta\sqrt{n_v}}{2\sqrt{n_v}}\right)-en_vV_v=+\left(\frac{e\hbar}{\alpha}\right).\mathrm{Re}\left(n_{cv}^{*}K.u_c\right)-2e\,\mathrm{Re}\left(\hat{n_{cv}}K.u_cu_v^{*}\right) \tag{108b}$$

where the $V_c=E_c+V$ and $V_v=E_v+V$. In order to close the above system of hydrodynamic equation (97, 98, we may add the following constituent equation for δ:

$$\left(\frac{e\hbar}{\alpha}\right)\nabla\delta+\frac{J_v}{n_v}-\frac{J_c}{n_c}=0 \tag{109}$$

This system can be solved for n_c, n_v, J_c and J_v. Other variables (n_{cv}, n_{vc} u_c, u_v, can be expressed in terms of these variables and the phase difference δ.

Naturally, this 2-band QHDM can be extended to higher orders, using the same concept.

18. QUANTUM TRANSPORT IN ATOMISTIC AND MOLECULAR DEVICES

Atomistic and molecular devices represent one of the major fields of current development in nanotechnology. Development of the electronic devices at the molecular scale involves study of the electronic level structure, response and charge transport at atomic scale. These devices can be constructed of single molecules, small groups of molecules, carbon nanotubes, nanoscale metallic or semiconductor wires. Molecular electronics emerging applications are tremendously exciting; for example, molecules can function as transistors, switches, rectifiers, IC inter-connects, photovoltaic, memories and sensors.

The advent of experimental progress in molecular electronics has given rise to new challenges in developing theoretical tools to describe the electron transport in molecular devices. In the case of very small semiconductor devices, the transverse dimensions are small, and the band structure of large volumes is no longer appropriate. Typical cases of this situation are resonant tunneling diodes, nanowires, carbon nanotubes, and molecules stretched between metallic leads. Here, the analytic band structure and effective mass approaches, we used previously, are no longer applicable.

Figure 15. Band structure of an InGaAs quantum well by TB, EMA, and k.p, methods
From Pan (2015)

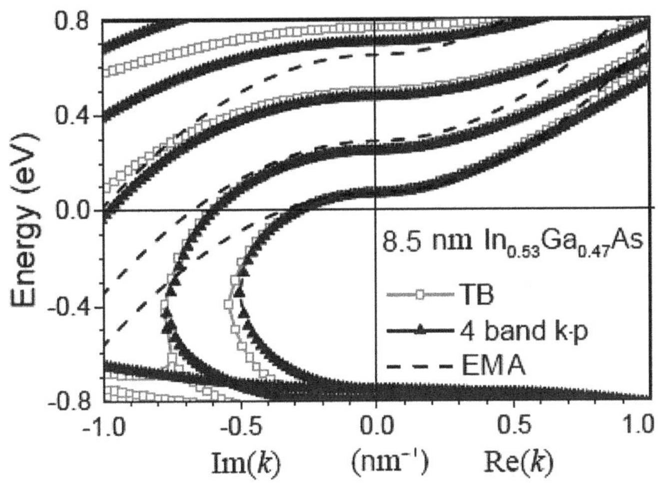

18.1 Quantum Transport Modeling from Atomic First Principles

In order to understand and model nanoscale devices, we need a detailed model of the interaction between the electrons and the individual atoms. At the atomic scale such a model must be based on a quantum mechanical model. The most fundamental description is through the Schrödinger equation involving all the electrons and the ionic cores of the atoms. The main numerical problem with the Schrödinger equation is that it couples the motion of the electrons and this makes a general solution intractable. A very popular strategy for avoiding solving the full Schrödinger equation is using a self-consistent field (**SCF**) model, where each electron is described as an independent particle interacting with the SCF from all the other electrons. The most successful SCF approaches are based on either Hartree-Fock (H-F) equation or the density functional theory (DFT).

In DFT each electron is influenced by a SCF determined from the total electron density through a classical electrostatic contribution (Hartree potential), and an exchange–correlation potential, which arises from the quantum mechanical nature of electrons. The exchange–correlation potential can only be calculated approximately, and there is a strong effort to develop improved exchange–correlation functionals for DFT. The local density approximation (LDA) and the generalized gradient approximation (GGA) to DFT, are popular approximations which proved very successful. Practically, the LDA and GGA methods are known to underestimate energy gaps in semiconductor materials and their results must be analyzed to judge their validity. Recent research seems promising for correcting the deficiencies of the DFT-LDA and DFT-GGA. However, if the energy gap or energy barriers are important in your study, the so-called GW approximation (Hedin, 1965) is more accurate, as shown in Figure 17.

The *Ab initio* methods can be also extended to study the transport of charge carriers in molecular and atomistic systems. In fact, the electronic transport properties of tiny systems are determined –to large extent- by their band structure. For instance, consider the atomistic system in the following figure where atoms are arranged in 2D layers. The electron and hole densities in each 2D layer are given by:

Figure 16. Example of a molecular device, consisting of a molecule stretch between two metal contacts
Values have been extracted from Tran and Blaha (2009).

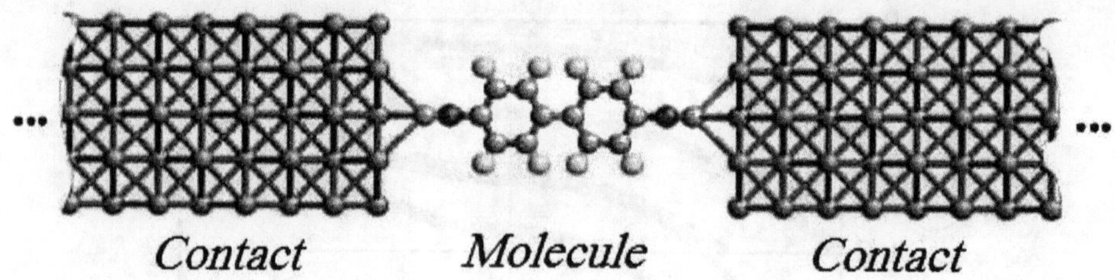

Figure 17. Experimental and theoretical bandgap energy of different materials
Values have been extracted from Tran and Blaha (2009)

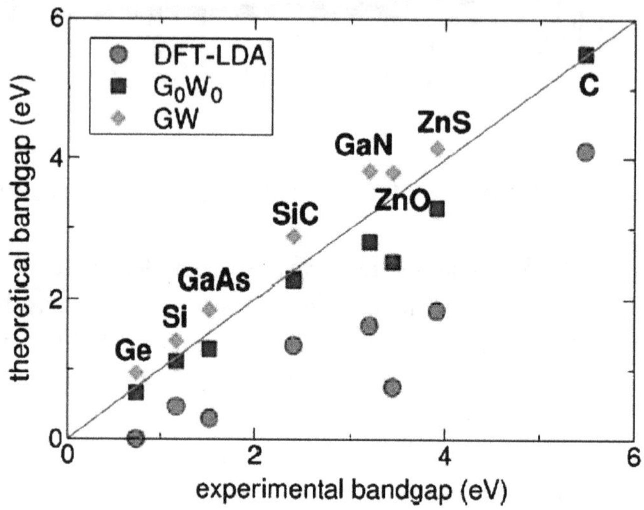

$$n(z) = \frac{1}{\left(2\pi\right)^2} \int_{BZ_{//}} d^2\mathbf{k}_{//} \sum_c \left|\left\langle z \middle| E_c \mathbf{k}_{//} \right\rangle\right|^2 f\left(E_c - F_n\right) \tag{111a}$$

$$p(z) = \frac{1}{\left(2\pi\right)^2} \int_{BZ_{//}} d^2\mathbf{k}_{//} \sum_v \left|\left\langle z \middle| E_v \mathbf{k}_{//} \right\rangle\right|^2 \left(1 - f\left(F_h - E_v\right)\right) \tag{111b}$$

The influence of free carrier charge redistribution and polarization fields can be included by solving the *Poisson equation:*

$$\frac{d}{dz} D(z) = \frac{d}{dz}\left(-\varepsilon \frac{d}{dz} V_H + P\right) = e\left(p - n + N_D^+ - N_A^-\right) \tag{112}$$

The Hamiltonian of the system can be then evaluated ($H = H_C+V_H$) and from this Hamiltonian, one can calculate energy levels, self-consistently, and substitute $E_i k_{//}$ in equation (111). In fact, the band structure and local strains in nanodevices should be taken into account, and therefore, the atomistic approaches, such as the DFT are important.

18.2 Energy Band Structure and Transport Calculation Methods

In the transport theory, band structure is first calculated and used as input for the calculation of the carrier distribution functions. There have been several methodologies developed to study the energy band structure of matter. In addition to the Hartee-Fock (H-F) method that we presented in the beginning of this chapter, we have the density function theory (DFT) as well. In addition, some approximate techniques, such as the tight-binding electron (TBE) model may be also used to build the material energy band structure.

Generally speaking, there exist two basic categories of computational methods, to develop the energy band structure of a matter, namely:

- Ab initio methods,
- Approximate (or empirical) methods.

The *ab initio* computational methods are based on first-principles quantum physics, to find out the energy band structure of a matter. *Ab initio* electronic structure methods have the advantage that they can be made to converge to the exact solution, when all approximations are sufficiently small. In this case, when all possible configurations are included, such methods tend to the exact solution of the electronic Schrödinger equation (in the Born-Oppenheimer approximation).

The H-F, technique is the simplest *ab inito* method, in conjunction with the self-consistence field methods. The computational load of the H-F method scales as N^4 (N being the number of basis wavefunctions). However in practice it can scale closer to N^3 as the program can identify and neglect zero

Figure 18. Flowchart of the Kohen-Sham method

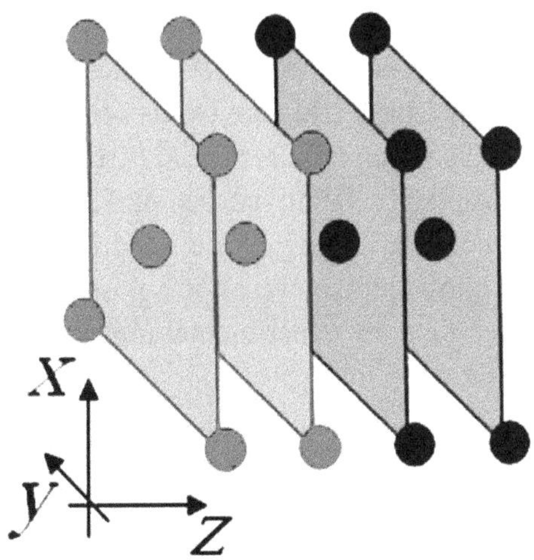

and small integrals. The Density functional theory (DFT) methods use functionals, which include the Hartree–Fock exchange potential (Kohn, Becke & Parr, 1996).

The dark side of *ab initio* methods is their high computational cost. They often take enormous amounts of CPU time and memory. For this reason, many researchers have resorted to the so-called approximate or empirical methods to develop the energy band structure of solid materials. In general, there exist two basic approximate methods, which are widely used for the calculation of energy bands in solids, namely:

- The quasi-free electron (QFE) model, and
- The tightly bound electron (TBE) model.

These approximate methods use the perturbation theory. But they differ in their zero approximation. The quasi-free electron (QFE) method (which represents the physicists approach) takes the free electron model as a zero-order approximation and the periodic field of the lattice as a perturbation. When the core contribution is replaced by a simplified pseudopotential, it results in the so-called pseudopotential methods (Hamann, Schlüter & Chiang, 1979).

Pseudopotential Methods

The pseudopotential methods are based upon the fact that the core electrons do not play a significant role in defining the chemical bond or the physical properties, which are mainly described by valence electrons. Therefore, core electrons are removed from the band structure calculation and the ionic potential is replaced with a soft pseudopotential. The overall potential function reproduces the true potential outside the core region ($r>r_c$), but is smooth inside the core ($r<r_c$), as shown in Figure 19. The pseudopotential V(r) can be expressed in terms of the structure factor SG and the form factors VG as follows:

$$V(r) = \sum_{G} V_G S_G \exp(jG.r) \tag{112a}$$

For diamond structures, there are two atoms in the primitive unit cell and the structure factor is given by: $S_G=cos(G.t)$, where $t = (a_o/8)(1,1,1)$. Assuming that the atomic pseudopotentials are spherically symmetric, the form factors VG only depend on the absolute value of the reciprocal lattice vector (G). Practically, the employed parameters in the empirical pseudopotential calculations consist of three local form factors (V3, V8 and V11), two parameters to model the nonlocal correction, and two parameters for the spin-orbit interaction term. In general, there are two approaches for calculating pseudopotential form factors for a specific material. They can be determined by fitting a small number of experimental data, such as the position of peaks in optical reflectivity spectra. This approach is known as the Empirical Pseudopotential Method (EPM). The disadvantage of the EPM is that it requires experimental inputs. With the availability of high-speed computers, however, it is possible to determine the pseudopotential form factors from first principles without any experimental input. The first-principles pseudopotential methods are known as self-consistent or ab initio pseudopotential methods.

After determining the pseudopotential, the energy dispersion relation E(k) and the properties of the material, under consideration, can be calculated by the following one-electron Schrödinger wave equation:

$$\left(\frac{p^2}{2m} + V_{ps}(r)\right)\phi_k(r) = E_k\phi_k(r) \tag{112b}$$

where Vps(r) is the pseudo-potential and fk(r) denotes the pseudo wavefunctions, which are good approximation to the true wavefunctions outside the core region. Since the pseudopotential is a weak perturbation on the free-electron band structure, a good initial solution for diagonalzing (112b) is to expand fk as a sum of plane waves (similar to those of nearly free electrons).

Tightly-Bound (TB) Model

The tightly-bound (TB) electron method (which represents the chemists approach) takes the electron atomic orbitals in isolated atoms as a zero approximation and the periodic potential of the lattice as perturbation (Slater and Koster in 1954). This results in the so-called linear combination of atomic orbitals (LCAO) methods (Chaney, Lin & Lafon, 1971).

Note 5: Linear Combination of Atomic Orbitals (LCAO)

The LCAO is a variant of the TB approximate method, that was introduced by Lennard-Jones, in 1929. In this method, we consider the trial wave-functions as linear combinations of the following functions:

$$\psi(r) = \sum_n d_n \chi_n(\boldsymbol{r})$$

Then, we form the basis of the Linear Combination of Atomic Orbitals (LCAO), as introduced by Lennard-Jones and others to compute the energies and wavefunctions of atoms and molecules.

The basis functions χn (which are not generally orthogonal) are chosen to achieve best accuracy with minimum computations. For instance, the Slater-type orbitals (STO) are using Hydrogen-like wavefunctions in the form:

Figure 19. Pseudopotential V_{ps} and its Fourier transform in the reciprocal k-space

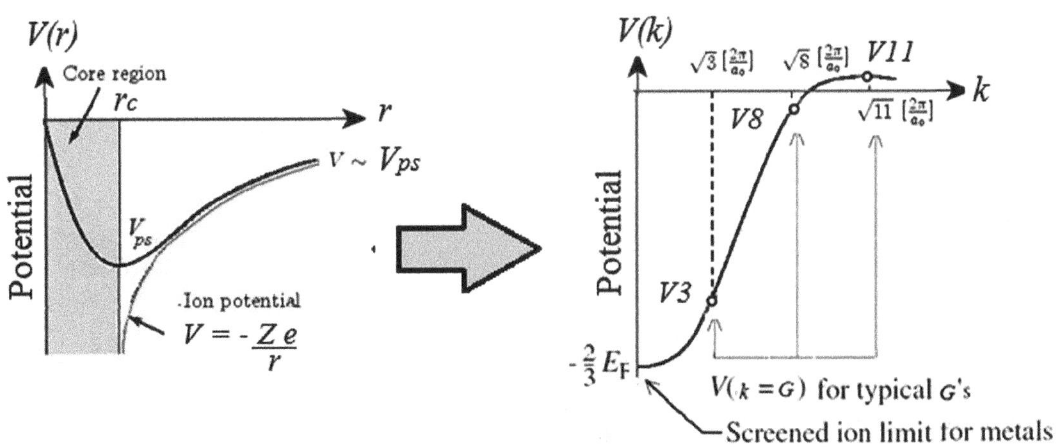

$$\chi(r) = Y_{\ell m}\left(\theta,\phi\right)e^{-\alpha r}$$

Alternatively, the Gaussian-type orbitals (GTO) may take the form

$$\chi(r) = Y_{\ell m}\left(\theta,\phi\right)e^{-\alpha r^2}$$

These are the most widely used forms. In particular, the Gaussian orbitals form the basis of many quantum chemistry computer codes.

The so-called Hubbard model is an improvement on the TB model, with two-terms Hamiltonian: a kinetic term allowing for tunneling (by hopping) of particles between lattice sites and a potential term expressing the Coulomb interaction, between electrons of the same atomic orbital. The Hubbard model is able to describe the interaction-driven transition from a metal to an insulator, commonly known as the *Mott transition*. This model is well suited for disordered structures, such as organic and molecular semiconductor devices. In particular, the so-called *sp3d5s** is a semiempirical tight-binding method, which is better at the nanometer scale. In this model, we take a number of nearest neighbor orbitals; each is twofold degenerate, if spin orbit coupling is considered (Jancu et al, 1998; Luisier et al, 2006).

In addition to the above approximate methods, one can cite the *k.p* method, which is a widely used for modeling the band structure near band edges. By this method we can obtain an analytical expression for the band structure near a given point k_0 in the k-space, for which the band structure is already known. This method has been already described in section 17.1

18.3 Density-Functional Theory (DFT)

The density-functional theory (DFT) is an *ab-inito* technique to solve the many-body Schrödinger equation, in terms of single-particle equations and an effective potential. The first-principles calculations based on DFT have become one of the most powerful approaches that allow one not only to calculate the energy band structure but also to model and predict the properties of emerging materials and nanodevices. In much the same way as the Hartree equation, the DFT provides a way to decouple the electron-electron interactions in the material, and solve the one electron equations by a self-consistent method to find out the electron wave-functions and the corresponding energies. However, the solution is performed starting from the ground-state electron density, like the self-consistent field method. Therefore, the many-electron Schrödinger equation is replaced by the so-called Kohn-Sham equation (Kohn & Sham, 1965).

$$\left[\frac{\hbar^2}{2m}\nabla^2 + V'_{eff}\right]\psi_i(r) = E_i\psi_i(r), i = 1,2,...N \tag{113a}$$

where N is the number of electrons in the system and the effective potentials, V_{eff}, is defined as follows:

$$V_{eff} = V(r) + V_H + V_{xc} \tag{113b}$$

The first term in equation $V(r)$ is the external potential and includes the potential originating from the nuclei as well as an external applied electric field if present. The term V_H corresponds to the classical

Coulomb potential (Hartree potential) for an electron density $n(r)$. This term is obtained by solving the Poisson equation for the scalar potential

$$\nabla^2 V_H(r) = e.(\rho/\varepsilon) = -(e^2/\varepsilon)\, n(r) \tag{114}$$

The exchange-correlation potential V_{xc} takes into account the many-body quantum effects, and is defined as the functional derivative of the exchange energy, E_{XC} with respect to the electron density $V_{xc} = \partial E_{xc}/\partial n$. The last term contains the remaining contributions to the potential that we do not know exactly. There exist N eigenfunctions ψ_i for each electron, from which the electron density can be calculated using the Thomas-Fermi model (Thomas & Fermi, 1927)[12]

$$n(\vec{r}) = \sum_i^{occupied} \left|\psi_i(\vec{r})\right|^2 = \sum_i \left|\psi_i(\vec{r})\right|^2 f_i \tag{115}$$

Here, he sum is taken over all the occupied states and $f_i = f(E_i - E_F)$ is the occupation probability (the Fermi-Dirac distribution in equilibrium). Also, the total energy is given by:

$$E_T = \Sigma_i E_i \,.\, f(E_i - E_F) \tag{86}$$

where the sum is taken over the total number of electrons (N).

18.4 DFT Procedure

The procedure to implement the DFT may be summarized as follows:

1. We first make use of a suitable set of basis functions to form the system of wavefunctions $\psi_i(r)$. This may be a set of orthogonal plane waves (OPW) or a linear combination of atomic orbitals (LCAO). The later choice is one of the best ways to make the iterative procedure for solving the K-S equation computationally accessible. Therefore, we consider the wavefunction expansion $\psi_i(r) = \Sigma_j c_{ij} \phi_j$, where c_{ij} is a set of coefficients $j=0 \rightarrow L$, for a set of L basis functions. Using the LCAO ansatz in the K-S equations, enables us to convert the nonlinear optimization problem into a linear one, with the expansion coefficients (c_{ij}) being the unknown variables. As we mentioned so far in Note 5, we try to choose the most accurate and fewest number of basis functions in the LCAO expansion. For instance, the basis functions may take the Slater-type orbitals (STO) or the Gaussian-type (GTO) form.

Using the LCAO ansatz with the K-S equations (82), we arrive at the following matrix equation:

$$[F^{KS}][C] = [S][C][E_i] \tag{117a}$$

where $[C]$ is the coefficient matrix and $[F^{KS}]$ is the K-S-Fock matrix, which is given by:

$$[F^{KS}]_{ij} = \int \phi_i f^{KS} . \phi_j \, d^3r \tag{117b}$$

Figure 20. Flowchart of the Kohen-Sham Method

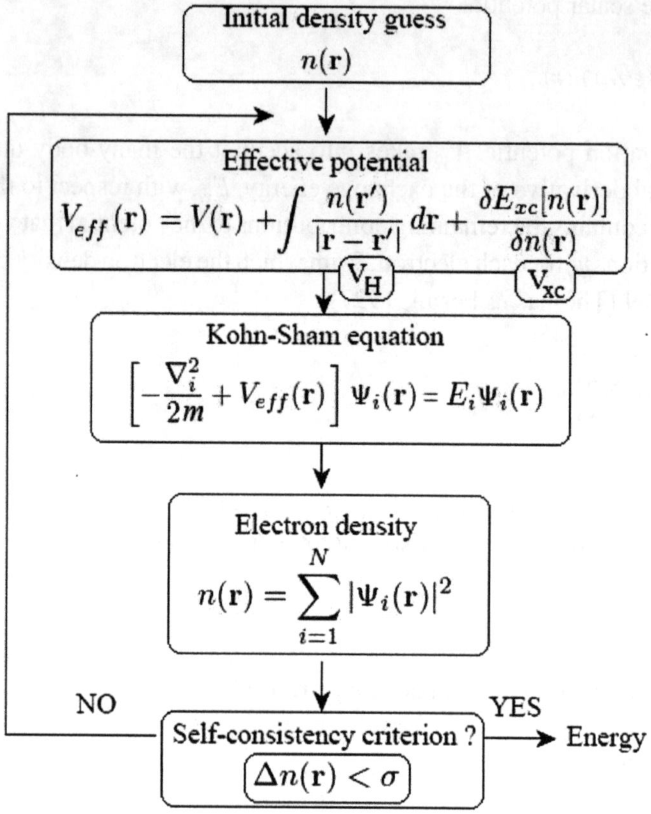

Also, the overlap matrix *[S]* is given by:

$$[S]_{ij} = \int \phi_i(r)\, \phi_j(r).\, d^3r \tag{117c}$$

And $[E_i]$ the electron allowed energies (eigenvalues) matrix:

$$[E]_{ij} = \delta_{ij}.E_i \tag{117d}$$

3. The above matrix equation may be prepared in the following form:

$$F_{ij}^{KS} = h_{ij} + J_{ij} + V_{ij}^{XC} \tag{118a}$$

This includes the following single-electron integrals:

$$h_{ij} \int \phi_i(r)\left[-\frac{\hbar^2}{2m}\nabla^2 - \sum_{i=1}^{M}\frac{e^2}{4\pi\varepsilon_0}\frac{Z_i}{\left|\boldsymbol{r}-\boldsymbol{r}_i\right|}\right]\phi_j(r)d^3r \tag{118b}$$

As well as the Coulomb interaction term:

$$J_{ij} = \sum_{n=1}^{L}\sum_{m=1}^{L} P_{nm} \iint \phi_i\left(r_1\right)\frac{1}{\left|r_1 - r_2\right|}\phi m\left(r_2\right)\phi_n\left(r_2\right)d^3r_1 d^3r_2 \tag{118c}$$

Here, r_1 and r_2 represent the coordinates of electrons and ions and the occupation probabilities P_{ij} are given by:

$$P_{ij} = \sum_{\mu=1}^{N} c_{i\mu} c_{j\mu} \tag{118d}$$

Finally, the exchange correlation term is given by:

$$V_{ij}^{XC} = \int \phi_i(r)V_{XC}(r)\phi_j(r)d^3r \tag{118e}$$

4. After solving the above matrix equation we get the coefficient matrix (c_{ij}) elements and the basis functions (ϕ_j). Therefore, we can construct the electron wavefunctions (ψ_i) and the density matrix (or electron density) can be calculated, as follows:

$$\rho(r) = \sum_{i=1}^{L}\left|\psi_i(r)\right|^2 = \sum_{i=1}^{L}\sum_{j=1}^{L}\psi_i(r)P_{ij}\psi_j(r) \tag{119}$$

The solutions of the Kohn-Sham (KS) equations in principle give the ground state properties of the system. However, the term E_{xc} is not known exactly and some kind of approximation is needed. The two types of approximations which are generally used are the local density approximation (LDA) and the generalized gradient approximation (GGA). In both LDA and GGA, the exchange-correlation potential V_{xc} is a local functional of the charge density. They are accurate when the charge density *n(r)* is a smoothly varying function. However, for systems where strong correlation effects are important, i.e., where the charge density *n(r)* changes sharply, these approximations most likely fail. DFT is used only for ground–state properties, as there is no practical scheme able to describe excited stated starting from the ground–state density. The excited state properties can be obtained from the extension of DFT to the time domain: This extension is known as the Time–Dependent DFT (TD-DFT).

$$\delta\rho(1) = \chi_{KS}(1, 1')\delta v_{KS}(1') \tag{120}$$

where $\chi_{KS}(1, 2)$ is the response function of the K-S non interacting system, while $\delta v_{KS}(1)$ is the variation of the TDKS potential needed to follow the evolution of the density when an external potential $\delta V_{ext}(1)$ is applied. The analogous version to the Hohenberg-Kohn theorem which includes time-dependent potentials is known as the Runge-Gross theorem.

18.5 Local Density Approximation (LDA)

The local density approximation (LDA) was proposed by Kohn and Sham. They showed that it could be applied to the limiting case of a slowly varying electron density. Therefore, one considers exchange energy E_{XC} to be that of a uniform electron gas of the partial density $n(r)$. In this case, the Kohn-Sham equation (90) may be written as follows:

$$\left[\frac{\hbar^2}{2m}\nabla^2 + V(r) + \int \frac{n(r')}{\left|r - r'\right|}dr' + V'_{LDA}\right]\psi_i(r) = E\psi(r), i = 1,2,...N \tag{121}$$

where $V_{LDA} = \partial E/\partial n$ is the local approximation to the Kohn-Sham exchange-correlation potential and $V(r)$ is the external potential energy. The Kohn-Sham LDA can be further extended to the spin dependent case by replacing the scalar external potential $V(r)$ by a spin dependent potential and replacing the electron charge density $n(r)$ by the density matrix. The DFT in the local density approximation (DFT-LDA) has been very successful in the analysis of interacting-electron ground states. Many efforts have also been devoted to DFT-based schemes for excited states.

18.6 Hybrid DFT Methods

Within the Kohn-Sham formulation, the total energy has the following form:

$$E_{KS} = V + \langle hP\rangle + ½ \langle PJ\rangle + E_X + E_C \tag{122a}$$

where, V is the nuclear repulsion energy $(H_{i\text{-}i})$, $\langle hP\rangle$ is the one-electron energy (kinetic plus potential) and $½\langle PJ\rangle$ is the classical Coulomb repulsion of electrons $(H_{e^-e^-})$. Also, E_x is the exchange energy due to the quantum (Fermion) nature of electrons and E_c is the correlation energy. The first four terms already exist in the Hamiltonian of the Hartee-Fock (H-F) equation (however, the exchange energy E_c in the H-F is given by an explicit exchange integral). The latter two terms are expressed as functionals (of carrier density or density matrix) in the DFT.

The functionals normally used in DFT are integrals of some function of the density and/or the density gradient. For instance, the exchange and correlation energy functionals may be written as follows:

$$E_X = \int f_x(\rho_\alpha(r), \rho_\beta(r), \nabla\rho_\alpha(r), \nabla\rho_\beta(r))dr \tag{122b}$$

$$E_c = \int f_c(\rho_\alpha(r), \rho_\beta(r), \nabla\rho_\alpha(r), \nabla\rho_\beta(r))dr \tag{122c}$$

The DFT implementation methods differ in which function f_x is used for E_X and which (if any) f_c is used for the correlation functional E_C. In fact, the proposed functionals in the literature lead to integrals which cannot be evaluated in closed form and are solved by numerical quadrature methods.

In addition to pure DFT methods and the approximation schemes to calculate the exchange and correlation terms (like DFT-LDA and DFT-GGA), the hybrid DFT methods have been introduced to provide more accurate results. In the Hybrid DFT methods, the exchange functional is a linear combination of the Hartree-Fock exchange and a functional integral of the above form.

The typical names for the pure and hybrid DFT models are usually combining the names for the exchange and correlation functionals, or sometimes they are synonyms by their inventors. For instance, the DFT method PBE is due to Perdew, Burke and Ernzerhof (1996). Also the hybrid DFT method B3LYP denotes the *Becke* Three Parameter Functionals (1993), with Lee, Yang, and Parr correlation functionals (2004).

$$A*E_X^{Slater}+(1-A)*E_X^{HF}+B*\Delta E_X^{Becke}+E_C^{VWN}+C*\Delta E_C^{non\text{-}local} \tag{122d}$$

where A, B, and C are the Becke constants (Becke, 1993, 1998).

18.7 NEGF-DFT Atomistic Method

The so-called NEGF-DFT description of electron transport is based on the Kohn–Sham equation and Keldysh nonequilibrium Green's function (NEGF). In the NEGF-DFT, the K-S Hamiltonian is typically utilized to derive the Green's functions, as shown in the following figure. The NEGF-DFT method may be initialized using LCAO basis functions.

The NEGF-DFT technique was first developed to allow parameter-free analysis of devices with up to 1000 atoms in the active region. It has been successfully applied to quantitatively predict transport properties in molecular, metallic, and carbon nanowires, and has led to understanding of electronic levels in molecular devices, vibrational excitations in transistors, tunnel junctions, and non-equilibrium charge distribution in nanocapacitors.

Here we calculate the electric current flow driven by a bias voltage (in non-equilibrium). Therefore, we use the non-equilibrium statistics (such as the NEGF method) in the device scattering regions. An

Figure 21. Comparison between the DFT and NEGF-DFT methods

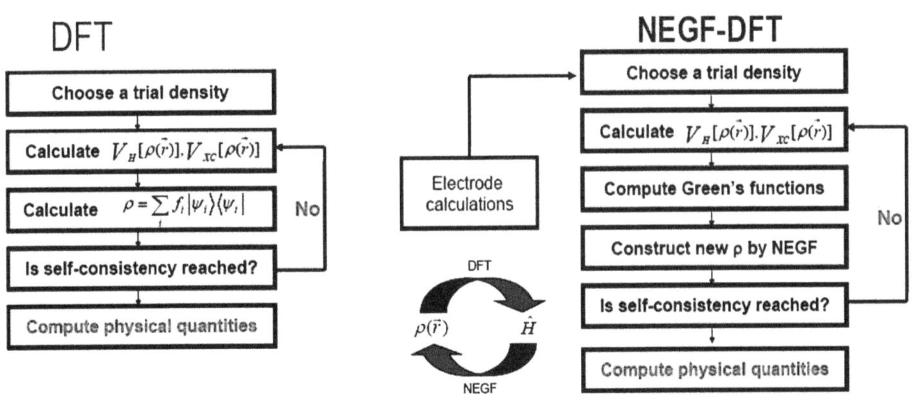

important consideration here is the calculation of the device Hamiltonian *H*, which determines the energy levels of the device. How to fill these levels is given by the non-equilibrium statistics. There exist many kinds and flavors of *H*, with different approaches, such as the TB, the *k.p*, the DFT, etc. In the end, one has to compare the final solution with experimental data to verify the validity of each approach. It should be noted that DFT in NEGF-DFT is not the ground state DFT: because density matrix of NEGF-DFT is constructed at non-equilibrium.

$$\hat{\rho}(E) = -\frac{i}{2\pi}\int_{-\infty}^{\infty}\hat{G}^{<}(E)dE \tag{123a}$$

Note that the above equation is similar to (77). The Lesser Green's function $G^{<}$ is related to the retarded and advanced Green's functions G^{r} and G^{a} and the lesser self-energy $\Sigma^{<}$ (which are already defined)

$$\hat{\boldsymbol{G}}^{<}(E) = \boldsymbol{G}^{r}(E)\Sigma^{<}(E)\boldsymbol{G}^{a}(E) \tag{123b}$$

The following table summarizes the steps involved in solving the self-consistent DFT Kohn–Sham equations for a two-probe system. For each step we summarize the input and output of the algorithm and its computational complexity when implemented using localized basis functions. The complexity is given in terms of the number of atoms *N* and the area of the electrode cell *A*. This formalism is well suited for the simulation of nanodevices and nanowires since the transport problem, including open boundaries and external bias can be solved fully self-consistently and since the calculation cost for the Hamiltonian operator scales only linearly with the number of atoms involved, making this approach ideal for the calculation of nanowires.

In order to simulate even bigger systems of up to one million atoms, we can use the tight binding (TB) model (Goedecker & Teter, 1995). In this case, the TB parameters are extracted from NEGF-DFT calculation of smaller devices at non-equilibrium. Thereby, they reflect the transport boundary condition and external fields.

18.8 GW Approximation

We have pointed so far that the DFT-LDA and DFT-CGA are known to underestimate energy gaps in semiconductor materials and the GW approximation (GWA) gives more accurate results. In the GWA, the Schrödinger equation can be written, within the Green's function formalism, as follows (Aryasetiawan & Gunnarsson, 1998):

$$\left[-\frac{\hbar^2}{2m}\nabla^2 + U_L(\boldsymbol{r}) + V_H(\boldsymbol{r})\right]\psi_i(\boldsymbol{r}) + \int \Sigma(\boldsymbol{r},\boldsymbol{r}')\psi_i(\boldsymbol{r}')d\boldsymbol{r}' = E_i\psi_i(\boldsymbol{r}) \tag{124}$$

where $\Sigma(r,r') = jG.W$ is the self-energy, *G* is the Green's function of the system and *W* is a specific function lumping the screened Coulomb interaction. The *W* potential can be calculated from the following sequence of Hedin's equations (Hedin, 1965):

$$\begin{aligned}
\Sigma(1,2) &= i\int G(1,3)\tilde{\Gamma}(3,2,4)W(4,1^{+})d(34)\\
G(1,2) &= G^{0}(1,2) + \int G^{0}(1,3)\Sigma(3,4)G(4,2)d(34)\\
\tilde{\Gamma}(1,2,3) &= \delta(1,2)\delta(1,3) + \int G(4,6)G(7,5)\frac{\delta\Sigma(1,2)}{\delta G(4,5)}\tilde{\Gamma}(6,7,3)d(4567)\\
\tilde{P}(1,2) &= -i\int G(1,3)G(4,1^{+})\tilde{\Gamma}(3,4,2)d(34)\\
W(1,2) &= v(1,2) + \int v(1,3)\tilde{P}(3,4)W(4,2)d(34)
\end{aligned} \tag{125}$$

The Figure 22 illustrates the screening potential *W*, and how it is weaker than the bare Coulomb interaction. The Figure 23 shows the outline of the GW approximation, to calculate the energy band structure starting from the DFT-LDA. The Figure 22 illustrates how to calculate the self-energy (Σ) in the GW approximation, as compared to the Hartee-Fock method.

It comes from the above discussion that the *GW* approximation is nothing but a *screened* version of the Hartree-Fock model, where *W* replaces V_{sc}. The following figure illustrates how the screening potential *W* is weaker than the bare Coulomb interaction V_{sc}.

18.9 Quantum Transport Simulators

There exist a variety of quantum device simulators, from the Academia and various companies. Among other techniques, these tools employ the nonequilibrium Green's function (NEGF) together with the TB and DFT calculations for atomistic modeling of nanodevices. For instance, the Vienna *ab initio* Simulation Package (VASP) is a program for atomic scale modeling, of electronic structures and molecular

Figure 22. Outline of the DFT-NEGF method

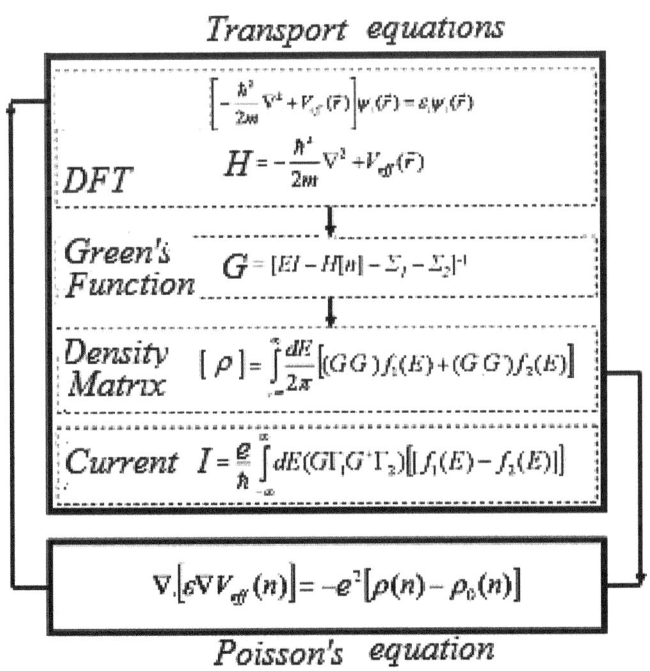

Table 3. Complexity vs. no. of atoms N and electrode area A

Step	Algorithm	Complexity
1	$n(\mathbf{r}) \to V^{\text{eff}}(\mathbf{r})$	$O(N \log N)$
2	$V^{\text{eff}}(\mathbf{r}) \to \bar{H}$	$O(N)$
3	$\bar{H} \to \bar{G}$	$O(N^3), O(NA^3)$
4	$\bar{G} \to \bar{D}$	$O(N)$
5	$\bar{D} \to n(\mathbf{r})$	$O(N)$

Figure 23. Schematic illustration of the screening potential W and how it is weaker than the bare Coulomb interaction V_{sc}

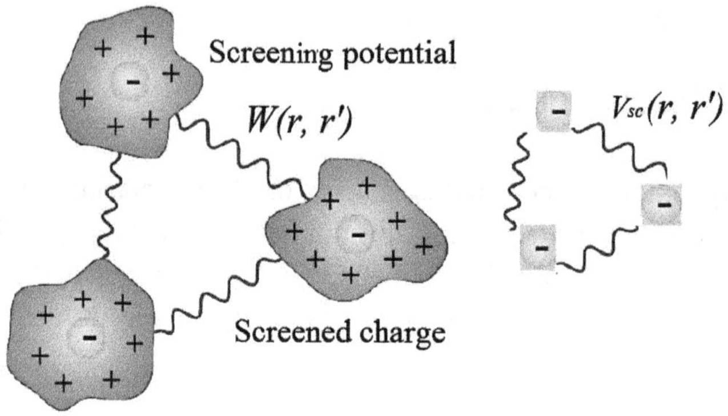

dynamics, from first principles (VASP, 2010). Also, Smeagol is an *ab initio* transport code based on a combination of DFT and NEGF (Rocha et al, 2005). In addition, Quantum ESPRESSO is an integrated suite of codes for band-structure calculations and materials modeling at the nanoscale. It is also based on DFT and pseudopotential method.

There are other software tools to perform DFT simulations, such as SIESTA (Spanish Initiative for Electronic Simulations with Thousands of Atoms). SIESTA is a program to perform band structure calculations and ab initio molecular dynamics of molecules and solids (Ordejón et al, 1996). This program has been extended to compute the conductance via Green's functions in a new package called: TranSIESTA (Brandbyge et al, 2002).

Another open tool, which illustrates the NEGF approach, is called NEMO (Nano-Electronic MOdeling), from Texas Instruments and Raythgeon. The NEMO simulator calculates the atomic structure from tight-binding (TB) method. The current version of NEMO (NEMO5) is intended to be a nanoelectronics simulation toolbox (Steiger et al, 2012). Also CP2K is a DFT package for studying transport phenomena in semiconductors from first-principles. CP2K is usually combined with OMEN, which is a quantum simulator for semiconductor structures and nanodevices (Villani et al, 2015). Another interesting open code that implements DFT and GWA is called MOLGW (Bruneval et al, 2016). MOLGW implements the many-body perturbtation theory to describe the excited electronic states in finite atomic and molecu-

Figure 24. Outline of the GW approximation, to calculate the energy band structure starting from the DFT-LDA. When no iteration is used we get the so-called G_0W_0

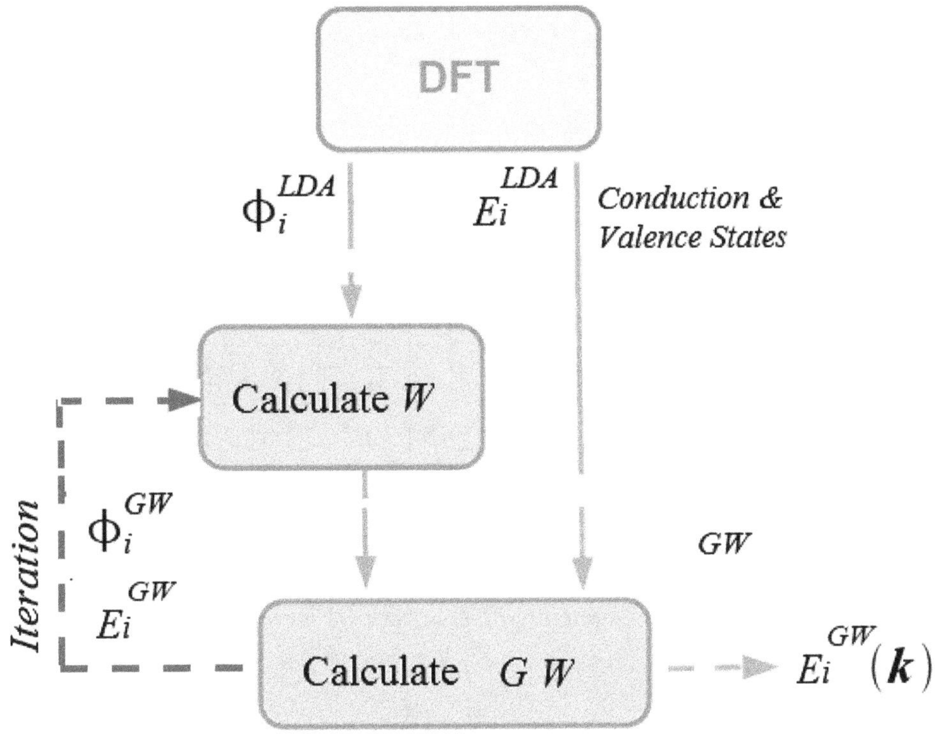

Figure 25 Expression of the self-energy in the GW approximation, compared to the Hartee-Fock method

Hartree-Fock	GWA
$\Sigma_{sc}(r,r') = \frac{j}{2\pi}\int_{-\infty}^{EF} G(r,r',\omega')V_{sc}(r,r')\,d\omega$	$\Sigma_{xc}(r,r',\omega) = \frac{j}{2\pi}\int G(r,r',\omega+\omega')W(r,r',\omega')\,d\omega'$

lar systems. It implements the *GW* approximation and the Bethe-Salpeter equation (Ziaei and Bredow, 2016) for calculating the optical excitations[13].

19. CASE STUDY 4: SIMULATION OF TUNNELING FET (TFET)

The following figure depicts the energy band structure of a tunneling FET, in both OFF and ON states. At low gate voltages, valence-band electrons in the source are reflected back at the heterojunction due to the long tunneling distance between source and drain. At higher gate voltages, instead, tunneling between the source valence band and the channel conduction band can occur due to the short tunneling distance (Baccarani et al, 2016).

The simulation of TFET has been carried out by several authors, using both ab initio and approximate quantum approaches (e.g., Shin, 2009).The following figure depicts the energy band (conduction and

Figure 26. Coupling between the DFT solver (CP2K) and the NEGF simulator (OMEN)

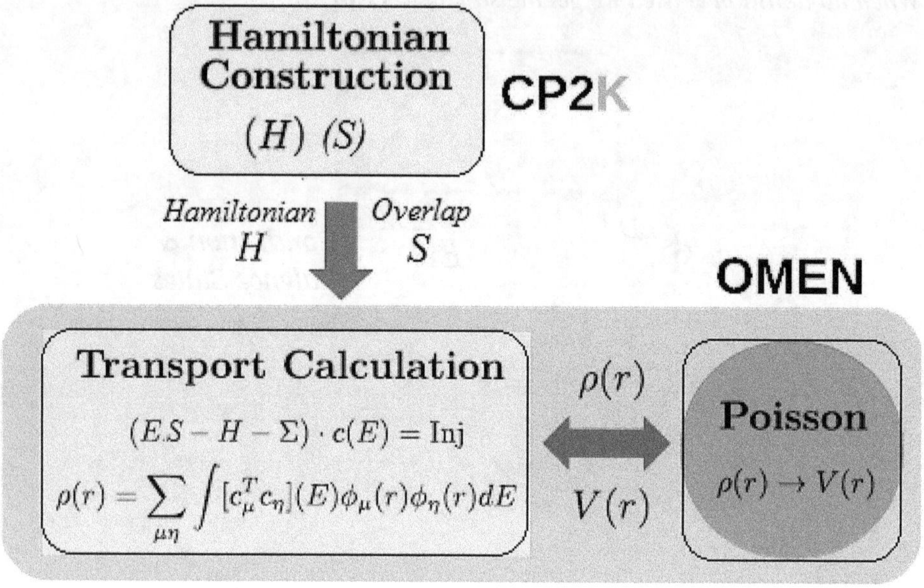

Figure 27. Energy band structure of a tunneling FET, in both OFF and ON states
After Baccarani et al (2016).

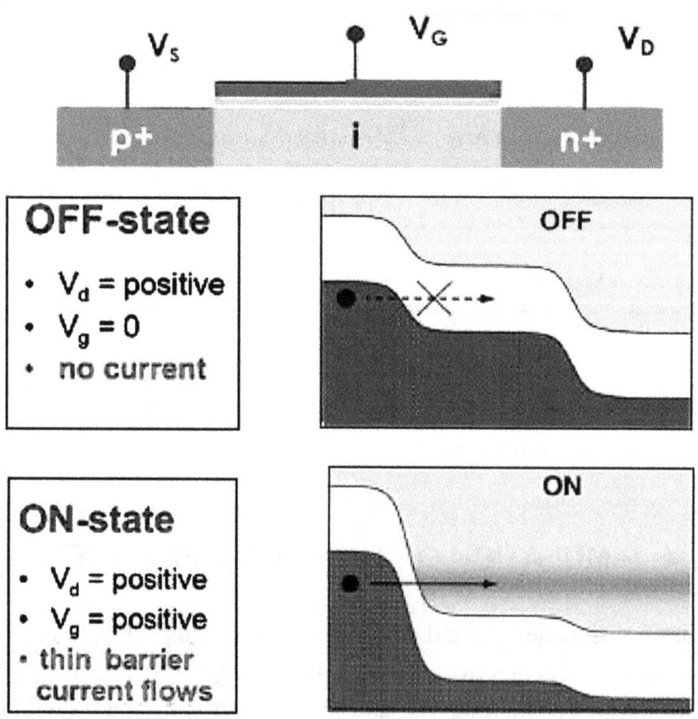

valence band edges) along the TFET, as obtained by the ***k.p*** method with Poisson's equation at different gate voltages (Luiser & Klemick, 2009). The analytical solution, according to (Villani et al, 2015) is also shown in symbols. The Figure 27 depicts I-V characteristics of TFET in the ON-state, with different gate lengths (left) and different widths (right).

According to Baccarani et al (2016), TFETs are still in the early phase of investigation, but the obtained results can indicate their upper performance limit. As for III–V based TFETs, they are should be affected by the density of interface traps, the effect of which can hardly be incorporated into quantum simulation tools. Besides, subthreshold swing is negatively influenced by trap-assisted tunneling, which may become important for heterojunction TFETs in the presence of defects.

20. SUMMARY

For ultra-small devices, whose active layers dimensions (at least one of them) are in the order of De Broglie wavelength (10 nm), the BTE should be replaced by one of the quantum transport models. The quantum transport theory is based on different approaches, such as the quantum Liouville equation and the Wigner-Boltzmann transport equation (WBTE). The quantum Liouville equation describes the temporal

Figure 28. Electrostatic potential and energy band (conduction and valence) edges along the TFET, as obtained by the k.p method with Poisson's equation at different gate voltages
After Luiser & Klemick,(2009).

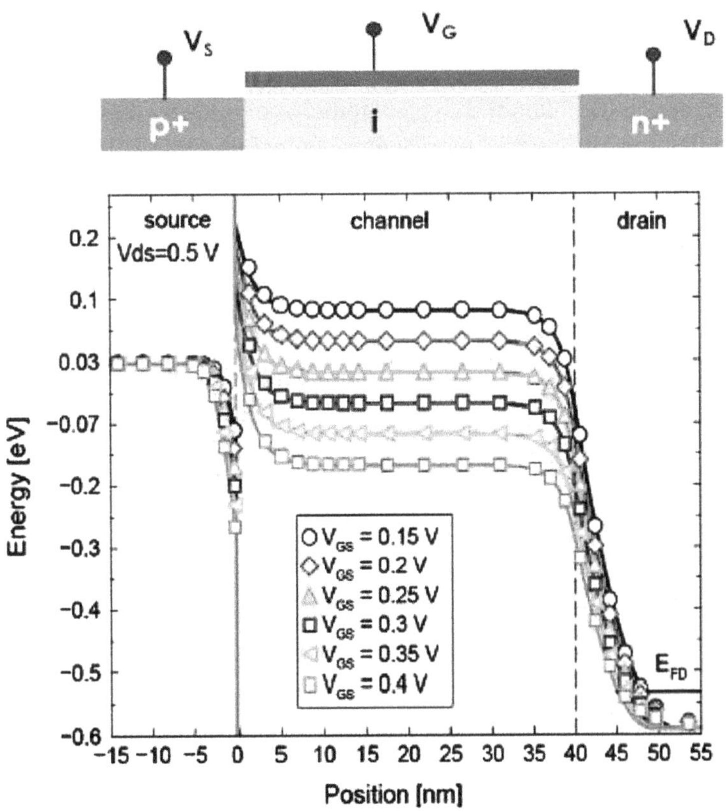

Figure 29. I-V characteristics of TFET in the ON-state, with different gate lengths (left) and different widths (right)
After Luisier & Klimeck (2009).

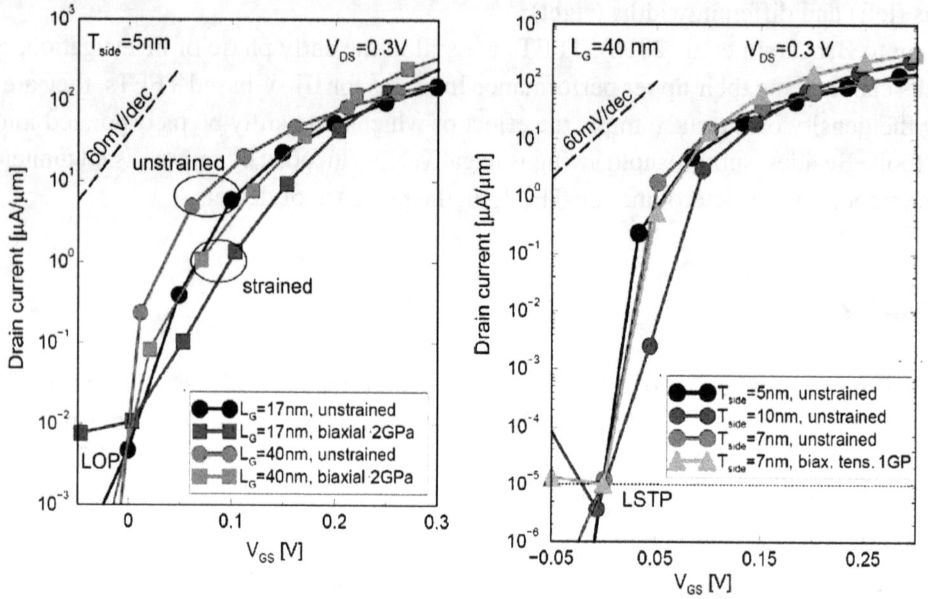

evolution of the density operator. The density *matrix* operator is the favorite mathematical instrument in quantum statistical physics. The *density* operator ρ *is* defined as $|\psi\rangle\langle\psi|$

In the real world, the interaction of any device with the surrounding is unavoidable. The study of real devices implies a many bodies problem which is extremely complex within any picture of the quantum mechanics. Therefore, there are two basic approaches to attack this problem:

1. The first approach consists in looking for a similar classical system and to get its associated Hamiltonian, then to proceed with the quantization of this phenomenological system.
2. The other approach makes use of the matrix density approach for the whole system and makes the trace over the surrounding variables, the resulting density matrix is called reduced density matrix, and its associated evolution equation is called master equation.

The quantum Liouville equation can be obtained by substituting the definition of the density operator, into the Schrödinger equation to obtain:

$$j\hbar\,\partial\hat{\rho}/\partial t = [\,\hat{H}\,,\,\hat{\rho}\,]$$

where $\hat{\rho}$ is the density operator and $\hat{H}$ is the system Hamiltonian.

The Wigner formulation of quantum mechanics allows the description of quantum mechanical systems without the need for wave functions. The kinetic equation for the Wigner distribution function including scattering effects is called the Wigner-Boltzmann transport function (WBTE). The Wigner Boltzmann transport equation (WBTE) may be written in the form:

$$\frac{\partial W}{\partial t}+\frac{\boldsymbol{p}}{m}.\nabla_x W+\theta.W=\left(\frac{\partial W}{\partial t}\right)_C$$

where θ is an operator acting on the Wigner distribution function $\boldsymbol{W}(r, p, t)$ as well as the potential energy of the system. A practically used approximation to incorporate scattering processes into the Wigner transport equation is to utilize the Boltzmann scattering operator, or by an even simpler scheme such as the relaxation time approximation. With this approximation, some quantum mechanical collisional effects, such as the collisional energy broadening, are neglected. After solving the WBTE, and calculating the Wigner distribution function $W(r,p,t)$, we can calculate the spatial density of carriers $n(r,t)$, as well as the average value of any microscopic physical variable.

$$n(r,t) = \int W(r,p,t)\, d^3p$$

Based on the WBTE, the quantum corrected Boltzmann equation, the quantum hydrodynamic model (QHDM), and the density gradient (DG) approximation can be obtained.

The Wigner distribution function may be also defined as the energy integral of the Green's function, $W(k,r,t) = \int G(k,\omega, r, t).d\omega$.

The Green's function gives the response of a system to a constant perturbation in the Schrödinger equation. The non-equilibrium Green's function (NEGF) formalism is a very powerful technique for evaluating properties of many-particle systems both in thermodynamic equilibrium and also in non-equilibrium situations. In coordinate representation the Green's function, $G(r_1, t_1; r_2, t_2)$, depends on two position arguments r_1, r_2 and two time arguments t_1, t_2, representing the non-locality in space and time. In quantum mechanics the non-equilibrium Green's function $G(r,r',t,t')$ depends on the continuous space variables r and r' through the creation and annihilation operators $\psi(r,t)$ and $\psi^{\dagger}(r',t')$.

Under steady state condition the Green's functions depend only on time differences. One can use the Fourier transform to change the time coordinate to energy $G(r,r.;E)$. The equation of motion for the Green's function is given by the integro-differential Dyson equation. Therefore, the retarded Green's function can be found by solving the following equation:

$$(E - H).G(E) = I$$

The NEGF allows for the calculation of scattering states in the quantum transport problem. When we take the coupling between the lesser and the retarded Green's function into account, we should solve a system of coupled PDE's. For the case of the electron-phonon interaction, these equations are:

$$G = (E.I - H_0 - \Sigma^r)^{-1},$$

$$G^n = G\, \Sigma^{in}\, G^{\dagger}$$

$$G^p = G\, \Sigma^{out}\, G^{\dagger}$$

where H_0 is the single-electron Hamiltonian, G (or G^r) is called the retarded Green's function, its Hermitian conjugate, G^+ (or G^a), is called the advanced Green's function, and Σ^{in} and Σ^{out} are the self-energy function related to electron-phonon and hole-phonon interactions. Also, G^n and G^p are the Green functions specifying the electron and hole density spectra. The following figure shows the self-consistence solution of the Green's function with Poisson's equation to obtain the electron density in a nanodevice. The electron density spectrum and the terminal current density spectrum can be evaluated, after self-consistent solutions are obtained for the correlation functions. In summary, we can underline the following notes.

Quantum corrections can accurately account for the size quantization effects. For modeling the nanodevices, which include tunneling and size quantization effects, one has better to utilize the Poisson-Schrodinger solvers or Quantum Monte Carlo or the NEGF methods. The NEGF is important when quantum interference effects are important. However, accurate and reliable multi-dimensional modeling of realistic nanoscale devices requires enormous computational efforts, yet the currently available NEGF algorithms are prohibitively expensive. For the case of nanodevices, in which the band-structure plays a significant role, the atomistic simulation using the *ab-initio* methods is necessary. The most famous *ab-initio* methods to calculate the band structure are the Hartee-Fock (H-F) theory and the Density-Functional Theory (DFT). The H-F method (or *mean field theory*) try to solve the Schrödinger equation, assuming that the wavefunction can be approximated by a single Slater determinant made up of one spin orbital for each electron. The actual goal of DFT method is the solution of the time-independent (non-relativistic) Schrödinger equation.

The DFT theory is very similar to the H-F theory, but its practical implementations depends on several approximations for the energy exchange-correlation (Exc) terms. A flagrant problem of the DFT approximations (such as DFT-LDA and DFT-GGA) is the systematic error for the energy gaps of semiconductors and insulators. Recently, new approximations for the exchange energy term have emerged (like the hybrid DFT methods and the GW approximation) to correct this error.

Figure 30. Selt consistent solution of NEGF and Poisson's equation

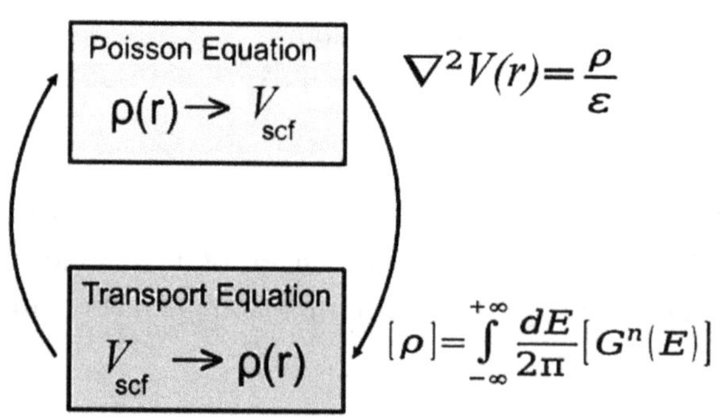

REFERENCES

Abdel Rahman, A. (2005). *Exploring New Channel Materials for Nanoscale CMOS* (PhD thesis). Purdue University.

Ancona, M. G., & Iafrate, G. J. (1989). Quantum correction to the equation of state of an electron gas in a semiconductor. *Physical Review B: Condensed Matter and Materials Physics*, *39*(13), 9536–9540. doi:10.1103/PhysRevB.39.9536 PMID:9947690

Ansgar, J. & Matthes, D. (2005). A derivation of the isothermal quantum hydrodynamic equations using entropy minimization. *Appl. Numer. Math.*

Aryasetiawan, F., & Gunnarsson, O. (1998). The GW Method. *Reports on Progress in Physics*, *61*(3), 237–312. doi:10.1088/0034-4885/61/3/002

Baccarani, G., Gnani, E., Gnudi, A., & Reggiani, S. (2016). Theoretical analysis and modeling for nanoelectronics. *Solid-State Electronics*, *125*, 2–13. doi:10.1016/j.sse.2016.07.020

Barker, J. (1973). Quantum transport theory of high-field conduction in Semi-conductors. *Journal of Physics. C. Solid State Physics*, *6*(17), 2663–2684. doi:10.1088/0022-3719/6/17/009

Barker, J. A. (1979). A quantum-statistical Monte Carlo method; path integrals with boundary conditions. *The Journal of Chemical Physics*, *70*(6), 2914–2911. doi:10.1063/1.437829

Barker, J. R. (1985). Quantum theory of hot electron tunneling in microstructures. *Physica B, Condensed Matter*, *134*, 22–31.

Barletti, L., Borgioli, G., & Frosali, G. (2014). *Semiclassical hydrodynamics of a quantum Kane model for semiconductors*. arXiv:1402.3429v2 [math-ph]

Barletti, L., Demeio, L., & Frosali, G. (2007). Multiband quantum transport models for semiconductor devices. In *Transport Phenomena and Kinetic Theory*. Birkhäuser Boston. doi:10.1007/978-0-8176-4554-0_4

Barletti, L., & Ben Abdallah, N. (2011). Quantum transport in crystals: effective mass theorem and k.p Hamiltonians. *Comm. Math. Phys.*, *307*, 567–607.

Bastard, G. (1988). *Wave Mechanics Applied to Semiconductor Heterostructures*. Paris: Hasted Press.

Becke, A. (1993). Density functional thermochemistry—III: The role of the exact exchange. *The Journal of Chemical Physics*, *98*(7), 5648–5652. doi:10.1063/1.464913

Ben Abdallah, N., & Degond, P. (1996). On a hierarchy of macroscopic models for semiconductors. *Journal of Mathematical Physics*, *37*(7), 3306–3333. doi:10.1063/1.531567

Birner, S. (2011). *Modeling of semiconductor nanostructures and Semiconductor–electrolyte interfaces* (Ph.D. Dissertation). Technischen Universität München.

Blount, I. (1962). Formalisms of Band Theory. Academic Press.

Blum, K. (1996). *Density Matrix Theory and Applications*. New York: Plenum. doi:10.1007/978-1-4757-4931-1

Blunt, K. (1981). *Density Matrix Theory and Applications*. Plenum.

Bogolyubov, N. N., & Tyablikov, S. V. (1960). Retarded and advanced Green functions in statistical physics. *Soviet Phys.*, *4*, 589–593.

Bohr, N. (1920). Über die Serienspektra der Element. *Zeitschrift fur Physik*, *2*(5), 423–478. doi:10.1007/BF01329978

Bonitz, M. (1998). *Quantum Kinetic Theory*. Stuttgart, Germany: Teubner.

Bonitz, M., & Semkat, D. (Eds.). (2003). *Progress in Nonequilibrium Green's Functions II*. World Scientific.

Borgioli, G., Frosali, G., & Zweifel, P. F. (2003). Wigner approach to the two-band Kane model for a tunneling diode. *Transport Theory and Statistical Physics*, *32*(3&4), 347–366. doi:10.1081/TT-120024768

Born, M., & Jordan, P. (1925). Zur Quantenmechanik. Zeitschrift für Physik 34: 858. Bibcode:1925. *ZPhy.*, *34*, 858B.

Born, M., & Oppenheimer, R. (1927). Zur Quantentheorie der Molekeln. *Ann. Phys. (Leipzig)*, *84*(20), 457–484. doi:10.1002/andp.19273892002

Boykin, T. B., Kharche, N., Klimeck, G., & Korkusinski, M. (2007). Projection method for approximate bandstructures of semiconductor alloys from supercell calculations. *Journal of Physics Condensed Matter*, *19*, 036203. doi:10.1088/0953-8984/19/3/036203

Brandbyge, M., Mozos, J. L., Ordejón, P., Taylor, J., & Stokbro, K. (2002). Density- Functional Method for Nonequilibrium Electron Transport. *Physical Review B: Condensed Matter and Materials Physics*, *65*(16), 165401–165417. doi:10.1103/PhysRevB.65.165401

Bruneval, F., Rangel, T., Hamed, S. M., Shao, M., Yang, C., & Neaton, J. B. (2016). MOLGW 1: Many-body perturbation theory software for atoms, molecules, and clusters. *Computer Physics Communications*, *208*, 149–161. doi:10.1016/j.cpc.2016.06.019

Buot, F., & Jensen, K. (1990). Lattice Weyl-Wigner formulation of exact many-body quantum-transport theory and applications to novel solid-state quantum-based devices. *Physical Review B: Condensed Matter and Materials Physics*, *42*(15), 9429–9457. doi:10.1103/PhysRevB.42.9429

Buot, F. A. (2009). *Nonequilibrium Quantum Transport Physics in Nanosystems: Foundation of Computational Nonequilibrium Physics in Nanoscience and Nanotechnology*. World Scientific. doi:10.1142/6042

Burghardt, I., Lorenz, S., Cederbaum, S., Moller, K. B., & Parlant, G. (2004). *Quantum Transport Atom to Transistor*. Cambridge, UK: Cambridge Univ. Press.

Burt, M. G. (1992). The justification for applying the effective-mass approximation to microstructure. *Journal of Physics Condensed Matter*, *4*(32), 6651–6690. doi:10.1088/0953-8984/4/32/003

Callaway, J. (1958). Electron Energy Bands in Solids. Academic Press.

Chai, L., Jin, S., Li, Q., & Morandi, O. (2015). A Multi-band semiclassical model for surface hopping quantum dynamics. *Multiscale Modeling & Simulation*, *13*(1), 205–230. doi:10.1137/140967842

Chaney, R. C., Lin, C. C., & Lafon, E. E. (1971). Application of the Method of Tight Binding to the Calculation of the Energy Band Structures of Diamond, Silicon, and Sodium Crystals. *Physical Review B: Condensed Matter and Materials Physics*, *3*(2), 459–472. doi:10.1103/PhysRevB.3.459

Connelly, D., Yu, Z., & Yergeau, D. (2002). Macroscopic simulation of quantum mechanical effects in 4-D MOS devices via the density gradient method. *IEEE Trans.*, *49*(4), 619–626. doi:10.1109/16.992871

Datta, S. (1990). A simple kinetic equation for steady-state quantum transport. *Journal of Physics Condensed Matter*, *2*(40), 8023–8052. doi:10.1088/0953-8984/2/40/004

De Broglie, L. (1924). *Recherches sur la théorie des quanta*. Paris: Masson.

De Broglie, L. (1926). *Ondes et Mouvements*. Paris: Gauthier-Villars.

Demeio, L., Barletti, L., Bertoni, A., Bordone, P., & Jacoboni, C. (2003). Wigner Function for Multiband Transport in Semiconductors, Journal. *Transport Theory and Statistical Physics*, *32*(3-4), 307–325. doi:10.1081/TT-120024766

Dirac, P. A. M. (1930). Note on exchange phenomena in the Thomas atom. *Proceedings of the Cambridge Philosophical Society*, *26*(03), 376. doi:10.1017/S0305004100016108

Dirac, P. A. M. (1931). The Principles of Quantum Mechanics. Clarendon Press.

Dyson, F. (1967). Ground-State Energy of a Finite System of Charged Particles. *J. Math. Phys.*, *8*(8), 1538–1545. doi:10.1063/1.1705389

Ehrhardt, M., & Koprucki, T. (Eds.). (2014). *Multi-Band Effective Mass Approximation*. Springer International Publishing. doi:10.1007/978-3-319-01427-2

Einstein, A. (1905). Über einen die Erzeugung und Verwandlung des Lichtes betreffenden heuristischen Gesichtspunkt, Concerning a heuristic point of view toward the emission and transformation of light. *Annalen der physik, 322*(6), 132-148.

El-Saba, M.H. (1993). *Modélisation et Simulation Hydrodynamique des phénomènes de Transport de porteurs chauds et de L'ionisation par Impact dans les Dispositifs à Semiconducteur* (Ph.D.). INSA Lyon, France, Order No.93 ISAL 0072.

Faraj, A., & Jin, S. (2015). *The Landau-Zener transition and the surface hopping method for the 2D Dirac equation for graphene.* arXiv preprint arXiv:1505.05988

Fermi, E. (1927). Un Metodo Statistico per la Determinazione di alcune Priopriętà dell'Atomo. *Rend. Accad. Naz. Lincei*, *6*, 602–607.

Ferry, D., & Goodnick, S. (1997). *Transport in Nanostructures*. Cambridge Univ. Press. doi:10.1017/CBO9780511626128

Ferry, D., & Grubin, H. (1996). Modeling of quantum transport in semiconductor devices. *Solid State Physics*, *49*, 283–448. doi:10.1016/S0081-1947(08)60300-8

Ferry, D., Ramey, S., Shifren, L., & Alkis, E. (2002). The Effective Potential in Device Modeling: The Good, the Bad and the Ugly. *Journal of Computational Electronics*, *1*(1), 9–65.

Ferry, D., & Zhou, J.-R. (1993). Form of the quantum potential for use in hydrodynamic equations for semiconductor device modeling. *Physical Review B: Condensed Matter and Materials Physics*, *48*(11), 7944–7950. doi:10.1103/PhysRevB.48.7944 PMID:10006981

Ferry, D. K., & Barker, J. R. (1980). Balance equation formulation and memory effects in retarded high-field semiconductor transport. *J. Appl. Phys. Chem. Solids.*, *3*, 106–141.

Feynman, R. P. (1948). Space–time approach to non-relativistic quantum mechanics. *Reviews of Modern Physics*, *20*(8), 367–387. doi:10.1103/RevModPhys.20.367

Fiolhais, C., Nogueira, F., & Marques, M. (2003). *A Primer in Density Functional Theory*. Springer. doi:10.1007/3-540-37072-2

Fischetti, M. V. (1998). Theory of Electron Transport in Small Semiconductor Devices using the Pauli Master Equation. *Journal of Applied Physics*, *83*(1), 270–291. doi:10.1063/1.367149

Fischetti, M. V. (1999). Master-equation approach to the study of electronic transport in small semiconductor devices. *Physical Review B: Condensed Matter and Materials Physics*, *59*(7), 4901–4917. doi:10.1103/PhysRevB.59.4901

Fischetti, M.V., & Vandenberghe, W.G. (2016). Overview of Quantum-Transport Formalisms. Chapter in Advanced Physics of Electron Transport in Semiconductors and Nanostructures. *Graduate Texts in Physics,* 361-380.

Fock, V. (1930). Näherungsmethode zur Lösung des quanten-mechanischen Mehrkörperproblems. *Zeitschrift fur Physik*, *61*(1-2), 126–148. doi:10.1007/BF01340294

Frensley, W. R. (1990). Boundary conditions for open quantum systems driven far from equilibrium. *Reviews of Modern Physics*, *62*(3), 745–791. doi:10.1103/RevModPhys.62.745

Frensley, W. R. (1994). Heterostructure and Quantum Well Physics. In N. G. Einspruch & W. R. Frensley (Eds.), *Heterostructure and Quantum Devices, a volume of VLSI Electronics: Microstructure Science*. San Diego, CA: Academic Press. doi:10.1016/B978-0-12-234124-3.50006-9

Garcia-Garcia, J., Martin, F., Oriols, X., & Sune, J. (1998). Quantum Monte Carlo simulation of resonant tunneling diodes based on the Wigner distribution function formalism. *Applied Physics Letters*, *73*(24), 3539–3541. doi:10.1063/1.122800

Garcia-Loureiro, A.J., Seoane, N., Aldegunde, M., & Vall, R. (2011). Implementation of the Density Gradient Quantum Corrections for 3-D Simulations of Multigate Nanoscaled Transistors. *IEEE Transactions on Computer-Aided Design Integrated Circuits And Systems, 30*(6), 841-851.

Gardner, C. (1994). The quantum hydrodynamic model for semiconductor devices. *SIAM Journal on Applied Mathematics*, *54*(2), 409–427. doi:10.1137/S0036139992240425

Gardner, C., & Ringhofer, C. (1996). Smooth quantum potential for the hydrodynamic model. *Physical Review E: Statistical Physics, Plasmas, Fluids, and Related Interdisciplinary Topics*, *53*(1), 157–166. doi:10.1103/PhysRevE.53.157 PMID:9964245

Gardner, C. L., & Ringhofer, C. (1996). Smooth quantum potential for the hydrodynamic model. *Physical Review E: Statistical Physics, Plasmas, Fluids, and Related Interdisciplinary Topics*, *53*(1), 157–167. doi:10.1103/PhysRevE.53.157 PMID:9964245

Gehring, A. (2003). *Simulation of Tunneling in Semiconductor Devices* (PhD thesis). Technische Universität Wien Fakultät für Elektrotechnik.

Gilbert, M. J., & Ferry, D. K. (2004). Full Quantum Mechanical Simulation of Ultra-Small Silicon Devices in Three-Dimensions: Physics and Issues. *Journal of Computational Electronics*, *3*(3-4), 355–358. doi:10.1007/s10825-004-7076-3

Goedecker, S., & Teter, M. (1995). Tight-binding electronic-structure calculations. *Physical Review B: Condensed Matter and Materials Physics*, *51*(15), 9455–9464. doi:10.1103/PhysRevB.51.9455 PMID:9977606

Green, M. S. (1954). Markov random processes and the statistical mechanics of time dependent phenomena. Irreversible processes in fluids. *The Journal of Chemical Physics*, *22*(3), 398–413. doi:10.1063/1.1740082

Grubin, H., & Kreskovsky, J.-P. (1989). Quantum moment balance equations and resonant tunnelling structures. *Solid-State Electronics*, *32*(12), 1071–1075. doi:10.1016/0038-1101(89)90192-5

Hamann, D. R., Schlüter, M., & Chiang, C. (1979). Norm-Conserving Pseudopotentials. *Physical Review Letters*, *43*(20), 1494–1497. doi:10.1103/PhysRevLett.43.1494

Hansch, W., & Mahan, G. D. (1983). Transport equations for many-particle systems. *Physical Review B: Condensed Matter and Materials Physics*, *28*(4), 1902–1922. doi:10.1103/PhysRevB.28.1902

Hartree, D. R. (1928). The wave mechanics of an atom with non-Coulombic central field. *Proc. Camb. Soc.*, *24*, 89. doi:10.1017/S0305004100011919

Haug, H., & Jauho, A. P. (2008). *Quantum Kinetics in Transport and Optics of Semiconductors*. Berlin: Springer-Verlag.

Havu, P., Havu, V., Puska, M.J. & Nieminen, R.M. (2004). Non-Equilibrium Electron Transport in Two-Dimensional Nano-Structures Modeled by Green's Functions and the Finite Element. *Physic Rev, B, 69*, 115325.

Hedin, L. (1965). New method for calculating the one-particle Greens function with application to the electron-gas problem. *Physical Review*, *139*(3A), A796–A823. doi:10.1103/PhysRev.139.A796

Heisenberg, W. (1927). Über den anschaulichen Inhalt der quantentheoretischen Kinematik und Mechanik. *Zeitschrift fur Physik*, *43*(3-4), 172–198. doi:10.1007/BF01397280

Heisenberg, W. (1958). *Physics and Philosophy*. Harper.

Hetmaniuk, U., Ji, D., Zhao, Y., & Anantram, M. P. (2015). A Reduced-Order Method for Coherent Transport Using Green's Functions. *IEEE Transactions on Electron Devices*, 62.

Hinckley, J. M., & Singh, J. (1990). Hole transport theory in pseudomorphic Si1-xGex alloys grown on Si(001) substrates. *Physical Review B: Condensed Matter and Materials Physics*, *41*(5), 2912–2926. doi:10.1103/PhysRevB.41.2912

Hirose, K., & Kobayashi, N. (2014). *Quantum Transport Calculations for Nano systems*. Pan Stanford.

Hohenberg, P., & Kohn, W. (1964). Inhomogeneous Electron Gas. *Physical Review*, *136*(3B), B864–B871. doi:10.1103/PhysRev.136.B864

Hove, L. V. (1955). Quantum-mechanical Perturbations giving Rise to a Statistical Transport Equation. *Physica*, *21*, 517.

Hydrodynamic methods for ultrafast quantum dynamics, quantum transport, and dissipation. (n.d.). Available: http://mail.math.ups-tlse.fr/~nanolab/Contents/Invited2.pdf

Jacoboni, C. (1992). Comparison between quantum and classical results in hot-electron transport. *Semiconductor Science and Technology*, *7*(3B), B6–B11. doi:10.1088/0268-1242/7/3B/002

Jacoboni, C. (2010). *Theory of Electron Transport in Semiconductors: A Pathway from Elementary Physics to Nonequilibrium Green Functions*. Springer Series Solid-State Sciences. doi:10.1007/978-3-642-10586-9

Jacoboni, C., Reggiani, L., & Ferry, D. K. (Eds.). (1992). *Quantum Transport in Semiconductors*. New York: Plenum.

Jancu, J. M., Scholz, R., Beltram, F., & Bassani, F. (1998). Empirical Spds* tight-binding calculation for cubic semiconductors: General method and material parameters. *Physical Review B: Condensed Matter and Materials Physics*, *57*(11), 6493–6507. doi:10.1103/PhysRevB.57.6493

Jerome, J. W. (2009). Mathematical Advances and Horizons for Classical and Quantum-Perturbed Drift-Diffusion Systems, Solid State Devices and Beyond. *Journal of Computational Electronics*, *8*(2), 132–141. doi:10.1007/s10825-009-0269-z

Jones, R. O., & Gunnarsson, O. (1989). The density functional formalism, its applications and prospects. *Reviews of Modern Physics*, *61*(3), 689–746. doi:10.1103/RevModPhys.61.689

Kadanoff, L. P., & Baym, G. (1962). *Quantum Statistical Mechanics: Green's Function Methods in Equilibrium and Non Equilibrium Problems*. New York: W.A. Benjamin Cummings.

Kane, E. O. (1956). Energy band structure in p-type Germanium and Silicon. *Journal of Physics and Chemistry of Solids*, *1*(1-2), 82–89. doi:10.1016/0022-3697(56)90014-2

Kane, E. O. (1966). The k•p Method. Semiconductors and Semimetals, 1, 75–100.

Keldysh, L. V. (1965). Diagram technique for non-equilibrium processes. *Soviet Physics, JETP*, *20*, 1018.

Klimeck, G., Ahmed, S. S., Bae, H., Kharche, N., Clark, S., Haley, B., & Boykin, T. B. et al. (2007). Atomistic simulation of realistically sized nanodevices using NEMO 3-D—Part I: Models and benchmarks. *IEEE Transactions on Electron Devices*, *54*(9), 2079–2089. doi:10.1109/TED.2007.902879

Klimeck, G., Fonseca, J. E., Rahman, R., Kharche, N., Lansbergen, G. P., & Rogge, S. (2016). *Metrology Studies with NEMO, a Multimillion Atom Simulation Tool*. Available at: http://nanohub.com

Kohn, W., Becke, A. D., & Parr, R. G. (1996). Density Functional Theory of Electronic Structure. *Journal of Physical Chemistry*, *100*(31), 12974–12980. doi:10.1021/jp960669l

Kohn, W., & Sham, L. J. (1965). Self-consistent equations including exchange and correlation effects. *Physical Review*, *140*(4A), A1133–A1138. doi:10.1103/PhysRev.140.A1133

Kosina, H., & Nedjalkov, M. (2006). Handbook of Theoretical and Computational Nanotechnology (Vol. 10). Los Angeles, CA: American Scientific Publishers.

Kosina, H., Nedjalkov, M., & Selberherr, S. (2003). A Monte Carlo Method Seamlessly Linking Quantum and Classical Transport Calculations. *Journal of Computational Electronics*, *2*(2-4), 147–151. doi:10.1023/B:JCEL.0000011416.93047.69

Kubis, T. & Vogl, P. (2007). Self-consistent quantum transport theory: Applications and assessment of approximate models. *J. Comput Electron.*, *6*, 183–186.

Kubis, T., Yeh, C., Vogl, P., Benz, A., Fasching, G., & Deutsch, C. (2009). Theory of nonequilibrium quantum transport and energy dissipation in terahertz quantum cascade lasers. *Physical Review B: Condensed Matter and Materials Physics*, *79*(19), 195323. doi:10.1103/PhysRevB.79.195323

Kurniawan, O., Bai, P., & Eli, E. (2009). Ballistic calculation of nonequilibrium Greens function in nanoscale devices using finite element method. *Journal of Physics. D, Applied Physics*, *42*(10), 105109–105120. doi:10.1088/0022-3727/42/10/105109

Landau, L. D., & Lifshitz, E. M. (1977). *Quantum mechanics: non-relativistic theory*. Oxford, UK: Pergamon Press.

Li, S., Ahmed, S., Klimeck, G., & Darve, E. (2008). Computing entries of the inverse of a sparse matrix using the FIND algorithm. *Journal of Computational Physics*, *227*(22), 9408–9427. doi:10.1016/j.jcp.2008.06.033

Lindblad, G. (1976). On the generators of quantum dynamical semigroups. *Communications in Mathematical Physics*, *48*(2), 119–130. doi:10.1007/BF01608499

Luisier, M., & Klimeck, G. (2009). Performance comparisons of tunneling field-effect transistors made of InSb, carbon and GaSb-InAs broken gap heterostructures. *IEDM Tech Dig*, *2009*, 913–916.

Luisier, M., Schenk, A., Fichtner, W., & Klimeck, G. (2006). Atomistic Simulation of Nanowires in the *sp3d5s** Tight-Binding Formalism: From Boundary Conditions to Strain Calculations. *Physical Review B: Condensed Matter and Materials Physics*, *74*(20), 205323. doi:10.1103/PhysRevB.74.205323

Lundstrom, M. (2000). *Fundamentals of Carrier Transport*. Cambridge Univ. Press. doi:10.1017/CBO9780511618611

Lundstrom, M. & Ren. (2002). Essential physics of carrier transport in nanoscale MOSFETs. *IEEE Trans. Electron Devices, 49*(1), 133-141.

Luttinger, J., & Kohn, W. (1955). Motion of Electrons and Holes in Perturbed Periodic Fields. *Physical Review*, *97*(4), 869–883. doi:10.1103/PhysRev.97.869

Maassen, J., Harb, M., Michaud-Rioux, V., Zhu, Y., & Guo, H. (2013). Quantum Transport Modeling from First Principles. *Proc. IEEE Trans*, *101*(2), 518-530.

Madelung, E. (1927). Quantentheorie in hydrodynamischer form. *Zeitschrift fur Physik*, *40*(3-4), 324–326. doi:10.1007/BF01400372

Madelung, O. (1978). *Introduction to Solid-State Theory*. Berlin: Springer-Verlag. doi:10.1007/978-3-642-61885-7

Mahan, G. D. (2000). *Many Particle Physics*. New York: Plenum. doi:10.1007/978-1-4757-5714-9

Majewski, J.A., Birner, S., Trellakis, A., Sabathil, M. & Vogl, P. (2004). Advances in the theory of electronic structure of semiconductors. *Physica st. solidi (c), 1.*

Mascali, G. (2015). A hydrodynamic model for silicon semiconductors including crystal heating. *European Journal of Applied Mathematics*, *26*(04), 477–496. doi:10.1017/S0956792515000157

Morandi, O. (2009). Multiband Wigner-function formalism applied to the Zener and transition in a semiconductor. *Physical Review B: Condensed Matter and Materials Physics*, *80*(2), 024301–024312. doi:10.1103/PhysRevB.80.024301

Morandi, O., & Schurrer, F. (2011). Wigner model for quantum transport in graphene. *J. Phys. A: Math. Theory.*, *26*(44), 265301. doi:10.1088/1751-8113/44/26/265301

Moyal, J. E., & Bartlett, M. S. (1949). Quantum mechanics as a statistical theory. *Mathematical Proceedings of the Cambridge Philosophical Society*, *45*(1), 99–124. doi:10.1017/S0305004100000487

Nelson, E. (1966). Derivation of the Schrödinger equation from Newtonian mechanics. *Physical Review*, *150*(4), 1079–1085. doi:10.1103/PhysRev.150.1079

Odanaka, S. (2007). A high-resolution method for quantum confinement transport simulations in MOSFETs. *IEEE Trans. Computer-Aided Design Integr. Circuits Syst.*, *26*(1), 80–85.

Ordejón, P., Artacho, E., & Soler, J. M. (1996). Self-Consistent Order-N Density- Functional Calculations for Very Large Systems. *Physical Review B: Condensed Matter and Materials Physics*, *53*(16), 10441–10444.

Pal, H. S., Nikonov, E. D., Kim, R., & Lundstrom, M. S. (2012). *Electron-Phonon Scattering in Planar MOSFETs: NEGF and Monte Carlo Methods*. Available at: https://www.researchgate.net/publication/230899595

Pan, A. (2015). *Physics and Modeling of Tunneling in Low Power Transistors* (Ph.D. Dissertation). University of California, Los Angeles, CA.

Parr, R. G. (1983). Density Functional Theory. *Annual Review of Physical Chemistry*, *34*(1), 631–656. doi:10.1146/annurev.pc.34.100183.003215 PMID:24341393

Pauli, W. (1921). *Über das Modell des Wasserstoff-Molekülions* (PhD thesis). Ludwig-Maximilians-Universität München.

Planck, M. (1900). Entropy and temperature of radiant heat. *Annalen der Physik*, *1*(4), 719–737. doi:10.1002/andp.19003060410

Quade, W., Scholl, E., Rossi, F., & Jacoboni, C. (1994). Quantum theory of impact ionization in coherent high-field semiconductor transport. *Physical Review B: Condensed Matter and Materials Physics*, *50*(11), 7398–7412. doi:10.1103/PhysRevB.50.7398 PMID:9974719

Querlioz, D. & Dollfus, P. (2010). *The Wigner Monte Carlo Method for Nano- Devices - A particle description of quantum transport and decoherence*. ISTE-Wiley.

Ravaioli, U., Osman, M., Potz, W., Kluksdahl, N., & Ferry, D. K. (1985). Investigation of ballistic transport through resonant tunneling quantum wells using Wigner function approach. *Physica B, Condensed Matter*, *134*, 36–40.

Richard, S., Aniel, F., & Fishman, G. (2004). Energy-band structure of *Ge, Si,* and *GaAs*: A thirty-band k•p method. *Phys. Rev.*, *B71*, 169901.

Rocha, A. R., García-suárez, V. M., Bailey, S. W., Lambert, C. J., Ferrer, J., & Sanvito, S. (2005). Towards Molecular Spintronics. *Nature Materials*, *4*(4), 335–339. doi:10.1038/nmat1349 PMID:15750597

Rossi, F., Brunetti, R., & Jacoboni, C. (1991). *An introduction to charge quantum transport in semiconductors and numerical approaches* (D. K. Ferry, Ed.). doi:10.1007/978-1-4899-3689-9_3

Schrödinger, E. (1926). Quantisierung als Eigenwertproblem. *Annals of Physics*, *385*(13), 437–490. doi:10.1002/andp.19263851302

Schrödinger, E. (1926). An Undulatory Theory of the Mechanics of Atoms and Molecules. *Physical Review*, *28*(6), 1049–1070. doi:10.1103/PhysRev.28.1049

Schwinger, J. (1961). Brownian Motion of a Quantum Oscillator. *J. Math. Phys.*, *2*(3), 407–432. doi:10.1063/1.1703727

Sellier, J. M. (2012). *NEMO5, a parallel, multiscale, multi physics nano-electronics modeling tool.* SISPAD.

Sellier, J. M. (2014). *Nano-archimedes*. Retrieved from www.nano-archimedes.com

Sellier, J. M., Nedjalkov, M., & Dimov, I. (2015). An introduction to applied quantum mechanics in the Wigner Monte Carlo formalism. In Physics Reports. Elsevier. doi:10.1016/j.physrep.2015.03.001

Shifren, L., & Ferry, D. K. (2002). A Wigner function based ensemble Monte Carlo approach for accurate incorporation of quantum effects in device simulation. *Journal of Computational Electronics*, *1*(1-2), 55–58. doi:10.1023/A:1020711726836

Shifren, L., Ringhofer, C., & Ferry, D. (2003). A wigner function-based quantum ensemble monte carlo study of a resonant tunneling diode. *IEEE Transactions on Electron Devices*, *50*(3), 769–773. doi:10.1109/TED.2003.809434

Shin, M. (2009). Full-quantum simulation of hole transport and band-to-band tunneling in nanowires using the k.p method. *Journal of Applied Physics*, *106*(5), 054505. doi:10.1063/1.3208067

Sho, S., & Odanaka, S. (2013). A quantum energy transport model for semiconductor device simulation. *Journal of Computational Physics*, *235*, 486–496. doi:10.1016/j.jcp.2012.10.051

Singleton, J. (2001). *Band theory and electronic properties of solids*. Oxford Univ. Press.

Siyuranga, O., Hasan, S., Lundstrom, M. S., Anantram, M. P., & Nikonov, D. E. (2007). Nonequilibrium Greens Function Treatment of Phonon Scattering in Carbon-Nanotube Transistors. *IEEE Transactions on Electron Devices*, *54*(9), 2339–2351. doi:10.1109/TED.2007.902900

Slater, J. C. (1937). Wave Functions in a Periodic Potential. *Physical Review*, *51*(10), 846–848. doi:10.1103/PhysRev.51.846

Slater, J. C. (1956). *Electronic Structure of Solids, Handbuch der Physik* (S.-V. Flagge, Ed.). Berlin.

Slater, J. C., & Koster, G. F. (1954). Simplified LCAO Method for the Periodic Potential Problem. *Physical Review*, *94*(6), 1498–1524. doi:10.1103/PhysRev.94.1498

Steiger, S. (2015). *NEMO5 User Manual*. Purdue University.

Stokbro, K. (2008). First-principles modeling of electron transport. *Journal of Physics Condensed Matter*, *20*(6), 064216. doi:10.1088/0953-8984/20/6/064216 PMID:21693878

Tarasov, V. E. (2008). *Quantum Mechanics of Non-Hamiltonian and Dissipative Systems*. Amsterdam: Elsevier Science.

Taylor, J., Guo, H., & Wang, J. (2001). Ab initio modeling of quantum transport properties of molecular electronic devices. *Physical Review B: Condensed Matter and Materials Physics*, *63*(24), 245407. doi:10.1103/PhysRevB.63.245407

Thomas, L. H. (1927). The calculation of atomic fields. *Proceedings of the Cambridge Philosophical Society*, *23*(6), 542–548. doi:10.1017/S0305004100011683

Tran, F., & Blaha, P. (2009). Accurate band gaps of semiconductors & insulators with a semilocal exchange-correlation potential. *Physical Review Letters*, *102*(22), 226401. doi:10.1103/PhysRevLett.102.226401 PMID:19658882

Tsuchiya, H., Horino, M., & Miyoshi, T. (2003). Quantum Monte Carlo Device Simulation of Nano-Scaled SOI-MOSFET. *Journal of Computational Electronics*, *2*(2), 91–95. doi:10.1023/B:JCEL.0000011405.54654.89

Vasileska, D., & Ahmed, S. S. (2004). Modeling of Narrow-Width SOI Devices. *Semiconductor Science and Technology*, *19*(4), S131–S133. doi:10.1088/0268-1242/19/4/046

Vasileska, D., Goodnick, S., & Klimeck, G. (2010). *Computational Electronics: Semi-Classical and Quantum Device Modeling and Simulation*. CRC Press. doi:10.1201/b13776

Villani, F., Gnani, E., Gnudi, A., Reggiani, S., & Baccarani, G. (2015). A quasi 2D semi-analytical model for the potential profile in hetero and homojunction tunnel FETs. *Solid-State Electronics*, *113*, 86–91. doi:10.1016/j.sse.2015.05.018

Vlasov, A. (1938). On Vibration Properties of Electron Gas,'. *Journal of Experimental and Theoretical Physics*, *8*, 444–470.

Vogl, P., Hjalmarsons, H. P., & Dow, J. D. (1983). A semi-empirical tight-binding theory of the electronic structure of semiconductors. *Journal of Physics and Chemistry of Solids*, *44*(5), 365–378. doi:10.1016/0022-3697(83)90064-1

Von Neuman, J., & Wigner, E. (1929). Uber merkwurdige diskrete eigenwerte. uber das verhalten von eigenwerten bei adiabatischen prozessen. *Zeitschrift fur Physik*, *30*, 467–470.

Von Neumann, J. (1955). *Mathematical Foundations of Quantum Mechanics*. Princeton University Press.

Von Neumann, J. (1996). *Mathematical Foundations of Quantum Mechanics*. Princeton, NJ: Princeton University Press.

Wagner, L., Grossman, J. C., Neaton, J. B., & Rousseau, I. (2016). *QWalk Quantum Monte Carlo Tutorial*. Available online: https://nanohub.org/resources/qwalk

Wannier, G. H. (1959). *Elements of Solid State Theory*. New York: Cambridge University Press.

Wannier, G. H. (1962). Dynamics of band electrons in electric and magnetic fields. *Reviews of Modern Physics*, *34*(4), 645–655. doi:10.1103/RevModPhys.34.645

Weyl, H. (1927). Quantenmechanik und gruppentheorie. *Zeitschrift fur Physik*, *46*(1-2), 1-46.

Weyl, H. (1931). *Theory of Groups and Quantum Mechanics*. New York: Dover.

Wigner, E. (1932). On the quantum correction for thermodynamic equilibrium. *Physical Review*, *40*(5), 749–759. doi:10.1103/PhysRev.40.749

Wilson, C. L. (1988). Hydrodynamic carrier transport in semiconductors with multiple band minima. *IEEE Transactions on Electron Devices*, *35*(2), 180–187. doi:10.1109/16.2438

Winstead, B., & Ravaioli, U. (2002). *Quantum Corrections for Monte Carlo Simulation* (PhD Dissertation). Beckman Institute, University of Illinois.

Yan, L., Voon, L. C., & Willatzen, M. (2009). *The k.p method - Electronic Properties of Semiconductors*. Springer.

Ziaei, V., & Bredow, T. (2016). GW-BSE approach on S_1 vertical transition energy of large charge transfer compounds: A performance assessment. *The Journal of Chemical Physics*, *145*(17), 174305. doi:10.1063/1.4966920 PMID:27825209

ENDNOTES

1 This is also called the Gorini–Kossakowski–Sudarshan–Lindblad equation (GKSL) equation

2 In quantum mechanics, particles are considered as indistinguishable, because of the uncertainty principle. Quantum particles are categorized as Bosons (like photons and phonons) or Fermions (like electrons).

3 The dephasing length is approximately given as: $\lambda_\phi \sim \hbar v^{th}/\Delta E$, where v^{th} and ΔE are the thermal velocity and energy broadening of the electron wave packet.

4 Since we cannot speak about momentum and space location simultaneously, in quantum mechanics, the Wigner distribution is not really a probability function. In fact $W(x, p)$ may have negative values. Therefore, it is called quasi-distribution function.

5 This combination is called the Wigner-Weyl transform of the density operator $\rho = |\Psi\rangle\langle\Psi|$

6 There are many QMC methods, which are based on other quantum transport approaches, such as the Liouville von Neumann equation (Jacoboni, 2010), the Pauli Master equation (Tsuchiya, Horino, & Miyoshi. 2003), the Feynman path integral (Shumway & Gilbert, 2015) and the stochastic Green function algorithm.

7 Generally, the LHS represent an N-particle Green's functions, which can be expanded in infinite chain of equations of motion. This chain is usually truncated to obtain an approximate solution. Here we consider, as an example, scattering due to collisions between electrons and holes in an interband system. Hence the two-points 1,2 and 3,4 are denoting electrons and holes in conduction and valence bands, respectively.

8 For computational reasons, an infinitesimal imaginary term may be added to E, such that $E \rightarrow E' = E + j\eta$, and $\eta \rightarrow 0$. Also, E' may be multiplied by an overlapping matrix O, so that $E \rightarrow E'.O$. Here, we take $O=I$.

9 Computer scientists talk about the scalability of algorithms in terms of computational complexity: Quicksort algorithm is O(n log n) in the size of the array; Big-O bounds describe the worst case performance of an algorithm as its inputs become large

10 The semiclassical HDM has been discussed in Ch3 and the QHDM in Section 4-9

11 The Madelung model has been already described in Section 4-11. Note that Φ is not actually a phase angle but a velocity field.

12 The Thomas-Fermi model was proposed, long time before Kohen-Sham equation (KSE), to compute the electronic properties of atoms with many electrons using the average electron density $n(r)$ as the basic variable instead of the many body Schrödinger wavefunction $\psi(r_1, r_2,..r_N)$.

13 The Bethe-Salpeter equation is presented in Chapter 8, in the context of photonic quantum transport

APPENDIX

An English translation of the original De Broglie article: “Ondes et Mouvements” can be found in “Phase Waves of Louis de Broglie”, American Journal of Physic, vol.40, No. 9, pp.1315-1420, 1972.
The Heisenberg uncertainty paper has been translated into English by John Archibald Wheeler and Hubert Zurek, in Quantum Theory and Measurement, Wheeler and Zurek, Editors, Princeton Univ. Press, 1983.
De Broglie, L. (1926). Ondes et Mouvements. Paris: Gauthier-Villars.
Heisenberg, W. (1927). Über den anschaulichen Inhalt der quantentheoretischen Kinematik und Mechanik. Zeitschrift fur Physik, 43(3-4), 172–198.

Chapter 5
Carrier Transport in Low-Dimensional Semiconductors (LDSs)

1. INTRODUCTION

Low-dimensional semiconductors (LDSs) are semiconductor structures in which the carrier transport is restricted in 2-dimensions (quantum well) or 1-dimension (quantum wires) or even zero-dimension (quantum dots).

A quantum well is a thin layer which can confine particles (typically electrons or holes) in the dimension perpendicular to the layer surface, whereas the movement in the other dimensions is not restricted. Quantum wells are formed in semiconductors by having a material, like *GaAs* sandwiched between two layers of a material with a wider bandgap, like *AlAs*. These structures can be grown by molecular beam epitaxy (MBE) or chemical vapor deposition (CVD) with control of the layer thickness down to monolayers. Thin metal films can also support quantum well states. Because of their quasi-two dimensional (Q2D) nature, electrons in quantum wells have a density of states as a function of energy that has distinct steps. Additionally, the effective mass of holes in the valence band is more closely matching that of electrons in the conduction band. These factors lead to better performance of quantum wells in optical devices such as laser diodes. They are also used to make HEMTs (High Electron Mobility Transistors), which are used in RF low-noise electronics. Quantum well infrared photodetectors are also based on quantum wells, and are used for infrared imaging. By doping either the quantum well, or the barrier with donor impurities, a two-dimensional electron gas (2DEG) may be formed. Such a structure forms the conducting channel of a HEMT, and has interesting properties at low temperature. Nowadays, the layered LDS materials with high carrier mobility are highly desirable in emerging nanoelectronic devices. For instance, the monolayers of germanium monosulfide (GeS) and molybdenum disulphide (MoS_2) are currently considered for the next generation of MOSFET transistors (Tomaneck et al, 2015).

Upon completion of this Chapter, the reader will be able to:

DOI: 10.4018/978-1-5225-2312-3.ch005

- Understand the notion of low-dimensional semiconductors (LDS) and define their main types.
- Differentiate between bulk (3-D) and low-dimensional structures (quantum wells Q2D, quantum wires Q1D and quantum dots Q0D).
- Explain the quantum band structure and density of states in LDS.
- Explain the charge carrier statistics in LDS.
- Explain the quantum transport mechanisms and models in LDS.
- Explain the concepts of a ballistic quantum transport and Landauer formulae.
- Explain the concepts of a quantum blockade and Kondo Effect.
- Describe the LDS-based devices, their properties and applications.
- Calculate the I-V characteristics of graphene FET transistor

2. INTRODUCTION TO LOW-DIMENSIONAL SEMICONDUCTORS (LDS)

We have learned so far that conduction electrons in the bulk of a semiconductor can move freely, in all the three dimensions of the physical space. However, in certain semiconductor devices, free electrons are only permitted to move in one or two dimensions. Such semiconductors, are called low-dimensional semiconductors (LDS's). For instance, the electrons filling the thin inversion layer in a conventional MOSFET (Metal-Oxide-semiconductor Field Effect Transistor) are only permitted to move freely in the 2 dimensions of the inversion layer. The motion of free electrons in the third dimension (which is normal to the thin layer plane) is confined and quantized.

The semiconductor structure is considered as a quasi two-dimensional Q2D, when one of its three spatial dimensions is getting so small (in the order of de-Broglie wavelength) compared to the other two dimensions. The gas of electrons filling such structures is sometimes called *two-dimensional* electron gas (2DEG). Thus, the electrons filling the thin inversion layer in conventional MOSFET devices act as quantum potential wells and form a 2DEG. Quantum wells are usual in hetero-junction semiconductor devices. On the other hand, the structure is considered as a quasi-one-dimensional Q1D, when two of its three spatial dimensions are so small compared to the third dimension. Such structures are sometimes called quantum wires. Many laser devices are built from 1-D structures, including nanotubes. If all the three dimensions are so small the structure is called a quasi-zero-dimensional Q0D structure or a quantum dot.

The idea behind confinement is all about keeping electrons trapped in a small area. The size we are talking about for confinement is in the order of 10 nm. One is probably wondering why confinement is so important. For one thing, it leads to new electronic properties that are not present in classical 3-D semiconductor devices. Consider the quantum dot. The typical quantum dot is anywhere between 3-60 nm in diameter. That's still 30 to 600 times the size of a typical atom. A quantum dot exhibits 0-D confinement, meaning that electrons are confined in all three dimensions.

2.1 0D Structures

The 0D structures (quantum dots) are semiconductor crystallite whose size is in the order of Brouglie wavelength or smaller than the size of the Bohr radius (1-100 nm for electrons). As we mentioned above, the quantum dots, are three-dimensionally constrained nanostructures. Figure 1 shows possible shapes of a quasi-zero-dimensional (Q0D) system or quantum dot.

Figure 1. Possible structures of quantum dots

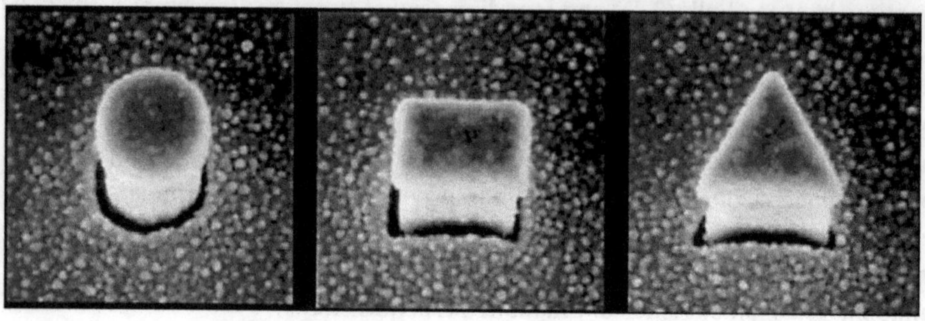

Quantum dots are usually fabricated from compound semiconductors (e.g., *InGaAs*), which are used in optical and fluorescent applications (e.g., Laser and solar cells). Quantum dots also have other applications in biological tagging and quantum computing. In fact the quantum dots are 2-D analogies for real atoms and suitable for atomic physics experiments, that cannot be done on real atoms. For instance, a magnetic flux quantum (fluxoid) in an atom requires a very high magnetic field of 1M Tesla, while for a quantum dot it is about 1Tesla, which is accessible. Therefore, quantum dots can be thought of as *artificial atoms*. As shown in Figure 2, the fabrication of quantum dots may be carried out by creating different semiconductor layers on top of each other, using one of the modern epitaxial techniques.

The energy levels of quantum dots can be modeled using the particle in a box model in which the energy of different states is dependent on the length of the box. The band gap can become larger in the strong confinement regime where the size of the quantum dot is smaller than the Bohr radius as the energy levels split up. The only things in nature that have true 0-D confinement are atoms. So a quantum dot can be loosely described as an *artificial atom*. One can easily deduce the density of states (DOS) and the density of electrons in LD's, from the $E(k)$ diagram. For simplicity, we may assume a parabolic $E(k)$ relation for electrons in such structures.

Besides confinement in all the three dimensions (in quantum dots), other quantum confined semiconductors include:

Figure 2. Array of quantum dots on top of AlGaAs substrate

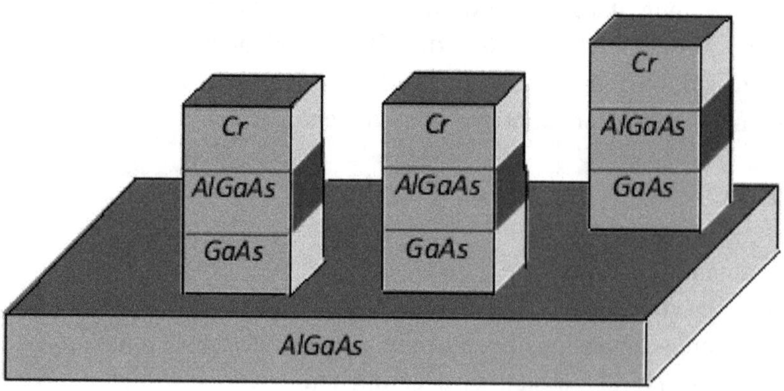

Figure 3. The density of states in zero-dimensional (quantum dot) semiconductors

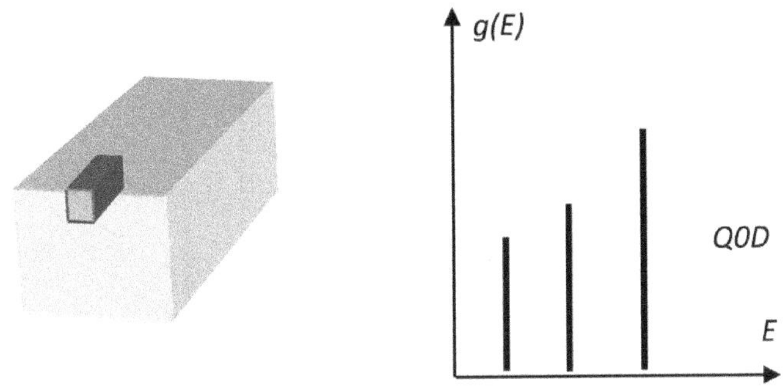

- **Quantum Wires:** Which confine electrons or holes in two spatial dimensions and allow free propagation in the third.
- **Quantum Wells:** Which confine electrons or holes in one dimension and allow free propagation in two dimensions.

2.2 1D Structures

The one-dimensional structures, such as quantum wires, are sometimes called quasi-one-dimensional (Q1D) devices or one-dimensional electron gas (1DEG). The motion of electrons in Q1D structures is free in one dimension and quantized in the other two dimensions. Therefore, the *E-k* is transformed to a series of sub-bands, as shown in figure 4. In the parabolic band approximation, the density of states (per unit length per unit energy) of one-dimensional structures $g_{c1}(E)$ is given by:

$$g_{c1}(E) = \frac{1}{\pi}\sqrt{\frac{2m_n^*}{\hbar^2}}\left(E - E_c\right)^{-1/2} \qquad (1)$$

2.3 2D Structures

The two-dimensional structures or quantum wells have thin sandwiched layer which can confine particles (electrons or holes) in the dimension perpendicular to the layer surface. In such structures, the density of states (per unit area per unit energy) $g_{c2}(E)$ is independent of energy and given by:

$$g_{c2}(E) = \frac{m_n^*}{\pi\hbar^2}.\Theta(E - E_{ci}) \qquad (2)$$

where Θ(E-Ec) is the unit step function. Note that, in Q2D structures, the motion of electrons is free in two dimensions and quantized in the third dimension. Therefore, the *E-k* is transformed to a series of sub-bands, as shown in Figure 4.

Figure 4. The E-k relation and density of states in low-dimensional semiconductors

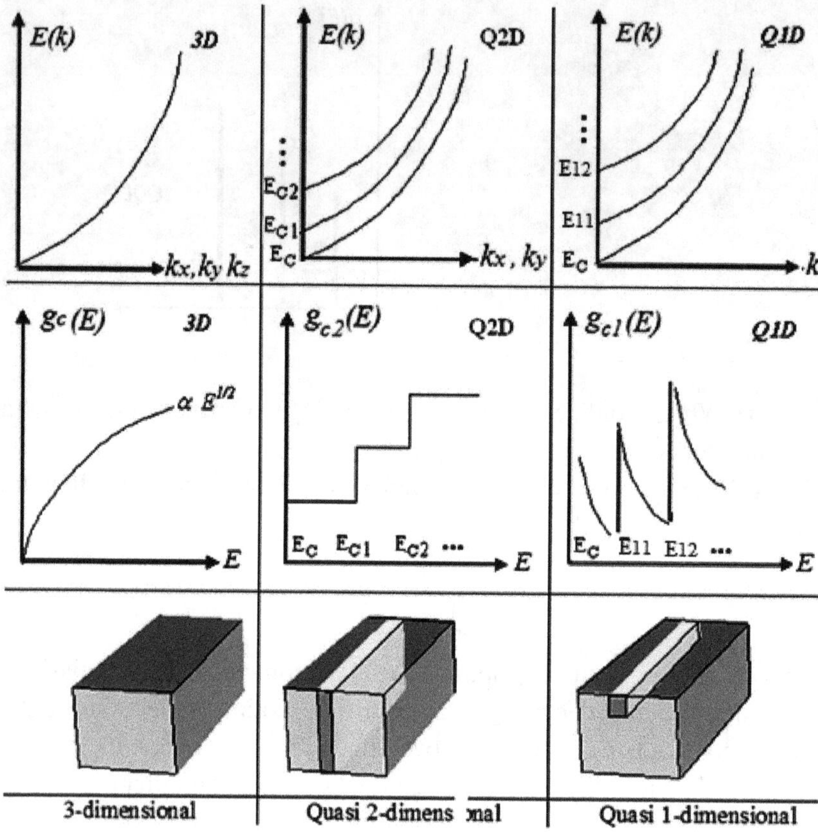

Table 1a summarizes the relations describing the dispersion relations and density of states (DOS) of parabolic semiconductors of different dimensions. Note the subband edges (E_{ci}, E_{cij} and E_{cijk}) are series of quantized values. For instance, the subband edges in Q2D may be expressed as follows:

$$E_{ci}(2D) = E_{kz} = (i + \tfrac{1}{2})\hbar\omega_i, \tag{3a}$$

where $\hbar\omega_i$ represents the transition energy (frequency) between adjacent subbands.

Similarly, the subband edges in Q1D may be expressed as follows:

$$E_{ij}(1D) = E_{k//} = (i + \tfrac{1}{2})\hbar\omega_i + (j + \tfrac{1}{2})\hbar\omega_j \tag{3b}$$

The *E-k* relation in certain valleys of some semiconductors cannot be accurately described by a simple parabolic relation. In this case, a non-parabolicity factor is usually introduced in the *E-k* relation of the semiconductor:

$$\frac{\hbar^2}{2m^*}\left(k - k_o\right)^2 = \gamma(E) \approx E\left(k\right)\ \left[1 + \alpha E(k)\right] \tag{3a}$$

Table 1a. Dispersion relations and density of states (DOS) in the conduction band of parabolic semiconductors of different dimensions. Θ(x) is the unit step function and δ(x) is the Dirac delta function. Also, E_c and E_{ci} denote the conduction band (or subband) edge

Dimension	Dispersion *E(K)*	DOS $g_c(E)$
3D (Bulk)	$(E - E_c) = \left(\frac{\hbar^2}{2m^*}\right)(k_x^2 + k_y^2 + k_z^2)$	$\left[\frac{\sqrt{2}(m^*)}{\pi^2\hbar^3}\right](E - E_c)$
2D (Well)	$(E - E_{ci}) = \left(\frac{\hbar^2}{2m^*}\right)(k_x^2 + k_y^2)$	$\left[\frac{m^*}{\pi\hbar^2}\right].\Theta(E - E_{ci})$
1D (wire)	$(E - E_{ij}) = \left(\frac{\hbar^2}{2m^*}\right)(k_x^2)$	$\left[\frac{(m^*)^{3/2}}{\pi\hbar^2\sqrt{2}}\right].\Theta(E - E_{ij})^{-1/2}$
0D (Dot)	-	$2\delta(E - E_{ijk})$

where α is the non-parabolicity factor. For the main conduction band of Si, the non-parabolicity factor is defined as follows:

$$\alpha = \frac{1}{E_g}.\left[1 - \frac{m^*}{m_o}\right] \tag{3b}$$

The nonparabolicity factor α a is 0.5 eV^{-1} in *Si*, *Ge* and 0.7 eV^{-1} in *GaAs*.

Table 1b depicts Dispersion relations and density of states (DOS) of non-parabolic semiconductors of different dimensions. Here, *E* is measured from the conduction band (or subband) edge.

Table 1b. Dispersion relations and density of states (DOS) in the conduction band of non-parabolic semiconductors of different dimensions

Dimension	Dispersion *E(K)*	DOS $g_c(E)$
3D (Bulk)	$E(1 + \alpha E) = \left(\frac{\hbar^2}{2m^*}\right)(k_x^2 + k_y^2 + k_z^2)$	$\left[\frac{\sqrt{2}(m^{*c3/2})}{\pi^2\hbar^3}\right](1 + \alpha E)^{1/2}(1 + 2\alpha E)$
2D (Well)	$E(1 + \alpha E) = \left(\frac{\hbar^2}{2m^*}\right)(k_x^2 + k_y^2)$	$\left[\frac{m^*}{\pi\hbar^2}\right].(1 + 2\alpha E)$
1D (wire)	$E(1 + \alpha E) = \left(\frac{\hbar^2}{2m^*}\right)(k_x^2)$	$\left[\frac{(m^*)^{3/2}}{\pi\hbar\sqrt{2}}\right](E)^{-1/2}$

2.4 Statistics of Low-Dimensional Carriers

The concentration of free carriers in the conduction band determines the location of the Fermi level. The carrier density for the d–dimensional case is

$$n = \int g_d(E)\, f(E).dE \tag{4a}$$

where $g_d(E)$ is the d–dimensional density of states and $f(E)$ is the carrier distribution function. At thermal equilibrium, the low-dimensional electrons density is related to the Fermi energy as follows:

$$n = \frac{1}{2^{d-1}}\left(\frac{2m_n^* k_B T}{\pi h^2}\right)^{d/2} \cdot F_{\frac{d}{2}-1}(\xi) \tag{4b}$$

where $\xi = -(E_c - E_F)/k_B T$ and $F_j(\xi)$ is the Fermi-Dirac integral of the jth order, which is defined as (Antia, 1993):

$$F_j(\xi) = \frac{1}{\Gamma(j+1)} \int_o^\infty \frac{x^j dx}{1+\exp(x-\xi)} \tag{5}$$

This relation is general for d-dimensional semiconductor. For the three-dimensional case, it reduces to the conventional form.

$$n_{3d} = 2\left(\frac{2\pi m_n^* k_B T}{h^2}\right)^{3/2} \cdot F_{\frac{1}{2}}(\xi) = N_c \cdot F_{\frac{1}{2}}(\xi) \tag{6}$$

where N_c is the 3-D effective density of states. Similarly, for the two-dimensional case, we get immediately

$$n_{2d} = \frac{m_n^* k_B T}{\pi \hbar^2}.\ln\left[1+\exp(\xi)\right] \tag{7}$$

Therefore, the two-dimensional electrons density in a given sub-band is related to the Fermi energy as follows:

$$n_{2d} = \frac{m_n^* k_B T}{\pi \hbar^2}.\ln\left[1+\exp\left(-\frac{E_c - E_F}{k_B T}\right)\right] \tag{8}$$

The total density of electrons (in all sub-bands) at equilibrium is given by the following sum:

Figure 5. The Fermi-integral of order 3/2 and its approximation limits
After Antia (1993).

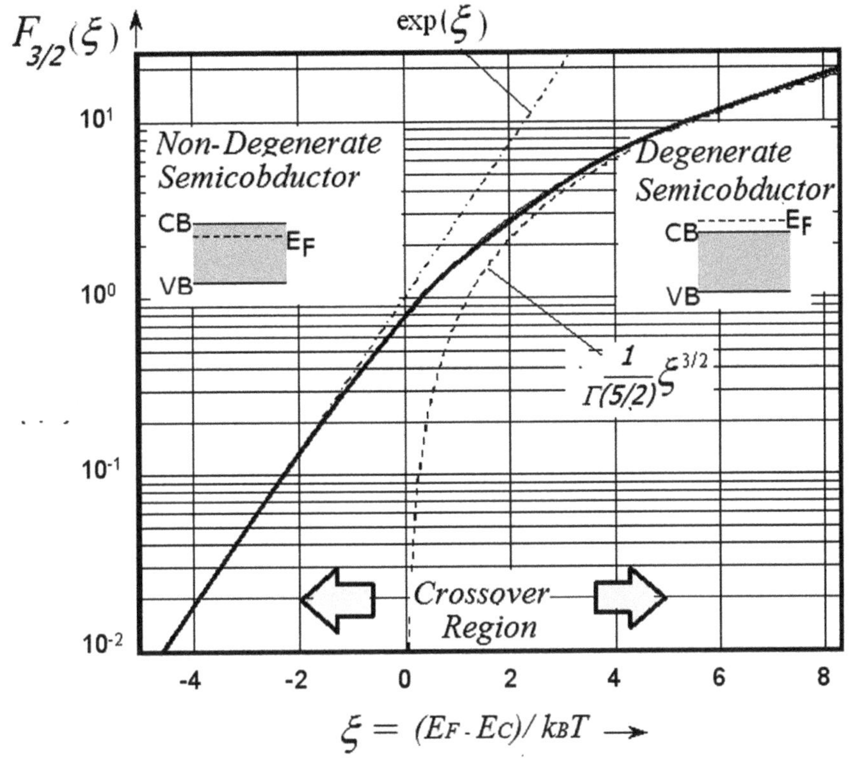

$$n_{2d} = \sum_i \frac{m_i^* k_B T}{\pi \hbar^2} . \ln\left[1 + \exp\left(-\frac{E_{ci} - E_F}{k_B T}\right)\right] \tag{9}$$

where the sum (for $i=0$ to ∞) is taken on all sub-bands and E_{ci} is the edge of the ith sub-band of the conduction band. If the conduction band of the semiconductor has several equivalent valleys, then:

$$n_{2d} = \sum_{j=1}^{Mc} \sum_i \frac{m_{ij}^* k_B T}{\pi \hbar^2} . \ln\left[1 + \exp\left(-\frac{E_{ij} - E_F}{k_B T}\right)\right] \tag{10}$$

where the first sum (for $i=0$ to ∞) is taken on all sub-bands and the outer sum (for $j=1$ to M_c) is taken on all equivalent valleys of the conduction band. Also, E_{ij} is the energy of the ith sub-band in the jth valley.

In the case of silicon, we have six equivalent valleys ($M_c = 6$). In silicon inversion layers, which are usually induced in MOSFET devices[1], we've j denotes the *4* transverse direction valleys (parallel to <100>) and the *2* longitudinal direction valleys (perpendicular to <100>).

$$n_{2d} = 4\sum_i \frac{m_t^* k_B T}{\pi\hbar^2} . \ln\left[1 + \exp\left(-\frac{E_{ci} - E_F}{k_B T}\right)\right] + 2\sum_i \frac{m_l^* k_B T}{\pi\hbar^2} . \ln\left[1 + \exp\left(-\frac{E_{ci} - E_F}{k_B T}\right)\right] \tag{11}$$

where $m_t^* = 0.19\ m_o$ and $m_l^* = 0.916\ m_o$. If the 2DEG potential well is triangular, we can express the energy of the ith sub-band, E_{ci}, in terms of the quantum number i by the following relation:

$$E_{ci} = E_c(0) + \left(\frac{h^2}{2m_\perp^*}\right)^{1/3} \left(\frac{3e\pi}{2}\zeta\right)^{2/3} \left(i + \frac{1}{4}\right)^{2/3} \tag{12}$$

where ζ is the applied electric field and $E_c(0)$ is the conduction band edge of the 2DEG structure (e.g., of the silicon inversion layer). Also $m_\perp^*$ denotes the effective mass in the perpendicular direction. When j denotes the transverse direction valleys, then $m_\perp^* = m_l^*$, and when j denotes the longitudinal direction valleys, then $m_\perp^* = m_t^*$.

In the above relations we assume the Fermi-Dirac statistics, which are valid in thermal equilibrium. We also assume that the electron motion is bound in the perpendicular directional to the 2-dimensional space. A more general relation of the electron density should take the probability of finding an electron along the perpendicular direction, so that:

$$n_{2d} = \sum_i \frac{m_i^*(z) k_B T}{\pi\hbar^2} \left|\psi_i(z)\right|^2 . \ln\left[1 + \exp\left(-\frac{E_i - E_F}{k_B T}\right)\right] \tag{13}$$

where $\psi_i(z)$ is the electron wavefunction in the ith subband and $|\psi_i(z)|^2$ is the probability of finding an electron, of the ith subband, along the z-direction. Note that the effective mass of the ith sub-band, m_i^*, is a function of z, in heterostructures of several 2-D layers of different semiconductors. Therefore, the electron wave-functions in each sub-band can be calculated by solving the Schrödinger equation in the normal direction.

$$-\frac{\hbar^2}{2}\frac{d}{dz}\left[\frac{1}{m_i^*(z)}\frac{d\psi_i}{dz}\right] + \left[eV(z) + \Delta E_{ci}(z)\right]\psi_i = E_{ci}\psi_i \tag{14}$$

where ΔE_{ci} is the band discontinuity, in the case of multiple semiconductor layers, along the z-direction. Also, the potential energy $V(z)$ can be calculated by solving the Poisson equation.

2.5 Examples of 2D Semiconductors (Quantum Wells)

Semiconductor nanostructures and superlattices, which are based on low-dimensional electron gas (2DEG) are expected to form future devices for sensing and information processing. Actually, there exist three well-known types of 2DEG systems, namely: Metal-Oxide-Semiconductor Field Effect Transistors (MOSFET's), superlattice and liquid helium surface. Figure 6 shows a heterojunction of a 2D semiconductor, sandwiched inside another semiconductor with wider energy gap. Note the discrete energy levels and the stair case density of states, inside the 2D semiconductor. The surrounding 3D semiconductor has a continuum of energy levels (energy bands), as shown in Figure 6.

In superlattices, like consecutive GaAlAs and GaAs layers, if the GaAlAs is n-type, the electrons migrate to fill the few holes on the top of the GaAs valence band but most of them will end up in states near the bottom of the GaAs conduction band. The transfer of electrons from GaAlAs to GaAs continues until a dipole layer is formed from the positive donors and the negative inversion layer. This dipole layer gives rise to a potential discontinuity which finally makes the Fermi level of the GaAs equal to that of the GaAlAs, as shown in the left of Figure 7. The electrical and optical properties of superlattice structures have been exploited in so many devices, such as quantum cascade laser (QCL) and terahertz (THz) emitters.

One of the important achievements in LDS systems is the concept of modulation doping, which led to the fabrication of high mobility semiconductor layers (up to 10^7 cm^2/Vs in GaAs). This very high mobility is achieved by selective doping of impurities far from charge carriers and hence reducing the carrier-impurity scattering. Many heterojunction devices, such as high-electron mobility transistor (HEMT), have been based on this concept.

Figure 6. Example of a 2-dimensional semiconductor structure (quantum well)

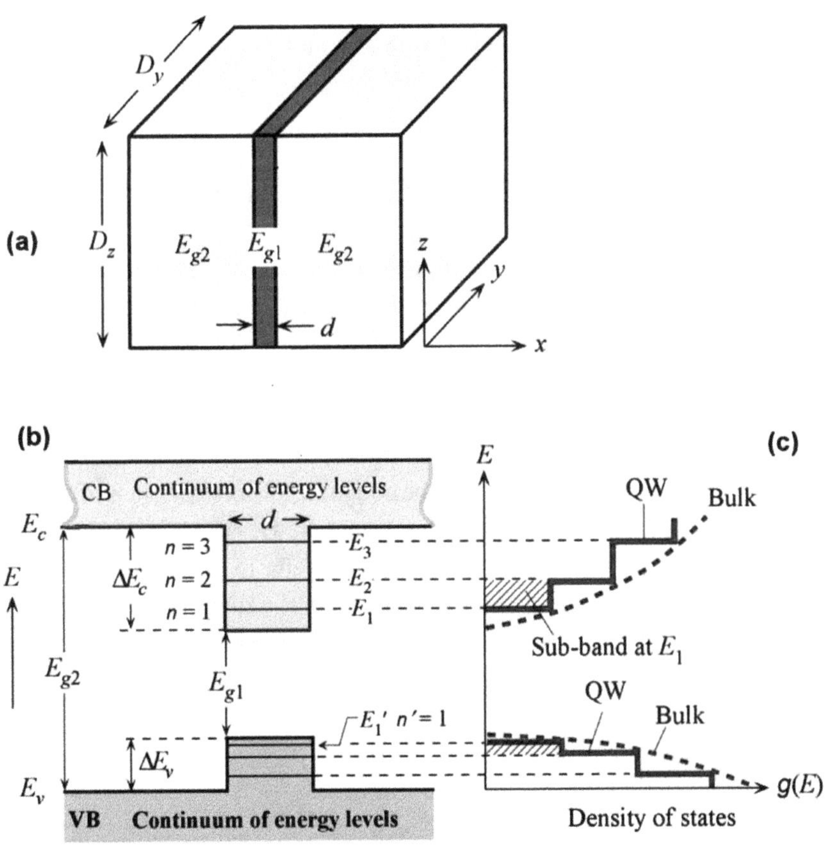

Figure 7. Illustration examples of energy band structure across 2-DEG systems: (a) At GaAs-GaAlAs interface, (b) At MOSFET inversion layer (SiO2-Si interface)

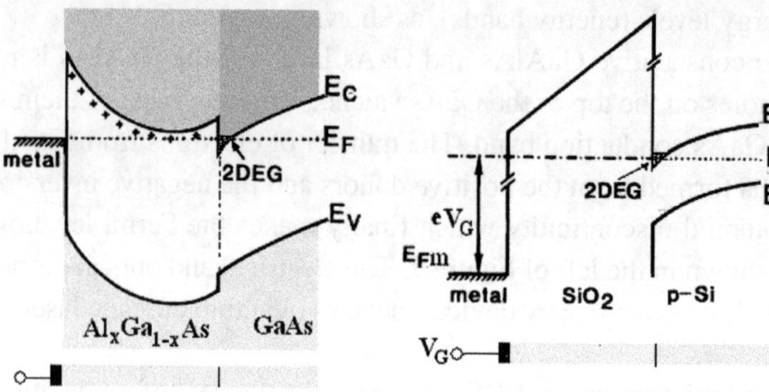

Another type of two-dimensional electron system is formed in MOSFET's, where an inversion layer is formed at the semiconductor/insulator (Si/SiO_2) interface. The principle of the electronic inversion layer is well-known. Assume an n-channel MOSFET, with p-type substratee, as shown in Figure 7. When a positive gate bias (V_G) is applied with proper amplitude, an electric field perpendicular to the interface attracts electrons from the semiconductor bulk and forms an inversion layer (rich sheet of electrons). The electrons can move in the plane sheet, but their perpendicular motion is quantized. The result is a 2D system of electrons (2DEG).

Another example of 2D systems is the utilization of semiconducting monolayers, such as germanium monosulfide (GeS) and Molybdenum disulphide (MoS_2) monolayers, as a channel in modern MOSFET devices.

3. TRANSPORT IN TWO-DIMENSIONAL ELECTRON GAS (2DEG) SYSTEMS

During the last three decades, an intensive research has been around the transport properties of 2DEG structures. The motivation for this trend emerged from the great demand on high speed MOSFETs and heterojunction microwave devices. This demand is still driven by the search for new devices and for higher integration levels in the electronics industry.

Electrical properties of 2DEG systems may be obtained from the solution of the Schrödinger and Poisson equations together with a suitable transport model. Classical transport such as drift diffusion model (DDM) or hydrodynamic model (HDM) may be adequate for specific class of devices. For instance, DDM is adequate for the electro-absorption modulators but the HDM or the direct solution of the BTE may be necessary for the simulation of hot carriers phenomena in sub-micron MOSFETs. However, in order to take some quantum transport effects into account, the quantum correction terms should be included. Optical properties, on the other hand, can be determined by a multiband quantum transport approach, such as the Maxwell-Bloch semiconductor equations (Stahl & Balslev, 1987).

3.1 Vertical and In-Plane Transport in 2-D Structures

The component of the electron momentum normal to narrow heterojunction interfaces (like 2DEG in a quantum well) is quantized and can take only discrete values. So, the energy associated with the motion along the direction normal to a 2DEG is quantized. This results in energy sub-bands. Hence, the vertical motion of charge carriers across a 2DEG is bound, but the in-plane motion is possible. Only electrons which have high kinetic energy can escape from the confining potential and behave like normal bulk electrons. Therefore, the semiclassical approximation breaks down in treating transport in the normal direction in heterojunctions. When transport along the in-plane (transverse) direction is assumed, one can still make use of the Boltzmann transport equation, but now in two dimensions, since electrons are free to move only on the plane of the heterojunction interface.The discrete energy levels (eigenvalues) at the interface of a heterojunction between two non-degenerate parabolic semiconductors can be obtained by solving the Schrödinger equation using the so-called Ben Daniel-Duke Hamiltonian:

$$\left[\frac{(p_x^2+p_y^2)}{2m(z)}+\frac{p_z}{2m(z)}\cdot p_z+\Delta E(z)_c\right]\psi(r)=E\psi(r) \tag{15}$$

where $\Delta E(z)$ is the discontinuity in the conduction band.

Schrödinger-Poisson (S-P) Model

The confinement effect appears in MOSFET devices with thin oxide layers where the barrier of potential at the interface SiO_2/Si is deep. This quantum confinement is well described by solving the single particle Schrödinger equation, self-consistently with the Poisson equation. This model provides the carrier eigenvalues and eigenfunctions along the three directions of the k-space. Considering m_l, m_{t1} and m_{t2} as the longitudinal and transverse effective masses of electrons, respectively, the electron density is given by:

$$n(x)=\frac{2k_BT}{\pi\hbar^2}\{S_{li}+S_{t1i}+S_{t2i}\} \tag{16a}$$

$$S_{li}=\sqrt{m_l m_{t1}}\sum\left|\psi_{li}\right|^2\ln\left[1+\exp\left(\frac{E_F-E_{li}}{k_BT}\right)\right] \tag{16b}$$

$$S_{t1i}=\sqrt{m_l m_{t2}}\sum\left|\psi_{t1i}\right|^2\ln\left[1+\exp\left(\frac{E_F-E_{t1i}}{k_BT}\right)\right] \tag{16c}$$

$$S_{t2i} = \sqrt{m_{t1}m_{t2}} \sum \left|\psi_{t2i}\right|^2 \ln\left[1 + \exp\left(\frac{E_F - E_{t2i}}{k_B T}\right)\right] \tag{16d}$$

where x is the position normal to the gate oxide, ψ_{li}, ψ_{ti}, E_{li} and E_{ti} are the ith longitudinal and transverse eigenvector and eigenvalues, respectively. For holes, a similar expression is obtained for the light and heavy holes. For a 2-D MOS device, the S-P equation is usually solved along a set of 1-D parallel slices under the gate. At the ends of each slice an infinite potential is set as a boundary condition. Unfortunately, this assumption is not physical at the SiO_2/Si interface. Therefore, the S-P model should be better designed to include the gate oxide so that the carriers and their eigenvectors could penetrate into the insulator layer (SiO_2), as shown in Figure 9. Figure 8 depicts the formation of energy sub-bands due to the confinement of electrons near the Si-SiO_2 interface (left) and the resultant electron density (right). Note that the classical carrier density rises towards the semiconductor–oxide interface, while the quantum (S-P) solution results in a more correct carrier density, which decreases towards the interface.

Figure 10 depicts the carrier density at SiO_2- Si interface of n-MOS and p-MOS structures, as obtained by semiclassical (SHE with DG corrections) and a quantum (S-P) solver. Figure 11 depicts the C-V characteristics of a MOSFET, according to the semiclassical, quantum and D-G approaches. We note the shift of the threshold voltage near 0.5V and the reduction of the quantum capacitance in inversion mode ($V_G > 0.5$ V). The difference observed between S-P approach and DG model in strong accumulation is explained by the fact the charge is treated in a quantum scheme in the S-P solver whereas a part of the charge should be treated semiclassically.

Example 1

The electric field at the surface of a semiconductor in the inversion layer of a MOSFET is $\zeta = 5\times10^4$ V/cm.

Figure 8. Bound states in a triangular quantum well at SiO_2- p-type Si interface

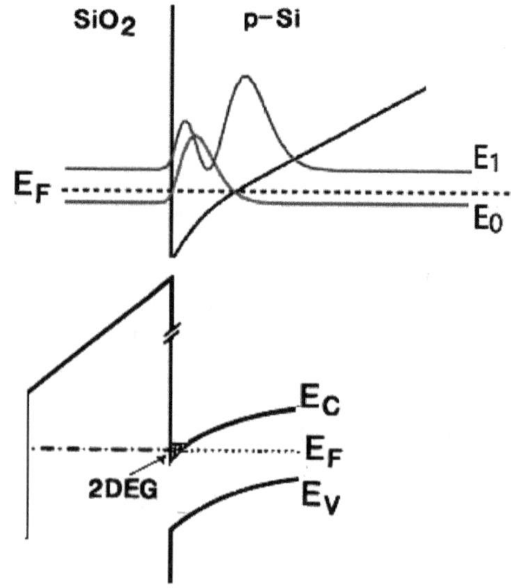

Figure 9. Formation of energy sub-bands due to well confinement of electron gas near the Si-SiO2 interface (left) and the resultant electron density (right)

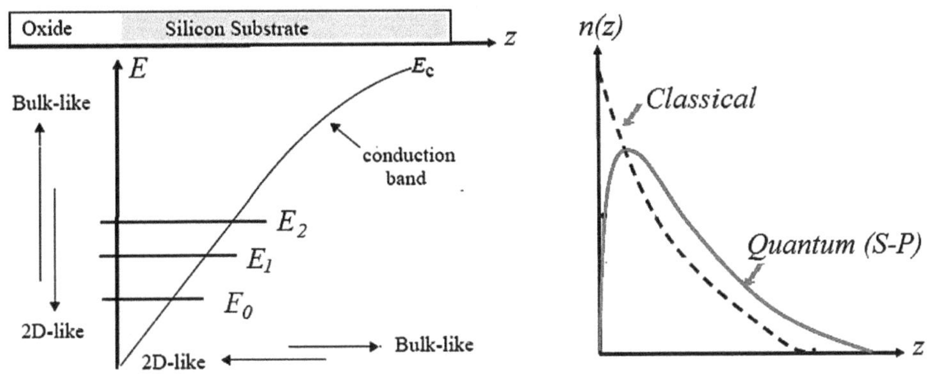

Figure 10. Carrier density at SiO_2- Si interface of n-MOS and p-MOS structures

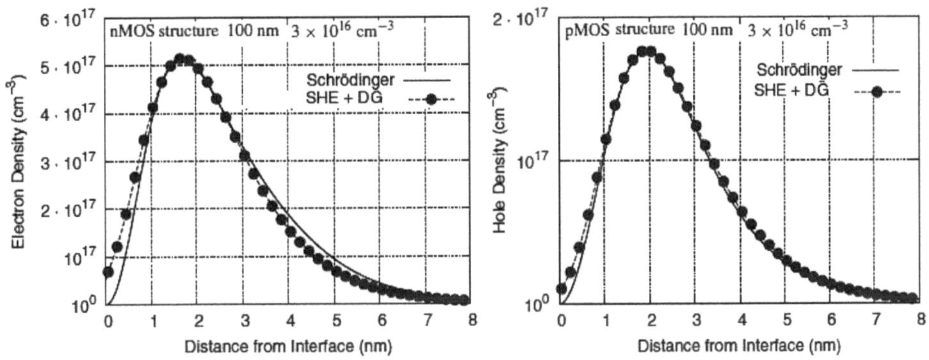

Using a trial wave function $\psi \sim z\, exp(-z/a)$, estimate the lowest energy of an electron in the triangle potential well (V = 0 for $z \leq 0$ and $V = e.\zeta.z$ for $z > 0$) formed by the electric field. Take the effective mass of the electron is $m^* = 0.063m_0$.

Solution

The energy of the ground state is a minimum of the function:

$$E = \frac{\langle \psi \mid H \mid \psi \rangle}{\langle \psi \mid \psi \rangle}$$

So, we have to solve the Schrödinger equation for an electron in a triangle potential,

$$H\psi = \frac{\hbar^2}{2m^*}\psi'' + ez\zeta\psi = E\psi$$

Figure 11. C-V characteristics of an MOS, according to the semiclassical model (dashed line), Schrodinger-Poisson model (dotted line) and density gradient method (solid line)
After Mamaluy et al (2005).

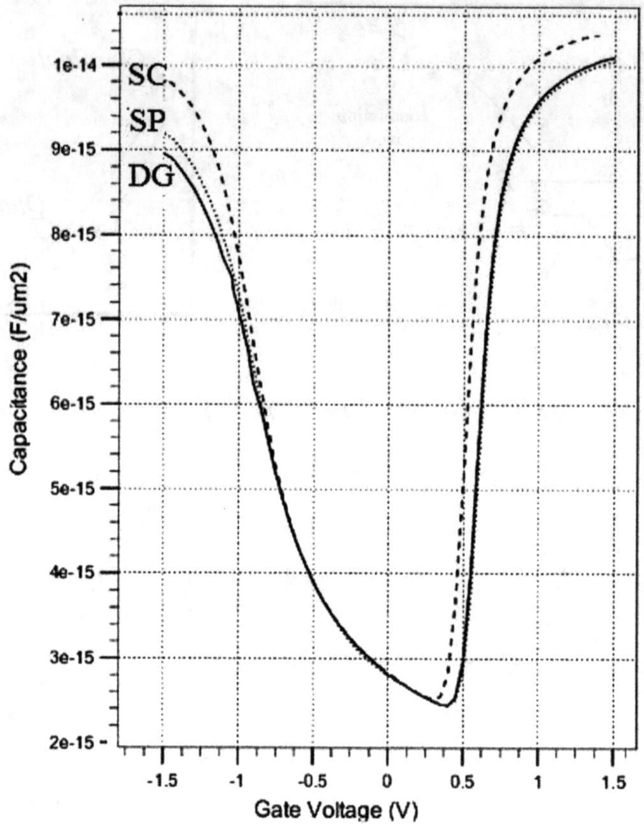

to find

$$\langle \psi \mid H \mid \psi \rangle = \frac{\hbar^2}{2m^*} \int_0^\infty \psi'' dz + e\zeta \int_0^\infty z \mid \psi \mid^2 dz$$

and

$$\langle \psi \mid \psi \rangle = \int_0^\infty \mid \psi \mid^2 dz$$

With the trial wave function $\psi = Az\, exp(-z/a)$ we obtain

$$\langle \psi \mid H \mid \psi \rangle = A^2 \left(\frac{\hbar^2}{2m^*} a + \frac{3e\zeta}{8} a^4 \right),$$

and

$$\langle\psi \mid \psi\rangle = \frac{1}{4}a^3A^2$$

Therefore, the energy of the ground state is a minimum of

$$E = \frac{\langle\psi \mid H \mid \psi\rangle}{\langle\psi \mid \psi\rangle} = \frac{\hbar^2}{2m^*a^2} + \frac{3}{2}ea\zeta$$

From $E' = dE/da = 0$, we get E_{min} occurs at $a = (2\hbar^2/3m^*e\ \zeta)^{1/3}$ and equals

$$E_{\min} = \tfrac{9}{4}\left(\frac{2\hbar^2e^2\zeta^2}{3m^*}\right)^{1/3}$$

Substituting, we get E_{min}=61 meV, which is bigger than k_BT 300K

Under the effect of external field, charge carriers can drift in-plane. There exists also a possibility of conduction across the vertical direction of 2DEG systems, under the influence of high electric fields. This vertical conduction may be associated with a negative differential conductivity, as has been shown by Esaki and Tsu (2000). In the following subsections we discuss both in-plane and vertical conduction mechanisms in 2DEG systems, with emphasis on super-lattices.

In-Plane Conductivity of a 2DEG

We can calculate the in-plane conductivity starting from the semiclassical BTE of electrons in the ith sub-band of the 2DEG:

$$-\frac{e\zeta_{//}}{m_n}.\hbar k_{//}\frac{\partial f_i}{\partial \varepsilon_i} = \sum_{k'_{//}} S(k_{//},k'_{//}).f'_i(1-f_i) - S(k'_{//},k_{//})f_i(1-f'_i) \tag{17}$$

where $k_{//} = [k_x, k_y]^T$ is the 2-dimensional wave vector, which characterizes the in-plane motion, $f_i(k_{//})$ is the electron distribution function, m_i is the effective mass and $E_i(k_{//}) = E_i + \hbar^2 k_{//}^2/2m_i$ is the electron energy in the ith sub-band of the 2DEG. The scattering rates $S(k_{//},k'_{//})$ can be calculated by the Fermi golden rule, with the proper matrix element and final density of states in two-dimensions. Figure 12 shows a comparison between scattering rate in the bulk and in 2DEG

The above coupled system of BTE equations (in different sub-bands) can be simplified at low fields, by adopting the relaxation time approximation:

Figure 12. Scattering rate in the bulk and in 2DEG
After Bastard (1990).

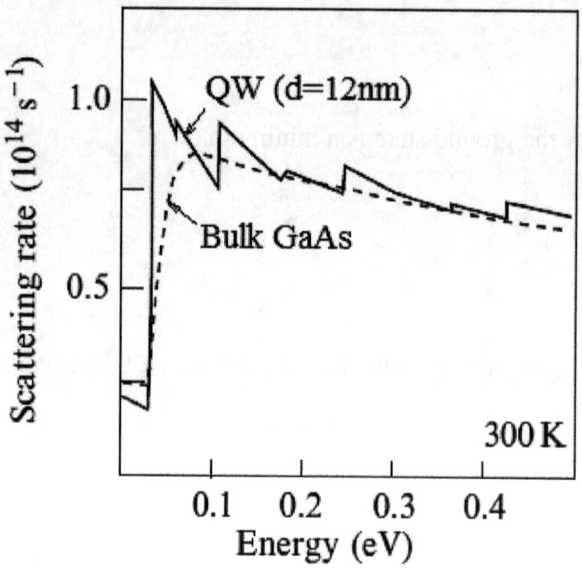

$$f_i(k_{//}) = f_{io}(k_{//}) + \frac{e\zeta_{//}}{m_i}.\hbar k_{//}\frac{\partial f_{io}}{\partial E_i}\tau_i(E_i) \tag{18}$$

with τ_i is the microscopic relaxation time in the ith subband of the 2DEG. In so doing, the plane current density averaged over all the sub-bands is given by $J_n = \sigma_n \zeta$, where $\sigma_n = \Sigma \sigma_{ni}$ is the 2DEG conductivity and

$$\sigma_{ni} = e^2\, n_{(i)}\tau_i / m_i \tag{19}$$

Here, $n_{(i)}$ is the electrons density in the ith subband. At 0K, $n_{(i)}$ is given by:

$$n_{(i)} = (m_i/\pi\hbar^2)[\,E_{Fi} - E_i] \tag{20}$$

where E_{Fi} is the Fermi energy level in the ith sub-band. Also, the electron mobility μ_n of the 2DEG is given by the average over mobility terms μ_{ni}:

$$\mu_n = \Sigma_i\, n_{(i)}\, \mu_{ni} / \Sigma_i\, n_{(i)} \text{ and } \mu_{ni} = e\tau_i / m_i \tag{21}$$

The total mobility is calculated from other mobilities by Matthiessen's rule

$$(\mu)^{-1} = (\mu_{imp})^{-1} + (\mu_{ac})^{-1} + (\mu_{op})^{-1} + (\mu_{polar})^{-1} + (\mu_{alloy})^{-1} \tag{22}$$

Figure 13. Illustration of calculated & experimental electron mobility of a GaAs 2DEG with $3.5x10^{11}$ cm^{-2} sheet doping
After Briner (2011).

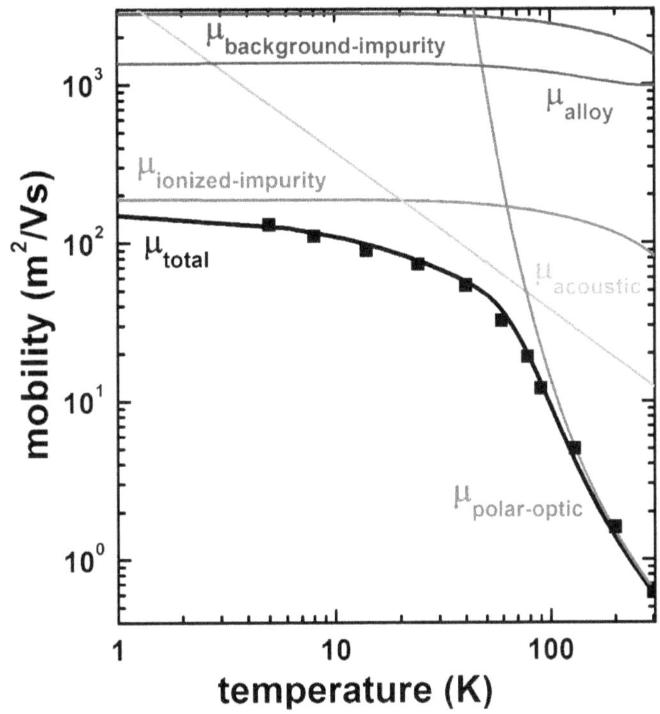

3.2 Vertical Transport

While in-plane motion is free (with plane electronic waves) the motion of carriers in vertical direction is quantized. When voltage is applied across a layered structure or a superlattice (SL), electrons will relax from higher to lower energy levels emitting either light or LO phonon. Assuming no transitions between the different sub-bands of an SL, it can be shown that electrons will have an oscillatory behavior (Bloch oscillations) with frequency:

$$f = e\, \zeta\, d / 2\pi \hbar \qquad (23)$$

where d is the spacing between repeating layers. For a static field ζ of 10kV/cm and d=100Å, we get f =*2.5* THz. Such Tera Hertz oscillations have never been reported experimentally. This is due to electron scattering events. The vertical transport in SLs is mainly due to hopping of electrons between localized states in adjacent quantum wells. Other parameters such as impurity density and interface roughness also influence the vertical mobility of an SL (Safa & Asgari, 2015). Figure 14 illustrates *the interband transitions in a q superlattice structure, called quantum cascade laser QCL. QCL* is, a semiconductor laser in which an electron passes through a series of quantum wells. In each quantum well, the electron emits a photon on an inter-subband transition before tunneling through to the next quantum well.

Figure 14. Schematic of interband transitions in a quantum cascade laser (QCL) superlattice structure

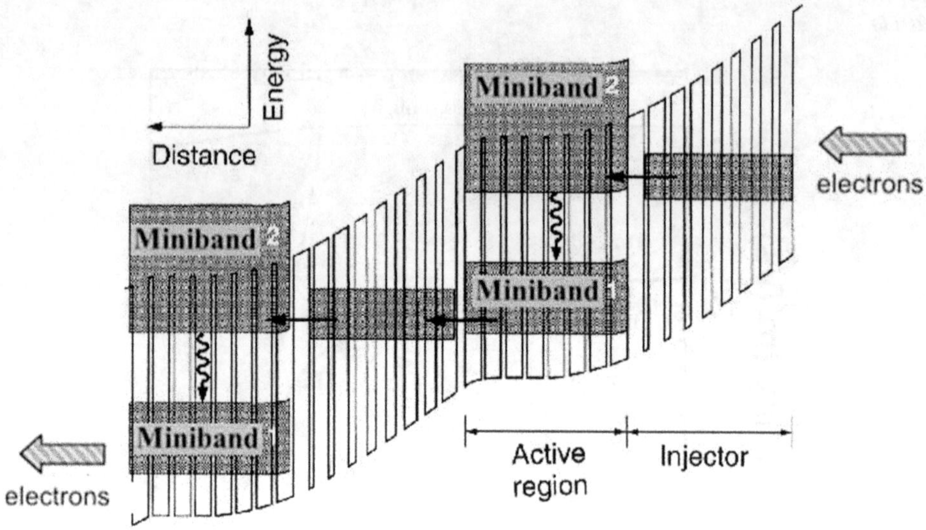

3.3 Tunneling and Resonant Tunneling

The tunneling across potential wells or thin layered structures is another mechanism of vertical transport. Tunneling refers to particle transport through a classically forbidden region, where the total energy of a particle is less than its potential energy. This is illustrated in figure 15, where a particle of energy E is incident on an arbitrary-shaped potential energy barrier of height $V>E$. In classical mechanics a particle is completely reflected. However, the quantum mechanics shows that there exists a probability of tunneling, for thin barriers. Here, we present the tunneling across quantum wells and layered semiconductor structures. Note that, tunneling is different from scattering limited conduction and phonon-assisted hopping between quantum wells (Ilatikhemeneh et al, 2015).

In case of vertical transport across a single quantum well (of thickness a) between two barriers (of thickness b), the tunneling probability is given by:

$$T = \frac{1}{1+\frac{1}{4}(t+t^{-1})^2 sh^2(k_b \cdot b)} \tag{24}$$

where $t = (m_b / m_w).(k_w / k_b)$ is the ratio of electron masses and wave vectors inside the well and the barrier regions, respectively.

$$k_w = \sqrt{(2m_w\varepsilon_1)}/\hbar,\ k_b = \sqrt{2m(V_b - \varepsilon)}/\hbar \tag{25}$$

where V_b is the barrier height and $\varepsilon_1 \leq V_b$ is the energy of first sub-band, which we only consider here.

The tunneling resonance happens when the transmission coefficient is equal to unity ($T=1$) and the transmitted and reflected waves interfere in a constructive manner inside the intermediate quantum well. This happens at certain energies, which satisfy the condition:

Figure 15. Example of transport by tunneling across SiO_2 dielectric barrier in a metal-oxide-semiconductor (MOS) structure, and the associated wavefunction

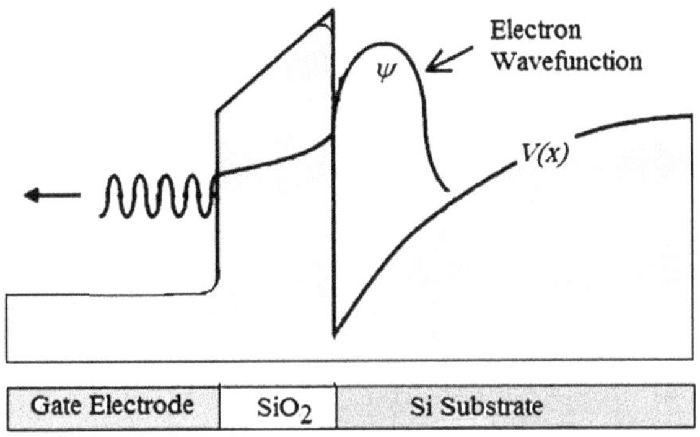

$$Cos(k_w a) = ½ (r + r t^{-1}) \sin (k_w a) \tag{26}$$

That is, an electron with this resonant energy can cross the double barrier without being reflected. Figure 16 depicts the resonant tunneling phenomenon. This resonance phenomenon is similar to that taking place in the optical Fabry-Perot resonator. This phenomenon is exploited in the resonant tunneling devices (RTD's). The RTD's possess fast response times and have been employed as solid-state oscillators at frequencies approaching 1 THz.

3.4 Case Study 5: Transport across a Resonant Tunneling Diode

A resonant-tunneling diode requires a band-edge discontinuity at the conduction band or valence band to form a quantum well and, thus, necessitates heteroepitaxy. The most common combination used is *GaAs-AlGaAs*. The middle quantum-well thickness is typically around 50A, and the barrier layers about 30A. Symmetry of the barrier layers is not required so their thickness can be different. A typical resonant tunneling diode structure with analytical band edge model as used in simulations is depicted in Figure 17. The figure depicts the practical structure of a double-barrier resonant tunneling (DBRT) diode, with engineered emitter.

The tunneling problem is usually treated by solving the Schrödinger equation with Ben Daniel-Duke Hamiltonian. However, the Wigner distribution function can be also used to provide an adequate quantum mechanical description of the electron transport through tunneling nanostructures. For a resonant tunneling device, like the double-barrier resonant tunneling (DBRT) diode shown below, we can write:

$$\frac{\partial W}{\partial t} + \frac{\hbar \boldsymbol{k}}{m^*} \cdot \nabla_x W + \theta \cdot W = \left(\frac{\partial W}{\partial t} \right)_C \tag{27a}$$

where

$$\theta \cdot W(x,k) = \int_{-\infty}^{+\infty} T(x,k,k') \cdot W(x,k') dk' \quad {}_C \tag{27b}$$

Figure 16. Illustration of the resonant tunneling phenomenon, which has a unity probability at certain carrier energy

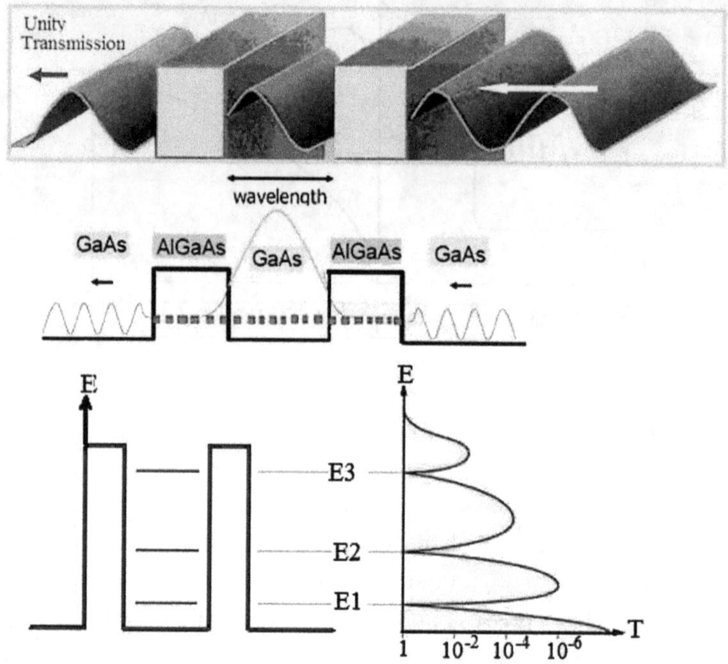

Figure 17. Structure of a double-barrier resonant tunneling (DBRT) diode, with double barriers of thickness b=30 Å and a single quantum well of thickness a=50Å

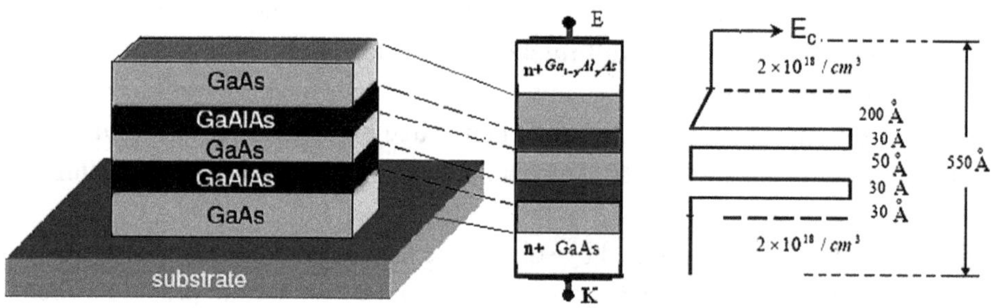

and

$$T(x,k,k') = \int_0^L \left[V(x+x') - V(x-x')\right] \sin\left[2x'(k-k')\right] \cdot dx' \tag{27c}$$

Here L is the length of the tunneling structure under consideration. The RHS term in the WBTE is due to scattering and may be modeled using the relaxation time approximation. The boundary conditions on the Wigner distribution function $W(x, k)$ at the emitter ($x=0$) and cathode ($x=L$) can be approximated to flat-band transport and equilibrium electron-distribution. In order to calculate the potential energy $V(x) = e\phi(x) - \Delta E(x)$, the WBTE should be coupled with the Poisson equation:

$$\frac{\partial}{\partial x}\left(\varepsilon\frac{\partial \varphi}{\partial x}\right) = -e.\left(n(x) - N_d^+\right) \tag{28a}$$

This equation may be also written as follows:

$$\frac{\partial}{\partial x}\left(\varepsilon\frac{\partial V}{\partial x}\right) = -e^2.\left(n(x) - N_d^+\right) \tag{28b}$$

3.5 Magneto-Transport in a 2DEG

The effect of magnetic on low-dimensional semiconductor devices such as superlattices and quantum cascade (QC) structures has been a fruitful subject during the last two decades.

In a magnetic field B, perpendicular to the 2DEG, the current is no longer in the direction of the electric field due to the Lorentz force. Consequently, the conductivity is no longer a scalar but a tensor σ, related to the diffusion tensor D by the Einstein relation (Tsui, Stormer & Gossard, 1982):

$$\boldsymbol{\sigma} = e^2\, g_c(E_F)\boldsymbol{D} \tag{29}$$

Between scattering events the electrons at the Fermi level execute circular orbits, with cyclotron frequency $\omega_c = (eB/m)$ and cyclotron radius $l_{cyc} = (mv_F/eB)$. Taking the 2DEG in the x−y plane, and the magnetic field in the positive z-direction, one can write in complex number notation:

$$v(t) = v_x(t) + j\, v_y(t) = v_F \exp\,(j\phi + j\omega_c t) \tag{30}$$

The diffusion tensor, $\boldsymbol{D}$, is obtained from the following relation:

$$D_{xx} + jD_{yx} = \frac{1}{2\pi}\int_0^{2\pi}\left[\int_0^{\infty} v(t)v_F\cos(\varphi)\exp\left(-\frac{t}{\tau}\right)\right]d\varphi = D_o\cdot\frac{1+j\omega_c\tau}{1+(\omega_c\tau)^2} \tag{31}$$

where D_o is the zero-field diffusion constant. One can easily verify that $D_{yy} = D_{xx}$ and $D_{xy} = -D_{yx}$. From the Einstein relation, we can obtain the conductivity tensor as follows:

$$\sigma = \frac{\sigma_o}{1+(\omega_c\tau)^2}\begin{pmatrix} 1 & -\omega_c\tau \\ \omega_c\tau & 1 \end{pmatrix} \tag{32a}$$

where σ_o is the zero-field conductivity. The resistivity tensor $\rho \equiv 1/\sigma$ has the form:

$$\rho = \frac{\rho_o}{1+(\omega_c\tau)^2}\begin{pmatrix} 1 & \omega_c\tau \\ -\omega_c\tau & 1 \end{pmatrix} \quad (32b)$$

with $\rho_o = \sigma_o^{-1} = (m/n_s e^2\tau)$ the zero-field resistivity. The off-diagonal element $\rho_{xy} \equiv R_H$ is the classical Hall resistance of a 2DEG:

$$R_H = \frac{B}{en_s} = \frac{1}{g_s g_v}.\frac{h}{e^2}.\frac{\hbar\omega_c}{E_F} \quad (33)$$

Note that in a 2D channel geometry there is no distinction between the Hall resistivity and the Hall resistance, since the ratio of the Hall voltage $V_H = W\zeta_x$ across the channel to the current $I = W J_y$ along the channel does not depend on its length and width (provided transport remains in the diffusive regime). The diagonal element ρ_{xx} is referred to as the longitudinal resistivity. Equation (28) tells us that the magnetoresistivity is zero (i.e., $\rho_{xx}(B) - \rho_{xx}(0) = 0$), in the classical limit. This counterintuitive result can be understood by considering that the force from the Hall voltage cancels the *average* Lorentz force on the electrons. As a general conclusion, the classical effects of magnetic fields are important only if $\omega_c\tau \geq 1$. In such fields an electron can complete several cyclotron orbits before being scattered out of orbit. In a high-mobility 2DEG this criterion is met at weak magnetic fields (note that $\omega_c\tau = e\mu B$).

In the application of the Einstein relation we have used the zero-field density of states. Moreover, we have assumed that the scattering time τ is B-independent. Both assumptions are justified in weak magnetic fields, for which $E_F/h\omega_c \geq 1$, but not in stronger fields. The deviations from the semiclassical result (33) appear when the magnetic field is increased. These deviations take the form of oscillatory magnetoresistivity (Shubnikov-De Haas effect) and plateaux in the Hall resistance (quantum Hall effect). The origin of these two phenomena is the formation of Landau levels by a magnetic field that leads to the B-dependent density of states.

Integer Quantum Hall Effect

The quantum Hall effect was first observed in 1980 by K. Klitzing who was awarded the Nobel Prize in physics in 1985. The integer quantum Hall effect (IQHE) was discovered while investigating the conductance of 2DEG structures at high magnetic fields and at very low temperatures. It was found that the Hall conductance of such a system, shows quantized plateaus at integer multiples of (e^2/h). For a 2DEG system, the density of states, in absence of magnetic field, is $g_{c2}(E) = (m_n/2\pi\hbar^2)$ states per unit energy per unit area (cm^2). After applying a magnetic field, the energy states contract into separate Landau levels.

Each Landau level is degenerate, including $(g_{c2}.\hbar\omega_c) = (eB/h)$ states per unit area (cm^2). We may designate this number of states per unit area for a filled Landau level by n_B, such that $n_B = (g_{c2}.\hbar\omega_c) = eB/h$.

The filling factor v is defined as the ratio between n_B and the electron density of the 2DEG:

$$v = n_{2D}/n_B = n_{2D}/(eB/h) \quad (34)$$

Therefore, $n_{2D} = (eB/hv)$. From the above equations (34, 33 we can get the Hall resistance (per unit area):

Figure 18. Quantum Hall measurement sample

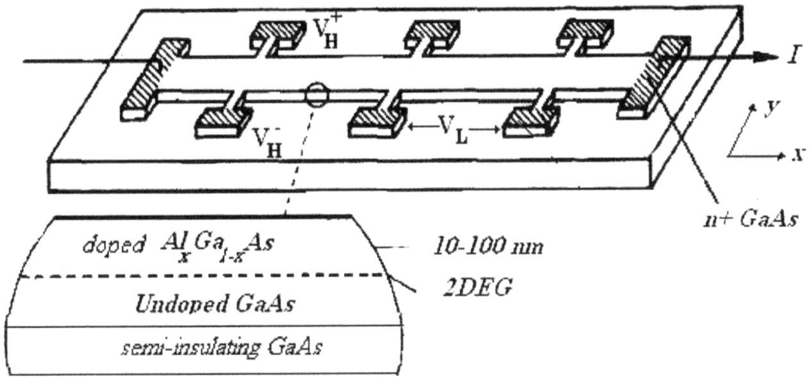

$$R_H = 1/G_H = (h/e^2)/v \tag{35}$$

When we substitute $\omega_c\tau \rightarrow \infty$ into the classic Hall conductivity tensor expression (32), then we obtain $\rho_{xy} \rightarrow en/B = (h/e^2v)$. Nevertheless, the above equation (35) is only correct in certain specific values of n_{2D}.

For the inversion layer of a MOSFET, n_{2D} is proportional to the gate voltage V_G. Hence, the Hall resistance should be inverse proportional to the gate voltage. Actually, Klitzing and his colleagues discovered the quantized Hall plateaus in the $R_H(V_G)$ relation of a high-mobility MOSFET. Using a Hall voltage method they found that $R_H = (h/e^2)/v$ to high accuracy. This result does not depend on material and has been verified in *Si* and *GaAs*. So, the quantum *Hall resistance* is given by:

$$R_{xy} = (V_H^+ - V_H^-)/I = (h/e^2)/n_{2D} = -(h/e^2)/\nu. \tag{36}$$

Figure 19. Transverse (R_{xy}) and longitudinal (R_{xx}) Hall resistivity of a 2DEG of density n_{2D} = 2.6 x10^{15} nm^{-2} as a function of the magnetic field B
After Yoshioka (2002).

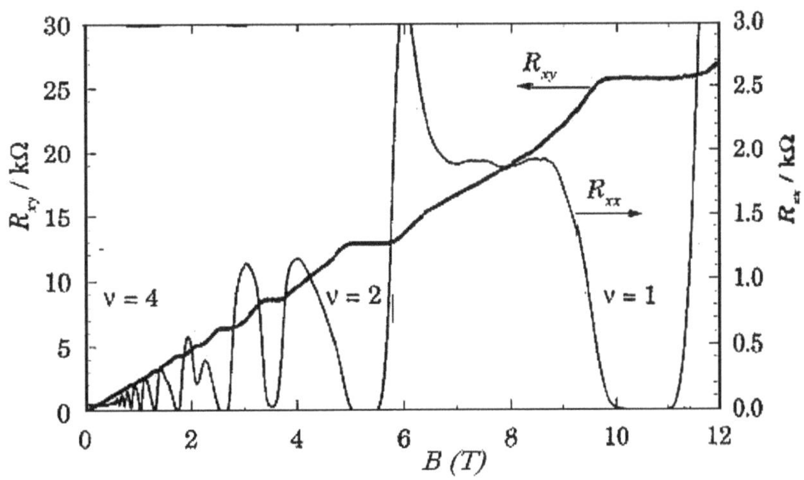

Note that the filling factor ν is equal to the number of filled Landau levels ($\nu = 1, 2, 3, ..$). In these levels, the motion of electrons is confined in two dimensions. Figure 18 shows the experimental setup. The Hall transverse and longitudinal resistance are defined as $R_{xy} = (V_H^+ - V_H^-)/I$ and $R_{xx} = V_L/I$. As shown in figure, the transverse Hall resistance R_{xy} has plateaus in a certain quantized values between adjacent Landau levels, while the longitudinal resistance R_{xx} almost vanishes at these values. This means that the electrons are localized in these regions. In fact the existence of the localized states explains the appearance of plateaus in the IQHE. As the electron density n_{2D} is increased (or $\boldsymbol{B}$ is decreased) the localized states fill up gradually, without any change in the Hall resistance. For these densities the Hall resistance is on a step and the longitudinal resistance vanishes.

The ratio $R_k = h/e^2 = 25.812807\ k\Omega$. is a fundamental unit in quantum Hall effect and has been adopted as a standard unit for the resistance measurement. In fact this unit is closely related to the fine structure constant $\alpha = \mu_o.c.e^2/2h = 1/137$.

Fractional Quantum Hall Effect (FQHE)

In the fractional quantum Hall effect (FQHE), the conductance is quantized in *fraction* multiples of (e^2/h). The fractional quantum Hall effect was discovered in 1982, by Tsui, Gossard and Stormer, in pure high electron mobility samples at very low temperatures. Tsui and Stormer were awarded the Nobel Prize for this discovery in 1998. Figure 20 depicts the fractional quantum Hall effect in *Si*. Tsui and his colleagues found that the filling factor of Landau level may take fractional values, such that:

$$\nu = n_{2D} / n_B = n_{2D} / (g_{c2} .\hbar\omega_c) = N/M \tag{37}$$

with N and M are integers. As shown in figure 20, the plateaus of ρ_{xy} correspond to minima in ρ_{xx}.with filling factor ν is indicated, particularly for $\nu = 2/3,\ 2/5$ and $3/5$. The interpretation of the fractional quantum Hall effect has been developing rapidly. One of the widely accepted theories (formulated by Jain, in 1989), attributed this effect to the Coulomb interaction forces between electrons.

According to this theory, when electrons are confined in two dimensions, cooled to very low temperatures and subjected to a strong magnetic field, their kinetic energy is quenched due to Landau level quantization. The behavior of electrons under such conditions is governed by the Coulomb repulsion alone. Therefore, electrons are arranging themselves in particular configurations (bound states called composite Fermions) in order to minimize their energy at certain values of ν. Such composite Fermions are supposed to move at an effective magnetic field $\boldsymbol{B}_{\text{eff}}$, which is much smaller than the applied magnetic filed, $\boldsymbol{B}$, and that vanishes at certain ν values.

$$\boldsymbol{B}_{eff} = \boldsymbol{B} - \boldsymbol{B}_{½} = \boldsymbol{B} - 2\hbar\, n_{2D} / e \tag{38}$$

Spin Quantum Hall Effect (SQHE)

The integer and fractional quantum Hall effects were the first topological states to be discovered in the 1980s, but they exist only in the presence of large magnetic fields. The search for topological states of matter that do not require magnetic fields for their observation led to the theoretical prediction in 2006

Figure 20. Longitudinal (ρxx) and transverse (ρxy) resistivity of a high mobility 2DEG at 0.15K, showing the fractional quantum Hall effect
After Yoshioka (2002).

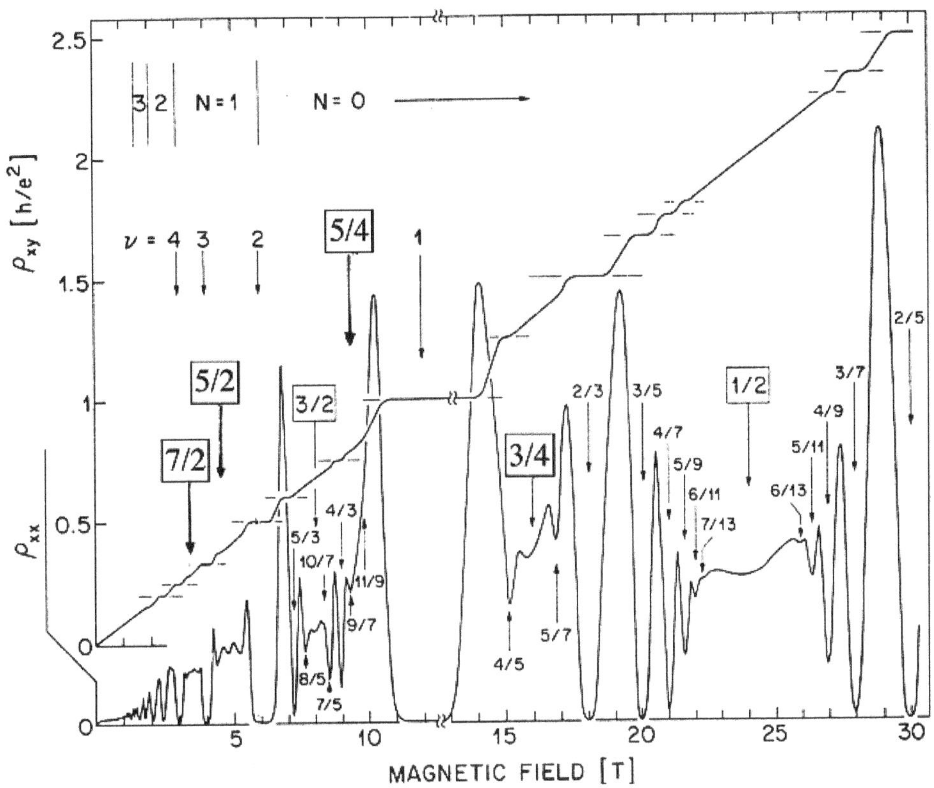

and experimental observation in 2007 of the so-called quantum spin Hall Effect (SQHE) in *HgTe* quantum wells, a new topological state of quantum matter (Maciejkoet al., 2011).

The quantum spin Hall Effect is the cousin of the integer quantum Hall Effect. It is a state of matter which exists in special, two-dimensional, semiconductors that have a quantized spin-Hall conductance. As shown in Figure 21, the charge current flows from left to right through a Hall bar made from a conductor such as aluminum. Spin–orbit interactions cause a separation of electron spins — the spin Hall Effect. If the charge current is unpolarized (with equal numbers of spin-up and spin-down electrons), the spin imbalance does not induce a charge imbalance or transverse voltage at the Hall cross. If electrons, which are polarized in the direction of magnetization *M*, are injected from a ferromagnetic electrode while a circuit drives a charge current (I) to the left, a spin imbalance is created. This produces a spin current (I_S) without a charge current to the right of the electrode. Spin–orbit interactions again separate spin-up and spin-down electrons, but now the superiority of one spin type means that a transverse charge imbalance and a spin Hall voltage, V_{SH}, is created. As the distance, *L*, between the electrode and the Hall cross increases, the voltage signal decreases, allowing the decay length of spin currents (spin diffusion length λ_{sf}) to be measured._

Figure 21. Illustration of the spin quantum Hall effect
After Maciejkoet al. (2011).

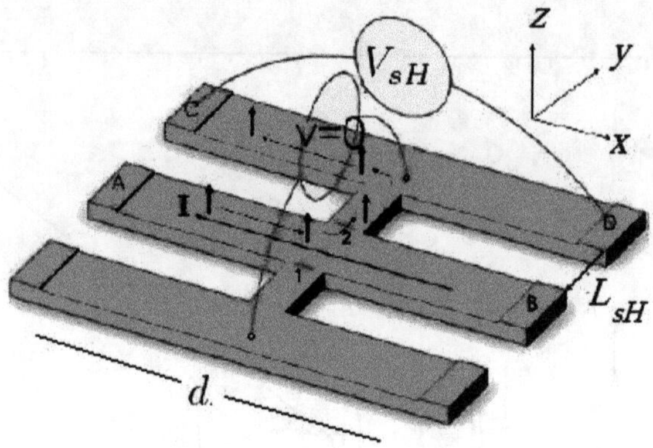

$$V_{sH} = -E_x\, d. = \frac{\sigma_{sH}}{\sigma^2}\, j_y^{z} . d. \exp\left(\frac{-L_{sH}}{\lambda_{sf}}\right)$$

Shubnikov - de Hass Effect in a 2DEG

The quantum oscillations of conductivity, in presence of magnetic field, is called the Shubnikov-de Haas effect. We have pointed out, so far, that the free energy of electrons orbiting in a magnetic field ($E- E_c- \hbar^2k^2/2m^*$) shows an oscillatory behavior with increasing magnetic field. Figure 22 depicts the Shubnikov-de Haas effect in a two-dimensional electron gas, which manifests itself as the oscillation of the longitudinal voltage, V_x, at different values of magnetic fields. The main effect of magnetic field is on the scattering rate τ^{-1}, which is proportional to the density $g_c(E_F)$:

$$\tau^{-1} = (\pi/\hbar)\, g_c(E_F)\, A_i\, u^2 \tag{39a}$$

Here g_c is the doping concentration per unit area (of impurities), and the impurity potential is modeled by a 2-D delta function of strength u. The diagonal element of the resistivity tensor is then:

$$\rho_{xx} = (m / e^2 n_s).\tau^{-1} \alpha\, g_c(E_F) \tag{39b}$$

Oscillations in the density of states at the Fermi level due to the Landau level quantization are therefore observable as an oscillatory magneto-resistivity. One expects the resistivity to be minimal when the Fermi level lies between two Landau levels, where the density of states is smallest. In view of equation (39b), this occurs when the Landau level filling factor $\nu \equiv (n_s/g_s g_v)(h/eB)$ equals an integer $N = 1, 2, \ldots$. The resulting Shubnikov-De Haas oscillations are periodic in $1/B$, with spacing $\Delta(1/B)$ given by:

$$\Delta\left(\frac{1}{B}\right) = \frac{e}{h} \cdot \frac{g_s g_v}{n_s} \tag{40}$$

This is a simple, but correct explanation of the Shubnikov-de Haas effect.

As shown in figure 22, the Shubnikov de Haas effect has smooth maxima. The basic pattern of the Shubnikov-de Haas oscillations around zero magnetic field repeats itself around each half integer filling factor (*ν=1/2, 3/2, 5/2, 7/2*, ..etc) and then again around each quarter integer filling factor (*ν=1/4, 3/4, 5/4*, .. etc). If we interpret this structure as a Landau level structure of a new type of quasi-particles (composite Fermions) then such quasi-particles should feel no magnetic field at filling factors of even denominator and an effective magnetic field ($\boldsymbol{B}_{eff}$) less than the external magnetic field (***B***) away from these points.

The symmetrical form of the oscillations around even denominator filling factors implies that the effective magnetic field felt by these quasi-particles can be either positive or negative with respect to the external magnetic field

Aharonov–Bohm Quantum Effect in LDS's

A fine oscillation was observed superimposed on peaks of the magneto-resistance of some LDS's called *antidots*. The antidot lattice is a 2D system modulated by a periodic repulsive potential. The transport in this system is ballistic, i.e., electrons are scattered from the antidote potential rather than impurities. Figures 21 depicts the **A**haronov–**B**ohm (**AB**) type oscillation.

The AB effect is an interference effect that arises from the vector-potential dependence of a wave function. The period is given by $\Delta B \approx \Phi_0/a^2$ where $\Phi_0 = ch/e$ is the magnetic flux quantum and a is the antidote lattice period.

Figure 22. Shubnikov-de Haas and quantum Hall voltages vs. magnetic field
After Singleton (2001).

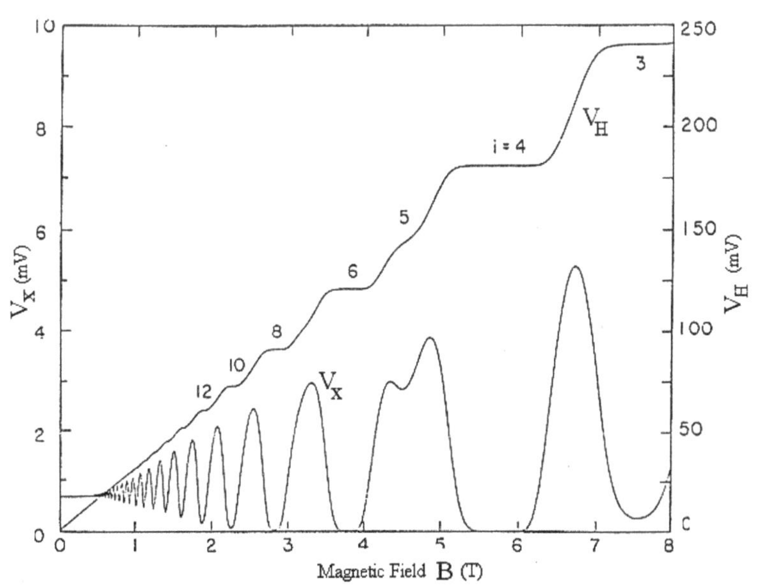

Figure 23. Example of observed quantum oscillations superimposed on the fundamental commensurability peak in an antidote lattice
After Weiss et al (2002).

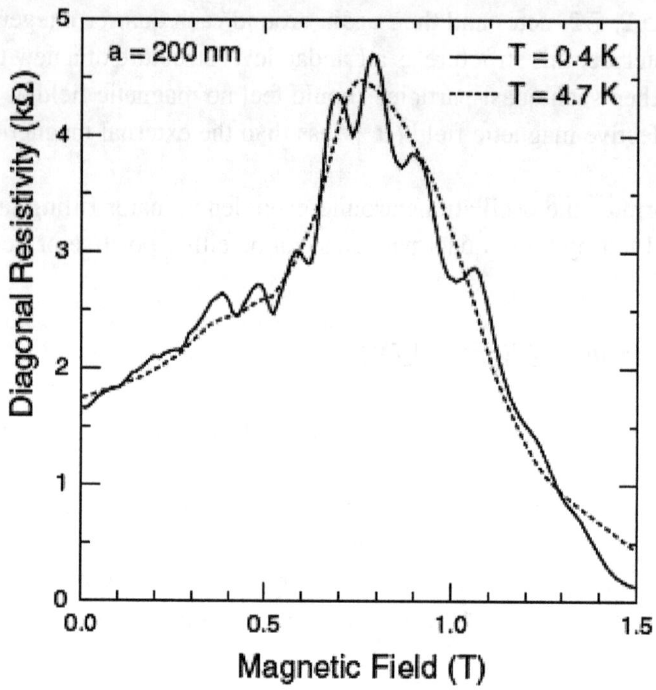

3.6 Effect of Magnetic Fields on Quantum Cascade Structures

When a magnetic field is applied parallel to the layers of a superlattice, it shifts the *E(k)* dispersion curves, as shown in Figure 25. The shift in k-states is proportional to the magnetic field density, and given by:

$$\Delta k = \frac{eB}{\hbar}. < z > \quad (41)$$

When a magnetic field is applied perpendicularly to the layers of a super-lattice, as shown in figure 24, we can separate the Hamiltonian in perpendicular and in-plane components. The in-plane states are further localized in Landau levels. A perpendicular magnetic field offers the possibility of simulating a quantum box structure, with a confining potential that depends upon the intensity of magnetic field.

The magnetic size of the quantum box is then given by:

$$l_c = [h/(eB)]^{1/2} \quad (42)$$

The above magnetic effects are very important when they are coupled with optical phenomena in superlattices, such as photo absorption and photo luminescence. It has been observed that parallel magnetic fields brought a quenching of the intensity of electro-luminescence signals, together with a broadening and a blue-shift of the photo-emission in superlattices On the other hand, a perpendicular magnetic field

Figure 24. Effect of (a) parallel and (b) perpendicular magnetic fields on the dispersion curves of a superlattice

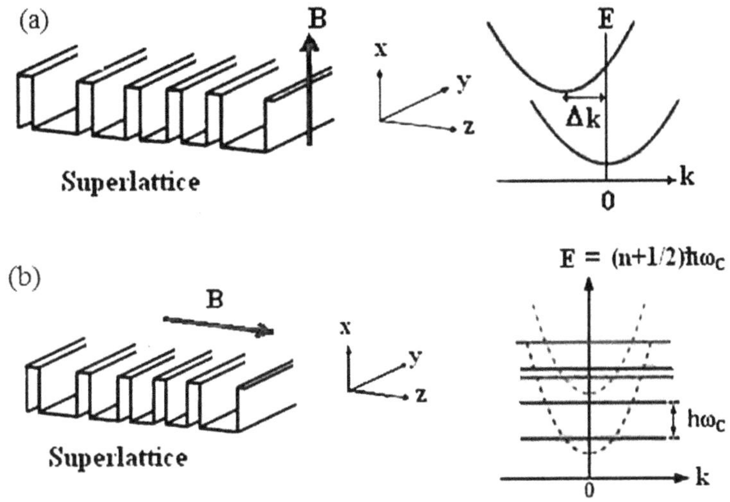

Figure 25. Effect of magnetic field on the spectra of a quantum cascade LASER, as detected by InSb cyclotron resonance detector
After Elliot (1998).

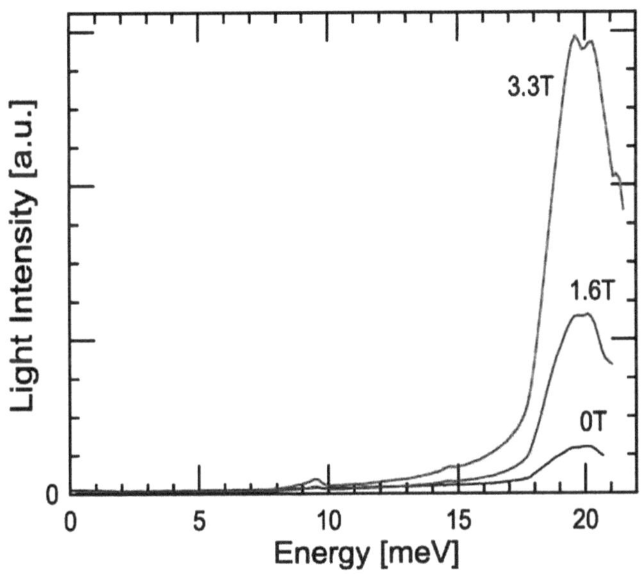

will allow transitions only between Landau levels and optical phonon emission is strongly decreased. Figure 26 shows the effect of magnetic field on the spectra of a quantum cascade LASER, as detected by *InSb* cyclotron resonance detector

3.7 Optical and Excitonic Effects

The energy required to create a free electron-hole pair in a semiconductor is simply equal to the energy gap. Hence we expect some optical absorption at photon energies just below the energy gap of the semiconductor. Figure 26 shows the absorption spectra for a typical *GaAs/AlGaAs* quantum well at room temperature. As shown in figure, there are many peaks in the spectra. The threshold absorption edge is shifted by ($E_{hh1} + E_{e1}$) compared to the bulk, such that: $\hbar\omega = E_g + E_{hh1} + E_{e1}$. Therefore, the energy (or frequency) of absorption is: $\hbar\omega = E_g + E_{hh1} + E_{el,} + \hbar^2 k_x^2/2m_h + \hbar^2 k_x^2/2m_e$

In order to understand the absorption peaks, we need to recall the concept of excitons. In fact, the optical properties of low-dimensional semiconductor nanostructures originate from excitons and exciton complexes such as *biexcitons* and *trions* (charged excitons). In this case, some electrons and holes are attracted to one another because of their Coulomb attraction. In order to include excitonic effects and their transport phenomena we need a specific formalism, which is different from the single particle picture. We can start with the Schrodinger equation for e-h pair (Wannier equation), which has the following form in the bulk of a parabolic semiconductor:

Figure 26. Absorption spectra for a typical GaAs/AlGaAs quantum well. The double peak at each quantum number (n) correspond to heavy and light holes
After Elliot (1998).

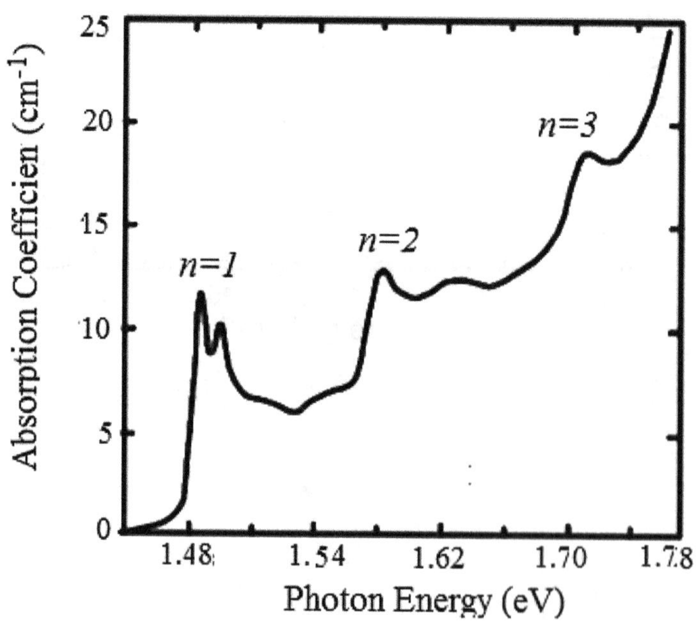

$$\left[-\frac{\hbar^2}{2m_e^*}\nabla_e^2-\frac{\hbar^2}{2m_h^*}\nabla_h^2-\frac{e}{4\pi\varepsilon\mid r_e-r_h\mid}\right]\psi_{ex}=E\psi_{ex} \tag{43a}$$

This equation can be split into two parts, for center of mass and relative motion parts, and solved to find out the exciton energy. For the case of an LDS, such as 2D quantum wells, we should take the confinement energy into account. The eigenvalues (energies) of the Hamiltonian of an exciton in a quantum well may be written as follows:

$$E_n^{2D}=E_g+E_{nze}+E_{nzh}-\frac{R_y^{*==}}{(n-\frac{1}{2})^2}+\frac{\hbar^2(k_x^2+k_y^2)}{2M} \tag{43b}$$

where $R_y=(13.6/\varepsilon^2).(m_r/m_o)$ is the exciton Rydberg binding energy (2-20 meV), $m_r = m_e m_h/(m_e+m_h)$ is the reduced mass of exciton and n is the principal quantum number. Also, $M = m_e+m_h$ and $K=k_e+k_h$ are the exciton mass and wavevector. Note the shift of these energies by the quantum well confinement energies of electron and holes ($E_{nz,e}$ and $E_{nz,h}$). The subband index nz should be chosen equal for electron and hole constituents of the exciton, according to the selection rules of optical transitions. Note that the excitonic effects are more pronounced in LDS, such as quantum wells, than in bulk semiconductors. For instance, the binding energy in 3D bulk is simply equal to R_y, while it is about $4R_y$ for the lowest excitonic state ($n=1$) in a quantum well. As excitons are e-h pairs, they form electric dipoles which can couple to electromagnetic wave, forming new quasi-particles, called exciton-polaritons.

3.8 Optoelectric Effects and Electroabsorption

When electric fields are applied to quantum wells, their optical absorption spectrum near to the band-gap energy can be changed substantially. Figure 27 shows the absorption spectra for a quantum well sample at room temperature with various electric fields applied in the plane of the quantum well layer. Note that the peaks of absorption (excitons) broaden as electric field increases. In fact, the high energy, supplied by electric field, may ionize excitons and cause them to disappear. The broadening of absorption peaks is hence due to the shortening of the exciton lifetime, due to electric field ionization effect. This is called the electroabsorption effect.

Other consequence of in-plane electric field is the appearance of a weak absorption tail at lower photon energies. The appearance of the decreasing tail is referred to as the Franz-Keldysh effect, which is explained by the variation of the material refractive index with applied electric field.

When the electric field is perpendicular the exciton absorption peaks are strongly shifted by the field (instead of being broadened). In this case, electrons are attracted to the positive bias side and holes are attracted to the negative side, but QW walls prevent the exciton separation (ionization) and the electron-hole pair is still close and bound.

3.9 Case Study 6: Transport across Graphene

We know that graphene is a two-dimensional crystalline form of carbon: a single layer of carbon atoms arranged in hexagons. In 2004 physicists, in different institutes, first isolated individual graphene planes

Figure 27. Absorption spectra for a quantum well at room temperature with various electric fields applied in the plane of the quantum well layers
After Gibson (2002).

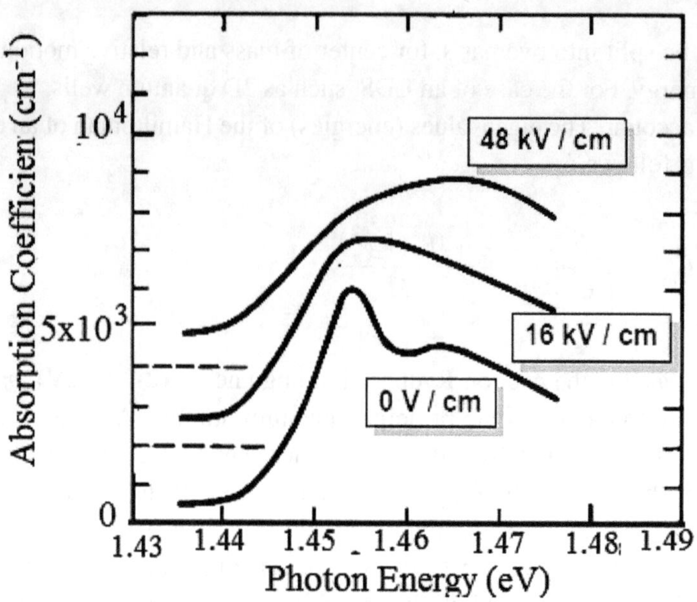

by using adhesive tape. A few years later, the researchers simplified the technique and started using dry deposition (Li et al, 2009). Since then, exfoliation procedures have been scaled up, and now companies sell graphene in large quantities. The graphene material has been subject of intensive research worldwide mainly due to its exceptional physical properties and various potential applications in electronics. The successful isolation of graphene with reasonable cost led to the current research boom. In fact, graphene has very high carrier (electrons and holes) drift mobility even at room temperature. The Graphene unusual electronic properties arise from the fact that the carbon atom has four electrons, three of which are tied up in bonding with its neighbors. But the unbound fourth electrons are in orbitals extending vertically above and below the plane, and the hybridization of these spreads across the whole graphene sheet. The hybridized electrons interact with the periodic field of the hexagonal crystal lattice and form Dirac Fermions, which are described by cone-like energy bands. Actually, the energy band diagram of grapheme is linear near the Fermi point, such that $E(k)=\hbar v_F \mathbf{k}$. This linear relation means a linear group velocity and zero effective mass[2].

One interesting consequence of this unique band structure is that the electrons in graphene are almost free. Unlike electrons in other solids, the electrons in graphene move almost ballistically (by hopping) over great distances, even at room temperature. As a result, the ability of the electrons in graphene to conduct electrical current is 10 to 100 times greater than those in a normal semiconductors at room temperature. The electron mobility of graphene is about 200,000 cm^2/Vs at room temperature, compared to about 1,400 cm^2/Vs in *Si*, and 77,000 cm^2/Vs in *InSb*, which is the highest semiconductor mobility known. However, the free electrons in graphene, though having such high mobility, are few in number compared to other metals. Therefore, the overall conductivity of graphene is limited.

The surface conductivity of an infinite graphene layer can be calculated by means of the Kubo formalism. The surface conductivity can be then represented in a local form with the Drude-like intraband contribution:

$$\sigma(\omega) = \frac{2ek_BT}{\pi\hbar^2}\left(\frac{j}{1 + j\omega\tau}\right)\ln\left[2\cosh\left(\frac{E_F}{2k_BT}\right)\right]. \tag{44a}$$

As well as the interband contribution, which is given by:

$$\sigma_i(\omega) = \frac{e^2}{4\hbar}\left(H(\tfrac{1}{2}\omega) + \frac{4j\omega}{\pi}\int_0^\infty \frac{H(\varpi) - H(\frac{1}{2}\omega)}{\omega^2 - 4\varpi^2}d\varpi \tag{44b}$$

Here $\tau = 10^{-13}$s is the relaxation time, T is the temperature, E_F is the Fermi level (chemical potential) and $H(\varpi)$ is defined as follows:

$$H(\varpi) = \frac{sh(\hbar\varpi / k_BT)}{\cosh(E_F / k_BT) + \cosh(\hbar\varpi / k_BT)} \tag{45}$$

Since graphene is a 2-dimensional structure, it can be modeled by a 2nd order conductivity tensor, as follows:

$$\sigma = \begin{bmatrix} \sigma_{xx} & \sigma_{xy} \\ \sigma_{yx} & \sigma_{yy} \end{bmatrix} = \begin{bmatrix} \sigma_d & \sigma_o \\ -\sigma_o & \sigma_d \end{bmatrix} \tag{46}$$

The conductivity depends on the band structure of graphene and other parameters including temperature and doping, as well as the applied electrical and magnetic fields. When applying an electric field, it injects more electrons in graphene and thereby allows the control of both real and imaginary part of the electrical conductivity. While graphene is mostly isotropic in the absence of magnetic fields, it shows its anisotropy in its conductivity when a magnetic field is applied, as shown in the figure 27. As shown in figure, the graphene conductivity is frequency-dependent under AC fields, and can have different behavior at RF, μW and THz spectra.

Among the most spectacular phenomena in graphene transport, one can cite the chiral quantum Hall effect (CQHEs). Figure 29 shows the CQHEs observed in single-layer graphene. It shows up as an uninterrupted ladder of equidistant steps in the Hall conductivity σ_{xy} which persists through the Fermi point, where charge carriers change from electrons to holes. The sequence is shifted with respect to the standard QHE sequence by ½, such that $\sigma_{xy} = \pm 4e^2/\hbar\ (N + ½)$ where N is the Landau level index and the factor 4 is due to double valley and double spin. The CQHE may be considered '*half-integer*' fractional QHE (FQHE) to reflect the shift. However, the grapheme CQHE is not a new FQHE, nor the standard integer QHE (IQHE). The unusual sequence is now understood as arising from the Dirac-like quantization of

Figure 28. Copennets of the conductivity tensor of grapheme as function of frequency, at various values of magnetic field (from 0T to 7T)
After Alison (2002).

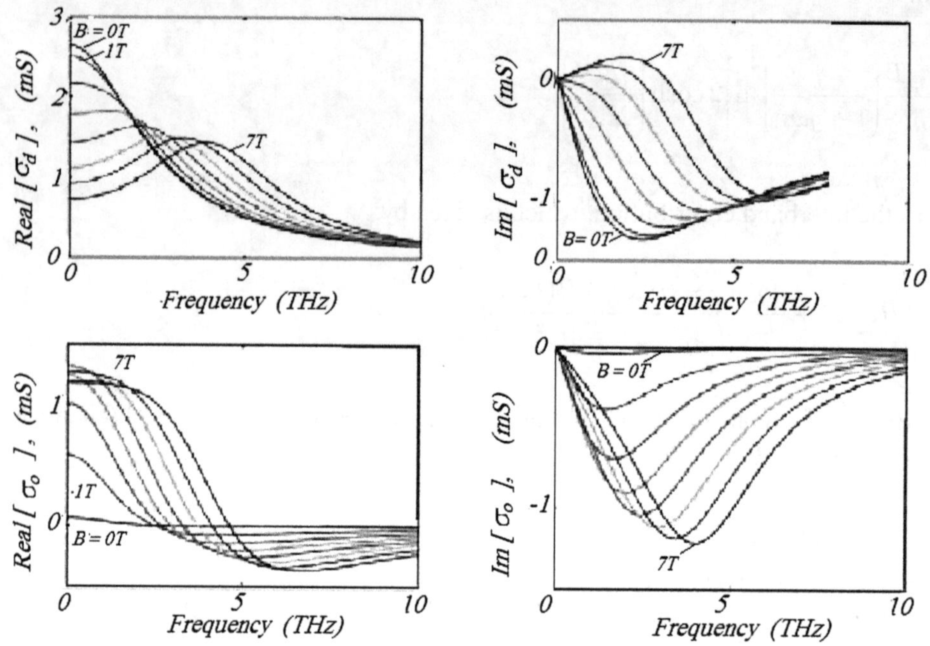

Figure 29. Chiral quantum Hall effect in grapheme
Adapted from Novoselov et al (2005).

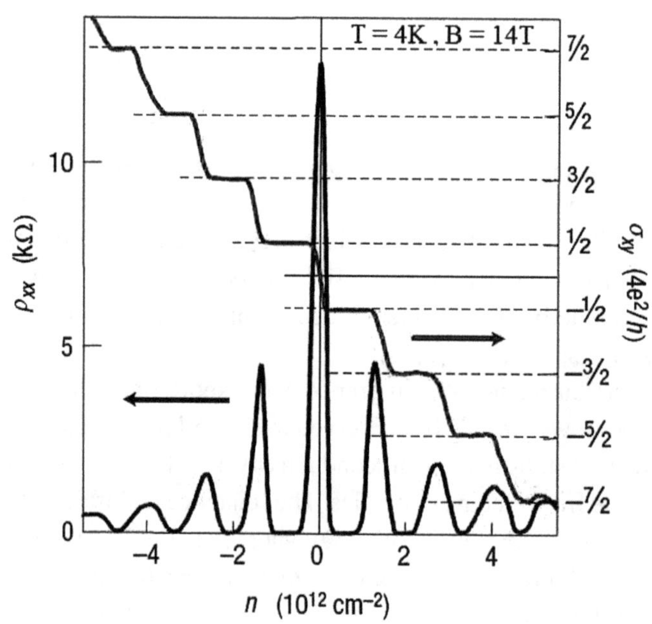

graphene's electronic bandstructure in the presence of magnetic field. The existence of a quantized level, shared by electrons and holes at zero energy, explains the anomalous behavior of CQHE in graphene.

3.10 Case Study 7: 2D Transistors

The so-called 2D transistors are similar to the conventional MOSFET devices, except for the channel material, which is formed of 2D semi-conductor monolayers or graphene. Monolayer molybdenum disulphide (MoS_2) is the second most-studied 2D material after graphene. The monolayers of transition metal dichalcogenide (TMD) materials, such as MoS_2, WTe_2 and GeS are very promising for implementing 2D transistors, because they have reasonable energy gap (about 2eV) and high carrier mobility (Wickramaratne, Zahid &, Lake, 2014). In fact, these characteristics are very critical for the implementation of high speed FET devices, with high ON-state/Off-state current ratio and threshold current swing.

Graphene Transistor

We know that graphene is a single sheet of graphite (carbon), which has a zero gap and high electron mobility (~200,000cm^2/Vs at 300K). The structure of a graphene nanoribbon (GNR) FET is shown below in figure. The figure depicts the structure of a back-gated graphene FET. The graphene sheet serves as the FET channel and sits on top of a stack of dielectric and conducting material. The graphene sheet, the dielectric and conducting material form a parallel plate capacitor with the conducting material acting as the gate (G). The graphene sheet is connected to conducting Source-Drain (S-D) contacts at the edges. The advantage of having a back-gated device is that the top surface of graphene makes conduction modulation via gas exposure possible. As a result, it can be used as a gas sensor.

Unfortunately, the graphene FET suffers from direct tunneling from source to drain. Figure 31 shows the energy band structure of a GNR FET, and how the introduction of a barrier (doped strip) in the middle of graphene can prevent the tunneling current.

The next figure depicts the drain-to-source current (I_{DS}) as a function of the applied gate voltage (V_{GS}), with and without a barrier. In the case of no barrier (only graphene flake), we observe no significant variation in $I_{DS}(V_{GS})$, because the tunneling current is dominant in the OFF state. The ON-OFF current ratio is far below what is needed for logic circuits. On the other hand, when a barrier is present (by

Figure 30. Schematic structure of a GNR FET

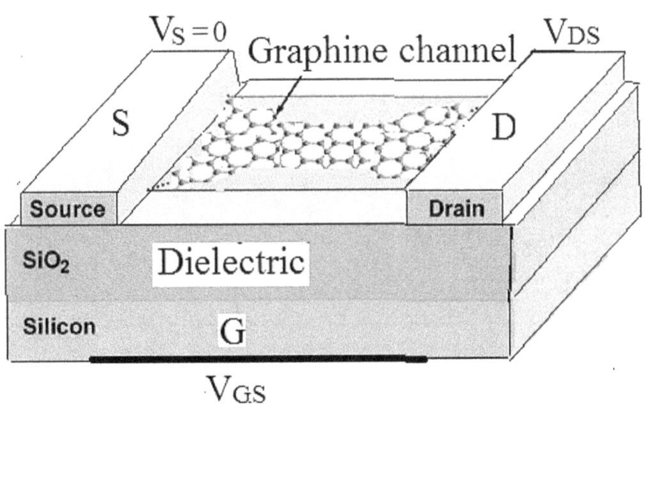

Figure 31. Schematic structure of a GNR FET

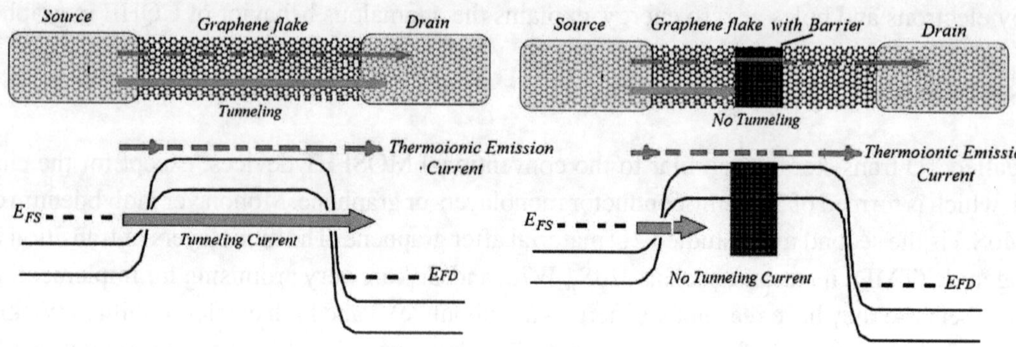

doping a middle strip of graphene with BN or BCN) we observe the switching of drain current, under the effect of gate bias. The figure shows the effect of different barrier doping constituents on the I-V characteristics of the GNR FET.

Many researchers are working hard to improve the performance of GNR FETs, with different configurations (e.g., Schwierz, 2010). Although the low ON–OFF current ratio in conventional GNR FET (without barriers) limits its use in logic circuits, this device may be promising candidate for RF applications. Also, the introduction of tunneling barriers can get rid of such inconvenience and extends the applications of GNR in new horizons.

Figure 32. Current-voltage characteristics of GNR FET, with and without a barrier
After Fiori and Iannacone (2013).

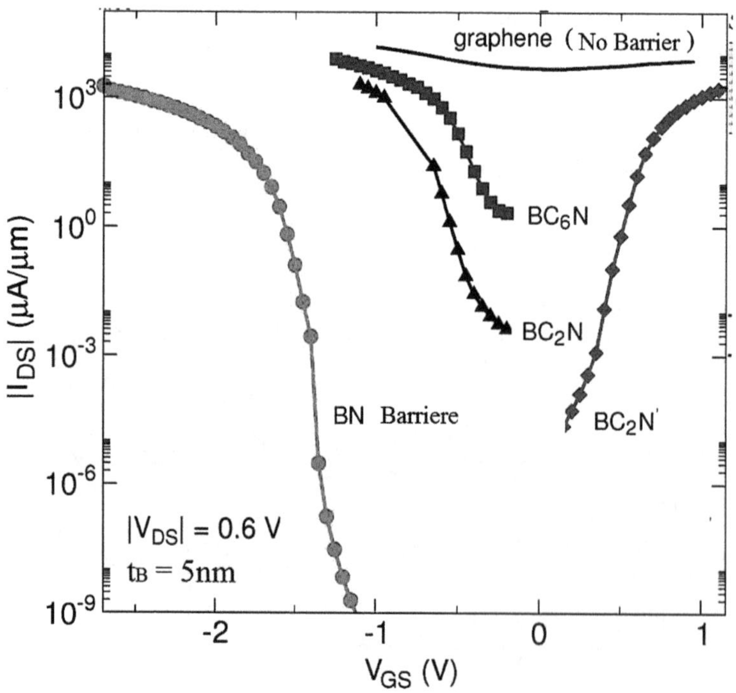

MoS2 Transistor

The so-called molybdenum disulfide (MoS_2) transistor has been introduced as an alternative to conventional MOS devices, with down to 1nm physical gate length. The MoS_2 transistor utilizes a layered MoS_2 as a channel and has been shown theoretically to be superior to Si at the sub-5-nm scaling limit (Liu, Lu & Guo, 2013). In fact, the modern deposition techniques allow the growth of high-quality 2D layers with no dangling bonds. This reduces the semiconductor surface states and enhances the carrier mobility in the channel of a 2D FET.

The Figure 34 depicts a comparison between Si MOSFET and MoS_2 FET, showing the direct source-to-drain tunneling leakage current, as a function of channel thickness and channel length. The MoS_2 is biased with $V_{DS} = V_{DD} = 0.43V$ with $V_{GS}=V_{FB}$ (OFF state).

4. CONDUCTANCE OF A QUASI-ONE-DIMENSIONAL (Q1D) SYSTEM

The properties of electron transport in quasi-one-dimensional (Q1D) systems have attracted considerable interest. By inspecting the momentum distribution function around the Fermi surface, Hu and Das Sarma have clarified that a clean Q1D electron system in a semiconductor wire shows the Luttinger-liquid behavior.

Figure 33. Comparison between conventional MOSFET and 2D transistors

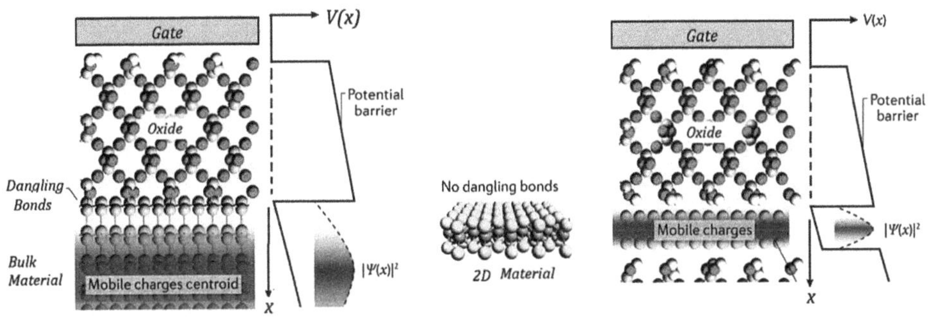

Figure 34. Schematic of the MoS2 FET, showing the MoS_2 monolayer

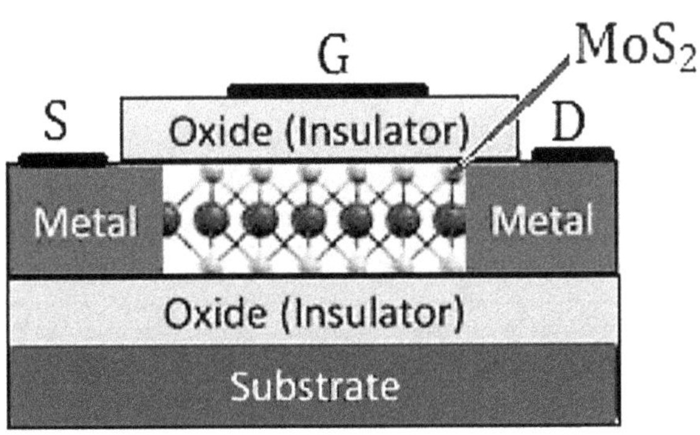

Figure 35. Comparison between Si MOSFET and MoS_2 FET, showing the direct source-to-drain tunneling leakage current as a function of channel thickness and channel length
After Desai et al (2016).

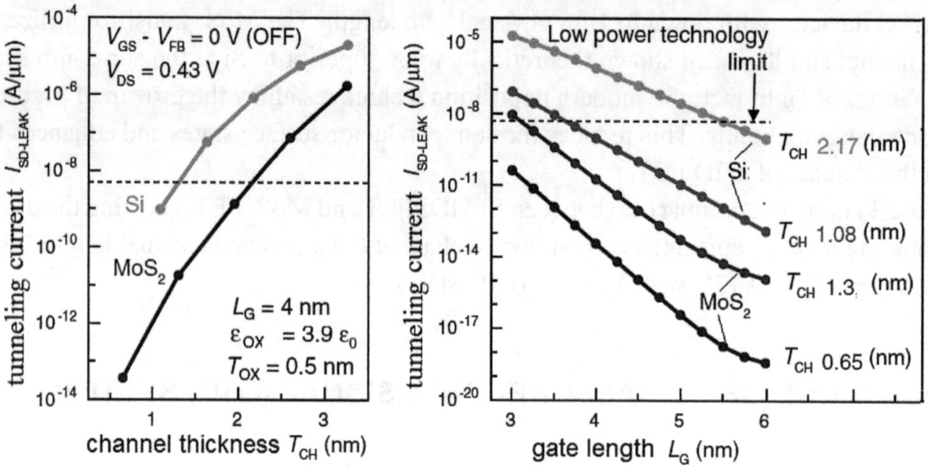

A Luttinger liquid (Luttinger & Kohn, 1955) is a theoretical model describing interacting electrons in a one-dimensional conductor (like carbon nanotubes). Such a model is necessary as the commonly used Fermi liquid model breaks down for one dimension. Even at zero temperature, the momentum distribution function of a Q1D does not display a sharp jump, in contrast to the Fermi liquid, where this jump indicates the Fermi surface. However, the existence of impurities in Q1D systems restores the Fermi-liquid behavior. The phenomenological theory of Fermi liquids was introduced by Lev Davidovich Landau in 1956 to explain why some interacting electron systems behave like non-interacting electrons of the Fermi gas. Actually, there exist two basic methods to calculate the two-terminal conductance of a quasi one–dimensional (Q1D) device, namely: the Kubo formalism and the Landauer formalism.

4.1 Landauer Formalism of Quantum Conductance

The Landauer formulism (Landauer, 1988) gives an accurate expression for the carrier current (energy or particle current) in non-interacting quantum systems, which is coupled to reservoirs at different temperatures and chemical potentials. The Landauer theory was generalize to multi-terminal devices in magnetic fields by Buttiker (Buttiker, 1988) and is referred to as the Landauer-Buttker (L-B) formalism.

The first step in deriving the Landauer formula is to convert the quantum mechanical problem into a scattering matrix problem for carriers, which are incident on a specific device. For an electronic device, let's consider the situation shown in Figure 35 in which two semi-infinite Q1D systems are connected by a region of electron gas with arbitrary effective potential V.

The corresponding 2D Schrödinger equation has the form:

$$[p^2/2m^* + V(x,y)]\Psi = E\Psi \tag{47}$$

In the left of the electron gas region the wavefunction has the form:

Figure 36. Open quasi-one-dimensional system connecting a region of electron gas with two semi-infinite contacts via arbitrary effective potential V

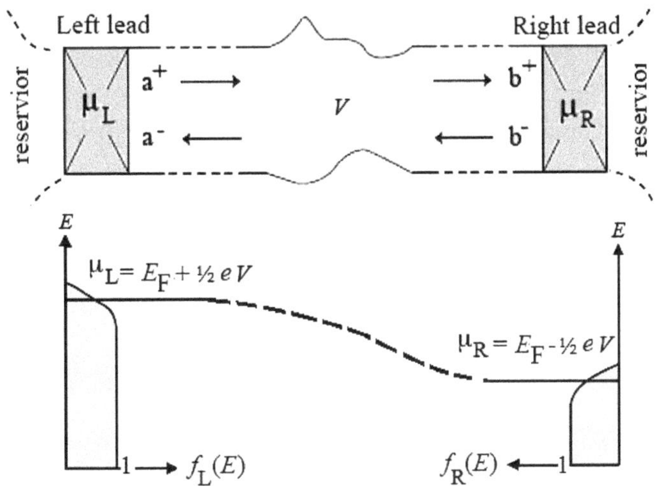

$$\Psi_L(x,y) = \Sigma_i\, exp(jk_x x)\phi_i^+(y)\, (b_i^+/\sqrt{J_i}) + \Sigma_i\, exp(-jk_x x)\phi_i^-(y)\, (b_i^-/\sqrt{J_i}) \tag{48a}$$

In the right of the electron gas region, it has the following form:

$$\Psi_R(x,y) = \Sigma_i\, exp(-jk_x x)\phi_i^+(y)\, (b_i^+/\sqrt{J_i}) + \Sigma_i\, exp(jk_x x)\phi_i^-(y)\, (b_i^-/\sqrt{J_i}) \tag{48b}$$

The quantities $\phi_i^+(y)$ and $\phi_i^-(y)$ are the transverse eigenfunctions in the Q1D systems evaluated at energy E. They form a complete basis at any chosen energy E only if all subbands are included. States with imaginary k^i_x are referred to as `evanescent modes' in analogy with their counterparts in optics. Also, J_i is the current in mode i. The coefficients a_i^+, a_i^-, b_i^+ and b_i^- are the current amplitudes for forward and reverse going states on the left and right of the connecting region. If we substitute equation (48a) and equation (48b) into equation (47) we can solve for the relation between the current amplitudes on the left and right of the inserted region. Typically this is done by discretizing (48) to achieve a finite transverse basis set and then solving numerically. The result is written in matrix form as

$$\begin{bmatrix} b^+ \\ a^- \end{bmatrix} = \begin{bmatrix} t^+ & r^+ \\ r^- & t^- \end{bmatrix} \begin{bmatrix} a^+ \\ b^- \end{bmatrix} = S \begin{bmatrix} a^+ \\ b^- \end{bmatrix} \tag{49}$$

where a^+, a^-, b^+, b^- (written now without subscripts) are vectors containing the amplitudes a_i^+, a_i^-, b_i^+ and b_i^-. The matrices t^+; r^+; t^-; r^- are the transmission and rejection matrices of the arbitrary region. The elements $t^+_{i;j'}$, $r^+_{i;j'}$, $t^-_{i;j'}$, $r^-_{i;j}$ are the current transmission and rejection probability amplitudes. The elements of the matrix S must be such that the current passing into the system via propagating modes is equal to the current passing out of the system via propagating modes. We now connect the quasi-one-dimensional leads to perfect Ohmic contacts with, as before, chemical potential $\mu_L = \mu - eV$ on the left and chemical potential $\mu_R = \mu$ on the right.

On a microscopic level, the currents flowing into each of the sub-bands of the Q1D systems from the Ohmic contacts are random due to the discreteness of electronic charge. In a simple transport measurement we detect the time averaged current. The average current flowing to the right in Figure 31(a) in a small energy range dE about energy E is given by:

$$dI^{+} = \Sigma_i <|\, b_i^{+}|^2>_{time} = <b^{+*}b^{+}>_{time} = <a^{+*}\, t^{+*}\, t^{+}a^{+}>_{time} = \Sigma_{i,j}\, (t^{+*}\, t^{+})_{i,j} <a_i^{+*}a_i^{+}>_{time} \quad (50)$$

The time average $<|a_i^{+*}a_j^{+}|^2>_{time}$ simplifies because the amplitudes for $i \neq j$ for are uncorrelated in time so that

$$<|a_i^{+*}a_j^{+}|^2>_{time} = \delta_{ij} <|\, a_i^{+}|^2>_{time} \quad (51a)$$

The current entering each subband $<|a_i^{+}|^2>_{time}$ will be exactly the same as was derived for the perfect case above

$$<|\, a_i^{+}|^2>_{time} = -(e/h)\, f(E + eV)dE \quad (51b)$$

Substituting, we get:

$$dI^{+} = (-e/h)\ Trace\ [t^{+\ne}\, t^{+}].f(E).dE \quad (52a)$$

and

$$dI^{-} = (-e/h)\ Trace\ [t^{+\ne}\, t^{+}].f(E).dE \quad (52b)$$

Therefore the total current will be

$$I = (-\, e/h) \int Trace\ [t^{+\ne}\, t^{+}].\ [f(E+eV) - f(E)].dE \quad (52c)$$

In the limit of zero temperature the conductance becomes

$$G = I\, /\, V = (e^2/\, h).Trace\ [t^{+\ne}\, t^{+}] \quad (53a)$$

This expression is evaluated at the chemical potential. This equation is called the Landauer formula. The Landauer formula is sometimes written as:

$$G = (e^2/\, h)\ T \quad (53b)$$

The above equation gives the conductance for an arbitrary effective potential sandwiched between two semi-infinite quasi-one-dimensional systems attached to perfect Ohmic contacts. Note that the Landauer formula (53) is similar to the Einstein relation $\sigma = e^2 g_c(E_F)D$. Both relations relates the conductivity and Fermi level properties. In the Einstein relation, the density of states $g(E)$ and the diffusion constant D, are both evaluated at E_F. The Landauer formula is another relation in terms of the transmission probability at the Fermi level rather than in terms of the diffusion constant.

4.2 Simple Derivation of the Landauer Formula

The Landauer formula can be derived simply as follows. Assume low bias is applied such that the difference between Fermi levels (or chemical potentials) across the conductor $\mu_L - \mu_R = e\Delta V$ is small. The emitted current is then given by:

$$I = e\, n\, v_g = e\, v_g\, (\partial n/\partial E)\, \Delta E = e\, v_g\, (\partial n/\partial E)\, (\mu_1 - \mu_2) \tag{54}$$

Also,

$$\partial n/\partial E = (\partial n/\partial k)\, \partial k/\partial E = (\partial n/\partial k) / (v_g\, \hbar). \tag{55}$$

In addition, for a Q1D system we have: $\partial n/\partial k = 1/2\pi$, where n is the electron density per unit length. Therefore,

$$I = (e/\hbar).(\mu_1 - \mu_2) = (e^2/\hbar)\, \Delta V \tag{56}$$

If an electron has probability T to be transmitted, then the conductance is:

$$G = I/\Delta V = (e^2/\hbar)T \tag{57}$$

The conductance of a quantum wire is determined by the number of occupied transverse modes or sub-bands. If we vary the number of occupied sub-bands, the conductance will jump discontinuously in values of $2e^2/h$. In order to vary the number of occupied sub-bands, we should change the Fermi Energy, by applying a gate bias, as shown in figure 37. Alternatively, we may change the width of the applied potential (W), so that the spacing between energy states (Modes) changes.

4.3 Ballistic Transport by Contact Block Reduction (CBR) Method

The CBR method is a variant of the nonequilibrium Green's function formalism, for ballistic transport, using an equilibrium distribution of the carriers (Mamaluy et al, 2005). By the CBR method, quantum transport in the ballistic limit in one, two and three dimensions can be calculated self consistently, with

Figure 37. Other representation of a quasi-one-dimensional system

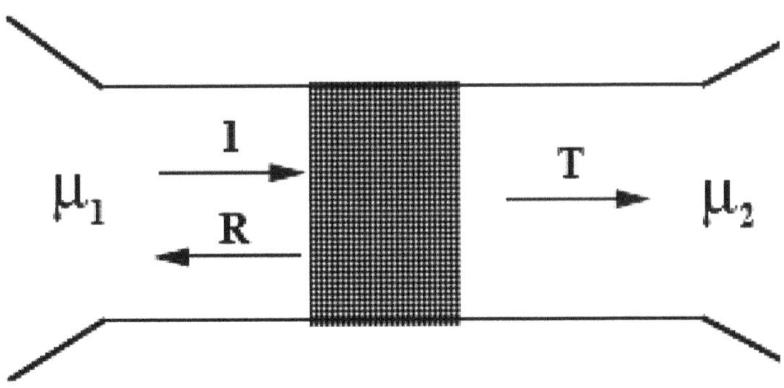

Figure 38. Conductance of a biased nanotube

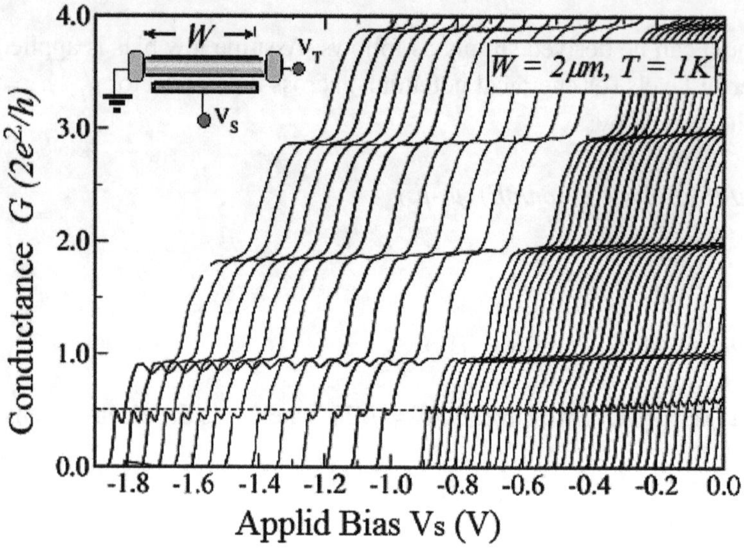

the Poisson equation (Briner, 2011). In fact, the self-consistent solution of the ballistic transport properties of an open device requires the iterative solution of the Schrödinger and Poisson equations due to their coupling through the electrostatic potential and the electron density. The solution convergence of the discretized forms of these equations is not easy[3], unless a suitable damping scheme is adopted. However, this problem can be solved using the so-called predictor-corrector approach. Using the CBR method, the energy resolved carrier density for a certain predictor potential

$$n_E^{\lambda}(x, E) = g_s\, g_v\, \rho^{\lambda}(x, E).f_{dFD}(E, \mu_{\lambda}) \tag{58a}$$

where $\rho(x,E)$ is the local density of states and f_{dFD} is the shifted Fermi-Dirac distribution, with the appropriate dimension (d = 1,2,3). Also, $g_s = 2$ is the spin degeneracy and g_v is the valley degeneracy. The predictor-corrector approach can be easily be applied by modifying this equation to get the energy resolved carrier density for a certain predictor potential ϕ, as follows:

$$n_E^{\lambda}(x,E) \rightarrow g_s g_v \rho^{\lambda}(x, E).f_{dFD}(E - e\Delta_{\phi}, \mu_{\lambda}) \tag{58b}$$

The local carrier density $n(x)$ for each lead (λ) is obtained by integrating over the energy E:

$$n^{\lambda}(x) = \int n_E^{\lambda}(x, E)dE \tag{58c}$$

Consequently, the total electron density, $n(x)$, can be calculated as follows:

$$n(x) = \sum_{\lambda=1}^{L} n^{\lambda}(x) \tag{58d}$$

Therefore, the nonlinear Poisson equation can be solved iteratively, using this predictor density. Once the new electrostatic potential for the predictor density has been obtained, the new quantum density, i.e. the new local density of states for this potential can be evaluated again and so on. Finally, the ballistic current from one lead to another can be calculated using the Landauer-Buttiker formula.

5. CONDUCTANCE OF A QUANTUM DOT (Q0D)

We have seen that a quantum dot (QD) is a small electronic island connected to metallic leads via two tunnel barriers. Figure 38 shows a possible structure fabricated from semiconductor materials. This structure is called the vertical resonant tunneling diode. The quantum dot forms between the two *AlGaAs* barriers. Conduction occurs vertically via heavily-doped *GaAs* contact regions above and below the quantum dot. The number of electrons trapped in the quantum dot is controlled by metal gate.

The figure 39 depicts the different transport mechanisms across quantum dot structures. As shown in figure, when, the QD radius is smaller than the exciton Bohr radius (a), strong confinement occurs. For weak confinement, charge carriers can move within the confined space of the quantum dot. In Q0D arrays, the electronic coupling between adjacent dots due to the finite overlap of charge carrier wave functions, allows for finite conductivity over macroscopic distances (Ulbricht et al, 2011).

The conductance of any mesoscopic system, including a quantum dot, may be calculated using the Landauer formalism (for 2-terminal devices) or the generalized Landauer-Büttiker formalism (for multi-terminal devices). The problem of calculating the conductance of a quantum dot therefore reduces to finding its transmission coefficients as a function of external parameters. We investigate here the energy dependence of the transmission coefficients of a quantum dot. As shown in figure, the two barriers of

Figure 39. Structure and energy band diagram of a quantum dot

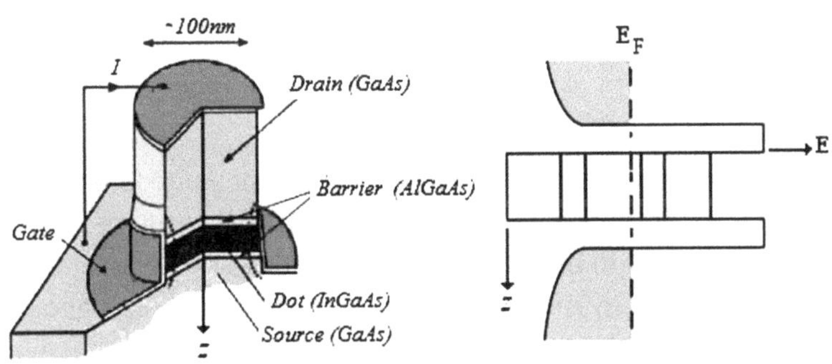

Figure 40. Different transport mechanisms of charge carriers in quantum dot

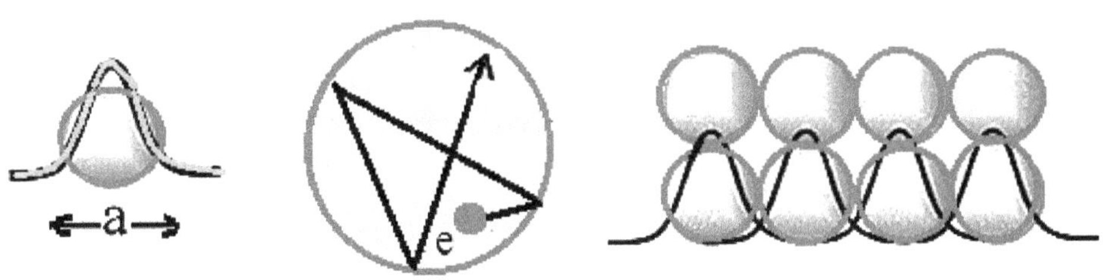

the quantum dot are assumed to be identical. If the two barriers are assumed to have reflection and transmission coefficients r and t, then the conservation of particles implies:

$$|r|^2 + |t|^2 = 1 \qquad (58e)$$

These coefficients are arbitrary complex numbers, but we will assume that the reflection coefficient is real - with zero or π phase change on reflection. When an electron enters the quantum dot it will scatter back and forth between the two barriers so that the total transmission and reflection coefficients are composed of the sum of all possible scattering paths through the dot. The total transmission coefficient of the quantum dot has the form:

$$T = t^2.\exp(\tfrac{1}{2}j\phi).\{1 + r^2.\exp(j\phi) + [r^2.\exp(j\phi)]^2 + [r^2.\exp(j\phi)]^3 + \ldots\} \qquad (5\text{-}59a)$$

The above series expansion sums up to:

$$T = t^2.\exp(\tfrac{1}{2}j\phi) / [1 - r^2.\exp(j\phi)] \qquad (59b)$$

The two-terminal conductance (G) of the quantum dot is then given by:

$$G = (e^2/\hbar).\ |T|^2 = (e^2/\hbar).\ t^4 / [1 + r^2 - 2r^2\cos(\phi)] \qquad (60)$$

Figure 41 depicts the conductance variation with the phase change on reflection ϕ. At first, the Figure 40 depicts $|T|^2$ as a function of ϕ (for t= ½). Note that the resonant peaks in this figure have unity height due to the symmetrical two barriers of the quantum dot. Figure 41 shows the conductance G as a function of both ϕ and t. As shown, the transmission resonances are broad for large values of t and narrow for small values of t. Obviously, from equation (60), the condition for a resonant peak of conductance is given by:

$$\phi = 2\, n\, \pi \qquad (61)$$

where n is an integer. This condition relates the spectrum of a quantum dot and its transmission coefficient. Therefore, the quantum dot transmits electrons resonantly when an eigen-energy passes the contact chemical potential. Interference in multiple scattering suppresses tunneling far from this condition. It should be noted that there are many mechanisms that can reduce the resonance peak height such as inelastic scattering, temperature; and asymmetry of barriers heights.

5.1 Classical Coulomb Blockade

Consider a quantum metallic dot in weak electrical contact with two contacts (reservoirs) by two high resistance contacts and capacitively coupled to a gate with a DC bias V_G, as shown in figure 41. Assume that the dot is large that its energy spectrum forms a continuum. The electrostatic potential of confined electrons inside the dot is given by:

Figure 41. Transmission and conductance of a quantum dot
From (Barnes, 2008).

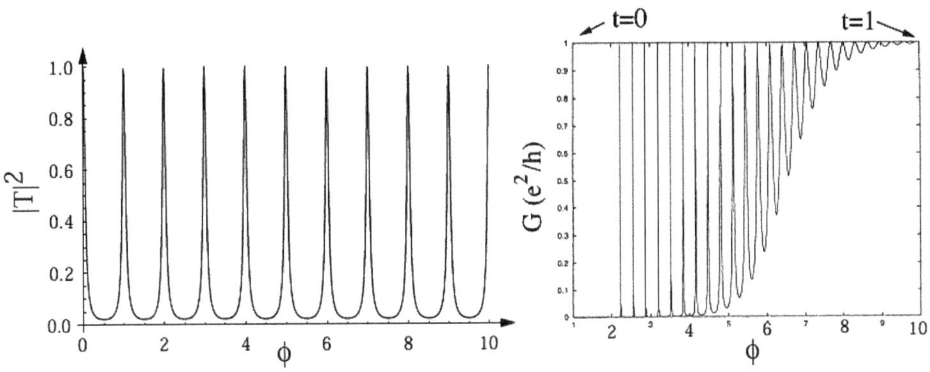

$$\phi(Q) = Q/C + \phi_{ext} \tag{62}$$

where Q is the charge on the dot, C is the capacitance between the dot and the rest of the system, and ϕ_{ext} is the external potential due to the external charge of the system so that $Q_{ext}=C\phi_{ext}$. This external charge is distributed between the surface gate, ionized donors and reservoirs. Integrating equation (62) the on the dot is:

$$U(Q) = \int_0^Q \varphi(q)\ dq \tag{63a}$$

As the dot is weakly coupled to the system, the total number of electrons on the dot should be integer multiples of e or $Q = Ne$. Therefore, (63a) yields:

$$U(N) = e^2N^2/2C - eN.\phi_{ext} \tag{63b}$$

The external charge Q_{ext} may be varied continuously by means of the gate V_G (in contrast to Q which is restricted to integer multiples of e). Rearranging equation (63b) and substituting for ϕ_{ext}, the electrostatic potential $U(N)$ takes the following parabolic form:

$$U(N) = (eN - Q_{ext})^2/2C - Q_{ext}{}^2/2C \tag{63c}$$

Figure 42 shows the electrostatic potential $U(N)$ as depicted by equation (63c) and plotted for a series of different values of Q_{ext} as a function of N. The values of $U(N)$ for which N is an integer are marked with black dots and labeled by the number of electrons trapped in the quantum dot. It can be seen from equation (63) for $Q_{ext} = Ne$, and from Figure 39, that when the dot contains N electrons the electrostatic potential is zero and $\phi(N)$=0. Hence, the electrostatic potential $U(N)$ is a minimum and the corresponding minimum energy to add an extra electron is:

$$U(N+1) - U(N) = e^2/2C \tag{64}$$

Figure 42. Schematic of a quantum metal dot, which is weakly coupled to two reservoirs (right and left) via tunnel barriers and capacitively coupled to metallic gate (below). Electrons can tunnel only at V_G for which $U(N+1) - U(N) = e^2/2C$

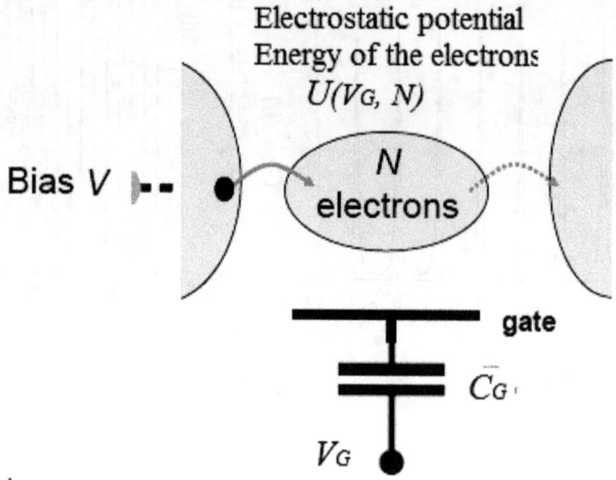

This energy should be supplied by the surrounding reservoirs (electrode bias) so that electrons can hop onto the dot and overcome its resistance. This phenomenon is referred as the Coulomb blockade because it is the Coulomb interaction that is preventing electrons from hopping onto the dot. The energy difference $\Delta U = U(N+1)-U(N) = e^2/2C$ is called the *Coulomb blockade energy* (or minimum addition energy).The addition spectrum of the dot is shown in Figure 43. As shown in figure, as more positive applied V_G, the Coulomb blockade energy decreases. Therefore, Q_{ext} increases and aproches zero when:

$$Q_{ext} = e\,(N + ½) \tag{65}$$

At this point, it takes as much energy to have N electrons in the dot as it does to have $N+1$ electrons and electrons may hop onto or off the dot without facing any Coulomb blockade. The resistance of the dot is then the sum of the resistances of the two contacts and the dot itself (classical resistance value). A current will flow through the dot once the DC bias between reservoirs is larger than the minimum addition energy. Once Q_{ext} increases near the value $Q_{ext} = e(N+1=2)$ there will be $N+1$ electrons trapped in the dot and again it will be Coulomb blockaded since there will be a gap allowing the $N+2^{nd}$ electron. The sequence then repeats itself and the resistance becomes very large until $Q_{ext} = e(N+1=2)$, where it is again equal to the classical resistance of the dot. Therefore, the conductance of a small metallic dot, as a function of an external gate voltage, consists of a series of equally spaced isolated peaks occurring when there is no Coulomb blockade. The external potential ϕ_{ext} has a linear dependence on the surface-gate potential V_G so that for some change in gate voltage ΔV_G: $\Delta\phi_{ext} = \alpha\Delta V_G$, between two peaks in conductance, one extra electron enters the dot and $\Delta Q_{ext} = e$. Therefore the spacing between conductance peaks is given by:

$$\Delta V_g = e / \alpha\, C = e / C_g \tag{66}$$

where C_g is the capacitance between the gate and the quantum dot.

Figure 43. Potential energy U(N) of electrons in a quantum dot as a function of N and V_G for a series of Q_{ext}. The figure below shows the addition spectrum of the quantum dot. Upper bands means addition of electron, Lower bands means removal of electron
From Quantum Electronics in Semiconductors (Barnes, 2008).

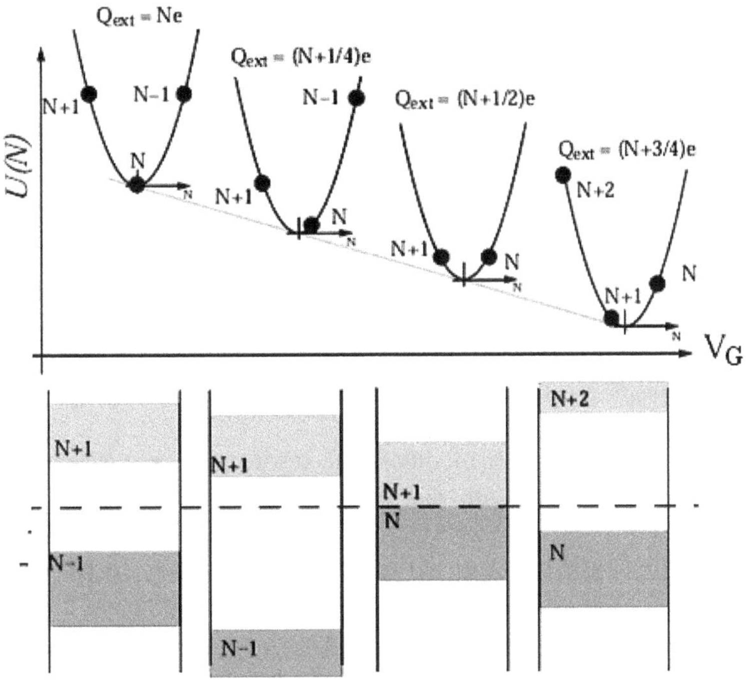

5.2 Quantum Coulomb Blockade

We have seen that the Coulomb blockade is originally a classical effect. However this phenomenon still appears in quantum dots. In these systems, the potential energy $U(N)$ is a sum of the Coulomb charging energy and the energies of discrete levels in the dot as follows:

$$U(N) = \frac{(e.N - Q_{ext})^2}{2C} - \frac{Q_{ext}^{\ 2}}{2C} + \sum_{i=1}^{N} E_i \tag{67}$$

Here E_i represents the single-particle energy of the ith state measured from the minimum effective potential of the quantum dot. Thus, as the dot fills, its effective potential changes and the single-particle spectrum changes. The minimum addition energy E_{min} of a quantum dot containing N electrons is:

$$E_{min} = U(N+1) - U(N) = [e^2(N + ½) - e\, Q_{ext}]/C + E_{N+1} \tag{68}$$

which depends on the single-particle energy of the $(N+1)^{th}$ state (E_{N+1}). In the classical case, the addition energies above E_{min} form a continuum. In the quantum case they form a discrete set of energies $\boldsymbol{E}_i^{'} = E_{min} + E_i - E_{N+1}$. Figure 43 shows the addition spectrum at a series of different Q_{ext}. Once the DC bias between the electrodes is larger than E_{min} electrons can tunnel through the dot. As the DC bias is increased and successive single-particle states pass beneath the chemical potential of any electrode, the conductance increases in a series of steps. At zero bias, a peak in the conductance of a quantum dot occurs when its addition energy is zero. However, away from resonant peaks, interference due to multiple scattering may aid the Coulomb blockade and makes it more resistive than the classical case. Therefore, the external charge at a conductance peak is given by:

$$Q_{ext} = e(N + \tfrac{1}{2}) + (C/e).E_{N+1} \tag{69a}$$

Therefore, between the Nth and $(N+1)^{th}$ resonant peaks, we have:

$$e\,\Delta Q_{ext} = e^2 + C.(E_{N+2} - E_{N+1}) \tag{69b}$$

The spacing between peaks in gate voltage is given by:

$$e\Delta Vg = e^2/C_g + a^{-1}(E_{N+2} - E_{N+1}) \tag{70}$$

The single-particle energies therefore modulate the resonant peak positions confirming that the oscillatory structure in the spectra is actually due to the single-particle spectrum of the quantum dot.

5.3 Effect of Magnetic Field on Quantum Dot (Fock-Darwin Model)

Quantum dots (QDs) that are subject to a magnetic field are an interesting research subject since many years. A popular approach to study the energy spectrum of such systems is the assumption of a two-dimensional parabolic confinement potential that is subject to a perpendicularly oriented magnetic field which introduces a further parabolic confinement. This approach makes sense for self-assembled quantum dots that have a strong confinement in the growth direction and a weak (parabolic) confinement in the plane perpendicular to it. We can then solve the single-particle 2D Schrödinger equation in the transverse (x-y) plane for a 2D parabolic confinement potential with a uniform magnetic field applied along the z-direction. The Hamiltonian which includes the magnetic field effect reads:

$$H = \tfrac{1}{2}\hbar^2 \{(\nabla + jgA).M(x).(\nabla + jgA)\} \tag{71}$$

where the g-factor $g = e.l.h$ and $\mathbf{A}$ is the vector potential. The parabolic confinement is typically chosen so that the energy separation between states $\hbar\omega_o$ is a few meV (without magnetic field) where ω_o is the oscillator frequency of the parabolic confinement.

Without magnetic field, and neglecting the twofold spin degeneracy, the ground state is not degenerate, the second level is twofold degenerate, the third level threefold, and so on.

$$E_{n,l} = (2n + |l| - 1)\hbar\omega_o \tag{72}$$

where $n = 1, 2, 3\ldots$ is the radial quantum number and $l = 0, \pm1, \pm2, ..$ the angular momentum quantum number.

If the magnetic field is present, then the eigenvalues (energies) are

Figure 44. Calculated the particle energy levels of a 2-D parabolic confinement potential ($\hbar\omega_o$=3meV) as a function of magnetic field. States labeled by (n; l) refer to radial quantum number n and angular momentum quantum number l
After (Harrison &, 2011).

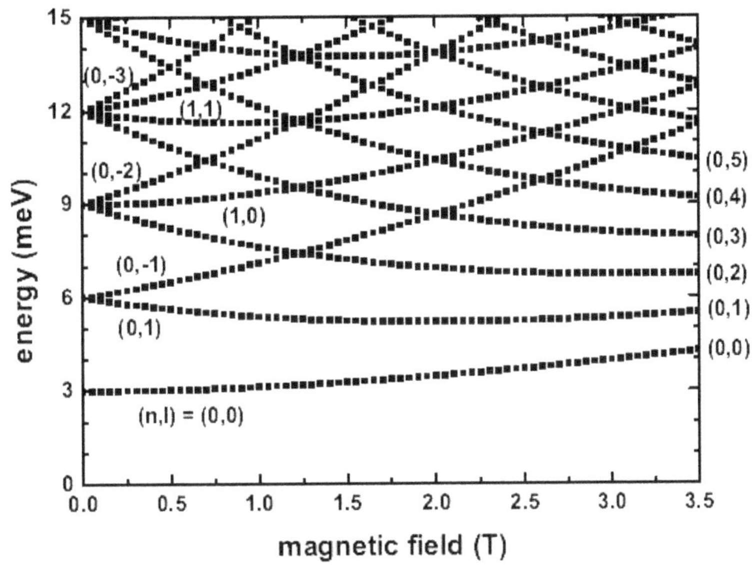

$$E_{n,l} = (2n+ |l| -1.\hbar.[\omega_o^2 + \tfrac{1}{4} \omega_c^2]^{1/2} - \tfrac{1}{2} l.\hbar\omega_c \qquad (73)$$

where $\omega_c=eB/m$ is the cyclotron frequency. Thus the degeneracy of the 2D harmonic oscillator can be lifted as seen in the calculated energy spectrum in figure 44. Here the energy levels are plotted as a function of the magnetic field magnitude. This relation is called the Darwin-Fock spectrum.

6. KONDO EFFECT

The Kondo effect is one of the most important effects of strongly correlated electrons. The Kondo effect appears as a dramatic increase of the conductivity of metals and certain quantum structures, when the temperature is decreased below a characteristic temperature, called the Kondo temperature. In particular, the Kondo effect has attracted great interest in semiconductor quantum dots. In fact, the simplest Kondo system consists of a localized, spin-½ electron coupled to a Fermi sea of electrons and this system can be realized with a quantum dot (QD). Below a certain temperature, the so-called Kondo temperature (T_K), a many-body singlet state is formed between the QD spin and the surrounding conduction electrons. This state adds a resonant level at the Fermi energy of the electrodes enabling the tunneling of electrons across the QD. Such a Kondo resonance can overcome the Coulomb blockade and enhance the conductance. Generally speaking, if the number of electrons confined in a quantum dot is odd, then the measured conductance increases due to the Kondo effect at low temperature. Otherwise, the conductivity will continue to decrease as temperature is decreased.

For metals with magnetic impurities the Kondo effect is also seen as an increase in the resistance (not conductance) for decreasing temperature. The phenomenon has been first explained by Jun Kondo in

1964. It was discovered in diluted magnetic alloys as a resisitance minimum in temperature dependence. The Kondo effect in both metals and semiconductor quantum dots is illustrated in figure 45. As shown in figure, the Kondo effect appears below the so-called Kondo temperature, T_K, which corresponds to the binding energy of the Kondo singlet state. It can be expressed in terms of the dot parameters as follows (Kouwenhoven & Glazman, 2001):

$$T_K = [(h.\Gamma.E_c)/2k_B].exp[\pi.E_o(E_o+E_c)/h.\Gamma.E_c] \quad (74)$$

where Γ is the tunnel rate to and from the quantum dot, h is Planck's constant and E_o is the energy level on the dot relative to the Fermi energy of the surrounding leads.

The Kondo effect may also arise from the interactions between a single magnetic atom and the many electrons in a non-magnetic metal. This leads to a Kondo effect caused by the interaction between a localized moment of doping atoms and conduction electrons. In this case, the Kondo effect can be understood as a magnetic exchange interaction between a localized impurity spin and free conduction electrons. In order to minimize the exchange energy, the conduction electrons tend to screen the spin of the magnetic impurity and the ensemble forms a spin singlet. In fact, we have seen so far that a quantum dot (with poor/tunneling leads) has Coulomb blockade peaks in the conductance as a function of gate voltage every time an energy level of the dot is aligned with the Fermi level of the leads.

In addition to spin-degeneracy, Kondo effect may be also attributed to other degrees of freedom of a singlet state. The degeneracy may also arise from degenerate orbitals. If the orbital degree of freedom is conserved during tunneling, then the orbital quantum number can behave as a spin, and we call this pseudospin. The orbital magnetic moment can be then thought of as a pseudospin, because it behaves similar to the electron spin. The orbital pseudospin leads to the so-called orbital Kondo (O-Kondo) effect, which appears at high magnetic fields.

Actually, there are many types of Kondo effect that arise in the presence of orbital degeneracy, such as:

- Two-level spin Kondo effect (TLS-Kondo),
- Orbital Kondo effect (O-Kondo),
- Singlet-triplet Kondo effect (ST-Kondo) and
- SU(4) Kondo effect.

Figure 45. Illustration of the Kondo effect in metal and quantum dots

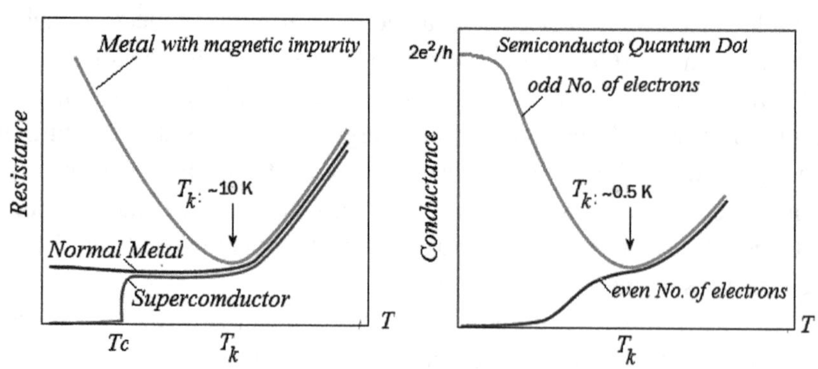

Note that the presence of degeneracy in the ground state is essential to all Kondo effects. The so-called SU(4) Kondo effect, is due to the combined spin and pseudospin degrees of freedom, which manifested as a multiple splitting of the Kondo resonance at magnetic finite fields. The Kondo effect is a well-defined quantum phenomenon, so that it has been an attractive testing tool for the new analytical models of quantum carrier transport. In addition, the Kondo effect helps us to understand the electronic properties of a wide variety of materials where the interactions between electrons are particularly strong. Previously, physicists could infer the role of the Kondo effect from the measurement of resistance and magnetic susceptibility. Nowadays, the scanning tunneling microscope (STM) has been used to image and characterize magnetic impurities on the surface of metals, opening a new avenue of research into the Kondo effect.

7. EFFECT OF STRAIN ON LDS

A crystal experiences strain when it undergoes some stress which raises its internal energy. Strain can be used to alter and optimize the electronic and optical properties of quantum wells, wires and dots by varying the extent of the electronic wave-functions and energy levels. Consequently, the conduction and valence band edges change as well. Therefore, an exact knowledge of the strain fields is of great importance for the design and development of electronic and optoelectronic devices.

The components of the strain tensor ε, are defend as

$$\varepsilon_{ij} = ½ (\partial u_i/\partial x + \partial u_j/\partial x) = \varepsilon_{ji} \tag{75}$$

where i, j = 1, 2, 3. The vector $u(x)$ describes the displacement due to lattice deformations. The diagonal elements of the strain tensor measure the extensions per unit length along the coordinate axes (positive values mean tensile strain, negative values compressive strain), i.e. the lengths of the considered volume element change while the angles remain constant. In contrast, the off-diagonal elements measure the shear deformations where the angles change and the volume remains constant. As the ε matrix is real and symmetric, it can always be diagonalized by an appropriate orthogonal transformation. The resulting diagonal matrix must not necessarily coincide with the crystal or simulation coordinate system.

The trace of the strain tensor Tr(ε) is equal to the hydrostatic strain, i.e. the change in volume. The strain tensor components are obtained by minimizing the elastic energy E

$$E = ½ \int V\, C_{ijkl}\, \varepsilon_{ij}\, \varepsilon_{kl}\, dV \tag{76}$$

where C_{ijkl} is the forth-ranked stiffness tensor. A derivation of the calculation of the strain tensor for arbitrary three-dimensional heterostructures can be found in (Hackenbuchner, 2002). The strain tensor components in the plane parallel to the substrate plane are independent of growth direction for cubic crystals and are given by the lattice mismatch

Figure 46. Schematic illustration showing how different materials with different lattice constants show strain

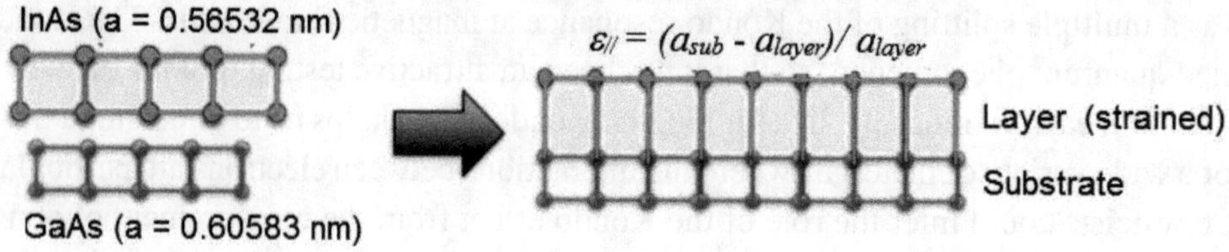

$$\varepsilon_{//} = (a_{sub} - a_{layer}) / a_{layer} \quad (77)$$

where a_{sub} and a_{layer} are the lattice constants of the substrate and the strained layer. Figure 46 shows how a deposited layer of *InAs* is strained when deposited on a *GaAs* substrate.

The strain analysis in nanostructures is usually implemented in quantum transport simulation as a pre-processing step. In the strain calculation, we calculate the total energy of the crystal and find the atoms position that minimizes the total elastic energy. The figure 46 depicts the effect of strain on band structure of heterojunction structure (*GaAs-InAs-GaAs* quantum well), as obtained by NEMO simulator (Steiger, et al, 2011). Note how the surface passivation got rid of surface states, which appear in the band gap (at the left figure).

8. SUMMARY

There is a current growing technological interest in low-dimensional systems (LDS's), driven mainly by the push for new devices and for ever-greater integration in the electronics industry. In LDS, free electrons are only permitted to move in one or two dimensions. For instance, the electrons filling the thin inversion layer in a conventional MOSFET are only permitted to move freely in the 2 dimensions of the inversion layer. The motion of free electrons in the third dimension (normal to the layer plane)

Figure 47. Effect of strain on band structure of heterojunction structure, as obtained by NEMO simulator. Note how the surface passivation got rid of surface state (in the left)
After Steiger, et al, (2011).

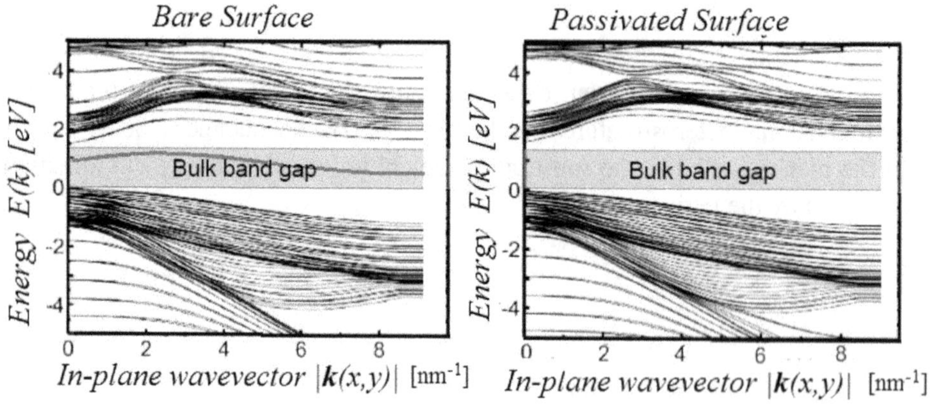

Figure 48.

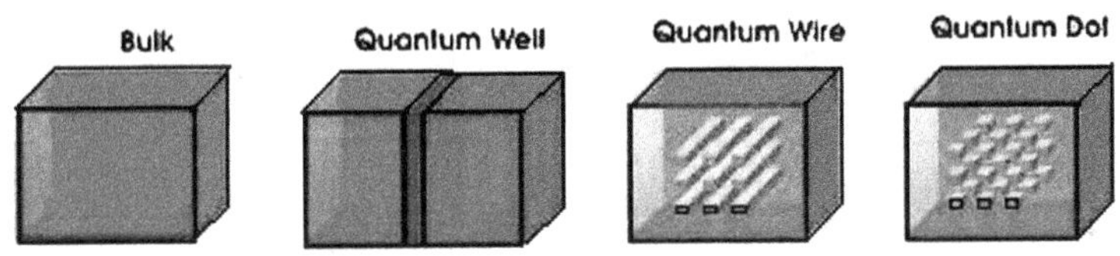

quantized. The same applies to *graphene*, which is a two-dimensional crystalline form of carbon. Unlike electrons in other materials, the electrons in graphene move almost ballistically over great distances, even at room temperature. As a result, the ability of the electrons in graphene to conduct electrical current is 10 to 100 times greater than those in a normal semiconductors at room temperature. A quantum dot may be considered as a 3-D quantum well, with no degrees of freedom at all and with quantized levels in three directions.

Table 2 and Figure 49 summarize the dispersion relations *E(k)* and density of states (DOS) of parabolic semiconductors.

In this chapter we investigate the carrier transport phenomena in low dimensional semiconductor structures and nanodevices. We describe different structures (quantum wells, quantum wires and quantum dots). We also describe various quantum effects that take place within such nanostructures, such as quantum Coulomb blockade, Aharonov–Bohm, Shubnikov-De Haas oscillations and Kondo effects.

The conductance of an LDS system, including a quantum dot, may be calculated using the Landauer formalism (for 2-terminal devices) or the generalized Landauer-Büttiker formalism (for multi-terminal devices). This formula, was initially developed for electronic transport in 1-D ballistic conductors. According to this formula, the electrical current from node *A* to node *B* of a ballistic conductor is given by

Table 2. Dispersion relations and density of states of low-dimensional semiconductors

Degrees of freedom	Dispersion (kinetic energy)	Density of states	Effective density of states
3 (bulk)	$E=\frac{\hbar^2}{2m^*}(k_x^2+k_y^2+k_z^2)$	$g(E)_{3D}=\frac{1}{2\pi^2}\left(\frac{2m^*}{\hbar^2}\right)^{\frac{3}{2}}\sqrt{E-E_C}$	$N_c^{3D}=\frac{1}{\sqrt{2}}\left(\frac{m^*kT}{\pi\hbar^2}\right)^{\frac{3}{2}}$
2 (slab)	$E=\frac{\hbar^2}{2m^*}(k_x^2+k_y^2)$	$g(E)_{2D}=\frac{m^*}{\pi\hbar^2}U(E-E_C)$	$N_c^{2D}=\frac{m^*}{\pi\hbar^2}kT$
1 (wire)	$E=\frac{\hbar^2}{2m^*}(k_x^2)$	$g(E)_{1D}=\frac{m^*}{\pi\hbar}\sqrt{\frac{m^*}{2(E-E_C)}}$	$N_c^{1D}=\sqrt{\frac{m^*kT}{2\pi\hbar^2}}$
0 (box)	-	$g(E)_{0D}=2\delta(E-E_C)$	$N_c^{0D}=2$

Figure 49. The dispersion relations and density of states (DOS) of low-dimensional semiconductors

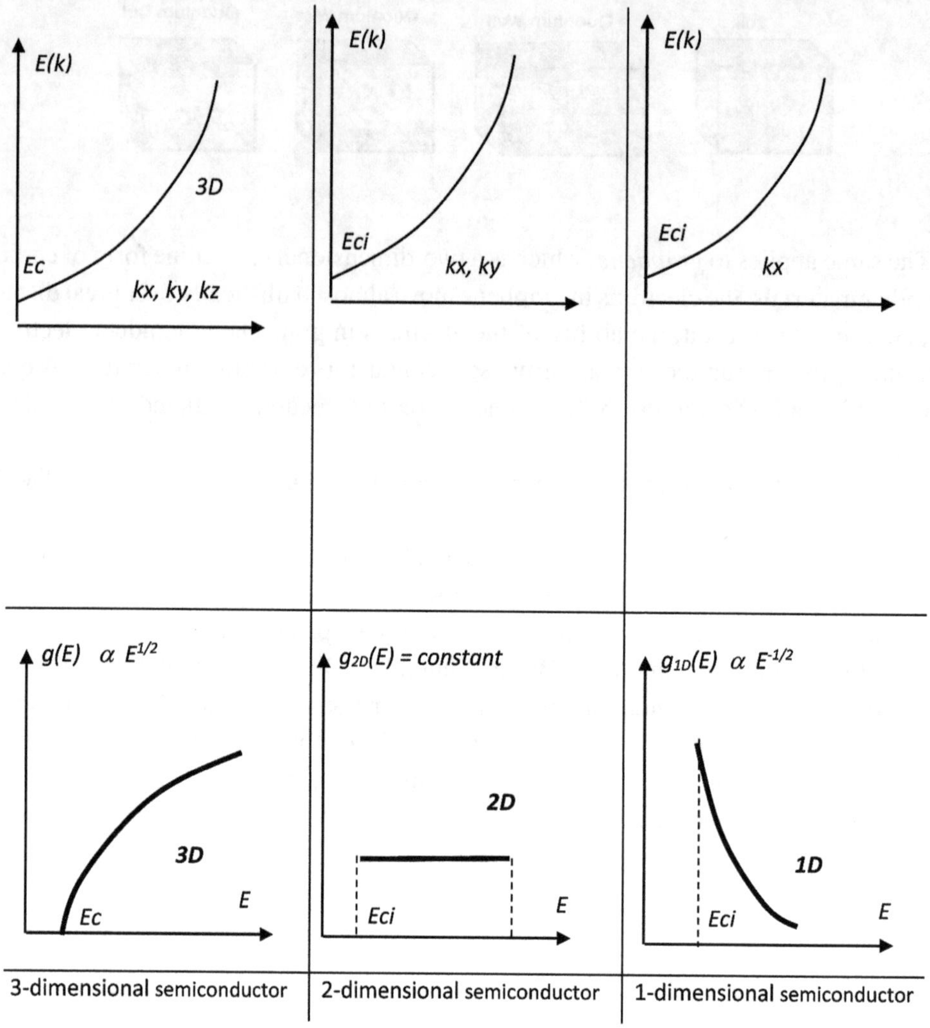

$$I_{AB} = \frac{g_S e}{h} \int_{E_{F_B}}^{E_{F_A}} M(E) f(E) T(E) dE$$

where e is the electron charge, g_s=2 due to spin degeneracy, h=Planck's constant, E_{FA} and E_{FB} are the Fermi levels of node A and node B, $M(E)$ is the number of propagating modes in the channel, f' is the deviation from the equilibrium electron distribution (perturbation), and $T(E)$ is the transmission probability (T=1 for ballistic transport).

REFERENCES

Aharonov, Y., & Bohm, D. (1959). Significance of electromagnetic potentials in quantum theory. *Physical Review*, *115*, 485–491.

Ando, T., Fowler, A. B., & Stern, F. (1982). Electronic properties of two-dimensional systems. *Reviews of Modern Physics*, *54*(2), 437–672. doi:10.1103/RevModPhys.54.437

Andreev, A. D., & OReilly, E. P. (2000, December). GaN / AlN hexagonal quantum dots. *Physical Review B: Condensed Matter and Materials Physics*, *62*(23), 15851–15870. doi:10.1103/PhysRevB.62.15851

Antia, H. M. (1993). Rational Function Approximations for Fermi-Dirac Integrals. *The Astrophysical Journal*, *84*(Supplement), 10–108.

Ashcroft, N. W., & Mermin, N. D. (1976). *Solid State Physics*. Saunders.

Bastard, G. (1988). *Wave Mechanics Applied to Semiconductor Hetero-structures*. Les Ulis, France: Les Editions de Physique.

Birner, S. (2011). *Modeling of semiconductor nanostructures and Semi-conductor–electrolyte interfaces* (Ph.D. Dissertation). Technischen Universität München.

Buttiker, M. (1988). Symmetry of electrical conduction. *IBM Journal of Research and Development*, *32*(3), 317–334. doi:10.1147/rd.323.0317

Davies, J. H. (1998). *The Physics of Low-Dimensional Semiconductors*. Cambridge, UK: Cambridge University Press.

Desai, S. B., Madhvapathy, S. R., Sachid, A. B., Llinas, J. P., Wang, Q., Ahn, G. H., . . . Javey, A. (2016). MoS_2 transistors with 1-nanometergate lengths. *Science Mag.*, *345*(6308), 99. Available at: https://www.researchgate.net/publication/309456228

Deslippe, J., Samsonidze, G., Strubbe, D. A., Jain, M., Cohen, M. L., & Louie, S. G. (2012). BerkeleyGW: A massively parallel computer package for the calculation of the quasiparticle and optical properties of materials and nanostructures. *Computer Physics Communications*, *183*(6), 1269–1289. doi:10.1016/j.cpc.2011.12.006

Ferry, D. K., Goodnick, S. M., & Bird, J. P. (2009). *Transport in Nanostructures* (2nd ed.). Cambridge, UK: Cambridge Univ. Press. doi:10.1017/CBO9780511840463

Fiori, G., & Iannaconi, G. (2013). Multiscale Modeling for Graphene-Based Nanoscale Transistors. *Proceedings of the IEEE*, *101*(7), 1653–1669. doi:10.1109/JPROC.2013.2259451

Fowler, R. H., & Nordheim, L. (1928). Electron emission in intense electric fields. *Proceedings of the Royal Society of London A: Mathematical, Physical and Engineering Sciences, 119*(781), 173-181. doi:10.1098/rspa.1928.0091

Geim, A., & Grigorieva, I. (2013). Van der Waals heterostructures. *Nature*, *499*(7459), 419–425. doi:10.1038/nature12385 PMID:23887427

Geim, A. K. (2009). Graphene: Status & prospects. *Science*, *324*(5934), 1530–1534. doi:10.1126/science.1158877 PMID:19541989

Geim, G. K., & Novoselov, K. S. (2007). The rise of graphene. *Nature Materials*, *6*(3), 183–191. doi:10.1038/nmat1849 PMID:17330084

Gilbert, T. L. (1975). Hohenberg-Kohn theorem for nonlocal external potentials. *Physical Review B: Condensed Matter and Materials Physics*, *12*(6), 2111–2120. doi:10.1103/PhysRevB.12.2111

Hackenbuchner, S. (2002). Selected Topics of Semiconductor Physics and Technology. Instituts der Technischen Universitat Munchen, Munchen, Germany.

Harrison, P. (2011). *Quantum Wells, Wires and Dots: Theoretical and Computational Physics of Semiconductor Nanostructures*. John Wiley & Sons.

Haug, H., & Jauho, A.-P. (1996). *Quantum Kinetics in Transport and Optics of Semiconductors*. Berlin: Springer.

Heitzinger, C., & Klimeck, G. (2007). Computational aspects of the three dimensional feature-scale simulations of silicon-nanowire field-effect sensors for DNA detection. *Journal of Computational Electronics*, *1*(6), 3, 387–390.

Ilatikhameneh, H., Ameen, T., Novakovic, B., Tan, Y., Klimeck, G., & Rahman, R. (2016). *Saving Moore's law down to 1nm channels with anisotropic effective mass*. arXiv:1605.03979

Ilatikhameneh, H., Tan, Y., Novakovic, B., Klimeck, G., Rahman, R., & Appenzeller, J. (2015). Tunnel field-effect transistors in 2d transition metal dichalco-genide materials. *IEEE Exploratory Solid-State Computational Devices & Circuits*, *1*(1), 12–18. doi:10.1109/JXCDC.2015.2423096

Ilatikhemeneh, H., Salazar, R. B., Klimeck, G., Rahman, R., & Appenzeller, J. (2015). *From Fowler-Nordheim to Non-Equilibrium Green's Function Modeling of Tunneling*. arXiv:1509-08170v1

Imry, Y. (1997). *Introduction to Mesoscopic Physics*. Oxford University Press.

Kane, E. O. (1961). Theory of Tunneling. *Journal of Applied Physics*, *32*(1), 83–91. doi:10.1063/1.1735965

Kane, E. O. (1969). *Tunneling Phenomena in Solids* (E. Burstein & S. Lundqvist, Eds.). New York: Plenum.

Kelly, M. J. (1995). *Low-dimensional semiconductors*. Oxford, UK: Clarendon Press.

Kittle, C. (1976). *Introduction to Solid-State Physics* (2nd ed.). New York: John Wiley & Sons.

Kouwenhoven, L. & Glazman, L. (2001). Revival of the Kondo effect. *Physics World*.

Landauer, R. (1988). Spatial variation of currents and fields due to localized scatters in metallic conduction. *IBM Journal of Research and Development*, *32*(3), 306–316. doi:10.1147/rd.323.0306

Li, X., Cai, W., An, J., Kim, S., Nah, J., Yang, D., & Ruoff, R. S. et al. (2009). Large-Area Synthesis of High-Quality and Uniform Graphene Films on Copper Foils. *Science*, *324*(5932), 1312–1314. doi:10.1126/science.1171245 PMID:19423775

Liu, L., Lu, Y., & Guo, J. (2013). On Monolayer MoS_2 field-effect transistors at the scaling limit. *IEEE Transactions on Electron Devices*, *60*(12), 4133–4139. doi:10.1109/TED.2013.2284591

Luttinger, J. M., & Kohn, W. (1955). Motion of Electrons and Holes in Perturbed Periodic Fields. *Physical Review*, *97*(2), 869–883. doi:10.1103/PhysRev.97.869

Maciejko, J. et al.. (2011). The Quantum Spin Hall Effect. Annual Rev. *Condensed Matter Physics*, *2*, 31–53.

Mamaluy, D., Sabathil, M., & Vogl, P. (2003). Efficient method for the calculation of ballistic quantum transport. *Journal of Applied Physics*, *93*(8), 4628–4633. doi:10.1063/1.1560567

Mamaluy, D., Vasileska, D., Sabathil, M., Zibold, T., & Vogl, P. (2005). Contact block reduction method for ballistic transport and carrier densities of open nano-structures. *Physical Review B: Condensed Matter and Materials Physics*, *71*(24), 245321. doi:10.1103/PhysRevB.71.245321

Mani, R. G., Smet, J. H., von Klitzing, K., Narayanamurti, V., Johnson, W. B., & Umansky, V. (2002). Zero-resistance states induced by electromagnetic-wave excitation in GaAs/AlGaAs heterostructures. *Nature*, *420*(6916), 646–650. doi:10.1038/nature01277 PMID:12478287

Miao, K., Sadasivam, S., Charles, J., Klimeck, G., Fisher, T., & Kubis, T. (2016). Buttiker probes for dissipative phonon quantum transport in semiconductor nanostructures. *Applied Physics Letters*, *108*(11), 113107. doi:10.1063/1.4944329

Miller, D. (1990). Optoelectronic applications of quantum wells. *Optics & Photonics News.*, *2*(1), 7. doi:10.1364/OPN.1.2.000007

Novoselov, K. S. (2005). Two-dim gas of massless Dirac fermions in graphene. *Nature*, 438.

Novoselov, K. S., Falko, V. I., Colombo, L., Gellert, P. R., Schwab, M. G., & Kim, K. (2012). A roadmap for graphene. *Nature*, *490*(7419), 192–200. doi:10.1038/nature11458 PMID:23060189

Pollak, F. H., & Cardona, M. (1966). Energy band structure of Ge and GaAs: The k.p method. *Journal of Physics and Chemistry of Solids*, *27*(2), 423–425. doi:10.1016/0022-3697(66)90049-7

Safa, S., & Asgari, A. (2015). *Vertical transport in InAs/GaSb superlattice: role of ionized impurity and interface roughness scatterings*. arXiv:1504.02871 [cond-mat.mtrl-sci]

Schwierz, F. (2010). Graphene transistors. *Nature Nanotechnology*, *5*(7), 487–496. doi:10.1038/nnano.2010.89 PMID:20512128

Seeger, K. (1999). *Semiconductor Physics: An Introduction*. Berlin: Springer. doi:10.1007/978-3-662-03797-3

Singleton, J. (2001). *Band theory and electronic properties of solids*. Oxford University Press.

Stahl, A., & Balslev, I. (1987). *Electrodynamics of the Semiconductor Band Edge*. Springer. doi:10.1007/BFb0045248

Steiger, S., Povolotskyi, M., Park, H.-H., Kubis, T., & Klimeck, G. (2011). NEMO5: A parallel multiscale nanoelectronics modeling tool. *IEEE Transactions on Nanotechnology*, *10*(6), 1464–1474. doi:10.1109/TNANO.2011.2166164

Störmer, H., & Tsui, D. (1995). Composite Fermions: New particles in the fractional quantum Hall Effect. Physics News, 33.

Takagi, S., Takayanagi, M., & Toriumi, A. (1999). Characterization of Inversion-Layer Capacitance of Holes in Si MOSFET. *IEEE Transactions on Electron Devices*, *46*(7), 1446–1450. doi:10.1109/16.772489

Tan, Y. P., Povolotskyi, M., Kubis, T., Boykin, T. B., & Klimeck, G. (2015). Tight-binding analysis of Si and GaAs ultrathin bodies with subatomic wave-function resolution. *Physical Review B: Condensed Matter and Materials Physics*, *92*(8), 085301. doi:10.1103/PhysRevB.92.085301

Thiele, S., Schaefer, J. A., & Schwierz, F. (2010). Modeling of graphene metal–oxide–semiconductor field-effect transistors with gapless large-area graphene channels. *Journal of Applied Physics*, *107*(9), 094505. doi:10.1063/1.3357398

Tsui, D. C., Stormer, H. L., & Gossard, A. C. (1982). Two-Dimensional Magneto-transport in the Extreme Quantum Limit. *Physical Review Letters*, *48*(22), 1559–1562. doi:10.1103/PhysRevLett.48.1559

Ulbricht, R., et al. (2011). *Reviews Carrier dynamics in semiconductors studied with time-resolved terahertz spectroscopy of Modern Physics APS*. Academic Press.

Vienna Schrödinger–Poisson Solver, V. S. P. 1.0. (2007). *User's Guide*. Institut fur Mikroelektronik, Technische Universitat Wien, Austria. Retrieved from www.iue.tuwien.ac.at/software

Vogl, P., & Kubis, T. (2010). The non-equilibrium Greens function method: An introduction. *Journal of Computational Electronics*, *9*(3-4), 237–242. doi:10.1007/s10825-010-0313-z

Von Klitzing, K., Dorda, G., & Pepper, M. (1980). New Method for High-Accuracy Determination of Fine-Structure Constant Based on Quantized Hall Resistance. *Physical Review Letters*, *45*(6), 494–497. doi:10.1103/PhysRevLett.45.494

Walukiewicz, W., Ruda, H. E., Lagowski, J., & Gatos, H. C. (1984). Electron mobility in modulation-doped heterostructures. *Physical Review B: Condensed Matter and Materials Physics*, *30*(8), 4571–4582. doi:10.1103/PhysRevB.30.4571

Weisbuch, C., & Vinter, B. (1991). *Quantum Semiconductor Structures: Fundamentals and Applications*. Academic Press.

Wickramaratne, D., Zahid, F., & Lake, R. K. (2014). Electronic and thermo-electric properties of few layer transition metal dichalcogenides. *The Journal of Chemical Physics*, *140*(12), 124710. doi:10.1063/1.4869142 PMID:24697473

Willatzen, M., Lassen, B., Yan Voon, L. C., & Melnik, R. V. N. (2006). Dynamic coupling of piezoelectric effects, spontaneous polarization, and strain in lattice-mismatched semiconductor quantum-well heterostructures. *Journal of Applied Physics*, *2*(100), 024302. doi:10.1063/1.2214361

Yoshioka, D. (2002). *The Quantum Hall Effect*. Springer. doi:10.1007/978-3-662-05016-3

Zagoskin, A. M. (1998). *Quantum Theory of Many-body systems: techniques and applications.* New York: Springer. doi:10.1007/978-1-4612-0595-1

ENDNOTES

1 Metal-Oxide-semiconductor (MOSFET) devices are usually fabricated on <100> silicon wafers.

2 The band structure of graphene will be presented all together with carbon nanotube in the next Chapter. However, we discuss the transport across graphene, here, because it is actually a 2D structure.

3 In order to accelerate the convergence of such highly nonlinear system, the Newton method is usually adopted. But this requires the numerical evaluation of the Jacobian terms, which is too expensive, since they cannot be derived analytically.

Chapter 6
Carrier Transport in Nanotubes and Nanowires

1. OVERVIEW AND CHAPTER OBJECTIVES

The interest in carbon nanotubes (CNTs) and their transport models is motivated by their many interesting electronic properties. The search for new high-speed devices has moved CNTs into the focus of current research. In fact, nanotubes have become at the heart of nanoelectronic devices. In this chapter we study the transport of charge carriers (electrons and holes) in nanotube structures. Nanotubes and nanowires with dimensions on the nanometer length scale cannot be treated as classical conductors because their dimensions are much smaller than the mean free path length (between successive collisions of electrons). The length of these structures is also large for the full quantum mechanical treatment as they consist of thousands of atoms. For these reasons, nanotubes are sometimes called quasi-one-dimensional (Q1D) structures or quantum wires.

As a starting point, we present the notion of carbon nanotubes, which are rolled sheets of graphene. We know that graphene is a two-dimensional crystalline form of carbon: a single layer of graphite carbon atoms arranged in hexagons. Graphene has unusual electronic properties, which arise from the fact that the carbon atom has four electrons, three of which are tied up in bonding with its neighbors (forming sp^2 'σ-bonds'). But the fourth electrons are in orbitals (p_Z -orbitals) extending vertically above and below the plane, and the hybridization of these electrons mix together forming delocalized electron states ('π-bonds'). These states are responsible for the electrical conductivity of graphene.

In the following sections, we present the basic properties of nanotubes and nanowires, such as silicon nanowires (SiNW) and carbon nanotubes (CNT's) and describe the physical transport mechanisms of charge carriers along them, with ample examples of real devices.

Upon completion of this chapter, students and readers will be able to

- Differentiate between nanotubes and nanowires.

DOI: 10.4018/978-1-5225-2312-3.ch006

Figure 1. Schematic of the transport of a quasi-free electrons in a graphite sheet

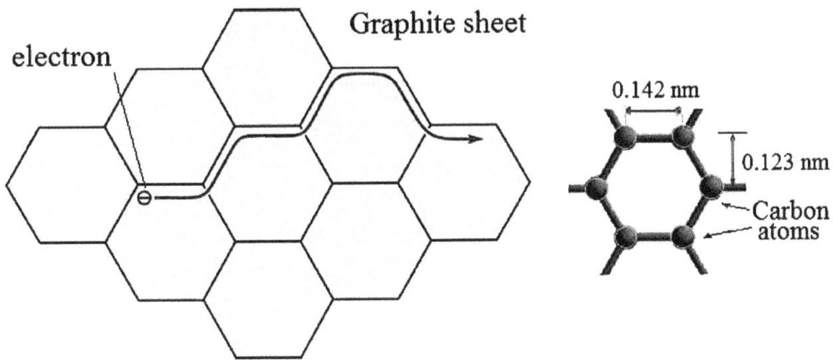

- Understand the notion of carbon nanotubes (CNT's) and define their main types and basic properties.
- Explain the concepts of a quantum wires and their transport mechanisms.
- Describe the nanotubes-based devices, their properties and applications.
- Calculate the I-V characteristics of carbon-nanotube FET transistor.

2. NANOTUBES AND NANOWIRES

Nanotubes are hollow one-dimensional form of carbon or other materials with nanometer diameter. Nanowires are solid materials in the form of wire with diameter smaller than 100 nm. Their name is derived from their size, since the diameter of a nanowire or nanotube is on the order of a few nanometers. Nanowires and nanotubes are the most confining electrical conductors. Nanotubes could be either single walled (SWNT) or multi-walled (MWNT) consisting of nested tubes with outer diameters ranging from 5 to 100 nm. A large percentage of researchers attribute the discovery of hollow, nanometer-size carbon tubes (CNT) to the Japanese Sumio Iijima of NEC in 1991. CNT's can be formed from graphene sheets which are rolled up to form tubes.

Currently, nanotubes are synthesized by different techniques. CNT synthesis can be performed near the focus of a high-power laser, in between two arcing graphite electrodes, or in a hot furnace full of hydrocarbon gas.

However, there are three famous methods for the synthesis of SWNTs: pulsed laser vaporization (Laser Ablation), arc discharge growth, or chemical vapor deposition (CVD) on supported or gas phase catalysts. The basic prerequisites for the formation of SWCNTs are an active catalyst, a source of carbon, and adequate energy. Most of these methods take place in vacuum or with gases. For the matter of completeness of this Chapter, we review these techniques and their growth methods.

3. CARBON NANOTUBES

Carbon nanotubes (CNT's) are long, thin cylinders of carbon or graphite sheets. The mono-atomic layers of graphite are sometimes called graphene sheets.

Figure 2. SEM picture of grown nanotubes
After Levchenko et al., (2013).

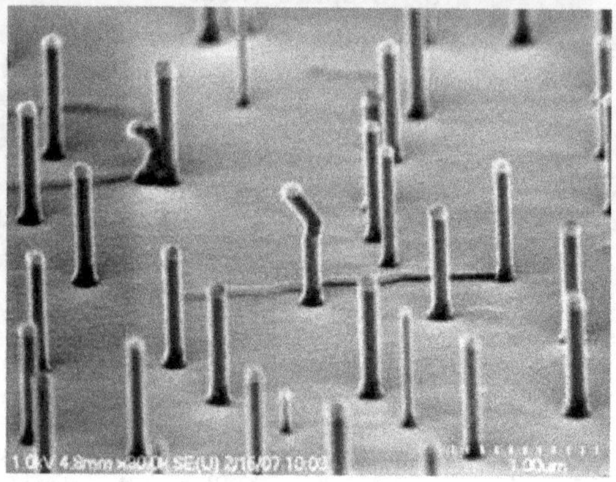

Figure 3. Schematic of a carbon nanotube (CNT) and how it may be formed from a rolled graphene (graphite sheet)

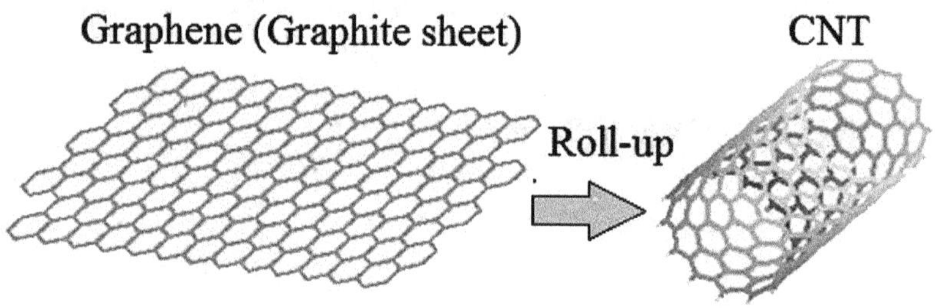

Nanotubes have different electronic, thermal, and structural properties that change depending on their different kinds. Nanotubes may have single cylindrical wall and can also have multiple walls.

The nature of the bonding of a nanotube may described by quantum chemistry, specifically, orbital hybridization. The chemical bonding of nanotubes is composed entirely of sp^2 bonds, similar to those of graphite. This bonding structure, is stronger than the sp^3 bonds found in diamonds, and provides molecules with a unique strength.

3.1 Single-Wall Nanotubes (SWNTs)

SWNTs are important type of CNTs because they exhibit electric properties that are not shared by other nanotube variants. SWNT's can be conceptualized by wrapping a one-atom-thick layer of graphite (graphene) into a seamless cylinder. The way the graphene sheet is wrapped is represented by a pair of indices (*n, m*) called the chiral vector indices. After rolling up a sheet of graphite in a certain direction we can determine the parameters characterizing the formed tube. Most SWNTs have a diameter close to 1nm, with a tube length millions of times longer. Carbon nanotubes form different types, which can be described by a chiral vector ***R***.

$$\boldsymbol{R}= n.\underline{a}_1 + m.\underline{a}_2 \tag{1}$$

where n and m are integer numbers and $\underline{a}_1$ and $\underline{a}_2$ are unit vectors of the hexagonal lattice. The integer indexes n and m denote the number of unit vectors along two directions in the honeycomb crystal graphene lattice. The unit vector a_1 lies along the zigzag line. The other vector ma_2 has a different magnitude than na_1, but its direction is a reflection of a_1 over the armchair line. When added together, they equal the chiral vector $\boldsymbol{R}$, as shown in Figure 4

In order to understand the meaning of chirality, imagine that the nanotube is unraveled into a planar sheet, and draw two lines along the tube axis. Next, find any point on one of these two lines that intersects one of the carbon atoms (point ***A***). Then, draw the armchair line, which travels across each hexagon, separating them into two equal halves. Once the armchair line is drawn, we find a point along the other tube axis that intersects a carbon atom nearest to the armchair line (point ***B***). Now connect A and B to get the chiral vector, $\boldsymbol{R}$. The wrapping angle ϕ is the angle between $\boldsymbol{R}$ and the armchair. It is given by:

$$\phi = 30^0 - \theta, \tag{2a}$$

where $cos\,(\theta) = \boldsymbol{R}.\boldsymbol{a}_1 / |R|\, a_1$, or

$$\theta = cos^{-1}\left[\frac{(2n + m)}{2\left(m^2 + nm + m^2\right)}\right]. \tag{2b}$$

If $\boldsymbol{R}$ lies along the Armchair line, then $\phi = 0°$ $(n=m)$ and the nanotube is an armchair nanotube. If $\phi = \pm 30°$,$(n=0$ or $m=0)$ then the tube is of the zigzag type. Otherwise, if $0° < |\phi| < 30°$ then it is a chiral tube.

The chirality affects the properties of the nanotube, like its conductance. A SWNT is considered metallic if n=m. If the value n-m is divisible by 3, the nanotube is semiconducting, with very small bandgap. Otherwise, the nanotube is moderate semiconducting. Thus, two-thirds of fabricated nanotubes are expected to be semiconducting, while the other third is metallic. Given a chiral vector (n,m), the length of chiral vector is given by:

Figure 4. Miller indices and chirality of single wall nanotubes

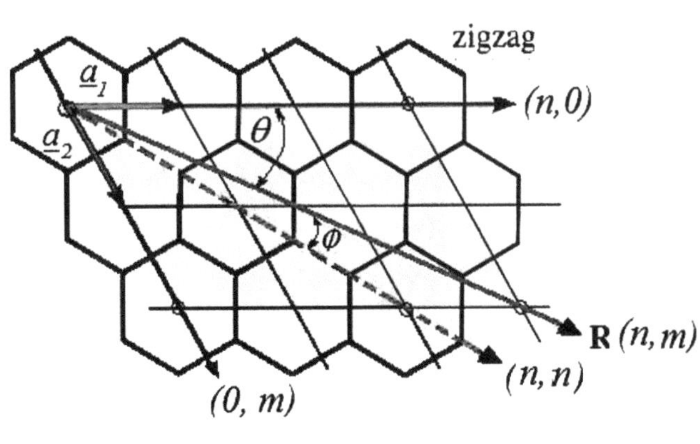

$$L = |R| = a\sqrt{n^2 + m^2 + nm} \tag{3}$$

where $a = \sqrt{3}a_o$ and $a_o = 1.42nm$ is the distance between neighboring carbon atoms in the flat graphene. In addition, the diameter of a carbon nanotube can be determined using the relationship:

$$d = \frac{L}{\pi} = \frac{a}{\pi}\sqrt{n^2 + m^2 + nm} \tag{4}$$

Example 1

Draw a schematic showing the CNTs with *R(11,11)*, *R(18,0)*, *R(14,7)*, and mention their type

Solution

The 1st CNT has $n = m = 11$ (armchair). The length of the chiral vector

$L = |R| = a\sqrt{(n^2 + m^2 + n.m)} = 11\sqrt{3}\ a$

The 2nd CNT has $n = 18$ and $m = 0$ *(Zigzag)*. The length of the chiral vector

$L = |R| = a\sqrt{(n^2 + m^2 + n.m)} = n\ a = 18\ a$

The 3rd CNT has $n = 14$ and $m = 7$ (Chiral*)*. The length of the chiral vector

$L = |R| = a\sqrt{(n^2 + m^2 + n.m)} = a\ \sqrt{(196 + 49 + 98)} = 7a\ \sqrt{7}$

Table 1.

Armchair SWNT	Zigzag SWNT	Chiral SWNT
R = (11,11)	*R = (18,0)*	*R = (14,7)*
$n = m$	$m = 0$	$n \neq m$
$\theta = 30^o$	$\theta = 0^o$	$0^o < \theta < 30^o$
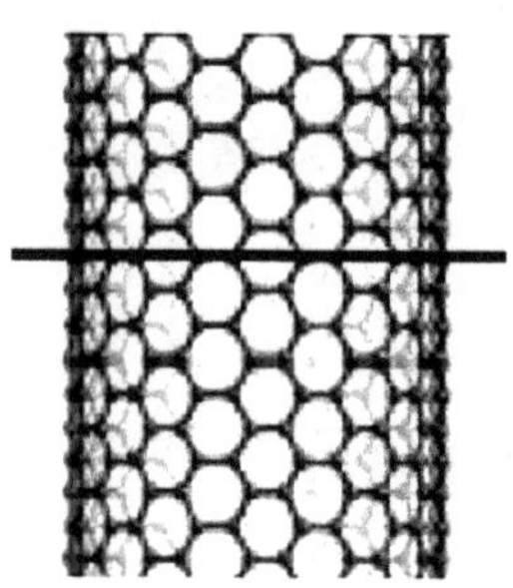	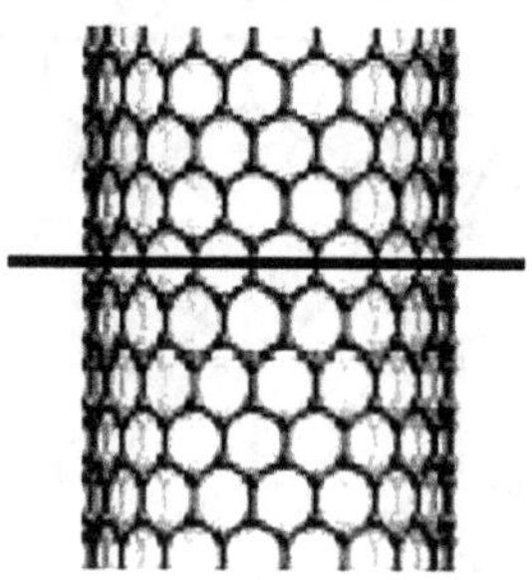	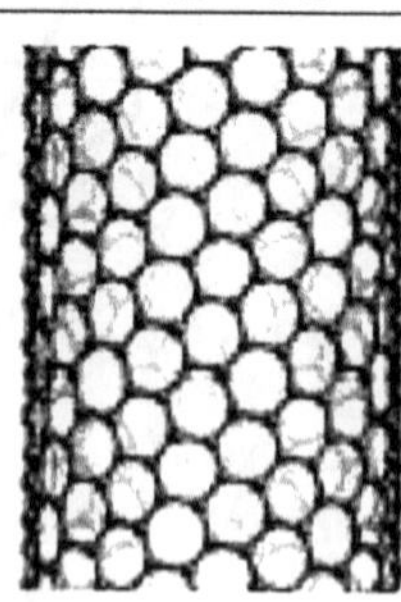

In conclusion, the electronic characteristics of a SWNT is essentially independent of length but strongly dependent on its transverse physical structure, which is denoted by a pair of integers (*n*, *m*). Figure 5 depicts the basic types of SWNTs, according to their chirality.

Unfortunately, the specific growth of distinguished semiconducting or metallic carbon nanotubes (CNT's) is a challenge (although some post-synthesis separation methods exist, according to our current knowledge). Also, the well-controlled doping of CNTs is not easy. However, the semiconductor nanotubes and nanowires (like SiNW) overcome these limitations.

SWNTs are unique materials for several reasons: they are prototypical quasi one-dimensional quantum wires composed of a single element (e.g., *C* or *Si*) or compound materials (like *SiGe*), with walls only one atom thick and tens of atoms in circumference.

Figure 5. Types of nanotubes, according to their conductivity

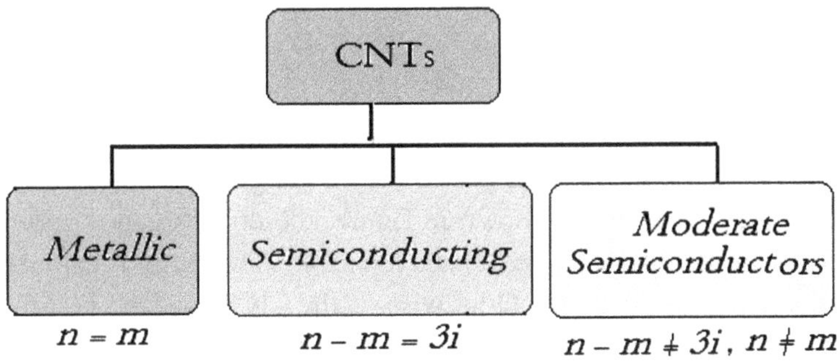

Figure 6. Single-wall gold nanotube (SWGNT), with chiral vector (5, 3)

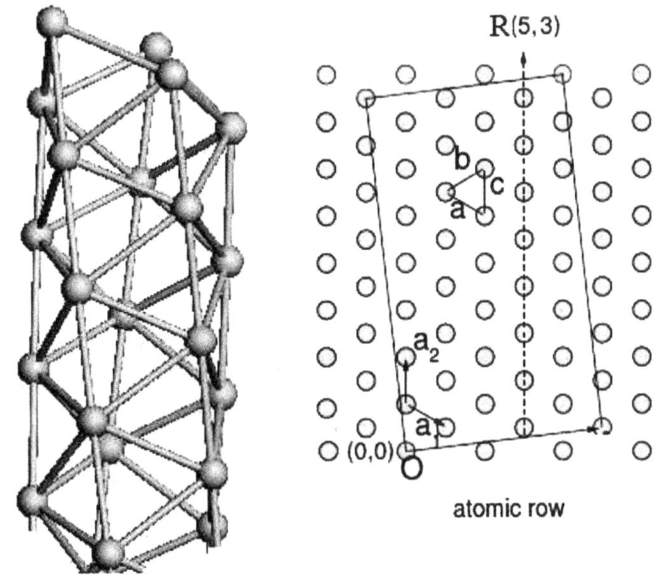

3.2 Energy Band Structure of Graphene and Carbon Nanotubes

The electronic structure of carbon nanotubes can be derived from the band structure of graphene. The band structure of graphene is shown in Figure 7. In graphene, the energy of an electron with wavenumber **k** is given by:

$$E(k) = \frac{E_p \mp \gamma_o \cdot w(k)}{1 \mp s_o w(k)} \tag{5a}$$

Here, γ_0 is the energy overlap integral[1] between nearest-neighbor carbon atoms, which is typically -3 eV < γ_0 < -2.5 eV. Also, E_p is the site energy (orbital energy) of the $2p_z$ atomic orbital, and s_0 is the overlap of the electronic wave function on adjacent sites. The geometrical function *w(k)* is given by (Saito and Kataura, 2001):

$$w(k) = \sqrt{1 + 4\cos\left(\frac{3}{2}a_o k_y\right) . \cos\left(\frac{\sqrt{3}}{2}a_o k_x\right) + 4\cos^2\left(\frac{\sqrt{3}}{2}a_o k_x\right)} \tag{5b}$$

where $a_o \approx 1.42$ nm is the lattice constant (interatomic distance).

Figure 7 depicts the calculated energy dispersion $E(k_x, k_y)$ of graphene using the parameters of Saito and Kataura (2001), with $s_0 = 0.129$. As shown in figure, the energy gap is zero at the K-point (the Fermi point), where the wavenumber $k = (4\pi/3a)$ and $a = \sqrt{3}a_o$. This point is called the Fermi point and makes two thirds of the nanotubes metallic. Otherwise, if the CNT miss the Fermi point, the nanotube is semiconductor. Near the Fermi points, the *E(k)* is conic and the slope of the cones is: $a_o\gamma_o\sqrt{3/2}$.

Near the Fermi level, the dispersion relation for electrons and holes is linear. Since the effective masses are given by the curvature of the energy bands ($m^* = \hbar^2/\partial^2E(k)/\partial k^2$), this corresponds to zero effective mass. The equation describing the excitations in graphene is formally identical to the Dirac equation for massless fermions which travel at a constant speed. Therefore, the connection points of the cones (Fermi points) are sometimes called the Dirac points. This gives rise to interesting analogies between graphene and nuclear particles.

Figure 7. Energy bands structure of graphene, the upper part refers to the conduction band, the lower part to the valence band
According to (Saito & Kataura, 2001)

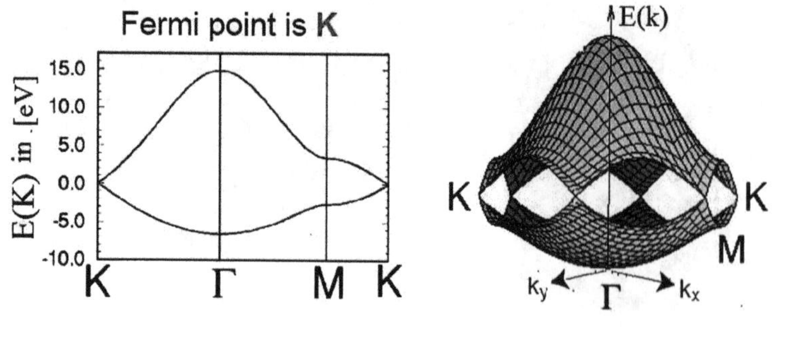

Tight Binding Model for Graphene

The graphene plane is a hexagonal lattice with two atoms per unit cell (say A, B) and a basis defined by two unit vectors ($\boldsymbol{a}_1$, $\boldsymbol{a}_2$). Applying the condition $\boldsymbol{a}_i \cdot \boldsymbol{b}_j = 2\delta_{i,j}$ allows us to obtain the reciprocal lattice vectors ($\boldsymbol{b}_1$, $\boldsymbol{b}_2$). Every carbon atom possesses four valence electrons (two 2s and two 2p electrons). When the atoms are placed onto the graphene hexagonal lattice the electronic wave functions from different atoms overlap. However, the overlap between the p_z orbitals and the s or p_x and p_y electrons is strictly zero by symmetry. Consequently, the p_z electrons, which form the π bonds in graphene, can be treated independently of other valence electrons. Within this π-band approximation, the A atom (or B atom) is uniquely defined by one orbital per atom site $p_z(r-r_A)$ or $p_z(r-r_B)$.

To derive the $E(k)$ relation, the corresponding Schrödinger equation has to be solved, and by applying the Bloch theorem, the wave functions can be decomposed as follows:

$$\psi(k) = C_A(r)p_{kz}^A(r) + C_B(r)p_{kz}^B(r) \tag{6a}$$

where

$$p_{kz}^J(r) = \frac{1}{N^{1/2}}\sum\nolimits_l p_z(r - r_J - 1)\exp(jkl) \tag{6b}$$

with k is the crystal momentum, l is the cell position index, N is the number of unit cells in the graphene sheet and the superscript $J = A,B$

The $E(k)$ relation is then derived by solving the Schrödinger equation by diagonalizing the following 2×2 Hamiltonian matrix:

$$\begin{pmatrix} H_{AA} - E & H_{AB} \\ H_{BA} & H_{BB} - E \end{pmatrix} \tag{6c}$$

Neglecting the overlap $<A|B>$, restricting interactions to nearest neighbors only and setting $H_{AA} = H_{BB} = 0$ as energy reference the dispersion relation are then:

Figure 8. Unit cell in real and reciprocal space of graphene

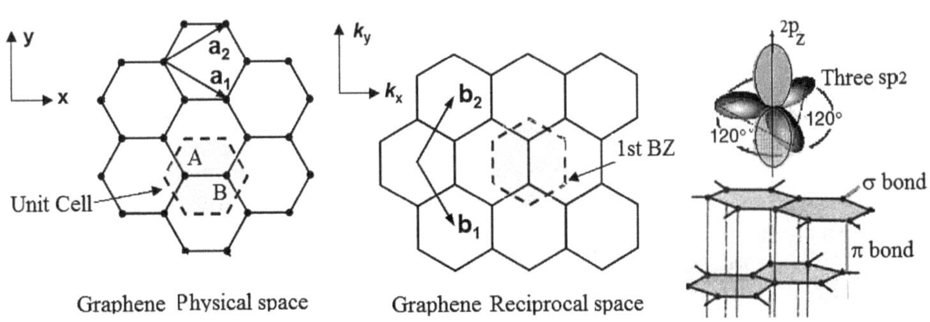

$$E(k) = \pm\, \gamma_o \sqrt{1 + 4\cos\left(\frac{3}{2}a_o k_y\right)\cdot \cos\left(\frac{\sqrt{3}}{2}a_o k_x\right) + 4\cos^2\left(\frac{\sqrt{3}}{2}a_o k_x\right)} \tag{6d}$$

where γ_0 is the nearest-neighbor hopping energy and a_o is the lattice constant. These are the π and the $\pi*$ bands in the TB model (Wallace, 1947). One of the above two bands (which represent the valence and the conduction bands), is completely filled and the other completely empty. Moreover they intersect only in two points, known as Dirac points, in the BZ. For this reason graphene is a semi–metal with a one–dimensional Fermi surface. For the same reason CNTs are predicted to be either metallic or semiconducting according to whether or not the Dirac points belong to the set of allowed k–points. The predictions of the TB model are partially confirmed by ab–initio calculations even if some differences appear. The limitations of the TB model of SWCNTs has been discussed in details in (Jamal & Mominuzzamanl, 2015)

Note 1: Carbon Bonds

All carbon forms (including graphite) have some sort of tetravalent bonds. Since the *2s* and the *2p* orbitals are very close in energy, one electron from the *2s* orbital jumps to the $2p_z$ orbital. The *2s* and three *2p* orbitals mix together and give rise to four new different types of orbitals, called sp, sp^2 and sp^3 hybridizations. The most basic forms of carbon structures are: graphite, with sp^2 bonding, and diamond, with sp^3 bonding. Note that carbon in graphite sheets (graphene) is different from diamond carbon. Carbon in graphite is bonded in such a way that one electron per carbon atom is freed up to move freely, rather than stay near its base atom. This is the situation in metals, where some electrons are not bound to their atom and can easily be pulled under the influence of an electric field.

Therefore, graphite conducts electricity due to the presence of such delocalized electrons between layers. Actually, there are two types of covalent bonds in carbon: *sigma (σ)* and *pi (π)* bonds:

- **Sigma (σ) Bond:** Is a covalent bond, which is linear aligned along the plane containing the atoms. Sigma bonds are strong and the electron sharing is maximum.
- **pi (π) Bond:** Where electron orbitals overlap laterally. The resulting overlap is not maximum. It follows that the π-bonds are weak. Example: Graphite has 3 σ-bonds and 1 π-bond.

Energy Band Structure of a Carbon Nanotube

We know that carbon nanotubes are rolled sheets of graphite sheets (graphene), and may be metallic or semiconducting according to their geometry and chirality. In fact, the unique energy band structure of graphite allows for it to behave as both a metal and a semiconductor. In order to know the electronic structure and properties of carbon nanotubes one must apply specific quantization conditions, since electrons are bound in the circumferential direction of the tube (Chiral vector ***R*** *(n,m)* direction). When the graphene sheet is rolled in this direction to form a CNT, the resulting CNT is metallic. This result can be obtained from a variety of considerations, starting from the so-called Zone Folding Approach (ZFA), based on knowledge of the electronic properties of the graphene, to the direct study of nanotubes using Tight–Binding (TB) or Density functional theory (DFT) approaches.

Zone Folding Approach (ZFA)

In the ZFA the wave–functions $\psi_n = \exp[j(k_r.r+k_z z)].u_n(r)$ of the graphene sheet are used to describe electrons in CNTs assuming that the curvature of the sheet has negligible effects. We also assume the tube direction is the z-axis. Then the only difference with respect to graphene, is the quantization of the angular momentum $L_z = R.k_r$. Accordingly, only the wavefunctions $\psi_n(r)$ are considered in the ZFA model.

When the boundary conditions are applied, we get specific k–points in the reciprocal space ($\boldsymbol{R.k_r=\pi}$ $dk_r=2\pi$). Therefore, each of the graphene bands (π, π^*, σ, σ^*) splits into a number of one-dimensional subbands (equal to the number of graphene hexagons in the nanotube unit cell or $N=(n^2+nm+m^2)/d$). The k-points that respect the boundary conditions are represented in Figure 9. The shown energy surface of the valence-π band and conduction−π^* band of graphene, are cut by the allowed k–points which can be used to construct the CNT band structure. As shown, the π-π^* bands are joint at six points lying at the Fermi points. In other directions in k-space, as in the Γ-M direction, the band structure has a semiconductor-like band gap.

Although the predictions of the ZFA give a good description of the CNT properties, but some corrections have to be considered. For instance, the tube curvature in SWNTs and the tubes interaction in MWNTs should be included. Also, the position of the so-called Dirac points has to be shifted in the Brillouin Zone due to curvature effects. In addition, the ZFA results depend on the method used to compute the band structure of graphene. The energy surfaces in the above figure have been calculated within the TB model, for example. The limitations of the TB model to calculate the dispersion relation of SWNT has been discussed in the literature, for instance in Reich et al (2000) and Jamal & Mominuzzaman (2015). Within the TB+ZFA scheme, we can show that all *(n,n)* and *(3n,0)* CNTs are metallic, with *n* any integer, while all the remaining *(n, 0)* tubes are semiconducting. The general rule is that a CNT is metallic if $n-m$ is a multiple of 3. Figures 10, 11 show the *E(k)* relation in 2-dimensional k-space and the density of states for metallic and semiconducting nanotubes.

The spikes in the density of states are called van Hove singularities (vHSs). They arise from the quasi one-dimensional character of electronic motion in systems with very high aspect ratios. Each van Hove singularity is labeled with the index of the subband to which it belongs. These singularities are of great interest for optical transitions, in a variety of optical phenomena.

Figure 9. Illustration of the the zone-folding approach (ZFA)

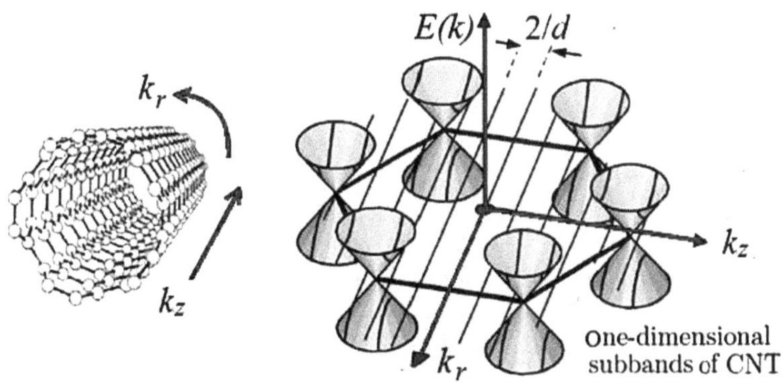

Figure 10. Energy band diagram in 2-dimensional k-space of metallic and semiconductor nanotubes

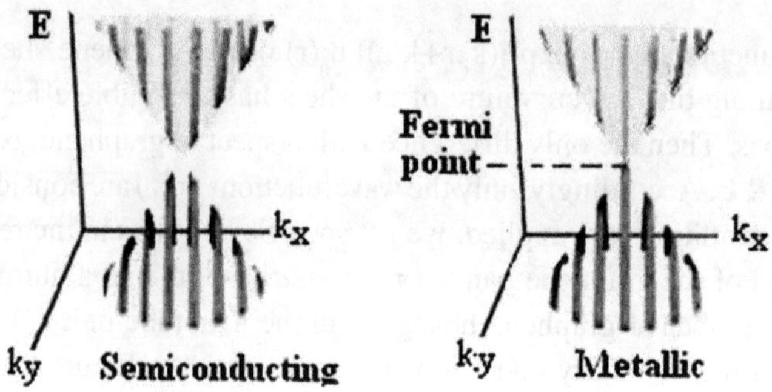

Figure 11. Energy-band structure and density of states of metallic nanotubes

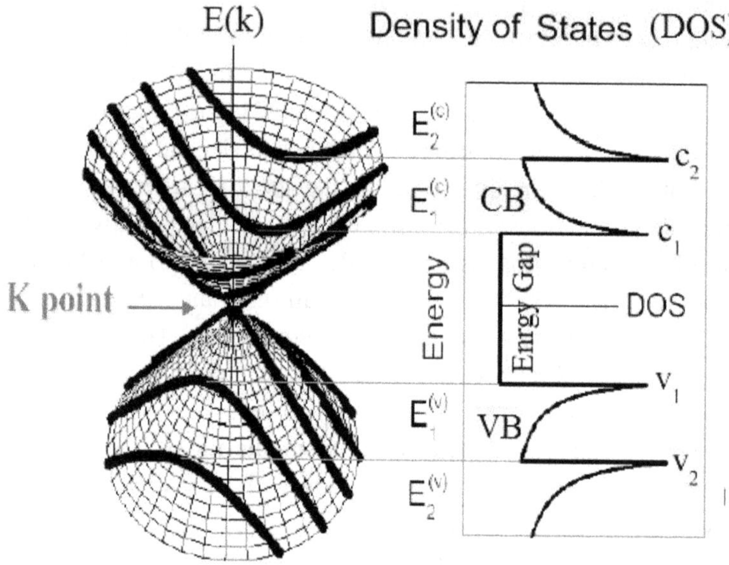

Energy Gap of Carbon Nanotubes

Carbon nanotubes do not all have the same band gap. For each diameter there is a set of allowed valences and conduction states. As nanotube diameters increase, more states are allowed. The relation between the energy gap (E_g, in eV) and CNT diameter (d, in nm) is approximately given by the following relation:

$$E_g \approx 2E_o\left(\frac{a_o}{d}\right), E_g(eV) \approx \frac{0.9}{d(nm)} \tag{7}$$

where a_o = *1.42 nm* is the distance between neighbor carbon atoms in a flat graphene sheet and E_o =2.7 eV is their binding energy. In this way, different-size nanotubes can have bandgaps as low as zero (like a metal), as high as the band gap of silicon, and maybe in between.

Figure 12a and 12b illustrates the energy gap of a SWNT as a function of its diameter. The figure shows the Katuara plots for the same parameter. The Katuara plot is a theoretical graph that was designed by Hiromichi Kataura in 1999 to rationalize some experimental findings. It relates the nanotube bandgap energy and its diameter. The oscillating of the of the Kataura plot branches reflects the strong dependence of the SWNT properties on the chirality (*n*, *m*) index, as well.

3.3 Effect of Magnetic Field on CNT Band Structure

At zero magnetic field, the low energy band structure of carbon nanotubes is doubly degenerate. This orbital degeneracy plays a crucial role in the transport properties of carbon nanotubes. In the presence of a magnetic field parallel to the CNT axis, B_z, the quantization condition becomes:

$$R.k + 2\pi(\frac{\phi}{\phi_o}) = 2\pi i \quad (8)$$

where $\phi = ¼\ \pi.d^2B_z$ is the magnetic flux, $\phi_o = h/e$ is the flux quantum and $2\pi\phi/\phi_o$ is the Aharonov-Bohm phase acquired by the electrons while travelling around the nanotube circumference. In fact, the Aharonov–Bohm effect shows that the energy dispersion (E-k) changes periodically with magnetic flux.

Applying an axial magnetic field B_z will shift the CNT subbands away from the cone vertices, and hence opening a bandgap. Therefore, a metallic nanotube can be transformed to a semiconducting nanotube by applying a magnetic field, and vice versa.

$$\Delta E_g = \pm ¼ ev_F dB_z \pm ½ g\mu_B.B \quad (9)$$

where ***B*** isb the total magnetic field, Bz is the magnetic field components in the z-direction, *g* is the electron g-factor, and μ_B is the Bohr magneton. For metallic CNTs, a finite field B_z doesn't change the subband degeneracy because the two subbands passing through the equivalent k-points (K1, K2) are shifted in the same direction. The case of semiconducting nanotubes is different, since the equivalent k-points are located on opposite sides of the dispersion relation cones ($K_2 = -K_1$). Because B_z shifts both subbands in the same direction, one subband gets closer to the K2 point, and its band gap decreases, while the other subband shifts away from the K_1 point, thereby increasing its band gap. The subband splitting can be thought of as an orbital splitting due to electrons with opposite orbital magnetic moments, like the Zeeman splitting for electrons with opposite spin magnetic moment.

3.4 Energy Band Structure of Silicon Nanowires

The silicon nanotubes (**SiNT**s) are some sort of hollow SiNWs, with tube-wall-thickness of several nanometers. SiNTs are expected to have more potential applications than SiNWs. Recently, crystalline silicon nanotubes (*c*-SiNTs), have been successfully synthesized. Figure 14 shows the shape of SiNW and its unit cell.

Silicon nanowires (SiNWs) are direct bandgap materials with a Brillouin zone half the length of that of bulk *Si* along the *Δ* line. In the conduction band, valley splitting reduces the averaged mobility mass

Figure 12. (a) Energy gap of carbon nanotube, as a function of tube diameter; (b) Katwara plots for the energy gap of carbon nanotube
After (Antram, 2006) and Saito & Kataura (2007).

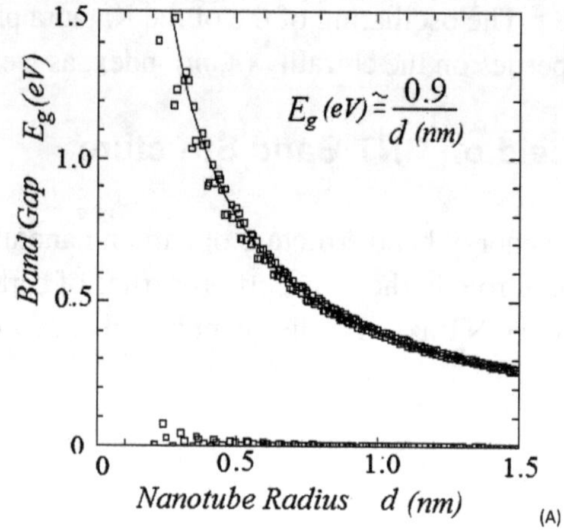

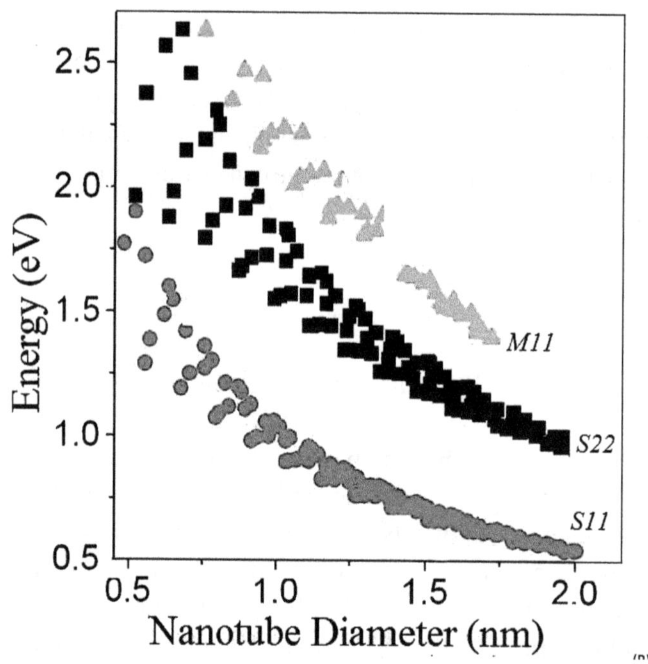

along the axis of the wire, but quantum confinement increases the transverse mass of the conduction band edge. For the wire thickness range up to 3nm, the effective mass at the conduction band edge is at least 35% heavier than that of transverse mass of bulk *Si*.

Quantum confinement has the largest effect on the carrier effective masses in the valence band. The effective mass at the valence band edge is at least 6 times heavier than that of the bulk in the [100] direction. The effective mass of the next highest band is even heavier. Small energy splitting occurs at the conduction band minimum. Figure 16 depicts the energy bands of a 1.54 nm SiNW. For wires greater than 1.54 nm thick, the 4 bulk valleys which compose the conduction band minimum are split into 3 energies.

Figure 13. Effect of axial magnetic field on the CNT band structure. The vertical lines represent allowed k_t values intercepting the dispersion cones at K1 and K2. The subband splitting in a semiconducting nanotube is shown in d.

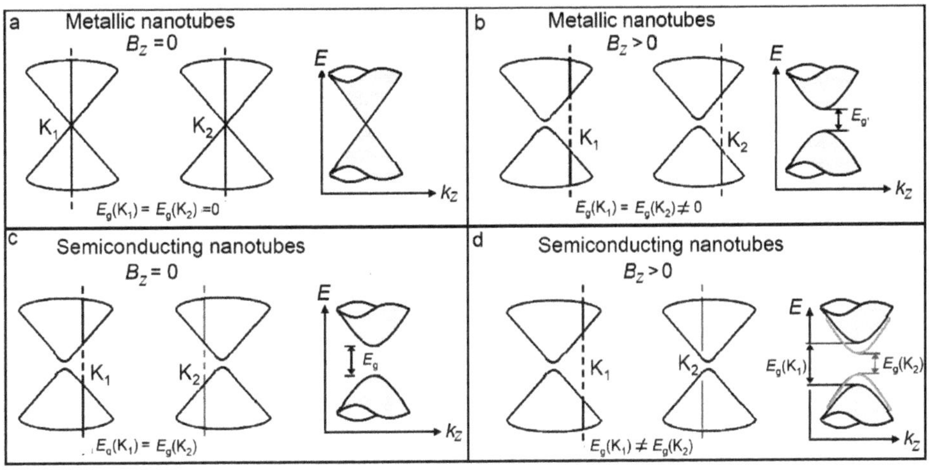

Figure 14. Structure of silicon nanowire (SiNW) and its unit cells

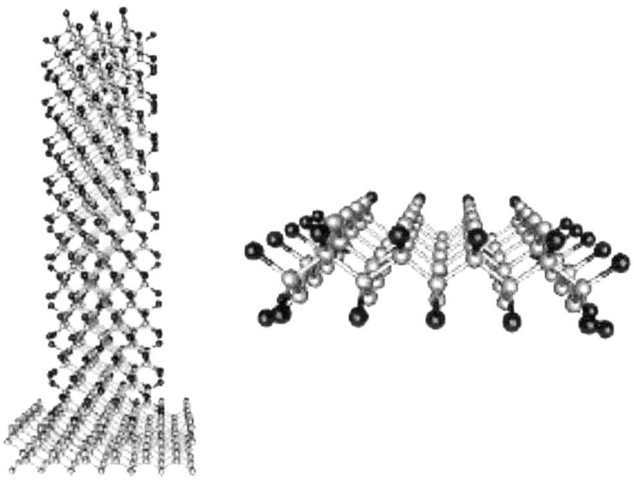

4. ELECTRONIC TRANSPORT IN NANOTUBES AND NANOWIRES

Nanotubes and nanowires with dimensions on the nanometer length scale cannot be treated as classical conductors and should be considered as quantum wires. In fact, the cross section dimension of nanotubes and nanowires is comparable to the mean free path length (between successive collisions of electrons) λ_m, and the phase coherence length (of electron wavefunction) λ_{ϕ}. Indeed, many quantum effects appear in nanotubes, especially at low temperatures. In fact, nanotubes are a nearly ideal realization of a Luttinger liquid, where the Fermi surface is not abrupt, even at zero temperature. Quantum transport effects in nanotubes include the Coulomb blockade of conduction (due to interference effects of coherent waves

Figure 15. Schematic of energy bands in 1.54nmx1.54nm SiNW
After Zheng et al., (2005).

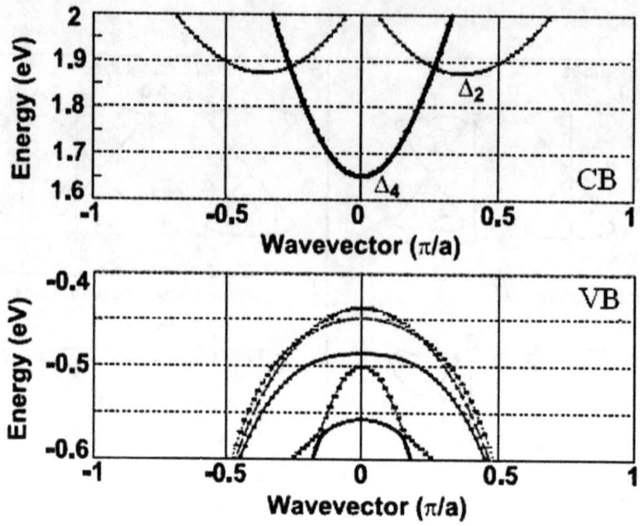

Figure 16. Effect of tube thickness (σ) on the energy band structure and energy gap of a uniform c-SiNT of diameter D= 1.8 nm

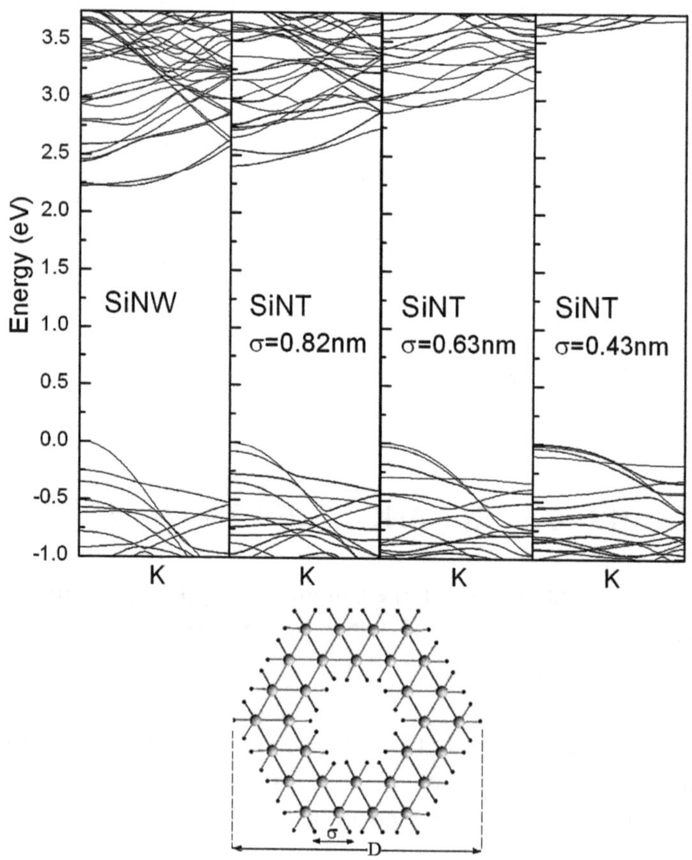

of carriers), resonant tunneling through energy levels, Kondo effects, and effects of ferromagnetism and superconductivity. However, the length scale of nanotubes is typically too large for the full quantum mechanical treatment. Therefore they are sometimes called mesoscopic or quasi one-dimensional (Q1D) structures (Datta, 1996).

There are several semiclassical and quantum approaches to study the charge carrier transport in carbon nanotubes. The semiclassical approaches include the solution of the Boltzmann transport equation (BTE) or its hydrodynamic moments for one or two fluids (π and σ electrons). The quantum approaches include the solution of the Schrodinger equation for ballistic transport or the utilization of the Landauer–Büttiker (LB) formalism or the nonequilibrium Green's functions (NEGF) method or the Wigner transport equation (Adessi et al, 2009). The atomistic simulation of CNTFET, using NEGF has also been carried out by several authors, e.g., Guo & Lundstrom, (2007) and Kordrostami & Sheikhi (2010).

In the following subsections, we examine quantum transport in carbon nanotubes, including the observation of quantized conductance as well as diffusion transport and scattering mechanisms. We also discuss the magnetic field effects and spin-dependent ballistic transport in carbon nanotubes.

4.1 Ballistic Transport of Charge Carriers in SWNT's

The electron transport in CNT's has been extensively studied since their discovery in 1990's. The quantum ballistic transport in nanotubes has been demonstrated by several researchers. Actually, the nanotube acts sometimes as a ballistic conductor. Some researchers succeeded to make contacts of nanotube fibers with mercury surface and the measurement revealed that the nanotube behaved as a ballistic conductor with quantized conductance (Frank et al, 1998). The coefficient of the quantum conductance is found to have integer quantized values, such as G_0. Some researchers (Sanvito, Kwon, Tománek, and Lambert) used a scattering technique to interpret the ballistic quantum conductance of nanotubes and found that some of the quantum conductance channels were blocked by inter-wall reactions. As shown in Figure 17, the nanotube conductance jumps by increments of $G_0 = 2e^2/h$ as additional nanotubes are touched to the mercury surface[2]. The value of G_0 was found to be $1/12.9k\Omega$.

Figure 17. Quantized conductance (multiples of $G_0=2e^2/\hbar$) of a multi-wall nanotube, as a function of tube thickness
Adapted from Frank et al (1998).

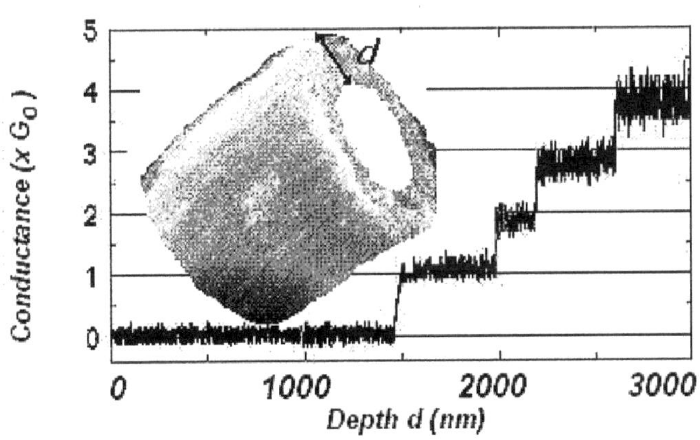

In order to understand the transport characteristics of carbon nanotubes, we start with the Landauer formula, which describes the conductance of a quasi-1D system. Therefore, the two-terminal conductance of a SWNT may be given by (Javey et al., 2003):

$$G = N(\frac{e^2}{h}).T \tag{10}$$

where N is the number of conduction channels (transverse modes or subbands) and T is contact transmission coefficient.

For a SWNT, the band structure gives rise to two propagating channels and taking into account the spin of electrons, there are 4 channels in total. Therefore, $N=4$ and the conductance is given by:

$$G_o = (\frac{4e^2}{h}).T \tag{11}$$

This formula gives the quantized conductance in the ballistic transport case. The conductance of a nanotube is corresponding to the following two-terminal resistance

$$R_o = \frac{1}{G_o} = (\frac{h}{4e^2}).\frac{1}{T} \tag{12}$$

If the contact is perfect ($T = 1$) and the CNT is scattering free, charge carriers can move through the nanotube ballistically, we have then

$$R(total) \approx R_o = \frac{h}{4e^2} \approx 6.5\ k\Omega \tag{13}$$

Figure 18 show a circuit diagram of an SWNT connected, between two metal leads, to an external bias voltage. Figure 19, depicts the quantized conductance of a silicon nanowire (**SiNW**), as a function of the width-to-length (W/l) ratio.

4.2 Diffusive and Ballistic Transport in Nanotubes

In a ballistic conductor, the electrons entering the conductor from one side will be transported to the other side without any scattering. Therefore, the conductance of a ballistic conductor that has M subbands at zero temperature is given by the Landauer formula (10).

In real experiments, ballistic transport can only be achieved if the tube length is much smaller than the electron mean free path (between successive collisions and diffusion events). Therefore, transport in semiconducting SWNT's has been previously interpreted as diffusive, with a low mobility (~20 cm^2/Vs). However electrostatic force microscopy (EFM) and low temperature transport measurements indicate that the conduction is limited by a series of large transport barriers.

Figure 18. Nanotube as a two terminal conductor

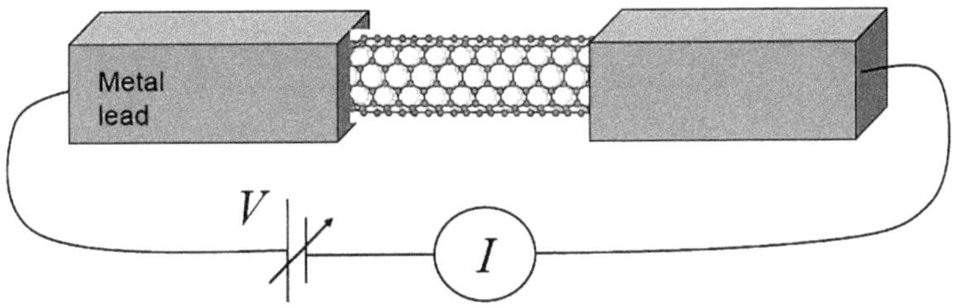

$$G = I/V = \frac{2e^2}{h}$$

Figure 19. Measurement setup and values of the conductance of a nanowire

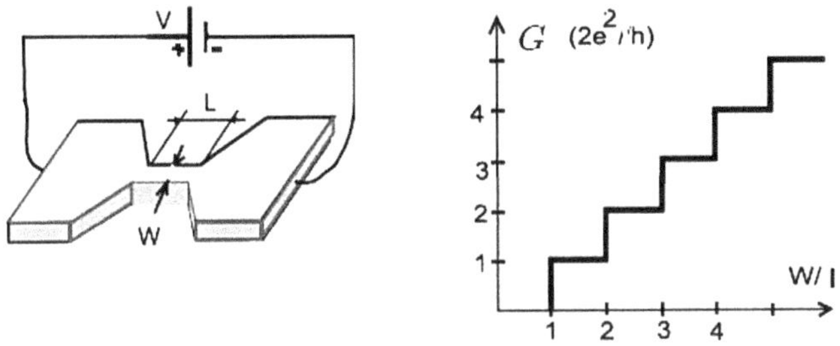

In very long semiconducting nanotube, where the resistance is dominated by intrinsic scattering processes, and not the contacts, an extremely high mobility is observed (20,000 cm^2/Vs at room temperature), which is greatly exceeding the carrier drift mobility of known semiconductors. Actually, the nanotube acts sometimes as a ballistic conductor. Some researchers succeeded to make contacts of nanotube fibers with mercury surface and the measurement revealed that the nanotube behaved as a ballistic conductor with quantum behavior.

Researchers could measure the resistivity of ropes of metallic SWNT's with a four-point probe technique and found it in the order of 10^{-4} Ω.cm at 300K. For this reason, the SWNT's are currently considered for revolutionary applications in nanoelectronics. While metallic SWNT's have been shown to conduct electrons ballistically at room temperature, the picture of electron transport in semiconducting SWNT's is less clear.

There are two types of barriers can form at the metal/tube interface and increase the contact resistance. The first is a barrier created by an imperfect interface between the contact metal and the nanotube. Its resistance is a function of the cleanliness of the interface and the overlap of the metal–nanotube electronic states. Both *Au* and *Pd* have proven to make good contacts to nanotubes, with near-perfect transmission coefficient.

The Schottky barrier can also form at the interface of a metal and a semiconducting nanotube. This contact gives rise to nonlinear resistance, which depends on the Schottky barrier height.

Figure 20. Ballistic transport across a nanotube conductor

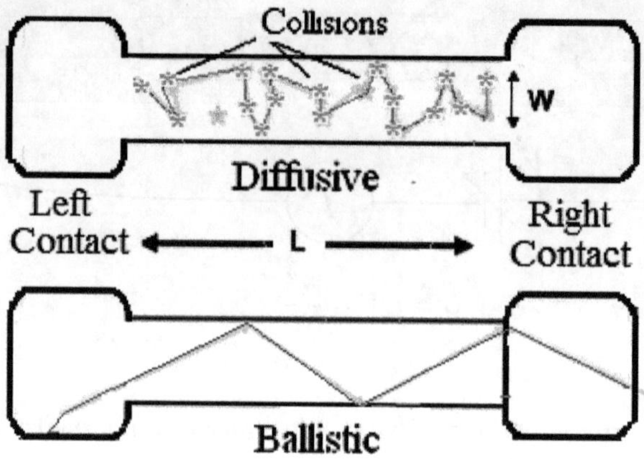

In addition, the scattering of charge carriers along the length of CNTs results in a Drude-like resistance, $R_D \sim L$, where L is the length of the CNT. Thus, the total resistance of a SWNT contacted by metal leads on both ends is the sum of these contributions: $R_t = R_o + R_c + R_D$.

Figure 21 depicts the conductance of a carbon nanotube, in the so-called field effect transistor (FET). The subsequent figure shows the Nanotube differential conductance at zero DC-bias voltage, as a function of applied gate voltage, measured at various temperatures. The nanotube has a length of ~20 microns between electrodes, and a diameter of 2.2 nm.

The measurement setup, using atomic force microscope (AFM), is shown in Figures 21 and 22.

Figure 21. Conductance of metallic and semiconducting nanotubes, in a field effect transistor (FET) configuration. The measurement setup, using atomic force microscope (AFM), is shown above
Zhou et al., (2005).

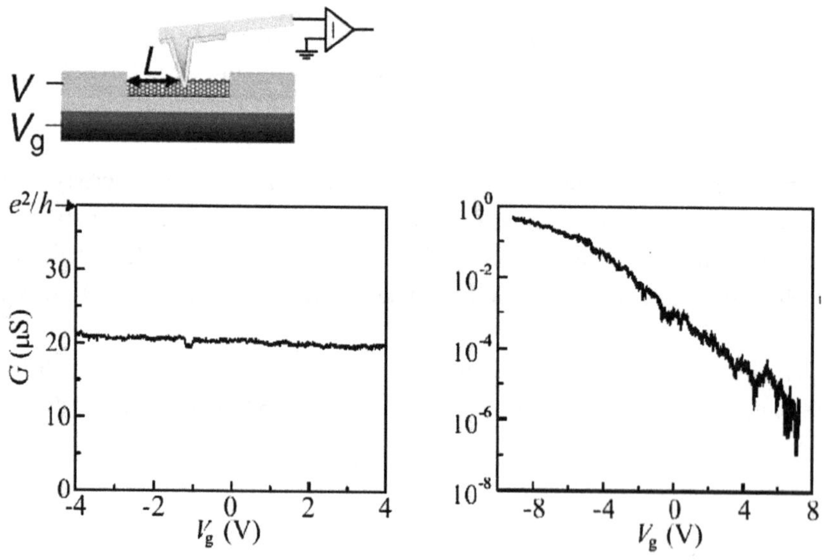

Figure 22. Nanotube differential conductance vs. gate voltage, in a FET configuration
Zhou et al, (2005).

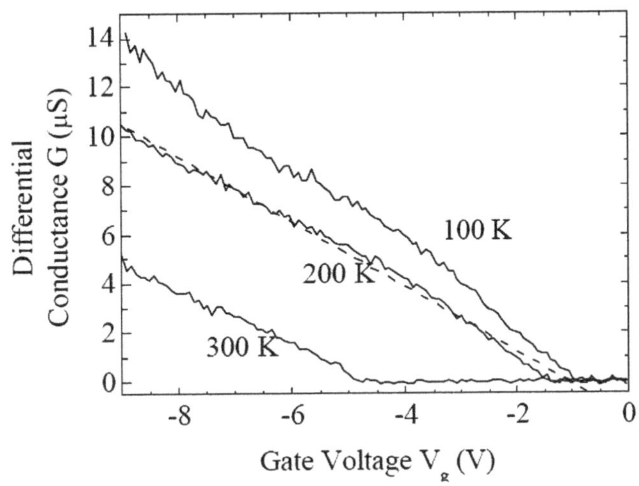

4.3 Effect of Lattice Vibrations (Phonons)

At room temperature, the dominant scattering mechanism in, high-quality metallic SWNTs is well known to be electron–phonon scattering. Recent measurements have shown also that phonon scattering limits the performance of semiconducting nanotube.

At low effective fields, the carrier mobility is dominated by Coulomb scattering, which is effectively screened at higher effective fields. At moderate effective fields, phonon scattering determines the mobility. As shown in Figure 23, the drift mobility of 79000 $cm^2/V{\cdot}s$ has been measured from a semiconducting CNT. This value exceeds those for all known semiconductors. In theory, metallic nanotubes can carry an electrical current density of $4{\times}10^9$ A/cm^2 which is more than 1,000 times greater than copper. Actually the sp^2 hybridized bonds of CNTs are typically delocalized, and this explains the extraordinary carrier mobility.

Measurements on many semiconducting CNTs yielded the following approximate expression for peak mobility (Perebeinos, Tersoff & Avouris, 2005)

$$\mu_t(cm^2 / Vs) \approx 1000 d_t \left(nm\right) . \left(\frac{300K}{T}\right) \tag{14}$$

Nanotubes have shown to exhibit strong electron-phonon resonance, and under certain bias and doping conditions the electron concentration on the tube oscillate at terahertz frequencies. These resonances could potentially be used to make terahertz (THz) sources or detectors.

4.4 Effect of Magnetic Field and Spin-Dependent Transport

We have seen in a previous section that applying an axial magnetic field, along the nanotube axis, changes the band structure from a metal to a semiconductor or vice versa. The energy gap of a semiconducting

Figure 23. Electron mobility in semiconducting SWNT's as function of drive voltage (Vth-Vg), at different temperatures
After Dürkop, Brintlinger and Fuhrer (2002).

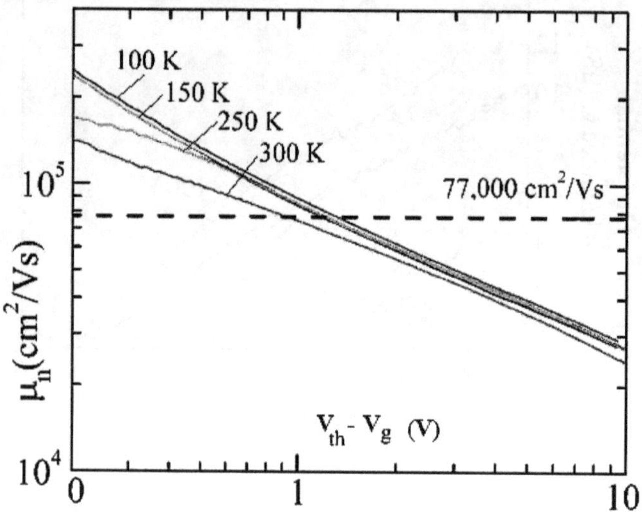

CNT is actually reduced as axial magnetic field increases, due to the formation of 2D Landau levels. This effect has important consequences on the quantum transport properties of carbon nanotubes

Magneto-resistance measurements can be carried out on individual MWNTs, which exhibit pronounced resistance oscillations as a function of magnetic flux. The quantum oscillation of conductivity with magnetic field is called the Shubnikov-de Haas effect. Such quantum oscillations of transport coefficients are widely used for investigation of the properties of metals and semiconducting nanotubes.

Spin and orbital magnetic moments lead to four-fold degeneracy in band structure and unusual Kondo effect in nanotubes (Jarello et al, 2005). Experiments on multiwall carbon nanotubes with ferromagnetic contacts demonstrate the capability of nanotubes to sustain coherent spin transport (Sahoo et al, 2005). These effects will be discussed in Chapter 9.

4.5 Hydrodynamic Modeling of Carrier Transport in Nanotubes

We have seen that at very low temperatures (below a few Kelvin), the quantum properties of CNTs have a strong influence in their transport behavior. However, due to the large size of complex nanotube structures, the full quantum methods, such as NEGF, are of limited use for describing the CNT devices. Therefore, the semiclassical modelling by the BTE or its hydrodynamic moments may be more adequate to capture the electrical behavior of CNT devices (Salahuddin, Lundstrom, & Datta, 2005).

The two-fluid version of the hydrodynamic model was originally developed by Cazaux to model σ-and π-plasmons in graphene (Cazaux, 1970) and has been lately used for modelling of SWNTs (Mowarby, 2008). However, the last work used the time-dependent density functional theory (TD-DFT) to derive a two-fluid two-dimensional hydrodynamic model describing the collective electronic response of a multiwall carbon nanotube.

In this model, the four carbon valence electrons are sp^2 hybridized, with three electrons per atom in planar σ-hybridized orbitals, and one electron per atom in out-of-plane 2pz hybridized π orbitals.

$$L_K = \frac{h}{2e^2 v_F} \quad (20a)$$

2. Electrostatic capacitance per unit length:

$$C_R = \frac{2\pi.\varepsilon}{\cosh^{-1}(2L / d)} \approx \frac{2\pi.\varepsilon}{\ln(L / d)} \quad (20b)$$

3. Quantum capacitance per unit length:

$$C_Q = \frac{2e^2}{hv_F} \quad (20c)$$

Substituting all the physical constants (with a rough estimation of the Fermi velocity $v_F \approx 10^8$ cm/s) gives the order of magnitude for the three parameters: $L_k \approx 10$ nH/ µm, $C_E \approx 50$ mF/µm and $C_Q \approx 100$ mF/µm.

The quantization of kinetic energy in the SWNT results in a significant inductance which may be exploited in RFICs. The capacitive coupling between CNT and metal plates is actually very small. To bring the resonance to GHz frequencies, CNTs of lengths about 100µm are needed. Measurement of the nanotube capacitance can directly probe the subband structure (Ilani et al., 2006). On the other hand, the deviation of the measured wave velocity of the CNT transmission line from the Fermi velocity directly determines the electron-electron interaction in 1-D systems.

6. OPTICAL PROPERTIES OF CARBON NANOTUBES

The band structure of carbon nanotubes is direct, which enables optical emission and the implementation photonic and optoelectronic devices. The optical properties of carbon nanotubes include their absorption (excitonic) spectra, photoluminescence (fluorescence), and their Raman spectroscopy. For instance, the optical selection rules of direct-gap SWNTs allow light polarization parallel to the tube axis to excite optical transitions between corresponding subbands in the valence and conduction bands. Figure 25 depicts the density of states in a semiconducting SWNT and the Van Hove singularities, as well as the possible optical transitions. Here, E_{ii} corresponds to the interband transitions. The nearest hopping parameter, γ_o, can be obtained from the relation: $\gamma_o \approx ½\, d(E_{ii}/a_o)$, where d is the SWNT diameter.

6.1 Absorption Spectra in Semiconducting SWNTs

Each semiconducting nanotube (with specific diameter and chiral vector) shows a set of well-defined absorption features. Typically, one of these absorptions lies in the near-infrared, one in the visible, and another in the near-UV. As is the case for molecules, rapid internal relaxation suppresses emission except from the lowest energy (near-IR) transition, which is labeled E_{11}. The metallic SWNTs also show

Figure 25. (a) Schematic illustration of a capacitively contacted CNT. (b) AC equivalent circuit of a CNT

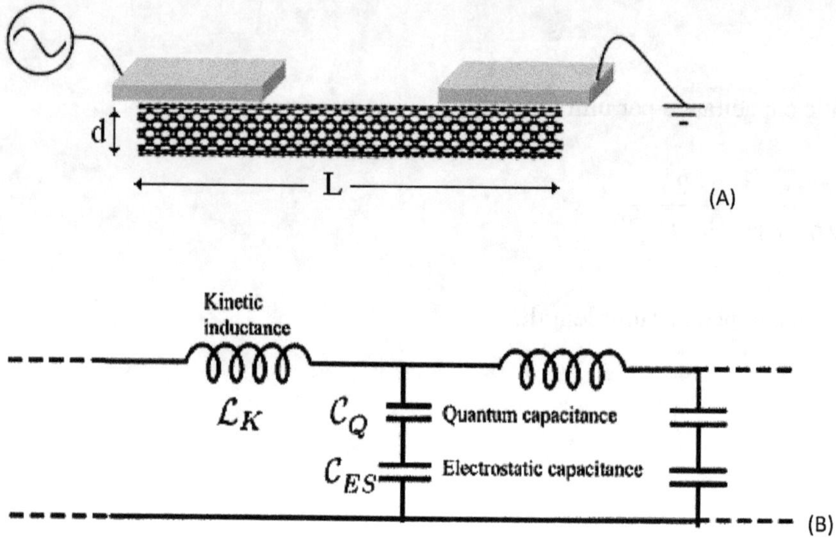

Figure 26. Schematic density of states and possible transitions in semiconducting SWNTs

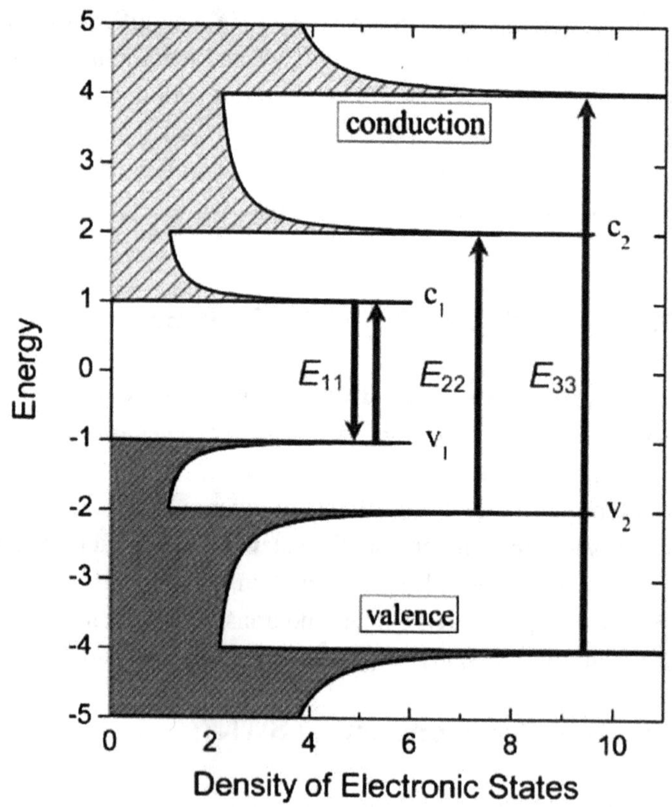

intense absorption features associated with transitions between Van Hove singularities in conduction and valence bands. However, the lowest energy optical transition in metallic nanotubes lies at shorter wavelength than the first two transitions of semiconducting tubes having similar diameter.

6.2 Excitonic Complexes in Semiconducting Nanotubes

CNTs exhibit well-pronounced optical properties. Consisting of a single layer of carbon atoms, the intrinsic many-particle screening is relatively weak resulting in strong Coulomb effects. Several theoretical and experimntal stuudies showed the formation of bound electron hole-pairs (excitons) in semiconducting and metallic nanotubes. While for semiconducting CNTs, we find excitonic binding energies in the range of 300-400 meV, metallic tube exhibit smaller energies around 50 meV.

Indeed, the optical properties of low-dimensional semiconductor nanostructures originate from excitons and exciton complexes such as *trions* (triplet charged excitons) and *biexcitons* (doublet excitons). The optical properties of semiconducting carbon nanotubes are also depending on the dynamics of excitons, and can be tuned by doping or by means of the quantum confined Stark effect (Maultzsch et al., 2006). Trions are generally more stable (has greater binding energy) than biexcitons in strongly confined quasi-1D structures with small reduced electron-hole masses, while biexcitons are more stable than trions in less confined quasi-1D structures with large reduced electron-hole masses. First experimental evidence for the trion presence in CNTs was reported by Matsunaga et al. (2000) and Santos et al. (2000) on p-doped and undoped CNTs. The nonlinear optics experiments were able to prove the presence of both trions and biexcitons in the same CNT sample, where the trion binding energy exceeds that of the biexciton. In all cases, the trion-to-biexciton binding energy ratio is greater than unity, and decreases as the CNT diameter increases. SWNTs have very complicated excitonic structures that comprise a total of 16 excitonic states, among which only one state is radiative and all others are non-radiative (dark). Figure 27 shows the excitonic absorption spectra in CNTs, with different chirality (Ando, 2004). The spectrum is analyzed in terms of the intensities of absorption peaks.

The configuration space model (Bondarev, 2014) can be used to evaluate biexciton binding energies in small diameter CNTs, and to obtain the asymptotic relations for the lowest energy trion and biexciton binding energies in quasi-1D semiconductors. The model operates in terms of the under barrier tunneling current between the equivalent configurations of the system in the configuration space and, therefore, allows for clear theoretical interpretation to uncover generic relative stability features of biexcitons and trions in quasi-1D semiconductors. The approach was originally pioneered by Landau, Gorkov and Pitaevski, as well as Holstein and Herring in their studies of molecular binding and magnetism (Tománek, 2014).

The problem is initially formulated for two interacting ground-state 1D excitons in a semiconducting CNT. Using the cylindrical coordinates with the z -axis along the CNT axis, and separating circumferential and longitudinal degrees of freedom of each of the excitons by transforming their longitudinal motion into their respective center-of-mass coordinates, as shown in Figure 28.

6.3 Measurement of Absorption Spectra in CNTs

There exist so many methods to analyze the optical properties of SWNT (Zhou et al., 2013). The Raman spectroscopy was the first optical method used for giving the detailed structural information of nanotubes. When the wavelength of the incident Laser is close to an optical absorption feature of a nanotube, relatively intense resonance Raman scattering is generated.

Figure 27. Absorption spectra of semiconducting SWNTs, for (6,4) and (6-5) tubes. The absorption spectrum of (6,4) SWNT is shifted upward for comparison. Also, PS peaks correspond to phonon side-band transitions
After Spataru et al (2004).

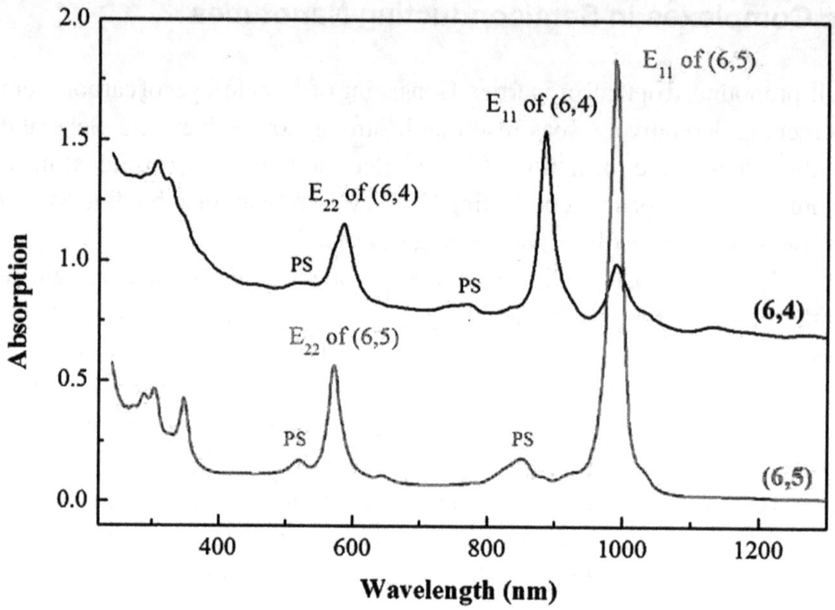

Figure 28. Absorption spectra of some metallic and semiconducting CNTs. The solid lines exhibit the excitonic absorption, and the dashed lines correspond to renormalized free-particle calculations. The arrows illustrate the excitonic binding energy. After Spataru et al (2004)
After Bondarev (2014).

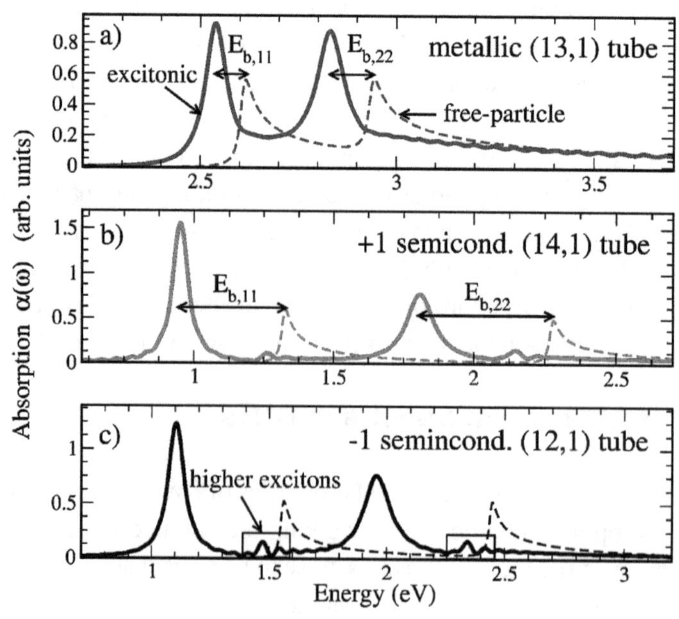

This Raman spectrum shows a G-band vibration mode, which arises from tangential and longitudinal stretching motions, and the radial breathing mode (RBM), which represents totally symmetric expansion and contraction of the nanotube cylinder. The RBM frequency varies inversely with nanotube diameter and thus provides valuable structural information for SWNTs of either metallic or semiconducting type. Although the G-band frequency is not sensitive to nanotube structure, its shape differs significantly for metallic and semiconducting species. It has been shown that the absorption, Rayleigh spectra, and Raman excitation profiles of the longitudinal and transverse G modes are best fitted by considering the nature of the E33 transition as excitonic (Tran et al., 2016).

6.4 Photoluminescence (Fluorescence)

Semiconducting single-walled carbon nanotubes emit near-infrared light upon photoexcitation, described interchangeably as fluorescence or photoluminescence (PL). The PL excitation usually occurs as follows: an electron in a nanotube absorbs excitation light via E22 transition, creating an electron-hole pair (exciton). Both electron and hole rapidly recombines from c2 to c1 (in conduction band) and from v2 to v1 states (in valence band). Then they recombine through a c1−v1 transition resulting in light emission.

Note that there is no excitonic luminescence in metallic tubes. Their electrons can be excited, thus resulting in optical absorption, but the holes are immediately filled by other electrons out of the many available in the metal. Therefore, no excitons are produced.

Figure 29. Illustration of the configuration space model showing two ground-state excitons sharing the same hole to form a negative trion state. The collinear axes, z1 and z2, represent independent relative electron-hole motions in the 1st and 2nd exciton, and their origins are shifted by DZ, the inter-exciton center-of-mass distance. After Bondarev (2014)
After Tran et al., (2016).

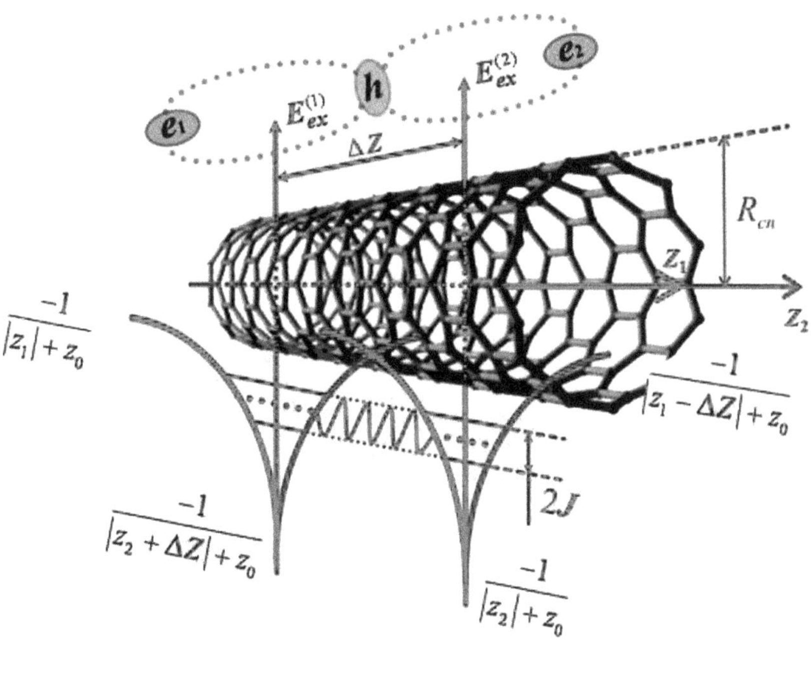

Figure 30. Raman spectra in semiconducting SWNTs. After Tran et al., (2016)

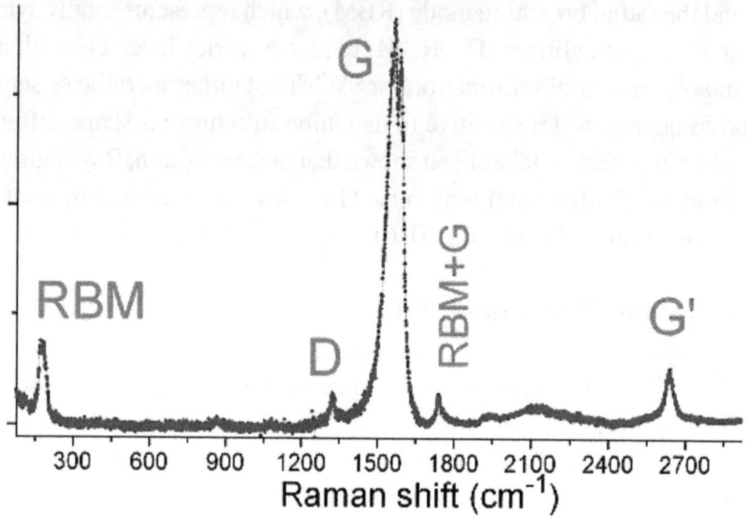

7. CASE STUDY 8: SILICON NANOWIRE TRANSISTOR (SiNWT)

The silicon nanowire transistor (SiNWT) is one of the promising nano-devices that attracted broad attention. In fact, Silicon nanowires (SiNWs) are direct bandgap materials with heavier effective mass at the conduction band edge than bulk Si. SiNW's should be distinguished from silicon nanotubes (SiNTs), which are some sort of hollow SiNWs, with tube-wall-thickness of several nanometers. Figure 31 depicts the structure of a cylindrical SiNWT. Of course the SiWNT may have other cross sectional shapes, such as square or triangular structures, as shown in Figure 31.

In order to explore the performance of SNWTs, understanding carrier transport in Si nanowires is important. The reader surely heard about the MOSFET and the role of the Si/SiO_2 interface. For bulk

Figure 31. Photoluminescence (PL) from SWNTs and the PL intensity as a function of excitation power

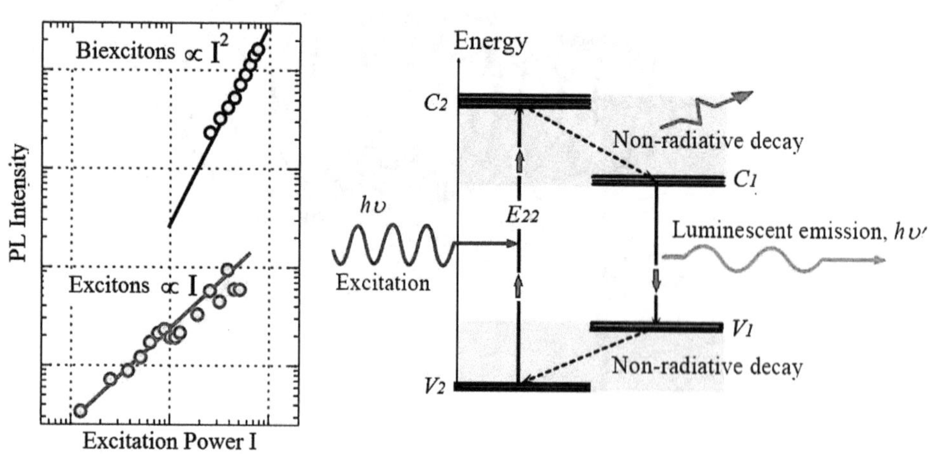

MOSFETs, electrons are confined at the Si/SiO_2 interface by an electrostatic potential well. The gate-all-around silicon nanowire transistor (SNWT) is a kind of multiple-gate MOS transistor

As shown in Figure 33, this device has a unique structure: the gate material surrounds the channel region. Because of the extreme length-to-width ratio, the nanowire channel can be considered as quasi-one-dimensional (Q1D) structure. Figure 33 depicts the dynamic characteristics (I_{DS} versus V_{GS}) of both n-type and p-type SiNWT.

8. CASE STUDY 8: CARBON NANOTUBE FET (CNTFET)

The exploration for new channel materials and device architectures to surmount the Si MOSFET scaling limit of about 5nm gate length, has led to the invention of carbon nanotube field-effect transistor (CNTFET). In fact, the direct source-to-drain tunneling at such gate lengths drastically affect the Off-state current and the whole performance of the device. The CNTFET is a field-effect transistor that utilizes a single (or array of) carbon nanotube as the channel material instead of the silicon in conventional MOSFET devices. CNTFET was first demonstrated by Dekker et al. (Tans, et al., 1998), by IBM (Martel, et al., 1998) and in Stanford University (Soh, et al., 1999). Compared to graphene nanoribon transistors, the CNTFETs provide better properties for building field-effect transistors (Schroter et al, 2013). As shown in Figure 33, there are two basic types of CNTFET devices, namely:

- Top-gated NTFET and

Figure 32. Schematic of Silicon nanotube transistor (SiNT) compared to the double-gate MOSFET

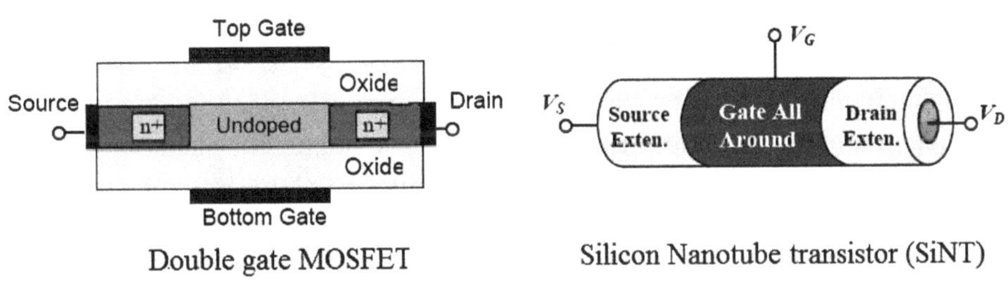

Figure 33. Schematic of gate-all-around silicon nanowire transistor (SNWT)

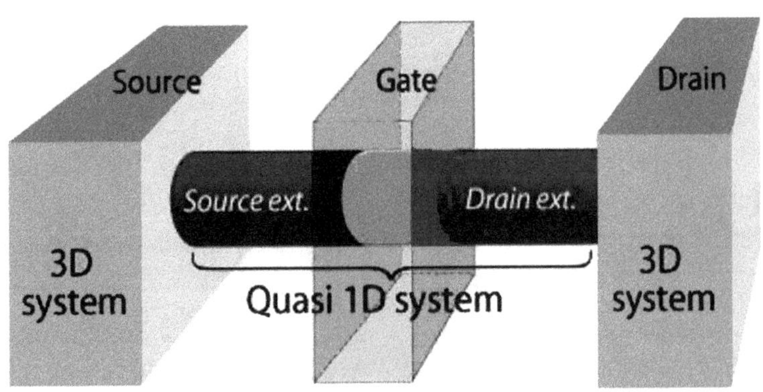

Figure 34. Simulated IDS-VGS characteristics of an SiNWT with circular cross section, 3nm diameter, and 8nm gate length. The channel orientation is [100]. Two drain biases, VDS=0.4V and VDS=0.05V, are shown in each type. After Jin, Tang & Fischetti (2008)

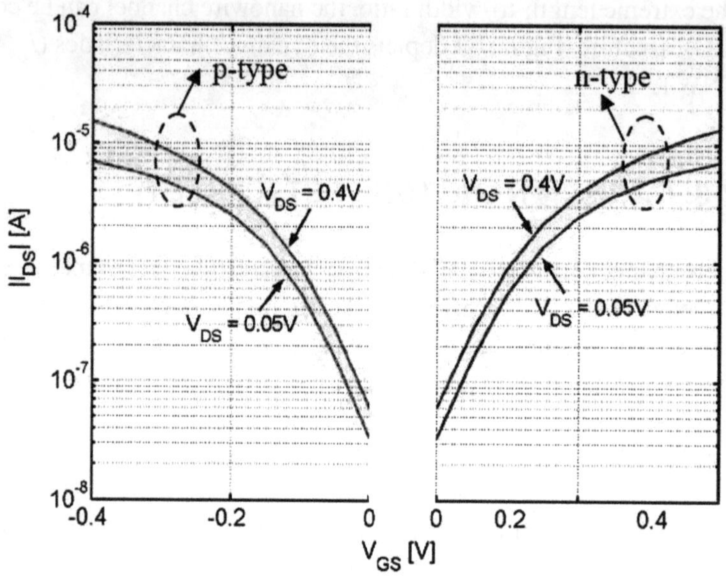

- Bottom gated. NTFET

Electrodes contacts may be metallic (Ohmic or Schottky barrier) or semi-conducting materials. Of course, the type of the contact determines the dominant mechanism of current transport and device output characteristics. For instance, the MOSFET-like CNTFET suffers from band-to-band tunneling and this leads to ambipolar current transport. Other common problems in the fabrication of CNTFETs include positioning of tubes across electrodes and the presence of metallic nanotubes mixed with semi-conducting ones. In order to get around these problems, dielectro-phoresis can be used to deposit and align tubes, while semiconducting electrodes (such as antimony telluride Sb_2Te_3 or bismuth telluride selenium Bi_2Te_2Se) can be used to turn off metallic nanotubes.

Figure 35. Basic structures of a nanotube field-effect transistor (NTFET)

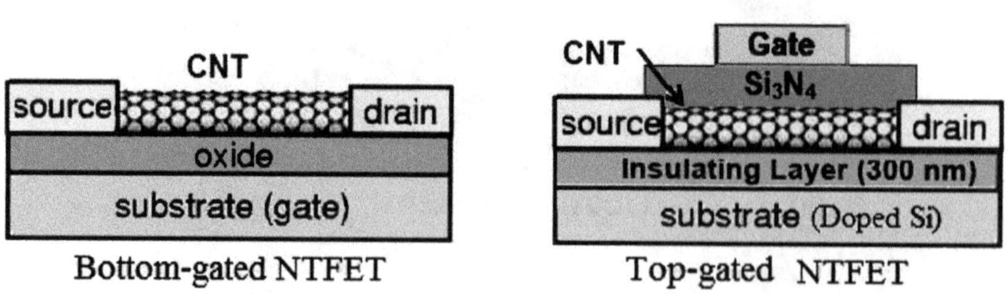

8.1 NEGF Transport Approach

The NEGF approach for CNTFET can be summarized as follows:

1. Write down the Hamiltonian matrix $\boldsymbol{H}$ in the isolated channel of the FET and identify a suitable basis set (e.g., of p_z orbitals). The size of the NxN Hamiltonian matrix is determined by the total number of basis functions in the channel region. The self-consistent potential, which appears in the Hamiltonian matrix, is included in the diagonal components of H.
2. Compute the self-energy matrices, Σ_S, Σ_D, and Σ^S, which describe how the channel couples to the source contact, the drain contact, and the dissipative processes (e.g., phonon scattering, electron-photon coupling), respectively. The source and drain self-energies can be computed using a recursive relation. At the ballistic limit, $\Sigma^S = 0$. The dissipative processes can be treated by adding a scattering contact, which in steady state takes carriers away from the initial states and put an equal number of carriers back to the final states. The carrier statistics of the scattering contact are determined by the distribution function of the channel, so that Σ^S needs to be iteratively solved with the Green's function using the Born approximation.
3. Compute the retarded Green's function, using the equation:

$$G^r(E) = [\, E.\boldsymbol{I} - \boldsymbol{H} - \Sigma_S - \Sigma_D - \Sigma^S]^{-1} \tag{21}$$

where E is the electron energy (plus an infinitesimal imaginary part in the order of 10^{-6}), $\boldsymbol{I}$ is the identity matrix and $\boldsymbol{H}$ is the Hamiltonian of a subband. As inverting this matrix equation is computationally expensive, especially for a large number of energy grid points, we usually make use of recursive algorithms.

4. Determine the physical quantities of interest from the Green's function matrix. For example, the electron density can be computed by integrating the diagonal entries of the following electron correlation function:

Figure 36. Top-gated CNTFET with semiconducting electrodes

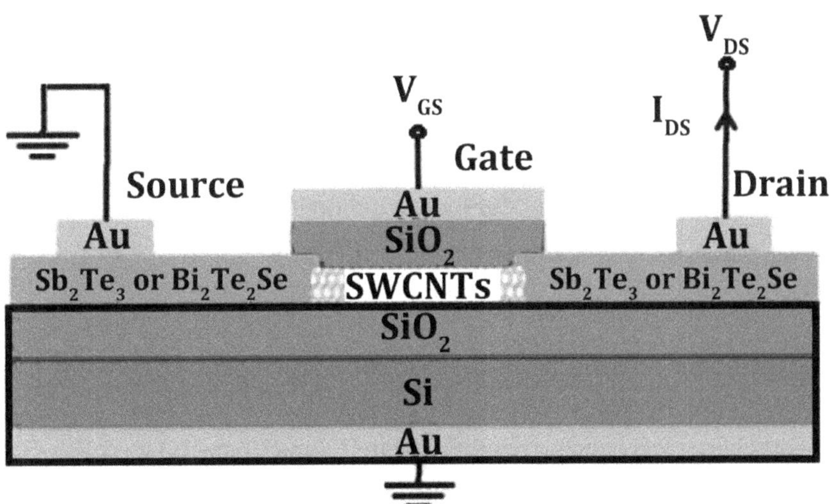

Figure 37. Simulation region of a top-gated CNTFET according to NEFG

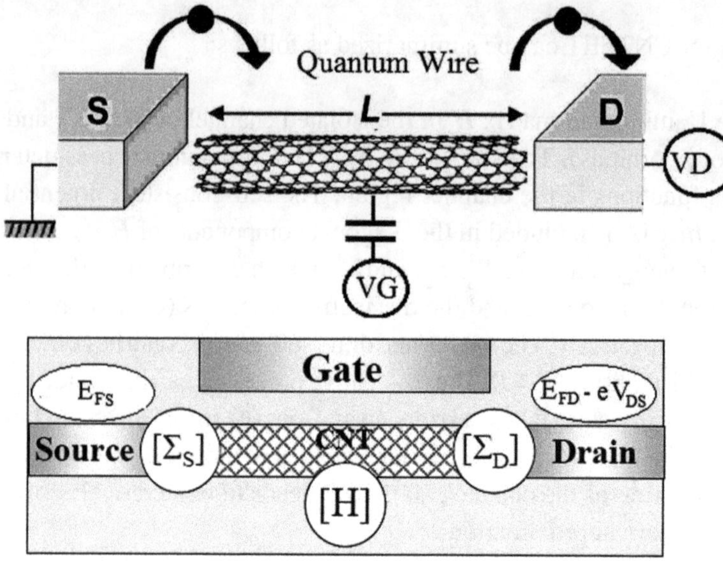

$$G^n(E) = G^r\,[\,\Gamma_S f_S(E) + \Gamma_D f_D(E) + \Sigma^S\,]^{-1} \tag{22}$$

where f_S, f_D are the equilibrium Fermi-Dirac functions at the source and drain contacts, $\Gamma_S = j(\Sigma_S - \Sigma_S^{\dagger})$ and $\Gamma_D = j(\Sigma_D - \Sigma_D^{\dagger})$ are the broadening functions of source and drain contacts. Also, Σ^S is the in-scattering function of the dissipative processes.

5. For a self-consistent solution, solve the NEGF transport equation (14) iteratively with Poisson's equation until self-consistency is achieved. The Poisson equation can be written as follows:

$$\nabla\,(\varepsilon \nabla V) = Q_t \tag{23}$$

The accumulated charge on the CNT can be calculated using the relation:

$$Q_t = -e\int [D_S f\left(E - E_{FS}\right) + D_D f\left(E - E_{FD}\right)]dE \tag{24}$$

where E_{Fs} and E_{FD} are the Fermi energy in the source and drain, respectively. Also D_s and D_D are local density of states (LDOS) in the source and drain, respectively. LDOS, in source and drain, can be written as follows:

$$D_S = G^r \Gamma_S G^{r\dagger}, D_D = G^r \Gamma_D G^{r\dagger} \tag{25}$$

The potential V has Dirichlet boundary condition at the Ohmic contacts ($V_D = V_{DS} - E_{FD}/e$, $V_S = -E_{FS}/e$) and Neumann boundary condition ($\nabla V.n=0$) at the remaining boundaries.

After the self-consistent solution is achieved, calculate the terminal currents. The source current, for example, can then be computed as

$$I_S = (\frac{4e}{h})\int Tr[\Sigma_S f_S A_S - \Gamma_S G^n]dE \tag{26}$$

where $A = j(G^r - G^{r\dagger})$ is the spectral function, and the factor of 4 comes from a spin and valley degeneracy in the energy band structure of the CNT

Note that using the NEGF formalism to treat the Schrödinger equation allows the truncation of the infinite interaction submatrix of source and drain contacts in the Hamiltonian using self-energies.

The above NEGF approach is limited to a single particle picture and the mean field theory. In order to improve this model, it is necessary to include the electron-electron correlations as well as the excitonic interactions. In addition, the transient simulation is necessary to obtain the RF parameters

8.2 Ballistic Transport Approach

According to the CNT ballistic transport theory, the drain current caused by the transport of the non-equilibrium charge across the nanotube can be calculated, as follows:

$$I_{DS} = (\frac{2ek_BT}{\pi\hbar})\left[F_o\left(U_{SF} / k_BT\right) - F_o\left(U_{DF} / k_BT\right)\right] \tag{27}$$

Here k_B is the Boltzmann constant, T is the ambient temperature, and $\hbar$ is Planck's constant divided by 2π, and F_0 is the Fermi–Dirac integral of order 0. Also, the potentials U_{SF} and U_{DF} are related to a self-consistent voltage V_{sc}, by the relations:

Table 2. Definitions of the MEGF terms, as applied to CNTFET simulation

Physical Quantity	Matrix Computation	Description
Broadening function	$\Gamma_S = j[\Sigma_S - \Sigma_S^+]$, $\Gamma_D = j[\Sigma_D - \Sigma_D^+]$	Related to the channel DOS
Spectral function	$[A_S(E)] = G\Gamma_S G^+$, $[A_D(E)] = G\Gamma_D G^+$	Diagonal elements give DOS $D(E) = \frac{1}{2\pi} Trace[A(E)]$
Correlation function	$[G^n(E)] = [A_S]f_S + [A_D]f_D$	Matrix From of electron density per unit energy
Density Matrix	$[\rho] = \frac{1}{2\pi}\int [G^n(E)]dE$	Diagonal elements give electron density $n(x) = Diag[\rho]$
Transmission Coefficient	$T(E) = Tr.[\Gamma_S G \Gamma_D G^+]$	Transmission probability

$$U_{SF} = E_F - e\,V_{SC} \tag{28}$$

$$U_{DF} = E_F - e\,V_{SC} - e\,V_{DS} \tag{29}$$

The above equations can be solved as long as the self-consistent voltage V_{SC} is known. This self-consistent voltage can be calculated from the iterative solution of Poisson's equation together with the quantum transport equation (24), across the nanotube. However the calculation can be achieved, by the calculation of the density of states at the source and drain sides, as follows:

$$C_\Sigma V_{sc} + Q_t = e\left(Ns + N_D - N_o\right) \tag{30}$$

Here, Q_t represents the charge stored in CNTFET capacitances, and the total terminal capacitance C_Σ is the sum of the gate, drain, source, and substrate capacitances. Also, the source and drain density of carriers are given by:

$$N_S = \tfrac{1}{2}\int D_S\left(E\right) f\left(E - U_{SF}\right) dE \tag{31a}$$

$$N_D = \tfrac{1}{2}\int D_D\left(E\right) f\left(E - U_{DF}\right) dE \tag{31b}$$

And their equilibrium value is

$$N_o = \tfrac{1}{2}\int D\left(E\right) f\left(E - E_F\right) dE \tag{31c}$$

The density of states at the channel $D(E)$ is given by:

$$D\left(E\right) = D_o\left[1 - \left(\frac{E_g}{2E}\right)^2\right]^{-1/2} \tag{32}$$

with $D_0 = 4/\pi\hbar v_F$. The factor 4 here comes from the degeneracy of each of the CNT subbands at zero magnetic field.

8.3 Simulation of CNTFET Characteristics

CNTFETs conduct electrons when a positive bias is applied to the gate and holes when a negative bias is applied, and drain current increases with increasing a magnitude of an applied gate voltage. Like other FETs, the drain current increases with an increasing drain bias unless the applied gate voltage is below the threshold voltage. For planar CNTFETs the FET with a shorter channel length produces a higher saturation current, and the drain saturation current becomes higher for the tube of smaller diameter.

In terms of RF operation, the CNTFET simulation shows the possibility of obtaining intrinsic cutoff frequency ($f_T = g_m/C_g$) in the THz frequency range. On the other hand, experimental demonstrations of individual CNTFETs have suffered from excessive parasitic effects. In order to mitigate the single tube limitations, CNT network-based transistors have been explored, as shown in Figure 38. The Figure 39 shows the design of a complementary CNTFET logic inverter and its transfer voltage characteristics.

8.4 MoS_2 CNTFET

We have already presented the basic structure of the molybdenum disulfide (MoS_2) FET, which utilizes a layered MoS_2 as a channel, as a case study of 2D transport in Chapter 5. We present here a modified

Figure 38. Output I-V characteristics of a CNTFET, with metallic electrodes at 300K
After Hassaniniat al (2008).

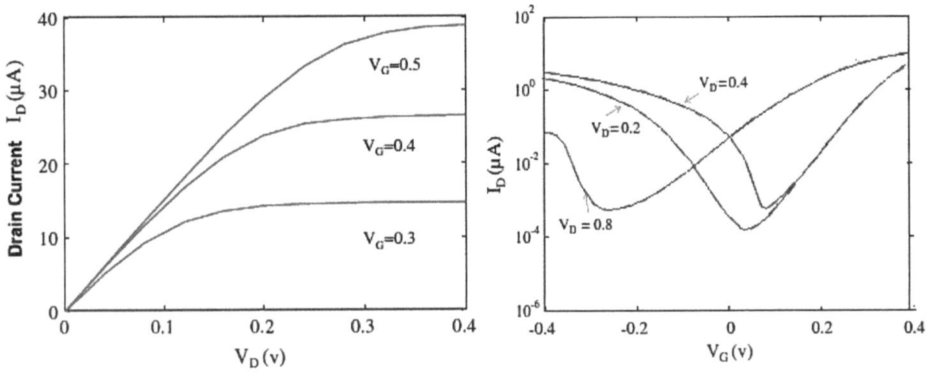

Figure 39. Cutoff frequency of a ballistic CNTFET with Schottky barrier contacts. The R-C parasitics are not included here
After Guo & Lundstrom (2007).

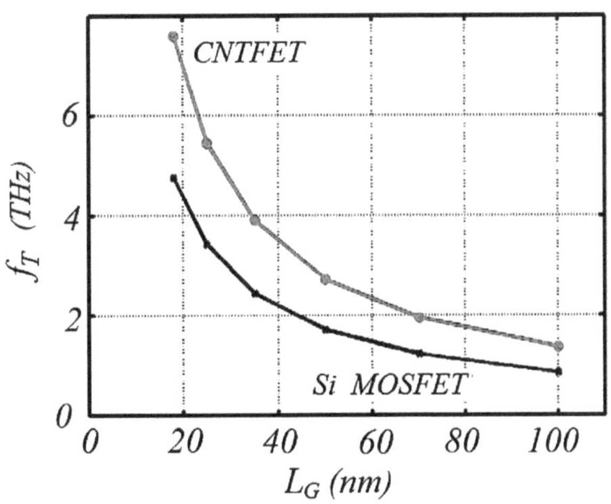

Figure 40. Layout of a CNT array-FET AFET (left) 3-D top view of AFET (middle) and printed AFET on a plastic substrate (right)

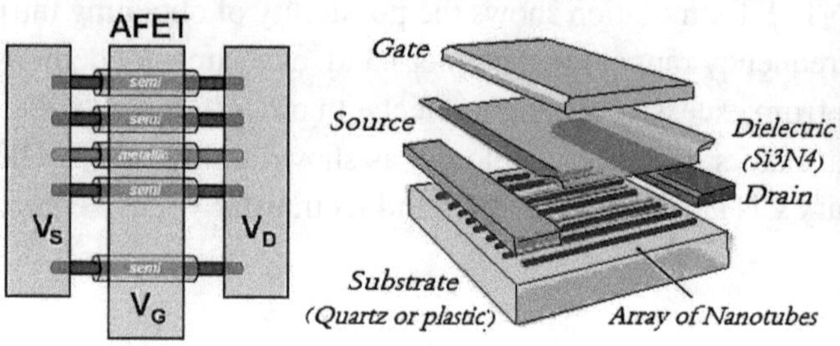

Figure 41. AFM Image of a complementary CNTFET and its transfer characteristics. Open circles are experimental data with V=±2V
After Avouris (2002).

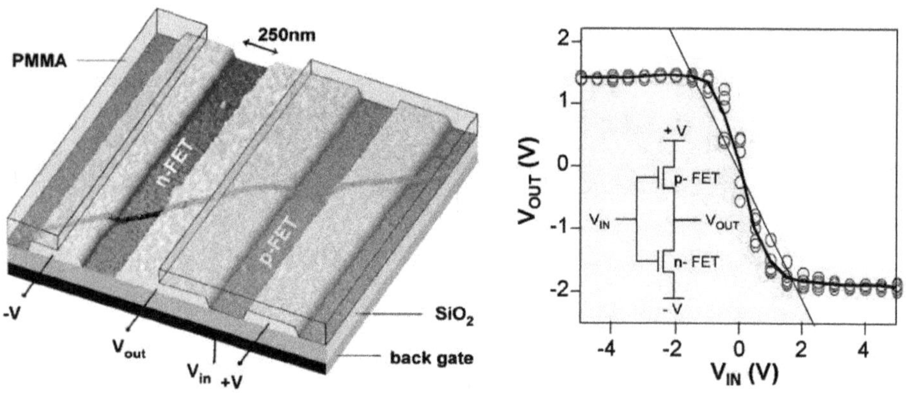

version the MoS_2 transistor which employs a single-wall carbon nanotube as the gate electrode (Desai et al, 2016). This device exhibits excellent switching characteristics with near ideal subthreshold swing of about 65mV per decade and an On/Off current ratio of about 10^6.

9. SUMMARY

Carbon nanotubes are one of the most interesting materials in nano-technology. Carbon nanotubes (CNTs) are rolled sheets of graphene (a single atomic layer of graphite). The nature of the bonding of a nanotube gives rise to their unique electrical, thermal and mechanical properties

CNTs are one-dimensional quantum systems and the geometry of a CNT can be described by a *chiral* vector $\boldsymbol{R} = n.\boldsymbol{a}_1 + m.\boldsymbol{a}_2$, where $\mathbf{a}_1$ and $\mathbf{a}_2$ are the unit vectors of the hexagonal grapheme lattice.

According to the chirality, CNTs are either Armchair ($n=m$), or Zigzag ($m=0$) or generally Chiral ($n \pm m$).

Theoretical studies have shown that a single-walled CNT (SWNT) can be either metallic or semiconducting depending on its chirality and diameter. Table 3 summarizes the carbon nanotube parameters.

Figure 42. Structure of MoS2 CNTFET
After Desai et al. (2016).

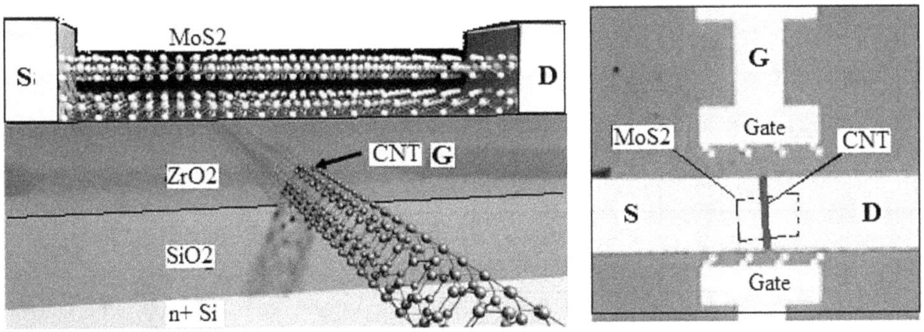

Figure 43. Chirality of a carbon nanotube (CNT)

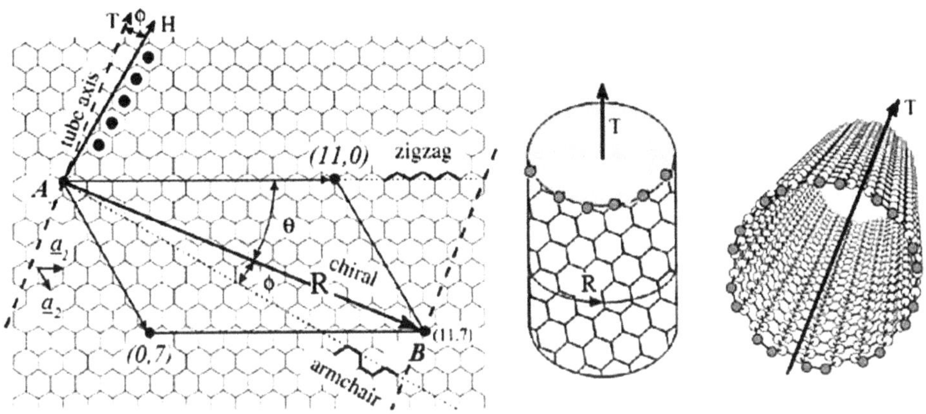

Figure 44. CNT types according to its geometry (zigzag, armchair and chiral)

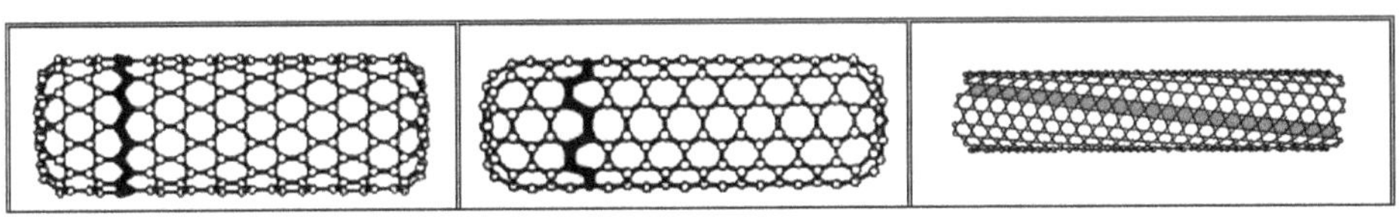

The (n, m) nanotubes with $n = m$ (armchair) are metallic, for $n - m \neq 3 \times$ integer, the nanotube are semiconducting with band gap proportional to $1/d$. For $n{-}m = 3\times$integer, the nanotubes is semiconducting, with a small band gap proportional to $1/d^2$. The transport properties of nanotubes and nanowires are intimately related to their band structure. The energy band structure of nanotubes can be obtained by the tight-binding (TB) model of graphene and the zone folding approxim-ation (ZFA) taking into account the tube curvature effects and the influence of mixing of in-plane σ and normal π orbitals. The $E(k)$ relation of CNT in the k-space can be described by the following equation:

Table 3. Physical parameters of a carbon nanotube

Parameter	Symbol	Expression
Chiral vector	*R(m,n)*	$R = na_1 + ma_2$
Length of ***R***	*L*	$L = a\sqrt{(n^2+m^2+nm)}$
Unit vector length	*a*	$a = \sqrt{3}.a_o = 2.49\,A$
Unit vectors (UVs)	$\boldsymbol{a}_1, \boldsymbol{a}_2$	$(\frac{1}{2}\sqrt{3}, \frac{1}{2}).a, (\frac{1}{2}\sqrt{3}, -\frac{1}{2}).a$
Reciprocal UVs	$\boldsymbol{b}_1, \boldsymbol{b}_2$	$(1/\sqrt{3}, 1).(2\pi/a), (1/\sqrt{3},-1).(2\pi/a)$
Diameter	*d*	L/π
Chiral angle	θ	$\cos\theta = (2n+m)/2\sqrt{(n^2+m^2+nm)}$

$$E(k) = \frac{E_p \mp \gamma_o \cdot w(k)}{1 \mp s_o w(k)}$$

where γ_0 is the nearest-neighbor C-to-C transfer energy ($3 < \gamma_0 < -2.5$ eV). Also, E_p is the energy of the $2p_z$ orbital, and s_0 is the overlap of the electron wave function on adjacent sites. The geometrical function *w(k)* is given by:

$$w(k) = \sqrt{1 + 4\cos\left(\frac{3}{2}a_o k_y\right) \cdot \cos\left(\frac{\sqrt{3}}{2}a_o k_x\right) + 4\cos^2\left(\frac{\sqrt{3}}{2}a_o k_x\right)}$$

where $a_o \approx 1.42$ nm is the lattice constant (interatomic distance).

Nanotubes and nanowires with dimensions on the nanometer length scale cannot be treated as classical conductors because their diameters are as small as to the mean free path length (between successive collisions) $l_{m'}$. They are sometimes called mesoscopic structures, because their dimensions lie between the macroscopic scale and the atomic scale where the full quantum mechanics should be applied. A metallic SWNT has two bands crossing the Fermi energy, and may show ballistic behaviors with ideal conductance $G = 4e^2/h$, according to the Landauer (collision-less) formula.

$$G_c = N \cdot e^2/h \cdot T,$$

where *N* is the number of conduction channels in parallel and *T* is contact transmission coefficient. Metallic SWNTs are ideal connectors for electronic devices due to their excellent electron transport behavior, optical transparency and flexibility. Semiconducting CNTs can have drift mobility as high as 79000 $cm^2/V{\cdot}s$ and are candidates for next-generation field-effect transistors (FETs). In the literature, one can find several modeling meyhods of charge carrier transport in nanotubes, some of them are based on quantum approaches and others are based on semiclassical approaches. The quantum transport approaches are usually based on the solution of Schrodinger equation, with the Poisson equation. On the other hand, the semiclassical approaches are based on the solution of the BTE or its moments for π and σ electrons, along the nanotube, together with the Poisson equation.

REFERENCES

Adessi, C., Avriller, R., Blase, X., Bournel, A., dHonincthun, H. C., Dollfus, P., & Zimmer, T. et al. (2009). Multiscale simulation of carbon nanotube devices. *Comptes Rendus Physique*, *10*(4), 305–319. doi:10.1016/j.crhy.2009.05.004

Anantram, M. P., & Leonard, F. (2006). Physics of carbon nanotube devices. *Reports on Progress in Physics*, *69*(3), 507–561. doi:10.1088/0034-4885/69/3/R01

Ando, T. (1997). Excitons in carbon nanotubes. *Journal of the Physical Society of Japan*, *66*(4), 1066–1073. doi:10.1143/JPSJ.66.1066

Ando, T. (2004). Excitons in carbon nanotubes revisited: Dependence on diameter, Aharonov–Bohm flux, and strain. *Journal of the Physical Society of Japan*, *73*(12), 3351–3363. doi:10.1143/JPSJ.73.3351

Ando, T. (2006). Effects of valley mixing and exchange on excitons in carbon nanotubes with Aharonov–Bohm flux. *Journal of the Physical Society of Japan*, *75*(2), 024707. doi:10.1143/JPSJ.75.024707

Ando, T. (2009). *Family Effects on Excitons in Semiconducting Carbon Nanotubes. Journal of the Physical Society of Japan*, *78*(10), 104703.

Ando, T., & Seri, T. (1997). Quantum Transport in a Carbon Nanotube in Magnetic Fields J. *Phys. Soc. Japan.*, *66*(11), 3558–3565. doi:10.1143/JPSJ.66.3558

Avouris, P. (2002). Molecular Electronics with Carbon Nanotubes. *Accounts of Chemical Research*, *35*(12), 1026–1034. doi:10.1021/ar010152e PMID:12484790

Berger, C., Poncharal, P., Yi, Y., & de Heer, W. (2003). Ballistic conduction in multiwalled carbon nanotubes. *Journal of Nanoscience and Nanotechnology*, *3*(476), 171–177. doi:10.1166/jnn.2003.180 PMID:12908247

Bondarev, I. V. (2014). *Relative Stability of excitonic complexes in quasi-one-dimensional semiconductors*. arXiv:1405.0777v2 [cond-mat.mes-hall]

Bosnick, K., Gabor, N. & McEuen, P. (2006). Transport in carbon nanotube *p–i–n* diodes. *Appl. Phys. Lett., 89*.

Buin, A. K., Verma, A., & Anatram, M. P. (2008). Carrier-phonon interaction in small cross-sectional silicon nanowires. *Journal of Applied Physics*, *104*(5), 053716. doi:10.1063/1.2974088

Cardona, M. (1996). *Fundamentals of Semiconductors*. Berlin: Springer-Verlag.

Cazaux, J. (1970). Sur lanisotropie des pertes denergie et des constantes optiques du graphite dans lultraviolet. *Optics Communications*, *2*(4), 173–175. doi:10.1016/0030-4018(70)90009-X

Cerda, J., & Soria, F. (2000). Accurate and transferable extended Huckel-type tight-binding parameters. *Physical Review B: Condensed Matter and Materials Physics*, *61*(12), 7965–7971. doi:10.1103/PhysRevB.61.7965

Chico, L., Benedict, L. X., Louie, S. G., & Cohen, M. L. (1996). Quantum conductance of carbon nanotubes with defects. *Physical Review B: Condensed Matter and Materials Physics*, *54*(4), 2600–2606. doi:10.1103/PhysRevB.54.2600 PMID:9986109

Cui, Y., Zhong, Z., Wang, D., Wang, W. & Lieber, C.M. (2003). .High performance silicon nanowire field-effect transistors. *Nano Lett.*, *3*(2), 14 6-152

Dai, H., Javey, A., Pop, E. & Lu, Y. (2006). Electrical Transport Properties and Field-Effect Transistors of Carbon Nanotubes. *Nano Brief Reports & Reviews, 1*(1).

Datta, S. (1996). *Transport in Mesoscopic Systems*. Cambridge Unniversity Press.

Datta, S. (2005). *Quantum transport from: atom to transistor*. Cambridge, UK: Cambridge University Press. doi:10.1017/CBO9781139164313

Dekker, C. (1999). Carbon nanotubes as molecular quantum wires. *Physics Today*, *52*(5), 22–28. doi:10.1063/1.882658

Desai, S. B., Madhvapathy, S. R., Sachid, A. B., Llinas, J. P., Wang, Q., Ahn, G. H., . . . Javey, A. (2016). MoS2 transistors with 1-nanometergate lengths. *Science Mag., 345*(6308), 99. Available at: https://www.researchgate.net/publication/309456228

Dresselhaus, M., Dresselhaus, G., & Avouris, Ph. (Eds.). (2001). *Carbon Nanotubes: Synthesis, Structure Properties &Applications*. Berlin: Springer-Verlag. doi:10.1007/3-540-39947-X

Durkop, T., Getty, S., Cobas, E., & Fuhrer, M. (2004). Extraordinary Mobility in Semiconducting Carbon Nanotubes. *Nano Letters*, *4*(1), 35–39. doi:10.1021/nl034841q

Forestiere, C., Maffucci, A. & Miano G. (2010). Hydrodynamic model for the signal propagation along carbon nanotubes. *J. Nanophoton., 4*(1). 10.1117/1.3424968

Frank, S., Poncharal, P., Walt, W., & de Hee, A. (1998). Carbon Nanotube Quantum Resistors. *Science*, *280*(5370), 1744–1746. doi:10.1126/science.280.5370.1744 PMID:9624050

Fu, Y. (1999). *Magnus Willander, Physical Models of Semiconductor Quantum Devices*. Springer. doi:10.1007/978-1-4615-5141-6

Gao, G., Cagin, T., & Goddard, G. A. (1997). *Energetics, Structure, Mechanical and Vibrational Properties of Single Walled Carbon Nanotubes (SWNT).* Available at: http://www.wag.caltech.edu/foresight/foresight_2.html

Geim, A., & Novoselov, K. S. (2007). The rise of graphene. *Nature Materials*, *6*(3), 183–191. doi:10.1038/nmat1849 PMID:17330084

Goldhaber-Gordon, D. (1998). Kondo effect in a single-electron transistor. *Nature*, *391*(6663), 156–159. doi:10.1038/34373

Guo, J., Koswatta, S., Neophytou, N., & Lundstrom, M. (2006). Carbon Nanotube Field-Effect Transistors. *International Journal of High Speed Electronics and Systems*, *16*(04), 897–912. doi:10.1142/S0129156406004077

Guo, J., Lundstrom, M., & Datta, S. (2002). Performance projections for ballistic carbon nanotube field-effect transistors. *Applied Physics Letters, 80*(17), 3192–3194. doi:10.1063/1.1474604

Guo, J., & Lundstrom, M. S. (2007). Device Simulation of SWNT-FETs. In A. Javey & J. Kong (Eds.), *Carbon nanotube electronics*. Springer.

Hassaninia, I., Sheikhi, M.H. & Kordrostami, Z. (2008). Simulation of carbon nanotube FETs with linear doping profile near the source and drain contacts. *Solid-State Electronics, 52*(6), 980-985.

Heinze, S., Tersoff, J., & Avouris, P. (2005). Carbon Nanotube Electronics and Optoelectronics. In G. Cuniberti & K. Richter (Eds.), *Introducing Molecular Electronics* (pp. 381–409). Springer.

Igami, M., Nakanishi, T., & Ando, T. (1999). Conductance of carbon nanotubes with a vacancy. *Journal of the Physical Society of Japan, 68*(3), 716–719. doi:10.1143/JPSJ.68.716

Iijima, S. (1991). Helical microtubules of graphitic carbon. *Nature, 354*(455), 56–58. doi:10.1038/354056a0

Ilani, S., Donev, L.A.K. Kindermann, M., & McEuen, P.L. (2006). Measurement of the quantum capacitance of interacting electrons in carbon nanotubes. *Nature Phys., 2*(458, 459, 474), 687–691.

Jamal, G. R., & Mominuzzaman, S. M. (2015). Limitations of Tight Binding Model in Describing Electronic Properties of Single Wall Carbon Nanotubes. *Journal of Nanoscience and Nanoengineering, 1*(2), 96–106.

Jarillo-Herrero, P. (2005). *Quantum transport in carbon nanotubes* (PhD Thesis). TU-Delft Univ. Retrieved from http://www.library.tudelft.nl/dissertations/

Jarillo-Herrero, P., Kong, J., van der Zant, H. S. J., Dekker, C. L., Kouwenhoven, P., & De Franceschi, S. (2005). Electronic Transport Spectroscopy of Carbon Nanotubes in a Magnetic Field. *Physical Review Letters, 94*(15), 156802. doi:10.1103/PhysRevLett.94.156802 PMID:15904173

Jarillo-Herrero, P., Kong, J., van der Zant, H. S. J., Dekker, C. L., Kouwenhoven, P., & De Franceschi, S. (2005). Orbital Kondo effect in carbon nanotubes. *Nature, 434*(7032), 484–488. doi:10.1038/nature03422 PMID:15791250

Javey, A., Guo, J., Wang, Q., Lundstrom, M. & Dai, H.J. (2003). Ballistic carbon nanotube field-effect transistors. *Nature, 424*, 654–657.

Jiang, J., Saito, R., Samsonidze, G., Jorio, A., Chou, S. G., Dresselhaus, G., & Dresselhaus, M. S. (2007). Chirality dependence of exciton effects in single-wall carbon nanotubes: Tight-binding model. *Physical Review B: Condensed Matter and Materials Physics, 75*(3), 035407. doi:10.1103/PhysRevB.75.035407

Jie, S. R., Jiang, L., Torres, F., & Saito, R. (2007). Charge transport in carbon nanotubes: Quantum effects of electron–phonon coupling. *Journal of Physics Condensed Matter, 19*(18). PMID:21690981

Jin, S., Tang, T.-W., & Fischetti, M. V. (2008). Simulation of Silicon Nanowire Transistors Using Boltzmann Transport Equation Under Relaxation Time Approximation. *IEEE Trans. Electron Dev., 55*(3), 727–736. doi:10.1109/TED.2007.913560

Jorio, A., Dresselhaus, G., & Dresselhaus, M. S. (2008). Carbon Nanotubes. *Topics in Applied Physics, 111*, 455–493. doi:10.1007/978-3-540-72865-8

Kaiser, A. (2001). Electronic transport properties of conducting polymers and carbon nanotubes. *Reports on Progress in Physics*, *64*(1), 1–49. doi:10.1088/0034-4885/64/1/201

Kataura, H., Kumazawa, Y., Maniwa, Y., Umezu, I., Suzuki, S., Ohtsuka, Y., & Achiba, Y. (1999). Optical properties of single-wall carbon nanotubes. *Synthetic Metals*, *103*(1-3), 2555–2558. doi:10.1016/S0379-6779(98)00278-1

Kelly, M. J. (1995). *Low-dimensional semiconductors*. Oxford, UK: Clarendon Press.

Kitagawa, M. (1988). Calculation of the stopping power at the surface of a solid. *Nuclear Instruments & Methods in Physics Research. Section B, Beam Interactions with Materials and Atoms*, *33*(1-4), 409–414. doi:10.1016/0168-583X(88)90595-2

Kordrostami, Z., & Sheikhi, M. H. (2010). Fundamental Physical Aspects of Carbon Nanotube Transistors. In *Carbon Nanotubes*. InTech. Available online: http://www.intechopen.com/books/carbon-nanotubes/

Ksenevich, V., Seliuta, D., Martunas, Z., Kasalynas, I., Valusis, G., Galibert, J., & Samuilov, V. et al. (2008). Charge carrier transport properties in single-walled carbon nanotube fibers. *Acta Phys. Pol. A Gen. Phys.*, *113*, 1039–1042.

Lanzani, G., & Luer, L. (2011). Carbon Nanotubes: Electronic Structure and Spectroscopy. *Comprehensive Nanoscience and Technology*, *1*, 23–39. doi:10.1016/B978-0-12-374396-1.00006-4

Levchenko, I., Han, Z.-J., Kumar, S., Yick, S., Fang, J., & Ostrikov, K. (2013). *Large Arrays and Networks of Carbon Nanotubes: Morphology Control by Process Parameters*. Intech.

Li, Q. & Li, L. (2012). *Selective separation of single-walled carbon nanotubes in solutions*. Academic Press.

Liu, B., Akis, R., & Ferry, D. K. (2014). Conductance fluctuations in graphene nanoribbons. *Journal of Computational Electronics*, *13*(4), 950–959. doi:10.1007/s10825-014-0613-9

Louie, S. G. (2001). *Carbon Nanotubes*. Berlin: Springer.

Malic, E., & Knorr, A. (2013). *Graphene and Carbon Nanotubes: Ultrafast Relaxation Dynamics and Optics*. Berlin: Wiley-VCH. doi:10.1002/9783527658749

Moradi, A., & Khosravi, H. (2007). Plasmon dispersion in metallic carbon nanotubes in the presence of of low-frequency electromagnetic radiation. *Physics Letters. [Part A]*, *371*(1-2), 1–6. doi:10.1016/j.physleta.2007.05.110

Mowbray, D. J. (2007). *Hydrodynamic Modelling of the Electronic Response of Carbon Nanotubes* (PhD Thesis). Univ. of Waterloo, Ontario, Canada.

Neophytou, N., Paul, A., Lundstrom, M., & Klimeck, G. (2008). Bandstructure effects in silicon nanowire electron transport. *IEEE Transactions on Electron Devices*, *55*(6), 1286–1297. doi:10.1109/TED.2008.920233

Pedersen, T.G. (2004). Exciton effects in carbon nanotubes. *Carbon Nanotubes, 42*, 1007–1010.

Perebeinos, V., Tersoff, J., & Avouris, P. (2005). Electron–phonon interaction and transport in semiconducting carbon nanotubes. *Physical Review Letters*, *94*(8), 086802. doi:10.1103/PhysRevLett.94.086802 PMID:15783915

Rafii-Tabar, H. (2008). *Computational Physics of Carbon Nanotubes*. Cambridge Press.

Rao, C. N. R., & Govindaraj, A. (2011). *Nanotubes and Nanowires* (2nd ed.). Royal Society of Chemistry.

Regis, E. (1995). *The Emerging Science of Nanotechnology*. Little Brown.

Reich, S., & Thomsen, C. (2000). Chirality dependence of the density-of-states singularities in carbon nanotubes. *Physical Review B: Condensed Matter and Materials Physics*, *62*(7), 4273–4276. doi:10.1103/PhysRevB.62.4273

Reich, S., Thomsen, C., & Ordejón, P. (2004). Electronic band structure of isolated and bundled carbon nanotubes. *Physical Review B: Condensed Matter and Materials Physics*, *65*(15), 155411. doi:10.1103/PhysRevB.65.155411

Ryu, H., Kim, J., & Hong, K. H. (2015). Atomistic study on dopant-distributions in realistically sized, highly p-doped silicon nanowires. *Nano Letters*, *15*(1), 450–456. doi:10.1021/nl503770z PMID:25555203

Sahoo, S., Kontos, T., Schonenberger, C. & Surgers, C. (2005). Electrical spin injection in multiwall carbon nanotubes with transparent ferromagnetic contacts. *Appl. Phys. Lett.*, *86*(478).

Salahuddin, S., Lundstrom, M., & Datta, S. (2005). Transport effects on signal propagation in quantum wires. *IEEE Transactions on Electron Devices*, *52*(8), 1734–1742. doi:10.1109/TED.2005.852170

Sanvito, S., Kwon, Y.-K., Tomnek, D., & Lambert, C. J. (2000). Fractional Quantum Conductance in Carbon Nanotubes. *Physical Review Letters*, *84*(9), 1974–1977. doi:10.1103/PhysRevLett.84.1974 PMID:11017674

M. Saraniti, & U. Ravaioli (Eds.). (2006). Nonequilibrium Carrier Dynamics in Semiconductors. *Proceedings of the 14th Int. Conf.* Springer.

Sattler, K. D. (2010). *Handbook of Nanophysics: Nanotubes and Nanowires*. CRC.

Schroter, M., Claus, M., Sakalas, P., & Wang, D. (2013). Carbon Nanotube FET Technology for Radio-Frequency Electronics. *State-of-the-Art Overview IEEE Journal of the Electron Devices Society*, *1*(1), 9–20. doi:10.1109/JEDS.2013.2244641

Sohn, L.L., Kouwenhoven, L.P., & SchÄon, G. (Eds.). (1997). *Mesoscopic Electron Transport*. Kluwer.

Somani, P. R., & Umeno, M. (2007). Carbon Nanotube: Multifunctional Material. Applied Science Innovations Ltd.

Spataru, C. D., Beigi, S. I., Benedict, L. X., & Louie, S. G. (2004). Excitonic Effects & Optical Spectra of Single-wall Carbon Nanotubes. *AIP Conference Proceedings*, *772*, 1061–1062. doi:10.1063/1.1994478

Tans, S. J., Devoret, M. H., Dai, H. J., Thess, A., Smalley, R. E., Geerligs, L. J., & Dekker, C. (1997). Individual single-wall carbon nanotubes as quantum wires. *Nature*, *386*(455), 474–477. doi:10.1038/386474a0

Tans, S. J., Verschueren, S., & Dekker, C. (1998). Room temperature transistor based on a single carbon nanotube. *Nature, 393*(6680), 49–52. doi:10.1038/29954

Tománek, D. (2014). *Guide through the Nanocarbon Jungle: Nanotubes, Graphene, and Beyond*. Morgan & Claypool in collaboration with IOP Science.

Tombros, N. (2008). *Electron spin transport in graphene and carbon nanotubes* (PhD Thesis). Zernike Institute for Advanced Materials.

Tran, H. N., Blancon, J.-C., Huntzinger, J.-R., Arenal, R., Popov, V. N., Zahab, A. A., & Paillet, M. et al. (2016). Excitonic optical transitions characterized by Raman excitation profiles in single-walled carbon nanotubes. *Physical Review B: Condensed Matter and Materials Physics, 94*(7), 075430. doi:10.1103/PhysRevB.94.075430

Tsukagoshi, K., Alphenaar, B. W., & Ago, H. (1999). Coherent transport of electron spin in ferromagnetic contacted carbon nanotube. *Nature, 401*(478), 572–574.

Wallace, P. R. (1947). The Band Theory of Graphite. *Physical Review, 71*(9), 622–634. doi:10.1103/PhysRev.71.622

Wang, L., Guo, W., & Hu, H. (2008). Group velocity of wave propagation in carbon nanotubes. *Proc. R. Soc., 464*, 1423–1438. doi:10.1098/rspa.2007.0349

Zheng, Y., Rivas, C., Lake, R., Alam, K., Boykin, T. B., & Klimeck, G. (2005). Electronic Properties of Silicon Nanowires. *IEEE Trans. Electron Devices, 52*(6), 1097-1103. Retrieved from http://docs.lib.purdue.edu/nanodocs/117

Zhou, J., Park, J.Y., Huang, S.M., Liu, J. & McEuen, P.L. (2005). Band structure, phonon scattering, and the performance limit of single-walled carbon nanotube transistors. *Phys. Rev. Lett., 95*.

Zhou, W., Sasaki, T., Nakamura, D., Saito, H., Liu, H., Kataura, H., & Takeyama, S. (2013). Survey of exciton-phonon sidebands by magneto-optical spectroscopy using (6,5) single-walled carbon nanotubes. *Applied Physics Letters, 103*(2), 021117. doi:10.1063/1.4813544

ENDNOTES

1 The γ_o parameter is sometimes called the transfer energy or the hopping energy between nearest-neighbor carbon atoms.

2 The conductance fluctuations in a semiconductor at low temperature may arise from the long-range disorder potential induced by random impurities and these fluctuations can be lifted by the application of a magnetic field (Liu, Akis and Ferry, 2014). The main sources of disorder in nanotubes are localized lattice defects, such as vacancies, substitutions and pentagon-heptagon defects, in addition to electrostatic potential fluctuations and mechanical deformations

Chapter 7
Phonon Transport and Heat Flow

1. OVERVIEW AND CHAPTER OBJECTIVES

The transport of heat energy is an old topic in physics, dated back to Joseph Fourier (1822). Different kinds of heat carriers have been proposed to explain this process, namely: molecules, electrons, phonons and photons. In this chapter, we concentrate our attention on the heat transport by phonons.

Phonons are quantized vibrations of the crystal lattice atoms. They are like electrons or photons (light particles), in that they carry energy. Phonons are actually the main heat carriers in semiconductors and insulators. Heat flows from the hot side to the cold side of a system whenever it is not in thermal equilibrium and temperature distribution is not uniform. Researchers are trying to answer fundamental questions about how phonons transport heat.

The energy transport and exchange between electrons and phonons in a crystalline solid is a vital topic in solid state physics. The interest of this topic has been raised by the miniaturization of electronic devices, where the heat removal imposes many restrictions on the development of semiconductor nanodevices. In fact, the heat transport and evacuation is related to the performance and the reliability issues, in a wide class of semiconductors devices, including MOSFET's and lasers. In addition, the thermoelectric applications motivate the studies of thermal transport in nanostructures and the emerging phononic nanodevices.

For long time, the heat transfer by conduction, in solid-state devices has been studied on the basis of Fourier's diffusion model ($Q=-k^{th}\nabla T$). Actually, the heat is carried predominantly by phonons in dielectrics and semiconductors. The relationship between the phonon mean free path of heat transfer and the device length scale determines whether thermal transport follows the classical thermal diffusion model or not. It is found that the optical and high frequency acoustic phonons provide the main contributions to the specific heat of bulk materials, where the mean free path of phonons is much smaller than the device length scale. However, for any length scale there will be some phonons of low enough frequency that propagate ballistically (without scattering) rather than diffusively (with scattering).

DOI: 10.4018/978-1-5225-2312-3.ch007

After a review of the basic physics and thermodynamics of heat and phonons we present the phonon transport theory. The phonon transport parameters such as thermal conductivity are introduced from the atomic-level properties using semiclassical and quantum approaches. We present the semiclassical and quantum schemes as well as phenomenological models, and examine their range of applications. In particular, we discuss the validity of different models of thermal conductivity for nanostructures and nanodevices. Figure 1 recapitulates the hierarchy of phonon transport approaches, which have been presented so far in the literature. Microscopic approaches may be quantum, semiclassical or based on molecular dynamics (MD). The microscopic semiclassical models, such as the Peierls-Boltzmann transport equation (phonon BTE) and Lattice Monte-Carlo simulation, can capture quasi-ballistic phonon transport.

Upon completion of this Chapter, the readers and students will:

- Review the notions of lattice temperature and heat transfer mechanisms
- Review the basics of lattice vibrations and phonon waves
- Describe the phonon scattering mechanisms and how they affect the heat conductivity of a semiconductor or insulator
- Understand the phonon transport mechanisms (diffusive, hydrodynamic and ballistic), and how they contribute to the thermal properties of solids
- Understand and know how and when to employ the phonon transport models, with molecular dynamics (MD) methods as well as semiclassical (phonon BTE) and quantum approaches (NEGF).
- Understand the phonon transport in low-dimensional structures, such as carbon nanotubes and graphene ribbons
- Understand and appreciate the concepts of the emerging phononic devices.

2. CONCEPT OF HEAT AND LATTICE TEMPERATURE

Heat may be defined as energy in transit from a high temperature object to a lower temperature object. In kinetic theory of gases, the temperature is considered as average kinetic energy of the chaotic motion

Figure 1. Hierarchy of phonon and heat transport models

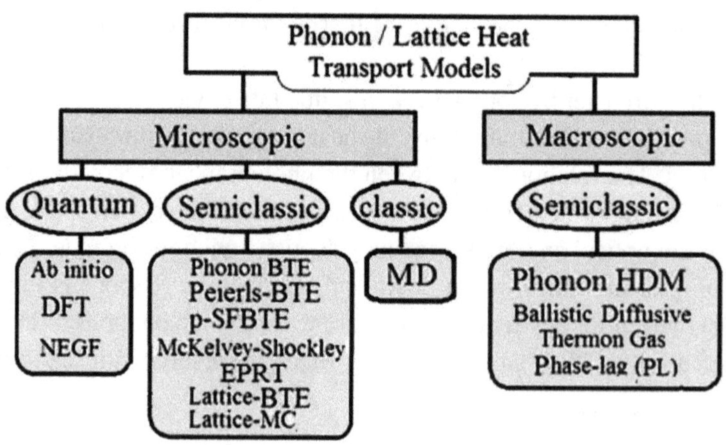

of molecules. The temperature of a gas of particles is then a quantity proportional to the average energy per particle at thermal equilibrium[1],

$$k_B T \propto \overline{E_i} \tag{1}$$

Similarly, the lattice temperature at thermal equilibrium is equal to the temperature of the gas of phonons filling the space between atoms. However, to describe a phonon system as a phonon gas we consider phonon as weakly interacting quasi particles, associated with wave packets, the size of which is much greater than the wavelength, λ, of concerned phonons.

2.1 Heat Transfer Methods

Heat is transferred to and from matter by the principal energy carriers, namely: phonons (lattice vibrations), electrons, photons and fluid molecules. In thermodynamics, heat transfer is usually classified as conduction, convection, and radiation, depending on the medium and the mechanism of heat transport across the medium. Conduction usually refers to heat transfer in solids, in which atoms and molecules are practically immobile. In such cases, heat is carried by vibration of atoms (phonons) and other mobile energy carriers such as electrons and holes. The convection mechanism expresses the heat transfer from a solid surface to fluids, mediated by diffusion or advection. When heat is carried by electromagnetic waves (photons), the heat transfer is called radiation. Unlike conduction and convection, no medium is required for radiation.

2.2 Heat Capacity of a Lattice

According to thermodynamics, when the lattice is held at a non-zero temperature its internal energy fluctuates about some mean value. The energy fluctuations are caused by lattice vibrations, which can be viewed as a gas of phonons. The temperature of a crystal lattice at thermal equilibrium is thus proportional to the average energy of the gas of phonons[2].

The heat capacity of a substance is the amount of heat required to change its temperature by one degree, and has units of energy per degree. The heat capacity of a gas of particles may be defined at

Figure 2. Conventional heat transfer methods

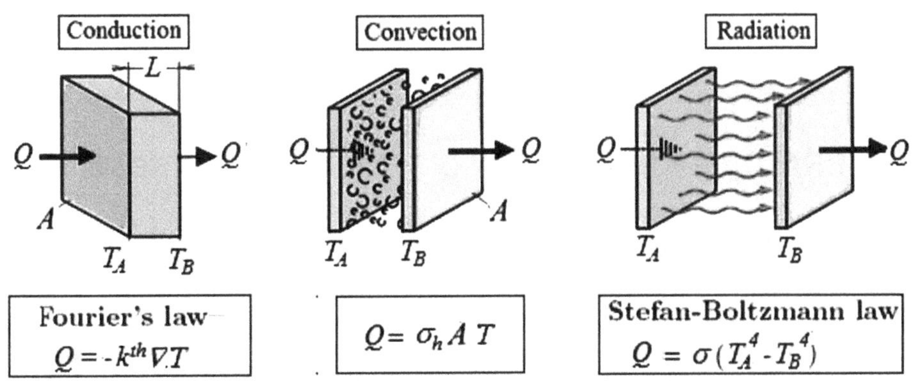

constant volume (C_v) or at constant pressure (C_p). However, the heat capacity of a solid is usually defined at a constant volume[3]. It is defined as the rate at which the total energy of the lattice system (U) increases as the temperature is raised:

$$C_v = \left[\frac{\partial U}{\partial T}\right], v = constant \tag{2a}$$

Therefore, the heat capacity is the solids ability to absorb and retain heat. The so-called *heat capacity per unit mole* or *molar heat ca*pacity is defined as ($c_v = \eta^{-1}\partial U/\partial T$). For one mole ($\eta = 1$), we have $C_v = c_v$.

According to the classical theory of gases $c_V = 3Nk_B$, which is a constant independent of temperature. Note that N is the total number of atoms, $3N$ is the number of crystal vibration (phonon) modes and k_B is the Boltzmann constant. This is known as the Dulong-Petit law. However, in solids this is only true above ambient temperatures. At lower temperatures the specific heat doesn't follow the Dulong-Petit law. Actually, the heat capacity may be due to other energy carriers in addition to phonons. For instance, in semiconductors, we should also consider the energy of conduction electrons. Therefore, the heat capacity may be expressed as follows:

$$C_v = C_v\,(phonons) + C_v\,(electrons) \tag{2a}$$

The experiment shows that $C_v(T)$ decreases as T decreases below ambient temperature according to the law:

$$C_v = a\,T^3 + b\,T \tag{2b}$$

The first term is due to phonons and the other is due to the contribution of electrons. Hence, we gave to apply the statistics of these quantum particles in order to interpret the correct behavior of C_v with temperature.

There exist two famous models which interpret the variation of the heat capacity at low temperature; namely: the Debye model (1912) and the Einstein model (1907). Figure 3 depicts the normalized specific heat, versus normalized temperature (T/T_D), according to Debye and Einstein models. The Debye temperature (T_D) is defined as follows (Vapaille, 1970):

$$T_D = E_m / k_B = h\omega_m / k_B \tag{3}$$

where E_m is known as the Raman (inelastic) phonon energy.

3. LATTICE VIBRATIONS AND PHONON DISPERSION RELATIONS

Heat conduction in crystalline solids (semiconductors and insulates) occurs primarily through lattice vibrations[4]. The crystal lattice of a solid may be viewed as a three dimensional structure of atoms held together by atomic bonds. The interatomic bonds act like springs that stretch and compress repeatedly about their equilibrium positions under the effect temperature.

Figure 3. Normalized specific heat ($c_v/3Nk_B$) versus normalized temperature (T/T_D)
After Glassenbenner (1964).

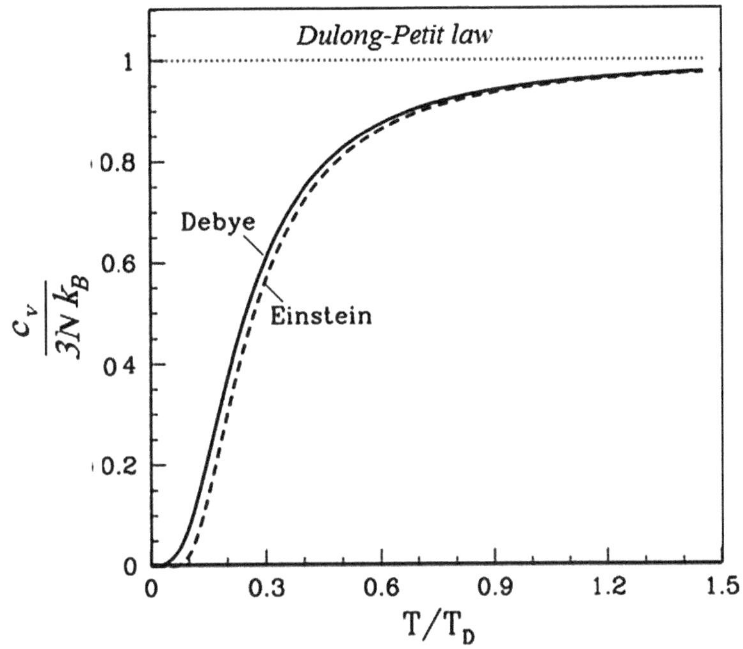

For the matter of illustration, the equations of motion of lattice atoms are presented here for a monoatomic chain, but it can be extended to the general case of polyatomic chains. Lattice vibrations are usually simplified as harmonic oscillators. The harmonic approximation is expressed as a Taylor expansion of the potential energy in terms of atomic displacements u, truncated after the second order terms (Ziman, 1972):

$$U = U_0 + \frac{1}{2}\sum_{i,j}\sum_{u,v}\frac{\partial^2 U}{\partial u_i\left(\boldsymbol{R}_\mu\right)\partial u_j\left(\boldsymbol{R}_\nu\right)} u_i\left(\boldsymbol{R}_\mu\right)u_j\left(\boldsymbol{R}_\nu\right) + \ldots \tag{4a}$$

Here $R_u \ldots R_v$ are the ideal atomic positions in the crystal lattice (at 0K!). The second derivatives of the potential energy U with respect to displacements $u_i \ldots u_j$ are called the force or stiffness constants ($C_{ij}^{uv} = \partial^2 U \partial u_i \partial u_j$)[5]. The stiffness matrix relates the acting stress ($\tau = \Delta F/\Delta A$) and the resultant strain ($\varepsilon = \Delta L/L$) in a solid as follows: $[\tau] = [C][\varepsilon]$.

Actually, $[C]$ is a fourth-order stiffness tensor of 81 elements, but due to crystal symmetry the number of coefficients can be reduced. For instance, the stiffness constant tensor has only 3 constants in cubic crystals (e.g., Si), which are termed C_{11}, C_{12}, C_{44}. The stiffness matrix elements can be obtained by the linear response or the small displacement methods (Kong, 2011). In these methods the crystal is considered in its ground state and the displacements of atoms are introduced, either by the perturbation theory or the supercell approach. The corresponding equation of motion of the ith atom in one dimensional chain is given by (Zimann, 1972):

Figure 4. Illustration of the interatomic forces in a 3-D lattice. The bottom figure illustrates the displacements of a 1-dimensional lattice with a monatomic basis.

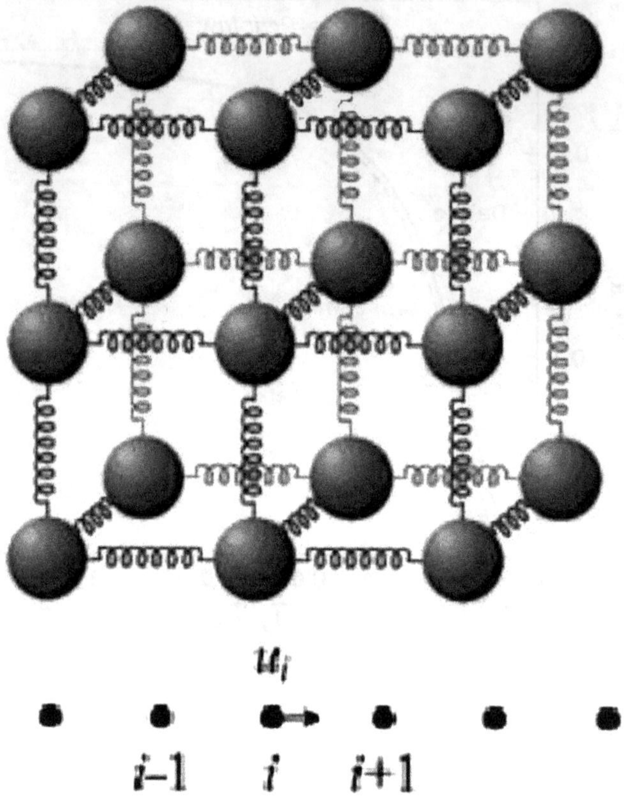

$$F_i = \Sigma_j C_{ij} . u_j \tag{4b}$$

After solving the above set of difference equations, we get the displacements u_i of each atom around its equilibrium position. The solution may be written in the following form of progressive waves:

$$u_i = A.exp[\, j\,(k\,n\,a - \omega\,t)\,] \tag{4c}$$

where $j = \sqrt{-1}$, ω is the angular frequency of vibrations, A is its amplitude and **k** is the wave vector. One can easily obtain the relation between ω and k by substituting equation (4c) into (4b). The result is that atoms vibrate together, and this collective vibration spreads throughout the crystal lattice.

3.1 Concept of Phonons

When an atom is vibrating around its ideal lattice position, the vibrational waves are associated with quasi particles called phonons. These phonons carry energy while propagating through the crystal. They propagate with certain frequency (ω), wave vector ($\boldsymbol{k}$) and polarization ($\boldsymbol{P}$). The quantum of energy contained by a phonon with frequency ω is $E_p=\hbar\omega$ and its momentum is $\boldsymbol{p}=\hbar\boldsymbol{k}$, where $\hbar$ is Planck's constant

Figure 5. Dispersion relation (ω -k diagram) of a 1-dimensional monatomic crystal

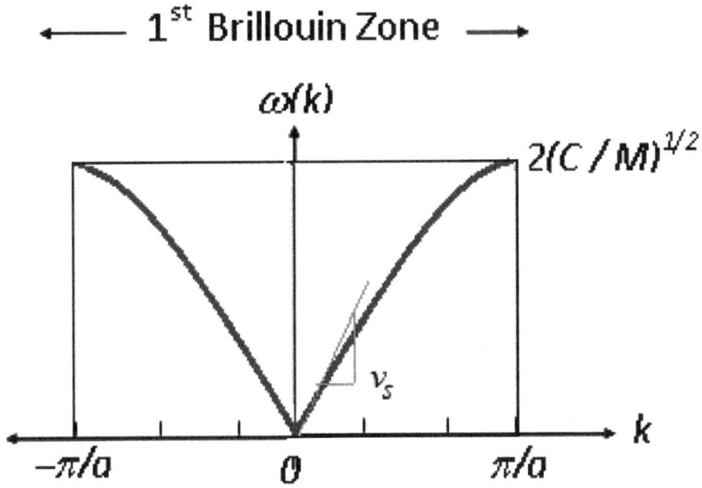

divided by 2π. Therefore, we treat phonons as having wavevector k, although this is not strictly correct. Actually $\hbar k$ is rather the crystal momentum (Ashcroft and Mermin, 1976). The thermodynamic properties of a solid material are closely related to its phonon spectrum. This is illustrated in the following sections.

3.2 Phonon Dispersion Relation

The displacement of atoms from their ideal positions in a crystal gives rise to a set of vibration waves propagating through the lattice. Every type of lattice vibration has its own characteristic frequencies of vibration, called *modes,* and the overall collective vibrational motion of the lattice is a superposition of many modes.

The $(\omega\text{-}k)$ relation, which describe the propagation of lattice waves, is called the material dispersion relation. The vibration modes are called normal (transversal) when the atomic planes vibrate in a direction perpendicular to the wave vector k (the direction of wave propagation). Alternatively, when the atomic planes vibrate in a direction parallel to the wave vector k, the modes of vibration are called longitudinal. The longitudinal modes are sometimes called *compression modes* because the vibrating atomic planes are stacked (or compressed) and released (decompressed) in the direction of wave vector. Also, the transverse modes are sometimes called *shear modes* because the vibrating atomic planes are slipped (as if they were subjected to shear stress) with respect to each other.

The $\omega(k)$ relation in real crystals has several branches, as shown in Figure 7. The upper branches are called optical modes and the lower branches are called acoustic modes. Some of optical branches are longitudinal (LO) and some are transverse (TO). Similarly, some of acoustic branches are longitudinal (LA) and some are transverse (TA).Generally speaking, a 3-D crystal with P atoms per unit cell, has 3P vibration modes, distributed as follows:

- *3* acoustic modes with the three directions of polarization,
- *3P-3* optical modes, from which *P-1* longitudinal & *2P-2* transverse modes.

Figure 6. Measured phonon dispersion relation in GaAs & Si at 300K
After Dollong (1963).

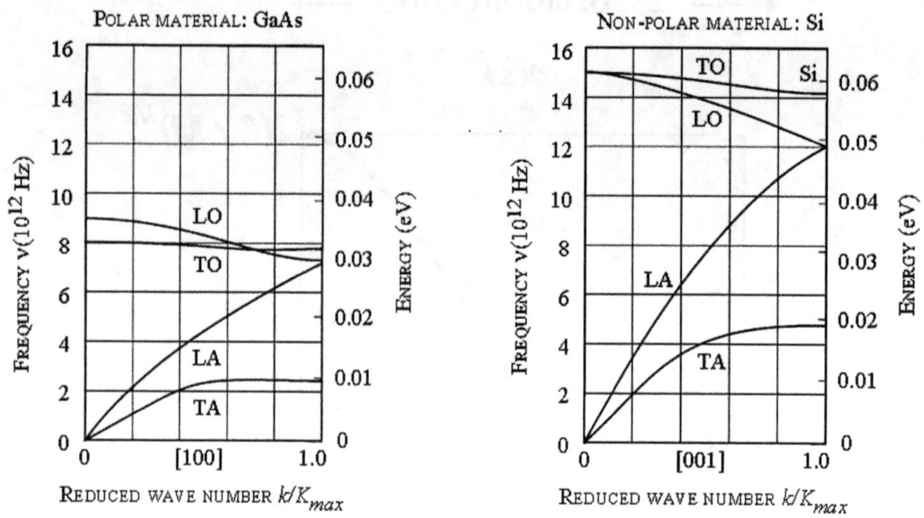

For example, a 3-dimensional lattice with one atom per unit cell, such as simple, body-centered, only three acoustic modes exist. For a 3-D lattice with two atoms per unit cell such as *Ge*, *Si* and *GaAs*, there exist three acoustic modes (one longitudinal acoustic *LA*, and two transverse acoustic *TA*) and three optical modes (one longitudinal optical *LO*, and two transverse optical *TO*). Note that all acoustic branches, are almost linear near $k=0$. In this case, the group velocity ($v_g = d\omega/dk$) is equal to the phase velocity ($v_p=\omega/k$) and both of them is equal to the sound velocity in the material (v_s).

3.3 Strain Effects

Stress inevitably arises in electronic devices due to mismatch in coefficients of thermal expansion, mismatch in lattice constants, and growth of different materials. Sometimes, stress is deliberately introduced to improve the electrical and optical device properties. For instance, the technology of strained silicon devices is utilized to increase the carrier drift mobility. Figure 7 depicts the phonon dispersion relation of both relaxed and strained *Si*, with 1 GPa of applied pressure (force).

The most effective way to induce tensile stress in semiconductors is through hetero-epitaxial growth on a substrate material that has larger lattice spacing (like *Si* on *SiGe*). As shown in Figure 8, if the *Si* is grown on *SiGe* below a critical thickness, it becomes strained with lattice symmetry change from cubic to tetragonal.

As we pointed out in Chapter 1, the application of strain on a semi-conductor shifts the energy levels of the conduction and valence bands and can remove the band degeneracy, when applied in certain directions. According to the deformation potential theory, which has been extended to strained semiconductors by Bir and Pikus (1974), the energy shift of a band extremum is expanded in terms of the components of the strain tensor ε_{ij}

Figure 7. Phonons dispersion curves for stressed Si under a pressure of 8.6 GPa

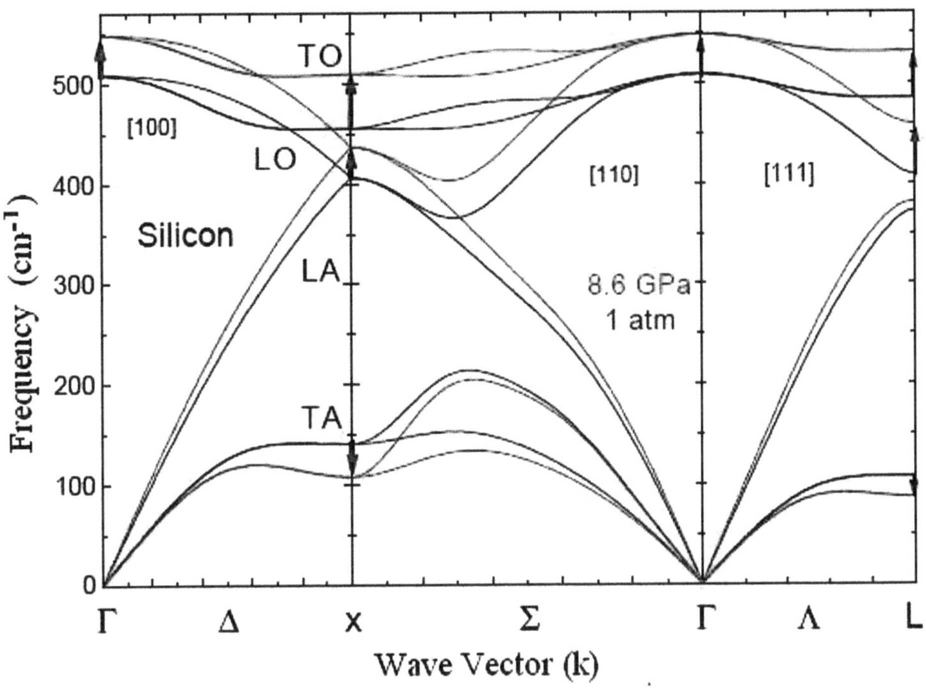

Figure 8. Epitaxial growth on a substrate with different lattice constant. For Si-on-SiGe, the grown Si epilayer deforms (strain) such that $a_{\perp} = a_o - 2(C_{11}/C_{12}).(a_s - a_o)$

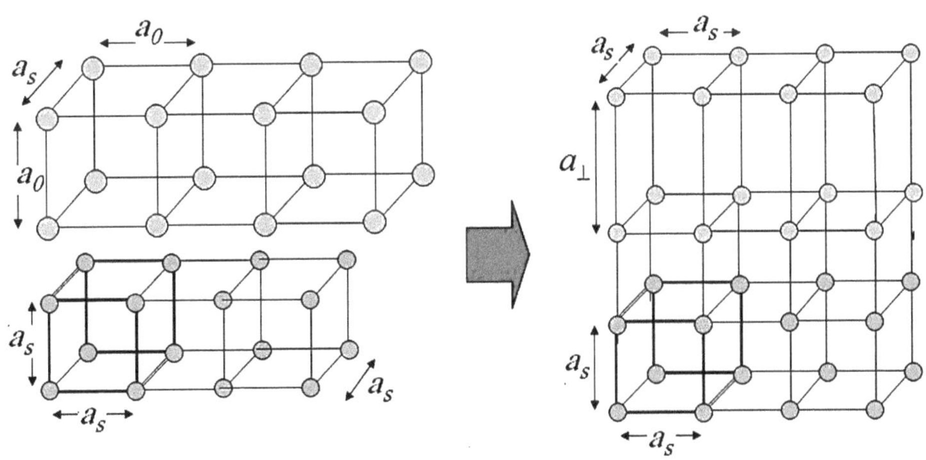

$$\Delta E = \Sigma_{i,j}\, \Xi_{i,j}\, \varepsilon_{i,j} \tag{5}$$

where $\Sigma_{i,j}$ is the deformation potential tensor of this band.

3.4 Phonon Energy Distribution

As phonons are indistinguishable (like electrons in their clouds), one has to recall quantum statistics describe their energy distribution. However, the number of phonons whose energies are equal (in a certain energy state) is not limited, which is not the case for electrons. Phonons are belonging to the family of *Bosons,* which obeys the statistics of Bose-Einstein at thermal equilibrium. The probability that a phonon can possess the energy $\hbar\omega$, at thermal equilibrium is given by (Kittle, 1980):

$$f_{BE}\left(E_i\right) = \frac{1}{A \cdot \exp\left(\frac{\hbar\omega}{k_B T}\right) - 1} \tag{6}$$

where A is a constant. At low temperatures, phonons can behave very differently than electrons and an unlimited number of them can condensate into one energy state. When $A = 1$, the occupation for low energy phonons is excessively high.

3.5 Phonon Density of States

The density of states (DOS) of phonons is the density of the normal modes of phonons in the frequency range ω to $\omega + d\omega$, divided by volume the crystal V. The phonon DOS is used to determine the total number of phonons N_p as well as the phonon-related properties. The density of normal modes for three-dimensional cubic lattices is given by the following relation for both acoustic and optical phonons (Kittle & Kroemer):

$$g_{ph}\left(\omega\right) = \frac{1}{\left(2\pi\right)^3} \sum_i \frac{4\pi k^2}{(d\omega_i(k) / dk)} \tag{7}$$

Obviously, the DOS is obtained, from the phonon dispersion relation $\omega(k)$, as shown in the Figure 9. The phonon DOS plots of different material can be found in the literature, e.g., Dollog (1999) or Brockhouse (2000).

Note that the phonon number depends on temperature and the vibration frequency. In the Debye limit (when $\omega = \upsilon.k$), the phonon density-of-states can be written as $g_{ph}(\omega) = a^3(\omega^2/2\pi^2\upsilon^3)$, where a is the lattice constant.

In order to understand heat conduction by phonons in crystalline dielectrics and semiconductors, it is important to realize that phonons do not carry same amounts of heat due to different polarizations (longitudinal or transverse) and their broad frequencies distribution. At high temperatures, most phonon modes are excited, but heat is carried mainly by a small fraction of these excited modes (long-wavelength

Figure 9. Phonon dispersion relation and density of states in Si, Ge and GaAs
After Henry & Chen, (2008).

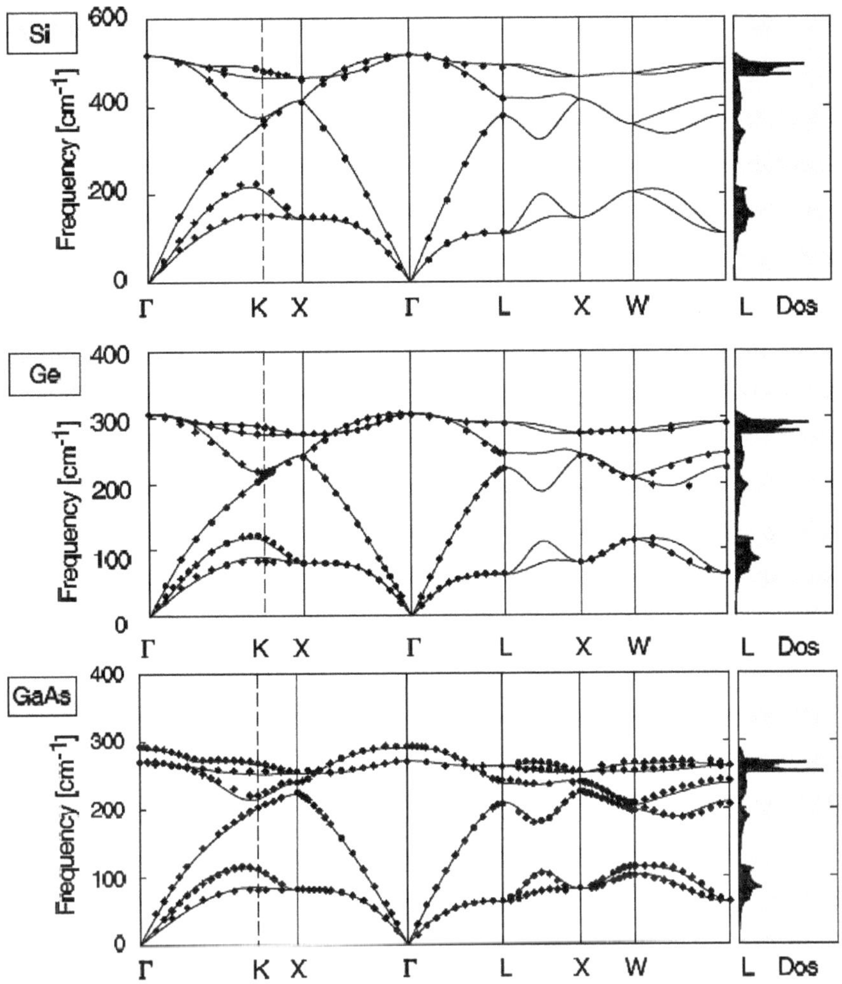

acoustic LA phonons). Acoustic phonons near the edge and the center of the Brillouin zone, and optical phonons are inefficient in carrying heat due to their low group velocity or heat capacity.

Note 1: Anharmonic Oscillations of Crystal Lattice

Note that in the above simple analysis, we neglected the higher order terms in the Taylor expansion of the lattice potential energy. This approximation results in a linear relation between force and displacements ($F = - C.u$). and sometimes called the harmonic approximation of lattice vibrations. However, if we considered higher order terms of the Taylor expansion of the crystal potential (at least the 3rd order term) we'd get nonlinear relations of force with atom displacements ($F = - C.u + \gamma.u^2$). The inclusion of nonlinear terms would reveal the anharmonic oscillations of the crystal lattice. It is amazing that, without these anharmonic oscillation terms, one cannot interpret properly some important phenomena, such as the thermal expansion of solids. This is because the phonon frequency does not depend on the amplitude of the oscillations and hence the pressure of a gas of harmonic phonons is independent of temperature. However, by introducing anharmonic terms, the phonon gas acquires a temperature-dependent pressure, which gives rise to the thermal expansion phenomena in a solid crystal lattice. However, the pressure of the electron gas should be also considered in the case of metals. The anharmonic phonons are also important in order to derive a correct expression of the thermal conductivity, with proper temperature dependence, especially in perfect crystal insulators. The only way to explain the experimental data is to admit that phonons can be scattered by other phonons, and this can only occur by including anharmonic phonons.

3.6 Phonon Scattering Mechanisms

In crystals, phonons are scattered by imperfections such as impurities and grain boundaries. The scattering of phonons is a main source of resistance for heat conduction (especially in non-metallic solids). The phonon scattering mechanisms include several types, among them one can cite:

- Phonon–phonon (*ph–ph*),
- Phonon–electron (*ph–e*),
- Phonon–impurity (*ph–i*), and
- Phonon–boundary (*ph–b*).
- Phonon-anharmonicity (three-phonon) scattering

The phonon scattering mechanisms can be also grouped into two categories:

- Normal (**N**) process and
- Resistive (**R**) process.

The energy of phonons is always conserved in any scattering process. However, ***N***-process conserves phonon momentum while ***R***-process does not. The ***R***-process includes three-phonon Umklapp process and phonon-boundary scatterings. The number of phonons is not conserved here, as they can be created or annihilated. For instance, in three-phonon scattering two phonons combine into one or one divides into two.

Each scattering mechanism can be characterized by a relaxation time (τ_i) and mean free path (λ_i), between collisions. The total scattering rate may be calculated using the Matissian rule:

$$\Gamma(\omega) = \tau^{-1}(total) = \Sigma_i \tau_i^{-1} \tag{8}$$

Phonon scattering rates are crucial to specify the collision term in carrier transport equations.

Phonon-Phonon Scattering

Like electron-phonon scattering, the phonon-phonon mechanism may be effectuated by normal process (conserving phonon momentum) or Umklap process (which involves translation to another Brillouin zone in the k-space and doesn't conserve phonon momentum). In the normal processes, the scattering rate $(\Gamma_i = 1/\tau_i)$ vary linearly with fequency (ω) while Umklapp processes vary with ω^2, as follows (Gurevich, 1986):

$$\tau_{t}^{-1}\,(Umklapp) = 2\gamma^2 .\omega^2 k_B T/(\mu.\omega_D.V_o) \tag{9}$$

where ω_D is Debye frequency, γ is Gruneisen anharmonicity parameter, μ is shear modulus and V_0 is volume per atom. The Grüneisen parameter is deduced from Grüneisen's law of thermal expansion and is a direct measure of anharmonicity of bonds (Grüneisen, 1912). Therefore, the Umklapp (U) scattering dominates at high frequency. Note that only Umklapp processes (in which the momentum of interacting particles is not conserved) are participating in the heat conduction of a solid crystal at high temperatures.

Phonon-Anharmonicty Scattering

The anharmonic oscillations of crystal lattice (anharmonic phonons) can be found out when we take the higher-order terms in the potential energy expansion of crystal lattice. Anharmonicity associated with finite temperature causes the normal modes of phonons to interact. In the phonon anharmonic interactions, a phonon decays into other phonons._

Three-Phonon Scattering

In the three-phonon interaction there are two types of allowed events. In a type I interaction, one phonon decays into two others. In a type II interaction, two phonons combine to form a third. In order to satisfy conservation of energy, processes in which three phonons are either created or destroyed are not allowed. There exist two selection rules for the allowed phonon interactions. First, from the translational invariance of the lattice potential energy, the wave vectors of the phonon modes in question must satisfy the following conservation rules:

$$k_1 = k_2 + k_3 + G\ (type\ I) \tag{10a}$$

$$k_1 + k_2 = k_3 + G\ (type\ II) \tag{10b}$$

where $\boldsymbol{G}$ is either equal to zero (corresponding to a Normal (**N**) process) or a reciprocal lattice vector (corresponding to an Umklapp (**U**) process. These criteria are valid in both the classical and quantum descriptions of the phonon system (Born, 1954).

The second selection rule, which only applies to the quantum system, is based on conservation of energy. For the type I and II interactions, conservation of energy then leads to

$$\hbar\omega_1 = \hbar\omega_2 + \hbar\omega_3\ (type\ I) \tag{11a}$$

$$\hbar\omega_1 + \hbar\omega_2 = \hbar\omega_3\ (type\ II) \tag{11b}$$

It should be noted that it is the anharmonic frequencies that should be used here. According to Herring (1954), the 3-phonons scattering due to longitudinal phonons under momentum conservation conditions at low temperatures is given by:

$$\tau_t^{-1}\,(3\ phonons) = A.T^3\omega^2 \tag{12}$$

where A is a fitting parameter which depends on the Gruneisen an-harmonicity parameter and the phonon velocity. Note that, the three-phonon normal process conserves phonon momentum and hence does not contribute to thermal resistance.

Phonon-Boundary Scattering

The phonon-boundary scattering is particularly important for low-dimensional semiconductors and nanostructures and its relaxation time is given by (Ju & Goodson, 1999):

$$\tau_t^{-1}\,(Boundary) = v_s(1-b)/L \tag{13}$$

where v_s is the average speed of sound, L is the characteristic length of the system and b is the surface roughness parameter. When $b=0$ we have a very rough surface and the transport is pure diffusive (the Casimir limit).

Phonon-Electron Scattering

When the material is lightly doped, the phonon-electron scattering processes is important. The electron–phonon scattering rates (for both the intravalley acoustic and the intervalley acoustic and optical scattering) are already treated in the context of charge-carrier transport in Chapter 2. They can be summarized as follows (Srivastava, 1990):

$$\Gamma_i = \frac{D_{a,s}^2 m_d}{4\pi p\hbar^2 k_s} \int_q \frac{1}{\omega_{q,s}} \left(N_{q,s} + \frac{1}{2} \mp \frac{1}{2} \right) I_q^2 q^3 dq \tag{14a}$$

$$\Gamma_{if} = \frac{\pi \Delta_{if}^2 Z_f}{2\rho\omega_{q,s}} \left(N_{q,s} + \frac{1}{2} \mp \frac{1}{2} \right) g_{df} \left(E_k \pm \hbar\omega_{q,s} \right) \tag{14b}$$

Here, $N_{q,s} = 1/[\exp(\hbar\omega_{q,s}/k_B T) - 1]$ is the equilibrium phonon distribution, which describes the average occupation of the phonon mode $\omega(q,s)$, with q and s represent the phonon momentum and polarization. The upper and lower signs ± correspond to phonon absorption and emission processes, respectively. Also, I_q is the overlap integral and g_{df} is the density of states. The values of intravalley and intervalley deformation potentials (D_a and Δ_{if}) are shown in Table 1. Other parameters have their usual meaning.

Phonon-Impurity Scattering

The phonon or host atom vibrations may be scattered on impurities with different mass. This is called *mass-difference impurity scattering* and given by (Asheghi, et al, 2002):

Table 1. The effective intravalley and intervalley deformation potentials D_a and Δ_{if} In Si

Electron Transport Parameters							
Δ_f^{TA} [eV/cm]	$\Delta_f^{LA/LO}$ [eV/cm]	Δ_f^{TO} [eV/cm]	Δ_g^{TA} [eV/cm]	Δ_g^{LA} [eV/cm]	Δ_g^{LO} [eV/cm]	D_{LA} [eV]	D_{TA} [eV]
$0.5x10^8$	$3.5x10^8$	$1.5x10^8$	$0.3x10^8$	$1.5x10^8$	$6.0x10^8$	6.39	3.01

Phonon Transport Parameters						
$A_{N,LA}$ [s-K^{-3}]	$A_{N,TA}$ [K^{-4}]	$A_{U,TA}$ [s^{-1}]	$A_{\delta M}+A_{\delta R}$ [s^3] n_0=1x10^{20} cm^{-3}	ω_1 [THz]	ω_2 [THz]	τ_{op} [ps]
$2.0x10^{-24}$	$9.3x10^{-13}$	$5.5x10^{-18}$	$1.3x10^{-44}(n/n_0)$	23.56	27.49	0.5-10

After Ju & Goodson (1999).

$$\tau_i^{-1}\,(Impurity) = V_o N_i\,\omega^2/4\pi v_g^{\,2} \tag{15}$$

where N_i is a measure of the impurity scattering strength and the group velocity $v_g = d\omega/dk$, can be obtained from the phonon dispersion relation._

3.7 Phonon Lifetime and Mean Free Path

Normally, the scattering process is quantified by the relaxation time (τ) or the mean-free-path (λ) of phonons. The mean-free-path (λ) of phonons, is the average distance phonons propagate without being scattered. For instance, the effective phonon mean free path of Si at room temperature is around 300nm. The relaxation time and mean-free-path are related by $\lambda = \upsilon_g \tau$, where υ_g is the group velocity of phonons. Both the relaxation time and the mean-free-path strongly depend on the frequency of phonon modes.

There are two main mechanisms which control the phonon lifetimes and mean free path (MFP) in a crystalline material, namely: the phonon an-harmonic interactions in which a phonon decays into other phonons, and the phonon scattering at impurity or defect centers. The mean-free-path of heat-carrying acoustic phonons is not constant but spans more than an order of magnitude in most crystals depending on phonon frequency.

The energy conservation law of the anharmonic decay requires that the energy of the decay phonon be equal to the sum of the energies of the created phonons. In fact, the recent results indicated that the classical heat diffusion equation significantly underestimates temperature distribution at nanoscales in the presence of external heat sources. Figure 11 depicts the phonon transport and energy exchange cycle, in semiconductor devices.

Generally speaking, describing phonon transport by a single MFP (the so-called gray body approximation) is an oversimplification of the problem. For any length scale there will be phonons of low enough frequency that propagate ballistically rather than diffusively. Most thermal energy sits in the high-frequency phonon modes with short MFP which can be modeled as a thermal reservoir with locally defined temperature and diffusive transport. Low-frequency phonons (with MFP on the order of or longer than ~1um in Si at room temperature) may contribute significantly to thermal conductivity but their contribution to the specific heat is very little due to their low density of states. The concept of

Figure 10. Phonon transport and energy exchange cycle
After Egley et al (2000).

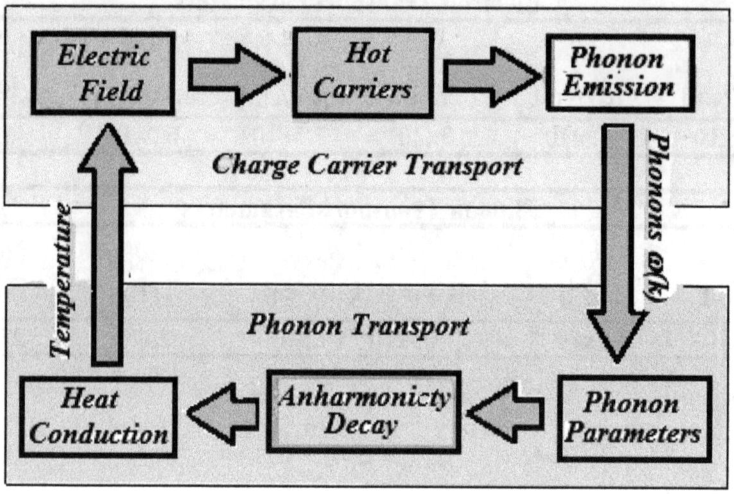

separating the phonon spectrum into low- and high-frequency parts goes back to the two-fluid model of thermal conductivity of Armstrong (1981). The Figure 11 depicts the experimental frequency dependence of the longitudinal phonon lifetime in Si at 300K.

As shown in figure, the phonon relaxation may be fitted to the Akhiezer relaxation model at lower frequencies (Kunal & Aluru, 2011). In sub-THz range, there is a transition in phonon-phonon interac-

Figure 11. Frequency dependence of the longitudinal phonon lifetime in Si at 300K. The experimental data (symbols) are taken from Asheghi et al (2002) and fitted by the Akhiezer relaxation model. The quadratic dependence extrapolation is obtained from first-principles calculations with three-phonon scattering (Landau-Rumer) model.

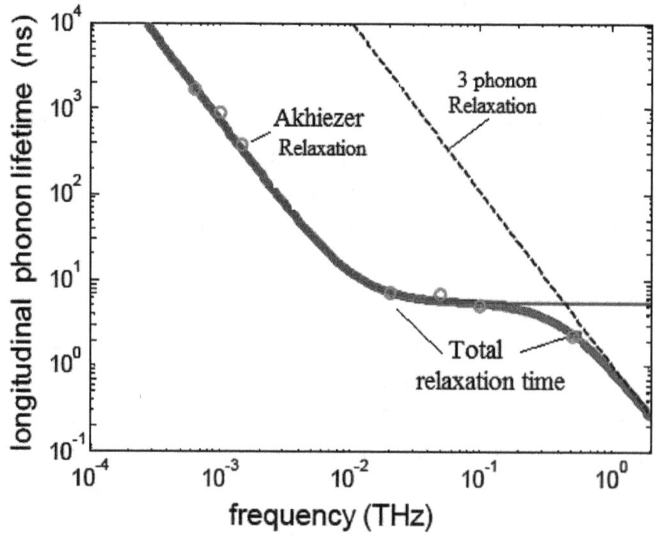

tion from the Landau-Rumer (3-phonon scattering) to the Akhiezer relaxation mechanism, and the frequency dependence of the phonon lifetime strongly deviates from the quadratic one. At frequencies below 100 GHz, the phonon lifetime is determined by the Akhiezer mechanism and can be modeled by a relaxation-type equation

$$\tau^{-1}\,(\text{Akhiezer}) = \tau_o(1 + 1/\tau_{th}^{2}\omega^{2}) \qquad (16)$$

where τ_{th} is on the order of the characteristic lifetime of dominant thermal phonons. In silicon at 300K, we may take τ_o = 5ns and τ_{th} = 14ps, to fit the experimental data, as shown in figure.

The phonon-lifetime in a certain material can be measured via the Raman spectral linewidth. The values of the Raman line widths at the zero slit width may be used to evaluate the phonons lifetime, τ, via the energy time uncertainty relation: $\Delta E = \hbar/\tau$. where ΔE is the Raman linewidth in units of cm^{-1} and $\hbar$ is Planck's constant divided by 2π.

4. PHONON TRANSPORT MECHANISMS

Phonon transport theory occupies a critical place in the hierarchy of information carrier transport especially at the nanoscale. The transport of phonons is usually considered as a diffusive process and described by Fourier's law of heat conduction. However, ballistic and hydrodynamic phonon transport mechanisms are also possible in bulk materials, especially at very low temperatures. Recent studies of low-dimensional materials have indicated the importance of ballistic phonon transport in thermoelectric materials and electronic devices.

4.1 Diffusive Phonon Transport

For heat conduction in solids, the rate of thermal energy transfer per unit time and unit area is assumed proportional to temperature gradient. Therefore, the heat flux term (***Q***) in the classical transport models is usually expressed by the following diffusion equation:

$$Q_n = -\,k^{th}\nabla T_L \qquad (17a)$$

Figure 12. Basic phonon transport mechanisms

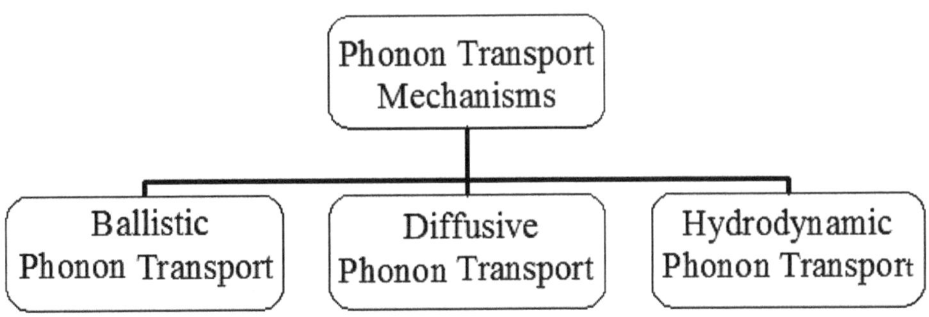

The above empirical relationship is known as the Fourier's law of heat conduction. The negative sign indicates that heat flows against temperature gradient, from the hot side to the cold side. The unit of heat flux is [W/m^2] and the unit of the thermal conductivity is [W/m.K]. For instance, *Al* (metal) has k^{th} >200 W/m·K and *InAs* (semiconductor) has k^{th} =27 W/mK.

The reciprocal of thermal conductivity is called the thermal resistivity. The thermal conductivity (k^{th}) is related to the electronic conductivity (σ_n) by the Wiedemann-Franz law, which was originally discovered for metals:

$$k^{th} = L\,\sigma_n\,T \tag{17b}$$

where $L = \gamma_n (k_B/e)^2$ is the Lorenz number (for metals γ_n = ⅓) and T is the ambient temperature.

In high thermal conductivity materials, the temperature distribution is almost homogenous since heat could be easily transported across the material. On the other hand, low thermal conductivity impedes heat transport. As shown in the Figure 14, a temperature jump may happen between dissimilar materials. Kapitza first encountered this phenomenon in 1941 when he submerged a solid in liquid helium (Kapitza, 1941). The Fourier diffusion equation has been the governing equation for heat transfer problems in conventional structures. The underlying assumption is that heat is transferred from one region to another subject to sufficient scattering events of phonons in the medium. If the size of the material or material constituents is much larger than the mean free path of phonons, many scattering events take place so that local thermodynamic equilibrium is restored in the material. However, it has been shown

Figure 13. Experimental thermal conductivity of different crystalline and amorphous solids, as a function of temperature
From (Kittel & Kroemer, 1980).

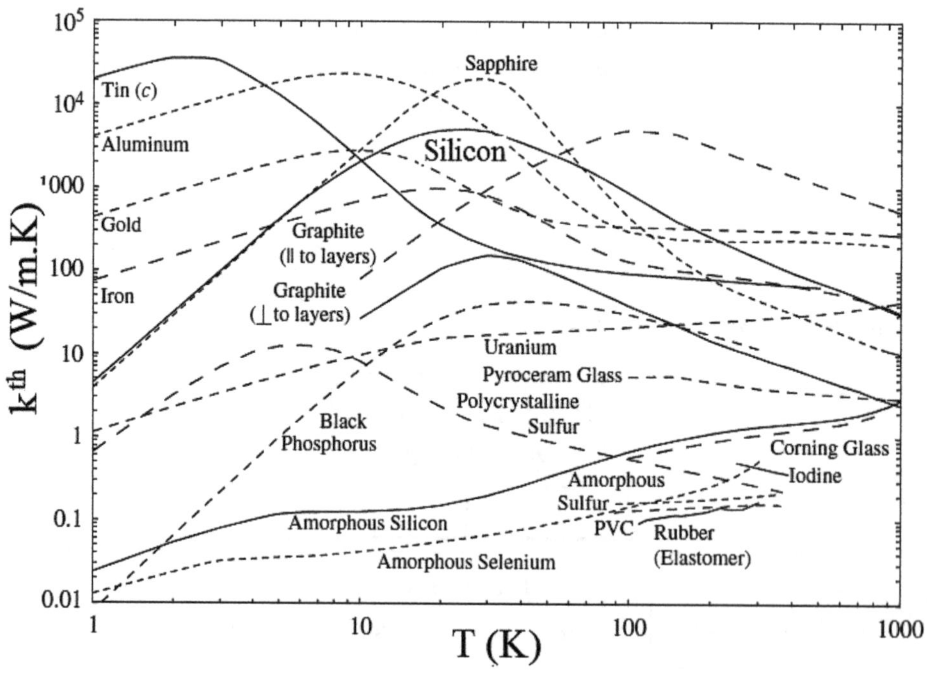

that the Fourier formula is not accurate for modeling the heat flux across thin semiconductor layers and interfaces (e.g., p-n junctions) and within nanodevices in particular.

At small length scales, the classical diffusive model for heat conduction begins to fail. When the phonon mean free path is greater than the characteristic size of the material, the heat flow process is in the ballistic regime where phonon-phonon scattering processes are rare and thermal conductivity of the material is governed by boundary and impurity scattering processes. For instance, SOI devices have dimensions smaller than the mean free path of phonons. In such a case, Fourier's law becomes invalid for describing the thermal transport process since the thermal transport is no longer diffusive.

4.2 Hydrodynamic Phonon Transport

The term hydrodynamic (HD) phonon transport came from the macroscopic transport phenomena in fluids. Like charge carrier transport, phonons in the HD regime move with an average (macroscopic) drift velocity. Therefore, HD phonon transport is different from the diffusive or ballistic phonon transport. The HD phonon transport has drawn less attention because it has been observed only at extremely low temperatures in bulk materials. However, several researchers have shown that the HD phonon transport can occur in certain two-dimensional nanostructures, like graphene at higher temperatures than in bulk materials (Lee et al, 2015). As shown in Figure 15, the phonons in the HD regime exhibit a non-uniform heat flux profile like the laminar fluid flow in a tube. In the HD regime, the main mechanism for momentum loss is boundary scattering. Therefore, the flux is small near the boundary. The HD transport of phonons has been demonstrated through drift motion, phonon Poiseuille flow and second sound.

Second sound is a quantum mechanical phenomenon in which heat transfer occurs by wave-like motion (phonon density wave), rather than by diffusion mechanism. Second sound refers to the propagation of a temperature wave, provoked by a heat pulse, analogous to ordinary sound in a fluid, which is pressure wave propagation[6]. This phenomena is observed in superfluids (like liquid *He*) and some dielectric materials.

Figure 14. Heat conduction through two layers of different materials

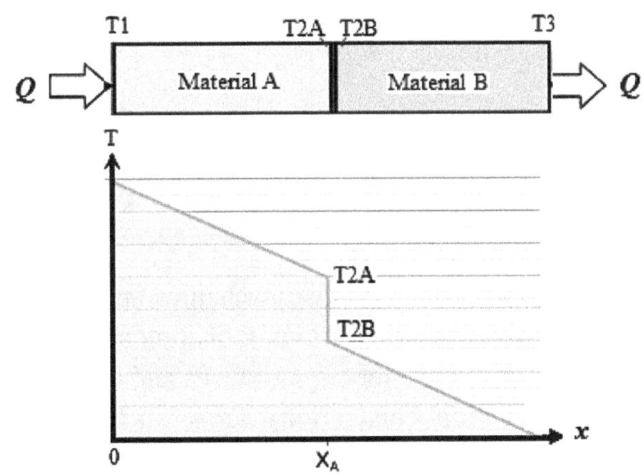

Figure 15. Steady-state heat flux profiles in diffusive and hydrodynamic regimes
After (Lee et al, 2015).

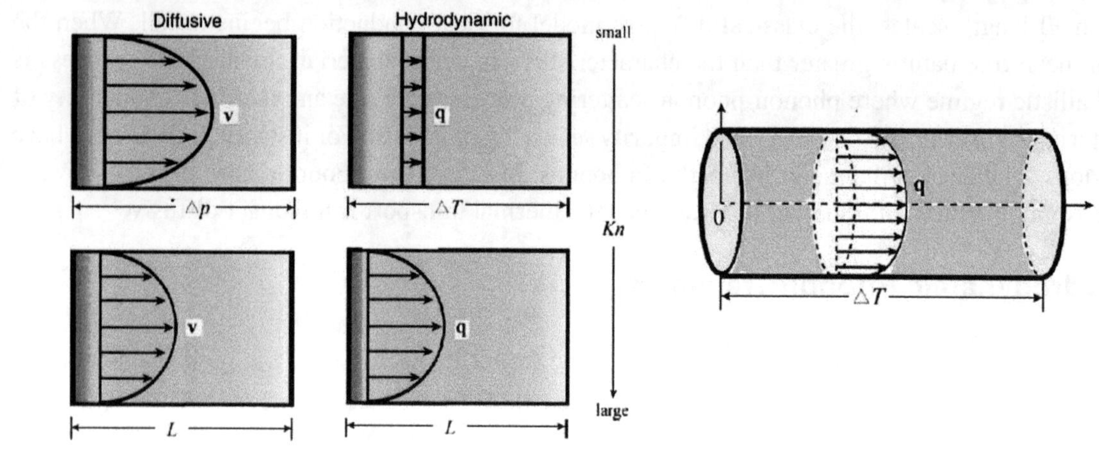

4.3 Ballistic Phonon Transport

Ballistic transport is the transport of carriers in a medium without scattering events. Recent studies of thermal transport in nano-materials have demonstrated the existence of ballistic transport (Buot, 2009). When the size of semiconductors is smaller than the phonon mean free path, phonons can carry heat with no internal scattering. In general, carriers will exhibit ballistic conduction when the device length $L < \lambda$., the mean free path between scattering events. Recent experiments demonstrate room-temperature evidence of ballistic phonon transport in various nano-materials. However, the thermal conductivity data for silicon in the length scale of 10-100 nm is still not available due to experimental challenges. It is now well known that ballistic phonon transport becomes important in structures with small feature sizes as well as in large structures in low temperature and under rapid transient conditions.

The ballistic thermal transport may be expressed by the Landauer-Büttiker formula (Büttiker et al., 1985), which was initially developed for electronic transport in 1-D ballistic channels. In spite of the big difference between electrons and phonons in nature and energy distribution functions, many researchers assume an equivalent Landauer formula for phonons. The heat current from node A to node B of a ballistic thermal channel may be given by the following relation (Munoz, Lu & Yakobson, 2010):

$$J^{th} = \int (\omega/2\pi)[\, N_A(\omega,T) - N_B(\omega,T)\,].\zeta(\omega).dw \tag{18}$$

where N is the phonon distribution and $\zeta(\omega)$ is the phonon transmission probability.

4.4 Coherent Heat Flow

The ballistic conduction is coherent in terms of wave mechanics and hence enables us to exploit the quantum mechanical properties of carrier wave functions. Coherent transport of phonons is particularly relevant in short-period heterostructures with smooth interfaces and long-wavelength phonons, such as two-dimensional superlattices. The thermal conductivity across such structure increases linearly with the repetition of stacked layers. The measured thermal conductivity increased linearly with increasing superlattice thickness over a temperature range from 30K to 150K (Ravichandran *et al.*, 2014).

Considerable research effort has been devoted to understanding the fundamental regime transition from incoherent to coherent in nanostructures. Recent experiments and atomistic simulations have shown that the transition to coherent phonon transport occurs in certain semiconductor nanostructures of 2-5 nm in thickness.

4.5 Quantized Thermal Conductance

The so-called thermal conductance quantum is a unit of heat conduction, which describes the rate at which heat is transported through a single ballistic phonon channel. It is given by:

$$G_o^{th} = (\pi k_B^2 / 6\hbar)\, T = g_o T \tag{19}$$

where the constant $g_o = 9.456 \times 10^{-13}\ W/K^2$. The thermal conductance of insulators that exhibits ballistic phonon transport is an integer multiple of G_o^{th}. The thermal conductance quantum was measured by Schwab and other researchers (Schwab et al, 2000). The measurement employed suspended Si_3N_4 nanostructures that exhibited a constant thermal conductance of $16G_o^{th}$ at 0.6K or below.

In ballistic electrical conductors, the electron contribution to the thermal conductance is also quantized due to the Wiedemann–Franz law (Wiedemann–Franz, 1853). This empirical law depicts the relation between the thermal and the electrical conductivities of metals, as: $k^{th}/\sigma = L.T$, where L is the Lorentz number (Lorentz, 1872) and σ is the electrical conductivity. As conductance and conductivity are related by geometry, the Wiedemann–Franz law may be written as $k^{th}/\sigma = G^{th}/G = L.T$, or $g_o/G = L$. In metals, we have $L = \frac{1}{3}\pi^2 k_B^2/e^2 \approx 2.44\times10^{-8}\ W\Omega/K^2$, therefore, we can write the corresponding quanta of electrical conductance as:

$$G = g_o/L = (\pi k_B^2 / 6\hbar) / \tfrac{1}{3}\pi^2 k_B^2/e^2 = \tfrac{1}{2}\pi\,(e^2/\hbar) \tag{20}$$

Although, we relied on an empirical law, the result is close to the quanta $2e^2/h$, which is a manifestation of collective behavior of electrons in 2-D structures at very low temperatures. However, note that L is constant because both heat and electrical currents are carried by electrons in metals, especially at low temperature. At room temperature, the contribution of phonons to the thermal conductivity cannot be neglected in solids, especially in insulators and semiconductors. For semiconductors, the Lorentz number (L) depends on the dominant collision mechanism, and other material parameters, as shown in Chapter 1 of this Book.

It should be noted that some researchers (Kane & Fisher, 1996) proved experimentally (Hussey et al, 2011), that the Wiedemann-Franz law could be violated in certain materials when electrons are confined in low-dimensional structures.

5. SEMICLASSICAL PHONON TRANSPORT

The lattice heat equation, which depicts the temporal and spatial distribution of lattice temperature along the semiconductor, has been already introduced in chapter 1. This equation is not only important

Figure 16. Quantized thermal conductance of Si_3N_4 thin layers of different thicknesses at very low temperature
After Rhyner, (2015).

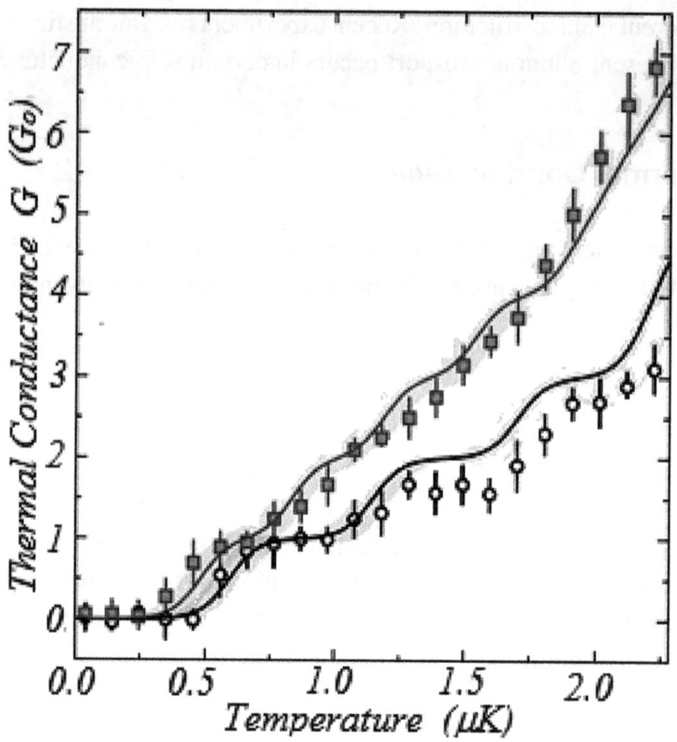

for power devices but also for nanodevices, including silicon-on-insulator (SOI) structures. The scale down in the size of electronic devices means lesser space for heat dissipation. In fact, the recent studies of SOI devices by the standard hydrodynamic model (HDM), showed certain anomalies due to lack of a heat evacuation mechanism at the insulator interface (Egley, 2000). Some authors suggested to fix this problem by considering a tensorial temperature and modifying the closure condition for energy flux term. Other authors suggested a more rigorous approach consists in solving the BTE for phonons (Chen, 2003). Therefore, we can add the phonon Boltzmann transport equation (BTE) to the set of transport equations to solve this problem. In this case, the semiclassical BTE gives the nonequilibrium energy distribution of phonons (in a specific system) in terms of its position and momentum coordinates $(\boldsymbol{r}, \boldsymbol{p})$. On the basis of the phonon BTE and its variants (such as the Equation of Phonon Radiative Transport, EPRT), researchers have successfully matched their solutions to the measured thermal conductivity data for bulk silicon, silicon thin films and silicon nanowires (Chen et al, 2008).

A similar approach to the phonon BTE, called the split-flux form of the phonon BTE (*p*-SFBTE), was proposed by Sinha *et al.* (2006). Here, the phonon distribution is split into two terms: $N_q = N_{qo} + n_q$. The first population term $N_{qo}(T_F)$ is a near-equilibrium component and has a Bose-Einstein distribution at a temperature T_F that follows the Fourier heat conduction law. The second population term n_q is a nonequilibrium component, which dominates the transport near hotspots and is determined by solving the phonon BTE, typically in the relaxation-time approximation. The *p*-SFBTE ensures macroscopic energy conservation and can capture the phonon ballistic conduction near hotspots while accounting for diffusive conduction far from hotspots.

The so-called McKelvey- Shockley flux method is another simple form of the BTE (McKelvey, Longini & Brody, 1961). This approach can provide accurate solutions for steady-state thermal transport from the ballistic to diffusive limits, with less computational efforts. The different solution methods of the phonon BTE and variant methods are discussed in details in (Chernatynskiy & Phillpot, 2010). A complete treatment of thermal transport phenomena in metals, with semiclassical approaches, can be found in Ashcroft and Mermin book (1976). We concentrate our attention here on the phonon transport models in semiconductor devices and nanostructures.

5.1 Phonon BTE

In equilibrium, phonons follow Bose-Einstein statistics and interact with each other via scattering processes. Using the Boltzmann transport equation (BTE), which was originally developed for studying the transport of classical particles, we can also study the phonon transport problems. The fundamental assumption in deriving the phonon BTE is that there exists a distribution function, $N_{q,s}(\boldsymbol{r},t)$, which describes the average occupation of the phonon mode (q, s) in the neighborhood of a location $\boldsymbol{r}$ at a time t. The equation assumes the simultaneous prescription of phonon position and momentum with arbitrary precision. As both position and momentum are conjugated in quantum mechanics (cannot be defined simultaneously, with precision), the phonon BTE considers phonons as wavepackets. Each phonon wave packet has a spread δq in the wavevector space and is localized in physical space within a region of size δr such that $\delta q \delta r \approx 1$.

The evolution (rate of change) of the phonon distribution in non-equilibrium can be described by the following phonon BTE (Pop, 2003):

$$(dN_{q,s}/dt) + v_g.\nabla N_{q,s} = (dN_{q,s}/dt)_{col.} \tag{21}$$

where $N_{q,s}$ is distribution function of an ensemble of phonons with wave vector q and polarization s, v_g is the phonons group velocity, $(dN_{q,s}/dt)_{col}$ is the rate of change of the phonon distribution function due to collisions. Phonon collision mechanisms include phonon–phonon, phonon–electron, phonon–impurity and phonon–boundary scattering mechanisms. The formal expression for the collision term in the phonon BTE has been developed by Peierls (1955), as an integral in wave vector space over all allowed scattering processes. The phonon BTE has shown promising results in predicting the ballistic-diffusive nature of heat transfer and temperature distribution. However, the collision term, which redistributes phonon energies in wavevector space, is non-linear and represents the primary source of computational difficulties in solving the phonon BTE.

In order to reduce complexity of solving the phonon BTE, scattering interactions are often approximated by the single mode relaxation time or assuming a3-fluid phonon transport model involving a stationary optical reservoir mode and a propagating acoustic mode. The latter approach can consider the heat evacuation by optical phonons as well as their anharmonic decay to *fast propagating* acoustic phonons, as shown in Figure 17.

Under the relaxation time approximation, the fundamental quantity associated with phonon scattering is the phonon mode lifetime. The single-mode relaxation time for a phonon system describes the temporal response of the system when that mode is excited and all other modes have their equilibrium

Figure 17. Energy flow in semiconductors
After Pop et al (2003).

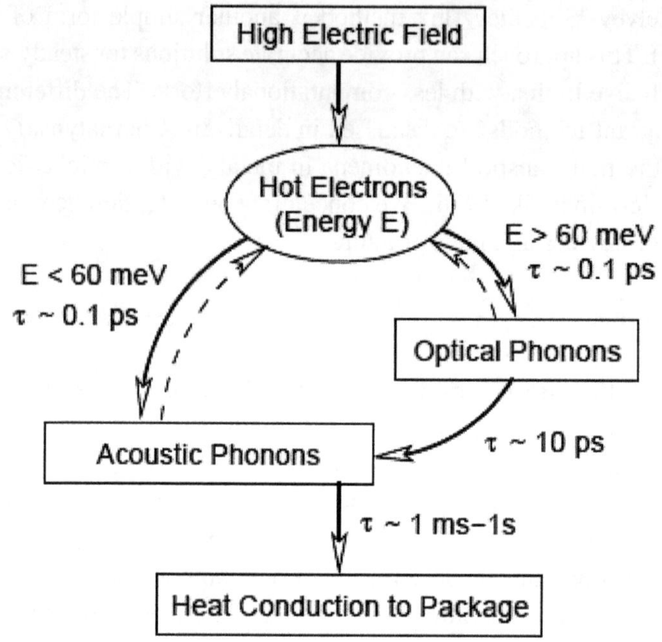

populations. The assumption (one mode activated and others at equilibrium) is never realized. This is because the natural decay of the energy in a mode occurs in the presence of other modes that are excited or diminished. In fact it is well known that the longitudinal optical (LO) phonons have small group velocities (~10^5 cm/s) and anharmonically decay into faster acoustic phonons which in turn transport the energy out of the semiconductor device.

5.2 Peierls -BTE

Peierls extended the Boltzmann transport equation which describes the thermodynamics of gas molecules to the study of phonons (Peierls, 1959). The Peierls–Boltzmann transport equation, which describes the evolution of the phonon distribution $N(x,t)$, has the same form as the phonon BTE[7]:

$$\partial N/\partial t + v_x (\partial N/\partial x) = (\partial N/\partial t)_{col.} \tag{22}$$

where $(\partial N/\partial t)_c$ represents the rate of change of the phonon distribution due to collisions. Using the moment method, one can obtain the average phonon energy and average momentum conservation along the phonon flow direction in the following macroscopic form (Guo &Wang, 2016):

$$\frac{\partial}{\partial t}\left(\sum_{\lambda}\int \omega N d\mathbf{q}\right)+\frac{\partial}{\partial x}\left(\sum_{\lambda}\int \omega v_x N d\mathbf{q}\right)=0 \tag{23a}$$

$$\frac{\partial}{\partial t}\left(\sum_{\lambda}\int q_x N d\mathbf{q}\right)+\frac{\partial}{\partial x}\left(\sum_{\lambda}\int q_x v_x N d\mathbf{q}\right)=0 \qquad (23b)$$

Here the sum over λ means all phonon polarizations. The right-hand side of equation (23a) is zero because energy is always conserved on scatterings. Also, the right-hand side of equation (23b) is also zero when we assume weak R-scattering and thus crystal momentum is conserved. The actual phonon distribution (N) under weak R-scattering can be approximated as a displaced distribution function (N_{BEd}). The displaced distribution can be linearized when the displacement in phonon distribution is very small:

$$N \approx N_{BEd} \approx N_{BE} + (\hbar/k_B T)\, N_{BE}\,(N_{BE}+1) q_x u_x \qquad (24)$$

5.3 McKelvey-Shockley Flux Method

The so-called McKelvey-Shockley flux method has been also used to treat transient phonon transport problem (McKelvey, Longini & Brody, 1961).

In comparison with other semiclassical phonon transport models, this approach has the following advantages:

1. It captures ballistic phonon effects, such as temperature jumps at the system boundaries,
2. It captures finite-velocity heat propagation, and
3. It can easily support full phonon dispersion relations, $\omega(k)$, and energy-dependent mean-free-path (MFP) of phonons.

In addition, the obtained results by this approach show good agreement with the phonon BTE, while requiring less computational effort. This technique categorizes phonons into two components, those that are forward moving ($v_x > 0$) and backward moving ($v_x < 0$). Therefore, the McKelvey-Shockley equations describe how the phonon flux (product of phonon density and phonon velocity), vary in space and time, as follows:

$$(v^+_x)^{-1}(dF^+/dt) + dF^+/dx = -\,(F^+ - F^-)/\lambda \qquad (25a)$$

$$(v^+_x)^{-1}(dF^-/dt) - dF^-/dx = (F^+ - F^-)/\lambda \qquad (25b)$$

Here, F^+ and F^- are the forward/backward phonon fluxes, $v^+_x(E)$ is the average x-projected phonon velocity and $\lambda(E)$ is the mean-free-path for backscattering. The right-hand sides of the above two equations describe how phonons scatter in/out of each flux type. Therefore, scattering is described by λ, which is the average distance traveled along x-axis before forward or backward scattering events.. Once these equations are solved under appreciate boundary and initial conditions, the heat current and heat flux can be obtained as follows:

$$I_Q(x, t, E) = I_Q^+(x, t, E) - I_Q^-(x, t, E) \qquad (26a)$$

$$Q(x,t,E) = [I_Q^+(x,t,E) + I_Q^-(x,t,E)] / v_x^+(E) \quad (26b)$$

where $E= \hbar\omega$ is the phonon energy and

$$I_Q^{\pm}(x,t,E) = E.F^{\pm}(x,t,E) \quad (27)$$

By integrating over energy, and hence all phonons, we obtain the total heat current (I_Q^{tot}) and heat density (Q^{tot}):

$$I_Q^{tot}(x,t) = \int_0^{\infty} I_Q(x,t,E).\, dE \; [W/m^2] \quad (28a)$$

$$Q^{tot}(x,t) = \int_0^{\infty} Q(x,t,E).\, dE \; [J/m^2] \quad (28b)$$

In addition to heat current and heat density, it is possible to calculate temperature profiles:

$$T(x,t) = \tfrac{1}{2}\,[\delta T^+(x,t) + \delta T^-(x,t)] + T_{ref} \quad (29a)$$

$$\delta T^{\pm}(x,t) = 2\,\delta I_Q^{\pm}(x,t) / C_v.v_x^+ \quad (29b)$$

where $\delta T^{\pm}$ is the correction in temperature relative to T_{ref} for each phonon component (forward and backward) and C_V is the heat capacity at T_{ref}. The reference temperature is an important issue in the temperature-dependent scattering process of phonons and heat transport. The Debye temperature, which is the temperature at which all phonon modes are excited, serves as the reference temperature. The Debye temperature (T_D) of silicon is 640 K.

5.4 Other Semiclassical Approaches

We presented so far the basic semiclassical models to study the phonon transport, which are based on the phonon BTE and variant methods. These methods are discussed and compared in (Chernatynskiy & Phillpot, 2010).

Due to the numerical implementation complexity and the computational cost, the phonon BTE is hard to solve in its generic form. Therefore, we can find other approximate approaches for solving the phonon BTE. Most of these approaches are based on the expansion of the phonon distribution function and the relaxation time approximations. For example, the so-called gray media (frequency independent phonons) is often assumed in the solution of the phonon BTE. Different orders of expansions of the phonon distribution function and phonon relaxation approximations give rise to different levels of phonons and heat transport equations. For instance, the transport of heat carriers can be described by the Chapman–Enskog (C-E) expansion to the phonon Boltzmann equation, with Callaway's relaxation approximation. This route yields the C–V equation (Cattaneo, 1958; Vernotte, 1958) as well as the G–K equation (Guyer, Krumhans 1966).

Chapman–Enskog (C–E) Equation

The Chapman–Enskog (C-E) method is an old solution technique to solve the BTE (Chapman, 1916; Enskog, 1917). It consists of the first order expansion of the BTE, around an equilibrium distribution (N_o), such that:

$$N = N_o + n_1 \tag{30a}$$

The deviation from equilibrium term (n_1) can be expressed as follows:

$$n_1 = -\tau_{q,s}\, v_{q,s}\, [\nabla T(\partial N_o/\partial T)] \tag{30b}$$

where N_o is the phonon equilibrium distribution at a given reference temperature. Obviously, the non-equilibrium components (n_1) reflects the Fourier diffusion mechanism. More accurately, n_1 can be written in terms of the equilibrium phonon distributions due to R and N processes as follows:

$$n_1 = -N_{No} - \tau_N\, [\partial N_{Ro}/\partial t + v_g(\partial N_{Ro}/\partial x)] \tag{30c}$$

where τ_N is the phonon relaxation time due to normal (N-process) scattering. The equilibrium distribution N_{Ro} takes the shape of Bose-Einstein (B-E) distribution and N_{No} takes the shapes of the displaced B-E distribution, as detailed in the G-K model below. The C–E method for the solution of phonon Boltzmann equation includes two derivatives, which express the continuous asymptotic expansion and the discrete lattice Boltzmann equation. The first-order Chapman–Enskog expansion can be utilized to develop the phonon hydrodynamic model (Jiaung & Ho, 2008). The lattice Boltzmann equation can be also derived starting from the Chapman–Enskog expansion and the phonon BTE (Guo & Wang, 2016).

Equation of Phonon Radiative Transport (EPRT)

The equation of phonon radiative transport (EPRT) expresses the evolution of the phonon intensity $I(\omega)$. The one-dimensional form of the EPRT reads:

$$(v)^{-1}(\partial I_\omega/\partial t) + \alpha(dI_\omega/dx) = -(I_\omega - I_{\omega o})/\nu\tau \tag{31a}$$

where α is the direction cosine and ν is the Debye velocity of phonons. The phonon intensity is given by following sum:

$$I_\omega = \Sigma\, g_{ph}(\omega).N(\omega_{q,s}).\nu.\hbar\omega \tag{31b}$$

Here $g_{ph}(\omega)$ is the phonon density of states. Note the resemblance between the EPRT and the McKelvey-Shockley flux method. The EPRT reproduces the expected radiative behavior in the acoustic 2D (thin film) limit and conforms to the Fourier heat flux law in the 3D (continuum) limit.

Guyer–Krumhansl (G–K) Equation

Guyer–Krumhansl (G–K) equation is a macroscopic heat transport equation, playing the same role as the Navier–Stokes equation in fluid dynamics. It, therefore, corresponds to the hydrodynamic description of a phonon system. Actually, the G–K can be derived from phonon Boltzmann equation, with Callaway relaxation time expression. As we'll see in the next section, the Callaway thermal conductivity is based on the following relaxation-time approximation of the phonon BTE:

$$(d N_q/dt)_{col.} = - (N_q - N_{No})/\tau_N - (N_q - N_{Ro})/\tau_R \quad (32)$$

where τ_N and τ_R are the phonons relaxation times due to normal (N) and resistive (*R*) phonon scattering processes, respectively. Note the different equilibrium distributions (N_{No} and N_{Ro}) in this expression. The first equilibrium distribution functions for N-processes (N_{No}) is a displaced Bose-Einstein (B-E) distribution function and the second equilibrium distribution functions for *R* processes (N_{Ro}) is the true B–E distribution[8]:

$$N_{No}\left(E_i\right) = \frac{1}{\exp\left((\hbar\omega - \hbar q)/k_B T\right) - 1},\ N_{Ro}\left(E_i\right) = \frac{1}{\exp\left(\hbar\omega / k_B T\right) - 1} \quad (33)$$

Here, the drift velocity $\boldsymbol{u}$ is a space-time dependent macroscopic velocity and not the same as the group velocity v_g of phonons. Therefore, the G–K transport equation reads:

$$\tau_R\,(\partial \boldsymbol{q}/\partial t) + \boldsymbol{q} + k^{th}\,\nabla T = 1/5 v_g^{\,2}\,\tau_N . \tau_R\,[\nabla^2 \boldsymbol{q} + 2\nabla(\nabla . \boldsymbol{q})] \quad (34)$$

where k^{th} is the bulk thermal conductivity

5.5 Solution Methods of Phonon BTE and Variant Methods

There exist several programs to solve the Phonon BTE and its variant methods. For instance; PhonTS is a lattice dynamics code that calculates thermal conductivity via the solution of the BTE for phonons. PhonTS has built in classical potentials (Lennard-Jones, Buckingham, Tersoff, etc.) and can interface to other codes to take advantage of other potentials or first principles band structure simulators, like VASP, and Quantum Espresso.

5.6 Lattice Monte Carlo (LMC) Method

Due to the implementation complexity and the computational cost, the phonon BTE is hard to solve in its generic form, using numerical techniques. Alternatively, one can solve the BTE using Monte Carlo (MC) simulation technique which is more convenient and efficient when frequency dependent phonons (non-gray media) are considered. Among the available solution methods to solve the Phonon BTE, the Lattice Monte Carlo (LMC) simulation is the best choice, although it may be expensive in terms of computer resources.

In LMC simulation, phonons are displaced inside the computational domain under the various boundary conditions and scattering effects. The phonons dispersion (frequency-dependent phonons) and their polarization effects can be then taken into consideration. The dispersion relationships for lattice vibrations $\omega(k)$ are available in the literature, for different materials. For instance, the $\omega(k)$ relations of phonon in Silicon may be taken, according to the measurements of Brockhouse (2000) or Dollog (1999).

When the dispersion relation is known, the phonon density of states (DOS) for a given polarization branch, $g_{ph}(\omega)$, is calculated as follows:

$$g_{ph}(w) = k^2/\left(2\pi^2 v_g(\omega)\right) \tag{35}$$

where the group velocity of phonon, v_g, is given by:

$$v_g(\omega) = \partial\omega/\partial k \tag{36}$$

Procedure of Lattice MC

The Lattice Monte Carlo (LMC) method for phonon transport can be used to find the phonons distribution function and the temperature distribution, in a certain device structure. It consists of the following steps. We initialize phonon ensembles, launch and trace them in terms of positions, frequency, temperature, and polarization. The temperature distribution varies with the location of these ensembles. Once the propagation stage is complete, the local temperature distribution can be calculated on the basis of the positions of ensembles. A general flowchart of the LMC is shown in Figure 18. The probabilities of scattering for the ensembles are calculated based on the scattering properties of phonons at a *pseudo-temperature*[9]. If an ensemble is scattered, its frequency and polarization are reset based on the equilibrium distributions. However, the local phonon distribution function after the propagation phase is different from the phonon distribution in equilibrium.

Phonon Equation of Motion

The propagation of phonons is governed by the following wave equation (in three-dimensions):

$$\rho.\omega^2.e_i = C_{ijml}\, k_j k_m e_i \tag{37}$$

where ρ is the crystal mass density, ω is the angular phonon frequency, $\boldsymbol{e}$ is the polarization vector, $\boldsymbol{k}$ is a wave vector and C_{ijml} is the elasticity tensor. For any given wave vector, k, there exist three eigenvalues for ω, and three eigenvectors for $\boldsymbol{e}$. These correspond to the three different polarization states: Longitudinal (*L*), Fast Transverse (*FT*) and Slow Transverse (*ST*).

The actual direction and velocity of propagation of phonons is given by the group velocity vector $v_g = d\omega/dk$.

Figure 18. Flowchart of LMC
After Wong e al. (2011).

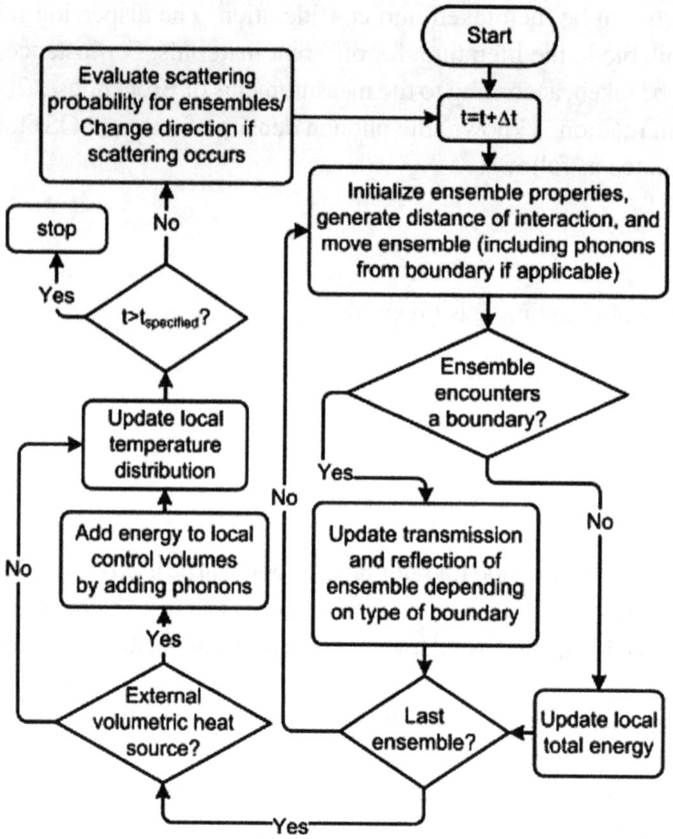

Application in Semiconductor Devices

The MC method has been used for phonons in silicon and compound semiconductor devices, For instance, Lugli and coworkers (Lugli et al, 1987, 1989) have performed self-consistent electron-phonon simulations in the bulk of *GaAs*, bulk *InP*, as well as $GaAs/Al_xGa_{1-x}/As$ heterostructures. In this work, only LO phonons were simulated. The results showed that the phonon population is driven into nonequilibrium during photoexcitation and impedes the relaxation of electrons and holes. Despite the subsequent rich applications, using MC in electrothermal device simulation, we still need to further investigations. The future simulations should respect the realistic phonon dispersion and address the ballistic phonon transport in nanodevices

5.7 Macroscopic Approaches

Although the microscopic approaches such as the phonon BTE and LMC are able to capture the statistical behavior of phonons at the nanoscale, we can resort to macroscopic models to provide a faster solution with reasonable physical insight into phonon transport. Actually, there are four basic macroscopic models for nanoscale phonon transport in the literature namely: ballistic-diffusive model, thermon gas model, phase-lag model and phonon hydrodynamic model.

The ballistic–diffusive model (BDE) divides the distribution function at any point into ballistic and diffusive parts, with *a priori* description of the phonon distribution function (Chen, 2001). The BDE is a mixed macro-scopic method as the heat flux of ballistic phonons is obtained from the kinetic theory.

The thermon gas model (Cao, Guo, 2002) describes heat transport through the classical fluid mechanics. In this model, heat conductions can be treated as thermon gas flows in solid driven by a temperature gradient. Thermon is defined as a quasi-particle carrying thermal energy. For solids, the thermon gas is the phonon gas of crystal lattice, attached to the electron gas or interacting between them. The phase-lag model (Tzou, 1995) suggests a causal response through adding one or two phase lags into Fourier's law.

Phase Lag Approaches (SPL & DPL)

It is well known that the utilization of the Fourier relation ($Q_L = -k^{th}\nabla T_L$) results in errors in the simulation of nanodevices (Pop, 2003). The single-phase lag (SPL) and dual-phase lag (DPL) approaches are mathematical techniques to avoid (actually to replace) the direct utilization of the Fourier diffusion relation. The phase lag techniques (which are sometimes called Jeffrey-type models) have been successfully employed to simulate SOI & Tri-gate nano MOSFET devices (Nasri et al, 2015).

The basic idea of the SPL model consists in expressing the lattice heat flux as follows (Cattaneo, 1958):

$$Q_L(r, t, \tau_L) = -k^{th}\nabla T_L(r, t) \tag{38}$$

where τ_L is a lag constant (roughly equal the phonon relaxation time τ_{ϖ}).

This allows us to write the heat continuity equation in the SPL model as follows (Cheng et al, 2008):

$$c_v\rho_L\frac{\partial T_L}{\partial t} + c_v\rho_L\tau_L\frac{\partial^2 T_L}{\partial t^2} - \nabla\cdot(k^{th}\cdot\nabla T_L) = \tau_L\frac{\partial H_s}{\partial t} + H_s \tag{39}$$

Hydrodynamic Phonon Transport Models

The hydrodynamic phonon transport model is another alternative approach to study phonon transport. This model is based on the macroscopic balance equations for phonons that can be derived from the phonons–BTE (PBTE). The balance equations of phonon energy and momentum are obtained through an integration of the PBTE, multiplied by the microscopic energy $\hbar\omega$ and crystal momentum $\hbar\mathbf{k}$:

$$\partial\varpi/\partial t + \nabla.\mathbf{q} = 0 \tag{40a}$$

$$\partial\mathbf{p}/\partial t + v^{-2}\nabla.\mathbf{Q} = -\mathbf{p}/\tau_R \tag{40b}$$

where $\varpi = \langle\hbar\omega\rangle$, $\mathbf{p} = \langle\hbar\mathbf{k}\rangle$, and the flux of heat flux $\mathbf{Q}$ is a symmetric tensor that can be defined from balance equation:

$$p= < \hbar k> = (1/2\pi)^3 \int \hbar k.N.d^3k \tag{41a}$$

$$\varpi = < \hbar\omega > = (1/2\pi)^3 \int \hbar\omega.\ N.d^3k \tag{41b}$$

$$q= < \hbar\omega.v_g > = (1/2\pi)^3 \int \hbar\omega.v_g.\ N.d^3k \tag{41c}$$

$$Q= < \hbar\omega.v_g v_g > = (1/2\pi)^3 \int \hbar\omega.v_g.v_g.\ N.d^3k \tag{41c}$$

Here, N denotes the phonon distribution function and $(1/2\pi)^3$ is the density of states in the momentum space. Substitution this definition yields the balance equation for heat flux (heat transport equation):

$$\partial q/\partial t + \nabla.Q= -q/\tau_R \tag{42a}$$

or

$$\partial q/\partial t + 1/3v^2\nabla\varpi + \nabla.Q= -q/\tau_R \tag{42b}$$

Lattice Heat Continuity Equation

As a compromise between the above approaches a lattice heat continuity equation can be added to the set of charge-carrier transport equations to form the so-called electrothermal transport model in semiconductors. The lattice heat continuity can be developed from the energy density form of phonon BTE in the relaxation time approximation (Sverdrup, Ju & Goodson, 2001):

$$\frac{\partial\varpi}{\partial t} + \vartheta\cdot\nabla\varpi = -\frac{\varpi-\varpi_o}{\tau_\varpi} + H_s \tag{43}$$

where ϖ is the phonon (excess) energy per unit volume per unit solid angle, υ the phonon velocity and τ_ϖ is the phonon energy relaxation time.

$$\varpi = \int \hbar\omega\cdot(N_q - N_{qo})g_{ph}(\omega)d\omega \tag{44}$$

The so-called phonon lattice Boltzmann equation (LBE) is a discrete form of the above equation.

In order to find out the macroscopic heat transport equation, Chapman–Enskog expansion can be applied to express the phonon distribution function. Assuming an isotropic distribution of phonons, the above equation may be integrated over the solid angle and written like the lattice energy conservation equation (Lai & Majumdar, 1996):

$$\frac{\partial W_L}{\partial t} + \nabla \cdot Q_L = -\frac{W_L - W_{Lo}}{\tau_\varpi} + H_s \tag{45}$$

where T_L is the lattice temperature, $W_L = \rho_L c_v T_L$ is the lattice energy density, W_{Lo} is the lattice energy density at equilibrium and $Q_L = -k^{th}\nabla T_L$ is the lattice heat flux. Also, H_s is the heat generation rate absorbed from hot electrons which can be extracted from a device simulator (usually, on the basis of DDM or HDM). In the bulk, the heat generation rate H_s is mainly due to the Joule heating effect ($\boldsymbol{J}.\boldsymbol{\xi}$) in addition to the net recombination cooling effect. In nanoscale devices, the phonon-boundary scattering effect should be added to these effects.

Lattice Boltzmann Equation (LBE)

The LBE is widely applied in classical hydrodynamics and has become a promising solver for phonon hydrodynamics in nanoscale phonon transport. The phonon lattice Boltzmann equation is obtained as a discrete form of the phonon Boltzmann equation

$$\frac{\partial \varpi_i}{\partial t} + \vartheta_i \cdot \nabla \varpi_i = -\frac{\varpi_i - \varpi_o}{\tau_{\varpi i}} \tag{46}$$

where υ_i (i = 1, 2, 3,...,n) are the discrete lattice velocities, and n is the number of them. The time and spatial derivatives of $\varpi_i(x,t)$ at the left side can be approximated by the finite difference method techniques. The phonon relaxation time may be approximated by, e.g., the Debye model, such that the equilibrium phonon energy density becomes $\varpi_i = C_v T/n$.

5.8 Thermal Conductivity Models

We now consider the modeling of thermal conductivity, which is a measurable macroscopic quantity, in the context of semiclassical phonon transport in insulators and semiconductors. The behavior of the lattice thermal conductivity in solids was qualitatively described by the early work of Debye and Peierls. The Debye and Peierls models described the lattice specific heat and thermal conductivity at very low temperatures (where phonons have long mean-free-path) on the basis of different assumptions. For instance, Debye assumed non dispersive phonon relation for all phonons ($\omega = \boldsymbol{k}.v_s$) and all phonons have the same velocity and mean free path (MFP).

Debye Model

Starting from the kinetic theory of gases, thermal conductivity of a gas of phonons is given by:

$$k^{th} = \Sigma_i c_{vi} v_{gi} v_{gi} \tau_i \tag{47a}$$

where the sum is taken over all phonon modes, c_{vi}, τ_i and $\boldsymbol{v}_{gi}$, are the phonon mode specific heat, lifetime and group velocity, respectively. Note that the group velocity is equal to the slope of the dispersion relation, $d\omega/dk$, and represents a very important quantity in heat transport. In fact, the group velocity in the optical branches of the dispersion relation is usually negligible, resulting in insignificant heat conduction by optical phonons. On the other hand, the acoustic branches of the phonon dispersion relation have large group velocities and contribute effectively in the heat transport. At the long wavelength (continuum) limit, the acoustic group velocity reaches the speed of sound ($v_s = \sqrt{(E/\rho)}$) in a crystal, where E is the elastic stiffness (Young's modulus) and ρ is the mass density. Assuming a non-dispersive phonon spectrum (linear), the above relation can be reduced as follows:

$$k^{th} = \tfrac{1}{3}\, c_v\, v_s\, \lambda \tag{47b}$$

Here, c_v is the average heat capacity per phonon, v_s is the average phonon speed and λ_p is the average phonon mean free path.

Indeed, λ depends on the phonon scattering mechanisms and is related to the phonon relaxation time τ by the relation ($\lambda = \tau . v_s$). Substituting the heat capacity, we get the Debye relations for heat capacity and thermal conductivity

$$c_v = \frac{dE_p}{dT}\Big|_v = \frac{9nk_B}{m}\left(\frac{T}{T_D}\right)^3 \int_0^{\frac{T_D}{T}} \frac{x^4 e^x}{\left(e^x - 1\right)^2} dx \tag{48a}$$

$$k^{th} = \left(48\pi^2\right)^{\frac{1}{3}} \frac{k_B^3 T^3}{a\hbar^2 T_D} \int_0^{\frac{T}{T_D}} \tau_p \cdot \frac{x^4 e^x}{\left(e^x - 1\right)^2} dx \tag{48b}$$

where T_D is the Debye temperature, m is atomic mass, n is the density of phonon modes and $x = \hbar\omega/k_B T$. Using this model, Debye found the T^3 dependence of heat capacity at low temperatures and the Dulong-Petit law at high temperature (about room temperature). However, this model leads to a thermal conductivity that declines from the experimental data as the temperature increases. In fact, the experimental data of thermal conductivity can only be explained by considering phonon-phonon scattering mechanisms, and the umklapp phonon scattering in particular. Following the above model, several theoretical models for the prediction of phonon thermal conductivity have been presented. Some of these models are based on the semiclassical BTE (like Callaway and Holland models) and others are based on *ab-initio* methods (like molecular dynamics MD). Though MD simulations account for anharmonicity, the classical nature of the MD makes it difficult to include quantum effects. Nevertheless, quantum correction terms can be added to correct the MD calculations.

Callaway Model

In1959, Joseph Callaway developed a solution of the phonon BTE on the basis of relaxation time approximation, and assumed an isotropic Debye-like phonon spectrum. In this model, there is no distinction between longitudinal and transverse phonons and the phonon branches are non-dispersive (linear ω-q

relation). Callaway considered only isotope impurity and boundary scattering mechanisms. Using these assumptions, Callaway developed a thermal conductivity model which is valid at low temperatures (up to 100K).

$$k^{th} = \frac{1}{2\pi^2 v_S}\int_0^{\omega_D} \frac{\hbar^2\omega^4 k_B T^{-2}}{\tau_p^{-1}} \cdot \frac{\exp\left(\hbar\omega / k_B T\right)^2}{\left[\exp\left(\hbar\omega / k_B T\right) - 1\right]} d\omega \tag{49a}$$

or

$$k^{th} = \frac{1}{2\pi^2 v_S}\int_0^{\omega_D} \frac{\hbar^2\omega^4 k_B T^{-2}}{A\omega^4 + BT^3\omega^2 + v_S L^{-1}} \frac{\exp\left(\hbar\omega / k_B T\right)}{\left[\exp\left(\hbar\omega / k_B T\right) - 1\right]^2} d\omega \tag{49b}$$

where A, B and L are constants of the phonon scattering rates. Note that the Callaway's model assumes a non-dispersive phonon spectrum and there is no distinction between phonon modes. Therefore, this model fails to explain the thermal conductivity of many solids like *Si* at high temperature (Ma, Li, & Luo, 2014).

Holland Model

In Holland's model (1963), the contributions of the transverse and longitudinal phonons are explicitly treated. In this model, Holland attempted to capture the thermal conductivity behavior at high-temperature, by modifying the expressions of the three-phonon normal and Umklapp processes. The final expression of Holland's model reads:

$$k^{th} = \frac{2}{3}\int_0^{\theta_T/T} \frac{C_T T^3 x^4 e^x \left(e^x - 1\right)^{-2}}{\tau_T^{-1}} dx + \frac{1}{3}\int_0^{\theta_L/T} \frac{C_L T^3 x^4 e^x \left(e^x - 1\right)^{-2}}{\tau_L^{-1}} dx \tag{50}$$

where $x = \hbar\omega/k_B T$, τ_T and τ_L are the total relaxation times for transverse and longitudinal phonons, respectively. Also, $\theta_T = \hbar\omega_T/k_B$ and $\theta_L = \hbar\omega_L/k_B$ are transverse and longitudinal temperature limits. Finally, the constants C_T and C_L are given by: $C_T = k_B^4/2\hbar^3\pi^2 v_T$ and $C_L = k_B^4/2\hbar^3\pi^2 v_L$.

A review of other models of thermal conductivity can be found in Ashcroft & Mermin (1976), Schroeder (2000) and Rowe (2006).

Empirical Models

The phenomenological models of thermal conductivity concentrated on the capture of the temperature dependence over a broader range with doping effect as well as the coherent phonon transport in nanostructures. The temperature dependence of kth in the bulk of many semiconductors may be simply modeled using the following phenomenological relation:

$$k^{th}(T_L) = k_{300}(T_L/300)^{\alpha}\ [W/K\ cm] \text{ for } T_L > 100K \quad (51)$$

where $\alpha = -1.3$ and $k_{300} = 1.45$ for Si. The parameter values when this model is applied to other semiconductors and insulators are summarized in Table 2.

At the interface, between different materials, the bulk thermal conductivity, k^{th}, should be replaced by the Kapitsa thermal conductivity, G^h, to express the temperature jump between dissimilar materials (Kapitsa, 1941). This abrupt change is due to the surface roughness at the interface between different materials and their different phonon properties (e.g., phonon sound velocity). Khalatnikov (1952) and Little (1959) developed the acoustic mismatch model (AMM) to explain the temperature jump by considering the sound velocity mismatch between different media.

6. QUANTUM THEORY OF PHONON TRANSPORT

When phase coherence effects cannot be ignored, the semiclassical approaches (like Boltzmann-Peierls equation) cannot be used to describe heat conduction and heat transfer in semiconductors and dielectrics. On the other hand, the molecular dynamics (MD) methods, which are based on classical mechanics, provide inaccurate results at low temperatures. This is especially true in nanostructures and nanodevices. Also, the concept of ensemble average in semiclassical distributions breaks down when the number of particles is small, which is the case in nanosystems (Tsalis, 1988).

Typically, heat in solids is transported incoherently because phonons scatter at interfaces and defects. Actually, heat is transported by a spectrum of phonon wave packets whose characteristic length scale, *i.e.* the thermal coherence length, determines the phonon transport regime. If the spatial coherence length of the wave packets is smaller than the feature size of the system, thermal transport is incoherent and phonons diffusively scattered at different defects. On the contrary, for wave packets with coherence lengths greater than the device length, transport becomes wave-like with coherent phonons. Recent experiments and atomistic simulations have shown that the transition to coherent phonon transport occurs in certain semiconductor nanostructures of 2-5 nm thickness. Considerable research effort has been devoted to

Table 2. Parameter values for thermal conductivity model

Material	Thermal Conductivity at $T_L = 300K$ K_{300} [W/K cm]	α
Si	1.48	-1.3
Ge	0.60	-1.25
GaAs	0.46	-1.25
AlAs	0.80	-1.37
InAs	0.273	-1.1
InP	0.68	-1.4
GaP	0.77	-1.4
SiO_2	0.0138	0.33
Si_3N_4	0.185	0.33

Source: Handbook Series on Semiconductor Parameters (Levinstein, Rumyantsev & Shur, 1996).

understanding the fundamental regime transition from incoherent to coherent in nanostructures. Recently, DFT combined with non-equilibrium Green's functions (NEGF) based on phonon interactions was proposed to describe thermal transport in microscopic systems.

6.1 Phonon Hamiltonian

Now we formulate the dynamic equations in a universal form to prepare them to quantum mechanical description. Let's first review the quantum-mechanical problem of a system of harmonic oscillators using creation/ annihilation operators. This system resembles the system of phonons in the harmonic approximation.

The Hamiltonian of such a phonon system is given as follows:

$$H_{ph} = \sum_{x} \frac{p^2(x)}{2m} + \frac{1}{2} \sum_{x,x'} u_i(x) \cdot C_{ij} \cdot u_j(x') \tag{52}$$

Notice the existence of the force constant tensor C_{ij} in the potential energy term. The solution of this can be carried out by several methods and leads to the dispersion relation, ω_α(k). The quantum mechanical Hamiltonian can be obtained from the classical Hamilton function by replacing classical momenta by operators. The corresponding quantized form of the solution may be written in terms of phonon annihilation and creation operators. These operators act directly upon the phonon occupation numbers. The phonon annihilation operator is given by

$$b_{\kappa,a} = \frac{1}{N^{1/2}} \sum_{x} e^{i(\kappa \cdot x)} s_\alpha(\kappa) \cdot \left[\left(\frac{m\omega_\alpha}{2\hbar} \right)^{1/2} u(x) + i \left(\frac{1}{2\hbar m\omega_\alpha} \right)^{1/2} p(x) \right] \tag{53a}$$

where N is the number of normal modes divided by α. Also, $\boldsymbol{s}_\alpha$ is the unit vector of mode α. Similarly, the phonon creation operator is defined as the adjoint of the phonon annihilation operator:

$$b^\dagger_{\kappa,\alpha} = \frac{1}{N^{1/2}} \sum_{x} e^{i(\kappa \cdot x)} s_\alpha(\kappa) \cdot \left[\left(\frac{m\omega_\alpha}{2\hbar} \right)^{1/2} u(x) - i \left(\frac{1}{2\hbar m\omega_\alpha} \right)^{1/2} p(x) \right] \tag{53b}$$

Therefore, the displacement $\boldsymbol{u}(\boldsymbol{x})$ and momentum $\boldsymbol{p}(\boldsymbol{x})$ can be expressed in terms of these operators $b_{k,\alpha}$ and $b_{k,\alpha}$, as follows:

$$u(\boldsymbol{x}) = \frac{1}{N^{1/2}} \sum_{\boldsymbol{\kappa},\alpha} \left(\frac{\hbar}{2m\omega_\alpha} \right)^{1/2} \left(b_{\boldsymbol{\kappa},\alpha} + b^\dagger_{\boldsymbol{\kappa},\alpha} \right) \boldsymbol{s}_\alpha(\boldsymbol{\kappa}) e^{i(\boldsymbol{\kappa} \cdot \boldsymbol{x})} \tag{54a}$$

$$p(\boldsymbol{x}) = -\frac{i^{1/2}}{N^{1/2}}\sum_{\kappa,\alpha}\left(\frac{\hbar m\omega_{\alpha}}{2}\right)^{1/2}\left(b_{\kappa,\alpha}+b^{\dagger}_{\kappa,\alpha}\right)\boldsymbol{s}_{\alpha}(\boldsymbol{\kappa})e^{i(\boldsymbol{\kappa}\cdot\boldsymbol{x})} \tag{54b}$$

Finally, the phonon Hamiltonian expressed in terms of these variables (the momentum and potential energies) becomes

$$\begin{aligned} H_{ph} &= \frac{1}{4}\sum_{\kappa,\alpha}\hbar\omega_{\alpha}\left(b_{\kappa,\alpha}-b^{\dagger}_{\kappa,\alpha}\right)\left(b^{\dagger}_{\kappa,\alpha}-b_{\kappa,\alpha}\right)+\frac{1}{4}\hbar\omega_{\alpha}\left(b_{\kappa,\alpha}-b^{\dagger}_{\kappa,\alpha}\right)\left(b^{\dagger}_{\kappa,\alpha}-b_{\kappa,\alpha}\right) \\ &= \frac{1}{2}\sum_{\kappa,\alpha}\hbar\omega_{\alpha}\left(b_{\kappa,\alpha}b^{\dagger}_{\kappa,\alpha}+b^{\dagger}_{\kappa,\alpha}b_{\kappa,\alpha}\right)=\sum_{\kappa,\alpha}\hbar\omega_{\alpha}\left(b^{\dagger}_{\kappa,\alpha}b_{\kappa,\alpha}+\frac{1}{2}\right) \end{aligned} \tag{55}$$

Note that applying the product $b^{\dagger}b$ to the wave function yields the phonon number N. This Hamiltonian corresponds to the solution of a system of harmonic oscillators in equilibrium, with quantized energies

$$E_n = \hbar\omega_n(n + ½) \tag{56}$$

Therefore, the normal vibrations behave as particles with the energy $\hbar\omega$ and quasi-momentum $\hbar k$. The quasiparticles are called phonons.

6.2 Phonon Green's Functions

The phonon transmission and thermal conductance can be obtained from the nonequilibrium Green's functions. The NEGF model for thermal transport is analogous to that for electron transport, which we described so far in Chapter 4. The Green's function is here the solution that results from the addition of a perturbation to the heat transport problem. This approach has been detailed in the literature, e.g., Zhang et al (2007) and Koswatta, et al. (2007).

Equilibrium Green's Functions for a Harminic Oscillator

At first, let's define the Green's functions (or correlation functions), for a single harmonic oscillator. We can define the Green's functions by the creation/annihilation quantum operators, for phonon systems. Better than this, we can use the vibrational displacement operators (Yamamoto, & Watanabe, 2006). Therefore, we define the greater Green's function as the position-position (r to r') correlation function in time (t to t') as follows:

$$g^{>}\left(t,t'\right) = -\frac{i}{\hbar}\left\langle u(t)u(t')\right\rangle \tag{57a}$$

Here $u_j(t)$ is the Heisenberg operator for the vibrational displacement where the time-dependence is according the Heisenberg evolution picture, and the angular bracket means the average (or trace) over the equilibrium density matrix ρ. Similarly, we define the lesser Green's function, as follows:

$$g^{<}\left(t,t'\right)=-\frac{i}{\hbar}\left\langle u(t')u(t)\right\rangle =g^{>}\left(t',t\right) \tag{57b}$$

Also, the time-ordered (causal) Green's function is defined as:

$$g^{t}\left(t,t'\right)=-\frac{i}{\hbar}\left\langle Tu(t)u(t')\right\rangle =\theta\left(t-t'\right)g^{>}\left(t,t'\right)+\theta\left(t'-t\right)g^{<}\left(t,t'\right) \tag{57c}$$

The Heaviside step function is defined *as* $\theta(t) = 1$ *if* $t \geq 0$ and 0 otherwise. In addition, the anti-time-ordered Green's function is given by

$$g^{\bar{t}}\left(t,t'\right)=-\frac{i}{\hbar}\left\langle \bar{T}u(t)u(t')\right\rangle =\theta\left(t'-t\right)g^{>}\left(t,t'\right)+\theta\left(t-t'\right)g^{<}\left(t,t'\right) \tag{57d}$$

It is useful to introduce also the retarded and the advanced Green's function

$$g^{r}\left(t,t'\right)=-\frac{i}{\hbar}\theta\left(t-t'\right)\left\langle\left[u(t),u(t')\right]\right\rangle =-\theta\left(t-t'\right)\frac{\sin\Omega\left(t-t'\right)}{\Omega} \tag{57e}$$

$$g^{a}\left(t,t'\right)=-\frac{i}{\hbar}\theta\left(t'-t\right)\left\langle\left[u(t),u(t')\right]\right\rangle \tag{57f}$$

The retarded Green's function appears in linear response theory, and it has the same meaning as that of Green's function in classical physics. In practical calculation, it is more convenient to work in the frequency domain. We thus define the Fourier transform of the Green's functions

$$g^{r}\left[\omega\right]=\int_{-\infty}^{+\infty}g^{r}\left(t\right)e^{i\omega t}dt \tag{58a}$$

and its inverse

$$g^{r}\left[t\right]=\int_{-\infty}^{+\infty}g^{r}\left[\omega\right]e^{-i\omega t}\frac{d\omega}{2\pi} \tag{58b}$$

The Fourier transform of the retarded Green's function for a single oscillator is

$$g^{r}\left[\omega\right]=-\int_{-\infty}^{+\infty}\theta\left(t\right)\frac{\sin\left(\Omega t\right)}{\Omega}e^{i\omega t-\eta t}dt=\frac{1}{\left(\omega+i\eta\right)^{2}},\left(\eta\to 0^{+}\right) \tag{58c}$$

Nonequilibrium Green's Functions for a Phonon System

Now, we generalize the definitions of the single harmonic oscillator and consider a general system with vibrational displacement u_j, where the index j runs over all degrees of freedom of the system. We define the greater Green's function $G^>$ as a matrix, whose elements are as follows:

$$G^>_{jk}\left(t,t'\right) = -\frac{i}{\hbar}Tr\left[\rho\left(t_0\right)u_j\left(t\right)u_k\left(t'\right)\right] \tag{59a}$$

where $u_j(t)$ is the Heisenberg operator for the displacement and the trace is taken over a complete set of states. Note that the lesser and the greater Green's function are interrelated as follows:

$$G^<_{jk}\left(t,t'\right) = G^>_{kj}\left(t',t\right) \tag{59b}$$

Also, the retarded Green's function is defined by:

$$G^r_{jk}\left(t,t'\right) = -\frac{i}{\hbar}\theta\left(t-t'\right)Tr\left\{\rho\left(t_0\right)\left[u_j\left(t\right)u_k\left(t'\right)\right]\right\} = \theta\left(t-t'\right)\left[G^>_{jk}\left(t,t'\right) - G^<_{jk}\left(t,t'\right)\right] \tag{59c}$$

Similarly, the advanced and retarded Green's functions can be obtained by swapping time arguments:

$$G^a_{jk}\left(t,t'\right) = G^r_{kj}\left(t',t\right) \tag{59d}$$

Nonequilibrium Phonon Density and Heat Current

In order to solve real systems, we need to incorporate real materials to the system and its contacts to the external world. We can do this through the incorporation of the retarded and advanced Green's functions $G^{r/a}$

$$G^{r/a} = [\ \omega^2 M - K - \Sigma_L^{\ r/a} - \Sigma_R^{\ r/a}\ ]^{-1} \tag{60}$$

where K is a coupling matrix, which can be constructed from the force constants between atoms. Also, M is a diagonal matrix of the mass of atoms of the involved material (e.g., Si or C atoms). The above matrix form of Green's functions includes self-energy matrices (Σ_L, Σ_R) that involve unperturbed Green's functions of contacts (boundaries) in a real system. These self-energies, as well as the coupling matrix elements, can be initially obtained from the atomistic analysis of contacts.

The equation of motion of such a system

$$\partial^2 G\left(\tau,\tau'\right) + KG\left(\tau,\tau'\right) = -\delta\left(\tau,\tau'\right)I \tag{61}$$

where I is the identity matrix and the coupling matrix K is given by:

$$K = \begin{pmatrix} K^L & V^{LC} & 0 \\ V^{CL} & K^C & V^{CR} \\ 0 & V^{RC} & K^R \end{pmatrix} \tag{62}$$

Note that $V^{LR}=V^{RL}=0$ as we assume that the contacts do not interact. The symmetric K matrix can be decomposed as ($K=D+\nu$) where:

$$D = \begin{pmatrix} K^L & 0 & 0 \\ 0 & K^C & 0 \\ 0 & 0 & K^R \end{pmatrix}, \nu = \begin{pmatrix} 0 & V^{LC} & 0 \\ V^{CL} & 0 & V^{CR} \\ 0 & V^{RC} & 0 \end{pmatrix} \tag{63}$$

with their corresponding equations of motion:

$$G(\tau,\tau') = g(\tau,\tau') + \int_C d\tau'' g(\tau,\tau'') \nu G(\tau'',\tau') \tag{64a}$$

$$\frac{\partial^2 g(\tau,\tau')}{\partial \tau^2} + Dg(\tau,\tau') = -\delta(\tau,\tau') I \tag{64b}$$

The heat current (heat flux in W) can be calculated, using the following Landauer-like formula (Das & Dhar, 2012):

Figure 19. Real systems of phonons between two contacts

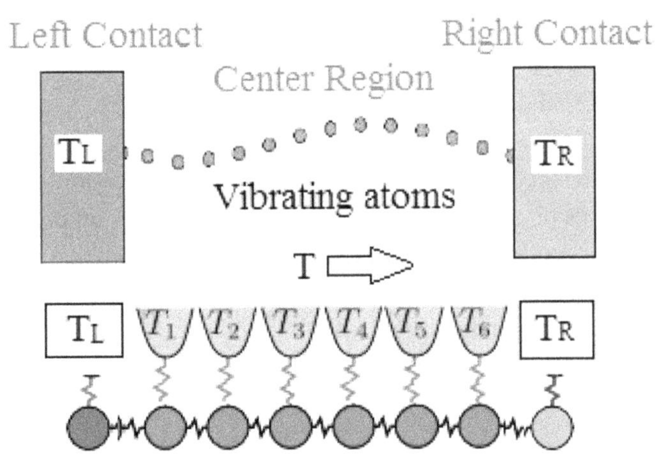

$$J = \int_0^\infty \frac{\hbar\omega}{2\pi} \Delta N(\omega)\zeta(\omega) \cdot d\omega \tag{65}$$

where the phonon distribution difference $\Delta N(\omega) = N_L(\omega) - N_R(\omega)$ may be calculated as follows

$$\Delta N = \frac{dN}{dT}\Delta T = \frac{\hbar\omega}{k_B T^2} \frac{e^{\hbar\omega/k_B T}}{\left(e^{\hbar\omega/k_B T} - 1\right)} \Delta T \tag{66}$$

Note that several authors contributed in the development of an expression of the thermal energy current in the form of a Landauer-like formula. For instance, Wu & Segal (2011) developed a Born-Oppenheimer-type formalism for the thermal energy current in the form of a generalized Landauer formula, which reduces to the standard Landauer form for heat transfer in the harmonic limit. Also, the phonon-energy transmission probability $\zeta(\omega)$ through the scattering region (from left to right) can be calculated as follows:

$$\zeta(\omega) = Tr\,[\,\Gamma_L(\omega).G^r(\omega).\,\Gamma_R(\omega).G^a(\omega)] \tag{67}$$

Here, $G^{r/a}$ are the non-equilibrium Green's function in the scattering region, as given by (59). Also, $\Gamma_{L/R} = Im[\Sigma^r_{L/R}(\omega) - \Sigma^a_{L/R}(\omega)]$ are the coupling constants. In the ideal ballistic limit without any scatterings, the transmission $\zeta(\omega)$ is equal to the number of phonon subbands[10]. The heat current can be then written as:

$$I_L = -\int_{-\infty}^{+\infty} \frac{d\omega}{2\pi} \hbar\omega \mathrm{Tr}\left[G^r[\omega]\Sigma_L^<[\omega] + G^<[\omega]\Sigma_L^a[\omega]\right] = \int_{-\infty}^{+\infty} \frac{d\omega}{4\pi} \hbar\omega \mathrm{Tr}\left[G^<\Sigma_L^> - G^>\Sigma_L^<\right] \tag{68}$$

Also, the thermal conductance (from left to right) is given by:

$$G(T) = \frac{k_B}{2\pi} \int_0^\infty d\omega \frac{\hbar\omega}{k_B} \frac{\partial n_{BE}}{\partial T} \zeta(\omega) \tag{69}$$

7. THERMAL TRANSPORT IN NANOSTRUCTURES

There are two information carrier types that can contribute to thermal conductivity in solids - electrons and phonons. In nanostructures phonons usually dominate and the phonon properties become of a particular importance for thermal conductivity. Over the past decade, researchers advanced the understanding of heat conduction in nanostructures through systematic measurements of the thermal conductivity of nanostructures. Our evolving understanding on nanoscale heat transport will facilitate the design of more efficient nanodevices and thermoelectric materials with reduced thermal conductivity.

The behavior of phonons in low-dimensional structures and their interaction with charge carriers have been extensively studied in the literature, for example, Yu & Cardona (2001). In this section, we show some theoretical and experimental studies on heat conduction by phonons in nanostructures, including thin films, and nanowires are briefly and selectively review.

Nano silicon wires and nano-composite dots have been shown experimentally to have tremendous potential as future thermoelectric materials (Li et al, 2003; Chen et al, 2008). This is due to their high electrical conductivity and low thermal conductivity which, when combined, yields a high figure-of-merit of thermo-electric power, and consequently, very high energy conversion efficiency. Also, the so-called phase-change memory (*PCM*), utilizes self-heating effects to switch nanostructured chalcogenide alloys (like *GST* and *GeTe*) between their crystalline and amorphous states (Malladi, 2013).

7.1 Phonon Transport in Thin Films

Thermal transport properties affect the performance and reliability of the thin films in transistors, lasers, sensors and nanoelectromechanical systems (NEMS). These devices often contain thin films of metals, silicon, oxides, nitrides and polysilicon. For instance, the silicon-on-insulator (SOI) devices contain a thin-layer of silicon (of 0.05 μm to 10 μm) above a buried silicon dioxide. Also, some types of laser devices contain superlattices of stacked thin-films such as *GaAs/AlGaAs* in lasers. The effective thermal conductivity of such thin films is reduced (50% in SOI) as compared to bulk materials, due to phonon-interface scattering and other defects.

In fact, the thermal conductivity of nanofilms is size dependent due to the contribution of ballistic transport of phonons. The ballistic transport has different effects on heat transport in the in-plane or cross-plane directions, which causes the anisotropy of thermal conductivity of nano-films.

In MOSFET devices, the gate electrodes are often made of polycrystalline silicon. When the polysilicon grain sizes are small, the heat conductivity is reduced due the internal scattering from grain boundaries. A simple approach to modeling the phonon scattering at grain boundaries of polycrystalline thin films is to include the following phonon scattering rate into account (Maiti et al, 1997):

$$\tau^{-1}\ (Grain\ boundaries) = B.(d/d_G) \tag{70}$$

Here, d_G is the characteristic grain size, v is the average phonon velocity in the material and B is a dimensionless constant which is correlated with the phonon reflection coefficient at grain boundaries.

7.2 Phonon Transport in Superlattices

A super lattice is typically consisting of a repeating sequence of two different thin films (film1, film2). The problem of heat conduction across heterojunctions of superlattices is more complex and less understood than the case of homogeneous thin films. A simple estimate of heat conduction of a superlattice is given by (Zhang et al, 2007):

$$k_n = \left(\frac{C_1 v_1 C_2 v_2}{C_1 v_1 + C_2 v_2}\right)\left(\frac{d_1 + d_2}{2}\right) \tag{71}$$

Figure 20. Thermal conductivity of Si nano-films
After Dong, Cao & Guo (2015).

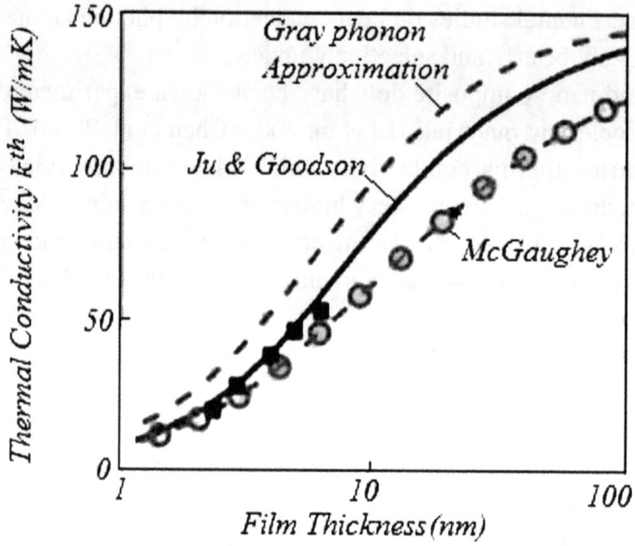

where C_1 and C_2 are the corresponding heat capacity of film1 and film2, respectively, v_1 and v_2 are the acoustic propagation velocities in film1 and film2, and d_1 and d_2 are the thicknesses of film1 and film2. This model neglects scattering in layers and assumes fully diffuse, inelastic scattering.

7.3 Coherent Heat Flow in Superlattices

Coherent transport is pertinent in short-period heterostructures with smooth interfaces and long-wavelength phonons, such as two-dimensional superlattices of graphene and semiconductor compounds. Considerable research effort has been devoted to understanding the transition from incoherent to coherent regime in order to control thermal conductivity in superlattices. Recent experiments and atomistic simulations have shown that the incoherent heat conduction shift to coherent phonon transport occurs for small periods of 2-5 nm in semiconductor and epitaxial oxide superlattices with atomically–smooth interfaces (Aaron et al, 2013).

It has been shown by Luckyanova *et al.* (2012) that thermal conductivity through sandwiched structure of *GaAs* and *AlAs* superlattices increases linearly with the number of superlattice repeats. This was attributed to the coherent transport of phonons in such nanostrustures. If the spatial coherence length of the wave packets is smaller than the period thickness of the superlattice, thermal transport is incoherent and phonons are scattered on different defects. On the contrary, for wave packets with coherence lengths greater than the superlattice period thickness, transport becomes wave-like with coherent phonons. Therefore, thermal conductivity is only limited by the total length of the superlattice.

Another example of short-period heterostructures with smooth interfaces and long-wavelength phonons, is shown in Figure 22. It depicts a short-period 2-D superlattice of graphene with boron nitride, that supports coherent phonon transport (da Silva et al, 2015).

Figure 21. Cross-plane thermal conductivity of twinning superlattices in single Si NWs
After Aaron et al. (2013).

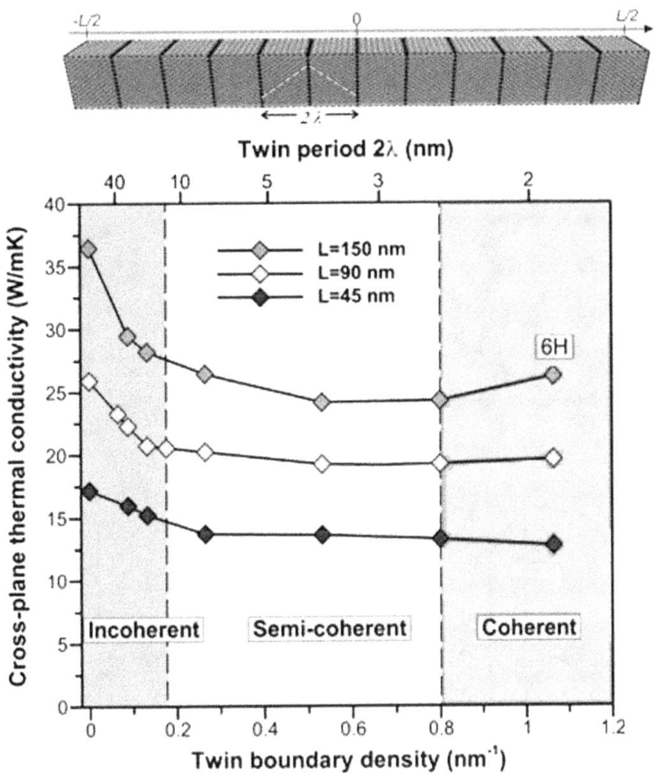

7.4 Phonon Transport in Nanowires

Due to the size confinement effects, the thermal properties of nanowires and nanotubes are different from the bulk. The first measurements of thermal conductivity in silicon nanowires were published in 2003 by Majumdar & Yang and Li et al. It was found that the thermal conductivities are much lower than that of the bulk and, decreases with wire diameter. In addition, the thermal conductivity decrease as the wire diameter is reduced, due to phonon boundary scattering rather than phonon–phonon Umklapp scattering processes. The observed thermal conductivity of single-crystalline silicon nanowires (c-SiNW) is less than 10 W/m.K for 20nm wire, which is two orders of magnitude smaller than the bulk value. This strong size-dependent thermal conductivity in these wires can be ascribed to the dominant role of boundary phonon (lattice vibration) scattering. The reduced thermal conductivity in semiconductor nanowires is greatly desired in thermoelectric generation and thermoelectric cooling, but is not preferable in conventional electronics and photonics devices.

At low temperatures, it was found that the thermal conductivity changes as T^3 for wide nanowires (about 100nm), or as T^2 for nanowires lengths widths about 50nm or as T for narrow nanowires (about 20nm), as shown in the Figure 22. For large diameter nanowires, theoretical models have been able to closely match the experimental results, assuming the nanowire diameters are comparable to the mean free path and that the mean free path is independent of phonon frequency. However, for very thin nanowires whose dimensions are comparable to the dominant phonon wavelength, a new model is required

Figure 22. Atomic structures of graphene layers wuth boron nitride (BN) in zigzag, and armchair interface and their superlattice cells

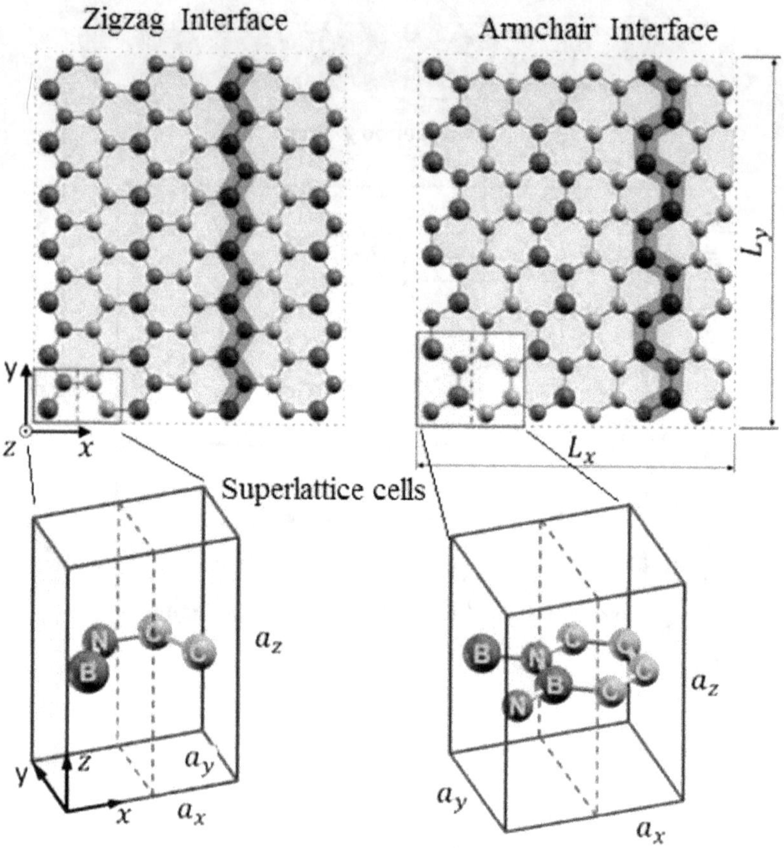

Figure 23. Thermal conductivity of SiNW with different diameters
After Majumdar & Yang, (Appl. Phys. Lett. 83, 3186, 2003).

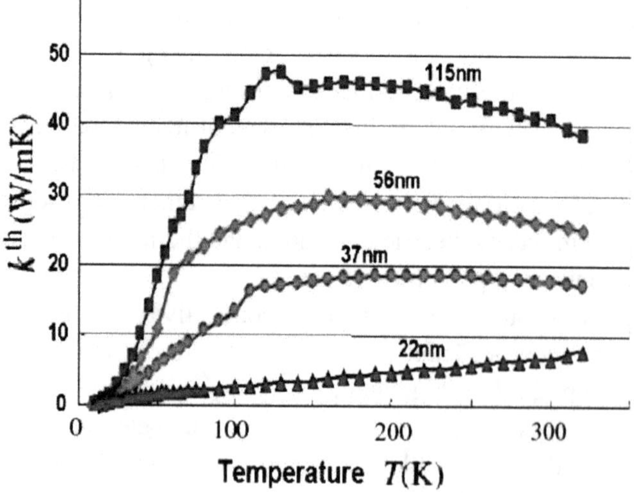

For large diameter nanowires the phonon-boundary scattering is frequency dependent. The mean free path is then given by:

$$\lambda^{-1} = \left(\frac{B}{d}\right)\left(\frac{h}{d}\right)^2\left(\frac{\omega}{\omega_D}\right)^2 N(\omega) \tag{72}$$

The parameter h is length scale associated with the disordered region, d is the nanowire diameter, $N(\omega)$ is number of modes at frequency ω, and B is a constant related to the disorder region. Thermal conductance is then calculated using Landauer-like formula:

$$G\left(T\right) = \frac{1}{2\pi}\int_0^\infty \left(\frac{\hbar^2\omega^2}{k_B T^2}\right)\left(\frac{N_1}{1+L/\lambda\left(\omega\right)} + \frac{N_2}{1+L/d}\right)\left(\frac{\exp(\hbar\omega/k_B T}{[\exp(\hbar\omega/k_B T)-1^2]}\right) \tag{73}$$

where d is the diameter, L is the length, λ is the frequency dependent mean free path, N_l is the number of modes with the mean free path λ, and N_2 is the number of modes with the mean free path limited to d, This formulism was developed by Chen *et al.* (2008) and has been successfully used to capture the thermal conductivity data of silicon nanowires.

7.5 Phonon Transport in Carbon Nanotubes

Thermal transport of carbon nanotube (CNT) and graphene become very important in their nano-electronic devices. The universal quantization of low-temperature thermal conductance in carbon nanotubes can be observed even in the presence of local structural defects such as vacancies and boundary defects, since the long wavelength acoustic phonons are not scattered by local defects. At room temperature, however, thermal conductance is critically affected by defect scattering since incident phonons are scattered by localized phonons around the defects. The low-temperature specific heat and thermal conductivity show direct evidence of 1-D quantization of the phonon band structure.

Modeling of the low-temperature specific heat allows determination of the on-tube phonon velocity, the splitting of phonon subbands on a single tube, and the interaction between neighboring tubes in a bundle. Measurements show a single-wall carbon nanotubes (SWNTs) room-temperature thermal conductivity about 3500 W/m·K. Thermal conductivity in CNT is mainly due to phonons rather than

Figure 24. Schematic artist of the heat transport across an MWNT

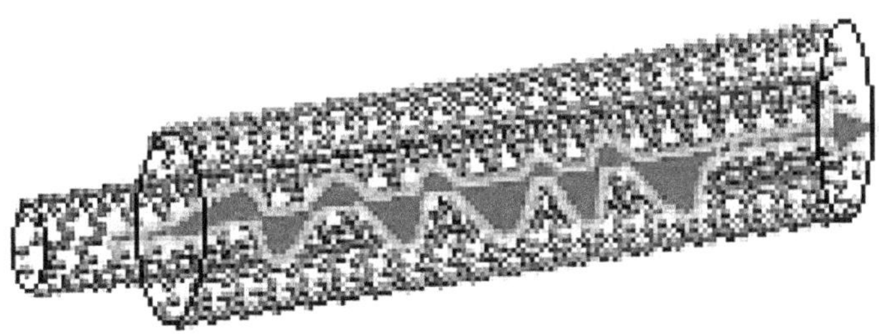

electrons so the Wiedemann–Franz law is not directly applicable. The diagonal elements of the thermal conductivity tensor, may be expressed as:

$$k_{zz} = \sum C v_z^2 \tau \tag{74}$$

where C_v is the specific heat, v_z is the group velocity and τ is th relaxation time of a given phonon. This definition is valid as long as the phonon wavelength is much smaller than the diameter of the SWNT.

At low temperatures (far less than Debye temperature), the relaxation time is determined by impurity scattering, sample boundaries, etc. and is roughly constant. Therefore, in ordinary materials, the low-temperature thermal conductivity has the same temperature dependence as the specific heat. However, in anisotropic materials, this relationship is not true. Because the contribution of each state is weighted by the scattering time and the square of the velocity, the thermal conductivity preferentially samples states with large velocity and scattering time. For instance, in graphite, the thermal conductivity parallel to the basal planes is only weakly dependent on the interlayer phonons. In SWNT bundles, it is likely that $k^h(T)$ depends only on the on-tube phonons, rather than the intertube modes. At high temperatures, three-phonon Umklapp scattering begins to limit the phonon relaxation time. Therefore the phonon thermal conductivity displays a peak and decreases with increasing temperature.

According to Berber *et al.* (2000), the phonon thermal conductivity of isolated nanotubes has a peak value of 37,000 W/m·K at about 100K, and decreases with increasing temperature. This value is comparable to the highest value ever measured (41,000 W/m·K for pure diamond at 104K). At room temperature, the thermal conductivity was also quite high (6600 W/m·K), twice the reported value of thermal conductivity of pure diamond at room temperature, as shown in Figure 26. The high thermal conductivity of nanotubes may be useful for a number of thermal management applications, such as heat sinking.

Figure 25. Calculated and measured thermal conductivity of isolated SWNT
After Berber et al (2006).

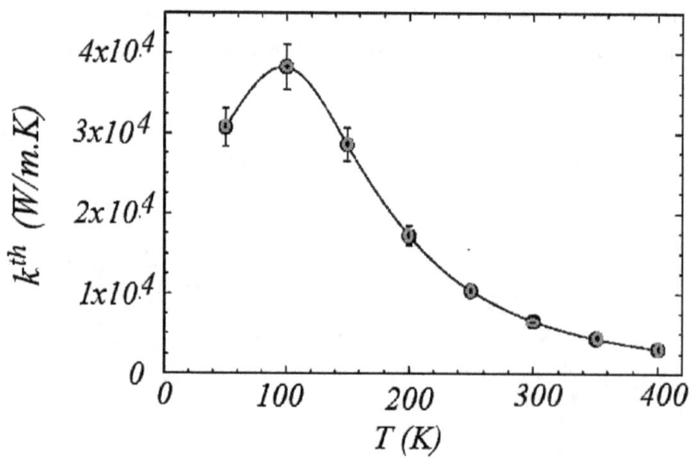

8. MOLECULAR DYNAMICS (MD) SIMULATION

For systems which are small enough, of the order of a micron or smaller, molecular dynamics (MD) can be used to simulate thermal transport. Molecular dynamics is a simulation technique where the time evolution of a set of interacting mass points (atoms) is followed by integrating their classical equations of motion. In MD we follow the Newton laws of mechanics. The forces between the mass points are defined as linear functions (difference equations) of the point displacements. Each atom is then treated as a point particle which interacts with other particles through an interaction potential. The Figure 26 depicts the flowchart of the MD simulationusing either empirical or quantum-based lattice potentials.

The required inputs to the MD simulation are initial conditions (atomic positions and velocities) and interatomic forces. The required forces between atoms may be obtained from interatomic potential energies. In order to find the interaction potential, empirical models (like Lennard-Jones potential model between two atoms) are used which are generally devised to fit experimental data in the bulk. However, force constants based on quantum *ab-initio* methods (like density functional theory DFT) or quantum approximate (like tight binding TB), are often used.

The position and velocity trajectory of every atom is calculated in the MD simulation for a set of discrete time steps. After obtaining the set of equations the next step is to integrate them. As atomistic vibrations occur at terahertz frequencies a time step on the order of 1fs is typically used. For bulk simulations, periodic boundary conditions (PBC) are usually applied to overcome the large scale problems. For simulating a nanostructure (e.g., a thin film or wire), free boundary conditions are applied in some directions. The resulting data can be analyzed and transport parameters such as thermal conductivity are extracted at the end of simulation.

In contrast to other anharmonic approaches, MD simulations include the full anharmonicity of the atomic interactions and all orders of phonon-phonon interactions. However, the vibrational modes in a MD simulation follow the classical Boltzmann statistics, which is the high-temperature limit of Bose-Einstein statistics.

8.1 MD Simulation Methods

There are three main types of MD simulation related to thermal transport: equilibrium, steady-state, and unsteady (Haile, 2001). In an equilibrium simulation, there are no spatial or temporal gradients in local temperature, allowing for averaging over space and time. In a steady-state, non-equilibrium simulation, spatial temperature variations may exist, but the system is steady in time, allowing for time averaging. Such simulations allow for direct observations of transport. In an unsteady simulation, spatial and temporal temperature variations may exist, allowing for observation of processes such as the propagation of a phonon wave packet. In all cases, the output of the MD simulation is the time histories of the positions and velocities of the constituent atoms. This output can be used to calculate forces, potential energy, kinetic energy, temperature, pressure, mechanical properties, and the thermal transport properties.

Figure 26. Flowchart of the MD simulation, for heat and phonon transport analysis

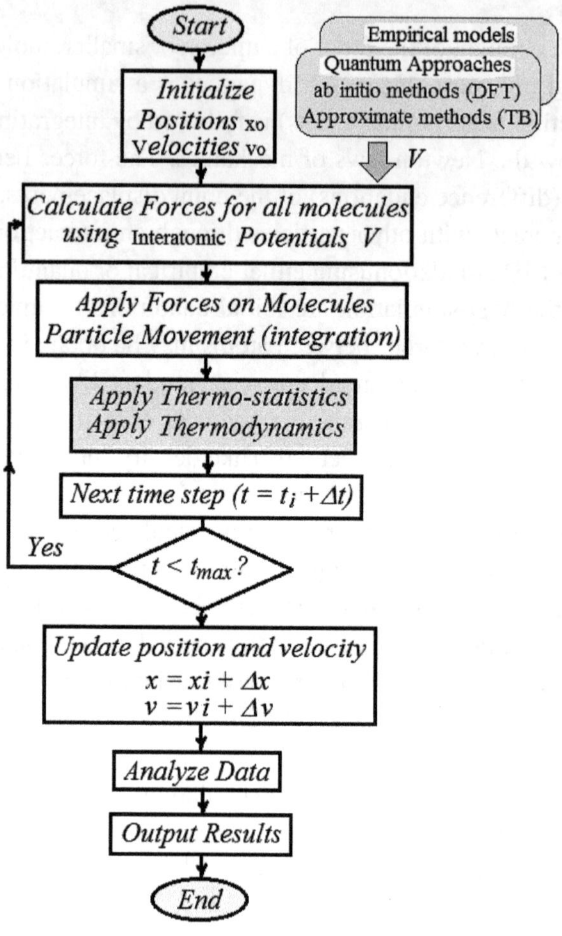

8.2 MD-Based Thermal Conductivity Models

We have already described the most famous models of thermal conductivity, which are based on the semiclassical approaches. Now we divert our attention to other models which are based on *ab-initio* (first principles) approaches, like the MD simulation. The models of thermal conductivity, which are based on molecular dynamics (MD) method, can be divided into two basic categories, namely:

- Equilibrium molecular dynamics (EMD) and
- Non-equilibrium molecular dynamics (NEMD).

In EMD, the system is initialized in a state which is non-equilibrium and is allowed to relax down to equilibrium. The rate of this relaxation is related to the thermal conductivity. In contrast, NEMD imposes a heat flux Q and calculates a temperature gradient $\Delta T/\Delta x$ and the thermal conductivity is computed as the ratio of the two $Q/(\Delta T/\Delta x)$.

Table 3. Comparison between MD-based thermal conductivity models

	Green-Kubo method	Direct method
Theory	Fluctuation-dissipation theorem	Fourier law
Type	Equilibrium **(EMD)**	Non-equilibrium **(NEMD)** Steady-state
Output	k^{th}	One component of kth
Length Scale	Phonon wavelength	Phonon mean free path

The so-called Green-Kubo method, uses equilibrium molecular dynamics (EMD) simulation. On the other hand, the so-called direct method, uses steady-state non-equilibrium molecular dynamics (NEMD) simulations. Table 3 shows a quick comparison between these methods

Green-Kubo Method

The Green-Kubo (G-K) method relates the equilibrium fluctuations of the heat flux, $\boldsymbol{Q}$, to the thermal conductivity tensor via the fluctuation-dissipation theorem. According to the G-K model, the thermal conductivity tensor components are given by following auto-correlation functions:

$$k_{ij}^{th} = \frac{1}{k_B V T^2} \int_0^{\infty} \left\langle Q_i(t) \cdot Q_j(0) \right\rangle dt, \quad i, j = x, y, z \tag{75}$$

Depending on the material system, the auto-correlation and averaging over symmetrical crystallographic directions may be possible. However, the Green-Kubo thermal conductivity shows a weak dependence on system size, and hence can capture the size effect in nano-structures and thin films.

Direct Method

In the direct method, a heat flow is applied across a simulation cell that is long in one direction and the resulting temperature gradient is used with the Fourier law, to predict the thermal conductivity in that direction.

8.3 Macroscopic, Atomistic, Ab initio Dynamics (MAAD)

The so-called macroscopic, atomistic, ab initio dynamics (MAAD) is a multi-scale approach that incorporates the molecular dynamics (MD) and the tight binding (TB) or density functional theory (DFT), to model nanostructures and nanodevices. In this method, a finite element method (FEM) is utilized to discretize the system domain until we approach the atomic spacing; then, the MD method is entered until we arrive in the concerned physical phenomenon. At this point, the TB or DFT quantum approach is utilized to calculate the band structure and coupling coefficients.

In MAAD, we have two overlapping regions, namely: FEM/MD and MD/TB/DFT. Such overlapping areas are termed *handshake* regions, where the contribution of each model to the total Hamiltonian is taken as the average of two separate Hamiltonians. The Lagrange multipliers are used to ensure the compatibility between the two models in the handshake regions.

Like other multiscale methods, the MAAD may be hierarchical or concurrent (or sometimes semi-concurrent), depending on the nature of the system and the phenomena to be simulated. In the hierarchical methods, the TB/DFT and MD models are executed consequently and the strain-stress relation is calculated first and used as a constitutive relation for the MD simulation. Alternatively, the two models are executed simultaneously in the concurrent methods. In the latter case, the quantum model (TB or DFT) is embedded in the MD algorithm and executed as frequently as needed in the fine-structure scale. The hierarchical methods are more efficient but the concurrent methods are sometimes preferable for nonlinear systems (Fish, 2010).

9. MEASUREMENT OF THERMAL TRANSPORT PARAMETERS

There are several ways to evaluate the thermal conductivity of a semi-conductor, including steady-state longitudinal heat flow, modified Angstrom's method, optical pump-probe, Laser flash, third-harmonic electrical method, and scanning thermal microscopy (SThM). Most of these approaches require either a destructive contacts and/or thick samples (>100μm), and do not have high spatial resolution

For measurements of heat conduction on submicron or nanometer length scales, temperature profile has to be monitored within the similar length scales (~1 μm). Also, the temperature decay profile should be monitored in nanosecond time scales, as well. Unfortunately, the conventional approaches to measure the thermal conductivity are incapable to reach such small length and time scales, and thus are not suitable for measurements of the thermal conductivity of nanostructures.

An important technique for measurements of heat conduction on nanometer scale is the time-domain thermoreflectance (TDTR). In addition, the cross-plane thermal conductivity of thin films can be measured by the so-called differential 3ω method.

9.1 Scanning Thermal Microscopy (SThM)

The Scanning thermal microscopy (SThM) mainly consists of a V-shaped probe, incorporated at the end of a flexible cantilever. The probe is made of a resistive element (Pt/10% Rh wire), with radius of curvature 1m. The probe tip is positioned using an atomic force microscopy (AFM) arrangement (Laser, mirrors, and detector), as shown in figure 26. The resistive element forms one leg (Rprobe) of a Wheatstone bridge. The bridge voltage should be controlled to keep the bridge balanced and thus to maintain a constant probe temperature. The SThM tip is operated at a preset value 405°C above the specimen temperature.

Figure 27. Illustration of the scanning thermal microscope (SthM) method

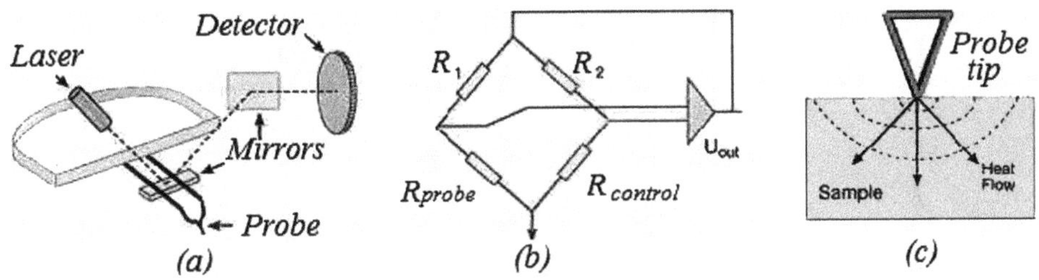

Figure 28. Illustration of the scanning thermal microscope (SthM) method

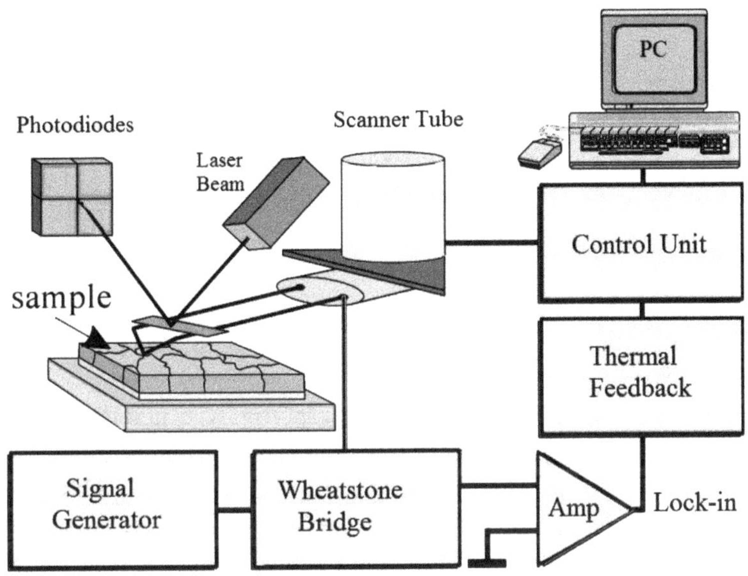

9.2 Time-Domain Thermoreflectance (TDTR)

The TDTR measurements should be performed using a two-tint configuration, as shown in Figure 29. Before TDTR measurements, the samples are usually coated with a thin layer (~80 nm) of metal with high thermo-reflectance (e.g., Al) by magnetron sputtering or thermal evaporation. The metal film serves as a transducer to absorb the heating pump beam and to convert the temperature excursions at the surface into changes in the intensity of the reflected probe beam. Usually, Al is used as the metal transducer due to the high thermo-reflectance.

In TDTR measurements, the output of a mode-locked Ti:sapphire laser oscillator is split into a pump beam and a probe beam, with the relative delay time between the pump and probe pulses being adjusted via a mechanical stage. Samples are heated by the pump beam, which is modulated by an electro-optic modulator at frequency f, $0.1 < f < 10$ MHz.

Cooling of the surface after being heated by pump pulses is then monitored through changes in the intensity of the reflected probe beam using a Si photodiode and a radio-frequency (RF) lock-in amplifier.

Figure 29. TDTR measurement apparatus

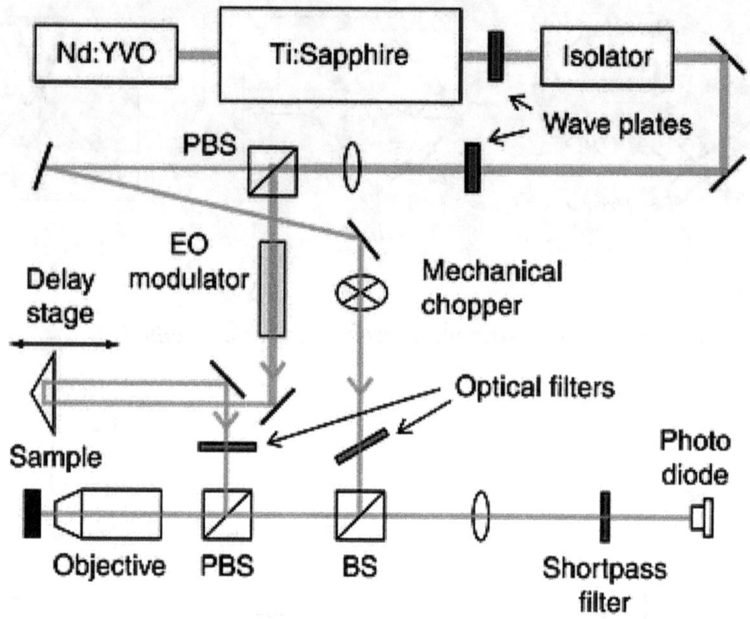

9.3 The 3-Omega Method

The 3-Omega technique is used to measure the thermal properties of materials very accurately. It is most commonly used to measure the through-plane thermal conductivity, but it can also measure the thermal diffusivity. As shown in Figure 30, an electrical current of angular frequency ω is driven through a metal heater line, causing Joule heating at a frequency 2ω. The periodic heating creates a thermal wave that penetrates the surrounding environment. The temperature oscillation then causes the resistance of the heater to oscillate at 2ω. Because the current is driven at a frequency ω and the resistance changes at a frequency 2ω, an RMS voltage at 3ω results. he 3ω voltage amplitude is directly measurable and provides information on the thermal properties of the desired material.

Figure 30. Standard heater line geometry for a thin film measurement

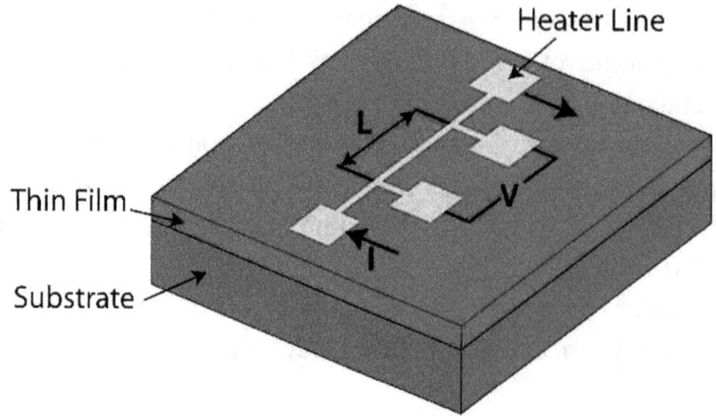

In particular, the cross-plane thermal conductivity of thin films can be measured by the differential 3ω method. A reference sample without the thin film of interest is prepared simultaneously with the sample containing the film of interest such that the metal line patterned on both samples has the same thickness and width. Both the thin film sample and the reference are measured using similar heating power and the same range of heater frequencies. The temperature drop across the thin film ΔT_f can then be derived from the difference in the amplitude of the temperature oscillation of the sample and the reference. Assuming one-dimensional (1D) heat conduction, the cross-plane thermal conductivity of the thin film is derived from ΔT_f. If the thin film is semiconducting, a thin dielectrics layer is required to electrically insulate the metal lines from the thin films. Figure 31 depicts the thermal conductivity of thin films of Si_3N_4 layers at room temperature. The measured thermal conductivity of Si_3N_4 thin films are smaller than that of the bulk materials found in literature, due to the scaling effects, and also strongly dependent on film thickness.

10. CASE STUDY 10: HEAT GENERATION AND TRANSPORT IN SOI MOSFETs

Phonons transport is very important in many thermal energy conversion applications, and heart evacuation from nanodevices and lasers. Heat conduction from transistors and interconnects has started to be a very critical design issue for the 22nm and lower technology nodes. In this example we look at the heat generation and transport mechanisms in silicon-on-insulator (SOI) MOSFET devices. In fact, the low thermal conductivity of buried oxide influences the heat transport and directly affects the performance of SOI and DG MOSFET devices. Whether the internal heat of a device is destructive depends on the rate at which heat is transferred away to its surroundings. The energy transport simulation of SOI devices, without taking proper physical models of heat transport presents several anomalies (Gritsch, 2002). In addition, the floating potential region in SOI devices, always presents a time consuming problem due to the slow convergence of the semiconductor equations with heat continuity equation. Therefore, a semi-

Figure 31. Measured heat capacity of a thin film of SiN

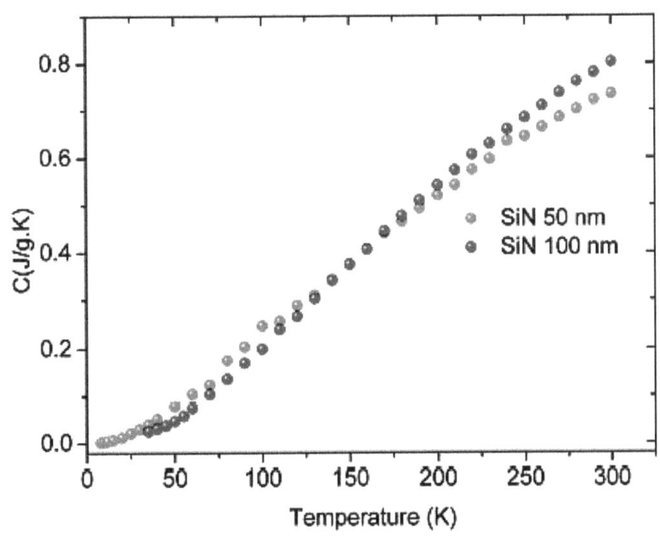

classical approach (BTE or complete HDM, with proper energy relaxation) should be adopted to find the electrical behavior and the Joule heating. A useful comparison of different phonon transport models for predicting heat conduction in SOI transistors can be found in (Sreekant et al, 2005) and (Nasri et al, 2015)

During the device operation, electrons are transferring their excess energy gained from the electric field to the phonon population by scattering. This results in heat generation throughout hot spots. The electron system cools in the drain end of the transistor through phonon emission. The phonon population then proceeds to evolve in a manner to bring its population back toward equilibrium. This evolution of the phonon distribution can be described by a phonon BTE, and solved by the methods we presented in previous sections of this Chapter, such as Monte Carlo method. Here, we present the simulation results with Monte Carlo method, which was developed by Pop *et al.* (2003) and later modified by Rowlette *et al.* (2008).

Figure 32. Energy flow mobility in Si at 300K
After Pop, Sinha & Goodson (2006).

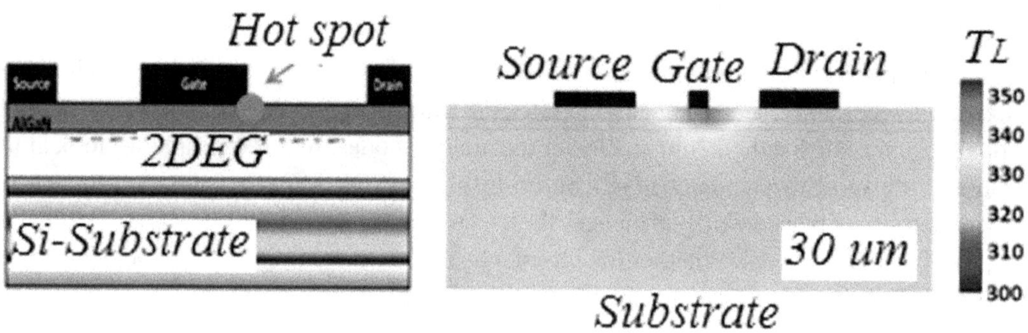

Figure 33. Temperature distribution in Si MOSFET, simulated with the phonon BTE (PBTE) and the conventional Fourier diffusive law
After Rowlette & Goodson (2008).

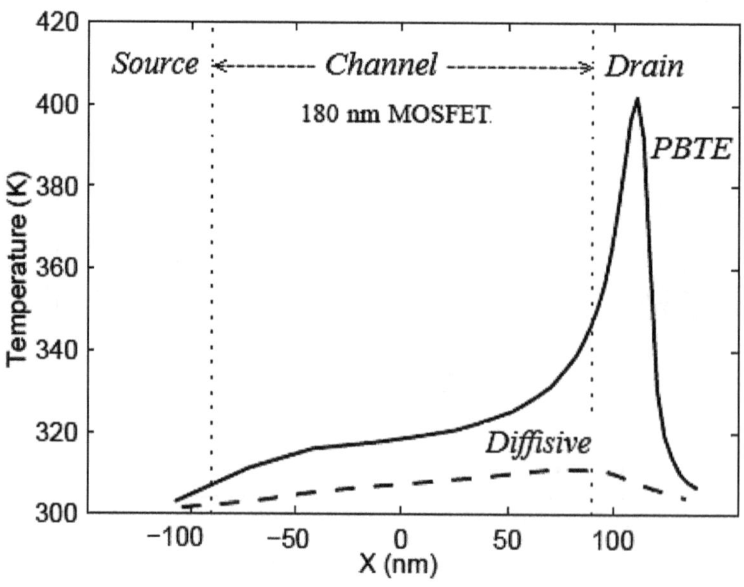

11. CASE STUDY 11: PHONON TRANSPORT ACROSS GRAPHENE

Single layer graphene (SLG) and multi-layer graphene (MLG) are promising semiconductor materials for nano-electronic devices due to their high carrier mobility, high thermal conductivity, and adjustable energy gap.

Experimental measurements of thermal properties of graphene encouraged researchers for more in-depth study of the heat conduction in graphene nanoribbons (GNR). Recently, a number of theoretical investigations about phonon transport and heat conduction were carried out in GNR with various lengths, widths and defect concentrations.

For instance, Evans, Hu & Keblinsky (2010) and Savin, Kivshar & Hu (2010), as well as Munoz, Lu & Yakobson (2010) used molecular dynamics (MD) simulations. Also, Zhai & Jin (2011) as well as Jinag & Wang (2011) used nonequilibrium Green's function (NEGF) method. In addition, Aksamija & Knezevic (2011) and Hu et al. (2011) made use of the phonon BTE approaches.

Keblinsky and co-workers (2010) found from the MD study that the thermal conductivity of a square graphene sheet is k^{th} ≈8000 - 10000 W/mK at 300K. This value was size independent for length L>5 nm . For the ribbons with fixed L=10 nm and width W varying from 1 to 10 nm, k^{th} increased from ~1000 W/mK to 7000 W/mK. The thermal conductivity in GNR with rough edges is reduced by orders of magnitude with respect to that in GNR with perfect edges (Savin, Kivshar & Hu, 2010). The isotopic defects also decrease the thermal conductivity (Jinag &Wang, 2011). The uniaxial stretching in the longitudinal direction enhances the low-temperature thermal conductance for armchair and zigzag GNR (Zhai & Jin, 2011).

Figure 34. Heat transport across graphene

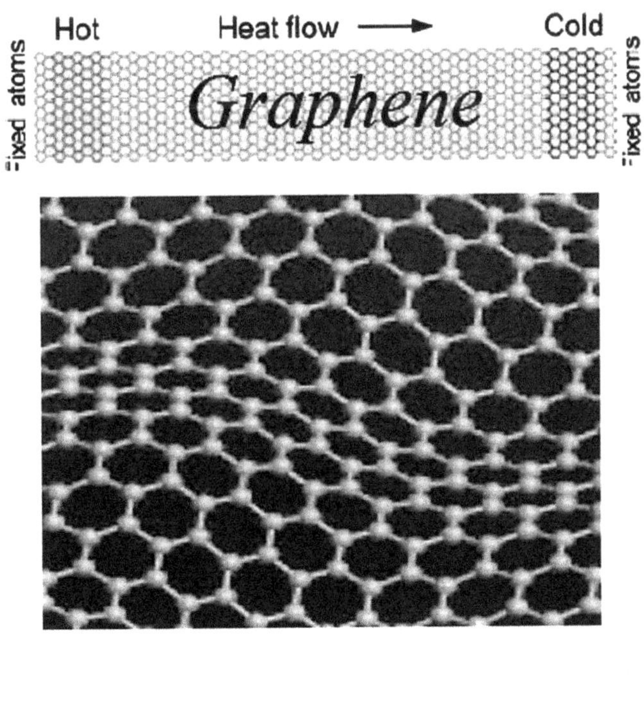

Aksamija and Knezevic (2011) calculated the dependence of the thermal conductivity of GNR, with 5nm width and 1nm edge roughness, on temperature. The thermal conductivity was calculated taking into account the three-phonon Umklapp and rough edge scatterings. The authors obtained thermal conductivity K ~5500 W/mK at T=300K for the graphene nanoribbon, as shown in Figure 35.

The study of the nonlinear thermal transport in rectangular and triangular GNRs under large temperature biases was reported by Hu et al. (2011). The authors found that in short rectangular GNRs (L~6nm), the negative differential thermal conductance exists in a certain range of the applied temperature difference. As the length of the rectangular GNR increases the effect weakens. Also, Ong and Pop (2011) examined thermal transport in graphene supported on SiO_2 using MD simulations. The approach employed by the authors utilized the reactive empirical bond order (REBO) potential to model the atomic interaction between the C atoms, Munetoh potential to model the atomic interactions between the Si and O atoms and Lennard-Jones potential to model the van der Waals type C-Si and C-O couplings.

Qiu and Ruan (2011) addressed the problem of relative contributions of ZA phonons to thermal transport in the framework of the equilibrium MD simulations. Their conclusion was that in suspended SLG out-of-plane ZA phonons are coupled with in-plane phonons due to the 3rd-order and higher-order anharmonic interactions, which results in about 25 – 30% contribution of ZA phonons to the thermal conductivity of graphene. Qiu and Ruan stated that the in-plane TA and LA phonons are the dominant heat carriers in supported SLG and make major contribution in suspended SLG.

12. PHONONIC NANODEVICES

The conventional thermoelectric devices are solid-state devices that convert heat into electricity on the basis of Seebeck effect or refrigerate objects by supplying an electrical current on the basis of the Peltier effect.

Figure 35. Thermal conductivity of GNRs of width W=5 nm and edge roughness Δ=1 nm, showing contributions from phonon branches TA, LA, ZA (flexural acoustic) and ZO (flexural optical). After Aksamija & Knezevic (2011).

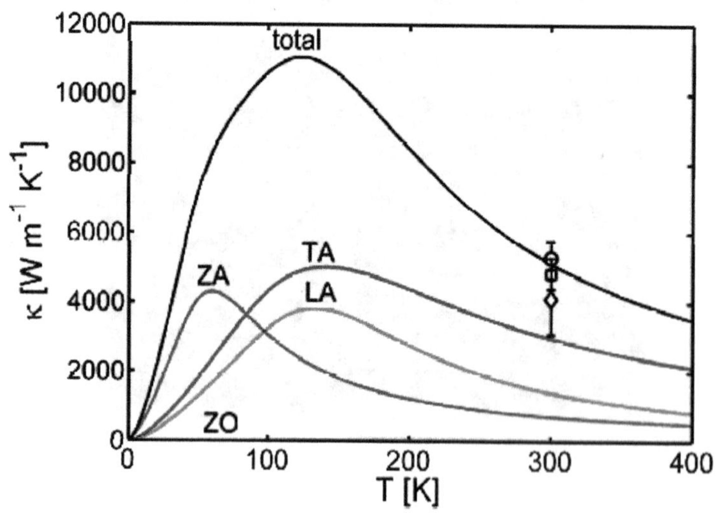

Figure 36. Schematic of a phononic device

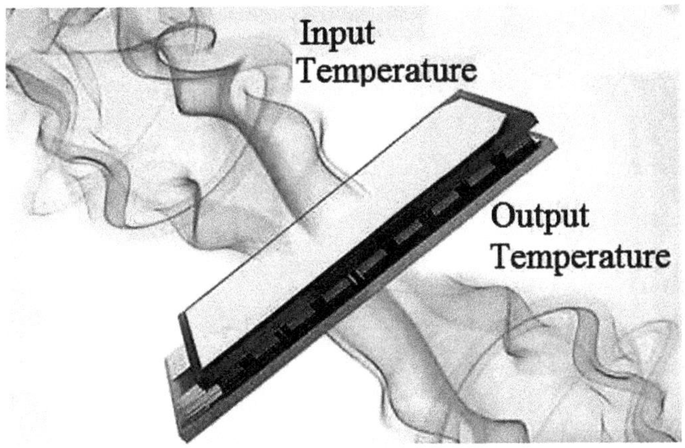

The so-called phononic nanodevices, such as nanoscale heat valves, thermal diodes and thermal transistors are potential candidates for controlling heat at the nanoscale level. They may be used for thermal insulation, as well as thermal storage and control in future microprocessors and computers. Therefore, we shouldn't think of heat as a waste by-product of conduction and switching processes. Instead, heat can be stored and maybe utilized again, resulting in better energy efficiency.

Several phononic devices such as diode, transistor and logic devices have been proposed in the literature. For instance, Chang et al. (2006) demonstrated the idea of solid-state thermal diode, which is formed by a non-uniform mass distribution of nanotubes. The resulting nanoscale device yields asymmetric axial thermal conductance with greater heat flow in the direction of decreasing mass density, and hence works as a heat rectifier. Also, Tian et al (2012) proposed a solid-state thermal rectifier based on reduced graphene oxide (*rGO*). Such thermal rectifiers have substantial implications for diverse thermal management issues, ranging from nanoscale calorimeters to macroscopic refrigerators

The phonon properties of some materials provide opportunities to thermally insulate buildings, reduce environmental noise, transform waste heat into electricity and develop earthquake protection (Goldsmid, 2010). Figure 38 illustrates the idea of the so-called phononic crystals, and their utilization in sound and heat isolation, by proper engineering of their phononic energy gaps.

Figure 37. Schematic of a thermal rectifier, using reduced graphene oxide (rGO) paper

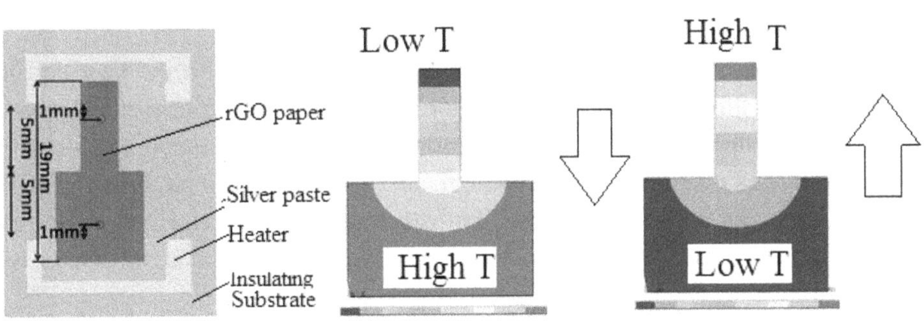

Figure 38. Schematic of phononic crystals and their potential in sound and heat isolation

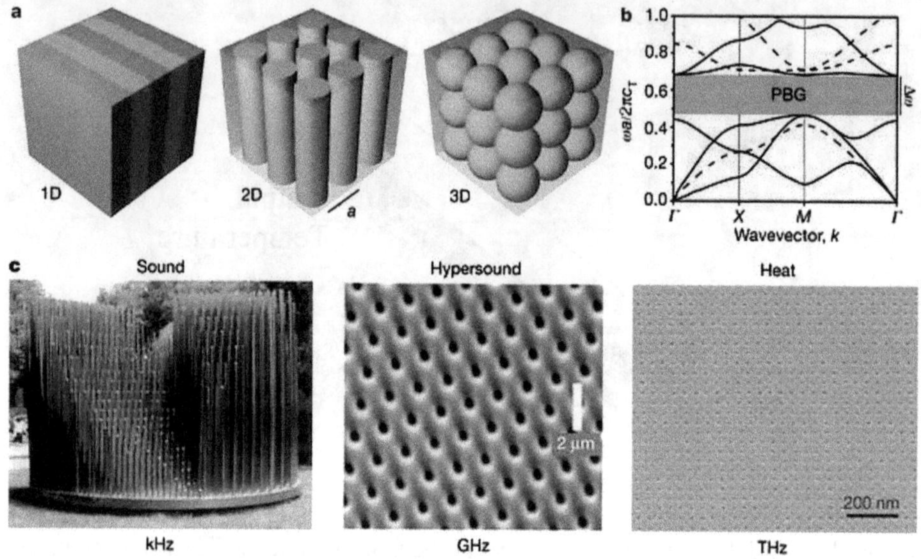

13. CHAPTER SUMMARY

The phonon is the quasi particle representing the mechanical vibrations of atoms and is responsible for the transmission of sound and heat. In semiconductors and dielectrics, energy transport occurs primarily through phonons. Phonon transport theory occupies a critical place in the hierarchy of information carrier transport. The heat conduction in solids is an extremely important physical process. It reveals the nature of lattice dynamics as well as the phonon scattering and transport mechanisms. Thermal conductivity of solids influences many technological issues, including thermal isolation and control of heat storage and conversion processes. In this chapter, we investigate phonon transport and thermal conductivity in semiconductor devices and nanostructures. Phonon transport modeling in semiconductor devices is still in the process of development, as compared to the mature state-of-the-art of electron transport modeling.

The recent experiments demonstrated room-temperature evidence of ballistic phonon transport in various nanomaterials. In particular, phonon transport properties in low-dimensional materials, such as carbon nanotubes (CNTs), have attracted much attention in the fields of nanodevices. The high thermal conductance opens the way to reduce the heating problems in nanometer-scale electron devices, while low thermal conductance leads to high efficiency of thermoelectric devices. In order to exploit the benefits of nano materials for energy applications, we must understand the physics of charge dynamics and heat transfer at the nanoscale.

Traditionally, heat transport has been described by Fourier's law with the heat equation, but an unphysical implication of this approach is that phonons can travel at infinite speed. The heat equation resolves this issue by adding a term to the heat equation that ensures a finite propagation velocity.

Microscopic approaches such as the Peierls-Boltzmann transport equation (phonon BTE) and Lattice Monte-Carlo (LMC) simulation can capture quasi-ballistic phonon transport. However, the phonon BTE is typically solved in the relaxation-time approximation. The phonon relaxation time, τ, as originally formulated in the BTE models of Callaway and Holland, gives an indication of how long it will take a system to return to equilibrium when one phonon mode has been perturbed. One can also think of the relaxation time as a temporal form of the phonon mean free path if the phonon-particle description is valid (i.e., $\tau = \lambda/v$), or as an indication of how long energy stays coherent in a given vibrational mode. Another approach, called the split-flux form of the phonon BTE (*p*-SFBTE), was proposed by Sinha *et al* to describe the phonon transport. Here, the phonon distribution is split into two terms. The first population term is a near-equilibrium component and has a Bose-Einstein distribution. The second population term is a nonequilibrium component, which dominates the transport near hotspots and is determined by solving the phonon BTE, typically in the relaxation-time approximation. The solution of the phonon BTE, particularly in the relaxation time approximation, is relatively faster than other microscopic approaches, such as lattice Monte Carlo (LMC) or the molecular dynamics (MD) methods. These approaches are valid only when heat transport is diffusive and the characteristic length scales are much larger than the phonon mean free path.

When phase coherence effects cannot be ignored, these semiclassical approaches fail and result is erroneous results. In fact, the recent results indicated that the heat diffusion equation significantly underestimates temperature distribution at nanoscales in the presence of an external heat source. The topic of ballistic (non-diffusive) phonon transport in nano structures has been subject of extensive theoretical work. However, the thermal conductivity data for Si and many semiconductors in the length scale of 10–100 nm is still rare due to the experimental challenges. For this reason, the researchers often recourse to modeling and simulation techniques. The simulations should respect the realistic phonon dispersion and address the quantum confinement and size effects in nanodevices. While first principles models and simulation techniques have contributed to our understanding of the phonon transport phenomena, there remains a need for simplified thermal transport approaches that capture the essential physics and that are computationally tractable.

Figure 39. Phonon transport models versus temperature and device scale size

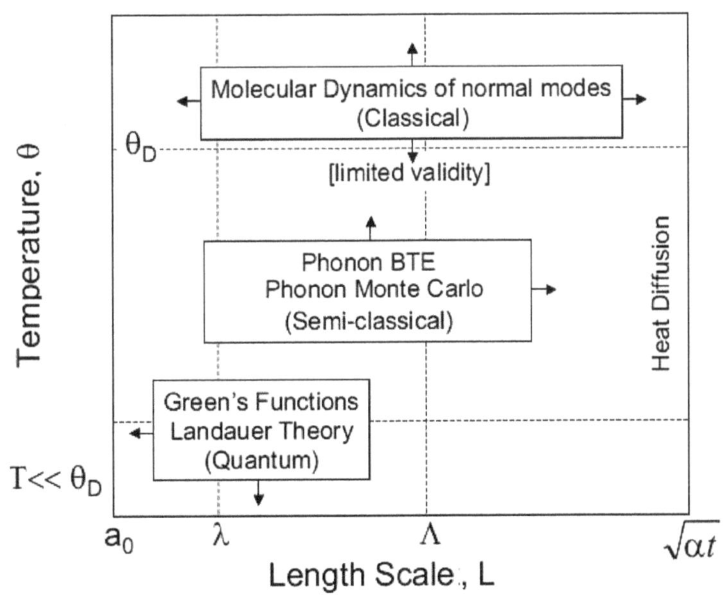

REFERENCES

Aaron, P., Tran, C., & Sansoz, F. (2013). *Semi-coherent Heat Conduction in Bulk and Single-Nanowire Twinning Superlattices*. Academic Press.

Aksamija, Z., & Knezevic, I. (2011). Lattice thermal conductivity of graphene nanoribbons: Anisotropy and edge roughness scattering. *Applied Physics Letters*, *98*(14), 141919. doi:10.1063/1.3569721

Armstrong, B. H. (1981). Two-fluid theory of thermal conductivity of dielectric crystals. *Physical Review B: Condensed Matter and Materials Physics*, *23*(2), 883–899. doi:10.1103/PhysRevB.23.883

Ashcroft, N. W., & Mermin, N. D. (1976). *Solid State Physics, HRW* (International Editions). CBS Publishing Asia Ltd.

Asheghi, M., Kurabayashi, K., Kasnavi, R., & Goodson, K. E. (2002). Thermal conduction in doped single-crystal silicon films. *Journal of Applied Physics*, *91*(8), 5079–5088. doi:10.1063/1.1458057

Bak, T. (Ed.). (1964). *Phonons and Phonon Interactions*. Benjamin.

Berber, S., Kwon, Y.-K., & Tománek, D. (2000). Unusually high thermal conductivity of carbon nanotubes. *Physical Review Letters*, *84*(20), 4613–4616. doi:10.1103/PhysRevLett.84.4613 PMID:10990753

Bir, G. L., & Pikus, G. E. (1974). *Symmetry and Strain Induced Effects in Semiconductors*. New York: Wiley.

Born, M., & Huang, K. (1954). *Dynamical Theory of Crystal Lattices*. Oxford, UK: Oxford University Press.

Brandt, D., et al. (2014). *Semiconductor phonon and charge transport Monte Carlo simulation using Geant*. Academic Press.

Brockhouse, B. N. (1959). Lattice Vibrations in Silicon and Germanium. *Physical Review Letters*, *2*(6), 256–258. doi:10.1103/PhysRevLett.2.256

Brown, E., Hao, L., Gallop, J. C., & Macfarlane, J. C. (2005). Ballistic thermal and electrical conductance measurements on individual multiwall carbon nanotubes. *Applied Physics Letters*, *87*(2), 023107. doi:10.1063/1.1993768

Buot, F. A. (2009). *Nonequilibrium Quantum Transport Physics in Nanosystems: Foundation of Computational Nonequilibrium Physics in Nanoscience and Nanotechnology*. World Scientific. doi:10.1142/6042

Büttiker, M., Imry, Y., Landauer, R., & Pinhas, S. (1985). Generalized Many-Channel Conductance Formula with Application to Small Rings. *Physical Review B: Condensed Matter and Materials Physics*, *31*(10), 6207–6215. doi:10.1103/PhysRevB.31.6207 PMID:9935492

Cahill, D., Ford, W. K., Goodson, K. E., Mahan, G. D., Majumdar, A., Maris, H. J., & Phillpot, S. R. et al. (2003). Nanoscale thermal transport. *Journal of Applied Physics*, *93*(2), 793–818. doi:10.1063/1.1524305

Callaway, J. (1959). Model for lattice thermal conductivity at low temperatures. *Physical Review*, *113*(4), 1046–1051. doi:10.1103/PhysRev.113.1046

Casimir, H. B. G. (1938). Note on the conduction of heat in crystals. *Physica*, *5*(6), 495–500. doi:10.1016/S0031-8914(38)80162-2

Cattaneo, C. (1958). A form of heat conduction equation which eliminates the paradox of instantaneous propagation. *CR (East Lansing, Mich.)*, *247*, 431–433.

Chang, C. W., Okawa, D., Majumdar, A., & Zettl, A. (2006). Solid-state thermal rectifier. *Science*, *314*(5802), 1121–1124. doi:10.1126/science.1132898 PMID:17110571

Chapman, S. (1916). On the law of distribution of molecular velocities, and on the theory of viscosity and thermal conduction, in a non-uniform simple monatomic gas. *Philosophical Transactions of the Royal Society of London*, *216*(538-548), 279–348. doi:10.1098/rsta.1916.0006

Chen, G. (2005). *Nanoscale Energy Transport and Conversion*. New York: Oxford University Press.

Chen, R., Hochbaum, A. I., Murphy, P., Moore, J., Yang, P., & Majumdar, A. (2008). Thermal Conductance of Thin Silicon Nanowires. *Physical Review Letters*, *101*(10), 105501–105505. doi:10.1103/PhysRevLett.101.105501 PMID:18851223

Chen, Y., Li, D., Lukes, J. R., & Majumdar, A. (2005). Monte Carlo Simulation of Silicon Nanowire Thermal Conductivity. *Journal of Heat Transfer*, *127*(10), 1129. doi:10.1115/1.2035114

Cheng, X., Xu, M., & Wang, L. (2008). From Boltzmann transport equation to single-phase lagging heat conduction. *International Journal of Heat and Mass Transfer*, *51*(25-26), 6018–6023. doi:10.1016/j.ijheatmasstransfer.2008.04.004

Chernatynskiy, A., & Phillpot, S. R. (2010). Evaluation of computational techniques for solving the Boltzmann transport equation for lattice thermal conductivity calculations. *Physical Review B: Condensed Matter and Materials Physics*, *82*(13), 134301. doi:10.1103/PhysRevB.82.134301

da Silva, C., Saiz, F. R., David, A., & Amon, C. H. (2016). Coherent phonon transport in short-period two-dimensional superlattices of graphene and boron nitride. *Physical Review B: Condensed Matter and Materials Physics*, *93*(12), 125427. doi:10.1103/PhysRevB.93.125427

Daly, B. C., Kang, K., Wang, Y., & Cahill, D. G. (2009). Picosecond ultrasonic measurements of attenuation of longitudinal acoustic phonons in silicon. *Physical Review B: Condensed Matter and Materials Physics*, *80*(17), 174112. doi:10.1103/PhysRevB.80.174112

Das, S. G., & Dhar, A. (2012). *Landauer formula for phonon heat conduction: relation between energy transmittance and transmission coefficient*. arXiv:1204.5595v1

Debye, P. (1912). The Theory of Specific Heat. *Annals of Physics*, *39*(14), 789–839. doi:10.1002/andp.19123441404

Dolling, G. (1963). Lattice vibrations in crystals with the diamond structure. *Symposium on Inelastic Scattering of Neutrons in Solids and Liquids*, 37-48.

Dong, Y., Cao, B.-Y., & Guo, Z.-Y. (2015). Ballistic–diffusive phonon transport and size induced anisotropy of thermal conductivity of silicon nanofilms. *Physica E, Low-Dimensional Systems and Nanostructures*, *66*, 1–6. doi:10.1016/j.physe.2014.09.011

Drabble, J. R., & Goldsmid, H. J. (1961). *Thermal Conduction in Semiconductors*. Oxford, UK: Pergamon Press.

Dresselhaus, M., Chen, G., Tang, M. Y., Yang, R. G., Lee, H., Wang, D. Z., & Gogna, P. et al. (2007). New directions for low-dimensional thermoelectric materials. *Advanced Materials*, *19*(8), 1043–1053. doi:10.1002/adma.200600527

Egley, J., Polsky, B., Min, B., Lyumkis, E., Penzin, O., & Foisy, M. (2000). SOI related simulation challenges with moment-based BTE solvers. Simulation of Semiconductor Processes & Devices, 241–244.

Einstein, A. (1907). The Planck Theory of Radiation and Theory of Specific Heat. *Annals of Physics*, *22*, 180–190.

Enskog, D. (1917). *Kinetische Theorie der Vorgänge in mässig verdünnten Gasen*. Uppsala Univertitet.

Evans, W. J., Hu, L., & Keblinsky, P. (2010). Thermal conductivity of graphene ribbons from equilibrium molecular dynamics: Effect of ribbon width, edge roughness, and hydrogen termination. *Applied Physics Letters*, *96*(20), 203112. doi:10.1063/1.3435465

Fan, X., Felipe, L., Pereira, C., Wang, H.-Q., Zheng, J.-C., Donadio, D., & Harju, A. (2015). *Force and heat current formulas for many-body potentials in molecular dynamics simulation with applications to thermal conductivity calculations*. arXiv:1503.06565v1 [cond-mat.mes-hall]

Fish, J. (Ed.). (2010). *Multi-scale Methods*. Oxford University Press.

Fourier, J. (1822). Théorie Analytique de la Chaleur. Gabay.

Franz, R., & Wiedemann, G. (1853). Ueber die Wärme-Leitungsfähigkeit der Metalle. [in German]. *Annalen der Physik*, *165*(8), 497–531. doi:10.1002/andp.18531650802

Geim G.K. & Novoselov, K.S. (2007). The rise of graphene. *Nature Matter, 6*, 183–191.

Glassenbenner, J. C., & Slack, G. A. (1964). Thermal Conductivity of Silicon and Germanium from 3K to the Melting Point. *Physical Review A.*, *134*(4A), 1058–1069. doi:10.1103/PhysRev.134.A1058

Goldsmid, H. J. (2010). *Introduction to Thermoelectricity* (Vol. 121). Springer. doi:10.1007/978-3-642-00716-3

Gritsch, M. (2002). *Numerical Modeling of Silicon-on-Insulator MOSFETs* (PhD Dissertation). TUW, Wein.

Grüneisen, E. (1912). Theorie des festen Zustandes einatomiger Elemente. *Annalen der Physik*, *344*(12), 257–306. doi:10.1002/andp.19123441202

Guo, Y., & Wang, M. (2015). Phonon hydrodynamics and its applications in nanoscale heat transport. *Physics Reports*, *595*, 1–44. doi:10.1016/j.physrep.2015.07.003

Guo, Y. & Wang, M. (2016). Lattice Boltzmann modeling of phonon transport. *Journal of Computational Physics Archive, 315*(C), 1-15.

Gurevich, V. L. (1986). *Transport in Phonon Systems*. Elsevier Science.

Guyer, R. A., & Krumhansl, J. A. (1966). Thermal conductivity, second sound, and phonon hydrodynamic phenomena in nonmetallic crystals. *Physical Review*, *148*(2), 778–788. doi:10.1103/PhysRev.148.778

Hamzeh, H., & Aniel, F. (2011). Monte Carlo study of phonon dynamics in III-V compounds. *Journal of Applied Physics*, *109*(6), 063511. doi:10.1063/1.3553409

Hellman, O., Abrikosov, I. A., & Simak, S. I. (2011). S.I. Lattice dynamics of an-harmonic solids from first principles. *Physical Review B: Condensed Matter and Materials Physics*, *84*(18), 180301. doi:10.1103/PhysRevB.84.180301

Henry, A. S., & Chen, G. (2008). Spectral Phonon Transport Properties of Silicon Based on Molecular Dynamics Simulations and Lattice Dynamics. *Journal of Computational and Theoretical Nanoscience*, *5*(2), 1–12. doi:10.1166/jctn.2008.2454

Herring, C. (1954). Role of low-energy phonons in thermal conduction. *Physical Review*, *95*(4), 954–965. doi:10.1103/PhysRev.95.954

Hirose, K., & Kobayashi, N. (2014). *Quantum Transport Calculations for Nanosystems*. Pan Stanford.

Hochbaum, A. I., Chen, R., Delgado, R. D., Liang, W., Garnett, E. C., Najarian, M., & Yang, P. et al. (2008). Enhanced thermoelectric performance of rough silicon nanowires. *Nature*, *451*(7175), 163–167. doi:10.1038/nature06381 PMID:18185582

Holland, M. G. (1963). Analysis of lattice thermal conductivity. *Physical Review*, *32*(6), 2461–2471. doi:10.1103/PhysRev.132.2461

Hu, J., Wang, Y., Vallabhaneni, A., Ruan, X., & Chen, Y. (2011). Nonlinear thermal transport and negative differential thermal conductance in graphene nanoribbons. *Applied Physics Letters*, *99*(11), 113101. doi:10.1063/1.3630026

Jiaung, W.-S., & Ho, J.-R. (2008). Lattice-Boltzmann modeling of phonon hydrodynamics. *Physical Review E: Statistical, Nonlinear, and Soft Matter Physics*, *77*(6), 066710. doi:10.1103/PhysRevE.77.066710 PMID:18643400

Jinag, J.-W., Wang, B.-S., & Wang, J.-S. (2011). First principle study of the thermal conductance in graphene nanoribbon with vacancy and substitutional silicon defects. *Applied Physics Letters*, *98*(11), 113114. doi:10.1063/1.3567768

Ju, Y. S., & Goodson, K. E. (1999). Phonon scattering in silicon thin films with thickness of order 100 nm. *Applied Physics Letters*, *74*(20), 3002–3007. doi:10.1063/1.123994

Kapitza, P. L. (1941). The study of heat transfer in helium II. *J. Phys (USSR)*, *4*, 181–210.

Kaviany, M. (2008). *Heat Transfer Physics*. Cambridge Academic Press. doi:10.1017/CBO9780511754586

Khalatnikov, I. M. (1952). Teploobmen mezhdu tverdym telom i geliem -II. *Journal of Experimental and Theoretical Physics*, *22*(6), 687–704.

Khan, A. I., Navid, I. A., Noshin, M., Uddin, H., Hossain, F., & Subrina, S. (2015). Equilibrium Molecular Dynamics (MD) Simulation Study of Thermal Conductivity of Graphene Nanoribbon: A Comparative Study on MD otentials. *Electronics*, *4*(4), 1109–1124. doi:10.3390/electronics4041109

Kittel, C., & Kroemer, H. (1980). *Thermal Physics* (2nd ed.). W. H. Freeman.

Kong, L. T. (2011). Phonon dispersion measured directly from molecular dynamics simulations. *Computer Physics Communications, 182*(10), 2201–2207. doi:10.1016/j.cpc.2011.04.019

Koswatta, S. O., Hasan, S., Lundstrom, M. S., Anantram, M. P., & Nikonov, D. E. (2007). Nonequilibrium Greens Function Treatment of Phonon Scattering in Carbon-Nanotube Transistors. *IEEE Transactions on Electron Devices, 54*(9), 2339–2351. doi:10.1109/TED.2007.902900

Koukaras, E. N., Kalosakas, G., Galiotis, C., & Papagelis, K. (2015). Phonon properties of graphene derived from molecular dynamics simulations. *Scientific Reports, 5*, 12923. doi:10.1038/srep12923 PMID:26316252

Kubo, R. (1966). Fluctuation–dissipation theorem. *Reports on Progress in Physics, 29*(1), 255–284. doi:10.1088/0034-4885/29/1/306

Kunal, K., & Aluru, N. R. (2011). Akhiezer damping in nanostructures. *Physical Review B: Condensed Matter and Materials Physics, 84*(24), 245450. doi:10.1103/PhysRevB.84.245450

Lai, J., & Majumdar, A. (1996). Concurrent thermal and electrical modeling of submicrometer silicon devices. *Journal of Applied Physics, 79*(9), 7353–7363. doi:10.1063/1.361424

Lee, J., Lim, J., & Yang, P. (2015). Ballistic phonon transport in holey silicon. *Nano Letters, 15*(5), 3273–3279. doi:10.1021/acs.nanolett.5b00495 PMID:25861026

Lee, J., Roy, A. K., & Farmer, B. L. (2011). Kapitza resistance in the lattice Boltzmann-Peierls-Callaway equation for multiphase phonon gases. *Physical Review E: Statistical, Nonlinear, and Soft Matter Physics, 83*(5), 056706. doi:10.1103/PhysRevE.83.056706 PMID:21728692

Lee, S. (2015). Hydrodynamic phonon transport in suspended graphene. *Nature Comm., 6*, 6290. Available online: www.nature.com/naturecommunications

Levinstein, M., Rumyantsev, S., & Shur, M. (Eds.). (1996). Handbook Series on Semiconductor Parameters (Vol. 1). Singapore: World Scientific.

Li, D., Wu, Y., Kim, P., Shi, L., Yang, P., & Majumdar, A. (2003). A. Thermal Conductivity of Individual Silicon Nanowires. *Applied Physics Letters, 83*(14), 2934–2936. doi:10.1063/1.1616981

Li, X. & Yang R. (2012). Size-dependent phonon transmission across dissimilar material interfaces. *Journal of physics. Condensed matter: Institute of Physics journal,* 24(15), 155302.

Lim, J. W., Hippalgaonkar, K., Andrews, S. C., Majumdar, A., & Yang, P. D. (2012). Quantifying Surface Roughness Effects on Phonon Transport in Silicon Nanowires. *Nano Letters, 12*(5), 2475–2482. doi:10.1021/nl3005868 PMID:22524211

Lindsay, L., Broido, D. A., & Reinecke, T. L. (2013). Ab initio thermal transport in compound semiconductors. *Physical Review B: Condensed Matter and Materials Physics, 87*(16), 165201. doi:10.1103/PhysRevB.87.165201

Little, W. A. (1959). The Transport of Heat Between Dissimilar Solids at Low Temperatures. *Canadian Journal of Physics, 37*(3), 334–349. doi:10.1139/p59-037

Luckyanova, M. N., Garg, J., Esfarjani, K., Jandl, A., Bulsara, M. T., Schmidt, A. J., & Chen, G. et al. (2012). Coherent Phonon Heat Conduction in Superlattices. *Science*, *338*(6109), 936–939. doi:10.1126/science.1225549 PMID:23161996

Lugli, P., Bordone, P., Reggiani, L., Rieger, M., Kocevar, P., & Goodnick, S. M. (1989). Monte carlo studies of nonequilibrium phonon effects in polar semiconductors and quantum wells. i. laser photoexcitation. *Physical Review B: Condensed Matter and Materials Physics*, *39*(11), 7852–7865. doi:10.1103/PhysRevB.39.7852 PMID:9947468

Lugli, P., Jacoboni, C., Reggiani, L., & Kocevar, P. (1987). Monte carlo algorithm for hot phonons in polar semiconductors. *Applied Physics Letters*, *50*(18), 1251–1253. doi:10.1063/1.97925

Ma, J., Li, W., & Luo, X. (2014). Examining the Callaway model for lattice thermal conductivity. *Physical Review B: Condensed Matter and Materials Physics*, *90*(3), 035203. doi:10.1103/PhysRevB.90.035203

Maassen, J., & Lundstrom, M. S. (2015). A Simple Boltzmann Transport Equation for Ballistic to Diffusive Transient Heat Transport. *J. of Applied Physics*. Available from http://www.researchgate.net/publication/271218501

Maiti, A., Mahan, G. D., & Pantelides, S. T. (1997). Dynamical simulations of nonequilibrium processes heat flow and the Kapitza resistance across grain boundaries. *Solid State Communications*, *102*(7), 517–521. doi:10.1016/S0038-1098(97)00049-5

Majumdar & Yang. (2003). Article. *Applied Physics Letters*, *83*, 3186. doi:10.1063/1.1619221

Malladi, M. (2013). *Phonon Transport Analysis of Semiconductor Nano-composites using Monte Carlo Simulations* (Master's Thesis). Clemson University.

Mandl, F. (2008). Statistical Physics (2nd ed.). John Wiley & Sons.

Maxwell, J. C. (1860). Illustrations of the dynamical theory of gases. Part I. On the motions and collisions of perfectly elastic spheres. *Philosophical Magazine*, *4*(19), 19–32.

Maxwell, J. C. (1860). Illustrations of the dynamical theory of gases. Part II. On the process of diffusion of two or more kinds of moving particles among one another. *Philosophical Magazine*, *4*(20), 21–37.

McKelvey, J. P., Longini, R. L., & Brody, T. P. (1961). Alternative Approach to the Solution of Added Carrier Transport Problems in Semiconductors Phys. *Rev.*, *123*, 51.

Mingo, D. A., Broido, D.A., Lindsay, L., & Li, W. (2013). Ab-Initio Thermal Transport. In Length-Scale Dependent Phonon Interactions. Springer-Verlag.

Mohammadi, S., Eftekhar, A. A., Khelif, A., & Adibi, A. (2010). Simultaneous two-dimensional phononic and photonic band gaps in opto-mechanical crystal slabs. *Optics Express*, *18*(9), 9164–9172. doi:10.1364/OE.18.009164 PMID:20588763

Munoz, E., Lu, J., & Yakobson, B. I. (2010). Ballistic thermal conductance of Graphene ribbons. *Nano Letters*, *10*(5), 1652–1656. doi:10.1021/nl904206d PMID:20402531

Murphy, P. G., & Moore, J. E. (2007). Coherent phonon scattering effects on thermal transport in thin semiconductor nanowires. *Physical Review B: Condensed Matter and Materials Physics*, 76.

Muscato, O., & Di Stefano, V. (2011). Hydrodynamic modeling of the electro-thermal transport in silicon semiconductors. *J. Physics A: Math. Theoretical*, *44*(10), 105501. doi:10.1088/1751-8113/44/10/105501

Narumanchi, V. S., Murthy, J. Y., & Amon, C. H. (2005). Comparison of Different Phonon Transport Models for Predicting Heat Conduction in Silicon-on-Insulator Transistors. *Journal of Heat Transfer*, *127*(7), 713–723. doi:10.1115/1.1924571

Nasri, F., Ben Aissa, M. F., Gazza, M. H., & Belmabrouk, H. (2015). 3D thermal conduction in a nanoscale Tri-Gate MOSFET based onsingle-phase-lag model. *Applied Thermal Engineering*, *91*(1).

Nasri, F., Echouchene, F., Ben Aissa, M. F., & Belmabrouk, H. (2015). Investigation of Self-Heating Effects in a 10-nm SOI-MOSFET with an Insulator Region Using Electrothermal Modeling. *IEEE Transactions on Electron Devices*, *62*(8), 2410–2415. doi:10.1109/TED.2015.2447212

Ong, Z.-Y., & Pop, E. (2011). Effect of substrate modes on thermal transport in supported graphene. *Physical Review B: Condensed Matter and Materials Physics*, *84*(7), 075471. doi:10.1103/PhysRevB.84.075471

Ouyang T, Chen Y P, Yang K K, & Zhong J X (2009). Thermal transport of isotopic superlattice graphene nanoribbons with zigzag edge. *EPL*, *88*, 28002.

Peirels, R. E. (1955). *Quantum Theory of Solids*. Oxford Clarendon Press.

Pohl, R. O. (1960). Influence of F centers on the lattice thermal conductivity in LiF. *Physical Review*, *118*(6), 1499–1508. doi:10.1103/PhysRev.118.1499

Pop, E. (2010). Energy dissipation and transport in nanoscale devices. *Nano Resourses*, *3*(3), 147–169. doi:10.1007/s12274-010-1019-z

Pop, E., Banerjee, K., Sverdrup, P., Dutton, R., & Goodson, K. (2003). *Localized Heating Effects and Scaling of Sub-0.18 Micron CMOS Devices*. IEEE.

Pop, E., Mann, D., Wang, Q., Goodson, K., & Dai, H. (2006). Thermal Conductance of an Individual Single-Wall Carbon Nanotube. *Nano Letters*, *6*(1), 96–100. doi:10.1021/nl052145f PMID:16402794

Pop, E., Sinha, S., & Goodson, K. E. (2006). Heat Generation and Transport in Nanometer-Scale Transistors. *Proceedings of the IEEE*, *94*(8), 1587–1601. doi:10.1109/JPROC.2006.879794

Potz, W., & Kocevar, P. (1992). *Hot Carriers in Semiconductor Nanostructures: Physics and Applications* (J. Shah, Ed.). Boston: Academic Press.

Qiu, B., & Ruan, X. (2011). *Molecular dynamics simulations of thermal conductivity and spectral phonon relaxation time in suspended and supported graphene.* arXiv:111.4613v1

Ravichandran, J. (2014). Crossover from incoherent to coherent phonon scattering in epitaxial oxide superlattices. *Nature Materials*, *13*(168). PMID:24317186

Rego, L. G. C., & Kirczenow, G. (1999). Fractional Exclusion Statistics and the Universal Quantum of Thermal Conductance: A Unifying Approach. *Physical Review B: Condensed Matter and Materials Physics*, *59*(20), 13080–13086. doi:10.1103/PhysRevB.59.13080

Ren, J. (2013). *Predicted Rectification and Negative Differential Spin Seebeck Effect at Magnetic Interfaces*. arXiv: 1305.0301

Rhyner, R. (2015). *Quantum Transport Beyond the Ballistic Limit*. Hartung-Gorre Verlag.

Romanov, D. A., Glavin, B. A., Mitin, V. V., & Stroscio, M. A. (1999). Stimulated decay of non-selectively pumped optical phonons in GaAs. *Physical Review B: Condensed Matter and Materials Physics*, *60*(7), 4771–4777. doi:10.1103/PhysRevB.60.4771

Rowe, D.M. (2006). *Thermoelectrics handbook: macro to nano*. CRC/Taylor & Francis.

Rowlette, J. A., & Goodson, K. E. (2008). Fully-Coupled, Nonequilibrium, Electron-Phonon Transport in Nanometer-Scale Silicon FETs. *IEEE Transactions on Electron Devices*, *55*(1), 220–232. doi:10.1109/TED.2007.911043

Savin, A. V., Kivshar, Y. S., & Hu, B. (2010). Suppression of thermal conductivity in graphene nanoribbons with rough edges. *Physical Review B: Condensed Matter and Materials Physics*, *82*(19), 195422. doi:10.1103/PhysRevB.82.195422

Schroeder, D. V. (2000). *An Introduction to Thermal Physics*. Addison-Wesley.

Schwab, K., Henriksen, E. A., Worlock, J. M., & Roukes, M. L. (2000). Measurement of the quantum of thermal conductance. *Nature*, *404*(6781), 974–977. doi:10.1038/35010065 PMID:10801121

Seol, J. H., Jo, I., Moore, A. L., Lindsay, L., Aitken, Z. H., Pettes, M. T., & Shi, L. et al. (2010). Two-dimensional phonon transport in supported graphene. *Science*, *328*(5975), 213–216. doi:10.1126/science.1184014 PMID:20378814

Shavit, A., & Gutfinger, C. (2009). Thermodynamics, From Concepts to Applications (2nd ed.). CRC Press.

Sinha, S., & Goodson, K. E. (2005). Review: Multiscale Thermal Modeling in Nanoelectronics. Int. *Journal for Multiscale Computational Engineering*, *3*(1), 107–133. doi:10.1615/IntJMultCompEng.v3.i1.80

Sinha, S., Pop, E., Dutton, R. W., & Goodson, K. E. (2006). Non-equilibrium phonon distributions in sub-100 nm silicon transistors. *Trans. ASME. Journal of Heat Transfer*, *128*(7), 638–647. doi:10.1115/1.2194041

Srivastava, G. P. (1990). *The Physics of Phonons, Adam Hilger*. Bristol: IOP Publishing Ltd.

Sun, L. & Murthy, J. (2010). Molecular Dynamics Simulation of Phonon Scattering at Silicon/Germanium Interfaces. *Journal of Heat Transfer-Trans ASME,* 132.

Sverdrup, P. G., Ju, S. Y., & Goodson, K. E. (2001). Sub-continuum simulation of heat conduction in silicon-on-insulator transistors. *ASME J. Heat Transfer*, *123*(1), 130–137. doi:10.1115/1.1337651

Tian, H., Xie, S., Ren, T.-L., Zhang, G., Wang, Y.-F., Zhou, C.-J., & Liu, L.-T. et al. (2012). A Novel Solid-State Thermal Rectifier Based On Reduced Graphene Oxide. *Scientific Reports*, *2*, 523. doi:10.1038/srep00523 PMID:22826801

Touloukian, Y. (1970). *Thermo-physical Properties of Matter* (Vol. 2). New York: Plenum.

Tsallis, C. (1988). Possible generalization of Boltzmann-Gibbs statistics. *Journal of Statistical Physics*, *52*(1-2), 479–487. doi:10.1007/BF01016429

Tzou, D. Y. (1995). The generalized lagging response in small-scale and high-rate heating. *International Journal of Heat and Mass Transfer*, *38*(17), 3231–3240. doi:10.1016/0017-9310(95)00052-B

Vapaille, A. (1970). *Physique des Dispositifs a Semiconducteur*. Paris: Masson & Cie.

Vernotte, P. (1958). Les paradoxes de la théorie continue de l'équation de la chaleur. *C. R. Acad. Sci.*, *246*, 3154–3155.

Ward, A., & Broido, D. A. (2008). Intrinsic lattice thermal conductivity of Si/Ge and GaAs/AlAs superlattices. *Physical Review B: Condensed Matter and Materials Physics*, 77.

Wilson, A. A. et al.. (2015). Thermal conductivity measurements of high and low thermal conductivity films using a scanning hot probe method in the 3ω mode and novel calibration strategies. *Royal Society of Chemistry*, *7*, 15404–15412. PMID:26335503

Wong, B. T., Francoeur, M., & Mengüç, M. P. (2011). A Monte Carlo simulation for phonon transport within silicon structures at nanoscales with heat generation. *International Journal of Heat and Mass Transfer*, *54*(9–10), 1825–1838. doi:10.1016/j.ijheatmasstransfer.2010.10.039

Wu, L.-A., & Segal, D. (2011). Quantum heat transfer: A Born-Oppenheimer method. *Physical Review E: Statistical, Nonlinear, and Soft Matter Physics*, *83*(5), 051114–0511148. doi:10.1103/PhysRevE.83.051114 PMID:21728497

Yamamoto, T. & Watanabe, K. (2006). Nonequilibrium Green's Function Approach to Phonon Transport in Defective Carbon Nanotubes. *Phys. Rev. Lett., 96*.

Yoder, P. D., Natoli, V. D., & Martin, R. M. (1993). Ab Initio Analysis of the Electron-Phonon Interaction in Silicon. *Journal of Applied Physics*, *73*(9), 4378–4383. doi:10.1063/1.352801

Zaitlin, M. P., & Anderson, M. C. (1975). Phonon thermal transport in non-crystalline materials. *Physical Review B: Condensed Matter and Materials Physics*, *12*(10), 4475–4486. doi:10.1103/PhysRevB.12.4475

Zhai, X, & Jin, G. (2011). Stretching-enhanced ballistic thermal conductance in graphene Nanoribbons. *EPL, 96*, 16002.

Zhang, W., Fisher, T. S., & Mingo, N. (2007). Simulation of Interfacial Phonon Transport in Si–Ge Heterostructures Using an Atomistic Greens Function Method. *Journal of Heat Transfer*, *129*(4), 483. doi:10.1115/1.2709656

Ziman, J. M. (1972). *Electrons and Phonons*. Oxford, UK: Oxford University Press.

Zwanzig, R. (2003). *Nonequilibrium Statistical Mechanics*. Oxford University.

ENDNOTES

1 This is true in ordinary systems where the kinetic energy of a gas dominates the energy of spin and the total average energy U is approximately equal to the average kinetic energy E.

2 Unlike particles of ordinary gas, phonons can be created or destroyed by energy fluctuations. In the statistical mechanics this means that the chemical potential for adding a phonon is zero.

3 Materials generally change their size when subjected to a temperature change while the pressure is held constant. In solid materials, the pressure does not appreciably affect the size of matter

4 In metals, the transfer of heat is strongly affected by electrons transport.

5 For simplicity, we'll write the tonsorial elements of force constant as C_{ij}. Actually, the i and j sum over all atoms in the system, and the u and v sum over the x, y, and z directions in 3-D crystals.

6 Normal sound waves are fluctuations in the density of atoms or molecules in a substance.

7 Actually, many authors name the Peierls-BTE as phonon BTE

8 Note that in the R scattering processes, the phonons exchange momentum such that each mode tends towards a B-E equilibrium distribution, whereas the N-scattering processes result in collective drift effects leading to a displaced B-E distribution.

9 Note that the temperature is strictly defined at the state of thermal equilibrium.

10 At small phonon energy, all Q1D structures have 4 acoustic modes, one longitudinal, two flexural, and one torsional modes, which do not depend on the structure details. In this case the thermal conductivity shows a universal quantization.

Chapter 8
Photon Transport

1. OVERVIEW AND CHAPTER OBJECTIVES

Photons are amazing information carriers that travel with the speed of light! The science of photonics is large and encompasses the generation, conversion, transport, guiding and detection of photons. Accurate modeling of photonic devices is essential for the development of high performance optical components required by current and future communication systems. In fact, the light speed and low noise properties make photons indispensable for telecommunications and information processing. Therefore, the modeling and simulation of photon transport, together with charge-carrier and phonon-heat transport, is of high interest for optoelectronic and photonic device design. In addition, the band structure calculation of the conveying medium and other quantization effects should be carried out, in order to obtain correct results. Figure 1 shows how transport models of charge carriers, phonons and photons should be integrated in the simulation process of a laser diode.

The description of photon transport in photonic semiconductor devices in general, and semiconductor lasers in particular, is important to generate improved designs. Laser Light is generated in a semiconductor laser by radiative recombination of electrons and holes. In order to generate more light by stimulated emission than is lost by absorption, the system has to be inverted. Inversion population is carried out electrically or by optical pumping. A laser is, thus, always a high carrier density system that entails many-body interactions. The operation of laser, as well as other photonic devices, involves an interaction between electrons and light waves (photons) as well as lattice vibrations (phonons). Also, the so-called excitons (bound electron-hole pairs) emerge in photo-excited semiconductors and strongly influence the carrier transport process. All these interactions cannot be taken into account exactly because of the high number of involved carriers and quasi particles. Therefore, the following approximations are usually adopted:

1. **Free Carrier Model:** In this model, many-particle interactions are neglected approximated phenomenologically. To take the effect of carrier interactions into consideration, a phenomenological

DOI: 10.4018/978-1-5225-2312-3.ch008

Figure 1. Simulation of carriers transport in laser devices

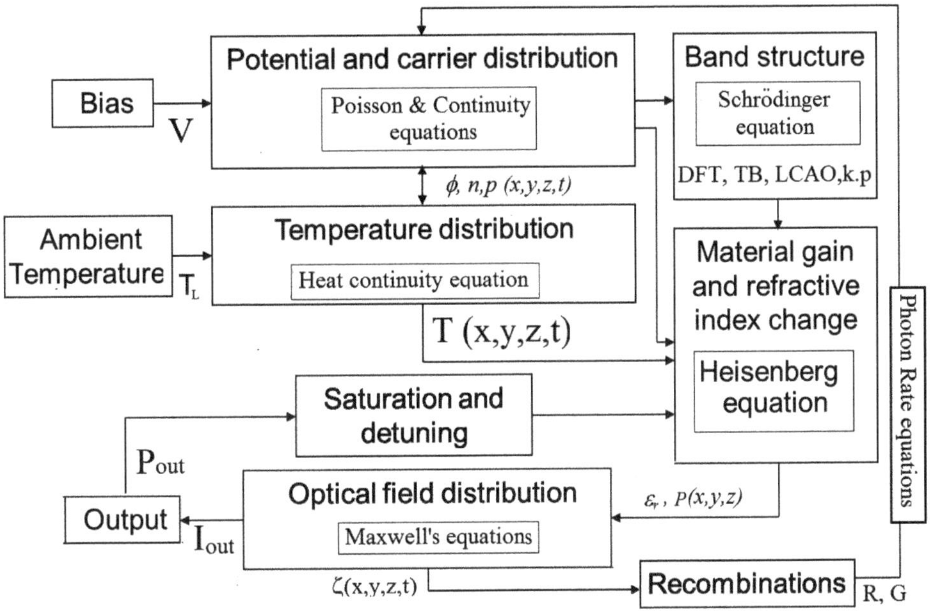

scattering time is introduced. Simple models for the gain coefficient are then used to obtain a system of laser rate equations, to calculate the time-dependent laser response.

2. **Hartree-Fock Approximation:** Here, the semiconductor Bloch equations (SBEs) are employed. To describe the carrier interactions, the Hartree–Fock approximation is utilized. In this case, carrier–carrier interaction leads to re-normalization terms for carrier energy and electric field. The collision terms, i.e., the terms describing carrier–carrier scattering have to be introduced phenomenologically using a relaxation time for the polarization field.
3. **Correlation Effects:** This microscopic approach takes the collision terms in the SBEs into account explicitly. Therefore, the collision terms in the SBEs can be included in the second-Born approximation. This model yields the correct laser linewidth for any excitation density or temperature. In the other models, the relaxation time has to be extracted from experiment.

Figure 2 depicts the different models of photon/exciton transport and electromagnetic field propagation. The electromagnetic field analysis of photonic devices may be carried out by classical or quantum models. The classical models are all about solving the Maxwell equations inside the device active region. The quantum approaches are based on some sort of dynamic wave equations (Schrodinger-like or Heisenberg-like) in the microscopic level or the SBEs in the macroscopic level. The so-called dynamics-controlled truncation (DCT) formalism is another successful microscopic approach that describes coherent correlations in optically excited semiconductors. On the other hand, the most successful approach to study incoherent effects and correlations in highly excited semiconductors is the nonequilibrium Green's functions (NEGF) approach.

Figure 2. Hierarchy of photonic and excitonic transport

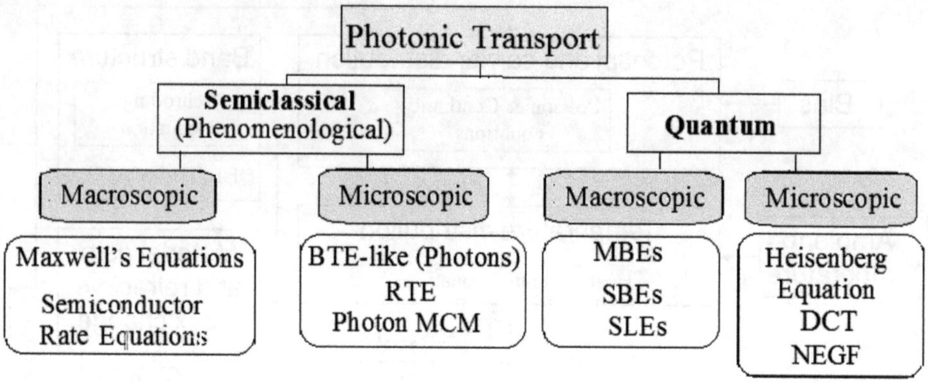

The purpose of this Chapter is to present the photon transport theory and to show how it is employed to simulate and design photonic and opto-electronic devices. In fact, the modeling approaches and techniques of photonic devices are not yet well standardized. Surprisingly the quantum theory describing photon interactions with nanophotonic structures, such as cavity resonators, has only recently been presented (Shen & Fan, 2009). In fact, the electronic and optical properties of semiconductor nanodevices depend strongly on the photon dynamics that take place in the femto- and picoseconds range.

Upon completion of this chapter, students and readers will be able to

- Review the semiclassical theory of electromagnetic waves and Maxwell's equations
- Review the basics of photons and photon waves
- Review the quantum theory of electromagnetic waves and second quantization operators
- Describe photon scattering mechanisms and their emission and absorption probabilities in semiconductors
- Understand the concept of excitons and exciton complexes (biexictons, trions) in low-dimensional structures.
- Understand the photon transport mechanisms, and how they affect the optical properties of solids
- Know how and when to employ the photon transport equations, in microscopic approaches, such as semiclassical (BTE-like) and quantum approaches, such as nonequilibrium Green's function (NEGF).
- Know how and when to employ the photon transport equations, in macroscopic, semiclassical and quantum approaches, such as Maxwell-Bloch equations (MBE's), Semiconductor Bloch equations (SBE's), Maxwell SBE's (MSBE's), Semiconductor luminescence equations (SLE's) and how they are coupled to charge carriers.
- Know how and when to employ the phenomenological rate equations in semiconductors and how they are coupled to the charge carrier transport in optoelectronic and photonic devices.
- Investigate the excitonic transport, in specific nanostructures such as carbon nanotubes and graphene ribbons
- Investigate the dynamics-controlled truncation (DCT) transport formalism, on different size scales of semiconductor devices.

2. FUNDAMENTALS OF OPTICAL WAVES AND INTERACTION WITH MATTER

To successfully build and optimize photonic devices and systems, you need to understand both the fundamentals of optics and electronics. In this section, we summarize and review the basics of the theory of light and light-matter interaction. The light -matter interaction is important for a quantum description of light gain in a medium and the rate equations, which describe the evolution of electrons and photons in optoelectronics. After this introduction, we present the quantum and phenomenological transport approaches for the description of optical emission and detection processes.

2.1 What Are Light Waves?

Light is electromagnetic wave (EMW). An electromagnetic wave is a traveling wave that has time-varying electric and magnetic fields (**ζ**, ***B***) that are perpendicular to each other and the direction of propagation.

The wave equation can have many solutions besides the traveling waves; depending on the boundary conditions. Actually, EMW's may be planar or spherical or diverted, as shown in the Figure 3. A surface over which the phase of a wave is constant is referred to as a wavefront. A wavefront of a plane wave is a plane perpendicular to the direction of propagation.

When we describe the interaction of a light wave with a nonconducting medium (with zero conductivity), we usually make use of the electric field component ζ_x rather than B_y. Hence, a monochromatic plane wave traveling along the positive z direction may be expressed as follows:

$$\zeta_x = \zeta_o \cos(\omega t - kz + \phi_o) \tag{1}$$

where ζ_x is the electric field along the x direction, ζ_o is the amplitude of the wave, k is the propagation (k=2π/λ), ω is the angular frequency (ω= 2πf) and ϕ_0 is Phase constant; at t = 0 and z = 0.

The direction of propagation is indicated with a vector $\boldsymbol{k}$, called the wave vector, whose magnitude is the propagation constant, (k = 2π/λ). Hence, $\boldsymbol{k}$ is perpendicular to constant phase planes.

The phase velocity $\boldsymbol{v}$ is defined as follows:

Figure 3. Representation of an electromagnetic wave

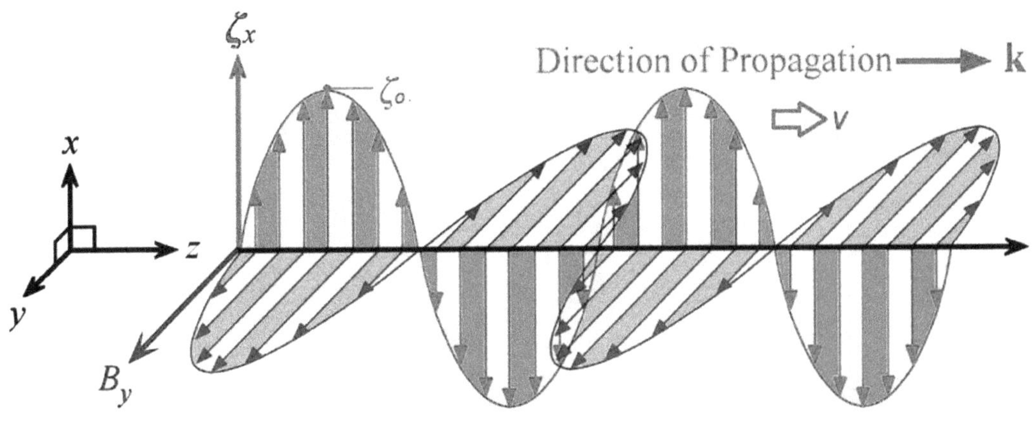

Figure 4. Propagation forms of electromagnetic waves
After Kassab (2001).

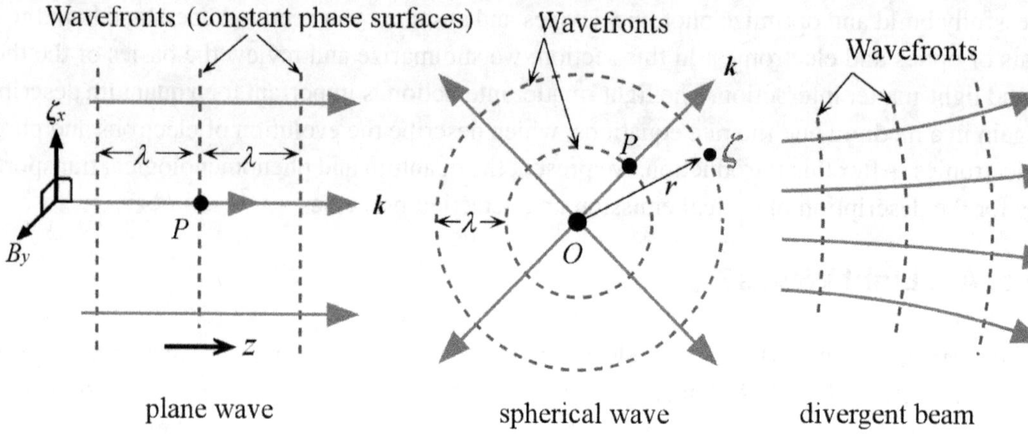

$$v = \frac{\delta z}{\delta t} = \frac{\omega}{k} = \upsilon\lambda \tag{2a}$$

The phase velocity of a plane wave in the medium is given by

$$v = \frac{\omega}{k} = \frac{1}{\sqrt{\varepsilon_o \varepsilon_r \mu_o}} \tag{2b}$$

The phase velocity in vacuum is

$$c = \frac{\omega}{k_0} = \frac{1}{\sqrt{\varepsilon_o \mu_o}} \tag{3}$$

The relative permittivity ε_r measures the ease with which the medium becomes polarized and hence it indicates the extent of interaction between the field and the induced dipoles.

When an EM wave is traveling in a medium, the oscillating electric field polarizes the molecules of the medium at the frequency of the wave. The stronger is the interaction between the field and the dipoles, the slower is the propagation of the wave. In non-magnetic materials, we usually have $\mu_r = 1$, so that the refraction index is given by:

$$n = \frac{c}{v} = \sqrt{\varepsilon_r} \tag{4}$$

Generally speaking, the refraction index is not constant, but rather is a function of the wavelength, such that $n = n(\lambda)$. When the medium has a finite conductivity σ, then its relative permittivity has an imaginary component: $\varepsilon_r = \varepsilon' - j\varepsilon'' = \varepsilon' - j(\sigma/\varepsilon_o\,\omega)$. This result can be obtained from the simple Drudé model for free electrons when we treat with sinusoidal fields. Therefore, the refraction index becomes a complex parameter, which is given by:

$$n = (\varepsilon' - j\,\varepsilon'')^{1/2} = n_o + j\,k_o \tag{5}$$

where n_o is called the real refraction index and k_o is called the extinction coefficient. Obviously, ε_r is real and $k_o=0$ for non-conducting media.

When the induced dipole oscillations in a material become large it is possible for the material to start absorbing energy from the applied field. When ε'' approaches zero, the absorption diminishes and the material becomes transparent.

2.2 Light Refraction and Refraction

Reflection, refraction and diffraction are all boundary behaviors of waves associated with the bending of the path of a wave. At the border of two materials usually both reflection and refraction appears. When light travels from one material to another it usually changes direction. The bending of light that occurs at the borderline of two materials is called refraction. If light travels from material with refraction index n_1 to another material with refraction index n_2, the Snell law determines the direction of the refracted ray: $n_1.\sin\theta_1 = n_2.\sin\theta_2$.

2.3 Light Polarization

Light waves are emitted spherically (omnidirectional), however EM waves may have polarization. Polarization of light means that the electric and magnetic fields are built up in a specific way, linear or circular. Light of a lincar polarization can be produced by sending un-polarized light through a polarizing

Figure 5. Reflection and refraction of light through a medium

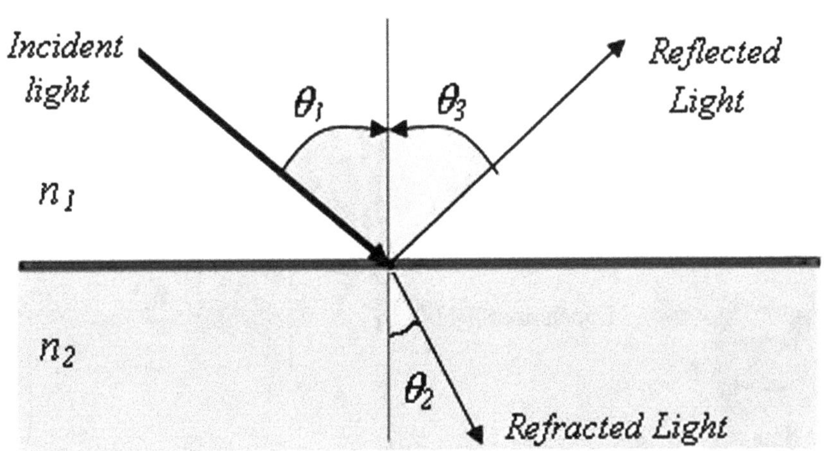

medium (Polaroid) whose polarizing axis is oriented along the direction of the desired polarization. A linear polarizer converts the un-polarized beam into one with a single linear polarization, as shown in Figure 6. The vertical components of all waves are transmitted, while the horizontal components are absorbed and reflected. The law of Malus (1809) describes the intensity of transmitted light via two linear polarizers (polarizer and analyzer), as $I(\theta)=I_o^2cos^2\theta$, where I_o is the initial unpolarized light intensity and θ is the angle between the axes of the polarizers.

2.4 Light Diffraction

Diffraction is the deviation and spreading of waves when they meet an obstruction. It can occur with any type of wave, including sound waves and light. Diffraction will not occur if the wave is not coherent, and diffraction effects become weaker as the size of obstruction is large compared to the wavelength. Diffraction from multiple slits is similar to what happens when waves are scattered from a periodic structure, such as atoms in a crystal lattice. Each atom (scattering center) acts as a point source of spherical wavefronts. Under certain conditions, the wavefronts undergo a constructive interference to form a number of diffracted beams. The direction of these beams is described by Bragg's law: $d \sin \theta = n \lambda$. The so-called Fraunhofer diffraction is the diffraction observed in the plane of an image formed by an optical system. The Fraunhofer diffraction pattern is the Fourier transform of the amplitude function of the diffracting aperture.

2.5 Light Coherence

Light waves may be coherent or incoherent. Coherence is associated with the wave-like nature of light. It manifests itself in interference and/or diffraction effects. Coherence is identified in two concepts; temporal coherence and spatial coherence. The temporal coherence is concerned with frequency content of a light wave, and whether they are in phase. For instance, a single wave like $f(t)=A\ exp(i\omega_0 t)$ has a perfect temporal coherence, because it has a single frequency and continuous in time. In practice the duration of a wave packet is finite (Δt), and the wave packet contains some frequency components and a bandwidth ($\Delta\omega=1/\Delta t$). The coherence depends on the bandwidth $\Delta\omega$ which in turn depends on the

Figure 6. Linear polarization of un-polarized light

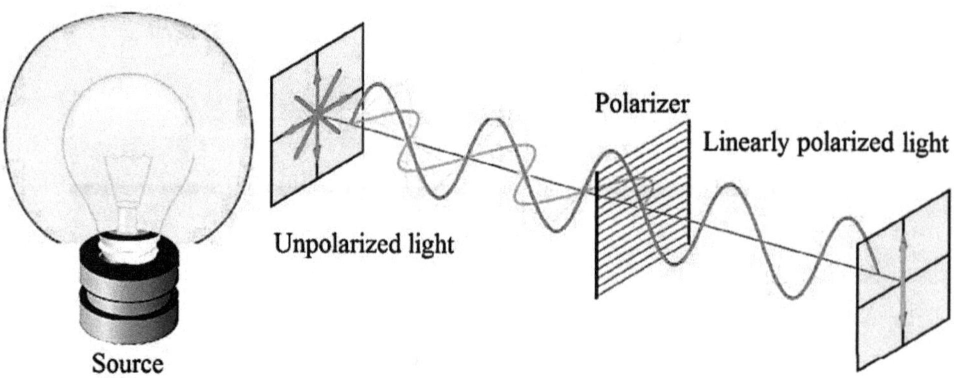

Figure 7. Light diffraction via two slits. The 2 rays interfere constructively when: d sinθ = nλ

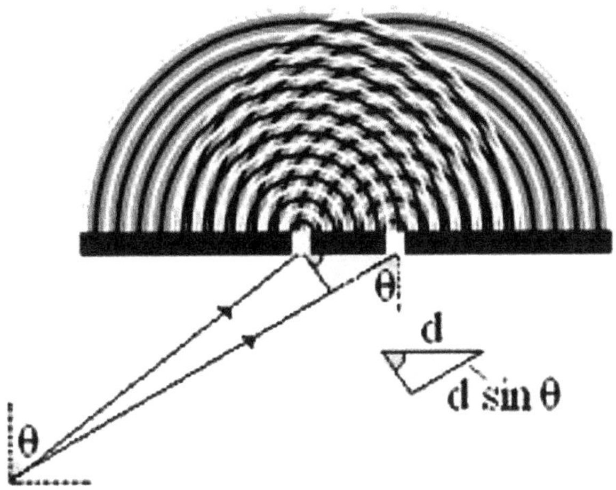

energy bandwidth ΔE of the wave packet. We can also observe the coherence of light perpendicular to the direction of travel. This is known as lateral coherence or spatial coherence.

Coherence may be characterized using correlation functions. Such correlations are found by taking time averages of the wave. Therefore, we combine the complex amplitudes arising from both the temporal (frequency) and spatial (wavelength) characteristics of the travelling wave. We can also get some information about the coherence of the light waves, when we measure interference or diffraction patterns of the light. Note that photons can only interfere with themselves. Different photons of different light sources never interfere.

2.6 Optical Power and Optical Attenuation

Optical power is the brightness (or intensity) of light. As the EM wave propagates in the direction of the wave vector ***k***, there is an energy flow in this direction. The wave brings with it electromagnetic energy. The energy densities in the ζ_x and B_y fields are the same,

Figure 8. Illustration of the light wave coherence

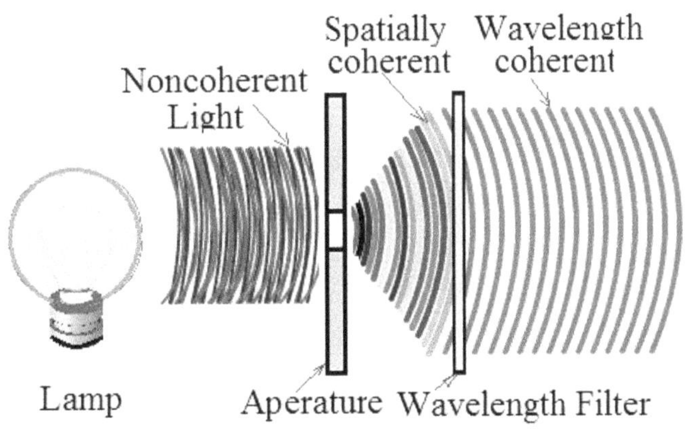

$$\frac{1}{2}\varepsilon_o\varepsilon_r\zeta_x^2 = \frac{1}{2\mu_o}B_y^2 \qquad (6a)$$

Note that we use the vectors **ζ** and ***B*** to represent the electric and magnetic fields in the EM wave. The total energy density in the wave is therefore $\varepsilon_o\varepsilon_r\zeta_x^2$. If *S* is the EM power flow per unit area, then $S = v\varepsilon_o\varepsilon_r\zeta_x^2 = v^2\varepsilon_o\varepsilon_r\boldsymbol{\zeta}_x\boldsymbol{B}_y$. The EM power flow per unit area can be written as

$$\widehat{S} = v^2\varepsilon_0\varepsilon_r\zeta x B \qquad (6b)$$

where $\widehat{S}$, called the Poynting vector, represents the energy flow per unit time per unit area in a direction of propagation (determined by $\boldsymbol{\zeta}\times\boldsymbol{B}$). The magnitude of power flow per unit area is called the irradiance (or intensity). The average irradiance is given by:

$$I = <\widehat{S}> = \frac{1}{2}v\varepsilon_o\varepsilon_r\zeta_o^2 = \frac{1}{2}c\varepsilon_o n\zeta_o^2 \qquad (6c)$$

where the refraction index $n = v/c$ and *c* is the speed of light in free space.

As light travels through a medium, some energy is lost; either absorbed by medium particles and converted to heat; or scattered by microscopic imperfections in the medium. This loss of intensity is called attenuation. In many cases, attenuation is an exponential function of the path length through the medium.

$$P_{out} = P_{in}\, exp(-\alpha L) \qquad (7)$$

where the attenuation coefficient α is defined as the fractional decrease in the optical power per unit distance. We typically measure optical power in dB (in reference to 1W) or in dBm (in reference to 1mW).

The signal intensity of light is inversely proportional to the distance squared (like sound). Attenuation decreases the intensity of electromagnetic radiation due to absorption or scattering of light quanta (photons). Attenuation does not include the decrease in intensity due to inverse-square law geometric spreading. Therefore, the total change in intensity involves both the inverse-square law and an estimation of attenuation over the path.

2.7 Classical Theory of Light and EM Waves

The classical theory of light is based on Maxwell's equations. The Maxwell's equations are second-order partial differential equations that describe the propagation of electromagnetic (EM) waves through a medium. An EM wave consists of two vector fields, namely, the electric field **ζ** and the magnetic field ***B***.

$$\nabla.\boldsymbol{B} = 0,\ \nabla \mathrm{x}\boldsymbol{\zeta} = -\,\partial\boldsymbol{B}/\partial t, \qquad (8a)$$

$$\nabla.\boldsymbol{D} = \rho,\ \nabla \mathrm{x}\boldsymbol{H} = \partial\boldsymbol{D}/\partial t + \boldsymbol{J} \qquad (8b)$$

With the constitutive relations

$$\boldsymbol{D} = \varepsilon \boldsymbol{E} = \varepsilon_o \boldsymbol{\zeta} + \boldsymbol{P} = \varepsilon_o (1+\chi)\boldsymbol{\zeta},\ \boldsymbol{B} = \mu \boldsymbol{H} \tag{8c}$$

Here, P and χ are the polarization and susceptibility of the medium. These equations are originated from a set of equations in electromagnetism, due to Gauss, Faraday, Coulomb and Ampere. Maxwell's contribution was to add the displacement current term, $(\partial D/\partial^2)$ in the last equation, and thus put the equations into a consistent form.

Macroscopically, Maxwell's equations show that in the absence of free charges ($\rho=0$) and currents ($J=0$), an electromagnetic field interacts with matter via the optical polarization $\boldsymbol{P}$. Therefore, the wave equation for the electric field $\boldsymbol{\zeta}$ reads

$$\left(\nabla \cdot \nabla - \frac{1}{c^2}\frac{\partial^2}{\partial t^2}\right)\zeta(r,t) = \mu_o \frac{\partial^2 P(r,t)}{\partial t^2} \tag{9}$$

Both electric and magnetic field are time-dependent vector fields that depend on the vector potential field $A(r,t)$, as well as a scalar field $\varphi(r,t)$:

$$B(r,t) = -\nabla \mathrm{x} A(r,t) \tag{10a}$$

$$\zeta(r,t) = -\nabla\varphi(r,t) - \frac{\partial A(r,t)}{\partial t} \tag{10b}$$

Choosing the Coulomb gauge[1], for which $\nabla \cdot \boldsymbol{A} = 0$, makes $\boldsymbol{A}$ into a transverse field. In a source-free region of space with a zero scalar potential, we only need to discuss the vector field, A, propagating through space. The direction of the vector potential $\boldsymbol{A}$ is parallel to the direction of the electric field $\boldsymbol{E}$. Therefore we only need to solve the wave equation in the Coulomb gauge

$$\nabla^2 A - \frac{1}{c^2}\frac{\partial^2 A}{\partial t^2} = 0 \tag{11}$$

The spatial-temporal solution of this wave equation, with periodic boundary conditions, consists of the traveling plane waves of the type:

$$A(r,t) = A_o.exp[\pm j(\boldsymbol{k}.\boldsymbol{r} - \omega t)] \tag{12}$$

The Fourier series expansion of the vector potential, in a finite cubic box of volume V, is then given by:

$$A(r,t) = \frac{1}{\sqrt{\varepsilon_o V}}\sum_k \sqrt{\frac{\hbar}{2\omega_k}}\left[b_k \exp[+j(k\cdot r - \omega_k t)] + b_k^* \exp[-j(k\cdot r - \omega_k t)]\right]\cdot e_k \tag{13a}$$

Note that the kth Fourier component of $\boldsymbol{A}$ is a vector perpendicular to the wavevector $\boldsymbol{k}$ and composed of a linear combination of two polarization vectors ($\boldsymbol{e}_k$ and $-\boldsymbol{e}_{k'}$) for left-hand and right-hand circularly polarized EM waves. Therefore,

$$\zeta = -\frac{\partial A(r,t)}{\partial t} = \frac{j}{\sqrt{\varepsilon_o V}} \sum_k \sqrt{\frac{\hbar \omega_k}{2}} \left[b_k \exp[+j(k \cdot r - \omega_k t)] - b_k^* \exp[-j(k \cdot r - \omega_k t)] \right] \cdot e_k \tag{13b}$$

$$B = \nabla x A(r,t) = \frac{j}{\sqrt{\varepsilon_o V}} \sum_k \sqrt{\frac{\hbar \omega_k}{2}} \left[b_k \exp[+j(k \cdot r - \omega_k t)] + b_k^* \exp[-j(k \cdot r - \omega_k t)] \right] (k x e_k) \tag{13c}$$

2.8 Quantum Theory of Light and EM Waves

Quantum theory of light shows us how light is described in terms of quantized photons, and how photons interact with matter. This is a field of research that uses both semiclassical and quantum-mechanical physics. The particle/wave duality nature of light can be included by quantizing the electromagnetic vector potential ($\boldsymbol{A}$). The quantum theory of the electromagnetic field starts by taking the Fourier expansion of the vector potential $\boldsymbol{A}(r,t)$ and then substituting operators for the amplitude terms. To do that, we replace the classical amplitudes (b_k, b_k^*) with the quantum creation and annihilation operators ($\boldsymbol{b}_k^\dagger$, $\boldsymbol{b}_k$) to produce the following quantum version of the vector potential. Note that the exponential time dependence, $exp(\pm j\omega t)$, is embedded into the quantum field operators.

$$A(r,t) = \frac{1}{\sqrt{\varepsilon_o V}} \sum_k \sqrt{\frac{\hbar}{2\omega_k}} \left[b_k(t).\exp(jk.r) + b_k^\dagger(t).\exp(-jk \cdot r) \right] e_k \tag{14}$$

It is now straight forward to write the quantized versions of ζ=-∂A/∂t and B=∇xA. The creation and annihilation operators must satisfy certain commutation relations. We can write the commutation relation as follows:

$$[\boldsymbol{b}_i, \boldsymbol{b}_j] = [\boldsymbol{b}_j^\dagger, \boldsymbol{b}_i^\dagger] = 0 \tag{15a}$$

$$[\boldsymbol{b}_i, \boldsymbol{b}_j^\dagger] = -[\boldsymbol{b}_j^\dagger, \boldsymbol{b}_i] = \delta_{ij} \tag{15b}$$

This procedure is called the field quantization or second quantization and was first done by Paul Dirac in 1927. In order to do so, Dirac took the phases of EM modes (Fourier components of the field) and the mode energies as dynamic variables to quantize (i.e., considered them as operators and postulated commutation relations between them)[2]. In this way, Dirac was able to align the photon concept with the new theory and hence to describe the interaction of photons with matter (Dirac, 1927).

2.9 Photon Wave Properties

In physics, a photon is the quantum of light and all other forms of EM radiation. It is also the force carrier for the EM force. The effects of this force can be observed at both the microscopic and macroscopic level, because the photon has no rest mass. If the photon is not a massless particle, it would not move at the speed of light. Because photons have no rest mass, they can interact at long distances. Like elementary particles, photons exhibit wave–particle duality. Unlike electrons, photons have no charge and belong to the category of Bosons that can condensate into single energy state at low temperature. The main features of photons are:

1. The photon has zero rest mass;
2. The photon energy is $E=h\nu=\hbar c|k|$ (k is wave vector, c is speed of light);
3. The photon electromagnetic momentum is $\boldsymbol{p}=\hbar k$;
4. The photon polarization $s = \pm 1$ is the eigenvalue of the z-component of the photon spin S_z.

Table 1 depicts some of the photon wave properties in 1 & 3-D in free space. The case of 2-D can be easily inferred from them. The photon radiation modes are plane waves characterized by a wavevector $\boldsymbol{k}$, a quasi-momentum $\boldsymbol{p}=\hbar\boldsymbol{k}$. In 3-D space, photons are characterized by the dispersion relation $E(p) = c|p|$. This relation can be derived from the relativity relation $E=(m_o c^2+c^2p^2)^{1/2}$, with the rest mass $m_o=0$. One can also write $E(k)=\hbar c|k|$ or $\omega(k)=c|k|$. In a material of refractive index n, we replace the speed of light c with $v=c/n$. In addition to carrying energy and having momentum, photons can possess a mechanical force. When an EM wave is absorbed by an object, the wave exerts a pressure (P in N/m^2) on the object equal to the wave irradiance (I in W/m^2) divided by the speed of light (c in m/s): $P = I/c$.

Photons have an angular momentum oriented mainly along the axis of light propagation. Measurements found that each light photon carries an *angular momentum* $L=\hbar$. This quantity is called helicity. We therefore also speak of the spin of a photon. Photons somehow 'turn' in a direction either parallel or antiparallel to their direction of motion. The magnitude of the photon helicity, or spin, confirms the classical relation $L = E/\omega$ between energy and angular momentum

Photons Density of States

The density of states for photons refers to the number of radiation modes per unit frequency interval per unit volume of real space. This quantity plays an important role in many optoelectronic devices. The density of states of photons can be highly modified in optical cavities because of photon confinement. For example, the number of photons in a laser field with intensity $I=1W/cm^2$ and photon energy of 1eV in a typical coherence volume $V= 1cm^3$ can be estimated as $N = (I/\hbar\omega)V/c = 2x10^8$ photons.

Photon Spin

In order to understand the photon spin, you have to bear in mind that the electric and the magnetic components are dipoles and have a direction. Photon spin is different from spin of other particles. If we were talking about a massive particle with spin 1, it'd have three possibilities for $S_z = (-\hbar, 0, \hbar)$. The fact that

Table 1. Photon wave equations. Here V denotes the volume of all photon radiation modes is in real space

	1-D	3-D
EM Wave Equation	$c^2\frac{\partial^2 u}{\partial x^2}=\frac{\partial^2 u}{\partial t^2}$	$c^2\left[\frac{\partial^2 u}{\partial x^2}+\frac{\partial^2 u}{\partial y^2}+\frac{\partial^2 u}{\partial z^2}\right]=\frac{\partial^2 u}{\partial t^2}$
Wave Solution	$u = \exp[j(kx - \omega t)]$	$u = \exp[j(\mathbf{k}.\mathbf{r} - \omega t)]$
Dispersion relation	$\omega = c\,k$	$\omega = c\,\lvert\mathbf{k}\rvert$
Density of states, $D(k)$	$D(k)=\frac{p}{\pi}$	$D(k)=\frac{pk^2}{2\pi^2}$
Density of states, $D(\omega)$	$D(\omega)=\frac{p}{\pi c}$	$D(\omega)=\frac{p\omega^2}{2\pi^2 c^3}$
Density of states, $D(E)$	$D(E)=\frac{p}{\pi\hbar c}$	$D(E)=\frac{pE}{2\pi^2\hbar^3 c^3}$
Intensity, $I(\lambda)$	$I(\lambda)=\frac{phc^2}{\lambda^3\left[\exp\left(\frac{hc}{\lambda k_B T}\right)-1\right]}$	$I(\lambda)=\frac{\pi phc^2}{\lambda^5\left[\exp\left(\frac{hc}{\lambda k_B T}\right)-1\right]}$
Wien's law	$\lambda_{max} = 0.0050994367\ T^{1}$	$\lambda_{max} = 0.0028977684\ T^{1}$
Stefan - Boltzmann law	$I=\frac{p\pi^2 k_B^2 T^2}{6h}$	$I=\frac{p\pi^5 k_B^4 T^4}{15c^2h^3}$
Internal energy	$U=\frac{p\pi^2 V k_B^2 T^2}{3hc}$	$U=\frac{4\pi^5 V k_B^4 T^4}{15h^3c^3}$

photon is massless, implies some mathematical conditions that rules out the 0 value. Then for photon, we don't talk about S_z, instead we talk about its helicity, which is either $-\hbar$ or $\hbar$. The two helicity values are related to the polarization states by associating one to the right-handed circular polarization and the other to the left-handed one.

Photon Polarization

The polarization of photons is carried out by the same orientation of the magnetic and electric component of the EM field. Figure 9 depicts the left- and right-handed circular polarization, and their associate angular momenta. note that the individual photons have spin either along or against their direction of motion, while the electric fields are perpendicular.

Figure 9. Illustration of left and right handed circular polarization, and their associate angular momenta

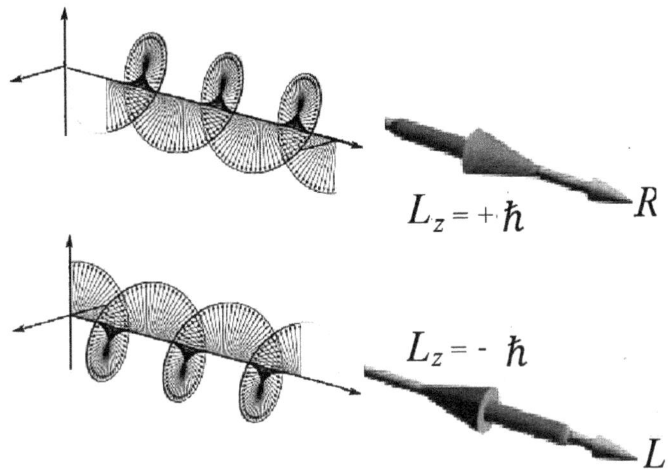

Photonic States

In optics, the polarization, wavelength, and the propagation vector specify the basic states (or modes) of the optical waves. The allowed wavelengths (frequencies) and polarization and spin of light characterize these modes. The fields have amplitude and phase, but these characteristics concern only the field and not the optical mode.

The Fock states, coherent states, and squeezed states represent three types of amplitude states. The Fock state represents one of the most fundamental notions of Quantum Electrodynamics (QED).

A Fock state has a definite number of photons in the mode (energy) but completely random phase. All Fock states have zero average electric field because of its random phase.

A coherent state has nonzero average electric field and well-defined phase. The laser emission has such coherent states. The electric fields for these states can take the form of sine and cosine waves. The coherent state actually consists of a linear combination of Fock states. However, coherent and Fock states are different. Fock state has an exact number of photons. On the other hand, the number of photons in the mode of a coherent state with given amplitude, has a Poisson probability distribution.

A squeezed vacuum state can be produced from the quantum vacuum state by reducing the noise (i.e., reducing the variance). Squeezing the vacuum state is equivalent to squeezing the coherent state since the vacuum and coherent states have the same type and amount of noise. The next figure is a schematic representation of coherent and squeezed state, with various possibilities, in the phase space.

Figure 10. Schematic representation of coherent and squeezed states in the phase space

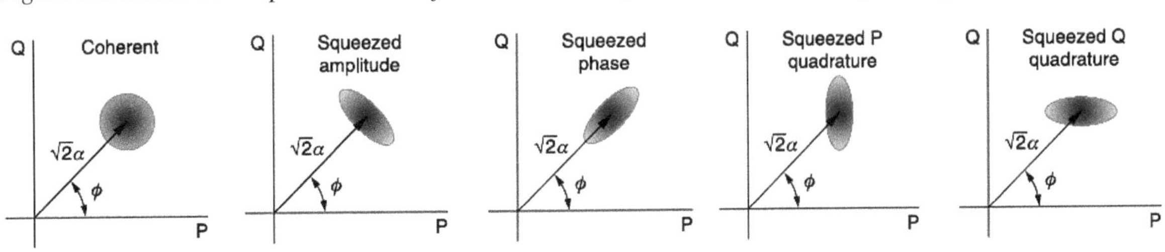

We can define a Wigner distribution (WD) that has the phase space coordinates (P and Q) as random variables. The WD is quantum analog to the classical probability distributions. However, the WD may be negative for some types of states (like Fock states). For coherent and squeezed states, the WD provides a near-classical picture. The so-called '*P*' and '*Q*' quasi-probability distributions can also be defined. The *P*-distribution helps to provide a mathematical representation, which is similar to the density operator of carriers.

2.10 Dirac Equation

In order to describe the emission and absorption of light quanta (photons), the basic Schrödinger equation has to be a little bit modified. The so-called Dirac's equation is equivalent to the Schrödinger equation, but with taking the special relativity into account. The Dirac equation states (Dirac, 1927):

$$j\hbar\, \partial\Psi/\partial t = H_{Dirac}\, \Psi \tag{16a}$$

where the Dirac Hamiltonian H_{Dirac} is given by:

$$H_{Dirac} = \beta\, m_o\, c^2 + c\, \Sigma\, \alpha \boldsymbol{p}_i \tag{16b}$$

Here m_o is the particle rest mass, c is the speed of light and p_1, p_2, and p_3 are the momentum components. In presence of external magnetic and electric fields, the Dirac Hamiltonian reads (Dirac, 1958):

$$H_{Dirac} = \beta\, m_o\, c^2 + \alpha.\, (\boldsymbol{p} - e\boldsymbol{A})c + e\phi \tag{16c}$$

where $\boldsymbol{A}$ is the vector potential and ϕ is the electrical potential. Also, the parameters β and α_i are 4x4 matrices, which are given by:

$$\alpha_1 = \begin{bmatrix} 0 & 0 & 0 & 1 \\ 0 & 0 & 1 & 0 \\ 0 & 1 & 0 & 0 \\ 1 & 0 & 0 & 0 \end{bmatrix}, \alpha_2 = \begin{bmatrix} 0 & 0 & 0 & -j \\ 0 & 0 & j & 0 \\ 0 & -j & 0 & 0 \\ j & 0 & 0 & 0 \end{bmatrix}, \alpha_3 = \begin{bmatrix} 0 & 0 & 1 & 0 \\ 0 & 0 & 0 & -1 \\ 1 & 0 & 0 & 0 \\ 0 & -1 & 0 & 0 \end{bmatrix}, \beta = \begin{bmatrix} 1 & 0 & 0 & 0 \\ 0 & 1 & 0 & 0 \\ 1 & 0 & -1 & 0 \\ 0 & 0 & 0 & 1 \end{bmatrix} \tag{16d}$$

These 4×4 matrices and are all Hermitian and their squares are equal to the identity matrix I: $\alpha_i^2 = \beta^2 = \text{I}$. These parameters mutually anticommute such that: $\alpha_i\, \alpha_j + \alpha_j\, \alpha_i = 0$ and $\alpha_i\, \beta + \beta\, \alpha_i = 0$

The signs of the operator β distinguish particles and antiparticles; it has two 1's and two -1's to take care of the 2 possible spin directions. With this Hamiltonian operator, a wavefunction for a particle has vanishing antiparticle components, and vice versa. The Hamilton operator yields the velocity operator v in the same relation that is valid in classical physics:

$$v = dx/dt = \beta\, p\, /\, (m_o^{\,2} c^4 + c^2 p^2)^{1/2}. \tag{16e}$$

As a consequence of the solution of this equation, spin and antiparticles could be mathematically explained. Therefore, the Dirac Hamiltonian describes how charged particles move by electromagnetic fields. A similar Hamiltonian (called Maxwell-Hamiltonian) can describe how fields move by charged particles (Dirac, 1966). Together, they form what is called quantum electrodynamics (QED).

3. LIGHT-MATTER INTERACTION

Matter and fields can interact by a variety of mechanisms. The material response to electromagnetic radiation may fall into three basic phenomena, namely: absorption, spontaneous emission and stimulated emission. Electric dipoles emit and absorb fields and therefore result the refractive index, gain and absorption properties. Therefore, both the classical and quantum theories incorporate dipoles at a fundamental level of their description of the matter–field interaction.

Microscopically, the optical polarization arises from quantum transitions between different states of the material system. For the case of semiconductors, electromagnetic radiation with optical frequencies is able to move electrons from the valence to conduction bands. When an electron is excited from the valence band, a hole is created. This electron and the corresponding hole create a microscopic polarization in the form of dipole (exciton) to which an incoming radiation field can couple. The polarization ***P*** is computed by summing over all microscopic transition dipoles (p_{cv})

$$P = \frac{1}{V}\Sigma_{cv}(d_{cv}p_{cv} + C.C.) \tag{17a}$$

where d_{cv} is the dipole matrix element which determines the strength of individual transitions between the valence and conduction states (v and c). The term CC denotes the complex conjugate of the first term, and $V$is the system volume. If E_c and E_v are the energies of the conduction and valence band states, the quantum evolution of dipole moments may be written as $p_{cv} = p_o\ exp\ [-jt(E_c - E_v)/\hbar)]$ so that

$$P = \frac{1}{V}\Sigma_{cv}\left(d_{cv}p_0 \exp[-j\frac{(E_c - E_v)t}{\hbar}] + C.C.\right) \tag{17b}$$

Therefore, ***P*** is given by a summation over the microscopic transition dipoles which all oscillate with frequencies $\hbar\omega_{cv} = (E_c - E_v)$. Thus, the optical polarization is a coherent quantity which has amplitude and phase.

In the classical theory, optical absorption occurs when an incident wave induces a dipole moment along the direction of the wave polarization and then the surrounding medium dissipates the energy. Emission occurs when an excited dipole synchronously radiates energy with its motion. Maxwell's wave equation incorporates the dipoles in terms of the susceptibility. The Maxwell's equations and the Poynting vector can explain emission, absorption and particle transport in terms of the classical theory of the microscopic dipoles. While quite successful, the description does not account for the quantum nature of matter and light, and does not explain basic phenomena such as the spontaneous emission of light from matter.

The quantum theory of the matter–field interaction uses the operator form of the dipole moment. The study of matter-light interaction usually begins with the time-dependent perturbation theory and the Fermi's golden rule for optical transitions. Fermi's golden rule relates the transition rate to the optical power, frequency and dipole moment The Fermi golden rule and density matrix formalism of quantum transport make use of the dipole moment as part of the interaction Hamiltonian.

In the quantum interaction representations, the state can evolve in time when material absorbs or emits light. These states are called Fock states, when the number of photons in the state is specified. Other types of photonic states have been already defined in §8-2.9 of this Chapter.

3.1 Light-Matter Interaction Hamiltonian

In order to understand the interaction of a particle with the electromagnetic fields we need to know the total energy of the system consisting of particle and field. The particles may borrow energy from the field (absorption) or it may give energy to it (emission), but the total energy remains conserved. In this section, we try to understand the absorption and emission phenomena of optical transitions. Our first step is to define the matter-light system Hamiltonian, which expresses the total energy of the system.

Classical Hamiltonian

The following equation depicts the interaction Hamiltonian of a single charged particle with an electromagnetic field

$$H = ½ m\text{-}1 (\boldsymbol{p}\text{-} qA)2 + V_o(r) \tag{18}$$

where q is the particle charge, m is its mass, V_o is the potential energy and $\boldsymbol{A}$ is the vector potential of the electromagnetic field. For a system of many charges, the classical Hamiltonian can be decoupled as follows:

$$H = H_0 - 1/2\Sigma_i[q_i m_i^{-1}(p_i.A - A.p_i)^2] + \boxed{1/2\Sigma_i[q_i m_i^{-1} \mid A \mid^2]} \tag{19a}$$

$$H_0 = 1/2\Sigma_i[m_i^{-1}p_i^2 + V_0(r_i)] \tag{19b}$$

where H_o is the Hamiltonian for a collection of particles in the absence of external field. For weak fields we can formulate this in the interaction picture as follows:

$$H = H_o + V(t) \tag{20a}$$

$$V(t) = 1/2\Sigma_i\, q_i m_i^{-1}(p_i.A + A.p_i) \tag{20b}$$

Quantum Electric Dipole Hamiltonian

Now we are in a position to replace the classical momentum with a quantum operator ($\dot{p} = -j\hbar\nabla$). Here the vector potential remains classical, and only modulates the interaction strength. Therefore, we get

$$V(t) = 1/2j\hbar\sum_i q_i m_i^{-1}(\nabla_i.A + A.\nabla_i) \tag{21a}$$

The first term is zero since we are working in the Coulomb gauge ($\nabla \cdot A = 0$).

For a single charge particle the interaction Hamiltonian, upon excitation by a sinusoidal EM field $E(t)=E_o.exp[j(k\cdot r - \omega t)]$ is given by:

$$V(t) = -qm^{-1}(A.p) = -qm^{-1}(A_0 \exp[j(k.r - \omega t)] + A_0^* \exp[-j(k.r - \omega t)]e_k.p \tag{21b}$$

where e_k is the polarization unit vector. In general, we can neglect the wave-vector dependence of the interaction potential. This is correct if the wavelength of the field is much larger than the molecular dimension such that ($\lambda \rightarrow \infty$ and $k \rightarrow 0$), then $exp(jk \cdot r) \approx 1$. This is known as the electric dipole approximation

$$V(t) = -(\zeta_0 / \omega)\sum_i [q_i m_i^{-1}(e_k.p_i)].\sin(\omega t) \tag{21c}$$

Substituting the perturbation $V(t)$, we get the electric dipole Hamiltonian .

3.2 Transition Dipole Matrix Elements

We are seeking to use the dipole Hamiltonian to evaluate the transition rates induced by $V(t)$ from the first-order perturbation theory expression. For a perturbation $V(t)=Vsin(\omega t)$ the rate of transitions induced by field is:_

$$\Gamma_{kl} = \left(\frac{\pi}{2\hbar}\right)|V_{kl}|^2 [\delta(E_k - E_1 - \hbar\omega)] \tag{22a}$$

Here, V_{kl} is the matrix elements in the eigenstates of a collection of particles

$$V_{kl} = -j\,\zeta_o\,(\omega_{kl}/\omega)\,d_{kl} \tag{22b}$$

where d_{kl} is the transition dipole matrix element. This expression allows us to write the interaction potential for a dipole in a field as follows:_

$$V(t) = -\text{đ}.\zeta(t) \tag{22c}$$

where đ is the dipole operator. Then the rate of transitions between quantum states induced by the electric field is_

$$\Gamma_{kl} = \left(\frac{\pi}{2\hbar}\right) | \zeta(\omega_{kl} - \omega) + \delta(\omega_{kl} + \omega)] \tag{23a}$$

This equation is an expression for the absorption spectrum since the rate of transitions can be related to the power absorbed from the field. Generally we can express the absorption spectrum in terms of a sum over all initial and final states, the eigenstates of H_0:

$$\Gamma_{kl} = \left(\frac{\pi}{2\hbar}\right) | \zeta_0 |^2 \, \Sigma_{kl} \, | d_{kl} |^2 \, [\delta(\omega_{kl} - \omega) + \delta(\omega_{kl} + \omega)] \tag{23b}$$

3.3 Optical Transitions

The study of matter-light interaction usually begins with the time-dependent perturbation theory and the approach of Fermi's golden rule for optical transitions. The Fermi's golden rule gives the rate of transition from a single initial state to a set of final states, which can be described by a density of state function.

The transition from a state to state may occur as a result of collision of electrons with lattice vibrations (phonon absorption or emission) or upon interaction with EM radiation (photon emission or absorption). Optical transition rate results from the Hamiltonian term $H = qp.A/m_o$, as shown in equation (8-22). By expanding the vector potential into Fourier series, we can arrive to the following expression. The probability of a transition from an initial state |*i*> to a final state |*j*> can be written, according to Fermi's golden rule, as follows (Parker, 2009):

$$probability(i \rightarrow j) = \left(\frac{d_{ij}\zeta_0}{\hbar}\right)^2 \frac{\sin^2[1/2(\omega_{ij} - \omega)t]}{(\omega_{ij} - \omega)^2} \tag{24}$$

where d_{ij} is dipole moment element, which is involved in the light-matter interaction process. By making a plot of this function, we see that it approaches a Dirac delta function as $\omega \rightarrow \omega_{ij}$. Therefore, the transition rate is given by:

Figure 11. Schematic illustration of an electromagnetic transition from an initial state i to one of final states j or vice versa

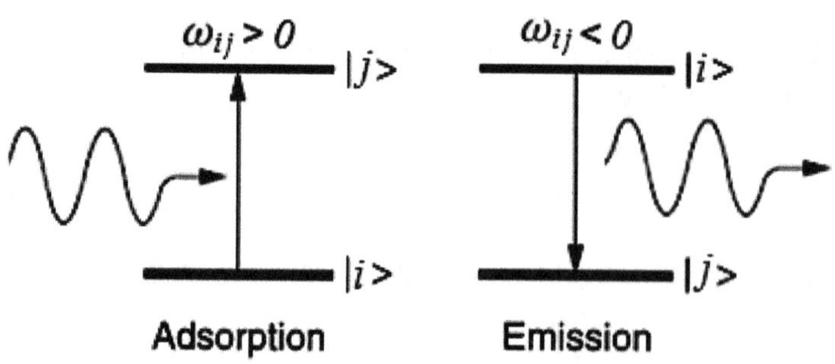

$$probability(i \rightarrow j) = \tau_{ij}^{-1} = \left(\frac{2\pi}{\hbar}\right)\left(\frac{d_{ij}\zeta_0}{\hbar}\right)^2 \delta(\omega_{ij} \pm \omega) \tag{25a}$$

This equation may be expanded as follows:

$$probability(i \rightarrow j) = \tau_{ij}^{-1} = \left(\frac{2\pi}{\hbar}\right)\left(\frac{d_{ij}\zeta_0}{\hbar}\right)^2 [\delta(\omega_i - \omega_j + \omega) + \delta(\omega_i - \omega_j - \omega)] \tag{25b}$$

Light Absorption Probability

We first consider the case for absorption where $\omega \cong \omega_{ji}$. Therefore, the transition probability for absorption by a field is given by:

$$\text{Absorption Probability}(i \rightarrow j) = \left(\frac{d_{ij}\zeta_0}{\hbar}\right)^2 \frac{\sin^2[1/2(\omega_{ij} - \omega)t]}{(\omega_{ij} - \omega)^2} \tag{26a}$$

Therefore,

$$\tau_{ij}^{-1}(Absorption) = \left(\frac{2\pi}{\hbar}\right)\left(\frac{d_{ij}\zeta_0}{\hbar}\right)^2 \delta(\omega_i - \omega_j \pm \omega) \tag{26b}$$

Light Emission Probability

When atom or molecule of a material is in excited state, it will eventually decay to a lower energy state by spontaneous emission. This case happens when $\omega \cong -\omega_{ji} > 0$. This gives the transition probability for emission:

$$\text{Emission Porbability}(i \rightarrow j) = \left(\frac{d_{ij}\zeta_0}{\hbar}\right)^2 \frac{\sin^2[1/2(\omega_{ij} - \omega)t]}{(\omega_{ij} - \omega)^2} \tag{27a}$$

Therefore,

$$\tau_{ij}^{-1}(Emission) = \left(\frac{2\pi}{\hbar}\right)\left(\frac{d_{ij}\zeta_0}{\hbar}\right)^2 \delta(\omega_j - \omega_i \pm \omega) \tag{27b}$$

3.4 Understanding Light Absorption and Emission Mechanisms

In order to understand how light is absorbed or emitted (generated) in semiconductor materials; we need first to investigate the energy transition phenomena in atoms or molecules. These phenomena include: spontaneous emission, stimulated emission/absorption and nonradiative decay.

According to quantum mechanics, the electrons of atoms can take different energy states, say E_1, E_2, E_3, such that, $E_1<E_2<E_3$. Lower energy level is more stable than higher energy levels, so electrons at high energy levels tend to decay to low energy levels, the energy difference between the two levels can be given out as electromagnetic radiation. This process is called Spontaneous Radiation. The relationship is:

$$E_2 - E_1 = h\nu_0 \tag{28}$$

where E_2 is the upper energy level, E_1 is the lower energy level, h is Plank's constant, ν_0 is frequency of the radiated EM wave. Note also that the energy difference between the two levels can decay in forms other than spontaneous radiation, which is called non-radiative decay. This is most likely happening in indirect-gap semiconductors. In this case, the energy difference can change into kinetic energy through collisions with atoms.

When external EM waves of frequency ν_0 are incident on the material whose atoms are initially at energy level E_2 and ν_0 is very near to the transition frequency between E_2 and E_1, there is a finite probability that the incident waves will force the atoms to undergo transition $E_2 \rightarrow E_1$. Such transition gives out a photon, while the incident photon still exists. Then we have two photons. This process is called Stimulated Radiation. In a group of such atoms, if the number of atoms in the excited state is given by N_j, the rate at which stimulated emission occurs is given by:

$$R_{ji} = - \partial N_j/\partial t = B_{ji} N_j \rho(\nu) \tag{29a}$$

Here, B_{ji} is proportionality constant for this particular transition. The constant B_{ji} is usually referred to as *Einstein **B** coefficient*, and $\rho(\nu)$ is the radiation density of photons of frequency ν. The corresponding rate R_{ij} for the total emission of photons of frequency ν and transition from a higher energy E_i to a lower energy E_j is

$$R_{ij} = - \partial N_i/\partial t = A_{ij} N_i + B_{ij} N_i \rho(\nu) \tag{29b}$$

where A_{ji} is the constant for spontaneous emission of a photon, and B_{ij} is the constant for stimulated emission in response to other photons. The rate of spontaneous emission A is also known as the Einstein A coefficient.

In thermodynamic equilibrium, the number of atoms in state i and that of atoms in state j must, on average, be constant; hence, the rates R_{ji} and R_{ij} must be equal. Also, from the Boltzmann statistics, the ratio of N_i and N_j is $exp(E_j - E_i) / k_BT$. From this, it is readily derived that:

$$A_{ji} = (8\pi h\nu^3 / c^2) B_{ji} \tag{29c}$$

On the above thermodynamic basis, Einstein postulated that the probability for spontaneous emission, ***A***, is related to the probability of stimulated emission, ***B***, by the relation (Einstein, 1917):

$$A / B = (8\pi\, h\upsilon^3 / c^3) \tag{29d}$$

Einstein argued that equilibrium would be possible, and the laws of thermodynamics obeyed, only if the ratio of the ***A*** and ***B*** coefficients had the value shown above. In recognition of Einstein's insight, the coefficients are called the Einstein *A* and *B* coefficients.

Note that the above Einstein relation does not give the values of the coefficients of spontaneous and stimulated emission. However, it is possible to calculate *A,* from which, the other coefficient can be calculated. According to the quantum electro-dynamics theory (QED), the coefficient of spontaneous emission, due to transition between two energy levels (say *m* and *n*), is given by:

$$A_{mn} = \frac{2\omega_{nm}^3}{3\varepsilon_o h c^3}\left|d_{nm}\right|^2 \tag{30a}$$

where ω_{nm} is the frequency of emitting radiation and d_{nm} is the transition matrix element (or dipole-moment matrix element). From quantum physics, the dipole moment matrix element is given by:

$$\left|d_{nm}\right|^2 = e\int \psi_m^{(0)*}\left[\sum_i r_i \psi_n^{(0)}\right]\cdot dV \tag{30b}$$

with $\psi_{n,m}^{(0)}$ are the zero-order solution of the wave functions associated with the *m* and *n* states. As indicated in equation (8-30a), the spontaneous emission coefficient is proportional to the cube of frequency of emitting radiation, in both the Einstein and quantum field theory. The critical detail of stimulated emission is that the emitted photon is identical to the stimulating photon in that it has the same frequency, phase, polarization, and direction of propagation. The two photons are totally coherent. It is this property that allows optical amplification to take place.

Figure 12. Absorption and emission of radiation

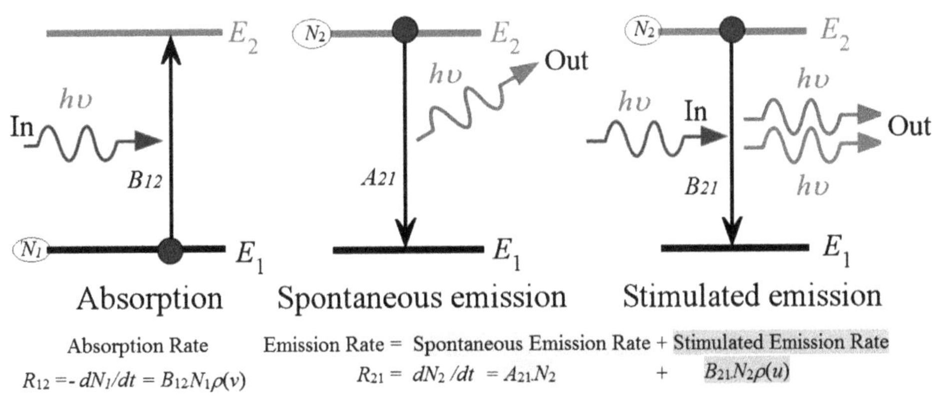

3.5 Optical Scattering

Photons interact with other particles such as electrons in an atom via several mechanisms. When a photon penetrates into a medium composed of particles whose sizes are much smaller than the wavelength of the incident photon, the scattering process is elastic and coherent. This means that the scattered photon has the same energy (and wavelength) of the incident photon. Otherwise, the photon may lose some of its energy and hence its frequency is changed.

Coherent (Rayleigh) Scattering

Rayleigh scattering, is elastic and coherent. In this scattering process, the energy (and therefore the wavelength) of the incident photon is conserved and only its direction is changed. In this case, the scattering intensity is proportional to the fourth power of the reciprocal wavelength of the incident photon. The condition of resonance (for Rayleigh scattering) means:

$$E_i - E_j = \hbar\omega - \Delta E \tag{31}$$

Raman Scattering

As shown in Figure 13, there exist other types of optical scattering mechanisms, which are not elastic. For instance, in the so-called Raman scattering, the scattered photon either gives energy to or takes energy from the medium. The Raman scattering means:

$$E_i - E_b = \hbar\omega_i \pm \hbar\omega_q - \Delta E = \hbar\omega_s - \Delta E \tag{32}$$

As we stated above, the photon scattering mechanisms (Raman and Rayleigh) may occur in dielectric materials. In this case, the incident photons may scatter over the dipoles of the dielectric atoms or molecules, and the photon frequency (*vi*) shifts up or shifts down, by the dipole frequency (ν). The Rayleigh scattering intensity for a single molecule (or particle) of radius *R*, is given by:

$$I= I_o\,(8\pi^4\alpha^2/\lambda^4R^2)\,(1+cos^2\theta) \tag{33}$$

The polarizability factor, α, is the proportionality factor between the applied field (electromagnetic field) and the induced dipole moment.

The so-called Raman spectroscopy employs the Raman scattering for materials analysis. The Raman spectroscopy is used to study vibrational, rotational, and other modes in solid materials and gasses. It relies on inelastic (Raman) scattering of monochromatic light, usually from a laser. The laser light interacts with phonons or other excitations in the material, resulting in the energy of the laser photons being shifted up or down. The shift in energy gives information about the phonon modes in the material.

Figure 13. Spectra and details of basic types of optical scattering mechanisms

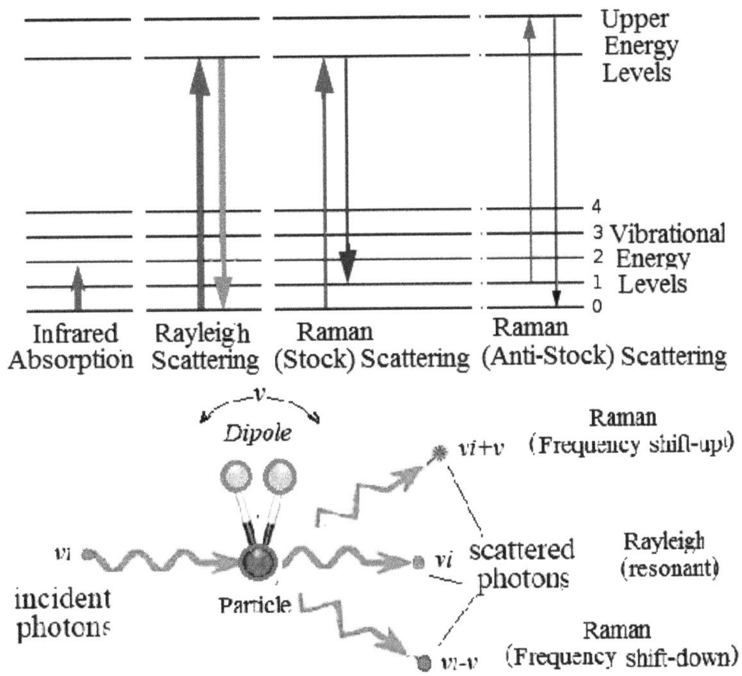

Compton (Incoherent) Scattering

The Compton (incoherent) interaction is a major mechanism for photon-produced electron excitation. In this case, a gamma-ray photon collides with an atom producing an energetic electron and a scattered photon of lower energy. The Compton interactions dominate other photon interactions in high energies (200 keV to 20 MeV).

Photoelectric Interaction (Photo-Ionization.)

Photons can interact with the inner shells of atoms in a solid and eject a photoelectron. Depending on the involved inner shell, the electrons are called K-, L and M-photoelectrons. Their energy is equal to the energy of the incoming photon less the binding energy of the inner shell. The K-shell photoelectric interaction varies with atomic number and photon energy and is dominant below 200 keV.

3.6 Nonlinear Optical Effects

It is well known that photons do not interact directly with themselves. However, photons can interact through their interactions with matter.

In a linear material the amount of charge displacement (which constitutes a dipole moment) is proportional to the instantaneous magnitude of the perturbing electromagnetic field or light. The charges oscillate at the same frequency as the frequency of the incident light. The oscillating charges may radiate light at the field frequency or the energy is transferred into nonradiative modes or other energy transfer mechanisms. Generally, the radiated light travels in the same direction as the incident light beam: The light is effectively bound to the material; the light excites charges that reradiate light that excites charges; and so on. As a result, the light travels through the material at a lower speed than it does in vacuum. If the motion of charges within the material decays without giving off light, some of the incident light is lost from the beam by scattering and absorption. The absorbance is defined as the ratio of light exiting a material to the light incident into the material divided by the material thickness. Both the absorbance and refractive index (ratio of speed of light in vacuum to the speed of light in the material) are linear optical properties of a material for low-intensity incident light.

In a nonlinear optical material, the displacement of charge from its equilibrium value is a nonlinear function of the electric field. All materials when exposed to a high enough light intensity (electromagnetic fields) show a nonlinear response such that the dielectric function becomes dependent on the electric field. Nonlinearity in optics occurs when the electromagnetic wave is large enough such that the medium responds not only at the fundamental driving frequency but also at higher harmonics. For small forces, the displacement of the charge is small and is approximated by a harmonic potential. When the displacement away from equilibrium is large, the harmonic approximation breaks down and the force is no longer a linear function of the displacement.

Nonlinear optical behavior is not observed when light travels in free space. The nonlinearity is a characteristic of the medium through which the light travels, rather than in the light itself. The response of a molecule is ***nonlinear*** if the charges are bound to the molecule by a non-harmonic potential. In this case, the dipole moment of the molecule is a nonlinear function of applied electric field. More generally, if a nonlinear molecule is exposed to light, the induced dipole moment is a nonlinear function of the time-dependent electric field. Therefore, when the intensity of the incident light to a material system increases to a large value the response of medium is no longer linear. Nonlinear optical phenomena can be described in terms of higher order susceptibilities. It is sometimes convenient to express the nonlinear effects in terms of the polarization density per unit volume $\boldsymbol{P}$ such that:

$$\boldsymbol{P} = P_1 + P_2 + P_3 + \ldots = x_1\zeta + x_2\zeta\zeta + x_3\zeta\zeta\zeta + \ldots \tag{34}$$

where P_1 is the linear (first order) polarization tensor, P_2 is the second order non-linear polarization, and so on. Also, x_1 is the linear susceptibility tensor, x_2 is the second order non-linear susceptibility, x_3 is the third order susceptibility and so on. Typically nonlinear optical phenomena happen at high fields. The high optical fields are generated by lasers: The coherence of laser light makes it possible to observe many nonlinear phenomena; when the molecules in the material respond coherently to the Laser,

EM Waves of different frequencies may exchange energy with one another via the nonlinear property of the medium, but their total energy is conserved. This class of nonlinear phenomena is known as parametric interactions. However, there are several nonlinear phenomena which are not parametric interactions. Examples of nonparametric nonlinear interactions include: laser interaction, multi-photon absorption, and nonlinear inelastic scattering (such as Raman and Brillouin scattering of light).

The second-order nonlinear polarization ($P_2 = x_2.\zeta(\omega_i).\zeta(\omega_j)$) gives rise to three-wave mixing processes, optical rectification and linear electro-optic effect. When $\omega_i = \omega_j$, we arrive at $\omega = 2\omega_o$, which corresponds to second harmonic generation. The third-order nonlinear polarization term ($P_3 = x_3\zeta(\omega_i).\zeta(\omega_j).\zeta(\omega_k)$) is responsible for four-wave mixing (FWM), stimulated Raman scattering, two-photon absorption, and Kerr-effect phenomena, including self-phase modulation (SPM). When $\omega_i = \omega_j = \omega_k$, we arrive at $\omega = 3\omega_o$, which corresponds to 3rd harmonic generation. All the above nonlinear phenomena have so many practical applications. In particular, the access to an optical non-linearity enables the processing of quantum information stored in light and generation of exotic states of light.

3.7 Coherent Optical Effects in Semiconductors

Recently, there have been several successful demonstrations of coherent light propagation effects in various semiconductor systems, including planar photonic crystals. The interaction of matter with electromagnetic fields is able to generate a coherent superposition of excited quantum states in the material. Coherent denotes the fact that the material excitations have a phase relation with the phase of the incident electromagnetic wave.

Macroscopically, the superposition state of the material results in an optical polarization; a rapidly oscillating dipole density. The optical polarization is a non-equilibrium quantity that decays to zero when the excited system relaxes to the equilibrium state after the electromagnetic excitation is switched off. Due to this decay or dephasing, coherent effects are observable only for a certain temporal duration after photoexcitation.

The field of nonlinear optics offers a host of fascinating phenomena, many of which are eminently useful. For instance, the nonlinear optical spectroscopy (using ultrafast laser pulses with durations of 10-100fs) reveals and interprets many coherent effects. We review here the phenomena of *quantum beats* and *photon echo* of excitons.

Quantum Beats

Quantum beats can be observed in systems in which the total optical polarization is due to a finite number of discrete and coupled transition frequencies. Assuming that all transitions have the same dipole matrix element, the optical polarization $P(t)$ of the system evolves as $\Sigma_i exp(-\Delta\omega_i t)$ after excitation with a short laser pulse. A finite number of frequencies results in temporal modulations of the squared modulus of the polarization $|P(t)|^2$ and thus of the intensity of the emitted electromagnetic field $|E(t)|^2$ with time periods $2\pi/(\Delta\omega_1 - \Delta\omega_2)$. For the case of two frequencies, the square modulus of polarization is proportional to $[1+cos(\Delta\omega_1 - \Delta\omega_2)t]$, i.e., due to the interference of two waves of same amplitude and different frequencies. Then the polarization varies between a maximum and zero. In semiconductors and semiconductor nanostructures, such as quantum wells, nonlinear optical quantum-beat spectroscopy has been widely used to investigate the temporal dynamics of excitonic resonance.

In particular, the consequences of coupling between different excitonic resonances via bi-excitons and Coulomb correlations, has been explored in many pump-probe and four-wave-mixing measurements. The theoretical analysis of such experiments in semiconductors requires a treatment on the basis of quantum mechanical many-body theory as is provided by the SBEs with many-body correlations incorporated on an adequate level.

Photon Echo

Recently, scientists have demonstrated the existence of magnetic-field-induced long-lived stimulated photon echoes. This coherent optical phenomenon results from resonant excitation of a medium by short optical pulses, which results in a delayed coherent optical flash response. This effect can be exploited to store optical information. In fact, it is possible to reverse the destructive interference of so-called inho-

Figure 14. Quantum beat in attosecond transient absorption
After (Chini et al, 2014).

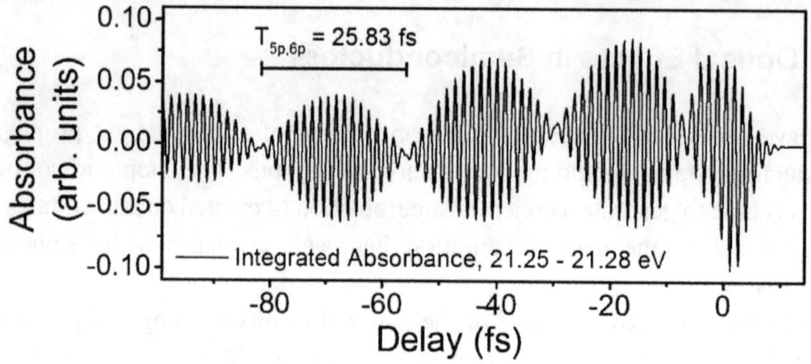

mogeneously broadened systems which contain a distribution of uncoupled subsystems with different resonance frequencies.

For example, consider a four-wave-mixing experiment in which the first short laser pulse excites all transitions at $t = 0$. As a result of the destructive interference between the different frequencies the overall polarization decays to zero. A second pulse arriving at $t = \tau > 0$ is able to conjugate the phases of the microscopic polarizations of the inhomogeneously broadened system. The subsequent unperturbed dynamical evolution of polarizations leads to rephasing and all polarizations are in phase at $t = 2\tau$ and results in a measurable macroscopic signal. This photon echo occurs when all polarizations add up constructively at $t = 2\tau$.

When photon echo experiments are performed in semiconductors with exciton resonances, it is essential to include many-body quantum effects in the analysis. For example, the solutions of the Semiconductor Bloch Equations (SBEs) have demonstrated that the reduction of the band gap due to the Coulomb interaction among the photo-excited electrons and holes is able to generate a photon echo even for a single discrete exciton with a pulse of sufficient intensity. Also, the spatial fluctuations of the energy can lead to a decay of the photon echo amplitude with increasing time delay. In order to consistently treat such dephasing phenomenon, we need to solve the SBEs, including biexciton correlations.

Figure 15. Schematic of the photon echo experiment

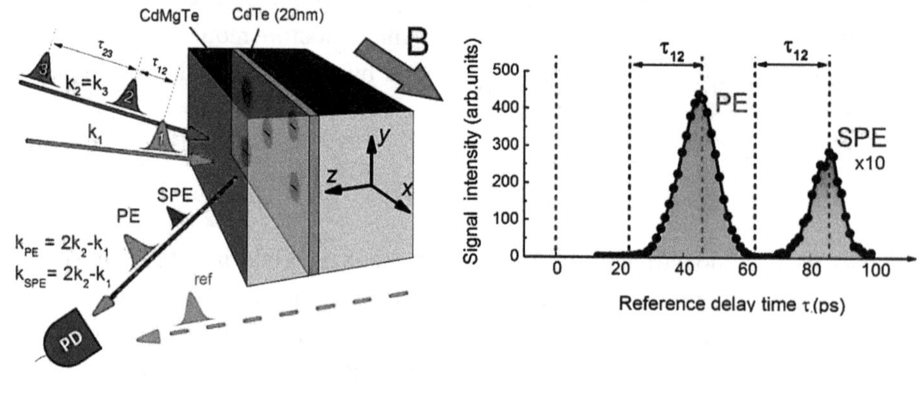

Note 1: Excitons, Bi-Excitons and Trions

An exciton is a bound state of an electron and hole, which are attracted to each other by the electrostatic Coulomb force, in a semiconductor (or insulator). The exciton has slightly less energy than the quasi-free electron and hole. The wave-function of the bound state is hydrogenic. However, the binding energy is much smaller and the size is much bigger than a hydrogen atom, just like electrons in the donor level. Technically speaking, excitons can be divided into two types, depending on the material properties. In semiconductors, the dielectric constant is large, and the Coulomb screening tends to reduce the interaction between excited electrons and holes. The result is weekly-bound electron-hole pair called a Mott-Wannier exciton, with a radius larger than the lattice spacing.

On the other hand, the excitations in materials with a small dielectric constant, the Coulomb interaction between an electron and a hole may be strong and the excitons tend to be small, of the same order as the size of the unit cell. This is called Frenkel exciton. Frenkel excitons have typical binding energy of 0.1 to 1eV. They are typically found in organic molecular crystals and alkali halides.

The existence of exciton states may be inferred from the absorption of light associated with their excitation. Excitons can be treated as Bosonic quasi particles. Excitons are quite stable and can have a relatively long lifetime (ps to ns). In addition, excitons are the main cause of light emission in semiconductors at low T (where $k_B T$ is less than the exciton binding energy E_B). However, excitons can dissociate into electron-hole (e-h) pairs.

An exciton can bind with other excitons to form a biexciton, like a hydrogen molecule. It should be noted that the optical properties of low-dimensional semiconductors and nanostructures originate from excitons and exciton complexes such as bi-excitons and tritons (charged triplet excitons). As shown in Figure 17, a trion is a localized excitation which consists of three charge carriers (electrons and holes). A positive trion consists of two holes and one electron and a negative trion consists of two electrons and one hole. Trions are long-lived and their transport is slower than singlet excitons. Trions were observed experimentally in optically excited low-dimensional semiconductors, such as quantum wells and quantum dots (Ganchevu et al., 2015), as well as nanotubes (Matsunaga et al, 2011).

Many optoelectronic devices are based on the exciton dynamics, including organic solar cells and light-emitting diodes. Therefore, an intensive research has been focused on the control of exciton transport in semiconductor devices and nanostructures.

Figure 16. Schematic illustration of the exciton (e-h) pair formation

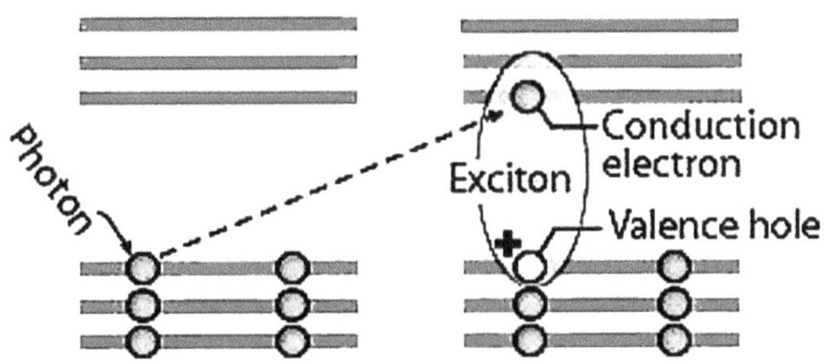

Figure 17. Schematic of excitons, biexcitons and trions

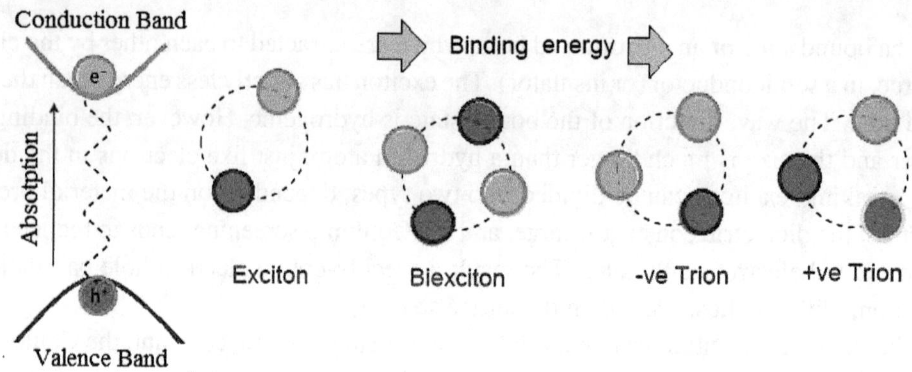

4. QUANTUM TRANSPORT MODELS FOR OPTOELECTRONIC DEVICES

A quick review of the published transport models in the domain of photonics, gives the impression that there is no standard framework, which encompasses all these models in a hierarchical manner. However, the wide family of the optoelectronic and photonic devices can be divided into two main classes: first one grouping semiconductor devices characterized by a quantum-mechanical behavior of their electronic subsystem and a second one which comprises low-dimensional nanostructures whose optoelectronic response in steady-state conditions may be safely treated within the semiclassical models. Based on this subdivision, it is possible to identify two distinct transport regimes, depending on the interaction conditions between photons and charge carriers. More specifically, we deal with two transport regimes, semiclassical and quantum, as summarized in Table 2.

Optical absorption in semiconductor nanostructures can be predicted from the classical theory of electromagnetism. This follows from the fact that the absorbing material size is much smaller than the wavelength of light. A semiclassical theory that treats light classically but treats atoms/molecules quantum mechanically can relate these phenomenological parameters to the atomic/molecular properties. Hence it seems not necessary to consider quantum theory of radiation to study absorption. For example the solutions to the classical Maxwell equations together with the Boltzmann transport equation may be an ideal tool to provide the spectral absorption and scattering rates in optoelectronic devices.

However, there is a general belief that classical physics cannot explain spontaneous emission[3]. This is sometimes true but it needs to be explained. In classical electromagnetism spontaneous emission is

Table 2. General classification scheme of the various approaches employed for the study of photonic semiconductor devices

Model	Semiclassical	Quantum
Macroscopic (Phenomenological)	Maxwell's Equations Semiconductor Rate Equations	MBEs SBEs SLEs
Microscopic	BTE-like (Photons) RTE Photon MCM	Heisenberg Equation, SBEs (density matrix) NEGF, Bethe-Salpeter Equation

shown as coming from accelerated charge carriers through radiation reaction of self-electromagnetic fields. For slowly varying (in space) optical excitations and confinement-potential profiles, it is possible to simulate the behavior of optical field using a distribution function (like the WDF). In this case, the distribution function can be described by semiclassical transport models. In contrast, in the presence of ultra-short spatial confinement and/or strongly inhomogeneous optical excitations, the utilization of a quantum Master equation (like the Heisenberg equation of motion) is necessary. In particular, the density-matrix formalism of the Maxwell-Bloch equations (MBEs) and the semiconductor luminescence equations (SLEs) are very useful in this concern. The NEGF approach has been also utilized to investigate the matter-field interactions in both linear and linear regimes.

4.1 Optical Bloch Equations (MBE's)

The optical Bloch equations, which are also called the Maxwell-Bloch equations, (MBE's) describe the dynamics of a two-state quantum system interacting with the electromagnetic field of an optical system. They are analogous to (but not at all equivalent to) the Bloch equations (1946), which describe the motion of spin systems in an electromagnetic field. These equations can be derived either semi-classically or with fully quantized field when certain approximations are made.

In order to describe phase-sensitive transport induced by optical excitation field U_{ab}, between two states or levels (a and b), we need to involve the inter-level polarization ($\boldsymbol{P}$). Combining the equations of motion of the level occupations f_a and f_b as well as of the inter-level polarization $\boldsymbol{P}$, results:

$$\frac{df_b}{dt} = -2\,\mathrm{Re}\left[\frac{U_{ab}P}{j\hbar}\right] \tag{35a}$$

$$\frac{dP}{dt} = \frac{(E_b - E_a)P}{j\hbar} + \frac{U_{ab}}{j\hbar}(f_b - f_a) \tag{35b}$$

$$dP/dt = (E_b - E_a)P/j\hbar + (U_{ba}/j\hbar)\,(f_a - f_b) \tag{35c}$$

In case of continuous optical excitation, which is resonant with the two-level system, we have: $U_{ab}(t) = U_o.exp(j\omega_o t)$, with $\hbar\omega_o = E_b - E_a$. The above treatment of light–matter interaction is based on an isolated two-level system and neglects the incoherence (phase-breaking) phenomena due to energy-dissipation by the host material. The incoherent phenomena may be incorporated within the two-level system by the following optical-Bloch-equations (Rossi, 2011):

$$\frac{df_a}{dt} = 2\,\mathrm{Re}\left[\frac{U_{ab}P}{j\hbar}\right] - \frac{(f_a - f_{ao})}{T_1} \tag{36a}$$

$$\frac{df_b}{dt} = 2\,\mathrm{Re}\left[\frac{U_{ab}P}{j\hbar}\right] - \frac{(f_b - f_{bo})}{T_2} \tag{36b}$$

$$\frac{dP}{dt} = \frac{(E_b - E_a)P}{j\hbar} + \frac{U_{ab}}{j\hbar}(f_b - f_a) - \frac{P}{T_2} \tag{36c}$$

Here, the incoherent phenomena are taken into account through the relaxation-times T_1 and T_2. After longtime, the populations f_a and f_b tend to their thermal-equilibrium values f_{ao} and f_{bo} via the relaxation time T_1, while the inter-level polarization P decays with the relaxation time T_2. This formulation is known as T_1/T_2 model. The derivation of the semiclassical optical Bloch equations is nearly identical to solving a system of two quantum states. Usually one casts these equations into a density matrix form. Therefore, the system can be described by the wave function:

$$\psi = c_b\psi_b + c_a\psi_a\,,\ |c_a|^2 + |c_b|^2 = 1 \tag{37}$$

The density matrix formulation of the above equations may take the following matrix form:

$$\rho = \begin{bmatrix} \rho_{aa} & \rho_{ab} \\ \rho_{ba} & \rho_{bb} \end{bmatrix} = \begin{bmatrix} f_a & p^* \\ p & f_b \end{bmatrix} = \begin{bmatrix} c_a c_a^* & c_a c_b^* \\ c_b c_a^* & c_b c_b^* \end{bmatrix} \tag{38}$$

This is the so-called density matrix for a two-level system. Recalling that the coefficients $c_a = \langle a|\psi\rangle$ and $c_b = \langle b|\psi\rangle$, the density matrix may also be regarded as a projection operator $\rho = |\psi\rangle\langle\psi|$.

One can now solve the Heisenberg equation of motion, or translate the results from solving the Schrödinger equation into density matrix form. One arrives at the following equations, including spontaneous emission:

$$d\rho_{aa}/dt = -\gamma\,\rho_{aa} + ½j(\Omega\rho_{ba} - \Omega^*\rho_{ab}) \tag{39b}$$

$$d\rho_{bb}/dt = \gamma\,\rho_{aa} - ½j(\Omega\rho_{ba} - \Omega^*\rho_{ab}) \tag{39a}$$

$$d\rho_{ba}/dt = -(½\gamma + j\delta)\,\rho_{ba} + ½j\Omega^*(\rho_{aa} - \rho_{bb}) \tag{39c}$$

$$d\rho_{ab}/dt = -(½\gamma - j\delta)\,\rho_{ab} - ½\Omega^*(\rho_{aa} - \rho_{bb}) \tag{39d}$$

In the derivation of these formulae, it was assumed that spontaneous emission follows an exponential decay of the coefficient ρ_{ab} with decay constant $\frac{1}{2}\gamma$. Also, the (generalized) Rabi frequency, Ω, is given by:

$$\hbar\Omega = 2U_0 = \hbar\sqrt{d^2 + |\chi_{ba}|^2} \tag{40}$$

where $\delta = \omega - \omega_o$ is the detuning which measures how far the light frequency, ω, is from the resonance transition frequency, ω_o. Also, $\boldsymbol{D}_o = \varepsilon.E_o$ is the amplitude of displacement vector (of electric field) including the polarization, $\chi_{ba} = \boldsymbol{p}_{ba}.\boldsymbol{D}_o/\hbar$ and $\boldsymbol{p}_{ba}$ is the transition dipole moment for the transition $b \rightarrow a$. The presence of spatial boundaries will modify the semi-conductor Bloch equations, whose open-system form may be written as:

$$\frac{d\rho_{aa}}{dt} = \left(\frac{d\rho_{aa}}{dt}\right)_{1e} + \left(\frac{d\rho_{aa}}{dt}\right)_{mb} + \left(\frac{d\rho_{aa}}{dt}\right)_{b} \tag{41}$$

Here, the last term describes the effect on the spatial boundaries on the time evolution of the density matrix ρ_{aa}.

4.2 Semiconductor Bloch Equations (SBE's)

For a correct description of optical processes in solid materials, it is necessary to go beyond the simple model of optical Bloch equations and to treat many-body interactions between material and optical excitations. For semiconductors the resulting system of equations is known as the semiconductor Bloch equations (SBE). The SBEs are most frequently used in situations where the optical properties are prominent, typically for excitations near the band gap, and for describing the dynamics of excitons and semiconductor lasers.

The SBEs are a set of integro-differential equations for the quantum dynamics of microscopic polarization and charge-carrier distributions. They describe the optical response of a semiconductor material, when it is excited by a coherent light source, such as laser. They also address the many-body quantum interactions among charge carriers and between carriers and phonons. Indeed, the SBEs form an excellent approach to describe optical properties of semiconductor, without phenomenological parameters. They describe the evolution of polarization, electron, and hole distributions as follows (Lindberg & Koch, 1988):

$$j\hbar\frac{dP_k}{dt} = \hat{E}_k P_k - [1 - f_k^e - f_k^h]\Omega_k + j\hbar\left(\frac{dP_k}{dt}\right)_{col.} \tag{42a}$$

$$\hbar\frac{df_k^e}{dt} = -2\,\mathrm{Im}[\Omega_k P_k^*] + \hbar\left(\frac{df_k^e}{dt}\right)_{col.} \tag{42b}$$

$$\hbar \frac{df_k^h}{dt} = -2\,\mathrm{Im}[\Omega_k P_k^*] + \hbar \left(\frac{df_k^h}{dt} \right)_{col.} \tag{42c}$$

where f_k^e and f_k^h are the distribution functions (densities) of electrons and holes and P_k is the microscopic polarization. Actually, P_k describes the transition amplitude for moving an electron from valence to conduction bands or vice versa (P_k*). Also, E_k is the energy of electron-hole pair, including the interaction corrections (due to Hartee-Fock):

$$\hat{E}_k = E_k - \Sigma_{k\neq k'} V_{k-k'}[1 - f_{k'}^e - f_{k'}^h] \tag{43a}$$

Note that the particle energies in the Hartree-Fock approximation include a part of carrier-carrier interactions. The role of the carrier-carrier interactions are given by the Coulombic potential $V_{k-k'}$. Other interaction terms are included in the collision terms. Also, Ω_k is the renormalized Rabi energy, which describes the interband dipole coupling to the radiation field:

$$\Omega_k = E_k d_{cv} + \Sigma_{k\neq k'} V_{k-k'} P_{k'} \tag{43b}$$

where $\boldsymbol{d}_{cv}$ is the dipole matrix element between the conduction and valence band and $E_k = E_e(\mathrm{k}) + E_h(k)$ is the energy of electron-hole pair. Remember that the Rabi frequency is a semiclassical concept representing the frequency of population oscillation between the levels of a resonantly illuminated two-level system. When the system is illuminated it will cyclically absorb photons and reemit them by stimulated emission. The Rabi frequency Ω_k is a measure of the coupling strength between the electronic system and the radiation field. The electron-hole system reacts to the combination of the applied field and the polarization field of the generated electron-hole pairs. Absorption of radiation and generation of electrons and holes are described by the factor $-2\, Im[\Omega_k P_k{}^*]$.

We dully note that the SBEs are particularly useful when solving the light transport in a semiconductor structure. In this case, we need to solve the SBEs with Maxwell's equations, driven by an optical polarization. The resulting set is called the Maxwell–SBE (MSBE) and they are often applied to simulate photonic devices (Haug & Koch, 2009).

For the simple case of a two-band model of a semiconductor, the SBEs can be written as follows (Kira &Koch, 2011):

$$j\hbar \frac{dP_k}{dt} = \Delta(\hat{E}_k P_k) - \Omega_k [n_k^c - n_k^v] + \left(j\hbar \frac{dP_k}{dt} \right)_{col.} \tag{44a}$$

$$j\hbar \frac{dn_k^c}{dt} = [\Omega_k^* P_k - \Omega_k^* P_k^*] + \left(j\hbar \frac{dn_k^c}{dt} \right)_{col.} \tag{44b}$$

$$j\hbar\frac{dn_k^v}{dt}=[\Omega_k^* P_k-\Omega_k^* P_k^*]+\left(j\hbar\frac{dn_k^v}{dt}\right)_{col.} \tag{44c}$$

Here, P_k is the microscopic polarization and n_k^c and n_k^v are the electron occupations in the conduction and valence bands (with superscripts c and v superscripts), respectively. As a result of the many-body Coulomb interaction and other scattering processes, the transition energy ΔE and the Rabi energy Ω_k depend on the state of the excited system. Therefore, they are expressed in terms of the polarizations P_k and the carrier occupations n_k^c and n_k^v, at any wavevector k'.

The original two-level SBE assume a scalar field and don't take the carrier spin or transverse exciting fields into account. Actually, the spin and the presence of transverse fields are interrelated. Optical excitation with circular polarization generates spin-polarized carrier populations (optical spin alignment) and can be used to investigate spin dynamics. Research on spin-dependent phenomena in semiconductors is rapidly progressing in many different directions, nowadays, and will be treated in Chapter 9.

The SBE's have many extensions in several directions. For instance, when the light field is quantized, one can treat several effects, such as photo-luminescence, in a quantum manner. The SBE's have been successfully utilized to study the impact ionization phenomena in coherent high-field semiconductor transport (Quade et al., 1994). Extension of the SBE formalism to treat holes including spin and polarization can be found in (Rössler, 2003). In this case, a six band model is used for the heavy hole, light hole, and split-off band, each with two spin degrees of freedom. Also, we can find specific SBE formulations for low-dimensional structures, such as quantum wells (Kira et al., 1999), carbon nanotubes (Hirtschulz et al., 2008) as well as quantum dots (Feldtmann, 2009).

4.3 Semiconductor Luminescence Equations (SLE's)

Luminescence is emission of light by a cold-body (not resulting from heat). It may happen due to chemical reaction of release of electrical energy or mechanical stresses in a crystalline solid. The semiconductor luminescence equations (SLEs) describe the spontaneous emission of radiation of semiconductors due to recombination of electrons and holes, which results in the luminescence phenomenon. These equations form a corner stone in semiconductor quantum optics as they include the quantized light–matter interaction as well as electronic excitations within a semiconductor. The SLEs are suited for modeling of semiconductor emission ranging from luminescence to laser effect.

Luminescence is related to the carrier and photon creation and annihilation operators (b_w, b_w^+). In optical transport the average intensity or photon number operator ($b_w^+b_w$), is a very important quantity. The total change in photon and carrier numbers are related as follows:

$$\frac{\partial[\sum_\omega \left\langle b_\omega^\dagger \right\rangle]}{\partial t}=-\sum_k f_k^{e,h} \tag{45}$$

where f^e_k and f^h_k are the electron and hole occupation probabilities (distribution functions) in the semiconductor. The above relation means that when an electron and a hole recombine, a photon is emitted. When the photon coherence, described here by the expectation value $<b_w>$, vanishes and the system becomes quasi-stationary, semiconductors emit incoherent light spontaneously, commonly referred to as luminescence (L). The luminescence flux L is proportional to the temporal change in photon number $<b^{\dagger}_{w}.b_{w}>$ as follows:

$$L(\omega) = \frac{\partial \left\langle b_{\omega}^{\dagger} b_{\omega} \right\rangle}{\partial t} = 2\,\mathrm{Re}[\Sigma_{k}(F_{\omega}^{*}\Pi_{k,\omega}] \tag{46a}$$

Here $\Pi_{k,\omega}$ describes a correlated emission of a photon of frequency ω when an electron with wave vector $\boldsymbol{k}$ recombines with a hole

$$\Pi_{k,\omega} = \Delta(<b_{\omega}^{\dagger} P_{\kappa}>) \tag{46b}$$

Note that the operator Δ means $\Delta(a.b) = <a.b> - <a>.<b>$. In general, the SLEs include all such single-particle and two-particle correlations, which can be used to calculate the luminescence spectrum.

SLEs are actually photon-number-like correlations and can be put in the following form (Kira et al., 1999):

$$j\hbar \frac{\partial \left(\Delta \left\langle b_{\omega}^{\dagger} b_{\omega} \right\rangle \right)}{\partial t} = \hbar(\omega' - \omega)\left(\Delta \left\langle b_{\omega}^{\dagger} b_{\omega} \right\rangle \right) + j\Sigma_{k}(F_{\omega}^{*}\Pi_{k,\omega} - F_{\omega}\Pi_{k,\omega'}^{*} \tag{47}$$

The diagonal form of this equation leads to the luminescence formula (8-46a). Also, the dynamics of photon-assisted correlations can be deduced from the following relation:

$$j\hbar \frac{\partial(\Pi_{k,\omega})}{\partial k'} = (\hat{E}_{k} - \hbar\omega)\Pi_{k,w} + \Omega_{k,\omega}^{sp} + T[II] - (1 - f_{k}^{e} - f_{k}^{h})(\Omega_{k,\omega}^{sp} + \Sigma_{k'} V_{k-k}\Pi_{k',\omega)} \tag{48}$$

The first contribution, $\hat{E}_{k}$, contains the Coulomb-renormalized single-particle energy that is determined by the band structure of the solid. The Coulomb renormalization are identical to those that appear in the semiconductor Bloch equations (SBEs), showing that *all* photon-assisted polarizations are coupled with each other via the Coulomb-interaction V_{k}. The three-particle correlations that appear are indicated by the $T[\Pi]$ term – they introduce excitation-induced dephasing, screening of Coulomb interaction, and highly correlated contributions such as phonon-sideband emission. Also, $\Omega_{k.\omega}^{spon}$ and Ω_{ω}^{stim} are the spontaneous-emission source and the stimulated emission contribution, simultaneously.

$$\Omega_{k,\omega}^{sp} = j\,F_{\omega}\,(f^{e}_{k} f^{h}_{k} + \Sigma_{k'}\,C_{x}^{k,k'}) \tag{49a}$$

$$\Omega_{\omega}^{st} = j\,\Sigma_{\omega}\,F_{\omega'}(\Delta<b_{\omega}^{\dagger} b_{\omega}>) \tag{49b}$$

where C_{x} denotes the exciton correlation. The electron and hole occupation probabilities, f^{e}_{k} and f^{h}_{k} modify the emission correlation, $\Pi_{k,\omega}$, through the Coulomb renormalizations and the Pauli- factor, $(1-f^{e}_{k} - f^{h}_{k})$. The occupation probabilities change with spontaneous recombination of electrons and holes, such that:

$$\partial f^e_k/\partial t\,|_L = \partial f^h_k/\partial t\,|_L = -2\,Re[\Sigma_\omega (F_\omega{}^*\Pi_{k,\omega})\,] \tag{50}$$

Under stationary optical pumping the analysis could become even simpler since the lost electron-hole pairs in recombination are replaced by a pump, keeping carrier occupations constant. But coherent pumping introduces additional correlation terms leading to a coupling of different modes.

In order to complete the set of SLEs, we should add the following equation to describes the quantum dynamics of exciton correlations (Jahnke, 2012)

$$j\frac{\partial(C^{k'})}{\partial t} = (\hat{E}_k - E_{k'})C_x^{k,k'} + (1 - f^e_{k'} - f^h_k)\Sigma_l V_{l-k'}C_x^{k,l} - (1 - f^e_k + f^h_k)\Sigma_l V_{l-k'}C_x^{l.k} + S_x^{k,k'} + D_x^{k,k'} + T_x^{k,k'} \tag{51}$$

The first term in the RHS contains the Coulomb-renormalized energy of electron–hole pairs. The summation terms define the main Coulomb sums that correlate electron–hole pairs into excitons. The last line contains one-, two- and three-particle correlations $S_x^{k,k'}$, $D_X^{k,k'}$ and $T_X^{k,k'}$, of singlet, doublet and triplet states, respectively.

4.4 Wannier Equation

The optical excitations and light emission in semiconductors can be described by the semiconductor Bloch equations (SBE) or the semiconductor luminescence equations (SLE). The so-called Wannier equation (WE) is a simplified version of these equations, which describes the optical excitations of electron-hole pairs (Wannier excitons). As shown in Chapter 5 of this Book, the WE resembles the Schrödinger equation for the hydrogen atom. Note that *Wannier exciton* are characterized by a radius that extends over several unit cells, in contrast to the *Frenkel excitons* whose size is about the unit cell. The generalized Wannier equation (GWE) includes the effect of many electrons and holes on the binding energy of excitons. It is equivalent to the homogeneous parts of SBEs or SLEs at low-density (Kira, 2011):

$$j\hbar\frac{\partial \hat{O}_H}{\partial t} = [\hat{O}_H, H_H] + j\hbar\left(\frac{\partial \hat{O}_H}{\partial t}\right)_H \tag{52a}$$

where, ϕ_λ is the e-h pair (exciton) wavefunction and $\hat{E}_k$ is the renormalized kinetic energy:

$$j\hbar\frac{\partial \hat{O}_H}{\partial t} = [\hat{O}_H, H_H] \tag{52b}$$

Here f^e_k and f^h_k, are the electron and hole occupations, respectively. Also, the effective Coulomb interaction ($V^{eff}_{k-k'}$) is given by:

$$\tilde{V}_{k-k'} = (1 - f^e_k - f^h_k)\,V_{k-k'} \tag{52c}$$

The term $(1-f_k^e - f_k^h)$ reduces the Coulomb interaction by the so-called phase-space filling factor (Haug & Koch, 2009). Due to the phase-space filling factor, the Coulomb attraction becomes repulsive for excitations levels where $f_k^e + f_k^h > 1$. Therefore, the GWE results only in a solution for unbound carriers, which follow the excitonic Mott transitions from bound to ionized electron–hole pairs. Figure 18 depicts the domain of applications of the excitonic transport models.

Note 2: Singlet, Doublet and Triplet (Spin) States

According to Pauli Exclusion Principle, the unexcited electrons are paired in their orbitals with opposite spins. However, after excitation, electrons will have different orbitals and their spin may be parallel or paired (with opposite spins). The spin angular momentum (S) of an electron (due to spin) is a quantized vector, which should precess around a certain axis, according to the uncertainty principle ($\Delta S.\Delta\phi > \hbar/2$). While precessing around the magnetic field axis, electrons are either spin-up or spin-down (with respect to the z-axis). In addition, electrons may precess in-phase or out of phase, according to their pointing direction in the x-y plane. In a two-spin system, this gives four possible vectors for the spin (1 vector for total spin S=0 and three vectors for total spin S=1). For S=0, the spin vector corresponds to magnetic momentum number 0 and it is called singlet state. For S=1, the three vectors correspond to magnetic momentum numbers (-1,0,+1) and they are called triplet states.

The above mentioned states of a 2-spin system (1 singlet and 3 triplets) may be visualized as spin (up-up), (up-down), (down-up) and (down-down).

The triplet states are more stable than singlet states due to the so-called exchange energy, which minimizes the binding energy between them. Therefore, A pair of spinning particles can be combined to form singlet of spin vector S=0 or one of three triplet states of total spin vector S=1. Note also, that upon rotation of a single electron with spin ½ (spin value = - ½ or ½), its state transforms as a doublet state. Almost all molecules in nature exist in a singlet state, except for molecular oxygen O_2, which exists in a triplet state at room temperature. Finally, it should be noted that singlet, doublet and triplet (spin) states may be attributed to some quasi particles such as excitons and polarons.

Figure 18. Domain of application of excitonic transport models (Maxwell SBEs & SLEs). Note that the emission correlation Π also describes photon assisted polarization and Ω is a measure of the coupling strength between the electronic system and the radiation field
After Haug & Koch (2011).

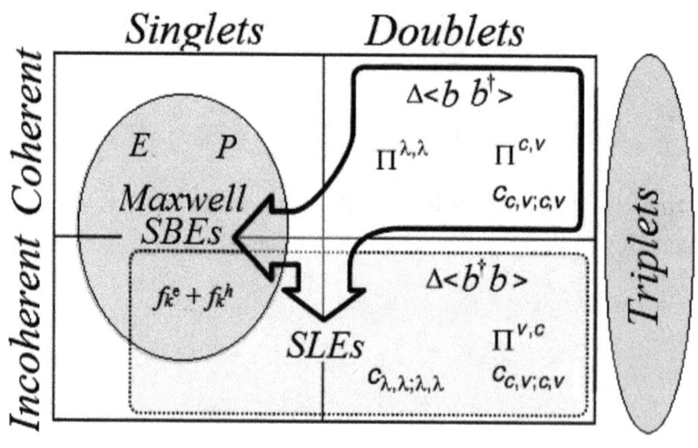

Figure 19. Singlets, doublets and triplets

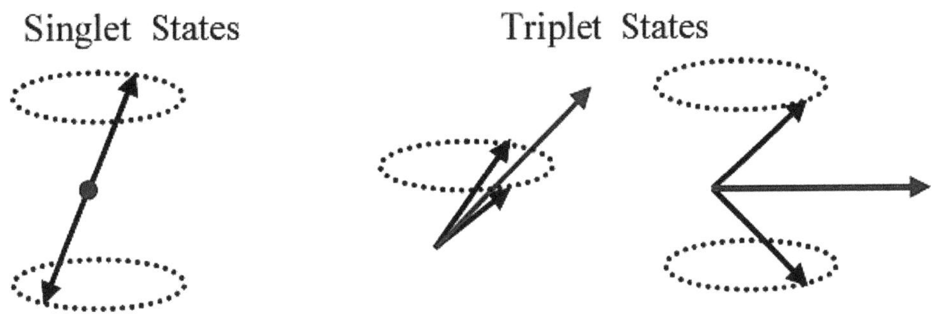

4.5 Heisenberg Equation of Motion (HEM)

We have seen, in Chapter 4, that in the Schrödinger picture the state vector evolve in time, $|\psi_s\rangle = |\psi_s(t)\rangle$, whereas the dynamic variables of the system are fixed during the time of undisturbed motion, $\hat{O}_s \neq \hat{O}_s(t)$. In the Heisenberg picture, we restrict the dynamic evolution to operators. This picture represents a scenario in which the state of the system has a fixed vector $|\psi_H\rangle \neq |\psi_H(t)\rangle$ and the dynamic variables are represented by moving linear operators $\hat{O}_H = \hat{O}_H(t)$. The transformation rule for a general observable O_s to O_H is obtained from the requirement that the expectation value should remain invariant. We then deduced the transformation relation: $O_H = \boldsymbol{T}^{\dagger} O_s \boldsymbol{T}$ where $\boldsymbol{T}$ is any unitary transformation matrix. The Heisenberg operators satisfy the following differential equation (Chow, Koch & Sargent, 1994):

$$H = \Sigma_k \left\{ \left[E_g + \left(\frac{\hbar^2 k^2}{2m_e^*} \right) \right] a_k^{\dagger} a_k + E_g + \left[E_g + \left(\frac{\hbar^2 k^2}{2m_e^*} \right) \right] b_k^{\dagger} b_k - [g_k a_k^{\dagger} b_k a_k + g_k^* a_k^{\dagger} b_k a_k] \right\} \tag{52a}$$

When the operator O_s has no explicit time dependence, the last term vanishes and the resultant equation is called the Heisenberg equation of motion (HEM)

$$j\hbar \, dO_H/dt = [O_H, H_H] \tag{52b}$$

The expectation value of the operator $O_H(t)$ in the Heisenberg picture is related to the density matrix (in the initial state of Schrodinger picture) as:

$$\langle O_H(t) \rangle = Tr\,[\rho_{so}\, O_H(t)] \tag{52c}$$

4.6 Heisenberg-Langevin Equation (HLE)

The so-called Heisenberg-Langevin equation (HLE) is a variant formula-tion of the Heisenberg equation of motion, for open quantum systems (like nano-electronic devices, with external electrodes). The Heisenberg equation of motion and its variants can be employed to observe the evolution of any average operator, such as the macroscopic polarization, of an optical system. The equation of motion for the effective dipole moment can be obtained by substituting the effective Hamiltonian for optical interactions in a semi-conductor as follows:

$$H=\Sigma_k\{[E_g+(\hbar^2k^2/2m_e^*)]a_k^\dagger a_k+[E_g+(\hbar^2k^2/2m_h^*)]b_k^\dagger b_k-[g_k a_k^\dagger b_k a+g_k^* a^\dagger a_k b_k]\} \quad (53a)$$

The equation of motion for the effective dipole operator ($b_k a_k$) is given by:

$$\frac{\partial b_k a_k}{\partial t}=-(g_k+jE_k)b_k a_k+jg_k(a_k^\dagger a_k+b_k^\dagger b_k-1)+F_k \quad (53b)$$

where γ is the damping, $E_k=E_g+(\hbar^2k^2/2m_{ec}^*)+(\hbar^2k^2/2m_{hv}^*)$ and F_k is the noise operator. Also, the equation of motion for the average number of photons may be written as follow (Scully & Zubairy, 1997)

$$\frac{da_k^\dagger a_k}{dt}=-\gamma\left\langle a_k^\dagger a_k\right\rangle+\int\left\langle F^\dagger(0),F(\tau)\right\rangle d\tau \quad (54)$$

where γ is the quantum Langevin drift coefficient and the noise force F is assumed stationary. This is a form of the famous fluctuation-dissipation theorem. It is actually a specific case of the quantum version of Langevin's equation (Langevin, 1923), which is useful for studying the time evolution of the current in presence of noise and fluctuation forces.

4.7 NEGF Modeling of Optoelectronic Devices

When we need to simulate real photonic devices, where the semiconductor active region has open boundaries to exchange charge carriers with the external environment, we can resort to the nonequilibrium Green's functions (NEGF) approach. The conventional Wigner-function quantum transport formalism to this problem leads to unphysical results, such as injection of coherent states from the contacts (Zaccaria & Rossi, 2003). The SBEs and their variants, on the other hand, have no special concern with carrier injection with open boundaries. Therefore, the NEGF is a useful tool to simulate nanoscale photonic devices, which involve both coherent transport (in the active region) and incoherent processes (in contact regions).

In the real space, the conduction band Green's functions (G^r, $G^<$) have four-parameters of positions z, z', energy E, and in-plane momentum value k. The retarded Green's function is the solution of the Dyson equation:

$$(E.I-H-\Sigma^r)\,G^r(E,k)=I \quad (55)$$

where the self-energy term Σ^r includes scatterings and couplings to the leads.

The valence part of the retarded Green's function is given by (Kolek, 2015):

$$G_{cv}^r=(E_z-E_v)^{-1}H_{cv}^\dagger G^r \quad (56a)$$

$$G^r_{cv} = (E_z - E_v)^{-1}[I + H^\dagger_{cv} G^\dagger_{cv}] \quad (56b)$$

Also, the differential Hamiltonian operator $\boldsymbol{H}_{cv}$ is defined as follows:

$$(\hbar^2/2m^*\gamma^{1/2})\, d/dz \rightarrow H_{cv} \quad (57)$$

and $\boldsymbol{E}_v = E_g(z) - E_c(z)$.

The lesser Green's functions can be obtained from the Keldysh equation:

$$\begin{bmatrix} G^< & G^<_{cv} \\ G^<_{vc} & G^<_{v} \end{bmatrix} = \begin{bmatrix} G^r \sum^< G^{r*} & G^r \sum^< G^{r*}_{vc} \\ G^r_{vc} \sum^< G^{r*} & G^r_{vc} \sum^< G^{r*}_{vc} \end{bmatrix} \quad (58)$$

where $\Sigma^<$ is the conduction band lesser self-energy. These equations should be iterated together with the Poisson equation. In this iteration, the Green functions valence components, $\boldsymbol{G}_{vc}{}^R$ and $\boldsymbol{G}_v{}^<$, should be included as they contribute to the density of electrons:

$$n(E, k, z) = -(j/2\pi a)\,.\,[G^< + G_v{}^<] \quad (59)$$

Once a self-consistent solution is achieved, the density of carriers and the optical absorption can be evaluated (Wacker, 2002). This model is particularly useful for a quantum cascade laser (QCL) devices. QCL laser operates on laser transitions not between different electronic bands, but instead on inter-subband transitions in semiconductor structures. The NEGF can treat quantum tunnelling and different scattering events, which are necessary for this type of devices. Figure 20 depicts the NEGF simulation of a quantum cascade laser (QCL) device. The population inversion of carriers at low *k*, is also shown in the right of the same figure.

Figure 20. Results of NEGF simulation of a QCL device. Note the population inversion of carriers at low k, in the right figure.
After Faist (2013).

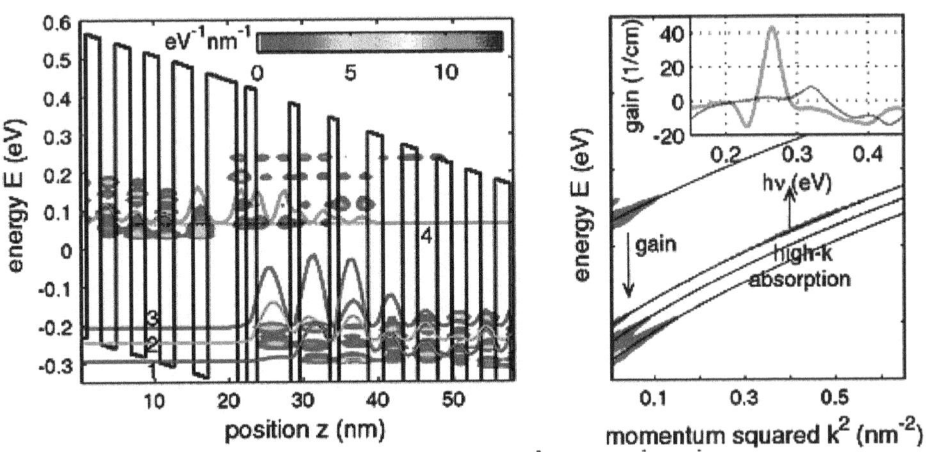

4.8 Bethe-Salpeter Equation (BSE) and Optical Excitations

The Bethe–Salpeter equation (BSE) is an equation of motion of bound states of a two-particle quantum system (Bethe & Salpeter, 1951). It is used to describe the coupled propagation of two-particle quantum systems, in a relativistic manner. For instance, the bound state of an electron–positron pair (positronium) and the bound state of excited electron–hole pairs (excitons). As the BSE can be derived by identifying bound-states of two particles, it can be related to the quantum description of scattering processes and Green's functions. Figure 21 depicts the coupled propagation of bound e-h pair (exciton) system. In this case the BSE may be written in the following form (4-point Bethe-Salpeter equation):

$$L = L_0 + L_0.K.L \tag{60a}$$

where L is denotes the electron-hole correlation function and, L_o is the free electron-hole pairs correlation, with the no interaction kernel ($K=0$). The set of variables (12,34, 56,78) comprises position, spin, and time coordinates: $(1)=(\mathbf{r}_1, \sigma_1, t_1)$, $(2)=(\mathbf{r}_2, \sigma_2, t_2)$ and so on. Actually, L depends on four time variables, related to two creation and two annihilation processes of electron and hole. The zero-interaction correlation may be written as follows:

$$L_o\,(12,34)= G(1,4).G(2,3) \tag{60b}$$

In optical excitations, we restrict ourselves to *simultaneous* creation and simultaneous annihilation, so only two of the four time variables are independent. In the absence of external fields we consider only the difference of the two time variables (of creation and annihilation) and the above equation is used to carry out a one-dimensional time-energy Fourier transform into $L(12,34,\omega)$ where the coordinates *12, 34* contains only position and spin of the creation and annihilation of an e-h pair. The BSE can be also written in the following irreducible form:

Figure 21. Schematic illustration of the coupled propagation of bound e-h pair (exciton)

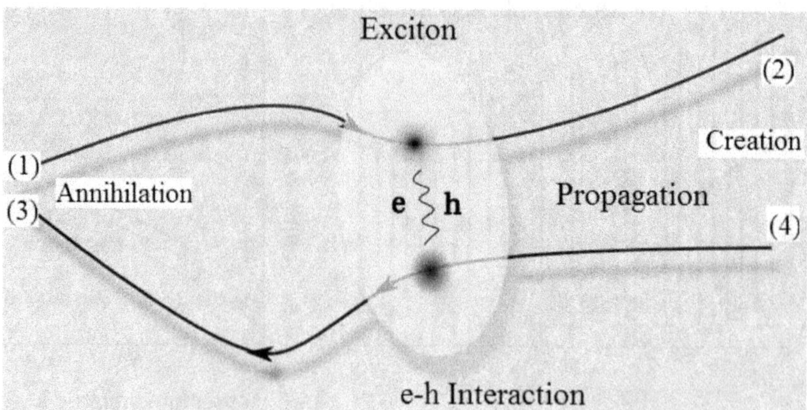

$$L(1234) = L_0(1234) + L_0(1234)\left[\frac{\delta\,\Sigma(56)}{\delta G(78)}\right]L(1234) \quad (60c)$$

where Σ and G are self-energy and Green's function of the system. The self-energy can be calculated using Hartee-Fock Coulomb potential ($\Sigma = jG.V_{sc}$) or the GWA using the screened Coulomb potential $W\,(\Sigma = jGW)$. The standard Bethe-Salpeter equation makes use of the time-dependent screened Hartree-Fock potential.

The solution if the BSE is difficult. However, if one of the particles is significantly more massive than the other, the problem is considerably simplified as one solves the Dirac equation for the lighter particle under the external potential of the heavier particle (Salpeter, 2008).

5. SEMICLASSICAL APPROACHES FOR PHOTON TRANSPORT

Semiclassical approaches help us to correlate optical material properties, such as absorption coefficient, to atomic properties. Light is treated classically (not quantized) but the transport medium is handled quantum mechanically. We have already presented the Boltzmann transport equation (BTE), as the basis of the semiclassical transport theory. The BTE accounts for changes of a carrier distribution function (or number density) in the phase space. We consider here photons as information carriers, and make use of a BTE-like equation to trace their evolution in the phase space.

5.1 Photon BTE (Radiative Transfer Equation)

Photon transport in scattering media can be modeled with BTE-like equations, such as the radiative transfer equation (RTE). The RTE can model the transfer of energy as photons move in a medium. The flow of radiation energy via a small area element is characterized by radiance or luminance. Luminance, *L*, is a measure of brightness of a radiating surface. It is defined as energy flux per unit time per unit normal area per unit solid angle (in $W/m^2.sterdian$) as follows:

$$L(\mathbf{r}, s, t) = d\Phi / d\Omega \quad (60a)$$

Here, Φ is the luminous flux (or photon fluence rate in W/m^2) and Ω is the solid angle. Also, $\mathbf{r}$ is position, $\mathbf{s}$ is unit direction vector and t is time.

The Photon fluence rate is defined as follows:

$$\Phi(\mathbf{r}, t) = \int_{4\pi} L(\mathbf{r}, s, t)\, d\Omega \quad (60b)$$

Photon current density (or energy flux) is defined as follows:

$$\mathbf{J}(\mathbf{r}, t) = \int_{4\pi} \mathbf{s} L(\mathbf{r}, s, t)\, d\Omega \quad (60c)$$

This is the vector counterpart of fluence rate pointing in the prevalent direction of energy flow.

The RTE is a differential equation describing radiance $L(\mathbf{r},s,t)$. It can be derived via conservation of energy. Briefly, the RTE states that a beam of light loses energy through divergence and extinction (including both absorption and scattering away from the beam) and gains energy from light sources in the medium and scattering directed towards the beam. Coherence, polarization and non-linearity are neglected.

Optical properties such as refractive index n, absorption coefficient μ_a, scattering coefficient μ_s, and scattering anisotropy g are taken as time-invariant but may vary spatially. Scattering is assumed to be elastic. The RTE (Boltzmann equation) is thus written as

$$c^{-1}\partial L(\mathbf{r},s,t)/\partial t + \mathbf{s}.\nabla L(\mathbf{r},s,t) + \mu_t L(\mathbf{r},s,t) = \mu_s \int_{4\pi} L(\mathbf{r},s,t)P(\mathbf{s}',\mathbf{s})d\Omega + S(r,s,t) \quad (61)$$

Here, c is the speed of light in the medium, as determined by the refractive index, $S(\mathbf{r},s,t)$ describes the light source., $\mu_t = \mu_a + \mu_s$ is the total attenuation coefficient and $P(s,\mathbf{s}')$ is the phase function, representing the probability of scattering of light with propagation direction $\mathbf{s}'$ into solid angle $d\Omega$ around $\mathbf{s}$. In most cases, the phase function depends only on the angle between the scattered $\mathbf{s}'$ and incident $\mathbf{s}$ directions (θ). The scattering anisotropy can be expressed as follows:

$$g = \int_{4\pi} P(\mathbf{s}',\mathbf{s})(\mathbf{s}'.\mathbf{s})d\Omega \quad (62)$$

The RTE is difficult to solve without introducing some approximations. By making appropriate assumptions about the behavior of photons in a scattering medium, the number of independent variables can be reduced. These assumptions lead to the so-called diffusion theory for photon transport. In this case, the radiance is expanded in spherical harmonics $Y_{n,m}$. Then, radiance is considered isotropic, so that only the isotropic and first-order anisotropic terms are used:

Figure 22. Schematic of energy flow through a differential area element dA at position r within a differential solid angle element dΩ

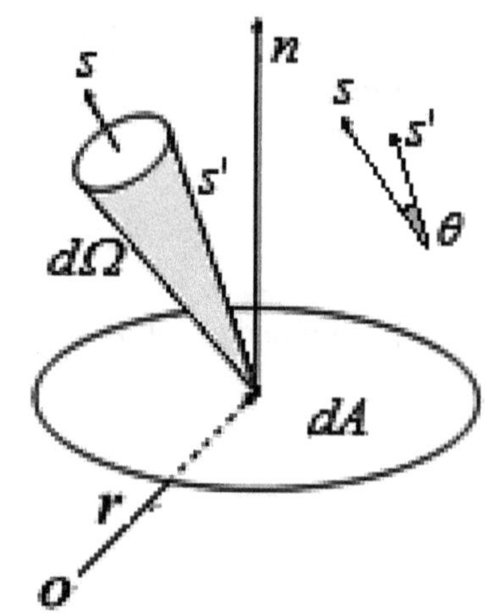

$$L(r,s.t) \approx \sum_{n=0}^{1}\sum_{m=-n}^{n} L_{n,m}(r,t)Y_{n,m}(s) \tag{63a}$$

where $L_{n,m}$ are the expansion coefficients. Radiance is expressed with 4 terms; one for $n=0$ (the isotropic term) and 3 terms for $n=1$ (the anisotropic terms). Hence we can approximate radiance as follows:

$$\sum_{m=-1}^{1} L_{1,m}(r,t)Y_{1,m}(s) = \frac{3}{4\pi} J(r.t).s \tag{63b}$$

Substituting the above expression for radiance, the RTE can be rewritten in scalar and vector forms as follows:

$$c^{-1}\partial\Phi(\mathbf{r},t)/\partial t + \mu_a\Phi(\mathbf{r},t) + \nabla.\mathbf{J}(\mathbf{r},t) = S(\mathbf{r}.t) \tag{64a}$$

$$c^{-1}\partial\mathbf{J}(\mathbf{r},t)/\partial t + \mu_t\mathbf{J}(\mathbf{r},t) + \tfrac{1}{3}\nabla\Phi(\mathbf{r},t) = 0 \tag{64b}$$

Here we assume that the fractional change in current density $\mathbf{J}$ over one mean free path is negligible. The second equation reduces to Fick's law,

$$J(r,t) = -\frac{\nabla\Phi(r,t)}{3(\mu_a + \mu_s^{'})} \tag{65}$$

which defines current density in terms of the gradient of fluence rate. Substituting Fick's law into the scalar representation of the RTE gives the diffusion equation:

$$c^{-1}\frac{\partial\Phi(r,t)}{\partial t} + \mu_a\Phi(r,t) - \nabla.[D\Phi(r,t)] = S(r,t) \tag{66a}$$

where $\mu'_s = (1-g)\mu_s$ is the reduced scattering coefficient and the diffusion coefficient D is

$$D = -\frac{1}{3(\mu_a + \mu_s^{'})} \tag{66b}$$

For the simple case of a short-pulsed point source

$$S(\mathbf{r},t,\mathbf{r}',t') = \delta(\mathbf{r}-\mathbf{r}')\,\delta(t-t') \tag{66c}$$

In infinite homogeneous material, the solution to the diffusion equation is

$$\Phi(r,t;r',t') = \frac{\exp[-\mu_a c(t-t')]}{4\pi D(t+t')^{3/2}} . \exp\left[-\frac{|r-r'|^2}{rDc(t-t')}\right] \tag{67}$$

The term $exp[\ -\mu_a c(t\text{-}t')]$ represents the exponential decay in fluence rate due to absorption in accordance with Beer's law. The other terms represent broadening due to scattering

5.2 Monte Carlo Method for Photon Transport

Monte Carlo method (MCM) is a stochastic technique that can be used to solve many physical problems. MCM can be used to keep track of the transport of multiple photons, in a scattering medium. The Monte Carlo photon simulation is equivalent to modeling photon transport by the radiative transfer equation (RTE), which describes the evolution of photons in a certain medium. Optical properties such as refractive index, absorption, scattering, and scattering anisotropy can be taken into account.

The first step of MC simulation, after initialization of all variables of interest, is to assign a step for the carrier motion in physical space. The step size, s, which is the distance the photon packet travels between interaction sites, may be expressed as $s = ln\xi/\mu_t$. where ξ is a random number and μ_t is the total direction cosine. Once a step size is selected, the photon packet is launched and propagated by a distance s in a direction defined by the direction cosines. We already know the Beer's Law, which describes the light intensity decay in an absorbing medium:

$$Relative\ intensity = I(d)/I(0) = exp(-\alpha_a\ d) \tag{68}$$

Figure 23. Diffuse reflectance vs. radius from an incident pencil beam as determined by Monte Carlo simulation and photon diffusion equation

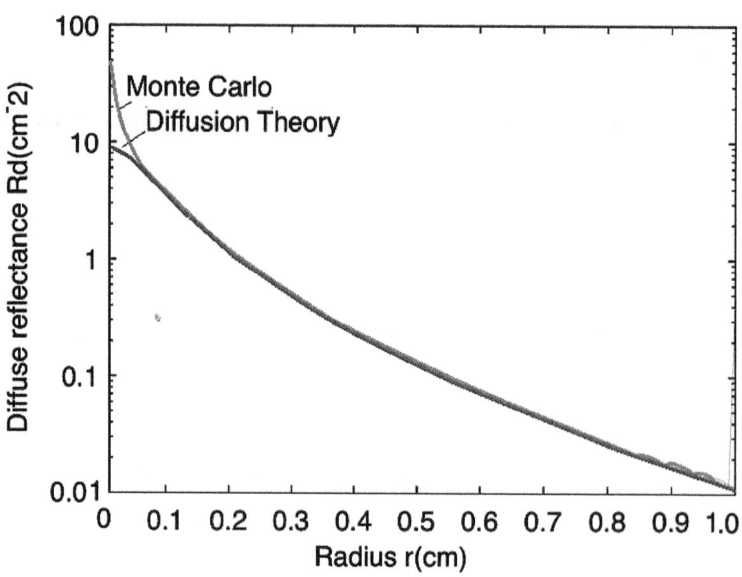

where α_a is the absorption coefficient [cm^{-1}] and d is the pathlength of photon travel [cm]. In a scattering medium, the photon path is not a straight line, but Beer's law still holds. Regardless of how curvy the path, the pathlength is given by $d = v_c.t$, where v_c is the speed of light in the medium ($v_c = c_o/n$) and t is time.

At any time point, one can calculate the probability of photon survival by $exp(-a_a d) = exp(-a_a ct)$. Therefore

$$P(t) = exp(- a_a d) = exp(- a_a v_c.t). \quad (69)$$

A portion of the photon weight is absorbed at each interaction site. The fraction of the weight is determined as $\Delta W = \mu_a/\mu_t$. The weight of the photon packet must then be updated as $W\text{-}\Delta W \rightarrow W$.

Following absorption, the photon packet is scattered. The weighted average of the cosine of the photon scattering angle is known as scattering anisotropy (g), which has a value between -1 and 1. If the optical anisotropy is 0, this generally indicates that the scattering is isotropic. If g approaches a value of 1 this indicates that the scattering is primarily in the forward direction.

In order to determine the new direction of the photon packet (and hence the photon direction cosines), we need to know the scattering phase function. Often the Henyey-Greenstein phase function is used to determine the scattering angle, θ, as follows:

$$Cos(\theta) = [1+g^2 - (1-g^2)/(1-g+2g\,\xi)] \text{ if } g \neq 0$$

$$= 2\xi - 1 \text{ if } g = 0 \quad (70)$$

The azimuthal angle φ is generally assumed to be uniformly distributed between 0 and 2π. Based on this assumption, we can set: $\phi = 2\pi\zeta$. Using these angles and the original direction cosines, we can find a new set of direction cosines. If a photon packet has experienced many interactions, for most applications the weight left in the packet is of little consequence. As a result it is necessary to determine a means for terminating photon packets of sufficiently small weight. A simple method is to use a threshold, and if the weight of the photon packet is below the threshold, the packet is considered dead. However, in time-resolved Monte Carlo photon simulation, we record the time-resolved spatial distribution of photons expressed as either an energy concentration (C in [J/cm^3]) or as fluence rate ($\phi = v_c.C$ in [W/cm^2]). Both photon energy concentration and fluence rate are usually expressed relative to the light impulse energy of the source at time zero (U_o in [J]). If one wishes to express C as photons/cm^3 or ϕ as photons/cm.s, then one simply multiplies the Monte Carlo result by U_o expressed in Joules divided by (hc/λ). For instance, a one joule beam at wavelength of 532nm is equivalent to $(1J)/(6.626 \times 10^{-34} Js)(2.998 \times 10^8 m/s) / (532 \times 10^{-9} m) = 2.7 \times 10^{18}$ photons.

Taking the absorption probability into account, the final expression of photon fluence rate is given by:

$$\phi(r,t) = v_c.C_o.exp(-a_a v_c.t). \quad (71)$$

where C_o reflects the spatial distribution, without absorption.

Monte Carlo method has been used to model electron/photon coupled transport, laser interactions, fluorescence and many other optical phenomena in radiation physics. Some of the MC simulators include the light polarization, before and after scattering events. A flow-chart that includes standard and polarized Monte Carlo is shown in the figure (Jessica et al, 2005). Beside well-established Monte Carlo

Figure 24. Effect of the g-factor on angular distribution of lighjt scattering

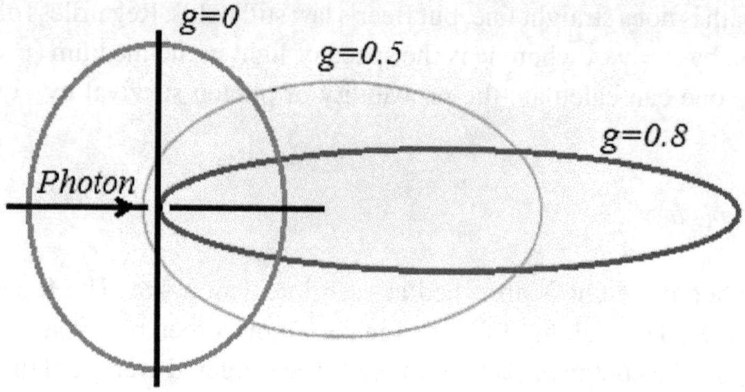

Figure 25. Flow chart of photon Monte Carlo method. Both standard and polarized methods (shaded blocks) are shown
After (Jessica et al., 2005).

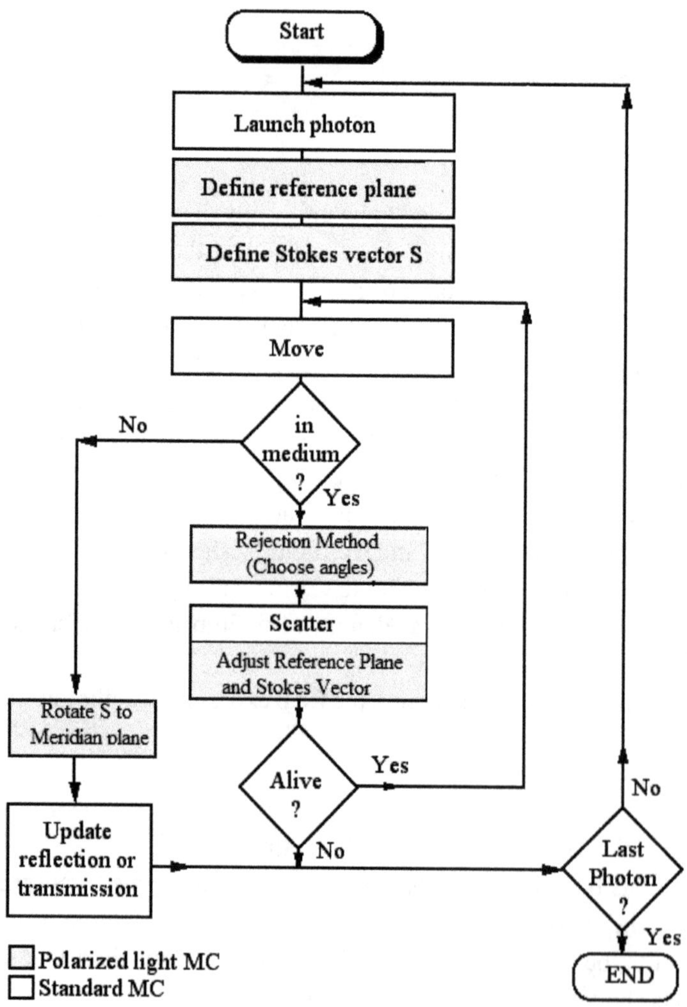

simulators, like EGS4, ITS and GEANT, one can find other programs in the literature, which are tailored for specific applications. For instance, the PENELOPE algorithm (Salvat, 2009) is specifically tailored for radiation therapy treatment.

6. PHENOMENOLOGICAL OPTICAL TRANSPORT MODELS

The dynamics of the optical field can be analyzed either by working with a space dependent dynamic equation for the complex field amplitude or using phenomenological rate equation for the scalar field energy together with a quasi-stationary wave equation to describe the evolution of the shape of the electromagnetic wave intensity distribution (Grondin, EL-Ghazaly and Goodnick, 1999). The main benefit of the latter approach is that it leads to simple results that are easier to interpret. However, the validity of approximating the shape of the instantaneous field distribution with that obtained from the quasi-stationary wave equation has not been yet proved (Vasileska, Goodnick & Klimeck, 2011).

6.1 Optoelectronic Modeling Using the Field Simulation

The electromagnetic field analysis of photonic devices may be carried out by either classical or full quantum models. The classical approaches are all about solving the Maxwell equations inside the device active region, while the quantum approaches are concerned with solving a sort of dynamic (Schrodinger-like or Heisenberg-like) wave equation. The classical electromagnetic approach may be represented by the Helmholtz equation:

$$\nabla x\left(\frac{1}{\varepsilon}\nabla xH\right)=\left(\frac{\omega}{c}\right)^2 H \text{ with the constraint: } \nabla.\boldsymbol{H}=0 \tag{72}$$

where $\boldsymbol{H}$ is the optical magnetic field vector and ε is the dielectric permittivity function of the material. The above equation can be obtained from the Maxwell equations in a transparent, time-invariant, source-free, and non-magnetic medium. This equation is sometimes called the Master Equation, and represents the basis of a Hermitian eigenvalue problem, which is not applicable if the wave equation were derived in terms of the electric field. If the eigen-operator in the LHS of the above equation is periodic (e.g., in a photonic crystal), then Bloch-Floquet theorem applies. The Bloch-Floquet theorem states that, due to the infinite periodicity in a photonic structure, the magnetic field will take the form:

$$H(\boldsymbol{r}) = exp(j\boldsymbol{k}.\boldsymbol{r}).h_k(\boldsymbol{r}). \tag{73}$$

where $h(r) = h(r+\boldsymbol{R})$ for all combinations of lattice vectors $\boldsymbol{R}$. Thus, we end up with the master equation in operator form:

$$(\nabla + j\boldsymbol{k})\{\varepsilon^{-1}\nabla + j\boldsymbol{k}\}h_k = (\omega^2/c^2)\, h_k \tag{74}$$

This is the fundamental equation, which needs to be solved. Different approaches can be explored to solve the final discretized problem, which may be put in the following matrix form:

$$[A]h = \omega^2[B]h \tag{75}$$

The result of solving the discrete problem is the dispersion relationship $\omega(k)$ between the frequencies of the modes and wave vector $\boldsymbol{k}$, usually plotted in the form of a band diagram.

6.2 Traveling Wave Model (TWM)

The traveling-wave model is based on the differential wave equations for counter-propagating optical fields in a laser cavity. These method was developed first by Homar and later by Serrat & Masoller (1999). The main component of the eclectic field of the fundamental transverse mode in a laser can be written as a superposition of forward and backward traveling waves:_

$$\zeta(r,t) = \boldsymbol{A}(r).[\psi^{+}.exp(j\pi.z/\Lambda) + \psi^{-}.exp(j\pi.z/\Lambda)].exp(j\omega_{o}\tau) \tag{76}$$

where Λ is the corrugation period of the laser waveguide and ω_o is an optical reference frequency, close to which the laser emits. The transverse field profile $A(r)$ of the waveguide can be normalized such that $|\psi^{+}(z,t)|^2$ and $|\psi^{-}(z,t)|^2$ are the powers guided by the forward and backward waves, respectively. The amplitudes are governed by the traveling wave equations

$$\partial\psi(z,t)/\partial t = j\boldsymbol{H}(z,t).\psi(z,t) \tag{77a}$$

with

$$\psi = [\psi^{+}, \psi^{-}]^{T} \tag{77b}$$

where $\boldsymbol{H}(z, t)$ is a Hamiltonian operator matrix, which is given by:

$$H = v_g \begin{bmatrix} j\dfrac{\partial}{\partial z} - \left(\beta - \dfrac{\pi}{\Lambda}\right) & -\kappa \\ -\kappa & -j\dfrac{\partial}{\partial z} - \left(\beta - \dfrac{\pi}{\Lambda}\right) \end{bmatrix} \tag{78}$$

The field energy can be expressed as

$$E = (1/v_g).\int_0^L |\psi|^2.\, dz \tag{79a}$$

with

$$|\psi|^2 = |\psi^{+}|^2 + |\psi^{-}|^2 \tag{79b}$$

Using the above equation, one can find easily the conservation relation

$$dE/dt = P_{gen} - P_{out} \tag{80}$$

where P_{gen} and P_{out} are the net power generated in the laser and the output power emitted from the two facets,

$$P_{gen} = 2 \int_0^L Im(\beta).|\psi|^2.\, dz \tag{81}$$

$$P_{out} = (1- |r_o|^2).|\psi^-(0)|^2 + (1- |r_L|^2).|\psi^+(L)|^2 \tag{82}$$

All terms in the energy balance equation (8-64) correspond clearly to those of the conventional photon rate equation (PRE)

6.3 Rate Equations in Semiconductors

The rate equations provide the most fundamental description of optical devices, such as lasers. Numerical analysis of a semiconductor laser indeed benefit from the set of rate equations, which are capable of describing the deterministic dynamics of the lasing modes as well as their mutual interactions The set of rate equations is a system of ordinary differential equations that relates the number or density of photons and charge carriers (electrons) in the device to the injection current and to device and material parameters such as the carrier lifetime, photon lifetime, and the optical gain. Such description is suited for analysis by existing simulation tools.

We can find at least three different sets of rate equations with three corresponding sets of variables. The first set of equations uses the variables describing the density of photons S (the number of photons per cm^3), the density of electrons n (carriers/cm^3) and the pump-current density $\boldsymbol{J}$ (carriers/s/cm^3). A second set of equations uses variables describing the optical power $P(W)$ and current $I(A)$. The first set using S and $\boldsymbol{J}$ must be equivalent to the second set with P and I since the number of photons correspond to the optical power. The phase does not enter into these equations since the optical power does not depend on phase. The third set involves the electric field $\boldsymbol{\zeta}$ (amplitude and phase), and carrier density n.

Let us consider the rate of change of a number of carrier pairs in the active region of a semiconductor-based device. We assume an intrinsic active region so that the density of electrons is equal to the number of holes per unit volume (n=p). For now, we assume that the semiconductor material comprising the laser as having two energy levels. These two levels correspond to the conduction and valence band edges obtained from the effective density of states. We want to know what physical phenomena can change the number of electrons in the conduction and valence bands. These changes must be related to the number of photons produced (for a direct bandgap semiconductor). We assume that the electrons and holes remained confined to the active region with volume V. The rate of change of the total number of electrons (or holes) $N = \boldsymbol{n}V$ is due to electron–hole generation and recombination. The rate equation has the basic form:

$$dN/dt = \text{Generation Rate} - \text{Recombination Rate} = G - R \tag{83}$$

Generation processes such as pumping and absorption increase the total number of electron–hole pairs (i.e., increases the number of electrons in the conduction band). Recombination processes such as stimulated and spontaneous emission reduce the total number of electrons in the conduction band. These facts can be taken into the basic rate equation:

G = Pump + (Stimulated Absorption –Stimulated Emission)

R = Nonradiative Recombination + Spontaneous Recombination (84)

This equation calculates the change in the number of carriers *N=**n**V* in the active region. Absorption and pumping increase the number while emission and recombination. Many of the processes that decrease the total number of carriers *N* must also increase the total number of photons *SV* in the modal volume *V*. We can therefore write a photon rate equation as

dS/dt = +Stimulated Emission + Fraction of Spontaneous Emission

- Stimulaed Absorption - Optical Loss (85)

The optical loss term accounts for the optical energy lost from the cavity. Some of the light scatters out of the cavity sidewalls and some passes through the mirrors. The light passing through the mirrors, although considered to be an optical loss, comprises a useful signal. Notice that the pump-current number density *J* does not appear in the photon equation since it does not directly change the cavity photon number.

The rate equations provide relations between the photon density, carrier density and the pump current density.

6.4 Single-Mode Rate Equations

In the resonant cavity of a laser, we get several standing wave patterns of electric field, corresponding to several axial modes (wavelengths).We may only consider one mode of field propagation, whose photon energy is closest to the gain peak. The intensity of this mode will grow faster than all others and eventually dominate. This simplification avoids the problem of finding the parameters and coefficients for other trivial modes.

The first rate equation (of electrons or excited atoms/molecules[4]) describes the change of excited photons as a function of time. The only gain term is the pumping term (*P=**J**/ed*), which increases the number of excited atoms by some external pumping sources. The two loss terms in this equation are due to stimulated and spontaneous emission/recombine-ation. Spontaneous emission/ recombination is a loss term ($-nf= -n/\tau_{sp}$) since this is unguided process that doesn't produce a net increase of photons in the cavity[5]. The stimulated emission/recombination part of this rate equation is represented by ($-n.S.G_o= -n.S/\tau_{ph}$), where G_o is the gain coefficient.

$$\frac{dn}{dt} = P - f \cdot n - G_o(n - n_o)S = \frac{J}{ed} - \frac{n}{\tau_{sp}} - \frac{(n - n_o)S}{\tau_{ph}} \tag{86}$$

where $G_o=1/\tau_{ph}$ is the gain coefficient, $f=1/\tau_{sp}$ is the decay rate due to spontaneous emission/recombination, and $P=J/e.d$ is the pump strength (e.g. due to pump current in a laser diode).

The second rate equation of photons describes the changing of laser photons as a function of time. The first term ($G_o.S = S/\tau_{ph}$) is the stimulated emission part of the equation. It is proportion to *n* and *S* since more excited atoms/cavity photons would obviously increase the rate of stimulated emission. The second term ($-A.S$) combines the photon loss term through the cavity by transmission (absorption S/τ_{ab}) and scattering or recombination (R_{sp}).

$$\frac{dS}{dt} = G_o S - A \cdot S = \frac{S}{\tau_{ph}} - \left(\frac{S}{\tau_{ab}} - R_{sp}\right) \tag{87}$$

where A is photon decay rate due to transmission/scattering, τ_{ph} is the photon lifetime and τ_{ab} is the photon absorption lifetime. Note that we substituted the gain coefficient $G_o = 1/\tau_{ph}$. It can be also expressed in terms of the modal $g(n)$, as follows:

$$G_o = v_g \Gamma .\ g(n) \tag{88}$$

Here, v_g is the group velocity of photons and Γ is the confinement factor (for $GaAs$ quantum well laser $\Gamma = 0.02$). If we assume that the gain is linear with the carrier density, n, then we can write the gain coefficient per unit length (the modal gain) as follows:

$$g(n) = \ell\ (n - n_o) \tag{89}$$

where ℓ is the differential gain coefficient and n_o is the transparency electron density. As for the spontaneous emission/recombination term (R_{sp}), it may be expressed as follows:

$$R_{sp} = \beta\, b\, n^2 \tag{90}$$

where b is the bimolecular recombination constant, β is a spontaneous emission coeffiucient. One can also add Auger recombination ($c.n^3$) to R_{sp}.

The rate equations may be solved by numerical integration to obtain a time-domain solution, or used to derive a set of steady state equations. Under steady state conditions, all parameters can be calculated from the carrier density (n) in a straight forward way. In this case, the rate equation for the photon density can be solved yielding:

$$S = \frac{\beta b n^2}{\tau_{ab}^{-1} - \tau_{ph}^{-1}} = \frac{\beta b n^2}{\tau_{ab}^{-1} - v_g \Gamma . g(n)} \tag{91}$$

The output optical power is given by the number of photons, which leave the semiconductor per unit time, multiplied by the photon energy:

$$P_{out} = hv\ (S/\tau_{ph})\ A = hv(S\ W)\ ln\ (1/\sqrt{R_1})\ v_g \tag{92}$$

where A is the active area of the device, and R is the reflectivity at the cavity interface edges.

$$R = [(n_1 - n_2)/(n_1 + n_2)]^2 \tag{93}$$

where n_1 and n_2 are the refraction indices at the interface (not to be confused with the electron density ***n***). For *GaAs*/Air interface, we have n_1=3.6, n_2=1 and $R = 0.32$. Finally one can find the current through the Laser device from the relation:

$$I = e.A\left(\frac{n}{\tau_{nr}} + bn^2 + \gamma n^3 + v_g \Gamma_g S\right) \tag{94}$$

where τ_{nr} (or τ_{ab}) is the non-radiative recombination time, b is the bimolecular recombination constant and γ is the Auger recombination constant. The power-current (*P-I*) curve of the Laser can be obtained, simply by plotting the values of the output power as a function of the Laser current. Note that laser action is induced above a certain threshold current density. Table 3 shows the typical values of the Laser diode parameters of a *GaAs/AlGaAs* edge-emitting single-quantum-well laser.

Table 3. Typical values of laser diode parameters (of a typical GaAs/AlGaAs edge-emitting single-quantum-well laser)

Laser and Material Parameters	Symbol	Typical Values
Cavity length	d	300 μm
Non-radiative recombination time	τ_{nr}	100 ns
Photon lifetime	τ_{ph}	2.58 ps
Bimolecular recombination constant	b	5 x 8^{-5} cm^2/s
Auger recombination constant	γ	6.25 x 8^{-18} cm^4/s
Group velocity	v_g	9 x 10^9 cm/s
Confinement factor	Γ	0.02
Differential gain coefficient		1.23 x 8^{-9} cm
Transparency carrier density	n_o	1.23 x 10^{12} cm^{-2}
Spontaneous emission factor	β	8^{-5}
Mirror #1 reflectivity	R_1	0.3
Mirror #2 reflectivity	R_1	0.3
Photon energy	$h\nu$	1.49 eV
Lasing wavelength	λ	830 nm
Waveguide losses	α	3 cm^{-1}
Threshold carrier (electron) density	n_0	2.99 x 10^{12} cm^{-2}
Threshold current	I_{th}	0.68 mA
Modal gain at threshold	$g(n_0)$	43 cm^{-1}
Gain slope	g_0	3x10^6 cm^{-1}/s
Gain saturation parameter	ε	3.4x10^{17} cm^3
Differential efficiency	η	0.47 mW/mA

According to (Parker, 2005).

6.5 Other Formulation for Large-Signal Single-Mode Laser

As just mentioned, the rate equations provide a primary description of light emission and absorption from a collection of atoms. We use them to describe the output optical power versus input current, the modulation response to a sinusoidal bias current, and the operating characteristics for laser amplifiers. The large signal rate equations in their modified form, due to Tucker (1999), read:

$$dS/dt = \Gamma g_o[\, n-n_o].S/(1+\alpha\, S) - S/\tau_{sp} + \beta\Gamma n/\tau_{sp} \tag{95a}$$

$$dn/dt = J/ed - g_o[\, n-n_o]S\,/\,(1+\delta\, S) - n/\tau_{sp} \tag{95b}$$

$$d\Phi/dt = \tfrac{1}{2}\,\alpha\,\{\Gamma g_o[\, n-n_o] - 1/\tau_{ph}\} \tag{95c}$$

Here n is the electron density, n_o is the electron density at the case of transparency, β is the fraction of spontaneous emission noise coupled into the lasing mode, g_o is the differential gain coefficient, ε is the gain saturation parameter,τ_{ph} is the photon lifetime,τ_{sp} is the carrier lifetime and α is the line width enhancement factor. Note that the term $(1+\varepsilon.S)$ is added in the denominator of the gain equation to account for gain saturation due to the high power densities in semiconductor lasers. Also, the output power is given by:

$$P(t) = \tfrac{1}{2}\,\eta.V.(\hbar\omega).S(t)\,/\Gamma.\tau_p \tag{96}$$

where η is the differential quantum efficiency, ω is the optical frequency and V is the volume of the active region,

6.6 Coupled Simulation of Electron/Photon Transport

During electron and photon transport in a medium (semiconductor or any other material) each particle produces the other. The rate equations can model the electrical and optical performance of optical semiconductor devices such a laser diode or a light-emitting diode (LED). Numerical analysis of a semiconductor laser indeed benefit from the set of rate equations, which are capable of describing the dynamics of the lasing modes as well as their mutual interactions. Such description in terms of first-order ordinary differential equations (ODEs) is well suited for an efficient analysis by existing simulation tools. In fact, the inclusion of spontaneous-emission noise in the modes and recombination noise in such frame is straightforward. Actually, in a multi-mode semiconductor diode laser several phenomena have their origin in the deterministic nonlinear dynamical evolution of the lasing modes. Starting from the solution of the Helmholtz equation:

$$\nabla^2\zeta + [\,\varepsilon_\omega(\omega^2/c^2) - \beta^2]\,\zeta = 0 \tag{97}$$

were ζ is the electromagnetic field and β is its propagation constant. We then solve the following continuity equation for each mode:

$$dS_m/dt = G_m(S_m+1) - S_m/\tau_{ph} \tag{98}$$

Here S_m is the photon density (occupation number) for the m-th mode and G_m is the mode gain. Also, τ_{ph} is the photon lifetime.

Poisson equation:

$$\nabla.(-\varepsilon \nabla\phi + \boldsymbol{P}) = e\,(p - n + Dop + N_t) \tag{100a}$$

Electron Continuity:

$$e\partial n/\partial t - \nabla.\boldsymbol{J}_n = -e(R - G) = -e(U_{rad} + U_{SRH}) \tag{100b}$$

Hole Continuity:

$$e\partial p/\partial t + \nabla.\boldsymbol{J}_p = -e(R - G) = -e(U_{rad} + U_{SRH}) \tag{100c}$$

$$\partial(n\,\omega_n)/\partial t + \nabla.\boldsymbol{S}_n - \zeta_n.\boldsymbol{J}_n - n(\omega_n - \omega_o)/\tau_{wn} = -\omega_n(U_{rad} + U_{SRH}) \tag{101a}$$

$$\partial(n\,\omega_n)/\partial t + \nabla.\boldsymbol{S}_p - \zeta_p.\boldsymbol{J}_p + - p(\omega_p - \omega_o)/\tau_{wp} = -\omega_p(U_{rad} + U_{SRH}) \tag{101b}$$

were $\boldsymbol{S}_n$, $\boldsymbol{S}_p$ are the carrier (electron and hole) energy fluxes and the carrier (electron and hole) current densities $\boldsymbol{J}_n$, $\boldsymbol{J}_p$ are given by the expressions:

$$\boldsymbol{J}_n = n\,\mu_n.m_n(\boldsymbol{v}_n.\nabla)(\boldsymbol{v}_n) - e\,n\,\mu_n.\zeta - k_B\,\mu_n.m_n\nabla.(n\,m_n^{-1}.\,T_n) \tag{102a}$$

$$\boldsymbol{J}_p = p\,\mu_p.m_p(\boldsymbol{v}_p.\nabla)(\boldsymbol{v}_p) - e\,p\,\mu_p.\zeta + k_B\,\mu_p.m_p\nabla.(p\,m_p^{-1}.\,T_p) \tag{102b}$$

And the carrier (electron and hole) energy fluxes $\boldsymbol{S}_n$, $\boldsymbol{S}_p$ are given by the expressions:

$$\boldsymbol{S}_n = -k_n\nabla T_n - (\omega_n + k_B T_n).(\boldsymbol{J}_n/e) \tag{103a}$$

$$\boldsymbol{S}_p = -k_p\nabla T_p + (\omega_p + k_B T_p).(\boldsymbol{J}_p/e) \tag{103b}$$

The conduction and valence band edges can be calculated from the electrostatic potential, as follows: $E_c = -e\phi - x$, $E_v = E_c - E_g$ where x is the electron affinity.

The net radiative recombination rate U_{rad} in the carrier continuity is given by:

$$U_{rad} \approx \Sigma_m B_m |E|^2 g_m \{S_m[f(E_c^m) - f(E_v^m)] + f(E^c_m)[1 - f(E^v_m)]\} \tag{104}$$

Here B_m *is* the Einstein coefficient of optical transitions for mode *m*. g_m is the equivalent density of states (DOS) for optical transitions between two levels (E_C^m in conduction band and E_V^m in valence band). Also, R_{cf} is the free carrier absorption rate.

In addition to the above system of six differential equations, we can add the heat flow equation:

$$-\nabla.k^{th}\nabla T + \nabla.H + \hbar\omega.(U_{rad} - R_{fc}) = 0 \quad (105a)$$

where k^{th} is the lattice thermal conductivity and the heat flux $\boldsymbol{H_s}$ may be calculated from the following phenomenological expression:

$$H_s = (\alpha_n k_B T + E_c)J_n + (\alpha_p k_B T - E_v)J_p \quad (105b)$$

$$\alpha_n = 2\,\digamma_1(\eta_n)/\digamma_0(\eta_n)\,,\ \alpha_p = 2\,\digamma_1(\eta_p)/\digamma_0(\eta_p), \quad (105c)$$

$$\eta_n = (E_{Fn} - E_c)/k_B T,\ \eta_p = (E_v - E_{Fp})/k_B T \text{ (8-105d)}$$

where $\digamma_m$ is the Fermi integral of order *m* and the quasi Fermi levels E_{Fn}, E_{Fp} are related to the density of carriers (near equilibrium), as follows:

$$n = N_c\digamma^{1/2}(\eta_n)\,,\ p = N_v\digamma^{1/2}(\eta_p) \quad (106)$$

7. NONLINEAR OPTICAL TRANSPORT IN SEMICONDUCTORS

Many optoelectronic and photonic devices are working near thermal equilibrium. The optical behavior of such devices can be modeled, by simple approaches, like the Kubo formula (Haug & Koch, 2012). However, the experiments demonstrate that some optical media exhibit nonlinear behavior, especially at high field excitations. For instance, when an intense laser beam passes through a material, the electric field can induce a change in the material refractive index that is proportional to the intensity of light. In addition, the high field may lead to electron tunneling from atoms.

7.1 Nonlinear Microscopic Approaches

In the presence of a time-dependent radiation field, the electrons will oscillate. When the radiation field is weak, there is mainly one oscillation frequency, that of the field. In a strong radiation field, the oscillatory motion becomes distorted and the dipole moment now includes a series of higher-order harmonics frequencies. In order to describe the response of the atom to strong laser fields, some models make use of the single-active-electron approximation, assuming that the interaction with the field involves essentially one active electron. Many methods have been proposed to solve the harmonic generation problems. The first approaches utilized the time-dependent Schrödinger equation, as proposed by ***Kulander*** (1989).

The interaction of intense laser pulse with matter may also lead to electron tunneling from atoms or molecules to the continuum. Tunneling of a particle through a potential barrier without having enough energy is a quantum phenomenon that challenges classical intuition. When the ionization of an atom by an electromagnetic field is considered, two different basic situations can happen. In the first one an atom absorbs photon with energy greater than the ionization energy of the atom. Hence, the atom is ionized and the electron is liberated. In the second situation, when the ionization potential is much greater than the photon energy, different ionization mechanisms such as Multi-Photon Ionization (MPI), Tunneling Ionization (TI) and Over-The-Barrier Ionization (OTBI), can occur in a non-linear regime. The direct numerical solution of the time-dependent Schrodinger equation for laser-induced processes is referred to as TDSE. The Ammosov-Delone-Krainov (ADK) theory is another alternative model used to study the tunneling ionization of atoms (Ammosov, 1986).

The above models and other variants have been used by several authors to interpret experimental results of laser-matter interaction (Barut, 2013).

Nonlinear Quantum Approaches

The optical excitations in semiconductors cannot be described on the level of isolated optical transitions but should be treated as an interacting many-body quantum system. Several quantum approaches and Master equations (ME) have been developed to study the nonlinear optical response of semiconductors (Hughes & Roy, 2012). Non-equilibrium Green's function (NEGF) and density matrix techniques are among the most common.

The semiconductor Bloch equations (SBEs) are able to theoretically describe coherent semiconductor optics on the basis of a fully microscopic many-body quantum theory. However, the standard SBEs have some drawbacks. For instance, the Coulomb interaction between free electrons and holes require the Hartree-Fock method, which readily treats excitons but cannot describe exciton-exciton interactions. In addition, the solution of coupled SBEs needs to huge computer resources. The SBEs, which are based on the density matrix, are an alternative to the Hartree-Fock treatment of the SBEs mentioned above. This is sometimes called the dynamics controlled truncation (DCT) equations.

The DCT equations are a successful microscopic approach that describe coherent effects in optically excited semiconductors. However, due to the computational difficulties, its application is limited to low-order nonlinearities in coherent transport. On the other hand, we know that the NEGF formalism can be successfully in examine the incoherent many-body transport, with carrier-carrier scattering and screening effects. For these reasons, researcher tried to combine the benefits of the advantages provided by the two approaches (NEGF and DCT) in a standard microscopic theory for nonlinear optical transport in semiconductors. Therefore, the DCT equations are written within the NEGF formalism (Ko, 2010). The Green's functions for optically-excited nonlinear systems (to third-order $\chi^{(3)}$) take the following form:

$$G_{ee}(i,t,j,t') = -j < \boldsymbol{T}[a_e(i,t)\, a_e^{\dagger}(j,t')] > \tag{107}$$

$$G_{eh}(i,t,j,t') = -j < \boldsymbol{T}[a_e(i,t)\, a_h(j,t')] > \tag{108}$$

$$G_{eehh}(i,t,j,t',k,t'',l,t''') = -< \mathbf{T}\, [a_e(i,t)a_e(j,t')a_h(k,t'')\, a_h(l,t''')] >$$

$$- G_{eh}(i,t,l,t''').G_{eh}(j,t',k,t'') + G_{eh}(i,t,k,t'').G_{eh}(j,t',l,t''') \quad (109)$$

The first two equations describe the Green's functions of electrons and holes densities, with polarization type functions. The third equation describes the coherent 4-particle (biexcitonic) functions. In these equations, T_C denotes the time ordering operator, along a Keldish contour and the angular brackets <.> means the average over the initial density operator. Also, $a_e(i,t)$ and $a_e^+(i,t)$ represent the annihilation and creation operators of electrons at the ith orbital (including band and spin) at a certain time t. In order to find the solution of this system, one can expand the Green's functions in diagrammatic series, using the Dyson theory.

Nonlinear Semiclassical Approaches

The semiclassical models provide us with a satisfactory degree of understanding of the physical interaction between atoms and laser fields. For instance, The Keldysh semiclassical strong field approximation (Keldish, 1967) was applied by Lewenstein (1999) to explain high-order-harmonic generation. Also, Van der Linden, Van der Heuvell, and Muller proposed a semiclassical model in the context of photoionization in 1999. This model was extended by Corkum and others to multiple ionization and high-order harmonic generation (Klingshirn, 2006). According to this model, the electron tunnels through the Coulomb energy barrier modified by the electric field of the laser, as illustrated in figure 8-21. It then undergoes (classical) oscillations in the field, during which the influence of the Coulomb force from the nucleus is negligible. If the electron comes back to the vicinity of the nucleus, it may be scattered or kicking out an electron. It may also recombine back to the ground state, thus producing a photon with energy, equal to the ionization potential plus the kinetic energy acquired from the field. We can investigate the harmonic generation properties from the classical calculations of the electron motion outside the binding potential. Assume that the electron to have zero velocity just after it has tunneled through the potential barrier at time $t = t_0$. Assume also that the laser field is given by $E = E_0 \sin(\omega t)$, then we get:

$$v(t) = - v_o \cos(\omega t) + v_o \cos(\omega t_o) \quad (8\text{-}110a)$$

$$x(t) = - (v_o /\omega)\sin(\omega t) + v_o (t-t_o).\cos(\omega t_o) \quad (110b)$$

where $v_0 = eE_0/m\omega$. As shown in Figure 22, the electron follows different trajectories depending on the time at which it was released into the medium (t_0). In the second half cycle, when the laser field changes its sign, electrons come back towards the core (at x=0) with a certain kinetic energy. This energy determines the emitted harmonic order. Except for the trajectory starting at 0.3T, which give the highest return energy (U_p), there are two main trajectories leading the same kinetic energy. This is illustrated in Figure 22, which shows the kinetic energy when the electron returns to the core, as well as the time spent in the medium, as a function of release time.

As shown in the figure, for each energy, and hence for each harmonic order, there are mainly two dipole strength intensity. Figure 23 shows for example the 35th harmonic generated in neon, calculated within the strong field approximation. The influence of the electron dynamics inherent to the harmonic-generation process is clearly visible on the intensity dependence of the harmonic components of the dipole moment.

Figure 26. Semiclassical model of nonlinear optical transport

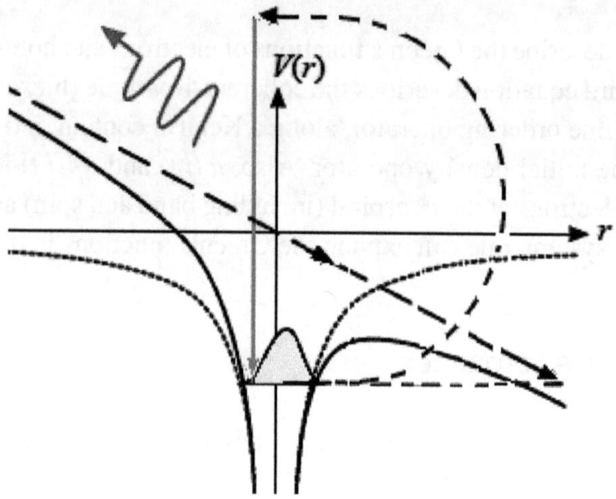

Figure 27. (a) Electron trajectories in the continuum at different release times. T is the laser period; (b) kinetic energy and time in continuum as a function of the release time.

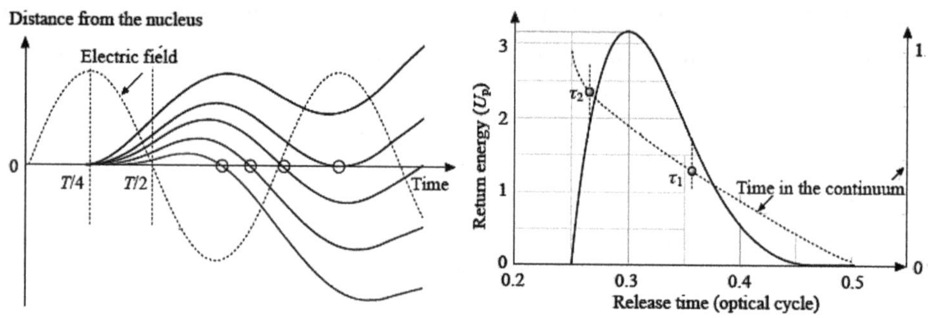

Figure 28. Single atom response within the strong field approximation. Intensity and phase of the 35-th harmonic in neon as a function of the laser intensity.

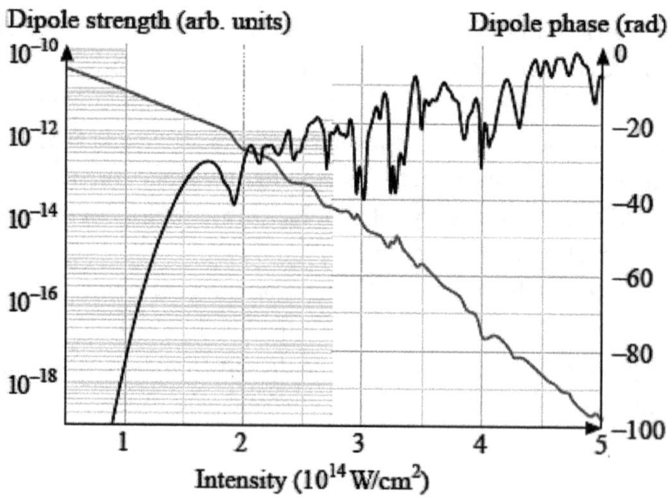

7.2 Nonlinear Macroscopic Approaches

We now turn to the macroscopic or measurable response of a medium to intense optical excitation. In a simple semiclassical model of a nonlinear medium consisting of non-interacting molecules, we can describe the interaction of light with molecules (atoms or ions) as follows:

$$\boldsymbol{P} = N\boldsymbol{p}\ , \boldsymbol{p} = \alpha\boldsymbol{\zeta} \tag{111a}$$

where p is the dipole moment of a molecule and E is the optical electric field strength N is the density of active molecules and α is the polarizability of molecules. The energy of a molecule in optical field is given by:

$$U = -\boldsymbol{p}.\boldsymbol{\zeta} = -\alpha(R)\ \zeta^2 \tag{111b}$$

Thus, the light-induced force driving molecular vibrations is given by

$$F = -\ \partial U/\partial R = (\partial\alpha/\partial R)\ \zeta^2 \tag{112}$$

where R is the generalized normal coordinate, which may be defined as the distance between the nuclei in a molecule. The force acting on a molecule in optical field may result in a resonant excitation. The propagation of light waves in a nonlinear medium is governed by the wave equation:

$$\Delta\boldsymbol{\zeta} - (n/c)^2\partial^2\boldsymbol{\zeta}/\partial t^2 = (4\pi/c^2).\ \partial^2\boldsymbol{P}_{nl}/\partial t^2 \tag{113}$$

where n is the refractive index, c is the speed of light, and $\boldsymbol{P}_{nl}$ is the nonlinear polarization of the medium.

7.3 Applications of Nonlinear Optical Transport

The so-called electro-optic effects are due to the change of the optical properties of matter in response to external electric field that varies slowly compared with the frequency of light. These effects include the Kerr and Pockels effects, which are very important in optical communications. The Pockels effect is a change in refractive index or birefringence that depends linearly in the field. Only certain crystalline solids (such as *LiNi* and *GaAs*) show the Pockels effect, as it requires inversion asymmetry. The Kerr effect is a change in index or birefringence that is quadratic in the field. We mean by birefringence, the double refraction, or the division of an incident ray of light into two rays (the *ordinary ray* and the *extraordinary ray*) when it passes through certain anisotropic materials, such as the cellophane paper. In both cases, the refractive index changes with applied field, according to the following expansion

$$n' = n + a_1\ \zeta + a_2\ \zeta^2 + \ldots$$

$$\Delta n = n' - n = \lambda\ K'\ \zeta + \lambda\ K''\ \zeta^2 + \ldots \tag{114}$$

where a_1, a_2, .. and K', K'',... etc. are constants.

Scintillation is another property of material luminescence, when excited by strong ionizing radiation. Luminescent materials, when struck by an ionizing particle, absorb its energy and scintillate, i.e. reemit the absorbed energy in the form of light, typically in the visible range. We mean by ionization particle radiation either charged particulate radiation (such as electrons and charged particles), or to uncharged radiation (like photons and neutrons), provided that they have enough energy to induce ionization. The first use of a scintillator dates back to an experiment in 1903 where Sir William Crookes observed a *ZnS* screen struck by α-particles. If the reemission occurs promptly, i.e. within the ~10ns required for an atomic transition, the process is called fluorescence· Sometimes, the excited state is metastable, so the relaxation back out of the excited state is delayed (from μsec to hours, depending on the material). The luminescence process then corresponds to either fluorescence or phosphorescence, depending on the type of transition and hence the wavelength of the emitted photon.If the energy levels are stacked, as shown in Figure 29, then the light emitted by fluorescence is of longer wavelength than the incident light.

8. THZ RADIATION FROM SEMICONDUCTORS

Terahertz (THz) waves are electromagnetic radiation in the frequency range from 0.1 to 10THz, between the range of electronic and photonic devices. THz radiation has several distinct advantages over other forms of spectroscopy: many materials are transparent to THz, THz radiation is safe for biological tissues because it is non-ionizing (unlike for example X-rays). THz emission can be obtained from quantum cascade laser (QCL), resonant tunneling diode (RTD) or down-converting a laser frequency, from the photonics side. In this section, I discuss the terahertz (THz) oscillations in semiconductor devices, upon the exposure to femto-second laser pulses. Optically-pumped THz laser (OPTL) is based on the switching of semiconductor structures, which are subjected to high electric field, or reverse-biased devices (usually P-I-N diodes) to femto-second laser pulses The interpretation of such THz radiation has been a subject of large debate (e.g., Xu et al, 1991; Tsen, 2004). In 2006, I published an article (El-Saba, 2006), which discussed the terahertz (THz) radiation from semiconductors at room temperature, upon the exposure to femtosecond laser pulses. In this work I studied the transient transport of hot carriers, in collision-limited regime, where coherence is completely destroyed. Starting from an analytical solution of the set of hydrodynamic equations in a reverse-biased P-I-N diode, I developed a formula to describe the evolution of hot carriers upon the exposure to femtosecond pulses:

Figure 29. Schematic illustration of the fluorescence emission process

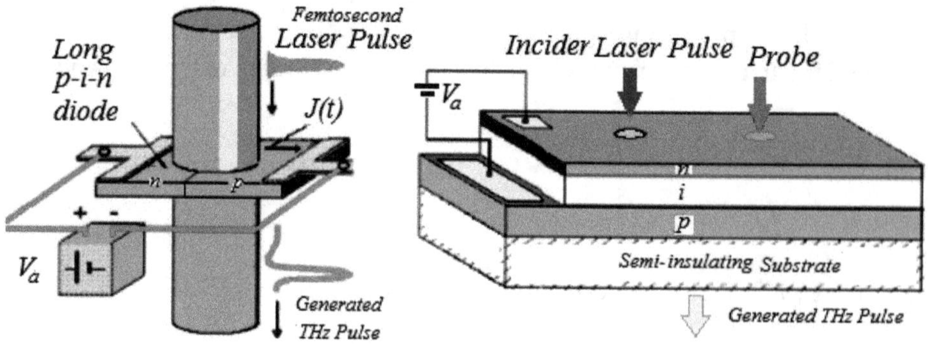

$$n(x,t) = \frac{N_o}{2v_{no}\sqrt{\pi \tau_{\omega n} t}} \exp\left[-\frac{(x - v_n t)^2}{2v_{no}^2 \tau_{\omega n} t} - \frac{t}{\tau_n} \right] \tag{115}$$

Here, τ_{wn} is the optical phonons relaxation time, τ_n is the carrier lifetime and v_n is the average velocity of carriers. Also, the pre-exponential factor, $N_{o,}$ is proportional to the density of photogenerated e-h pairs. It should be noted that Femtosecond laser pulses create dense, excited carrier plasma in an intrinsic semiconductor within a time shorter than the typical oscillation cycle of the bare LO phonon and plasmon resonance frequency. Phonon collisions and screened Coulomb interactions must therefore be considered. It is well known that the LO phonons have small group velocities (~10^5 cm/s) and anharmonically decay into faster acoustic phonons which in turn transport the energy out of the semiconductor device. Therefore, some phonons become hot and obtain much higher temperatures than the less active phonons. Consequences of having hot phonons are a substantially prolonged carrier cooling time (larger energy relaxation time). After cooling and return to equilibration of carrier and lattice temperatures, a final *condensation* process of e-h pair recombination completes the photoexcitation cycle. As the photoexcited carriers start to cool, the free electron-hole pairs begin to condensate into excitons. Both excitions and free electron-hole pairs have long lifetimes. Recombination can happen within a free electron-hole pair or via exciton recombination.

In 2011 Suzuki & Shimano have investigated the exciton Mott transition in Si at 30K, using optical pump and THz probe measurements and obtained very interesting information. They evaluated the time-dependent fraction of free carriers and excitons. The waveforms of the THz pulse transmitted after the sample with and without the optical pump were recorded by electro-optic sampling. The complex transmittance change was then obtained by Fourier transformation, thereby yielding a photo induced change in the complex dielectric function. To estimate the free *e-h* pair density at each delay time, the data were fitted with the free-carrier based Drude model (for lattice and electronic permittivity) in the low photon energy region (2.0–7.8 meV):

$$\varepsilon\ (Lattice) = \varepsilon_{\infty}\ (\omega^2 - \omega_{LO}^2)/(\omega^2 - \omega_{TO}^2),\ \varepsilon_e\ (Electronic) = \varepsilon_o - \omega_p^2/\omega^2 \tag{116}$$

Figure 30. Schematic representation of possible THz pulse generators. The femtosecond laser pulses are directed towards the space-charge region of the p-i-n diode.'

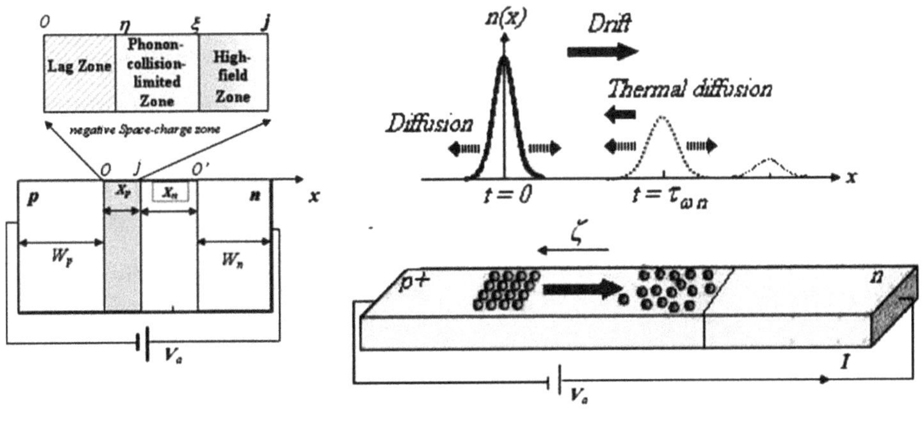

where, ω_{LO} and ω_{TO} are the longitudinal optical and transverse optical phonon frequencies and $\omega_p = ne^2/m^*\varepsilon_\infty$ is the plasma frequency. The Drudé electronic component decreases with delay time, and an exciton component emerges as a peak at about 10-12 meV, which corresponds to the 1*s*-2*p* transition of excitons in Si. The formation of excitons dynamics is thus revealed through the observation of the 1*s*-2*p* transition at about 3THz. This information is indeed important when we study THz radiation from Si structures at low temperature. For instance, when the laser pulse excitation is just above the energy gap threshold, we need to calculate the ratio of free e-h pairs to the total number of incident photons. This may be calculated by the aid of the **Saha** equation in the bulk (Kaindl et al., 2009, Suzuki):

$$N_{eh} = \tfrac{1}{2} N_{ex}^{\ 1/2} (2m_u k_B T_n/\hbar\pi)^{3/4} exp(-E_0/2k_B T_n) \tag{117}$$

where N_{eh} and N_{ex} are the density of e-h pairs and excitons that result in the semiconductor upon the exposure to the laser pulse and T_n is the carrier (electrons) temperature. Also, m_μ and E_0 are the exciton reduced mass and binding energy, respectively. For a given total e-h pair density $N=N_{eh}+N_{ex}$, The Saha equation gives the equilibrium ratio between the density of excitons N_{ex} and free carriers N_{eh} (when the two populations are reciprocally heated) in the Boltzmann approximation.

However, at room temperature, the dynamics of photon transport and photgenerated e-h plasma is quite different, especially at hot-carrier regime.

In our study, we considered THz radiation in a real device (P-I-N diode), where carrier heating is effectuated by the applied electric field. We attributed the THz in hot carrier regime to completely other reasons. The so called hot-carrier rebelling effect was introduced (EL-SABA, 2006)[6]. In this effect, we pointed out the role of back thermal diffusion of hot carriers, from hot to cold regions, which counteracts the sweep of cold carriers, which are mainly generated by femtosecond laser pulse, under the effect of electric field. This happens when the reverse bias voltage is properly tuned and the laser pulse is deliberately focused near the space charge region. Theretofore, the width of the phonon-collision-limited region (around which back and forth diffusion happen) becomes in the order of x_p:

$$x_p \sim v_{no}\, \tau_{\omega n} \tag{118}$$

which is the product of energy relaxation time and the saturation velocity of carriers (electrons in the p-side), as shown in Figure 32. Because holes have heavier mass (and slower response) compared to conduction electrons, we concentrated here on the electronic plasmon oscillation. Our simulation showed

Figure 31. Schematic representation of the transient behavior of photogenerated electrons, upon the exposure to laser pulses
After EL-Saba (2006).

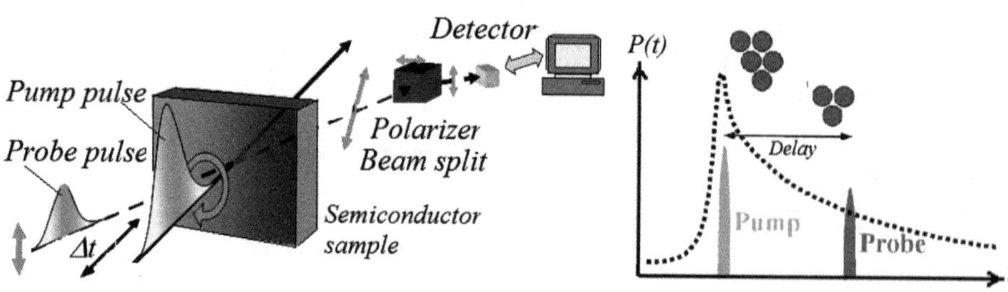

that this phenomenon can happen at room temperature, and far from breakdown region, where impact ionization is dominant (EL-SABA, 2012). Therefore, it may be used to generate low-noise high-power THz radiation at room temperature in low cost Si devices.

9. TIME-RESOLVED LASER SPECTROSCOPY OF SEMICONDUCTORS

The time-resolved laser spectroscopy is the study of dynamic processes in materials by means of laser spectroscopic techniques. Different methods are used to examine dynamics of charge carriers, spanning different time scales and photon energy ranges. We present in this section a short summary of this method, with emphasis on semiconductors and femtosecond laser excitation. In fact, the electronic and optical properties of semiconductors depend strongly on the electron dynamic that takes place in the femto- and picoseconds range.

When a semiconductor is subjected to ultrafast laser pulse, the absorbed photons create photoexcited hot carriers. The non-equilibrium dynamics of these hot carriers can be studied by ultrafast pump-probe spectroscopy. In this method, an ultrashort laser pulse creates a nonequilibrium distribution of free electrons-hole (e-h) pairs. The optically excited carriers then relax, eventually reaching thermal equilibrium with the crystal lattice.

Short pulses and short time delays between pump and probe provide a direct insight into carrier relaxation and hot phonon generation phenomena. On longer timescales the pump beam can give rise to luminescence from recombining excitons or electron-hole pairs. Output is gathered by either the reflected or the transmitted probe signal, as shown in the figure 24. The time-resolved laser spectroscopy is described in details in (Shah, 1999). Recent advances in this method can be found in (Brudevoll et al., 2011). The analysis of results of this method is usually carried out using the SBEs, probably with extensions to include spin and polarization.

10: CASE STUDY 12: SIMULATION OF A *SiGe* PHOTODETECTOR

The Photodetector converts an optical signal into an electrical signal. Optical detection is critical for defense and telecommunication applications, which requires near-infrared (NIR) detection in the 1300-1550nm wavelength range. In this example, we study the performance of vertical *SiGe* photodetector with FDTD optical simulator (like FDTD Solutions) and a device electrical simulator (like DEVICE). For the optical simulation, we model the guiding of light into the active area of the device, absorption of the optical signal and resulting generation of electrical charge.

Figure 32. Femtosecond pump-probe technique

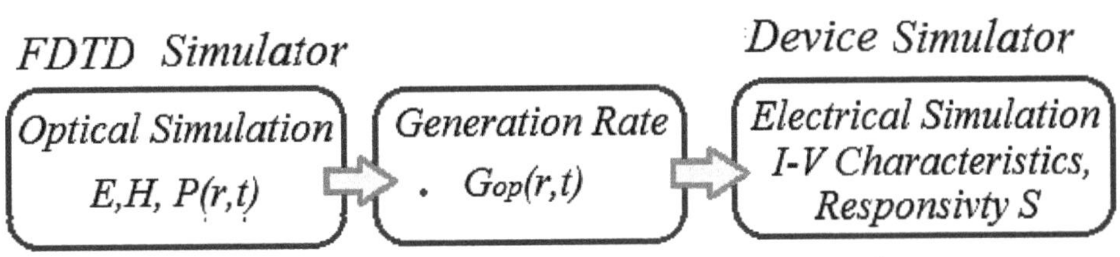

Figure 33. The photodetector simulation workflow

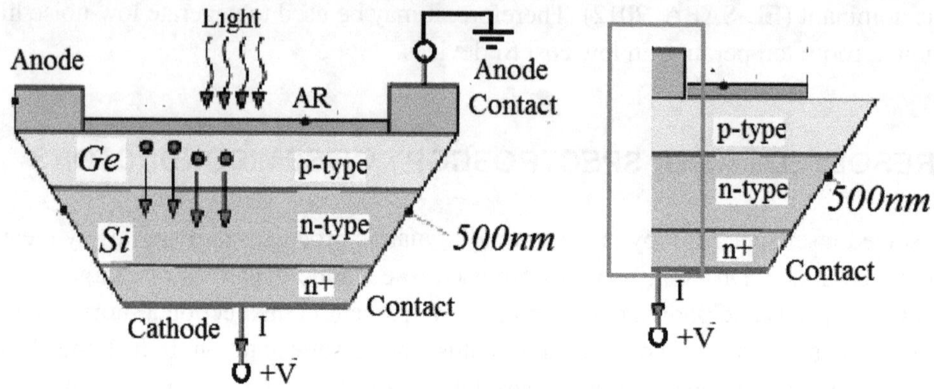

As a first approach, the interaction between optical and electrical effects can be modeled through the recombination-generation term. For instance, in a vertical photo diode, the supplied photons generate free charge carriers generating the photo current. We model this effect by adding - to the Shockley-Read-Hall term - the generation rate $G_{op}(x)$ of free carriers at depth x, caused by the optical irradiation power P-I-N with frequency ω:

$$G_{op} = \alpha_{ab}.I.\eta.(1\text{-}r).(W_{op}/h\omega).\alpha_{ab}\,exp(\alpha_{ab}\,x) \tag{119a}$$

Here, the physical parameters are the quantum efficiency η, which expresses the number of generated electron-hole pairs by one photon. (in Si, $\eta \approx 1$), the reflectivity r of the irradiated surface with area A, and the optical absorption coefficient α_{ab}. The optical generation rate G_{op} can be also calculated from the incident power density. Note that the optical power density, W_{op} (J/cm^3s), can be calculated from the light intensity field, which is equal to the time average of the Poynting vector, $S_{op} = ½\,Re(ExH)$,

$$W_{op} = -\,\nabla.S_{op} = ½\,\sigma\,|E|^2 \tag{119b}$$

Here E and H are the electrical and magnetic field intensities of the light electromagnetic wave and σ is the conductivity of the absorbing medium. Both E and H can be calculated from the solution of the Maxwell equations (or the separated Helmholtz equations). To calculate the absorption as a function of space and frequency, we only need to know the electric field intensity and the imaginary part of the permittivity. Both quantities are easy to measure in an FDTD simulation. The number of absorbed photons per unit volume can then be calculated by dividing this value by the energy per photon $(g = P_{abs}/\hbar\omega)$

The absorbed photons will generate electron hole pairs which will be separated out of the depletion region by the electric field and produce a flow of current. The photodetector is set up as a 3D simulation in *FDTD* with a length of 50um. Perfectly matching layer (PML) boundary conditions are used in all directions. To model the input light, we use a mode source with wavelength of 1.55um. The cathode and anode contacts set to simulate bias range of -5V to 0.5V. Once the sweep task is done, we can get a plot of current versus voltages and temperatures. The workflow starts with optical simulations in FDTD. Finite-Difference Time-Domain (FDTD) is one of the time-domain analysis tools that enable the simulation of passive and non-linear photonic components. The FDFT algorithm divides space and time into

Figure 34. The structure of the simulated photodetector

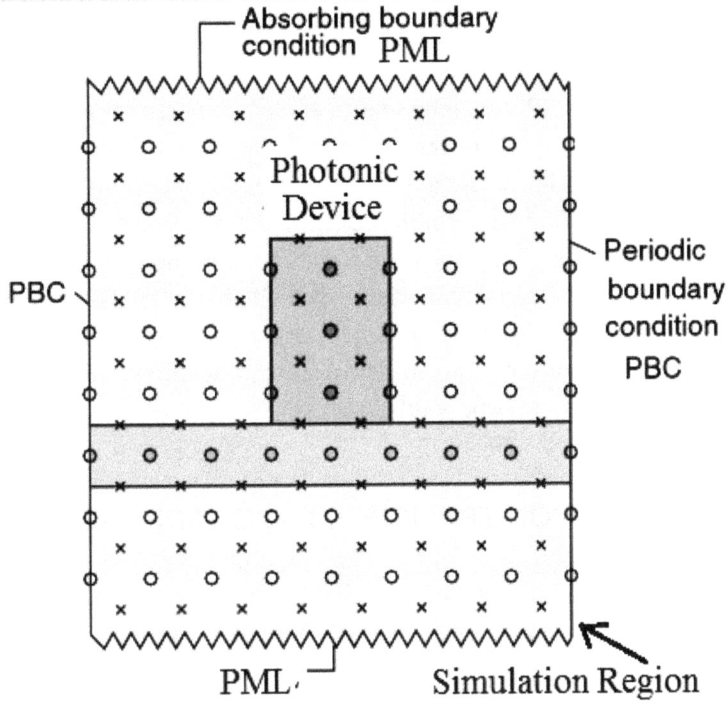

discrete grids. The time-dependent Maxwell's equations (in partial differential form) are discretized using central-difference approximations to the space and time partial derivatives.

The electric field vector components in a volume of space are solved at a given instant in time; then the magnetic field vector components in the same spatial volume are solved at the next instant in time; and the process is repeated. This results in the basic FDTD time-stepping relation that, at any point in space, the updated value of the E-field in time is dependent on the curl of the local distribution of the H-field in space and vice versa

$$\frac{\partial H}{\partial t} = -\frac{1}{\mu}\nabla x\zeta \Rightarrow \left.\frac{\partial H_z}{\partial t}\right|_{i+\frac{1}{2},j+\frac{1}{2}} = -\frac{1}{\mu}\left(\frac{\zeta_y}{\partial x} - \frac{\zeta_x}{\partial y}\right) \tag{120a}$$

And the explicit time-stepping is expressed as follows:

$$\left.\frac{\partial H}{\partial t}\right|_{t=n\Delta t} = -\frac{1}{\mu}\nabla x\zeta\Big|_{t=n\Delta t} \approx= \frac{H\left(n+\frac{1}{2}\right) - H\left(n-\frac{1}{2}\right)}{\Delta t} \tag{120b}$$

For stability, space and time resolutions are proportional. This scheme is known as Yee lattice, and has proven to be very robust, and forms the core of current FDTD software (Yee, 1967). The basic FDTD algorithm for photonic device simulation must be modified at the boundaries of the simulation region, to account for external contacts (electrical and magnetic conductors) and artificial boundaries, where suitable absorbing boundary conditions should be applied. Perfect Electrical Conductor (PEC) and Perfect Magnetic Conductor (PMC) boundary conditions are symmetric walls for the symmetric structure with symmetric wave propagation. PMC can be seen as the special case for Bloch's boundary condition (periodic boundary condition PBC) where the k-vector is set to zero in the corresponding direction. The PEC is the complement of PMC. There are many choices for absorbing boundary conditions. The Perfectly Matched Layer (PML) boundary conditions have the best performance. A perfectly matched layer (PML) is an artificial absorbing layer for wave equations, commonly used to truncate computation regions to simulate problems with open boundaries, especially in the FDTD methods.

The generation rate is calculated from the optical absorption and used as a source in the subsequent electrical simulation to calculate current. The electrical simulation allows one to extract key parameters such as dark current, responsivity and bandwidth.

11. CASE STUDY 13: PHOTON TRANSPORT IN GRAPHENE-FET

Recently, graphene, which is a single sheet of graphite, has attracted a great attention in field of photonics and nanoelectronics. The dispersion relation of the honeycomb lattice of graphene is linear in the vicinity of the Fermi energy. The quasiparticle nature of electrons in graphene can be described by a two-dimensional Dirac equation, whose speed of light is replaced by the Fermi velocity ($v_f = c/300$, where c is the speed of light). We present here an example describing the electromagnetics transport in a graphene device, namely graphene nanoribbon (GNR) FET.

The example shows the nonlinear optical response and quantum transport in graphene. It also highlights the role of time transients of hot carrier, upon the exposure to femtosecond laser. The GNR FET

Figure 35. Different types of boundary conditions around a photonic device

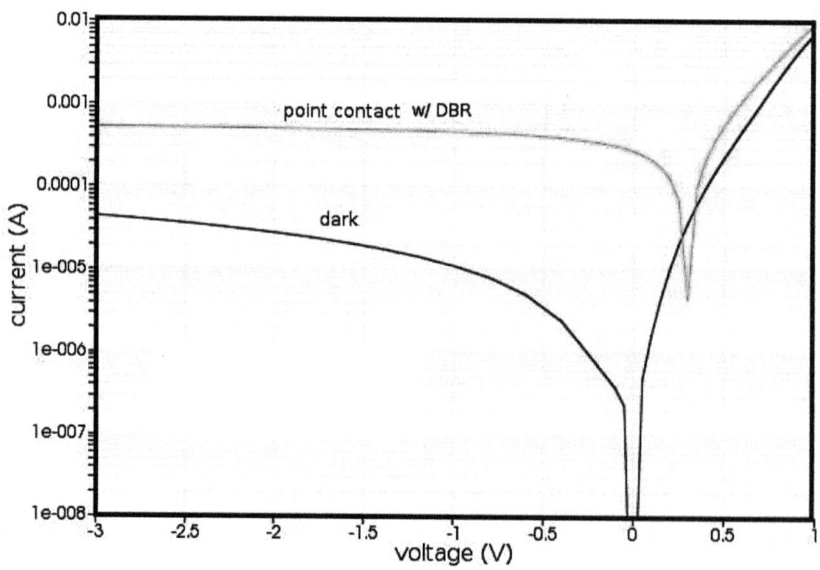

Figure 36. Photodiode I-V characteristics (dark and photo-current)
After Berrouth (2005).

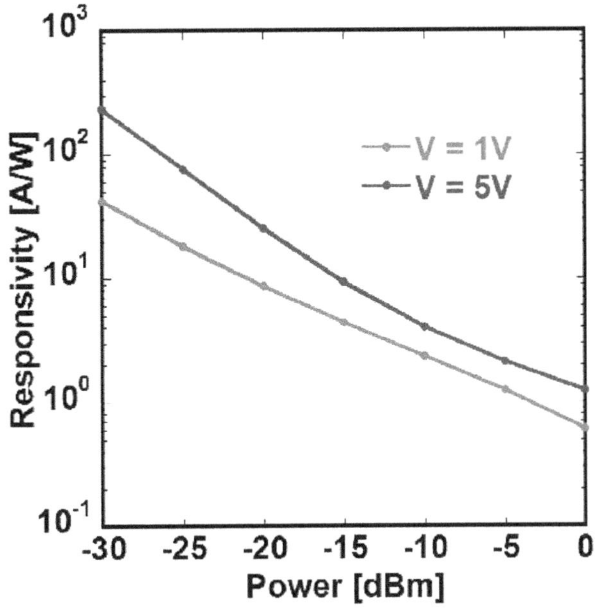

Figure 37. Responsivity of the photodetector at wavelength of 1,552 nm

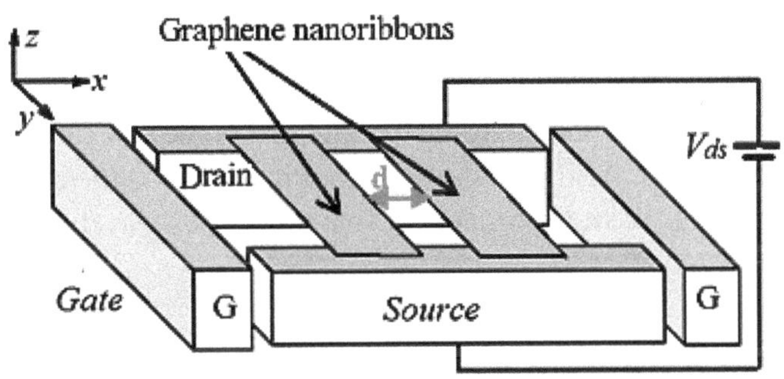

is shown in Figure 39. As shown, the device consists of two semiconducting GNRs connecting the source and drain of a FET device. A potential difference of 0.1 V is applied between drain and source; the source is assumed at 0 V, equipotential with the lateral gate (G). The nanoribbons are about 2.2 nm wide and the area of the square window delimited by the electrodes is 20x20 nm^2. For graphene, in the presence of an EM field, the Dirac equation reads (Ang, 2014):

$$j\hbar\left(\frac{\partial}{\partial t}+\frac{j}{\hbar}e\phi\right)\psi^{+}=\sigma.(p-eA)c.\psi^{-} \tag{121a}$$

Figure 38. Schematic of a GNR FET

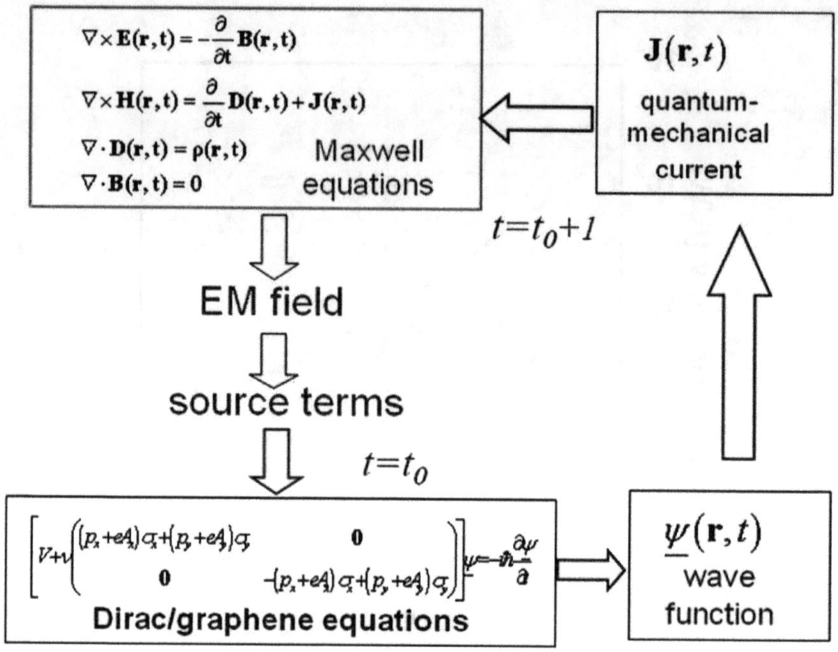

$$j\hbar\left(\frac{\partial}{\partial t}+\frac{j}{\hbar}e\phi\right)\psi^{-}=\sigma.(p-eA)c.\psi^{+} \quad (121b)$$

where $\boldsymbol{A}$ and ϕ are vector and scalar potentials, directly related to the EM field through the appropriate gauge and q is the electron charge. Also, $\boldsymbol{\sigma}$ is the Pauli matrices and $\boldsymbol{p}$ is the canonical momentum. The solution of the Dirac equation is the four component spinor complex wavefunction $\psi(r,t)$:

$$\psi(r,t)=[\psi_1\ \psi_2\ \psi_3\ \psi_4]^T=[\ \psi^{+}\ \psi^{-}]^T \quad (122)$$

The Pauli matrices $\boldsymbol{\sigma}$ and the canonical momentum $\boldsymbol{p}$ are related to the kinematic momentum $\boldsymbol{k}$, which includes the EM field contribution, as:

$$\boldsymbol{p}=-j\hbar\nabla, \boldsymbol{k}=\boldsymbol{p}-e\boldsymbol{A} \quad (123)$$

Figure 39 depicts the flowchart of the method: The computational scheme develops as follows.

At first, the EM field is discretized by the Transmission Line Matrix method. The Quantum phenomena are then introduced in a subregion of the 2D-domain, of graphene/nanoribbon region, described by the Dirac equation. At each time step, the Dirac equation is solved by accounting for the quantum device boundary conditions, initial conditions (e.g., injected charge), and additional source terms constituted by the EM field, sampled in the domain of the quantum devices. From the wavefunction (charge) solution of the Dirac equation, we derive the quantum mechanical current (J) over the device domain. This current is a distribution of local sources for the EM field that is injected into the TLM nodes, located only on

Figure 39. Flowchart of Maxwell-Dirac method
After Pierantoni & Mencarelli, (2012).

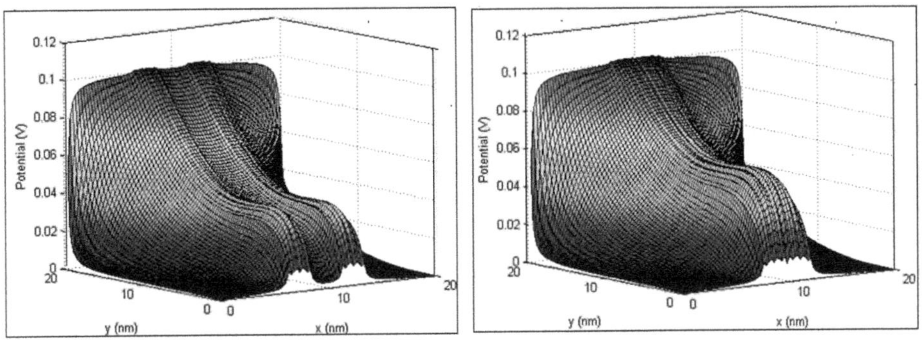

the grid points of the Dirac equation domain. At the next time step t+1, the TLM method provides a new updated distribution of field values that are, again, sampled over the device domain, and so on, iteratively.

Figure 40 shows the self-consistent potential distribution for different distances of the two coupled GNR.

12. OPTOELECTRONIC SOFTWARE PACKAGES

There exist various simulation tools as well as custom-made simulation software for simulating the behavior of optoelectronic devices. We give a summary of high-end simulation tools (such as APSYS,

Figure 40. Self-consistent potential for different distances of the two coupled GNR channels: a) d=2.4 nm, b) d=0.15 nm
After, Pierantoni & Mencarelli (2012).

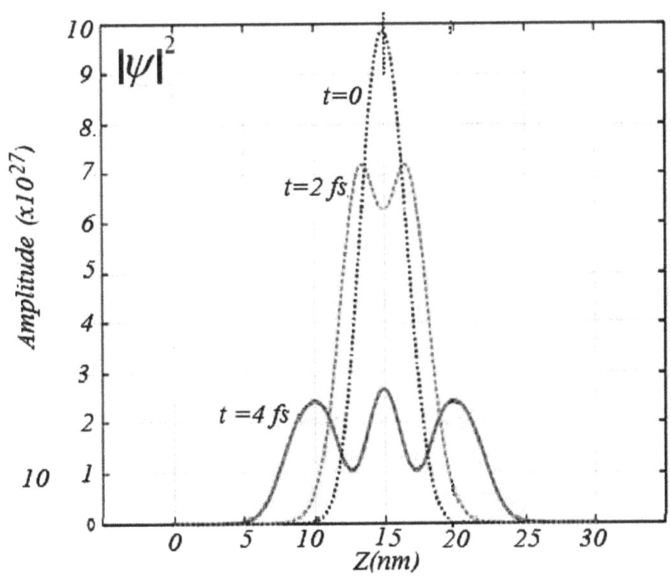

LASTIP, PICS3D, Lumerica, Liouminous2D). These software packages combine electrical, thermal, optical, and quantum models in 1- or 2- or 3-dimensions. They can be applied to a large variety of optical devices such as laser diodes (LD), light-emitting diodes (LED), photodetectors, and photonic amplifiers.

- **LASTIP:** 2D simulation of laser diodes. Considers competition of optical laser modes. Includes optical gain function for quantum well or bulk material with different models of spectral broadening, Coulomb interaction, and inter-band optical transitions integrated over ***k·p*** nonparabolic subbands. Physical models of various laser effects and material data base for many semiconductor compounds are available.
- **APSYS:** 2D/3D simulation of non-lasing devices. Features include: hot carrier transport, quantum mechanical tunneling, multi-quantum well structures, ***k·p*** band structure calculation, self-consistent Poisson and Schrodinger equations solving, wave guiding in media with arbitrary complex refractive index distribution, transient models, small ac-signal analysis, interface and bulk charge trapping, piezoelectric effect, impact ionization, optical absorption and emission with exciton, and other many-body phenomena, LED ray tracing model to optimize device structure and packaging for light extraction efficiency, photon recycling effect, different relaxation models. Broad range of semiconductor devices, including resonant tunneling diodes, bipolar and field effect transistors, LEDs, solar cells, detectors, semiconductor optical amplifiers, optical modulators.
- **PICS3D:** 3D simulation of edge-emitting lasers and vertical-cavity laser diodes (VCSELs). It can be used for waveguide photo detectors, semiconductor optical amplifiers, and coupling to external passive optical components. It can calculate longitudinal distribution of carrier density, gain, optical field, and surface emission modes. In addition to steady I-V characteristics, it can be used for AC, and transient analysis of laser diodes, mode emission power, and spectrum analysis. It may include 3D current flow, vector waves, Poisson and Schrodinger Equations self-consistent solution, and quantum-mechanical tunneling.
- **tiberCAD:** Is a software tool for numerical simulation of electronic and optoelectronic devices. It allows to model and design nanostructured devices, such as III/V LEDs, nanowire FETs, Dye Solar Cells (DSCs). With tiberCAD, quantum and classical descriptions can be used in different regions of a device/nanostructure within the same simulation.
- **DEVICE:** Is semiconductor TCAD device simulation software for the design, analysis and optimization of semiconductor-based optoelectronic components. DEVICE employs the drift-diffusion equations to model steady state and transient behavior of charge carriers in semiconductors. Using advanced finite-element mesh generation algorithms, DEVICE is capable of handling arbitrary geometries, enabling accurate modeling of integrated optoelectronic designs. DEVICE provides users with a complete set of tools to design and evaluate semiconductors, including integrated process simulation, comprehensive physical material models and robust numerical algorithms for device simulation. When combined with MODE simulator and Lumerical design environment, DEVICE provides a powerful tool for the design and analysis of optical components such as high speed optoelectronic modulators. Along with FDTD Solutions and MODE Solutions, DEVICE is a part of Lumerical's photonic design and analysis tools. All of Lumerical solvers provide support for data import/export and integration into EDA tools, for complete product design flows from component architecture to large scale device and system design.
- **OptiFDTD:** Provides comprehensive design, simulation, and post-analysis tools for photonic crystal (PhC) and photonic band gap (PBG) analysis:

- ◦ Layout designer provides the tools to define the lattice relation, atom properties, defects, wave path, and observation projects
- ◦ Band solver is an independent simulation engine that gives the band-diagram for the defined lattice structure
- ◦ Simulator can perform the plane wave simulation based on the periodic cell with a periodic boundary condition
- ◦ Analyzer can view the field pattern and extract the transmittance / reflection coefficients

- **OptiSPICE:** Is the first circuit design software for analysis of integrated circuits including interactions of optical and electronic components. It allows for the design and simulation of optoelectronic circuits at the transistor level, from laser drivers to transimpedance amplifiers, optical interconnects and electronic equalizers.
- **tMCimg:** Uses a Monte Carlo algorithm to model the transport of photons through 3D highly volumes with spatially varying optical properties and arbitrary boundary conditions.
- **MOLGW:** Is another interesting open code that implements DFT and GWA to describe the excited electronic states in finite atomic and molecular systems. It implements the *GW* approximation and **the Bethe-Salpeter** equation for the study of optical excitations (Ziaei and Bredow, 2016).

13. CHAPTER SUMMARY

Photons are the force carriers of the electromagnetic waves (e.g., light), with no charge and zero mass. The study of light-matter interactions and photon transport is a fundamental topic in quantum photonic devices and optical spectroscopy.

Accurate modeling of photonic transport is essential for the development of optical devices in fields like communications, sensing, biomedical instrumentation, consumer electronics and defense. The specific challenge of optoelectronic device simulation lies in the combination of photonics and electronics, including the sophisticated interaction of photons and electrons. The vast spectrum of optoelectronic device concepts and applications leads to a large diversity of modeling approaches. This is a field of research that uses semiclassical and quantum-mechanical physics to investigate phenomena involving light and its interactions with matter. In this Chapter, we review the predominant modeling techniques for photonic transport.

Quantum theory of light shows us how light is described in terms of quantized photons, and how the interaction of photons with matter is understood. Quantum transport models include Maxwell-Bloch equations (MBE's), semiconductor Bloch equations (SBE's) and semiconductor luminescence equations (SLE's), as well as nonequilibrium Green's functions (NEGF) techniques.

Semiclassical models for photon transport include BTE-like transport equations, the radiative transport equation (RTE), the Monte Carlo method (MCM) and other phenomenological or semi-empirical models,

The macroscopic radiation transport equation (RTE) is governed by an integro-differential equation, which is extremely difficult to obtain exact solutions. Macroscopic photonic transport analysis requires the consideration of seven independent variables: three spatial directions, two angular directions, frequency and time._

In order to study the carrier energy-relaxation processes, there are two different classes of experiments that have been employed: photoluminescence (PL) and pump-probe (PP) measurements. In addition, there are other techniques which rely completely on the phase coherence in the carrier system. The most popular technique which provides direct information on the carrier coherence in the semiconductor is four-wave-mixing (FWM) spectroscopy.

REFERENCES

Alsunaidi, M.A., Imtiaz, S.S., & El-Ghazaly, S.M. (1996). *Electromagnetic wave effects on microwave transistors using a full-wave time-domain model*. IEEE.

Ammosov, M. V., Delone, N. B., & Krainov, V. P. (1986). Tunnel ionization of complex atoms and ions in alternating EM field. *Soviet Physics, JETP*, *64*, 1191.

Ang, Y. S. (2014). *Nonlinear Optical Response and Transport properties in Graphene* (PhD Thesis). University of Wollongong. Retrieved from http//ro.uow.edu.au/theses/4117

Bandelow, U., Gajewski, H., & Hunlich, R. (2005). Fabry-Perot lasers: thermo-dynamic based modeling. In J. Piprek (Ed.), *Optoelectronic Devices. Advanced Simulation and Analysis* (pp. 63–85). Berlin: Springer. doi:10.1007/0-387-27256-9_3

Bandelow, U., Schatz, R. & Wunsche, H.J. (1996). A Correct-Single-Mode Photon Rate Equation for Multi section Lasers. *IEEE Photonics Tech. Letters, 8*(5).

Barut, A. (2013). *Foundations of Radiation Theory and Quantum Electro-dynamics*. Springer Science.

Bastard, G. (1990). *Wave Mechanics Applied to Semiconductor Hetero-structures*. Les Ulis, France: Les Editions de Physique.

Becerra, F. E., Fan, J., & Migdall, A. (2014). Photon number resolution enables quantum receiver for realistic coherent optical communications. *Nature Photonics*, 9.

Bethe, H., & Salpeter, E. (1951). A Relativistic Equation for Bound-State Problems. *Physical Review, 84*(6), 1232–1242. doi:10.1103/PhysRev.84.1232

Blanchard, F., Sharma, G., Razzari, L., Ropagnol, X., Bandulet, H. C., Vidal, F., & Hegmann, F. et al. (2011). Generation of Intense Terahertz Radiation via Optical Methods. *IEEE Journal of Selected Topics in Quantum Electronics*, *17*(1), 5–16. doi:10.1109/JSTQE.2010.2047715

Blasse, G., & Grabmaier, B. C. (1994). *Luminescent materials*. Springer Berlin. doi:10.1007/978-3-642-79017-1

Bloch, F. (1929). Űber die Quantenmechanik der Elektronen in Kristallgittern. *Zeitschrift fur Physik, 52*(7-8), 555–600. doi:10.1007/BF01339455

Born, M., & Wolf, E. (1993). *Principles of Optics* (6th ed.). Pergamon Press.

Born, M., & Wolf, E. (1999). *Principles of Optics* (7th ed.). Cambridge University Press. doi:10.1017/CBO9781139644181

Boyd, R. W. (2003). *Nonlinear Optics* (2nd ed.). San Diego, CA: Academic.

Brillouin, L. (1922). Diffusion de la Lumière et des Rayonnes X par un Corps Transparent Homogéne; Influence de l'Agitation Thermique [Diffusion of light and X-rays by a homogenous transparent body: Influence of thermal agitation]. *Annales de Physique*, *17*, 88.

Brudevoll, T., Storebo, A. K., Skaaring, O., Kirkemo, C. N., Norum, O. C., Olsen, O., & Breivik, M. (2011). *Time-Resolved Laser Spectroscopy of Semiconductors - Physical Processes and Methods of Analysis, Femtosecond-Scale Optics*. InTech. Available from: http://www.intechopen.com/books/femtosecond-scale-optics/time-resolved-laser-spectroscopy-ofsemiconductors-physical-processes-and-methods-of-analysis

Brunk M. & Jungel, A. (2007). *Numerical Simulation of Thermal Effects in Coupled Optoelectronic Device-circuit Systems*. Academic Press.

Brunk M. & J¨ungel, A. (2007). *Simulation of thermal effects in optoelectronic devices using energy-transport equations*. Unpublished.

Carrol, J. E. (1990). *Rate Equations in Semiconductor Electronics*. New York: Cambridge University Press.

Carrol, J. E. (1990). *Rate Equations in Semiconductor Electronics*. New York: Camridge University Press.

Chang, E., Bussi, G., Ruini, A., & Molinari, E. (2004). Excitons in Carbon Nanotubes: An Ab Initio Symmetry-Based Approach. *Physical Review Letters*, *92*(19), 196401. doi:10.1103/PhysRevLett.92.196401 PMID:15169423

Chen, L., Tang, Y., Bowers, J. E., & Theogarajan, L. (2012). CMOS enabled silicon photonics for data center packet switching. *IEEE Microwave Symposium Digest (MTT)*, 17-22. doi:10.1109/MWSYM.2012.6259474

Chini, M., Chen, Y., & Chang, Z. (2014). Probing electronic wavepackets with attosecond transient absorption, IOP Science. *Journal of Physics. B, Atomic, Molecular, and Optical Physics*, 10.

Chow, W., & Koch, S. W. (2011). *Semiconductor-Laser fundamentals*. Springer.

Chow, W., Koch, S. W., & Sargent, M. (1994). *Semiconductor-Laser Physics*. Springer. doi:10.1007/978-3-642-61225-1

Chrostowski, L., & Hochberg, M. (2015). *Silicon Photonics Design*. Cambridge. doi:10.1017/CBO9781316084168

Chrostowski, L., & Hochberg, M. (2015). *Silicon Photonics Design*. Cambridge. doi:10.1017/CBO9781316084168

Corkum, P. B., & Krausz, F. (2007). Attosecond science. *Nature Physics*, *3*(6), 381–387. doi:10.1038/nphys620

Csele, M. S. (2014). *Laser Modeling: A Numerical Approach with Algebra and Calculus*. CRC Press. doi:10.1201/b16770

Deng, J., Shur, M. S., Fjeldly, T. A., & Baier, S. (2000). CAD Tools And Optical Device Models For Mixed Electronic/Photonic VLSI, Numerical Methods in Electromagnetics, North Holland. *International Journal of High Speed Electronics and Systems*, *10*(01), 299–308. doi:10.1142/S0129156400000325

Dirac, P. (1958). *Principles of Quantum Mechanics. International Series of Monographs on Physics*. Oxford University Press.

Dirac, P. (1966). Article. *Lectures on Quantum Field Theory*.

Dirac, P. A. M. (1927). The Quantum Theory of the Emission and Absorption of Radiation. *Proceedings of the Royal Society of London. Series A, Containing Papers of a Mathematical and Physical Character*, *114*(767), 243–265. doi:10.1098/rspa.1927.0039

Drummond, P. D., & Hillery, M. (2014). *The Quantum Theory of Nonlinear Optics*. Cambridge University Press. doi:10.1017/CBO9780511783616

Einstein, A. (1916). Strahlungs-Emission und -Absorption nach der Quantentheorie. *Verhandlungen der Deutschen Physikalischen Gesellschaft*, *18*, 318–323.

El-Saba, M. H. (2006). Investigation of Hot Carrier Repelling Effect in Semiconductor Devices, Using an Analytical Solution of The Hydrodynamic Model. *IEEE Transactions on Electron Devices*, *53*(7), 1615–1622. doi:10.1109/TED.2006.876040

Faist, J. (2013). *Quantum Cascade Lasers*. Oxford, UK: Oxford University Press. doi:10.1093/acprof:oso/9780198528241.001.0001

Feldtmann, T. (2009). *Influence of Phonons on Semiconductor Quantum Emission* (Dissertation). Available at http://deposit.ddb.de/cgibin/ dokserv?idn= 999828991&dok_var=d1&dok_ext=pdf&filename=999828991.pdf

Feldtmann, T., Schneebeli, L., Kira, M., & Koch, S. W. (2006). Quantum theory of light emission from a semiconductor quantum dot. *Physical Review B: Condensed Matter and Materials Physics*, *73*(15), 155319. doi:10.1103/PhysRevB.73.155319

Ganchev, B., Drummond, N., Aleiner, I., & Falko, V. (2015). Three-Particle Complexes in Two-Dimensional Semiconductors. *Physical Review Letters*, *114*(10), 107401. doi:10.1103/PhysRevLett.114.107401 PMID:25815964

Geim, A. K. (2009). Graphene: Status and Prospects. *Science*, *324*(5934), 1530–1534. doi:10.1126/science.1158877 PMID:19541989

Geim, A. K., & Novoselov, K. S. (2007). The rise of graphene. *Nature Materials*, *6*(3), 183–191. doi:10.1038/nmat1849 PMID:17330084

Griffiths, D. J. (2004). *Introduction to Quantum Mechanics*. Prentice Hall.

Grondin, R. O., El-Ghazaly, S. M., & Goodnick, S. (1999). A Review of Global Modeling of Charge Transport in Semiconductors and Full-Wave Electromagnetics. *IEEE Transactions on Microwave Theory and Techniques*, *47*(6), 817–829. doi:10.1109/22.769315

Guo, S., & Albin, S. (2003). Simple plane wave implementation for photonic crystal calculations. *Optics Express*, *11*(2), 167–175. doi:10.1364/OE.11.000167 PMID:19461720

Hafezi, M., Chang, D., Gritsev, V., Demler, E. A., & Lukin, M. D. (2011). Photonic quantum transport in a nonlinear optical fiber. *Europhysics Letters*, *94*(5), 54006. doi:10.1209/0295-5075/94/54006

Haken, H. (1983). *Quantum Field Theory of Solids: An Introduction*. Elsevier Science Publishing Company.

Haug, H., & Jauho, A. P. (1996). *Quantum Kinetics in Transport and Optics of Semiconductors*. Berlin: Springer.

Haug, H., & Koch, S. (2009). *Quantum Theory of the Optical and Electronic Properties of Semiconductors* (5th ed.). World Science. doi:10.1142/7184

Haug, H., & Koch, S. W. (2009). *Quantum Theory of the Optical and Electronic Properties of Semiconductors* (5th ed.). World Scientific. doi:10.1142/7184

Heuvell, L., & Muller, H. G. (1993). Article. *Laser Physics*, *3*, 694.

Hirtschulz, M., Milde, F., Malić, E., Butscher, S., Thomsen, C., Reich, S., & Knorr, A. (2008). Carbon nanotube Bloch equations: A many-body approach to nonlinear and ultrafast optical properties. *Physical Review B: Condensed Matter and Materials Physics*, *77*(3), 035403. doi:10.1103/PhysRevB.77.035403

Hughes, S., & Roy, C. (2012). Nonlinear photon transport in a semiconductor waveguide-cavity system containing a single quantum dot: Anharmonic cavity-QED regime. *Physical Review B: Condensed Matter and Materials Physics*, *85*(3), 035315. doi:10.1103/PhysRevB.85.035315

Jackson, J. D. (1998). *Classical Electrodynamics* (3rd ed.). Wiley.

Jahnke, F. (2012). *Quantum Optics with Semiconductor Nanostructuresm Woodhead Publishing Ltd.* Springer.

Jam, S. A., & Kan, S. M. (1995). Transforming Tuckers Linearized Laser Rate Equations to a Form that has a Single Solution Regime. *Journal of Light Wave Technoloy*, *13*(9), 1899–1904. doi:10.1109/50.464741

Johnston, M. B., Whittaker, D. M., Corchia, A., Davies, A. G., & Linfield, E. H. (2003). Simulation of terahertz generation at semiconductor surfaces. *Physical Review B: Condensed Matter*, *65*(16), 165301. doi:10.1103/PhysRevB.65.165301

Jutzi, M., Berroth, M., Wohl, G., Oehme, M., & Kasper, E. (2005). Ge-on-Si vertical incidence photodiodes with 39-GHz bandwidth. *IEEE Photonics Technology Letters*, *17*(7), 1510–1512. doi:10.1109/LPT.2005.848546

Kaindl, R. A.; Hagele, D., Carnahan, M.A., & Chemla, D.S. (2009). Transient terahertz spectroscopy of excitons and unbound carriers in quasi-two-dimensional electronhole gases. *Phys. Rev. B, 79*(4), 0453.

Kalt, H., & Hetterich, M. (2004). *Optics of Semiconductors and their Nano-structures*. Springer. doi:10.1007/978-3-662-09115-9

Kasap, S. O. (2001). *Optoelectronics and Photonics, Principles and Practices*. Prentice Hall.

Keldysh, L. V. (1965). Ionization in the field of a strong electromagnetic wave. *Soviet Physics, JETP, 20*, 1307–1314.

Kira, M.; Jahnke, F.; Hoyer, W. & Koch, S.W. (1999). Quantum theory of spontaneous emission and coherent effects in semiconductor microstructures. *Progress in Quantum Electronics, 23*, 189.

Kira, M., & Koch, S. W. (2011). Maxwell-semiconductor Bloch equations. In Semiconductor Quantum Optics (pp. 521-549). Cambridge University Press.

Klingshirn, C. F. (2006). *Semiconductor Optics*. Springer.

Knox, R. S. (1963). *Theory of excitons, Solid state physics Edited by Seitz and Turnbul*. Academic.

Kolek, A. (2015). Nonequilibrium Greens function formulation of intersubband absorption for nonparabolic single-band effective mass Hamiltonian. *Applied Physics Letters, 106*(18), 181102. doi:10.1063/1.4919762

Kolek, A. (2016). Modeling of optoelectronic devices with one-band effective mass equation: Non-equilibrium Greens function approach. *Optical and Quantum Electronics, 48*(2), 118. doi:10.1007/s11082-016-0384-6

Korkin, A., & Rosei, F. (2008). *Nanoelectronics and Photonics: From Atoms to Materials*. Devices, Springer Science. doi:10.1007/978-0-387-76499-3

Kubis, T., Yeh, C., Vogl, P., Benz, A., Fasching, G., & Deutsch, C. (2009). Theory of nonequilibrium quantum transport and energy dissipation in terahertz quantum cascade lasers. *Physical Review B: Condensed Matter and Materials Physics, 79*(19), 195323. doi:10.1103/PhysRevB.79.195323

Kwong, N. H., & Binder, R. (2000). Greens function approach to the dynamics-controlled truncation formalism: Derivation of the χ(3) equations of motion. *Physical Review B: Condensed Matter and Materials Physics, 61*(12), 8341–8358. doi:10.1103/PhysRevB.61.8341

Langevin, P. (1908). Sur la théorie du mouvement brownien [On the Theory of Brownian Motion]. *C. R. Acad. Sci. (Paris), 146*, 530–533.

Lavrinenko, A. V., Lægsgaard, J., Gregersen, N., Schmidt, F., & Søndergaard, T. (2014). *Numerical Methods in Photonics*. CRC Press.

Lavrinenko, A. V., Lægsgaard, J., Gregersen, N., Schmidt, F., & Søndergaard, T. (2014). *Numerical Methods in Photonics*. CRC Press.

Leonhardt, U. (1997). *Measuring the quantum state of light*. Cambridge, UK: Cambridge University Press.

Lewis, G. N. (1926). The conservation of photons. *Nature, 118*(2981), 874–875. doi:10.1038/118874a0

Li, X. (2009). *Optoelectronic Devices: Design*. Cambridge, UK: Modeling, and Simulation. doi:10.1017/CBO9780511581144

Lindberg, M., & Koch, S. W. (1988). Effective Bloch equations for semiconductors. *Physical Review B: Condensed Matter and Materials Physics, 38*(5), 3342–3350. doi:10.1103/PhysRevB.38.3342 PMID:9946675

Mandel, L., & Wolf, E. (1995). *Optical Coherence and Quantum Optics*. Cambridge University Press. doi:10.1017/CBO9781139644105

Matsunaga, R., Matsuda, K., & Kanemitsu, Y. (2011). Observation of Charged Excitons in Hole-doped Carbon Nanotubes Using Photoluminescence and Absorption Spectroscopy. *Physical Review Letters*, *106*(1), 037404. doi:10.1103/PhysRevLett.106.037404 PMID:21405298

Meier, T., Thomas, P., & Koch, S. W. (2007). *Coherent Semiconductor Optics: From Basic Concepts to Na nostructure Applications*. Springer. doi:10.1007/978-3-540-32555-0

Mukamel, S. (1995). *Principles of Nonlinear Optical Spectroscopy*. Oxford, UK: Oxford University Press.

Nakanishi, N. (1969). A general survey of the theory of the Bethe–Salpeter equation. *Progress of Theoretical Physics*, *43*(Supplement. 43), 1–81. doi:10.1143/PTPS.43.1

Ning, C. Z., Indik, R. A., & Moloney, J. V. (1997). Effective Bloch equations for semiconductor lasers and amplifiers. *IEEE Journal of Quantum Electronics*, *33*(9), 1543–1550. doi:10.1109/3.622635

Obayya, S. (2010). *Computational Photonics*. Wiley. doi:10.1002/9780470667064

Parker, M. (2005). *Physics of Optoelectronics*. CRC Tylor & Francis. doi:10.1201/9781420027716

Parker, M. P. (2009). *Solid State and Quantum Theory for Optoelectronics*. Springer Science. doi:10.1201/9781420019452

PENELOPE2014. (2014). *A Code System for Monte-Carlo Simulation of Electron and Photon Transport*. NEA.

Phillips, R. T. (1993). *Coherent Optical interactions in Semiconductors*. Springer.

Pierantoni, L. & Mencarelli, D. (2012). *Numerical Techniques for the Analysis of Charge Transport and Electrodynamics in Graphene Nanoribbons*. Academic Press.

Piprek, J. (Ed.). (2005). *Optoelectronic Devices: Advanced Simulation and Analysis*. Springer. doi:10.1007/b138826

Quade, W., Scholl, E., Rossi, F. & Jacoboni, C. (1994). *Quantum theory of impact ionization in coherent high-field semiconductor transport*. The American Physical Society.

Ramella-Roman, J. C., Prahl, S. A., & Jacques, S. L. (2005). Three Monte Carlo programs of polarized light transport into scattering media. *OSA*, *13*(12), 4420. PMID:19495358

Rossi, F. (2011). Ultrashort Space- and Time-Scales:Need for a Quantum Description. In *Theory of Semiconductor Quantum Devices Microscopic Modelling and Simulation Strategies*. Springer. doi:10.1007/978-3-642-10556-2_2

Rössler, U., Tejedor, C., & Vina, L. (2003). *Semiconductor Bloch Equations including Spin and Polarization Degrees of Freedom. Institute of Physics Conference Series 171* (A. R. Long & J. H. Davies, Eds.). Bristol, UK: IOP Publishing.

Saleh, E. A., & Teich, M. C. (1991). *Fundamentals of Photonics*. New York: John Wiley & Sons, Inc. doi:10.1002/0471213748

Salpeter, E. (2008). Bethe–Salpeter equation (origins). *Scholarpedia.*, *3*(11), 7483. doi:10.4249/scholarpedia.7483

Salvat, F., & Fernandez-Varea, J. M. (2009). Overview of physical interaction models for photon and electron transport used in Monte Carlo codes, IOP Science. *Metrologia*, *46*(2), S112–S138. doi:10.1088/0026-1394/46/2/S08

Schäfer, W., & Wegener, M. (2002). *Semiconductor Optics and Transport Phenomena*. Springer. doi:10.1007/978-3-662-04663-0

Schäfer, W., & Wegener, M. (2013). Semiconductor Optics and Transport Phenomena. -. *Science*.

Schleich, W. P., Mayr, E., & Kraehmer, D. (1999). *Quantum Optics in Phase Space*. Weinheim: Wiley.

Scully, M. O., & Zubairy, S. M. (1997). *Quantum Optics*. Cambridge. doi:10.1017/CBO9780511813993

Shah, J. (1999). *Ultrafast Spectroscopy of Semiconductors and Semiconductor Nanostructures*. Springer. doi:10.1007/978-3-662-03770-6

Shen, J., & Fan, S. (2009). Theory of single-photon transport in a single-mode waveguide. I. Coupling to a cavity containing two-level atom. *Physical Review A.*, *79*, 1–11.

Sujecki, S. (2014). *Photonics Modelling and Design*. CRC Press.

Sultan, S., Ang, Y. S., & Zhang, C. (2012). Room-temperature strong Terahertz photon mixing in Graphene. *Journal of the Optical Society of America. B, Optical Physics*, *29*(3), 247–249.

Suzuki, T., & Shimano, R. (2011). Cooling Dynamics of photoexcited Carriers in Si studied using optical Pump and THz Probe Spectroscopy. *Physical Review B: Condensed Matter and Materials Physics*, *83*(8), 085207. doi:10.1103/PhysRevB.83.085207

Tai, C. (1993). *Dyadic Green functions in electromagnetic theory*. IEEE.

Tamai, Y., Ohkita, H., Benten, H., & Ito, S. (2015). Exciton Diffusion in Conjugated Polymers: From Fundamental Understanding to Improvement in Photovoltaic Conversion Efficiency. *The Journal of Physical Chemistry Letters*, *6*(17), 3417–3428. doi:10.1021/acs.jpclett.5b01147 PMID:26269208

Tamayama, Y., Nakanishi, T., Sugiyama, K., & Kitano, M. (2008). An Invisible Medium for Circularly Polarized Electromagnetic Waves. *Optics Express*, *16*(25), 20869–22087. doi:10.1364/OE.16.020869 PMID:19065225

Tromborg, B., Lassen, H. E., & Olesen, H. (1994). Traveling wave analysis of semiconductor lasers: Modulation responses, mode stability and quantum mechanical treatment of noise spectra. *IEEE Journal of Quantum Electronics*, *30*(4), 939–956. doi:10.1109/3.291365

Tsen, K. T. (2004). Ultrafast Dynamical Processes in Semiconductors. *Topics in Applied Physics*, *92*, 1–59. doi:10.1007/b13749

Vasileska, D., Goodnick, S. M., & Klimeck, G. (2011). *Computational Electronics*. CRC Press.

Wacker, A. (2002). Gain in quantum cascade lasers and superlattices: A quantum transport theory. *Physical Review B: Condensed Matter and Materials Physics*, *66*(8), 085326. doi:10.1103/PhysRevB.66.085326

Wang, L.-H., & Wu, H.-I. (2007). *Biomedical Optics: Principles & Imaging*. Wiley.

Wang, S. A., Marks, D. B., & Menyuk, C. (2013). Comparison of numerical methods for modeling laser mode locking with saturable gain. *Journal of the Optical Society of America. B, Optical Physics*, *30*(11), 3064–3074. doi:10.1364/JOSAB.30.003064

Wannier, G. H. (1937). The structure of electronic excitation levels in insulating crystals. *Physical Review*, *52*(3), 191–197. doi:10.1103/PhysRev.52.191

Wartak, M. S. (2013). *Computational Photonics*. Cambridge.

Wartak, S.M. (2013). *Computational Photonics, An Introduction with MATLAB*. Academic Press.

Wiedemann, E. (1888). Über Fluorescenz und Phosphorescenz, I. Abhandlung (On fluorescence and phosphorescence). *Annalen der Physik*, *34*, 446–463. doi:10.1002/andp.18882700703

Won, R. (2010). Integrating silicon photonics. *Nature Photonics*, *5*, 498–499.

Wu, L. A., Kimble, H. J., Hall, J. L., & Wu, H. (1986). Generation of squeezed states by parametric down conversion. *Physical Review Letters*, *57*(20), 691. doi:10.1103/PhysRevLett.57.2520 PMID:10033788

Xu, L., Zhang, X.-C., Auston, D. H., & Jalali, B. (1991). Terahertz radiation from large aperture Silicon p-i-n diodes. *Applied Physics Letters*, *59*(26), 23. doi:10.1063/1.105725

Yee, K.S. (1966). Numerical solution of initial boundary value problems involving Maxwell's equations in isotropic media. *IEEE Trans. Antennas and Prop.*, 302-307.

Ziaei, V., & Bredow, T. (2016). GW-BSE approach on S_1 vertical transition energy of large charge transfer compounds: A performance assessment. *The Journal of Chemical Physics*, *145*(17), 174305. doi:10.1063/1.4966920 PMID:27825209

ENDNOTES

1 Actually, there exist a number of different Gauge transformations, among which the Coulomb and Lorentz gauges are the most famous ones. We typically use the Coulomb gauge to develop the Poisson equation for a scalar electrostatic potential. This also admits the vector potential to become a transverse field and to quantize the electromagnetic fields. We avoid the use of Lorentz gauge that manifests the Lorentz invariance because it makes the electromagnetic fields more difficult to quantize.

2 In modern books of quantum physics, it is more common to quantize the Fourier components of the vector potential

3 The Einstein coefficients relate spontaneous and stimulated emission coefficients, on the basis of thermodynamics. However, the Einstein relation does not give their values.

4 In other types of laser, we may talk about the transition of atoms/molecules rather than electrons.

5 Sometimes this term is referred to as the non-radiative recombination rate and written as $(-n/\tau_{nr})$.

6 As far as I know, this Effect has been cited only once, through the last 10 years!, I'd prefer to rename it as *ElSab3 Effect*.

Chapter 9
Electronic Spin Transport

1. OVERVIEW AND CHAPTER OBJECTIVES

As well as mass and charge, an electron has another intrinsic property, called spin. The spin property of the electron was demonstrated experimentally by Stern and Gerlach (1922). The word '*spin*' was coined by Wolfgang Pauli (1926) to explain the fine-structure of atomic spectra, following the proposal of the two students Uhlenbeck and Goudsmith (1926). They proposed that the spin angular momentum, obeys the same quantization rules as those governing orbital angular momentum of atomic electrons. It had been shown later by Paul Dirac (1928) that electron spin arises naturally in the relativistic treatment of quantum mechanics.

Unlike position and momentum, which have classical analogs, spin does not. But if we think of spin in classical terms, we can think of a spinning charged particle as a loop of current. Thus, if a particle spins about the z-axis, then the spin vector S points along the z-axis. Since the spinning charge is negative, the left-hand rule can be applied. When the fingers of the left hand rotate (curl) in the direction of spin, the thumb will point in the direction of spin. Therefore, a spinning charge carrier produces a magnetic field similar to that of a tiny bar magnet. In this case, the spin vector S points to the south pole of the bar magnet. If the spinning particle is placed in a magnetic field, it tends to align the spin vector in the opposite direction to the magnetic field lines, as shown in the above figure.

Therefore, electrons have intrinsic angular momentum, which is composed of two components, namely angular orbital and spin angular momentum:

$$J = L + S \tag{1a}$$

Quantization of angular momentum had already known for orbital angular momentum, and if the electron spin behaves the same way, an angular momentum quantum number $s = \pm½$ is required to give just two states. The spin angular momentum $\boldsymbol{S}$ has the following magnitude:

DOI: 10.4018/978-1-5225-2312-3.ch009

Figure 1. Classical representation of the spin action of electrons. The electromagnetic field emanating from an electron can be considered to emanate from an idealized tiny bar magnet with north and south poles (dipole).

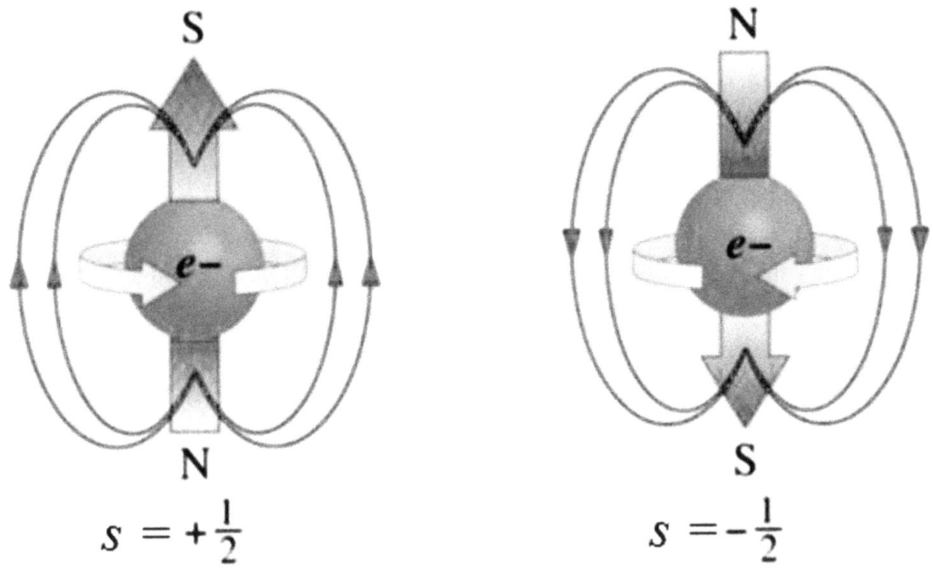

$$S^2 = S_x^2 + S_y^2 + S_z^2 = s\cdot(s+1)\cdot\hbar^2 \tag{1b}$$

Therefore, this intrinsic electron property gives an additional z-component for the angular momentum:

$$S_z = s\,\hbar, \text{ with } s = \mp 1/2 \tag{1c}$$

Spin electronics (or Spintronics) refers to the study of the role played by electron spin in solid state physics and devices. Spintronics utilizes the electron spin degree of freedom for information storage and transmission. The spintronic devices use the electron spin to carry and switch information, rather than the electron charge transport. The application of spintronics in information processing is a relatively new endeavor and is motivated by the belief that spintronics may offer a more power-efficient route compared to the traditional transistors. In fact, spin transport and spintronics have seen interesting developments in the past decade and have continued to attract attention.

In the past decades, the research on spin transport and magnetism has led to the discovery of giant magnetoresistance (GMR). The GMR prototype device is the read head of hard disks and the memory-storage cell, which consists of alternating ferromagnetic and nonmagnetic metal layers. Depending on the relative orientation of the magnetizations in the magnetic layers, the device resistance changes from small (parallel magnetizations) to large (antiparallel magnetizations). This change in resistance is used to sense changes in magnetic fields. The first use of spin-valve sensors in hard disk drive was in the IBM Deskstar, which was released in late 1994 with 16.8 GB of storage[1]. Recent efforts in GMR technology involves magnetic tunnel junction (MTJ) devices where the tunneling current depends on the spin orientation

Spintronic information transport can be achieved in three different ways, namely: (*a*) classical information transport by all-spin monolithic circuits, (*b*) classical information transport by hybrid (charge & spin) circuits and (*c*) quantum information transport by spin qubits

In order to boost spintronic research, a solid understanding of spin transport properties is essential. Current efforts in designing and manufacturing spintronic devices involve two different approaches. The first approach is concerned with finding novel ways of generation and detection of spin-polarized currents. The second approach is improving the existing giant-magnetoresistive (GMR)-based technology. Other Spintronic topics include the spin Seebeck effect, spin pumping in ferromagnet/normal metal heterostructures, magnon condensation, and phonon (heat) transport in magnetic insulators.

This chapter reviews the mainstream knowledge of semiconductor Spintronics, with emphasis on spin transport. The relevant theory of electron spin injection and detection is introduced in the beginning. Then we discuss the semiclassical and quantum spin transport theory in ferromagnet/semiconductor heterostructures.

The spin-polarized transport theory keeps track of the spin distribution, in addition to the charge distribution of electronic systems. In much the same way as charge transport, it is possible to describe transport of spin polarization by BTE-like equations. This can be also done within the density matrix formulation, or Wigner functions or the NEGF quantum approaches. The microscopic transport approaches (semiclassical or quantum) can be also reduced to simple macroscopic models, such as the spin drift-diffusion model (SDDM).

Upon completion of this Chapter, the readers and students will be able to:

- Review and understand the spin-induced phenomena.
- Identify quantum mechanics behind spin motion.
- Explore the characteristic of nanostructures made from diluted magnetic semiconductor (DMS).

Figure 2. Spin transport and spintronic devices and their overlapping with applications of other information carriers

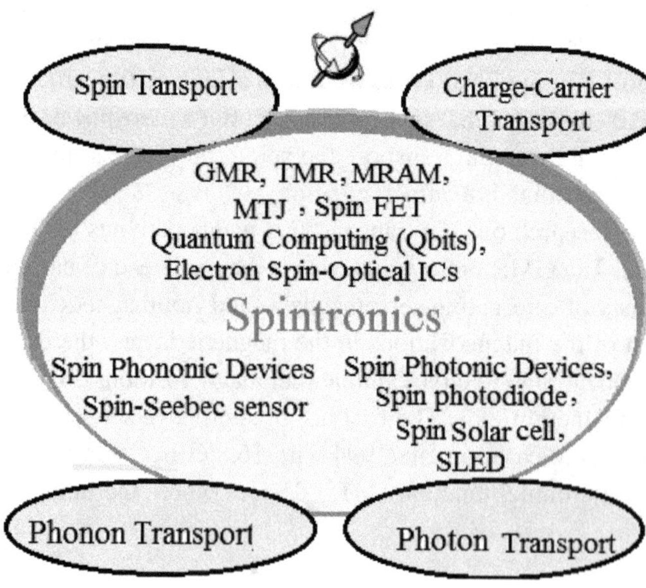

- Understand the semiclassical and quantum models of spin transport
- Describe the spin-dependant electron transport in magnetic devices and their microscopic and macroscopic models.
- Describe the phenomenon of spin-dependent tunneling, and magnetic tunnel junctions and their quantum models.
- Identify the Spin Magnetization Quantum Distribution Function (SMQDF), its equation of motion and its macroscopic moments.
- Know how and when to use the different spin-transport models.
- Analyze the operation of existing spin-based devices and spintronics.

2. CONCEPTS OF SPIN POLARIZATION, SPIN COHERENCE AND SPIN CURRENT

Although spin was originally thought of as particles spinning around their own axis, it had been shown by Paul Dirac (1928) that electron spin arises naturally within relativistic quantum mechanics[2]. The spin quanta that are associated with spin waves are called magnons. The term magnon was initially introduced by Felix Bloch in 1930 to explain the spontaneous magnetization in a ferromagnet.

The magnetic properties of matter are closely related to the spin motion of electrons. In fact, magnetism, at its root, arises from two sources, namely: electric current and internal magnetic moments due to spin motion of electrons. In magnetic materials, the most important sources of magnetization are, more specifically, the electrons' orbital angular motion around the nucleus, and the electrons' intrinsic magnetic moment. The component of electron spin measured along any direction is quantized and can only take on the values, $S = \hbar s_z$, $s_z = \pm\frac{1}{2}$. The intrinsic magnetic dipole moment associated with a spin of electron is very close to the Bohr magneton $\mu_B = e\hbar/2m$ =57.9 µeV/T or the free electron magnetic moment,

As we pointed out earlier, the electrons total angular momentum has two contributions, orbital and spin angular momentum ($J=L+S$). The corresponding magnetic dipole moment of an electron in solids is given by:

$$\mu = -(g_L\mathbf{L} + g_s\mathbf{S})\,\mu_B \tag{2a}$$

where the spin **g-factor** $g_s \approx 2$ to a good approximation[3], and the orbital g-factor g_L=1 for free electrons but changes with the electron effective mass in solids. In fact, the conduction electron in solids can have a g-factor, which is very different from the free electron case (Dyson, 1956). In metals the deviation is trivial (< 1%), but in semiconductors its value may be greater (g ≈-50 in *InSb*) or smaller (g≈-0.44 in *GaAs*). Actually, the g-factors of conduction electrons are affected by the spin-orbit interaction (SOI).

The so-called effective gyromagnetic ratio, γ, is defined as:

$$\mu = \tfrac{1}{2}\gamma\,\mu_B \tag{2b}$$

The magnetism originating from the nuclei is usually neglected, as the magnetic moment of protons and neutrons is 2000 times smaller than μ_B. If an external magnetic field $\boldsymbol{B}$ is applied, it will exert a torque on the magnetic dipole such that the magnetic potential energy is given by the Zeeman term:

$$V = \mu\sigma.\boldsymbol{B} \qquad (2c)$$

where σ is the Dirac spinor, which appears in the Dirac equation.

For most materials, ferromagnetism arises from an unbalanced occupation of the *d*-bands (transition elements) or the *f*-bands (rare earth elements). In metallic ferromagnets, such as the transition metals, *Ni, Fe, Co,* the conducting electrons are spin polarized, and a current of electrons drags a current of magnetization (Holstein, 1940)

Note 1: Origin of Ferromagnetism

The study of ferromagnetism has been carried out by several authors. In 1935 Mott predicted that the conductivity in the transition metal ferromagnets *Fe, Ni* and *Co* is different for electrons whose magnetic moment is aligned with the magnetization (majority electrons) and for electrons with anti-parallel magnetic moments (minority electrons). Thirty years later, the experimental evidence of this spin-polarized electron conduction in transition metal ferromagnets was found. As argued by Mott, the reason for this effect lies in the shifted band structure of majority and minority electrons, respectively, as shown in Figure 3.

In *Fe, Ni* and *Co* ferromagnets, the narrow 3d band features a large exchange splitting and moreover crosses the Fermi energy only for the minority electrons; it is thus responsible for the ferromagnetic character. On the other hand, the transport in these materials is dominated by the delocalized electrons in the much wider 4s-band, which has a small exchange splitting and crosses the Fermi energy for both spin-up and spin-down electrons. Nevertheless, the conductivity is spin-polarized, because the main transition induced by scattering is 4s- band to 3d- band.

As minority electrons have a density of states (DOS) of 3d-electrons which is much larger than DOS of majority electrons near the Fermi energy, the 4s-to-3d transition is more probable for minority electrons. This is because the probability of transition is proportional to the final DOS, according to Fermi golden rule.

2.1 Spin Polarization

In equilibrium state of conventional (nonmagnetic) materials, the spin of electrons is homogemeously distributed and equally populated for both spin-up and spin-down electrons. However, in ferromagnetic materials (such as *Co, Fe, Ni,* and some alloys), electron spins are aligned spontaneously, resulting in unequal density of spin-up and -down electrons. In such ferromagnetic materials, spin-resolved electronic

Figure 3. Energy band structure E(k) and DOS of Cu (nonmagnetic metal) and Co (ferromagnetic metal)

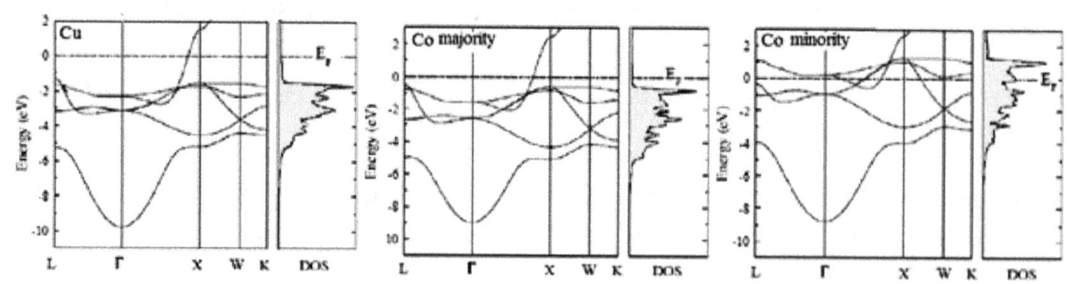

structures show a band difference between spin-up and spin-down electrons, as shown in Figure 4. The figure depicts the spin polarization of spin-up and spin-down electrons, as well as their total polarization (***P***) in different types of magnetic materials. The asymmetry in density of states (DOS) at Fermi energy gives rise to many spin related transport phenomena. As shown, the net magnetic polarization in paramagnetic materials is zero (in absence of external magnetic field). The so-called half-metal (HM) is a special class of metals that can produce 100% spin polarization (***P=1***).The polarization (***P***) is usually expressed in terms of the number (or density of states) of spin-up ($N_\uparrow$) and spin-down ($N_\downarrow$) electrons, resulting in a spin polarization

$$P = (N_\uparrow - N_\downarrow)/(N_\uparrow + N_\downarrow) \tag{3}$$

2.2 Spin Waves and Magnons

Magnons are the quanta of spin waves of excited spin states in magnetic materials. They arise when neighbor spins interact strongly in a magnetic material, making it favorable to excite the collectivity of spins, rather than flipping a single spin. Figure 5 shows some possible cases of spin polarization distributions of a ferromagnet. In the ground state of the ferromagnet (in equilibrium) almost all spins are aligned. In excited state of the ferromagnet (in non-equilibrium) some spins may be flipped. At lower energy of excitation, the spins are rotating (precessing) around their equilibrium, forming a spin wave (magnon). Magnons open up the possibility to transmit and manipulate information in a new class of materials. The magnons carry a spin current; a flux of angular momentum, that can be used to encode and transmit information. The field of magnon spintronics (or magnonics) studies magnon spin currents and aims to integrate this technology with conventional electronic devices

Figure 4. Spin polarization of up and down electrons in magnetic materials. All states up to the Fermi-level are occupied. In presence of external magnetic field, an excess of spins aligned parallel to the magnetic field will appear.

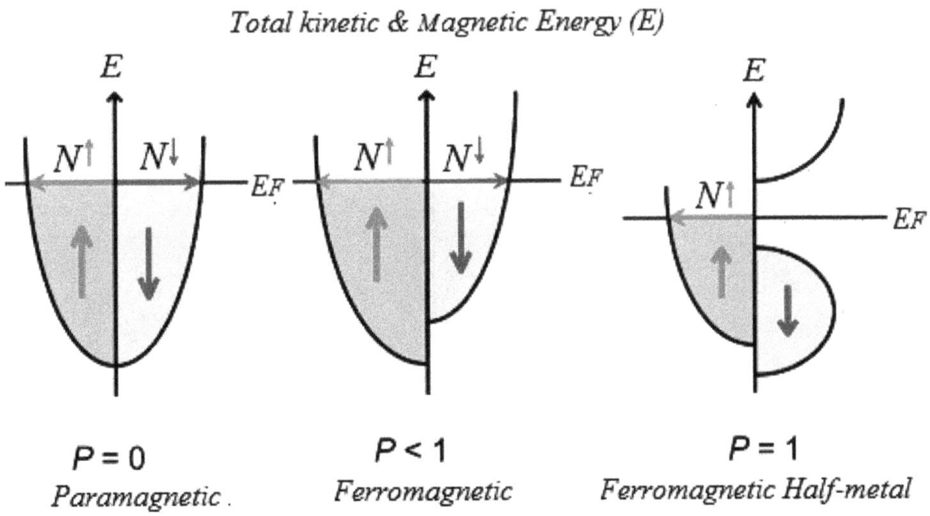

Figure 5. Possible states of spin distributions in a ferromagnet and spin waves

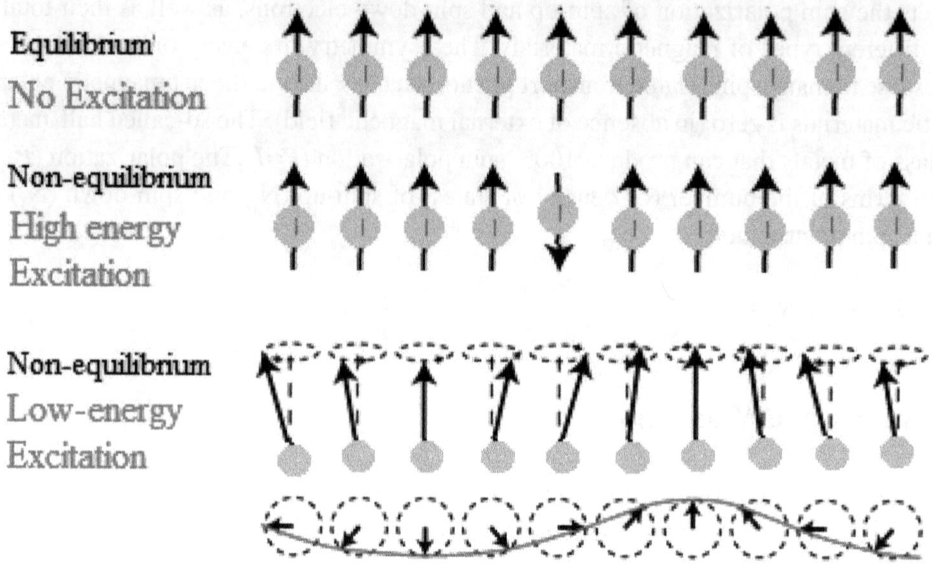

Note 2: What's Spin Precession?

Like a top, a rapidly spinning particle will precess in a direction determined by the torque exerted by its mass (weight). The precession angular velocity is inversely proportional to the spin angular velocity, so that the precession is faster as the particle slows down. The direction of the precession is visualized in Figure 6.

Figure 6. Spin precession

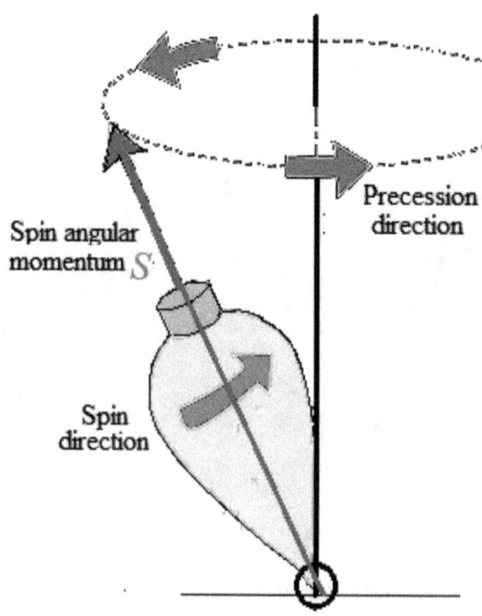

2.3 Spin Injection and Spin Currents

The design and implementation of spintronic devices requires an efficient source of spin-polarized carriers. In addition, injecting an excess magnetization in a non-magnetic conductor allows us to study properties of the spin dynamic properties, such as spin lifetime in this conductor. Spin injection and transport can be carried out in a variety of materials including metals, semiconductors, and carbon-based materials. In fact the recent experimental studies showed that the spin polarization can be maintained and controlled in certain semiconductor structures. For instance, the spin polarization of coherent injected electrons could be maintained in *GaAs* quantum wells, for up to nanoseconds and up to 100μm in homogeneous materials at room temperature (Ohno et al, 1999).

Spin Injection

A spin polarized current can be injected into a non-magnetic conductor or semiconductor by driving a current from a ferromagnetic metal. This creates an imbalance in the spin population if injection is faster than relaxation.

The basic idea of spin injection in a paramagnetic semiconductor is as follows. In a paramagnet, at equilibrium, the spin magnetic moments of charge carriers (electrons in the conduction band and holes in the valence band) point along random directions in space. Therefore the ensemble average of spin moment at any location inside the material is null and the spin polarization of electrons is zero. Under an applied voltage bias, we can inject a stream of spin-polarized electrons from a ferromagnet into the paramagnetic semiconductor. In fact, it has been proven theoretically and experimentally that spin-polarized electrons can be injected into a semiconductor from a metal ferromagnetic contact through a Schottky barrier (Albrecht & Smith, 2003).

Spin Currents

The spin polarization is transported by a spin-polarized current. There are two well-known types if spin currents, namely (Schmidt et al, 2000):

Figure 7. Injection of spin in a semiconductor, from a metal ferromagnet

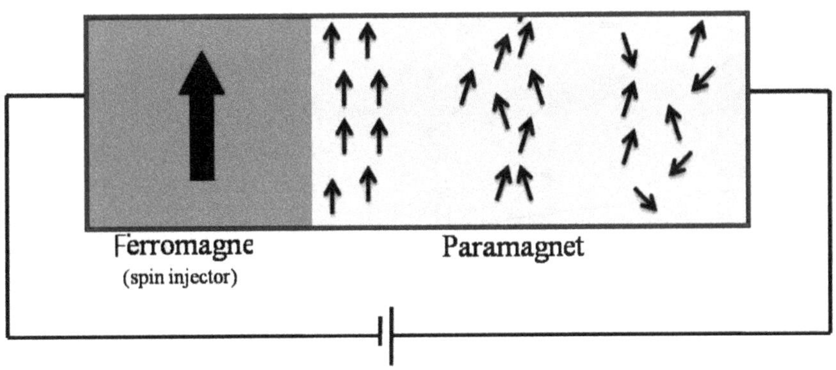

- Pure spin currents,
- Spin-polarized currents.

Pure spin currents carry only spins (spin angular momenta) unlike conventional spin-polarized currents which carry both charges and spins. One of well-known methods to generate the pure spin current is non-local spin injection. When the spin-polarized current is injected from a ferromagnet (FM) into a non-magnet (NM), spins are accumulated in the vicinity of the FM/NM interface. The accumulated spins diffuse in the NM, and thus the spin accumulation is formed in the place where no charge current is present.

2.4 Spin Detection and Extraction

In electrical spin injection we drive spin-polarized electrons from a ferromagnet into a nonmagnetic conductor. As a result, non-equilibrium spin accumulates in the nonmagnetic conductor. The opposite is also true: If a spin accumulation is generated in a nonmagnetic conductor that is in proximity of a ferromagnet, a current flows in a closed circuit, or an electromotive force (EMF) appears in an open circuit. This effect is called the Silsbee-Johnson spin-charge coupling[4]. The presence of the electron spin can then be detected electrically.

Other ways of detecting spin include the spin-valve effect, in which the injected spin-polarized electrons enter a detecting ferromagnetic electrode with an efficiency given by the relative orientation of the injecting and detecting electrodes, or optical detection in which spin-polarized electrons recombine with un-polarized holes and emit circularly polarized light.

A standard method to extract spin relaxation time in a paramagnetic material is to perform a spin-valve experiment, as shown in Figure 9.

A spin-valve is a tri-layered construct, in which the paramagnetic material of interest is sandwiched between two ferromagnetic electrodes of different coercivities. Unlike giant magnetoresistive (GMR) devices, the ferro-magnets are not magnetically coupled with each other. As a result, their magnetizations can be controlled independently by magnetic field. One of these ferromagnets acts as spin injector *i.e.*

Figure 8. Schematics of spin injection at Ferromagnet-Non-magnet interface

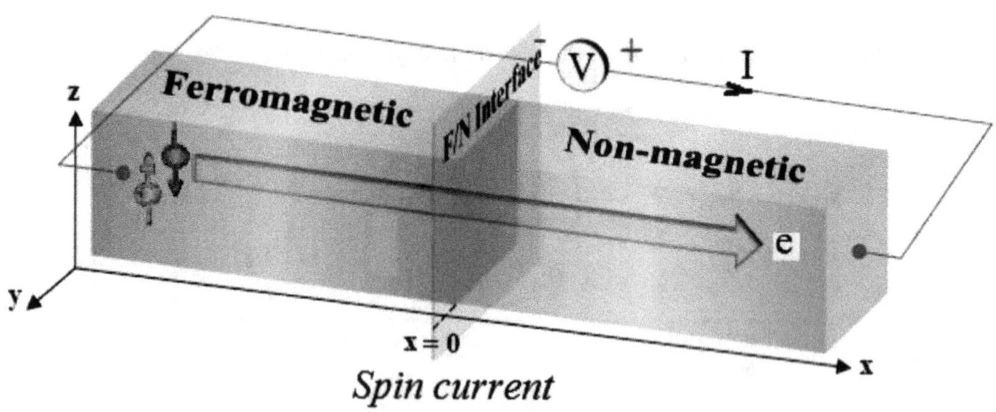

Figure 9. Spin valve experiment

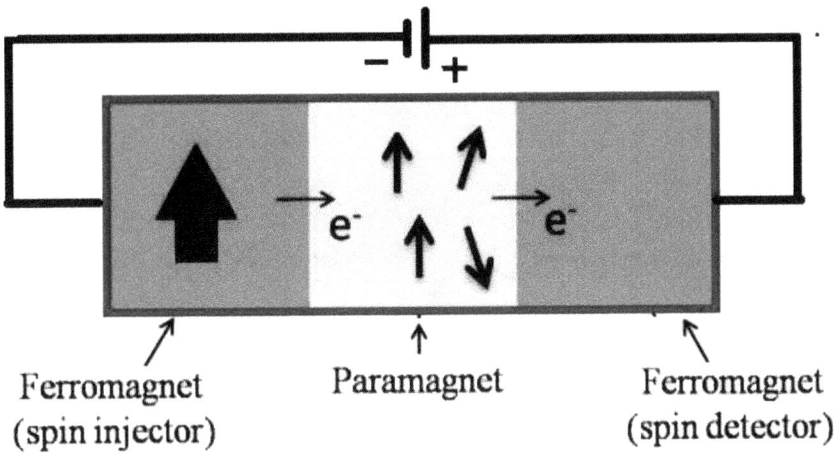

under an applied electrical bias it injects spins into the paramagnet. The second ferromagnet provides unequal spin-up and spin-down at the Fermi level and transmits spins of one particular orientation. Recent experiments allowed determining the spin dephasing time from all-electrical high frequency measurements. The inferred time is of the order of 1ns.

2.5 Spin Coherence and Spin-Orbit Interaction (SOI)

Coherence and phase interference phenomenon of electron spin waves are attractive quantum effects that are exploited in spintronic devices. The spin of electrons is affected by momentum scattering and other mechanisms, which leads to spin dephasing. The interaction mechanisms may be, for instance, due to:

- Dipole-dipole (Magnetic) interaction, between magnetic dipoles,
- Spin-orbit interaction (SOI), due to induced magnetic field,
- Exchange interaction due Coulombic fields, and
- Hyperfine magnetic field (between nuclei and electrons)

The spin–orbit interaction (SOI) is an interaction of electron spin with its electromagnetic motion in a crystal (Thomas, 1926). Although the spin–orbit interaction is a small perturbation to the crystal potential in solids (hundreds of milli eV), it may play an important role in the band structure. The *SOI* interaction for example splits bands which would be otherwise degenerate. The narrow-bandgap semiconductors (like *InSb* and *InAs*), and the semimetals (like *Bi*) have substantial spin-orbit interaction.

The *SOI* interaction energy (Hamiltonian) is proportional to the *L.S* product

$$H_{SO} = \lambda\,(\boldsymbol{L}.\boldsymbol{S}) \tag{4a}$$

where λ is spin-orbital coupling (SOC) constant and the product may be calculated from angular momentum numbers (orbital & spin), as follows:

Figure 10. Illustration of the spin-orbit interaction (SOI) and the effective magnetic field, which is controllable by electric field

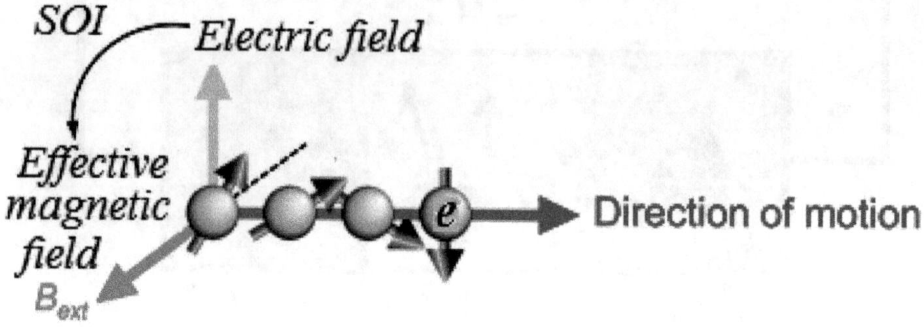

$$<L.S> = \frac{1}{2}(J^2 - L^2 - S^2) = \frac{1}{2}\hbar^2\,[(j(j+1) - l(l+1) - s(s+1)] \tag{4b}$$

The SOI tends to preserve spin coherence at lower temperatures and with an applied perpendicular magnetic field

3. SPIN RELAXATION IN SEMICONDUCTORS

During the electron transport, the spin polarization is gradually lost by spin flip and spin dephasing (spin decoherence) events. This loss typically follows an exponential decay that is characterized by a spin-diffusion length λ_s, such that $P=P_o.exp\,(-x/\lambda_s)$. In fact, the spin of electrons is affected by scattering events, which leads to spin dephasing and flip. Spin relaxation refers to the processes that bring an unbalanced population of spin states into equilibrium. Inside a metal (ferromagnetic or not), a polarized spin loses its original orientation due to scattering after a spin relaxation time τ_s.

There are several mechanisms in solids that are responsible for spin relaxation of conduction electrons. In case of semiconductors and metals, the most dominant mechanisms are

1. Elliott-Yafet (EY),
2. Dyakonov-Perel (DP),
3. Bir-Aronov-Pikus (BAP) and
4. Hyperfine interaction with nuclei (HF).

3.1 Elliott–Yafet Mechanism

The Elliott–Yafet (EY) mechanism is based on the fact that in real crystals Bloch states are not spin eigenstates. Elliott noticed that an ordinary interaction with impurities, boundaries, interfaces, and phonons can connect ‘up’ and ‘down’ electrons, leading to spin relaxation.

When the EY is the dominant scattering mechanism, a direct proportionality between the momentum scattering time and the spin relaxation time can be found. In the case of single-valley direct-gap semiconductors (*like GaAs*), the spin scattering time τ_s due to the EY mechanism is given by:

$$\tau_s(E) \sim [\Delta_{so}/(E_g+\Delta_{so})]^2.\ (E/E_g)^2.\ \tau(E) \tag{5}$$

where $\tau(E_k)$ is the momentum relaxation time. This may be important in narrow-gap semiconductors with large spin-orbit splitting Δ_{so} (like *InSb*).

3.2 Dyakonov and Perel Mechanism

Dyakonov and Perel showed that the lifting of the spin degeneracy by induced magnetic fields leads to spin relaxation (Dyakonov & Perel, 1972). The induced field arises in crystals that lack inversion symmetry, such as zincblende semiconductors (like *GaAs*, *ZnSe*), and lifts the spin degeneracy. Therefore, spin-up and spin-down electrons have different energies even when they are in the same momentum state. The interaction Hamiltonian, which accounts for the DP relaxation in the bulk, can be written as follows:

$$H_{DP}(k) = -\mu.B_i\sigma \tag{6a}$$

where $\mu = e\hbar/2m$ and $B_i(k)$ is called the effective (induced) magnetic field in the crystal and σ is the Dirac spinor (or Pauli vector). This spin-orbit interaction can be also written as follows

$$H_{DP}(k) = \alpha\ [\ \sigma_x k_x(k_y^2 - k_z^2) + \sigma_y k_y(k_z^2 - k_x^2) + \sigma_z k_z(k_x^2 - k_y^2)] \tag{6b}$$

where α is a coupling constant. The DP mechanism is a precessional spin relaxation because it describes how spins precess between the momentum scattering events. The Elliott-Yafet EY and the DP mechanisms can be distinguished from each other by the opposite dependences of their spin relaxation rates on mobility. In the former mechanism, the spin relaxation rate is inversely proportional to the mobility and in the latter mechanism; it is proportional to the mobility.

3.3 Bir, Aronov, and Pikus Mechanism

Another source of spin relaxation for conduction electrons was found by Bir, Aronov, and Pikus in the electron-hole exchange interaction. This interaction depends on the spins of interacting electrons and holes and appears in p-type semiconductors. It may be expressed as follows:

$$H_{BAP} = A.\boldsymbol{S}.\boldsymbol{J}\delta(r) \tag{7}$$

where $\boldsymbol{S}$ is the electron spin operator and $\boldsymbol{J}$ is the total angular momentum operator of holes. This mechanism is effective for the spin relaxation of electrons in *p*-type semiconductors, where overlapping of hole wave-functions is significant.

3.4 Hyperfine Interaction

The spin hyperfine interaction, happens for electrons bound on impurity sites or confined in a quantum dot.

The quantity $\lambda_s = \sqrt{(D_s.\tau_s)}$ is usually employed to define the characteristic length scale over which the spin relaxes. In a ferromagnet, this is typically 5–50 nm. For a non-magnetic metal, it is of the order of hundreds of nm. However, one may ask if the electron diffusion constant D_n is the same as the spin diffusion constant D_s which one would use to define the spin relaxation length.

Spin relaxation in semiconductors is a complex process, because there are two types of charge carriers to consider (electrons and holes) and both have spin and can carry spin-polarized currents. Also, some features in the polarization luminescence spectra imply to take excitons (bound electron-hole pairs) into account, because they can be also polarized. Furthermore, the spin relaxation is sensitive to temperature, doping and impurity content, dimensionality, strain, magnetic and electric fields as well as the material band structure. The variation of the relaxation lifetime with doping and temperature is illustrated in Figure 11, as obtained by the time-resolved Faraday rotation (TRFR) measurement (Kikkawa & Awschalom, 1998).

4. MAGNETIC SEMICONDUCTORS

Magnetic semiconductors are semiconductor materials (e.g. GaAs) which are converted into ferromagnetic material by adding magnetic dopants (e.g. manganese). The realization of materials that combine semiconducting behavior with robust magnetism has long been a dream of material physics. The following criteria should be met in such materials:

- Ferromagnetic transition temperature (T_c) must exceed room temperature
- Mobile charge carriers should respond strongly to changes in the ordered magnetic state, and
- The material should retain fundamental semiconductor properties, such as sensitivity to doping and light.

Figure 11. Spin relaxation mechanisms
After Fabian et al (2007).

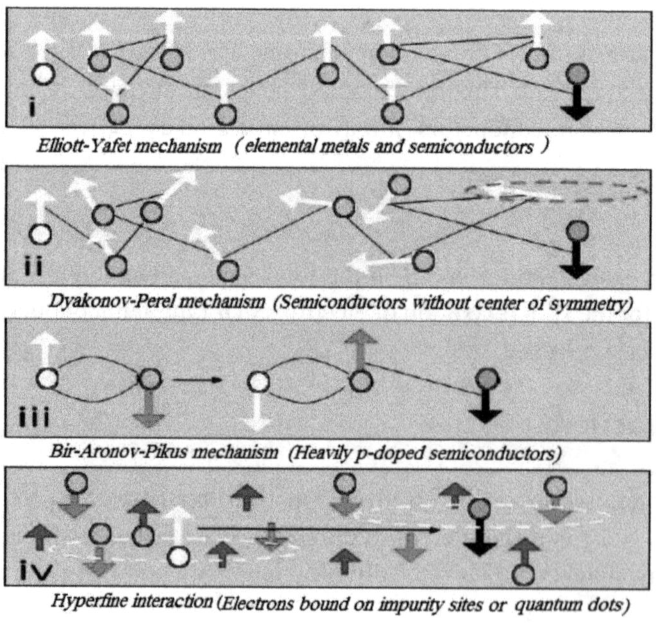

Figure 12. Measured spin lifetime, in GaAs, against temperature at different values of magnetic field (left), and against excitation density (right). The predictions of D-P and E-Y are also reproduced in the left figure.
After (Kikkawa & Awschalom, 1998).

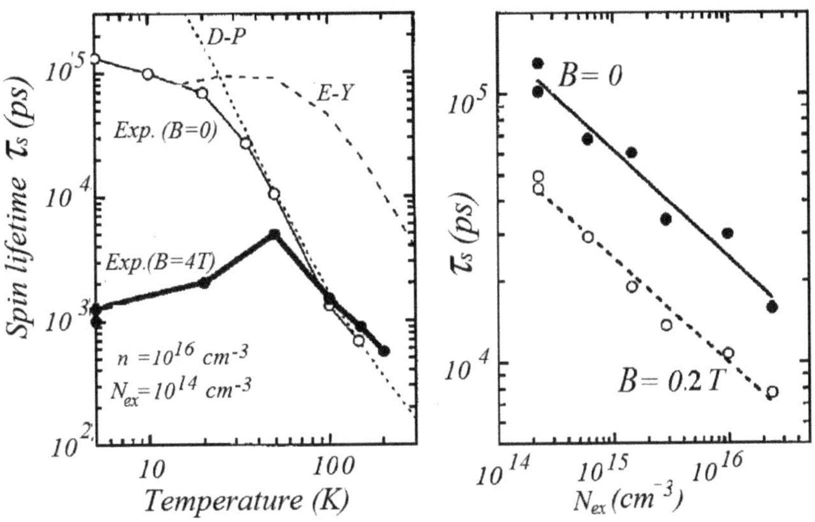

One strategy for creating such materials that are simultaneously semiconducting and magnetic, is to introduce local moments. The result is a new class of materials now known as diluted magnetic semiconductors (DMS). In such materials one of the semiconductor atoms is replaced by a magnetic atom. For instance, the *GaAs*, which is heavily doped with *Mn* acceptors shows a magnetic moment.

4.1 Dilute Magnetic Semiconductors (DMS)

The dilute magnetic semiconductors (DMS) are a class of semiconductor materials, which allow for a high spin polarization. The DMS materials, like *(Zn,Mn)Se* or other *Mn* doped semiconductors, are paramagnetic (non-magnetic) when no external magnetic field is applied. Such paramagnetic materials exhibit a giant spin splitting at low temperatures in the presence of magnetic field, which can be as large as 20meV in the conduction band. At a temperature of 4K the material conductivity is strongly affected by the magnetic polarization due to spin asymmetry.

In addition to the high spin polarization of the carriers in equilibrium, the *Mn* atoms in the material play another important role. They cause a strong spin scattering, which guarantees that non-equilibrium carriers that enter the material (when a current is flowing). Therefore, DMS materials are ideal candidates for perfect spin aligning devices. The most widely studied DMS in the past was the *Mn* ions embedded in the *CdTe*, which has the zincblende structure.

4.2 Magnetic Impurities in Semiconductors

It has been established that several semiconductors become ferromagnetic when they are heavily doped with ferromagnetic atoms, like *Ti, Cr, Mn, Fe, Co*, and *Ni*. For instance, some (III,V) compound

Figure 13. Crystalline structure and 3d orbitals of (Ga,Mn)AS
After Jungwirthet al (2006).

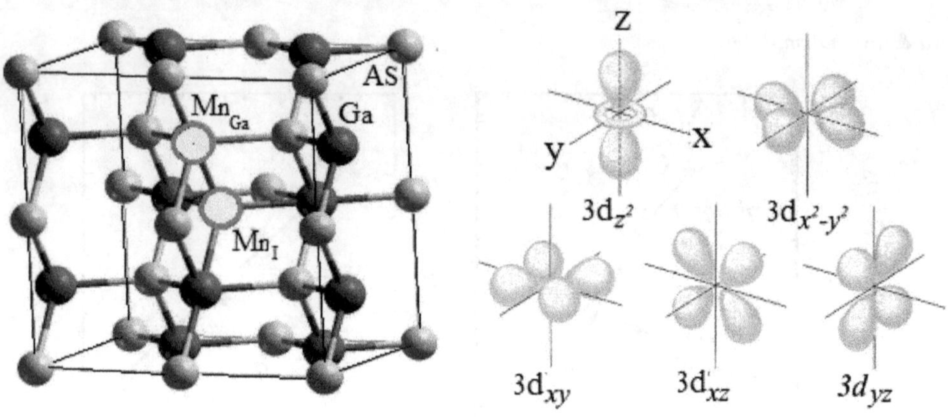

semiconductors become ferromagnetic when heavily doped with *Mn*, and the ferromagnetic transition temperatures may be well above 100K. In semiconductors like *GaAs* and *InAs*, *Mn* has been shown to act both as an acceptor and as a source of local magnetic moments.

The incorporation of *Mn* impurities in DMS causes a strong spin scattering and makes DMS materials ideal candidates for spin devices. For instance, the transport properties of *GaAs*, which is heavily doped with *Mn* acceptors reveal a phase transition at a critical temperature (50K) from a paramagnetic to a ferromagnetic material. This phase transition is due to the interaction between the *Mn*-hole complexes. The elements in the *(Ga,Mn)As* alloy have an atomic structure *[Ar]* $3d^{10}4s^2p^1$. The sub-stitutional MnG_a, and the interstitial Mn_I, positions are shown in Figure 13.

The local moments model of $Mn(d^5)$, which describes the behavior of *Mn*-doped IV-VI DMS in terms of momentum exchange to itinerant *sp*-band carriers, provides a good description. As shown n figure 14, the itinerant holes (solid small circles) couple antiferro-magnetically to the local *Mn* moments via the exchange interaction due to the overlap of the hole wave function with the *d*-orbitals of the local *Mn* electrons. For high hole densities a long range ferromagnetic order is established. The two ovals illustrate the flipping of a local Mn-moment through the kinetic exchange interaction with a passing-by hole.

Figure 15 depicts the band structure of *(Ga,Mn)AS* and shows the position of the 3d level for low Curie temperature T_C, as well as the positions of the majority-spin 3d bands, which lead to high T_C. Using conventional growth processes, the incorporation of magnetic *Mn* ions into III-As semiconductor crystals has been limited to 0.1%. In order to circumvent the solubility problems, a low-temperature molecular-beam-epitaxy (LT MBE) process is usually applied to grow *(In,Mn)As* and *(Ga,Mn)As* DMS ternary alloys with more than 1% *Mn* ions. The problems related to the critical temperature limits in *(III,Mn) V* DMS materials have been investigated in the literature. In 1998 the Tohoku University researchers announced a jump of T_c in *p*-type *(Ga,Mn)As* to 110K and pointed out that indirect coupling between Mn *d*-shell moments is mediated by induced spin polarization in a free-hole itinerant-carrier system. The main difference between (III,Mn)V materials like *(Ga,Mn)As* and IV-VI and II-VI compounds is that *Mn* substituting for the trivalent cation (*Ga*) is simultaneously an acceptor and a source of magnetic moments (Jungwirth et al., 2006).

The theoretical calculations of critical-temperature, which are based on the kinetic-exchange model, predict room-temperature ferromagnetism in *(Ga,Mn)As* with 10% *Mn* content. In spite of these predic-

Figure 14. Schematic of the itinerant holes, which couple antiferro-magnetically to local Mn moments via the exchange interaction

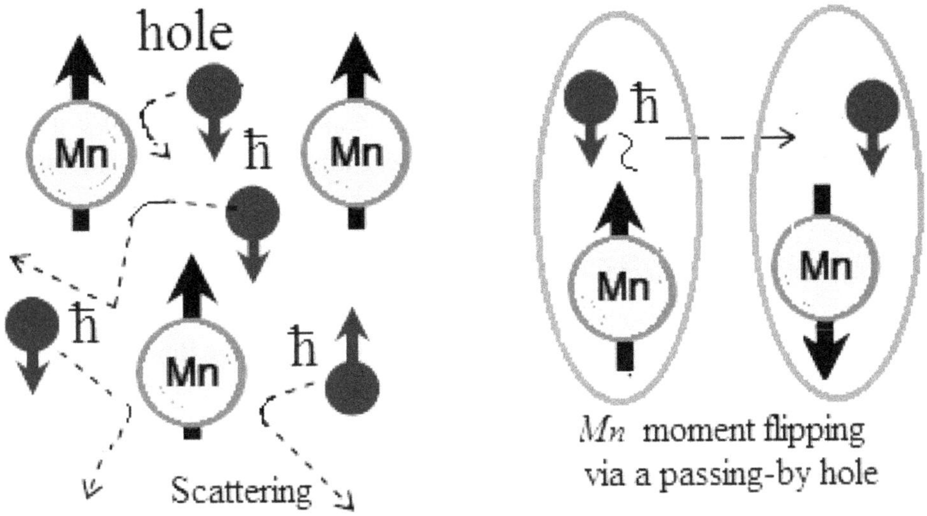

Figure 15. Band structure of (Ga,Mn)AS, showing the position of the 3d level for low Curie temperature T_C (a) and the positions of the minority (b) or majority-spin (c) 3d bands, which lead to high T_C.
After (Zhou & Schmidt, 2010).

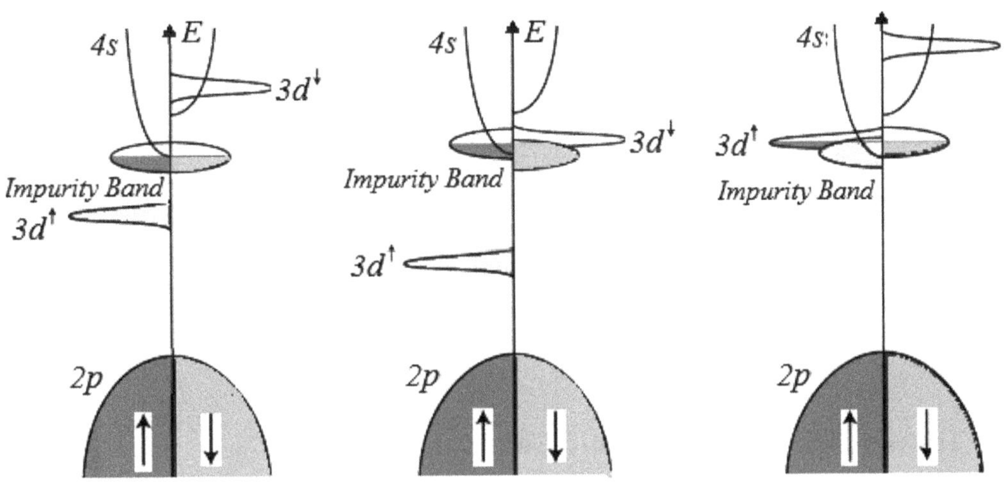

tions, the 110K milestne in *(Ga,Mn)As* was harsh for many years. However, with the advent of molecular beam epitaxy (MBE), it was possible to, push T_c in *(Ga,Mn)As* up to 173 K. The room temperature DMS started to appear in 2001, by the aid of the metal organic chemical vapor deposition (MOCVD).

On the basis of many experimental studies, the *(III,Mn)Sb* materials were predicted to follow the same route as *(Ga,Mn)As* and *(In,Mn)As* ferromagnetic semiconductors. It was also predicted that moving in the opposite direction of the periodic table toward *(III,Mn)P* and *(III,Mn)N* would lead to FMS materials with high critical temperature. Experimental samples of *p*-type and *n*-type ferromagnetic layers of *(Ga,Mn)N*, with high T_c (about1000K), are also developed (Zhou & Schmidt, 2010).

4.3 Ferromagnetic Semiconductors (FMS)

The so called ferromagnetic semiconductors (FMS) are displaying both semiconducting and ferromagnetic properties. We have seen that DMS materials yield high spin injection efficiency, but they require an external magnetic field and are thus not ideal for device applications. The ferromagnetic semiconductors are a promising alternative. The *(III,Mn)V* materials, such as *(Ga,Mn)As,* are examples of ferromagnetic semiconductors. Magnetism in such ferromagnets originates from *Mn* local moments. Note that the term *ferromagnetic* is reserved for magnetic materials in which ferromagnetism is due to coupling between magnetic moments that is mediated by conduction electrons or valence-band holes.

Experimentally, ferromagnetism in *(Ga,Mn)As* is observed when *Mn* doping reaches approximately 1% and the system is near the Mott insulator-to-metal transition, This happens when the average distance between *Mn* impurities ($r_c = (3/4\pi N_{Mn})^{1/3}$) is equal to the impurity effective Bohr radius ($a^{*}= \varepsilon\hbar/m^{*}e^{2}$). At larger *Mn* concentrations, the localization length of impurity-band states is extended to a degree that allows them to mediate ferromagnetic exchange interaction between *Mn* moments. At even higher *Mn* concentrations, the impurity band gradually merges with the valence band and impurity states become delocalized. However, most of FMS, like *(Ga,Mn)As*, are p-type and ferromagnetic appears at very low temperatures. Also, the short spin diffusion length of holes in semiconductors limits the spin injection efficiency from an FMS to nonmagnetic semiconductors. In 2004, Van Dorpe and others succeeded to generate a spin polarization from p-type ferromagnetic *(Ga,Mn)As* with high efficiency (80%) using an optimized tunnel (ESAKI) diode. In this device the spin polarized holes tunnel from the valence band of the FMS into the conduction band of the highly n-doped nonmagnetic semiconductor. FMS materials possess all properties that can be exploited in spintronics. For instance, the *(In, Mn)As* has been used to build field-effect transistors. It has been demonstrated that changes in the carrier density and distribution in thin-film FMS systems due to an applied bias voltage can reversibly induce the ferromagnetic/paramagnetic transition. Figure 16 depicts the reversible hysteresis curve of *ZnO-Mn.* As the energy levels of a magnetic ion split in the magnetic field, the magnetic susceptibility of the material can be measured, according to the relation:

$$\chi = \lim_{H \to 0} \frac{\partial M}{\partial H} \tag{8}$$

4.4 Spin Transport in Ferromagnetic Semiconductors

A ferromagnetic semiconductor exhibits strong magneto-transport effects, namely negative magnetoresistance (MR) and anomalous Hall Effect (AHE), and provide the possibility to control the spin by an external electric field. For ferromagnetic *GaMnAs*, usually the AHE is taken as a measure of its magnetization. The observation of AHE is considered as an important criteria for FMS materials. The MR and AHE have been reported in the *Ge:Mn* systems (which are prepared by LT-MBE or ion implantation at elevated temperatures). However, the correlation between magnetization, MR, and AHE, which is a hallmark of III-Mn-V and *ZnMnTe* FMS, has not been proven for *Ge:Mn* so far.

The anomalous Hall effect (AHE) shown in Figure 17, which completely dominates the low-field Hall response in *(Ga,Mn)As* and some other III-V DMSs, has become one of the key tools used to

Figure 16. Magnetization characteristics of ZnO: Sn, with 3% Mn
After (y, 2000).

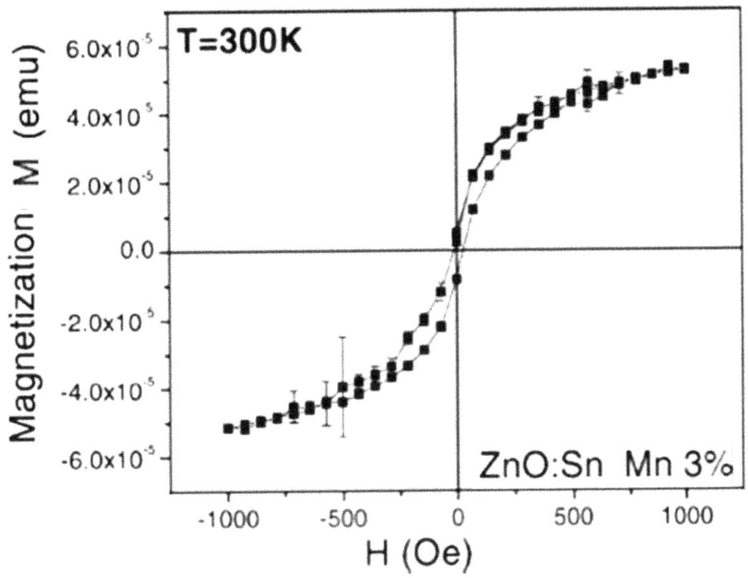

detect the paramagnetic/ ferromagnetic transition. The large value of Hall resistance is due to the spin-polarization of holes and provides strong evidence for the participation of mobile charge carriers in the ordered magnetic state of such DMS.

The electric field control of ferromagnetism in DMS materials, has been investigated using *(In,Mn) As*-based field-effect transistors. It has been demonstrated that changes in the carrier density and distribution in thin-film DMS systems due to an applied bias voltage can reversibly induce the ferromagnetic/

Figure 17. Hall resistance vs. external magnetic field for $Ga_{0.94}Mn_{0.06}As$ at different temperatures
After Edmonds et al. (2002).

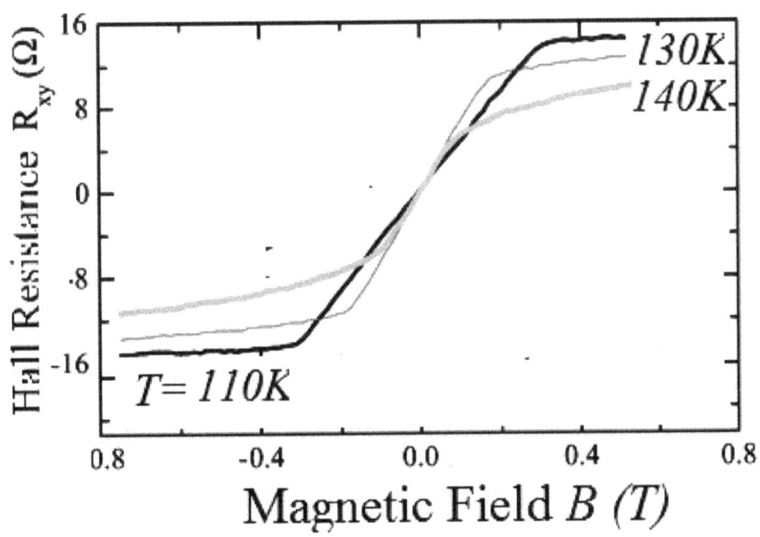

paramagnetic transition. Another remarkable effect observed in this magnetic transistor is electric-field-assisted magnetization reversal. This has been attributed to the dependence of the hysteresis loop width on bias voltage. Experiments in which ferromagnetism in a *(III,Mn)V* DMS system is turned on and off optically add to the list of functionalities that result from the realization of carrier-induced ferromagnetism in a semiconductor host material.

The Hamiltonian of the magnetic ion in the semiconductor can be written as follows:

$$H = \lambda\,(\boldsymbol{L.S}) + H_{\text{ion}} + Hc + (\boldsymbol{L} + 2\boldsymbol{S})\,\mu_{\text{B}}\boldsymbol{B} \tag{9}$$

where the first term in the RHS terms is the spin–orbit (*L.S*) coupling, λ being the spin-orbit coupling (SOC) constant. The second term (H_{ion}) is the free ion term, and the third term (H_c) is the crystal field term. The last term accounts for the interaction with external magnetic field, where μ_B is the Bohr magneton. However, the crystal field splitting energy is typically larger than the spin orbital coupling energy.

5. SPIN TRANSPORT MODELS

As the spin is a degree of freedom of charge carriers, it has been suggested to incorporate spin and spin-dependent interactions into the electron transport equations (quantum and semiclassical). However, the spin properties of information carriers have their own features, which turn the spin transport modeling more complex. For example, it has been proved that the direction of the electron spin is not conserved after a scattering. This fact has to be included in a model of spin transport in solids. Due to spin non-conservation the definition of the spin current has some phenomenological parameters. These parameters are typically used to express the spin relaxation time in the semiclassical transport equations.

The so-called two-component spin-drift-diffusion model (SDDM) is a simple semiclassical and straight-forward method for spin transport modeling. The semiclassical spin-transport approaches include the spin Boltzmann equation (spinor-BTE), the Bloch equations, and Monte Carlo methods (MCM). Some semiclassical models, (e.g., SDDM) are derived from quantum approaches, using the so-called semiclassical limit (EL Hajj, 2008). On the other hand, some semiclassical approaches are extended to include quantum transport features. For instance, the classical Bloch equations of spin magnetization transport have been utilized by Buot, Loberternos & Villarin (2012) to develop a transport equation for the nonequilibrium spin magnetization quantum distribution function (SMQDF). By taking the macroscopic moments of the equation of motion of the SMQDF, we can get a density-matrix based version of the SDDM. However, there exist some quantum models of spin-transport, which include the full quantum description, using the kinetic transport equation of density matrix as well as the NEGF.

One approach that is totally free of phenomenological parameters is density-functional theory (DFT), as well as its spin-density-functional (SDF) generalization. Like DFT, the SDF is based on Kohn-Sham theory (1965) and the many-body quantum effects are buried in an exchange-energy functional. Although DFT & SDF theories are theoretically exact, their application requires some approximation of the exchange energy. Approximate forms for this functional can be partially phenomenological and are normally based on microscopic calculations of correlation effects in the electron-gas model system. This is the case for the local-(spin-) density approximation (LSDA). In practice LSDA theory may be considered as a mean-field theory in which the exchange energy increases with the self-consistently determined local spin density.

It should be noted that both semiclassical (phenomenological) and full quantum approaches can be used to study ferromagnetism in magnetic semiconductors, and *(III,Mn)V* compounds in particular. For instance, LDA+*U scheme* (Anisimov et al. 1991) and self-interaction-corrected LSDA schemes have been used to obtain energy spectra and to understand the microscopic origin of ferromagnetism in (III,Mn) V alloys. There are also theoretical research on *(III,Mn)V* DMS materials based on simpler models in which holes are assumed to hop between *Mn* acceptor sites, where they interact with *Mn* moments via phenomenological exchange interactions. The so-called model Hamiltonian theories are also important because they sometimes provide more physical insight of ferromagnetism and enable predictions of transport properties that are sometimes beyond the reach of full quantum methods. The model Hamiltonian theories will be discussed in the following section, in addition to the semiclassical methods.

5.1 Quantum Description of Spin Transport

In general, the electron wavefunction has a spin-up and spin-down components, called spinors:

$$\psi(x) = [\psi\uparrow(x), \psi\downarrow(x)]^T \tag{10a}$$

Note that the spinor components should be normalized such that:

$$\int(|\psi\uparrow(x)|^2, |\psi\downarrow(x)|^2)dx = 1 \tag{10b}$$

A quantum state of an electron with spin can be described by the density matrix operator, ρ, which is related to the wavefunction components (or spinors $\psi\uparrow$ and $\psi\downarrow$) as follows:

$$\rho(x,x) = \begin{bmatrix} |\psi_\uparrow(x)|^2 & \psi_\uparrow(x).\psi_\downarrow^*(x) \\ \psi_\downarrow(x).\psi_\uparrow^*(x) & |\psi_\downarrow(x)|^2 \end{bmatrix} \tag{11}$$

The relative phase of the spinor components determines the spin direction, according to $\langle S\rangle = \frac{1}{2}\hbar \langle\psi(x)|\ \sigma\ |\psi(x)\rangle$, where the vector σ is called the Dirac spinor, which has three components, called Pauli matrices:

$$\sigma_x = \begin{pmatrix} 0 & 1 \\ 1 & 0 \end{pmatrix}, \sigma_y = \begin{pmatrix} 0 & -i \\ i & 0 \end{pmatrix}, \sigma_z = \begin{pmatrix} 1 & 0 \\ 0 & -1 \end{pmatrix} \tag{12}$$

Note 2: Pauli Matrices and Dirac Spinor

The Pauli matrices are a set of three 2×2 matrices usually indicated by the letter sigma (σ). They are given by:

$$\sigma_x = \begin{pmatrix} 0 & 1 \\ 1 & 0 \end{pmatrix}, \sigma_y = \begin{pmatrix} 0 & -i \\ i & 0 \end{pmatrix}, \sigma_z = \begin{pmatrix} 1 & 0 \\ 0 & -1 \end{pmatrix}$$

Pauli matrices must satisfy the following conditions: $\sigma_i \sigma_i = I$, $\sigma_i \sigma_j = - \sigma_j \sigma_i$.

In quantum mechanics, Pauli matrices occur in the Pauli equation which takes into account the interaction of the spin of a particle with an external electromagnetic field. The Pauli matrices obey the following commutation relations:

$\left[\sigma_a, \sigma_c\right] = 2i\varepsilon_{abc}\sigma_c$ and ant-commutation relations: $\left[\sigma_a, \sigma_b\right] = 2\delta_{ab} I$.

Spinors are elements of a complex vector space that can be associated with Euclidean space. Like vectors and tensors, spinors transform linearly when the Euclidean space is subjected to a small rotation. The most famous type of spinor, is the Dirac spinor. The Dirac spinor is sometimes called the Pauli spin vector or shortly spin vector. It is defined as: $\sigma = (\sigma_x, \sigma_y, \sigma_z)$

The interaction of the spin with the local magnetic field can be calculated by the Zeeman relation (2c), as follows:

$$H_Z = \mu\boldsymbol{\sigma}.\boldsymbol{B} = \tfrac{1}{2}\hbar\, g_s\, \mu_B \boldsymbol{\sigma}.\boldsymbol{B} \tag{13}$$

where μ_B is the Bohr magneton, g_s is the spin g-factor and $\sigma = (\sigma_x, \sigma_y, \sigma_z)$ is the Dirac spinor whose components are the Pauli matrices. A second interaction arises as relativistic correction to the motion of the electron in an electric field. This is the spin-orbit interaction, which has the general form:

$$H_{so} = \frac{\hbar^2}{4m_0^2c^2}\left(\nabla V x p\right)\cdot\sigma \tag{14}$$

where m_0 is the free electron mass, c is the speed of light at vacuum, V is the electric potential, $\boldsymbol{p}$ represents the canonical momentum and $\boldsymbol{\sigma} = (\sigma_x, \sigma_y, \sigma_z)$ is the Pauli spin vector. In presence of an external magnetic field $\boldsymbol{B}$ (where $B=\nabla xA$ and A is the magnetic vector potential), the momentum $\boldsymbol{p}$ should be replaced by the kinetic momentum $\boldsymbol{p} + e.\boldsymbol{A}$.

In the case of atoms in a specific solid, the spin orbit interaction refers to the interaction of the electron spin with the average Coulomb field of the nuclei and other electrons. In addition to the SOI resulting from the lattice-periodic crystal potential there is a SOI corresponding to external and built-in electric fields that may be present in the system. The SOI resulting from external or built in electric fields is usually referred to as the Bychkov-Rashba spin-orbit interaction. In what follows, we assume that the SOI due to the crystal potential has been incorporated into the band structure (by replacing m_0 with the effective mass m^*) and we focus on the effect of external and built in electric fields on the spins of the conduction electrons.

Spin Hamiltonian

Most of spintronic devices which are utilizing spin-orbit interaction, make use of low-dimensional structures. In this case, the motion of electrons is typically confined by an effective potential in the direction orthogonal to the active magnetic material interface. We consider here a two-dimensional gas (2DEG) of non-interacting electrons with effective mass m^*. The effective mass Hamiltonian for an in-plane electron motion in the one subband approximation can be written as follows (in the presence of magnetic field):

$$H = \frac{1}{2m^*}\left(p + eA\right)^2 + V(r) + \mu.B,\sigma \tag{15a}$$

where $(p+ eA)$ is the kinetic momentum and the electron spin is described by the Pauli spinor σ. Also $\mu = ½\gamma.\mu_B$ is the magnetic moment and γ is the effective gyromagnetic ratio. The electrostatic potential $V(r)$ defines the system confinement. This Hamiltonian may be also written as:

$$H = \frac{\mathbf{p}^2}{2m^*} + V(\mathbf{r}) + H_{SO} \tag{15b}$$

where H_{SO} is the spin-orbit interaction (SOI) Hamiltonian, which arise in Dirac's equation and is given by equation (14)

The potential, $V(\mathbf{r})$, corresponds to the interaction with in-plane electric fields. The spin-orbit interaction term, H_{SO}, can be written in a general dyadic form, which is linear in electron momentum. This term is assumed small in comparison with other terms in the Hamiltonian. However, H_{so} may contain terms like the Elliott-Yafet, D'yakonov-Perel or Rashba spin-orbit coupling mechanisms.

For motion in a two-dimensional electron gas (2DEG) of electrons, the effect of quantum well asymmetry (Rashba term) may be incorporated as follows:

$$H_R = \alpha\,(k_x\sigma_x - k_e\psi) \tag{16a}$$

where α is a coupling constant. This effect is isotropic with respect to crystallographic axes. The effect of crystal inversion asymmetry (Dresselhaus term) may be also added as follows:

$$H_D = \beta\,(k_y\sigma_x - k_x\sigma_y) \tag{16b}$$

This effect is anisotropic with respect to crystallographic axes. The constants α and β are generally electric field dependent.

Equation of Motion for Spin Density Matrix

We have seen so far that a quantum state of an electron with spin can be described by the density matrix operator, ρ, which can be related to the wavefunction components (or spinors $\psi\uparrow$ and $\psi\downarrow$) by equation (11). Alternatively, the density matrix operator, ρ, can be related to the electron spin polarization vector $\boldsymbol{P}$, as follows (Ivchenco et al., 1990):

$$\rho(t) = \begin{pmatrix} 1 + P_z(t) & P_x(t) - iP_y(t) \\ P_x(t)\_iP_y(t) & 1 - P_z(t) \end{pmatrix} \tag{17}$$

The equation of motion for spin density matrix is given by:

$$i\hbar\frac{\partial \rho}{\partial t} = -\frac{\hbar^2}{m^*}\sum_j \begin{matrix} \dfrac{\partial^2 \rho}{\partial R_j \partial \Delta r_j} + \left(V\left(R + \Delta r / 2\right) - V\left(R - \Delta r / 2\right)\right)\rho \\ + \dfrac{i}{2}\sum_{j,\alpha} A_{j\alpha}\left\{\sigma_\alpha, \dfrac{\partial \rho}{\partial R_j}\right\} + i\sum_{j,\alpha} A_{j\alpha}\left[\sigma_\alpha, \dfrac{\partial \rho}{\partial R_j}\right] \end{matrix} \tag{18}$$

Note that the density matrix $\rho = \rho(\mathbf{r},\mathbf{r}', s,s', t)$ is dependent on two coordinate variables and two spin variables. After the transformation to new spatial variables, $R= ½ (r+r')$ and $\Delta R= ½ (r++r')$, we get the above equation.

The effect of spin-orbit interaction is introduced by the last two terms. The square and curly brackets $[\sigma,\ldots]$ and $\{\sigma,\ldots\}$ are commutator, and anti-commutator with the Pauli spin matrixes, respectively

Spin Magnetization Quantum Distribution Function (SMQDF)

The spin magnetization quantum distribution function (SMQDF) should have a spinor correlation function which separates into charge and spin magnetization distributions as follows (Saikin, Pershin & Privman, 2005):

$$W = ½ (W_n I + W_{\sigma\alpha}.\sigma_\alpha) \tag{19a}$$

where the velocity

$$v_j = ½ (v_n{}^j I + v^j{}_{\sigma\alpha}.\sigma_\alpha) \tag{19b}$$

The spin Wigner distribution function can be obtained by the following Wiley transformation of density matrix to the Wigner function:

$$W_{ss'}(R, k, t) = \int \rho(R, \Delta r, s, s', t).\exp(-jk.\Delta r).d^2\Delta r. \tag{20}$$

SMQDF Transport Equation

Assuming that the potential, $V(\mathbf{r})$, varies slowly and smoothly with the position $\mathbf{r}$, we obtain the transport equation for a single electron with spin

$$\frac{\partial W}{\partial t} + \frac{1}{2}\left\{v_j, \frac{\partial W}{\partial x_j}\right\} - \frac{1}{\hbar}\frac{\partial V}{\partial x_j}\frac{\partial W}{\partial k_j} + jk_j\left[v_j, \frac{\partial W}{\partial x_j}\right] = \left[\frac{\partial W}{\partial t}\right]_{col} \tag{21}$$

The initial terms in the left-hand side of the above equation consist of the Boltzmann transport equation (BTE) with spin-orbit coupling in a magnetic field, in addition to the spin-dependent scattering effects.

At the right hand side of the above equation, we see the scattering term, $(\partial W\partial/t)_{col}$, which is responsible for interactions of an electron with phonons and non-magnetic impurities.

$$(\partial W\partial/t)_{col} = \int S(k,k').[W(R,k',t) - W(R,k,t)]d^2k' \tag{22}$$

where $S(\mathbf{k},\mathbf{k}')$ is the scattering rate for electrons without spin.

5.2 Spin Hydrodynamic Model (SHDM)

In order to get the macroscopic spin transport equations we take the moments of the SMQDF transport equation. Then we get relations for the particle density, spin density, particle current density, and spin current density (Saikin, 2004).

$$n_n = \int W_n d^2k \tag{23a}$$

$$n_{\sigma_\alpha} = \int W\sigma_\alpha d^2k \text{ (23 b)}$$

$$J_n^j = \int (v_n^j W_n + v_{\sigma_\alpha}^j W_{\sigma_\alpha}) d^2k \tag{24a}$$

$$J_{\sigma_\alpha}^j = \int (v_n^j W_{\sigma_\alpha} + v_{\sigma_\alpha}^j W_n) d^2k \tag{24b}$$

The particle and spin conservation equations are obtained by multiplying by the powers of $\boldsymbol{V}$ or $\boldsymbol{k}$ and integrating over the electron wave vector

$$\frac{\partial n_n}{\partial t} + \frac{\partial J_n^j}{\partial x_j} = 0 \tag{25a}$$

$$\frac{\partial n_\sigma}{\partial t} + \frac{\partial J_\sigma^j}{\partial x_j} - \frac{2m^*}{\hbar}(v_\sigma^j x J_\sigma^j) = 0 \tag{25b}$$

With the following constitutive equations

$$J_n^j = -\left[(k_B T)\frac{\partial n_n}{\partial x_j} + n_n \frac{\partial V}{\partial x_j}\right] \cdot \left(\frac{\tau}{m^*}\right) \tag{26a}$$

$$J_\sigma^j = -\left[(k_B T)\frac{\partial n_\sigma}{\partial x_j} + n_\sigma \frac{\partial V}{\partial x_j} - \left(\frac{2m^* k_B T}{\hbar}\right)(v_\sigma^j x n_\sigma)\right] \cdot \left(\frac{\tau}{m^*}\right) \tag{26b}$$

Note that the collision term is expressed using the relaxation time approximation. After solving these equations, we obtain the particle current density and spin current density

6. SEMICLASSICAL SPIN TRANSPORT MODELS

The semiclassical approaches for modeling the spin transport have been detailed in the literature, e.g., Pershin, Saiken and Privman (2008), Ben Abdalla & EL-Hajj (2009), EL-Hajj (2014), as well as Ansgar Jungel, Shpartko & Zamponi (2016). These semiclassical approaches are almost all based on the so-called spinor-BTE or the Bloch equations for spin transport and sometimes the Landau-Lifshitz equation (Landau & Lifshitz, 1935).

6.1 Spinor-BTE

The spinor-BTE is semiclassical equation (like the BTE) with special collision operators, to model the spin-polarized, electron transport in ferromagnetic structures. The collision operator is usually expressed as a 2x2 matrix that accounts for spin-up and spin-down scattering rates (Possanner & Negulescu, 2011).

6.2 Bloch Equations

The classical Bloch equations for spin transport are the analogue of the classical BTE for charge-carrier transport. They were originally invented (by Felix Bloch in 1946) as a spin and magnetization transport model, in the context of nuclear magnetic resonance (NMR). The Bloch equations are actually a set of phenomenological equations that are used to calculate the nuclear magnetization ***M****(t)* in the presence of an applied magnetic field ***B****(t)*. Sometimes, they are called the equations of motion of nuclear magnetization.

The classical Bloch equations for spin systems are usually expressed in the diffusive regime as follows (Buot, 2009):

$$\frac{\partial M_x}{\partial t} = -\left[\frac{M_x}{T_2} - D \cdot \nabla^2 M_x\right] + \gamma \cdot \left(M \times B\right)_x \tag{27a}$$

$$\frac{\partial M_y}{\partial t} = -\left[\frac{M_y}{T_2} - D \cdot \nabla^2 M_y\right] + \gamma \cdot \left(M \times B\right)_y \tag{27b}$$

$$\frac{\partial M_z}{\partial t} = -\left[\frac{M_z - M_o}{T_1} - D \cdot \nabla^2 M_z\right] + \gamma \cdot \left(M \times B\right)_z \tag{27c}$$

where $\boldsymbol{M}(t) = (M_x, M_y, M_z)$ is the nuclear magnetization, γ is the gyromagnetic ratio and $\boldsymbol{B}(t) = (B_x, B_y, B_{0z} + \Delta B_z)$ is the magnetic field experienced by the nuclei. Also, T_1 and T_2 are relaxation times. Actually, T_1 is called spin longitudinal relaxation time and T_2 is called spin decoherence (or dephasing) transverse relaxation time.

The T_1 relaxation occurs when a spin exchanges energy with lattice (spin-lattice interaction). Quantitatively, the decay of M_z is given by the Bloch equation (25c). Integrating this equation with $M_z(0) = -M_0$ gives:

$$M_z(t) = M_0\,[1 - 2\,exp\,(-\,t/T1)\,] \tag{27d}$$

The $\boldsymbol{T_2}$ relaxation is a spin-spin relaxation mechanism. It is the process by which the transverse components of magnetization (M_{xy}) decay or dephase. As originally described by Felix Bloch, T_2 relaxation follows a simple exponential decay with time constant T_2. Thus T_2 is the time required for the transverse magnetization to fall to about 37% ($1/e$) of its initial value. Figure 18 illustrates the spin relaxation times on the Bloch sphere[5]

Note that in solids $T_2 << T_1$. The z component of the magnetic field $\boldsymbol{B}$ is sometimes composed of two terms:

- A constant in time term, B_{0z},
- A time dependent, term $\Delta B_z(t)$.

Figure 18. Spin relaxation times on the Bloch sphere

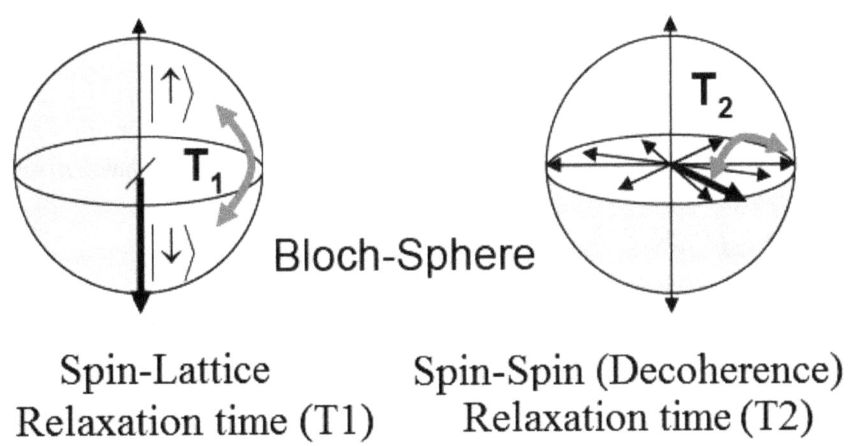

The later term is present in magnetic resonance imaging (MRI) and helps in spatial decoding of the nuclear magnetic resonance (NMR) signals. Also, $\boldsymbol{M}\times\boldsymbol{B}$ is the cross product of the two vectors. M_0 is the steady state nuclear magnetization (when $t \rightarrow \infty$); and it is usually in the z direction.

6.3 Spin Drift-Diffusion Model (SDDM)

The drift-diffusion approximation is the simplest and straightforward method for spin transport modeling. Like the drift-diffusion model (DDM), the spin transport is composed of the carrier continuity equations, but with two-components to account for spin-up and spin-down[6]. According to Shafir, Shen & Saikin (2004), the model can be formulated as follows:

$$e\,\partial n_{\uparrow}/\partial t = \nabla . J_{n\uparrow} + \tfrac{1}{2}\, e\,(n_{\uparrow} - n_{\downarrow})/\tau_{sp} + eS_{\uparrow} \tag{28a}$$

$$e\,\partial n_{\downarrow}/\partial t = \nabla . Jn_{\downarrow} + \tfrac{1}{2}\, e\,(n_{\downarrow} - n_{\uparrow})/\tau_{sp} + eS_{\downarrow} \tag{28b}$$

with the following current constituent equations:

$$J_n = J_{n\downarrow} + J_{n\uparrow} \tag{28c}$$

$$J_{n\uparrow} = e\, n_{\uparrow}\, \mu_n \zeta + e\, D_n\, \nabla n_{\uparrow} \tag{28d}$$

$$J_{n\downarrow} = e\, n_{\downarrow}\, \mu_n \zeta + e\, D_n\, \nabla n_{\downarrow} \tag{28e}$$

Here, $n_{\uparrow}$ and $n_{\downarrow}$ are the spin-up and spin-down electron densities, $J_{n\uparrow}$ and $J_{n\downarrow}$ are the corresponding electron current densities and τ_{sp} is the spin-relaxation time. Also $S_{\uparrow}$ and $S_{\downarrow}$ are the spin-up and spin-down polarization sources (spin net generation rates) and other terms have their conventional meaning. We assumed here that the mobility and diffusion coefficient as well as the spin-relaxation time of spin-up and spin-down electrons. The above equations should be coupled with the Poisson equation:

$$\varepsilon \nabla . \zeta = e\,(Dop - n) \tag{29}$$

where $n = n_{\uparrow} + n_{\downarrow}$ is the total electron density and Dop is the net doping concentration. This model is valid when the spin-dephasing time is shorter than studied times. An alternative approach for modeling the spin transport is to add the following polarization equation to the conventional drift-diffusion model.

$$\partial P'/\partial t = D\nabla^2 P' + (eD/k_B T)\nabla .(\zeta . P') + (S_{\uparrow} - S_{\downarrow}) - P'/\tau_{sp} \tag{30}$$

where the spin polarization density $P' = n_{\uparrow} - n_{\downarrow}$ can be added or subtracted from the total electron density, $n = n_{\uparrow} + n_{\downarrow}$, to obtain the density of spin-up and spin-down electrons.

In the literature, there have been extensive studies on the spin lifetime and spin diffusion length in both n- and p-type semiconductors under a wide range of doping concentrations, temperatures, with different ferromagnets and barriers. For instance, the spin relaxation time, τ_{sp} is temperature dependent ($\tau_{sp} \propto T^{1/2}$), as shown in Figure 19. Elliott has shown that the electron spin relaxation time τsf in semiconductors is related to the momentum relaxation time τ, as indicated in equation (5), for Elliott–Yafet spin relaxation mechanism.where E_n is the electron energy and Δ_{so} is the spin-orbit energy gap. Thus, τ_{sp} is important for small-gap semiconductor with large spin-orbit splitting (e.g. *InSb*). The spin relaxation time also depends on the magnetic field, according to the following relation:

$$\tau_{sp}(B) = \tau_{sp}(0).[1 + (\omega_c \tau_{mn})^2] \quad (31)$$

where $\omega_c = (e\ B/m_n^*)$ is the cyclotron resonance frequency.

6.4 Spin-Monte Carlo Approach

The spin transport can be incorporated into Monte Carlo simulation of charge carriers as spin polarization vector or spin density matrix for each simulated carrier (Saikin et al., 2003). This is illustrated in the figure 18.

In the latter case, the spin density matrix of the ith particle evolves coherently, during the free flight, as follows:

$$\rho_i(t+\delta t) = exp\ [-jH_s/\hbar]\ .\ \rho_i\ (t).\ exp\ [+jH_s/\hbar] \quad (32)$$

Figure 19. Spin relaxation time in Si as a function of temperature
After Fabian et al.

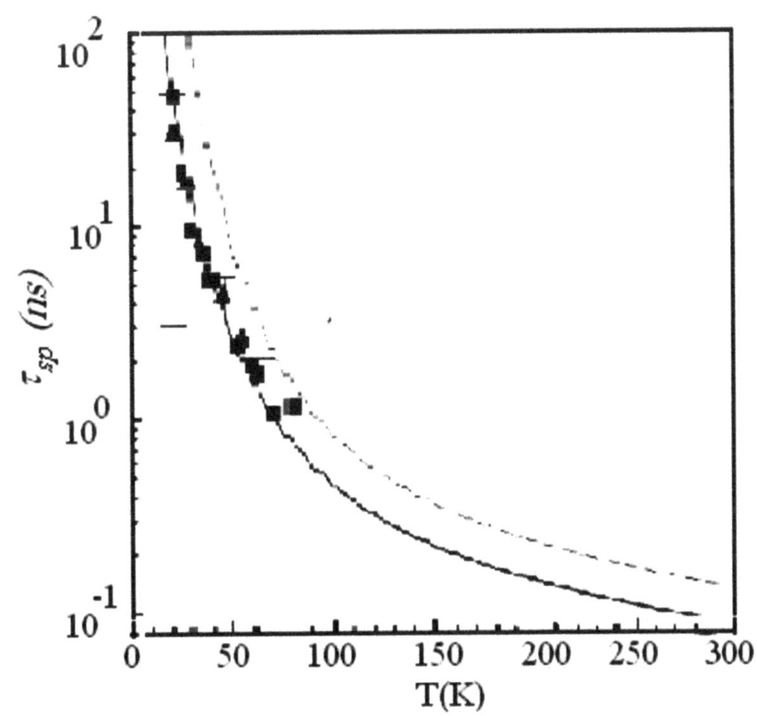

where $\delta_{ts} = -ln\ r/\Gamma$, and r is a random number between 0 and 1, while Γ is the total scattering rate including the self-scattering rate. Also, H_S is the (spin-dependent) Hamiltonian, which is assumed constant for short time steps, and it changes, instantaneously, during spin scattering events. Therefore, the spin dynamics of a single electron spin is coherent during the free flight. However, stochastic momentum fluctuations due to electron scattering events produce the distribution of spin states, thus causing effective dephasing at times $t > 0$.

The Monte Carlo simulation approach has been successfully applied for investigation of the spin polarized transport properties in 1D and 2D semiconductor structures in the presence of electric field. The flowchart of the Monte Carlo simulation including spin transport is illustrated in figure 19. Also, figure 20 depicts the spin relaxation process, upon the application of spin injection, with applied electric field of 1kV/cm at 77K.

7. SPINTRONIC DEVICES

Traditional electronic devices use the electronic charges for amplification and switching. An alternative way for information coding is to use the electron spin. The category of electronic devices which make

Figure 20. (a) Flowchart of Monte Carlo simulation including spin transport; (b) Monte Carlo simulation of spin polarization injection and relaxation via a Schottky contact
After Shen et al (2004).

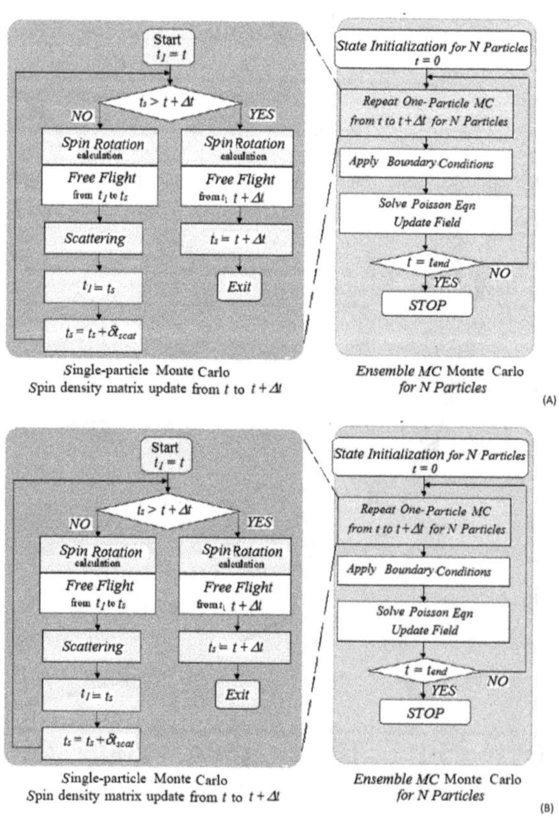

use of spin properties is called spintronic devices. Spintronics devices usually incorporate magnetic materials, as they possess more electrons with one spin direction than the other. The spintronic devices allow the control of functionality by spin-orbit and magnetic interaction. Different designs of transistors and spin-filtering devices utilizing control for the spin polarization in semiconductor structures have been proposed.

The first generation of spintronics, utilized giant magnetoresistance (GMR) and tunneling magnetoresistance (TMR) in ferromagnetic-metal structures such as GMR heads with high magnetic storage density. The 2nd generation of spintronic devices integrates magnetic materials with semiconductors to extend the benefits of spin to the wide electronic industry. One of the major challenges in developing the 2nd generation of spintronic devices was the synthesis of high quality spintronic materials with Curie temperatures above room temperature and similar electrical conductivity like semiconductors. In comparison with magneto-electronic devices utilizing GMR and TMR, semiconductor spintronic devices promise to be more universal in application due to the ability to adjust a potential variation and spin polarization in an active region of spin devices by external voltages and doping profiles. We present here a quick review of some demonstrated and actual spintronic devices.

7.1 Giant Magnetoresistance (GMR)

The giant magnetoresistance (GMR) is a large change in resistance that is observed in ferromagnetic/paramagnetic multilayer structures. GMR effect was first discovered in 1988 by the groups of Peter Grunberg and Albert Fert[7]. In the simplest case, GMR utilizes thin non-magnetic metal (typically *Cu*), sandwiched between two ferromagnetic metal layers (typically *Co* or *NiFe*). However, the ferromagnetic layers may be replaced with the so-called ferromagnetic semiconductors, such as *Mn*-doped *GaAs*. Initially, the GMR phenomenon has been interpreted in terms of spin-dependent scattering. A more accurate explanation is based on the density of states in ferromagnets, in which the density of states is different for spin-up and spin-down electrons (Parkin, 2004).

The GMR phenomenon is exploited in read heads of some hard disk drives, as shown in Figure 21. It should be noted that when a current is injected perpendicular to the planes of a ferromagnet metal multilayer system (spin valve), it produce dynamical magnetic states in the thin magnetic layers. Depending on the magnitude of the injected current, precession or complete reversal of the magnetization is observed. The underlying effect is called spin-transfer-torque (STT).

Figure 21. Illustration of the GMR effect and its use in read heads of hard disk drives

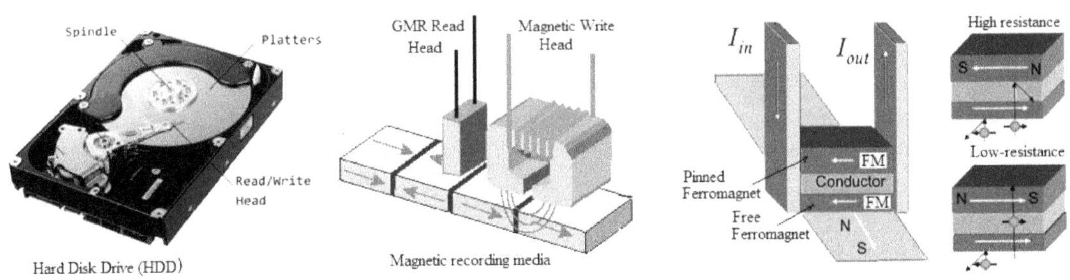

7.2 Magnetic Tunnel Junctions (MTJ)

The magnetic tunnel junction (MTJ) is based on the so-called tunneling magnetoresistance (TMR) effect. This is a quantum mechanical tunneling of spin-polarized electrons through a very thin dielectric (Binasch, 1989). The MTJ consists of a thin insulating layer (tunnel barrier) sandwiched between two ferromagnetic electrodes (FM1, FM2) and exhibits tunnel magneto-resistance (TMR). The MTJ is an important building block in many spintronic devices and usually exploited to study the fundamental physics of spin dependent tunneling. Since the 1995 discovery of room-temperature TMR, the MTJs with an amorphous Al_2O_3 tunnel barrier have been studied extensively. Such MTJs exhibit magnetoresistance (MR) ratios up to about 70% at room temperature. The figure 22 depicts TMR effect in an MTJ.

Note that. $D1\uparrow$ and $D1\downarrow$ are the density of states at the Fermi level (E_F) for the majority-spin and minority-spin bands in the first electrode (FM1), and $D2\uparrow$ and $D2\downarrow$ are the density of states at E_F for the majority-spin and minority-spin bands in the second electrode (FM2). Therefore, the polarizations in the two electrodes are $P_1 = (D1\uparrow - D1\downarrow)/(D1\uparrow + D1\downarrow)$ and $P_2 = (D2\uparrow - D2\downarrow)/(D2\uparrow + D2\downarrow)$. The TMR ratio is then defined as:

$$TMR = (R_P - R_{AP})/R_P \approx 2P_1P_2/(1 - P_1P_2) \tag{33}$$

where R_P and R_{AP} denote the parallel and anti-parallel polarization resistances, respectively. Recently, ferromagnetic *CoFe* alloys which are made amorphous by the addition of boron have gained special interest because MTJs incorporating them show high TMR values at room temperature. However, the thickness of layers in an MTJ are on the order of 1–2nm, which requires very precise characterization tools for their proper implementation. Nowadays, MTJ are used in read-heads of hard disk drives as well as magnetic sensors and magneto-resistive RAM (MRAM) devices.

Figure 22. Illustration of the TMR effect: (a) magnetizations in the two electrodes are aligned parallel (P-state); (b) magnetizations are aligned antiparallel (AP-state)

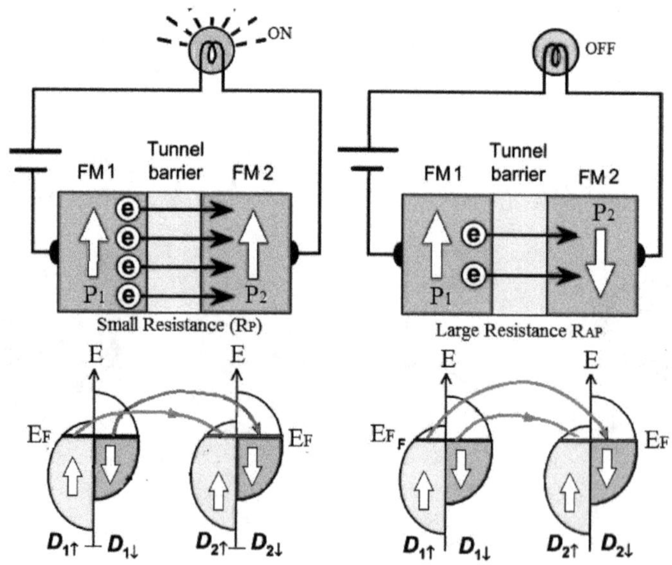

7.3 Magnetic RAM (MRAM)

Magnetic random access memory (MRAM) is a sort of nonvolatile memories, which stores information bits by the magnetization of small ferromagnets. MRAM typically employs magnetic tunnel junctions (MTJs) as storage elements (Akerman, 2005). Figure 23 depicts the structure of a typical MRAM. As shown, the MRAM is physically similar to the DRAM, although it does not require a transistor for the write operation. MRAM cells are designed to have two stable states that correspond to high or low resistance values and retain those values without external power. The current sent through the MRAM cell creates a magnetic fields and a corresponding torque; called the spin transfer torque (STT). If this torque is strong enough it can rotate the magnetization. In fact, the coupling between the magnetic and spin is achieved in the magnetic tunnel junction (MTJ) at the quantum level.The figure 24 shows the hysteresis loop in the magnetoresistance (R) of the MRAM cell, as the magnetic field (H) changes. The MRAM cells are read by sensing the resistance to determine if the state is high or low, while the writing is carried out by the magnetic field generated from the current flow in the bit and word lines. The above described memory is usualy called spin transfer torque MRAM or STT-MRAM. The STT-MRAM devices retain the state of the memory cell when power is turned off, in contrast to the volatile memory in standard DRAM devices. Also, MRAM should consume a negligible amount of power, in contrast to the steady drain of semiconductor DRAM.

7.4 Spin FET

The spin Field Effect Transistor (spin-FET) is a three terminal non-conventional charge spin-based, device which is similar to the semi-conductor MOSFET. Although the device structures are similar, their operating principles are quite different. The main feature of a spin-FET is its current drivability by the

Figure 23. Magnetic RAM (MRAM) cell and array structure (below) and circuit diagram (right)

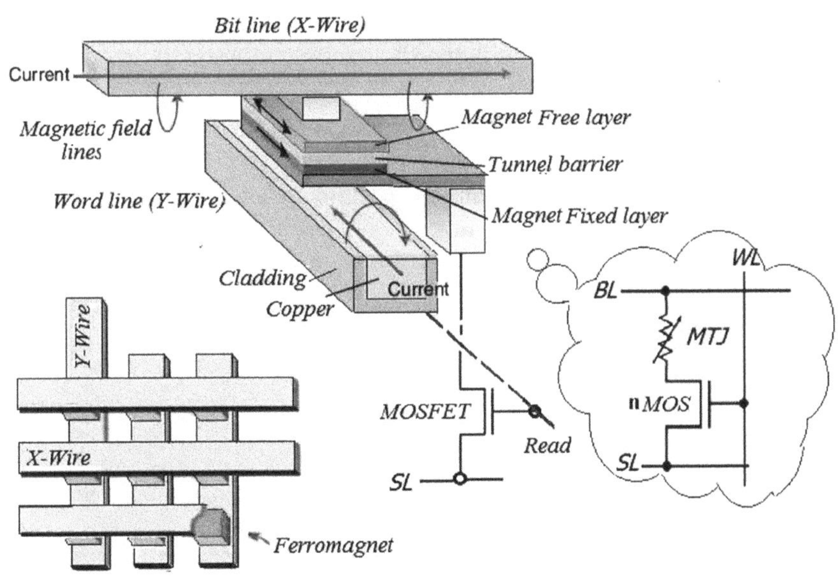

Figure 24. Hysteresis loop of the MRAM cell, with TMR effect

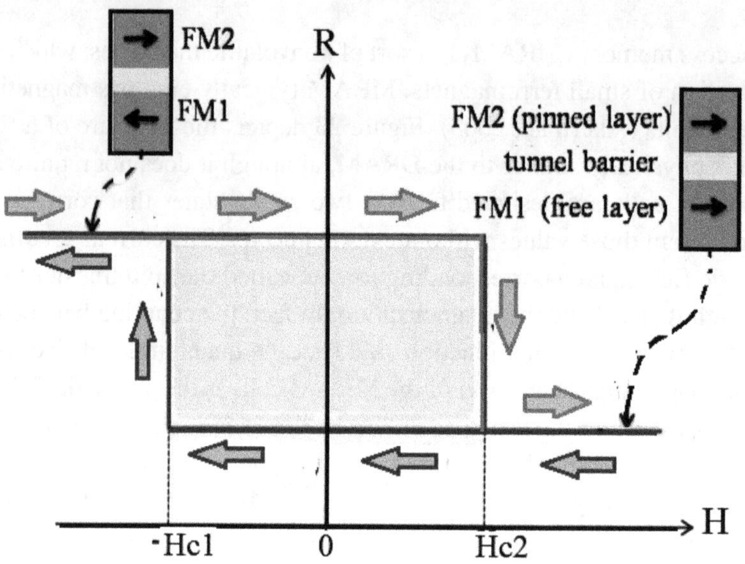

magnetization of the ferromagnetic electrodes. The spin-FET has passed many phases of development since its invention (Datta & Das, 1990).

Figure 25 depicts the basic structure of spin-FET. The source and drain are ferromagnetic and the transport channel is a two-dimensional electron gas (2DEG). When the source and drain magnetizations are parallel and gate voltage V_G=0, the channel resistance is low and the transistor is conducting (ON state). Applying a gate voltage V_G, rotates the spin via Rashba and Dresselhaus spin-orbit interactions. At a particular gate voltage, the spins are rotated by 180° resulting in high resistance and the transistor is not conducting (OFF state).

The spin drift-diffusion model (SDDM) can be applied to model the spin relaxation rate in non-ballistic spin-FET. During the operation of this device, the spinning electrons will suffer from the Rashba interactions in the channel, that is formed beneath the gate. Transport of spin density along the channel in this case can be described by the exponential decay

$$P(x) = P_o.exp\,[- x / \lambda_s\,] \tag{34a}$$

where the spin coherence length in the channel can be approximated by the following relation (Shafer,Shen & Saikin, 2004):

$$\lambda_s^{-1} = \left[\frac{\mu\zeta}{2D} + \sqrt{\left(\frac{\mu\zeta}{2D}\right)^2 + \left(\frac{2m^*(\alpha-\beta)}{\hbar^2}\right)^2}\right] \tag{34b}$$

Figure 25. Schematic illustration of the spin-FET operation

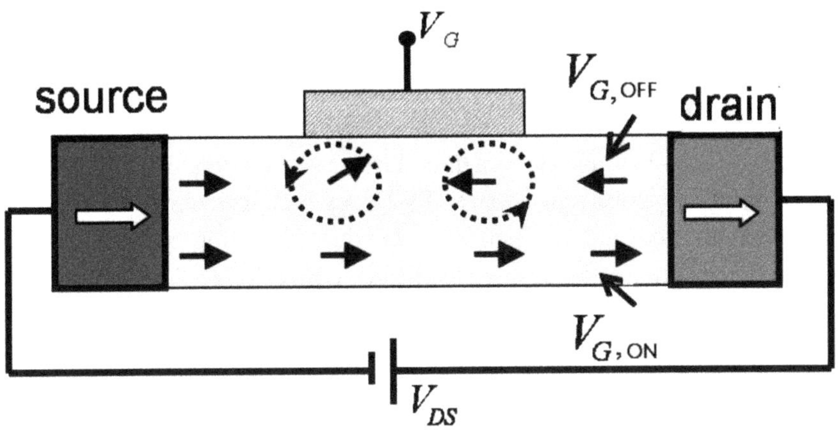

where ζ is the applied electric field, D is the diffusion constant of electrons and m^* is their effective mass. Also, α and β are the coupling coefficients of the spin-orbit interactions, due to Rashba and Dresselhaus, respectively. If these coefficients are not equal, then electron spin polarization will decrease with distance, along the channel. This can be controlled by the gate voltage, because the spin-orbit coupling coefficients depend on the perpendicular electric field in the channel. At the drain side, which is usually made of ferromagnetic material, the current will change according to the spin polarization value.

In addition to its utilization in electronic systems, the Spin-FET has been utilized as an analysis tool to study spin relaxation mechanisms and other phenomena in low-dimensional semiconductors. For instance, Parmanik, Bandyopadhyay & Cahay (2005) have studied the D'yakonov–Perel spin relaxation mechanism in the channel of a spin-FET and showed that it can be completely eliminated if the channel is a quantum wire and transport is strictly in a single mode.

Figure 26. Structure and operation of the heterojunction spin-FET

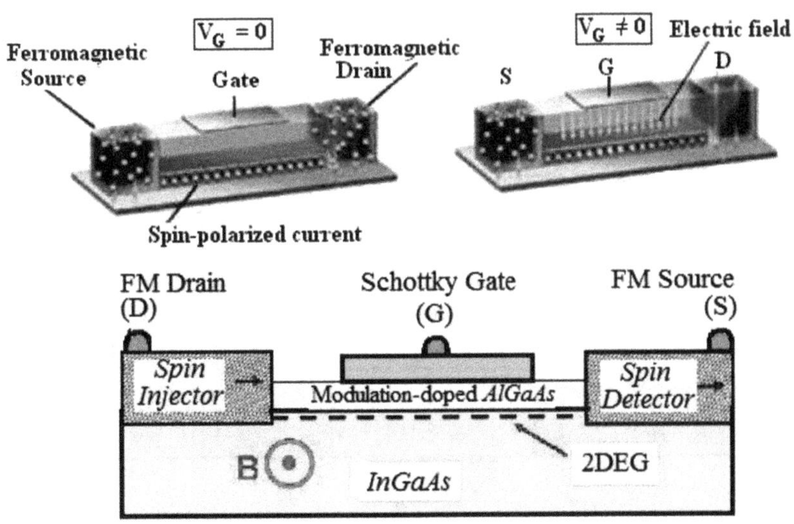

7.5 Spin Bipolar Transistor

Bipolar spintronic devices refer to semiconductor devices, which are based on the spin-polarized transport of both electrons and holes. The general theory of magnetic bipolar devices was developed by Fabian, Zutic, & Das Sarma (2002), Lebedeva & Kuivalainen, (2003) and Flatte et al., (2003) as well as Fabian and Zutec (2005). The spin bipolar transistor is a three-layer, three-terminal, device similar to some extent to the semiconductor bipolar junction transistor (BJT). The emitter and collector layers of the spin bipolar transistor are ferromagnetic films with magnetization *M* whose direction can be externally manipulated. The base layer is a nonmagnetic metal, such as gold (*Au*), copper (*Cu*) or silver (*Ag*). Figure 27 shows a schematic of the structure of the spin bipolar transistor and its energy band diagram in active mode.

As shown in figure, we first consider the active mode of operation, where a forward current flow across the emitter-base (E-B) junction. Part of the emitter current is returned, from the base layer, and another major part is injected to the collector side. The band structure of the ferromagnetic emitter is such that the emitter current is carried primarily by electrons of one spin subband. Thus, the electric current is a spin polarized current. There is an associated current of magnetization $I_M = \mu_B \cdot (I_e/e)$. You can think of the emitter as a reservoir of electrons that carry charge and spin. According to the law of detailed balance, the emitter can exchange particles with the nonmagnetic base layer only if those electrons are of specific spin (e.g., spin-up). Thus, the electric current carried into the base is also spin polarized. Also, current conservation requires that electrons be drawn out from the base at the same rate that other electrons are added at the E-B junction. Near this junction, there is a surplus of spin-up electrons. Therefore, the conduction electrons move diffusively, to the collector side, where they attracted by the base-collector bias. The ratio of injected spin density in the collector is much smaller than the density of the source spin in the emitter side, due to collisions and recombination in the base. However, the change in spin polarization at the collector may be significant, according to the following relation:

$$\Delta Pc = P_{ob} \cdot \gamma_{be} \cdot \gamma_{bc} \cdot (n_{oB}/N_C) \,.\, exp[V_{be} / V_T] \tag{35a}$$

Figure 27. Structure of the spin-bipolar transistor and its band diagram

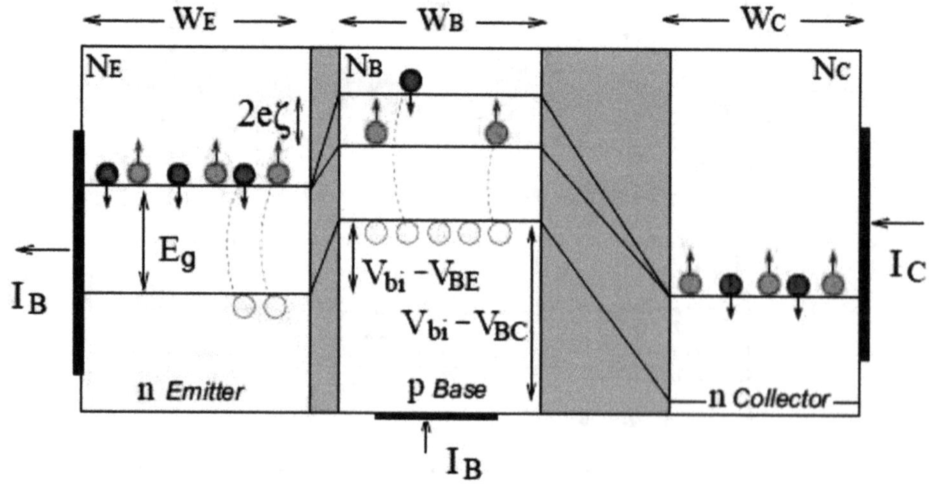

where $V_T = k_B T/e$ is the thermal voltage, P_{ob} is the equilibrium polarization at the base and γ_{be} and γ_{bc} are constants expressing the diffusion constants of the excess spin within the emitter (γ_{be}~W_E/λ_{se}) and at the edge of collector (γ_{bc}~λ_{sc}/W_B). Also n_{oB} is the equilibrium electron density in the base:

$$n_{oB} = (n_i^2/N_B)/\sqrt{(1-P_{ob}^2)}, \tag{35b}$$

Now, it is possible to prove that the magnetoamplification is proportional to $\sqrt{(1-P_{oe})}/(1-P_{ob}^2)$, and the current gain is given by (Lebedeva and Kuivalainen, 2003):

$$\beta = I_C/I_B = \beta_o.\sqrt{(1-P_{ob})}/(1-P_{oe}^2) \tag{35c}$$

where $\beta_o \approx (N_E/N_B).(\lambda_{sc}/\lambda_{se})$. The transistor current gain can be then controlled by the equilibrium magnetizations in the emitter or the collector.

It should be noted that there exist a variety of spin bipolar transistors. For instance, the Figure 28 depicts the structure of the magnetic tunnel transistor, which uses spin filtering in a spin-valve base layer to produce changes in collector current depending on the *NiFe* and *CoFe* base moments. Note the presence of an Al_2O_3 layer, which provides a potential barrier between the metal ferromagnet and the semiconductor. This layer solves the impedance mismatch problems, which prevent spin injection from a metal to a semiconductor.

The so-called magnetic bipolar transistor or MBT is entirely made of semiconductors, while the active (base) region is a magnetic semiconductor (Fabian, Zutic, & Das Sarma, 2004). In such spin bipolar transistors, the source of a nonequilibrium spin may be from electrical spin injection or optical spin orientation, and the magnetic control of magneto-amplification is typically achieved through the equilibrium spin. However, new effects may appear from the interaction of the equilibrium and non-equilibrium spins, when they are in an electrical contact. This interaction, gives rise to the spin-voltaic

Figure 28. Band structure of the magnetic tunnel transistor
After Parkin (2004).

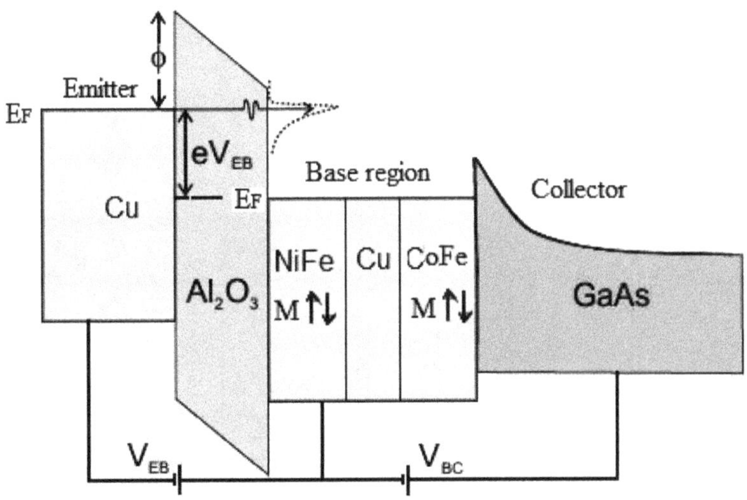

effect and the *giant-magneto-amplification* (GMA), which is controlled by the relative orientation of equilibrium to nonequilibrium spins (Fabian & Zutic, 2004).

7.6 Spin Photonic Devices

The prospects for new devices, based on DMS materials in which the spin and other ferromagnetic properties can be controlled by light have motivated research in the so-called spin-photonic devices. For instance, the spin injection and extraction can be exploited in such spin photonic devices. The Figure 29 depicts the structure of spin solar cell and spin photodiode. The spin solar cell generates electric current, when exposed to light radiation, by the spin tunneling effect, from an n-type semiconductor to a semiconductor ferromagnet. In the case of photodiode, the photocurrent is driven by spin injection from a ferromagnet. The Figure 30 depicts the structure and operation of the spin LED. Note that recombination of singlet[8] excitons (e-h pairs) results in radiative emission of photons (light).

In addition to the above mentioned spin-photonic devices, the magneto-optic coupling effects have helped a lot in studying DMS materials and their optical properties. For example, the absorption in *(Ga,Mn)As* occurs in the visible range and the position of its edge on the frequency axis depends on the circular polarization of incident light. Analysis of this magneto-optical effect provided information on the *p-d* exchange-induced band splitting and on doping in DMS materials (Dietl, 2002).

7.7 Plastic Spintronics

Organic (or plastic) electronics uses conducting organic materials in electronics applications. The materials we call plastics are organic materials, which consist mostly of carbon and hydrogen. Most of the studies in spintronic devices have been performed on metallic systems (GMR, metallic spin valves), tunneling insulators (MTJ) and inorganic semiconductors (e.g. *Si* and *GaAs*). Organic semiconductors are relatively new entrants in the domain of spintronics. During the last decade, a great effort has been exerted to understand the mechanism of spin injection and transport in organics. Organic semiconductors are believed to be very interesting candidates for Spintronics applications because of the predicted long spin lifetimes and low implementation cost.

Figure 29. Schematic diagram showing the operation of spin solar cell and spin photodiode

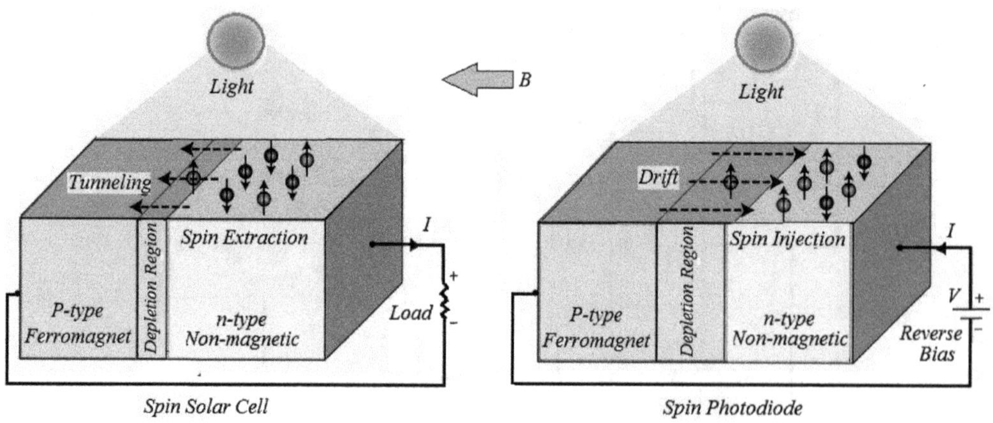

Figure 30. Structure and operation of the spin LED

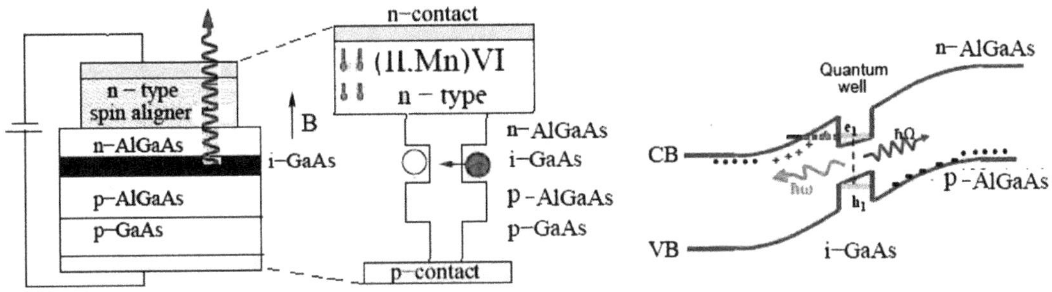

Note 3: History of Plastic Spintronics

In the late 1970's, it was shown by Alan Heeger, Kideki Shirakawa and Alan Macdiarmid that certain plastics or organic materials can conduct electricity. This discovery was rewarded with the Nobel Prize in Chemistry in 2000, and was the start of the huge research field of organic electronics.

A great boost came in 1987 when the first organic light emitting diode (OLED) was demonstrated, showing that organic materials can be used for producing light. Organic solar cells are another promising application. Although their efficiency is still much lower than that of their inorganic counterparts they are much cheaper and allow production in large volumes. As both organic electronics and spintronics offer many advantages and possibilities, the combination of the two fields seems to be a logical step. On the one hand, by using spins in organic electronics, new functionalities can be added to existing organic devices. On the other hand, organic materials can be employed in spintronics applications, benefitting from their low cost, as well as ease of processing. The first report on electrical spin injection and transport in organic semiconductor thin films appeared in 2002 and since then these materials have attracted significant interest

One of the main questions regarding spin transport in organic materials is about the origin of the processes that determine the spin relaxation time and spin diffusion length. Are they caused by the spin–orbit interactions or do other processes dominate. Recent experiments confirmed the dominant role of hyperfine fields in determining the spin-diffusion length in organic materials (Wu et al, 2007). The Figure 31 depicts the spin of a polaron (S) on a pentacene molecule and how it interacts with the hyperfine fields from H_2 nuclei. As shown, the polaron spin precesses around the hyperfine fields from surrounding H_2nuclei and the external magnetic field (B).

According to Geng (2016), the research on organic spintronic phenomena during the last few years has been concentrated on three axes, namely:

- Magnetic field effect in organic light emitting diodes (OLED), where spin mixing between singlet and triplet polarons can be influenced by an external magnetic field leading to organic magneto-resistive effect
- Magnetoresistance in organic spin valve, where spin injection, transport, and detection can be demonstrated, and
- Magnetoelectroluminescence or spin-OLED. Here, the spin polarized electrons and holes are injected into the organic semiconductor layer, leading to the dependence of electroluminescence intensity on relative magnetization of the ferromagnetic electrodes.

Figure 31. Illustration of the spin of a polaron on a pentacene molecule

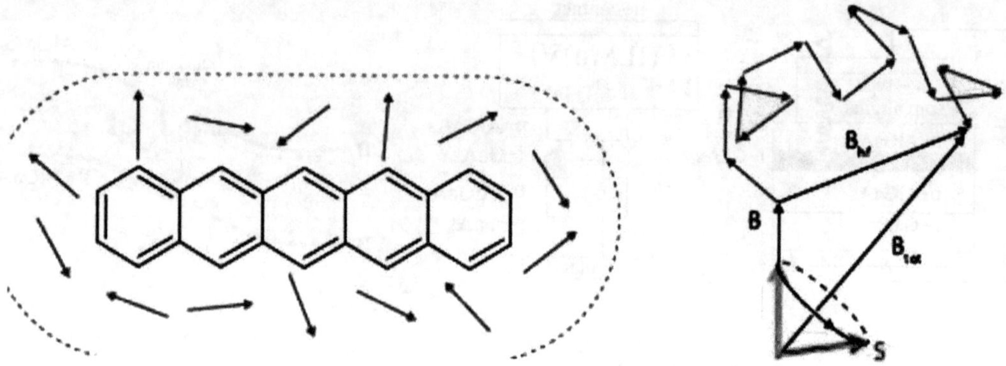

Figure 32 shows the structure and operation of an OLED. The OLED displays have already been launched in the market as displays for TV's and smart phones. An OLED is basically a p-n junction where the p- and n-layers are made of organic semiconductors. Under forward bias, the electrons injected from the n-layer and the holes injected from the p-layer form a cloud of excitons around the junction. We remember that excitons may be singlet (spin) or triplet (spin). The singlet excitons decay rapidly, emitting photons (light) but triplet excitons decay slowly, emitting phonons (heat). Therefore, the population ratio of singlet to triplet excitons determines the quantum efficiency of the OLED. Beside the exciton formation efficiency (η_r), there exist other factors that determine the external quantum efficiency (η_{ext}) of an OLED, namely:

1. Ratio of electron and hole injection, transport, and recombination (υ),
2. Internal electroluminescence quantum yield (φ_p), and
3. light out-coupling efficiency (η_p)

These parameters ae illustrated in Figure 32, as explained in (Adachi & Nakanotani, 2009).

The spin effects can be exploited to boost the internal quantum efficiency of the OLED. This is done, if spin polarized electrons and holes are injected in the OLED from a ferromagnetic electrode; and that is the spin-OLED. The spin polarized electrons and holes are injected into the energy levels of an organic emissive layer, causing electroluminescence (EL). The spin OLED are inexpensive with respect to conventional LEDs, and can be produced on flexible substrates. Figure 33 depicts the structure and operation of a spin OLED.

Note 4: Polaron-Pair (PP) Model

There are several polaron-pair (PP) models which are used to explain the evolution of singlet and triplet excitons in organic materials. For instance, the PP model of Ehrenfreund et al. (2012) is based on the evolution of PP spin subenergy levels, in presence of magnetic field. It is assumed that the PP excitations are immobile, hence PP diffusion is ignored, but the overall rate of PP decay (through exciton recombination and/or dissociation into free polarons that contribute to the device current) is taken into account. The steady state singlet fraction of the PP population is then calculated from the coherent time evolution of PP wavefunctions subject to the above interactions. The spin Hamiltonian includes

Figure 32. Structure and operation of OLED under bias. White dots are holes and dark blue dots are electrons. The arrows represent the light that penetrates the indium tin oxide (ITO)–glass substrate to air. Alq_3 is tris (8-hydroxyquinoline) aluminum; TPD is N,N′-diphenyl-N,N′-bis (3-methylphenyl)-(1,1′-biphenyl)- 4,4′-diamine

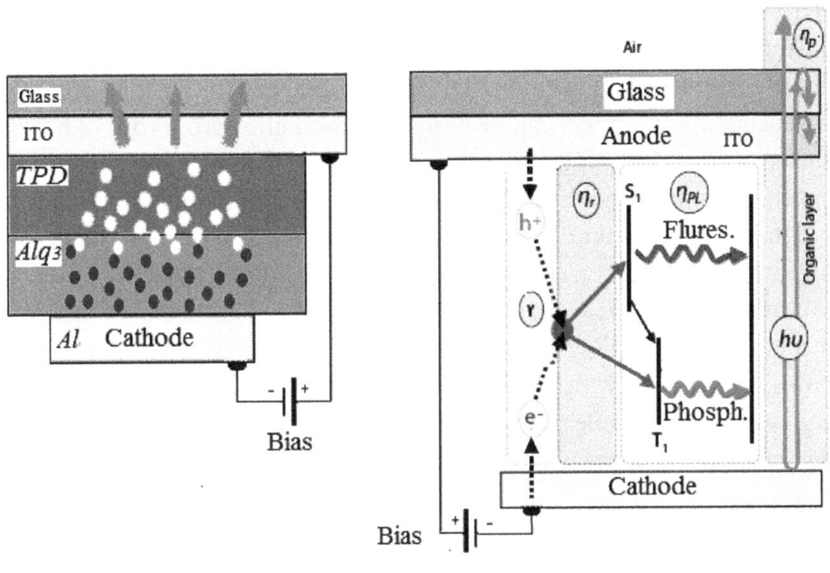

Figure 33. Structure and operation of the spin OLED

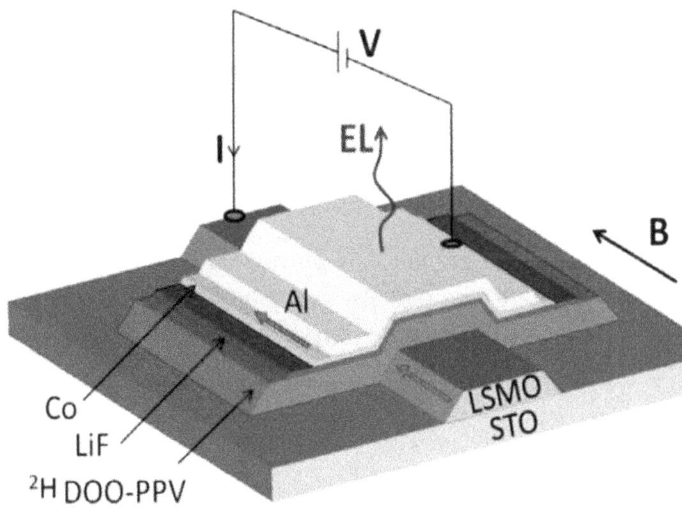

exchange interaction, hyperfine field and Zeeman terms. The calculated magneto electroluminescence (MEL) response is then expressed as a weighted average of the singlet and triplet PP. This may be carried out by Monte Carlo Simulation, as indicated in section 6.4. The relevant time evolution of the singlet-triplet inter-mixing that determines the steady state PP population is obtained in the model via the time dependent density matrix, $\rho(t)$.

7.8 Carbon-Based Spintronics

We know that carbon exists in many allotropes, among which *graphene* (graphite sheets) and carbon nanotubes CNTs (rolled sheets of graphene) are the most famous examples. These forms combine low-dimensionality (e.g., CNT is Q1D) and can be produced with high purity, allowing access to several exotic quantum phenomena.

From the point of view of Spintronics, both CNTs and graphene offer long spin coherence times, meaning that spin-encoded information can be easily retrieved after processing. The current diffusion length in graphene is in the order of 4 microns at 300K. Due to its long spin coherence length at room temperature, graphene is an ideal candidate for nanodevices. The possibility of spin transport at room temperature and to tuning the transport properties of graphene by electrostatic gates makes it an attractive material for spintronic applications (Geim, 2009).

In addition to the unusual electronic and optical properties of graphene, the low-energy electrons have two distinct spin states, namely: valley and pseudospin. This arises from the energy band structure of low-energy electrons in graphene, which is featured by two equivalent valleys (at the two equivalent K points). The emerging domains of Valleytronics and Pseudospintronics will offer unprecedented capabilities to exploit and manipulate spin transport in graphene nanostructures and nanodevices (Ang, 2014).

The electronic transport properties of graphene can be controlled by the application of external electric or magnetic fields, or via the setup of specific graphene topology. Figure 35 is a schematic illustration of a graphene spin FET transistor. The device consists of ferromagnetic metals as source and drain which are connected to the graphene channel. The ferromagnetic source and drain are for spin polarized injection and detection, respectively. However, the spin rotation in the graphene channel is dependent on the electron exchange interaction with the ferromagnetic dielectric (FMD) layer which is placed between the gate electrode and the channel in place of the Bychkov-Rashba effect, which is strongly dependent on spin-orbit coupling. This exchange interaction can be treated as an effective magnetic field. If an external electric field is applied normal to the graphene sheet via the gate, the exchange interaction can be adjusted, which in turn will influence the spin rotation angle thus, the drain current

Figure 34. Schematic of spin transport in graphene, deposited on a magnetic substrate of yttrium iron garnet (YIG), as a thermal spin pumping material. The spin pumping rate (S) versus temperature of YIG and a ferromagnet are shown in the right side. After Barker and Bauer (2016)

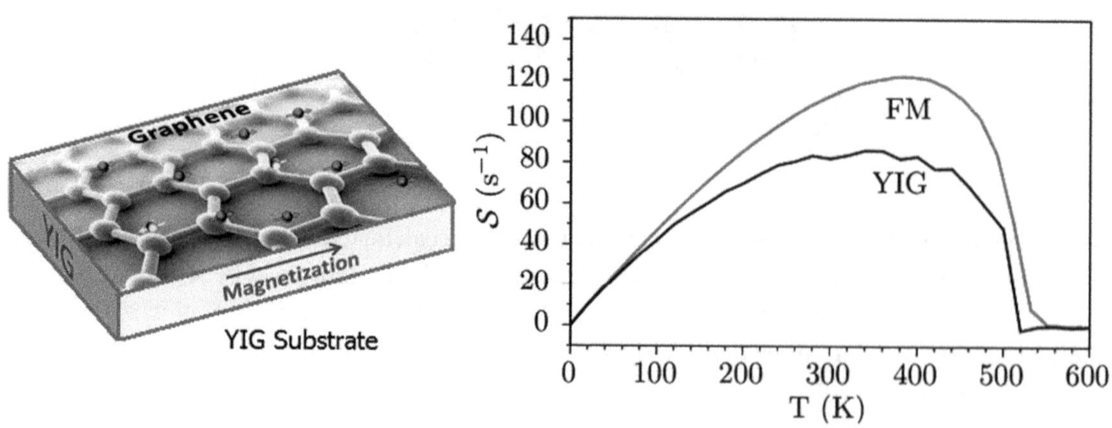

Figure 35. Schematic illustration of graphene spin FET transistor

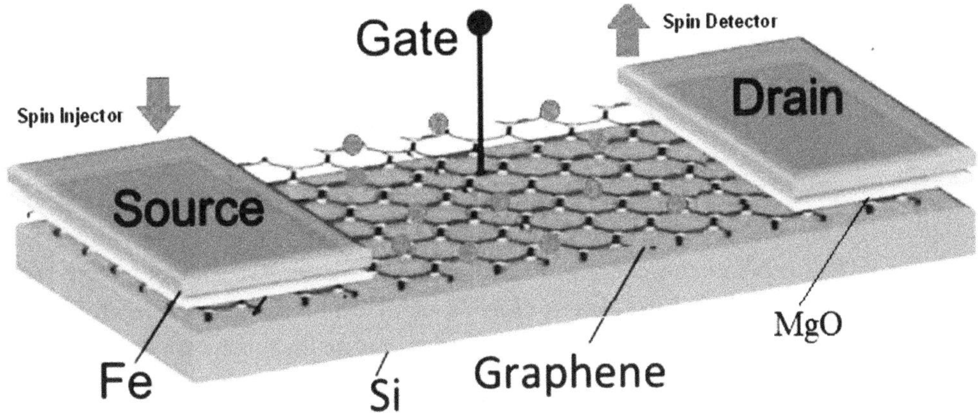

The discovery of magnetism in carbon structures containing zigzag edges has stimulated new directions in the development and design of spintronic devices (Dutta & Pati, 2010). Zigzag edges in sp^2 carbon have demonstrated spin ordered properties in many structures as graphene nanoribbons, open nanotubes and nanorings. It has been demonstrated by simulation, that the spin ordering on the zigzag of such nanostructures, can be changed by external magnetic fields or even static electric field (Yazyev, 2010).

One popular family of structures in the field of carbon spintronics is that of the triangular zigzag graphene nanoflake (TZGNF, or triangulene). Unfortunately, many of the proposed structures suffer from a phenomenon known as topological frustration, which leads to localized non-bonding states and instability. One of the proposed solutions to this problem is mixing between TZGNF with it mirror, as shown in Figure 36 (Bullard et al, 2016).

7.9 Spin Transport in Topological Insulators

Topological insulators are a new class of material that have gained attention in spintronics. These materials (like Bi_2Te_3) are electrically insulating in the bulk but have a conic linear dispersion at the surface. The surface states are excited by massless fermions. The spin of such fermions (or Kramers partners)

Figure 36. Some carbon-based zigzag structures: (a) Kekule structure of a triangular zigzag graphene nanoflake (TZGNF); (b) Bow-tie graphene nanoflake (Clar's goblet); (c) TZGNF with its mirror
After Bullard et al. (2016).

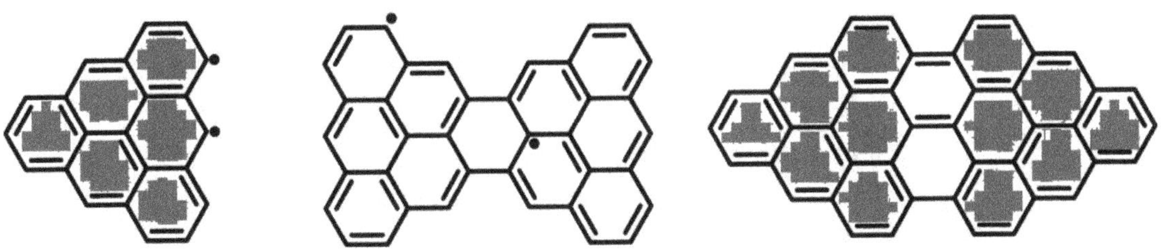

(a) TZGNF *(b) Clar's goblet* *(c) TZGNF with mirror*

is locked at a right angle to their momenta due to spin-orbit coupling, making them protected against non-magnetic impurity scattering and localization effects, and leading to non-dissipative transport. This makes this class of materials appealing for spintronic applications, where spins can be manipulated without the use of ferromagnetic contacts.

Topological insulators can be found in a subclass of narrow-gap semi-conductors and heterostructures where the band gaps are smaller than the relativistic corrections to the band structure. In fact, the heterostructures of some compound semiconductors such as $Bi_{1-x}\,Sb_x$ and $Pb_{1-x}Sn_xTe$ exhibit a strong topological phase. In $Bi_{1-x}Sb_x$, with increasing x the gap at the L point decreases and the L^+ and L^- bands cross at $x = 0{:}04$. After further increase in the mole fraction x, the gap reopens with inverted band. Some 2D topological insulators, such as $HgTe$ quantum wells, exhibit the quantum spin Hall Effect (SQHE). In the interface of the quantum well of such 2D topological insulatorss, the spin changes sign, thereby creating a pair of gapless helical edge states inside the bulk energy gap.

7.10 Spin-Caloritronics

Thermoelectric effects like the Seebeck and Peltier effects have been subject to extensive research for already more than fifty years. Renewed interest in this topic, driven by the progress in nanoscale Spintronics, has led to a new research field called spin Caloritronics. In this field the focus lies on how the spin degree of freedom may alter normal thermoelectric effects. The research in this area focusses on the thermal effects in non local spin valve Caloritronics devices at the nanoscale

8. MEASUREMENT OF SPIN TRANSPORT

Spin injection and transport is usually electrically characterized by nonlocal spin valve and Hanle precession measurements, from which the spin lifetime and spin diffusion length can be extracted. Similarly,

Figure 37. Band inversion of a topological insulator

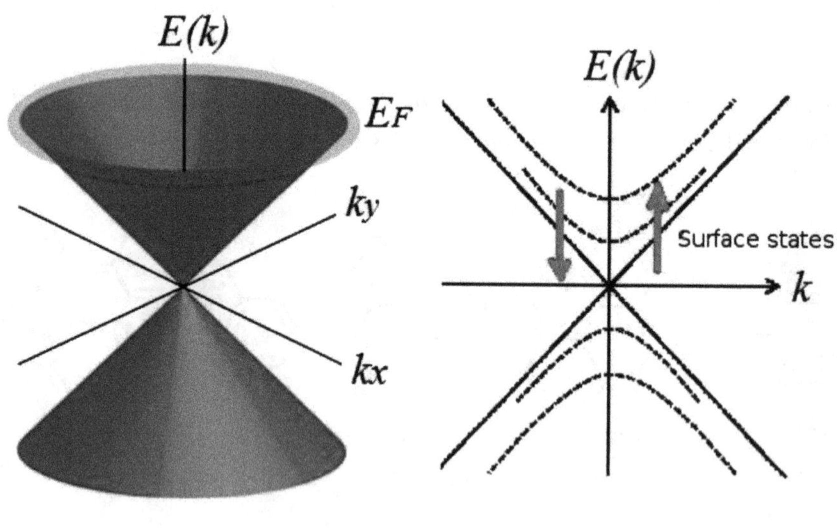

Figure 38. Concept of thermal spin injection

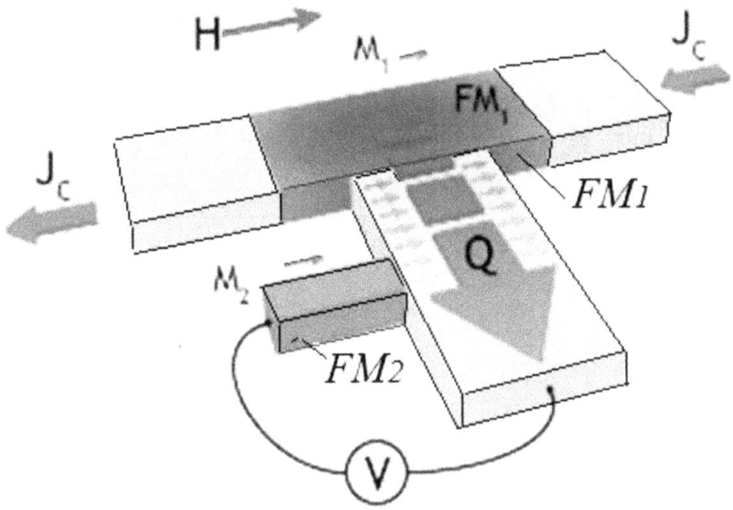

optical detection and visualization of the spin injection has also been reported in several structures, such as the spin-LED, by detecting the emitted polarized light (Crooker et al, 2007).

8.1 Nonlocal Spin Valve Measurement

In the nonlocal spin valve measurement, a spin-polarized current is injected from a ferromagnetic contact (injector) to create a spin accumulation in the semiconductor channel, while another ferromagnetic contact (detector), outside the charge current loop, probes the spin-dependent l potential of one spin (up or down) in relation to the reference contact.

There are two types of measurements which are typically made: spin valve in a magnetic field parallel to the plane of magnetization, and spin precession (Hanle experiment) in a magnetic field perpendicular to the plane of magnetization. Let's first consider the case of nonlocal spin-valve (NLSV) in a magnetic field parallel to the plane of magnetization. When the magnetization is swept from a negative field to a positive field one FM switches its magnetization before the other. Therefore, the direction of the spin injector and the detector switches from the parallel state to the anti-parallel state, and a bipolar nonlocal voltage V_H is sensed, such that $V_P > 0$ and $V_{AP} < 0$. Figure 40 shows an example of typical nonlocal spin valve signals measured from *Si* at 150K. In this measurement, the spin potential (voltage) sense loop is separated from the current loop and the nonlocal resistance is considered as an evidence of successful spin injection.

8.2 Hanle Experiment

In the presence of a uniform magnetic field, the magnetization precesses at constant rate while, at the same time, relaxing at the rate τ_s^{-1}. The magnetic time scale $\tau_B = 2\pi\hbar / g.\mu_B.|B|$ is the time for a complete precession of 2π. The magnetic length scale associated is $\lambda_B = \sqrt{D\tau_B}$. Spin precession is also called Hanle Effect. In a Hanle experiment, the manipulation of spins during transport is carried out with a perpen-

Figure 39. Schematic of the nonlocal spins valve and Hanle measurements

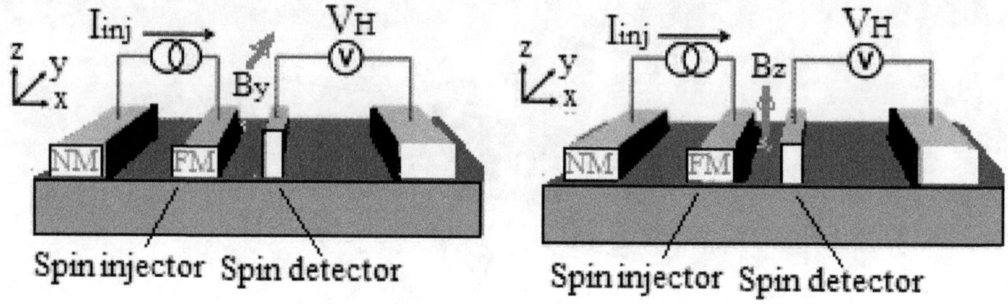

Figure 39

Figure 40. Typical nonlocal spin valve signals measured in n-Sie at 150K
Data from, (Appelbaum, 2011).

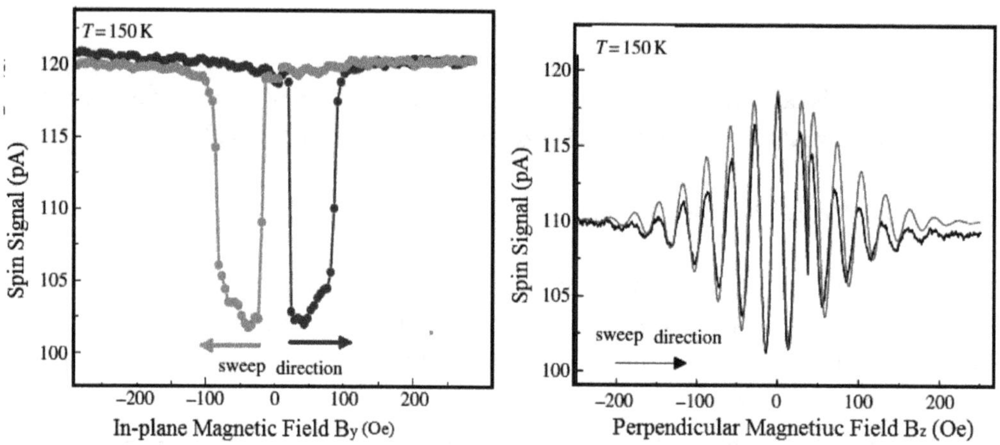

dicularly applied magnetic field. In this case, the polarization is electrically detected at a fixed length *d* while changing the magnitude of the perpendicularly applied magnetic field. In such measurements, the signal is proportional to the polarization at the detecting electrode. The Figure 41 shows the spin polarization *p* as a function of field *B* at different thicknesses *d*. Note that B_0 is the standard deviation of hyperfine field. As shown in figure, with $\theta = 0°$ (open symbols), an λ_s increase with field. For $\theta = 90°$, at small *d, p* slowly decays and no negative values are observed. With increasing *d, p* decays more quickly and a negative dip is visible.

8.3 Photoemission Spectroscopy

The two-photon Photoemission (TPPE) spectroscopy is one of the most popular and versatile methods for studying solid surfaces. In a typical photoemission process electrons are photoexcited from below the Fermi level to the vacuum level. The extracted photoemission spectrum contains energy distribution of the photoelectrons which provides quantitative information about the initial density of states. In

Figure 41. Spin polarization as a function of B at different positions d, with θ = 0○ (open symbols) and θ = 90○ (solid symbols)
Data from (Huang, 2008).

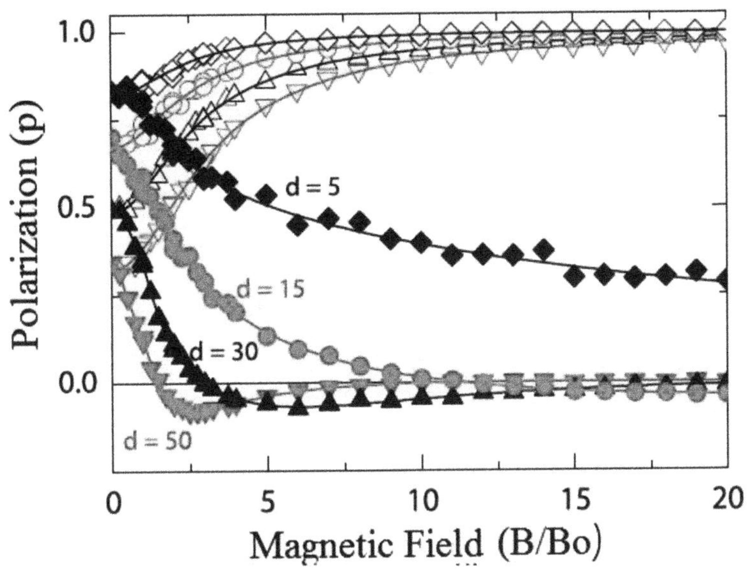

particular, two-photon photoemission (TPPE) has attracted special attention as it allows simultaneous detection of occupied and unoccupied excited electronic states in a single measurement. It is a second order process where the photon energy has to be less than the work function of the material under study.

The two photons (pump and probe photons) may have same or different energies. The pump photon excites an electron from an occupied level below the Fermi level to an unoccupied intermediate level in between Fermi level and vacuum. From this level electron is photo emitted after absorbing the second photon (probe photon). This method is used to measure spin injection efficiency and spin diffusion length. The spin polarized electrons lying above the lowest unoccupied orbital of a certain material are excited by the probe laser which gives rise to photoemission spectra. Information about the spin injection efficiency can be then extracted from these spectra.

8.4 Case Study 14: Spin Measurements in Graphene

We have shown that the interest in graphene spintronics is motivated by its weak spin-orbit interaction, which gives rise to a long spin relaxation time. Also, the fact that 99% of carbon isotopes do not have nuclear spins, leads to excellent spin transport properties. The spin injection and transport in single layer graphene can be investigated using nonlocal magnetoresistance (MR) measurements. Spin injection is performed using either transparent contacts (Co/SLG) or tunneling contacts (Co/MgO/SLG).

Typically, there are two geometries for electrical spin transport measurements (local and nonlocal). The conventional spin transport geometry, known as the local measurement, measures the resistance across two ferromagnetic electrodes. Spin polarized electrons are injected from one electrode, transported across the graphene, and detected by the second

Figure 42. Schematic of 4-terminal graphene spin-valve in non-local measurement. Current I is injected from contact 3 via the MgO barrier into graphene and extracted at contact 4. The voltage is measured between contacts 1 and 2. The non-local resistance is given by: $R=(V^{+}-V^{-})/I$.
After Tombros et al (2007).

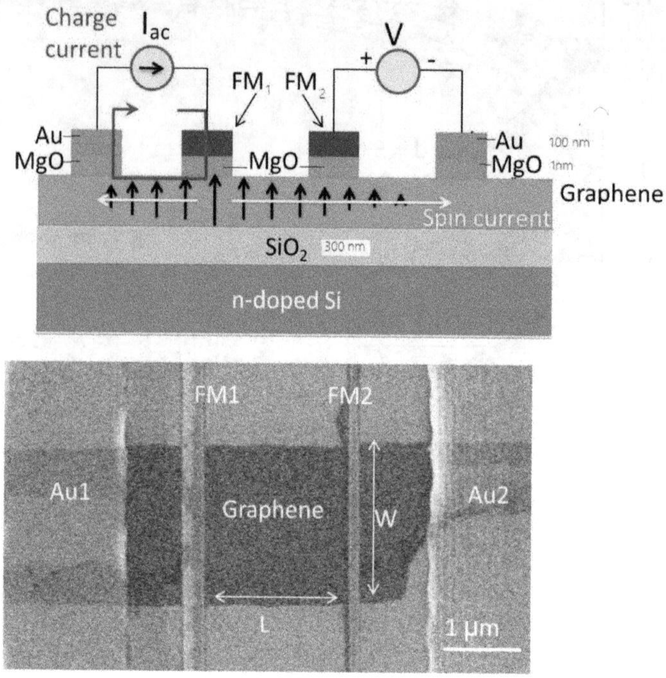

The second geometry is the nonlocal measurement which uses four electrodes is shown in Figure 41. Here, a current source is connected across *Co* electrodes E1 and E2 to inject spins at E2. For spin detection, a voltage is measured across Co electrodes E3 and E4, and the signal is due to the transport of spins from E2 to E3. This measurement is called *nonlocal* because the voltage probes lie outside of the current loop.

9. CHAPTER SUMMARY

The electron spin (spin-up and spin-down) is an intrinsic property of electrons. Conventional electronics devices use the electron charge to carry information. In contrast, the field of Spintronics relies on the electron spin. In spin-based electronics, information is encoded by the spin state of electron bunches. Spintronics, or spin based electronics, attempts to manipulate the spin degree of freedom, instead of or in addition to the charge of electrons to create new functionality. The advantages of spintronic devices include reducing the power consumption and overcoming the velocity limit of electric charge.

Transport in spintronic devices can be characterized by the creation of a non-equilibrium spin polarization in the device (spin injection), measurement of the final spin state (spin detection), external control of spin dynamics by the electric (gate modulation) or magnetic field. Figure 43 shows some building blocks of a spin polarization system comprising both magnetic and non-magnetic materials.

Figure 43.

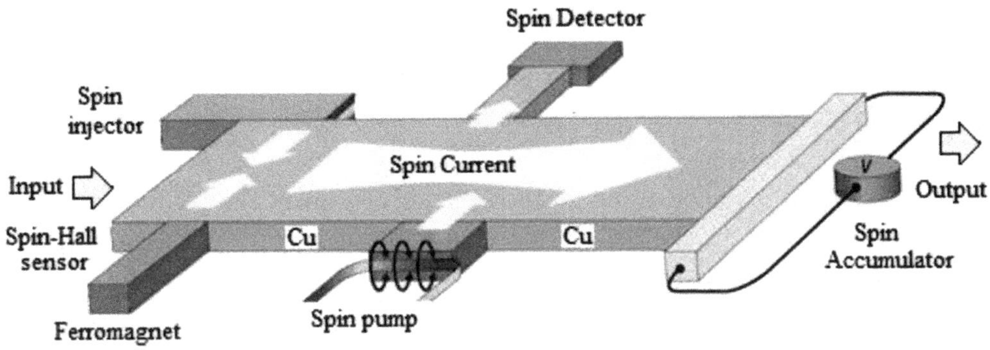

Applications of spintronics make use of one or combination of several phenomena:

- Giant magnetoresistance (GMR) (Hard disk drive),
- Magnetoresistive random-access memory (MRAM),
- Spin FET,
- Spin quantum computing.

The study of electron-spin transport through non-magnetic materials is an extremely active field, because of the rich physics involved and the important applications in the area of magnetic sensors. If the spin diffusion length is larger than or comparable to the distance between the electrodes, the current through such structures depends strongly on the mutual orientation of the magnetization directions of the electrodes, which is called the spin-valve effect, leading to Giant Magnetoresistance (GMR). Traditionally, non-magnetic metals are used as the spacer-layer material between electrodes in these structures. Spintronic devices utilizing spin injection and transport through a semiconducting spacer layer offer additional functionalities, such as spin transistors and the possibility to realize quantum computation logic.

Spin current has been found to play a central role in causing spintronics phenomena to occur not only in traditional magnetic metals but also in various materials including metals, semiconductors and oxides. Pure spin currents carry only spins (spin angular momenta) unlike conventional spin-polarized currents which carry both charges and spins. The so-called magnetic semiconductors are new promising materials, which are currently used to develop such devices. In such materials one of the semiconductor atoms is replaced by a magnetic atom. For instance, the *GaAs*, which is heavily doped with *Mn* acceptors shows a magnetic moment.

Topics in spintronics include magnetization reversal by spin torque, spin relaxation, spin injection in semiconductors and spintronic devices.

The spin relaxation mechanisms may be summarized as follows:

- Elliot-Yafet mechanism,
- D'yakonov-Perel mechanism,
- Bir-Aronov-Pikus mechanism,
- Hyperfine interaction.

In this chapter, we present a full quantum and a semiclassical description of spin transport, which explains how the motion of carriers gives rise to a spin current. The general drift-diffusion approach for the spin-polarized transport is based on the two, spin-up and spin-down, states model, originally developed for ferromagnetic metals. The phenomenological model for non-collinear spin transport including effects of spin-orbit coupling has been described for the regime where spin dynamics is significantly faster than a momentum scattering rate.

The semiclassical models can be useful for investigation of a broad class of transport problems in semiconductors, but they do not include effects of a spin phase memory. The quantum approach of spin density matrix with spin polarization vector of a spin state is more appropriate for this case.

The classical Bloch equations for spin transport are the analogue of the classical BTE for particle transport. They can be extended to time-dependent non-equilibrium quantum transport equations, using a suitable non-equilibrium quantum distribution function, like the Wigner distribution function (WDF).

REFERENCES

Adachi, C., & Nakanotani, H. (2009). *Organic Semiconductor Laser. Materia Material Matters*, *4*(3), 74.

Adelmann, C., Lou, X., Strand, J., Palmstrom, C.J. & Crowell, P.A. (2005). Spin injection and relaxation in ferromagnet-semiconductor heterostructures. *Phys. Rev. B, Condens. Matter Mater. Phys., 71*, 121301R/1–4.

Åkerman, J. (2005). Toward a Universal Memory. *Science*, *308*(5721), 508–510. doi:10.1126/science.1110549 PMID:15845842

Alam, K.M. & Pramanik, S. (2014). *Spin Transport in Organic Semiconductors: A Brief Overview of the First Eight Years*. Academic Press.

Albrecht, J. D., & Smith, D. L. (2003). Spin-polarised electron transport at ferromagnet/semiconductor Schottky contacts, Phys. Rev. B. *Condens. Matter Mater. Phys.*, *68*(3), 035340/1–14. doi:10.1103/PhysRevB.68.035340

Ang, Y. S. (2014). *Nonlinear optical response and transport in Graphene* (PhD. Thesis). Wollongong Iniv. Retrieved from http//ro.uow.edu.au/theses/4117

Appelbaum, I. (2011). Introduction to spin-polarized ballistic hot electron injection and detection in silicon. *Royal Society A.*, *369*, 3554–3574. PMID:21859721

Baibich, M. N., Broto, J. M., Fert, A., Van Dau, F. N., Petroff, F., Etienne, P., & Chazelas, J. et al. (1988). Giant magnetoresistance of (001) Fe/ (001) Cr magnetic superlattices. *Physical Review Letters*, *61*(21), 2472–2475. doi:10.1103/PhysRevLett.61.2472 PMID:10039127

Bakker, F. L., Slachter, A., Adam, J.-P., & van Wees, B. J. (2010). Interplay of Peltier and Seebeck effects in nanoscale nonlocal spin valve. *Physical Review Letters*, *105*(13), 136601. doi:10.1103/PhysRevLett.105.136601 PMID:21230794

Bandyopadhyay, S., & Cahay, M. (2009). Electron spin for classical information processing: A brief survey of. *Nanotechnology*, *20*(41), 412001. doi:10.1088/0957-4484/20/41/412001 PMID:19755729

Ben Abdallah, N., Degond, P., & Genieys, S. (1996). An energy-transport model for semiconductors derived from the Boltzmann equation. *Journal of Statistical Physics*, *84*(1-2), 205–231. doi:10.1007/BF02179583

Ben Abdallah, N., & El Hajj, R. (2009). *On hierarchy of macroscopic models for semiconductor spintronics*. Available at https://www.researchgate.net

Binasch, G., Grünberg, P., Saurenbach, F., & Zinn, W. (1989). Enhanced magneto-resistance in layered magnetic structures with antiferromagnetic interlayer exchange. *Physical Review B: Condensed Matter and Materials Physics*, *39*(7), 4828–4830. doi:10.1103/PhysRevB.39.4828 PMID:9948867

Bloch, F. (1930). Zur Theorie des Ferromagnetismus. *Zeitschrift fur Physik*, *61*(3-4), 206–219. doi:10.1007/BF01339661

Bloch, F. (1946). Nuclear induction. *Phys. Rev.*, *70*, 460-474.

Boross, P., Dóra, B., Kiss, A., & Simon, F. (2012). A unified theory of the Elliott-Yafet and the D'yakonov-Perel spin-relaxation mechanisms. *Scientific Reports*, *3*(11), 1–9.

Bullard, Z., Girao, E. C., Owens, J. R., Shelton, W. A., & Meunier, M. (2016). Improved All-Carbon Spintronic Device Design. *Scientific Reports*, *5*, 7634. doi:10.1038/srep07634 PMID:25579392

Buot, F. A. (2009). *Nonequilibrium Quantum Transport Physics in Nanosystems: Foundation of Computational Nonequilibrium Physics in Nanoscience and Nanotechnology*. World Scientific. doi:10.1142/6042

Buot, F. A., Loberternos, R. A., & Villarin, D. L. (2012). *Nonequilibrium Spin Magnetization Quantum Transport Equations: Spin and Charge Coupling*. arXiv:112.4220v2

Burch, W., Shrekenhamer, D. B., Singley, E. J., Stephens, J., Sheu, B. L., Kawakami, R. K., & Basov, D. N. et al. (2006). Impurity Band Conduction in a High Temperature Ferromagnetic Semiconductor. *Physical Review Letters*, *97*(8), 087208. doi:10.1103/PhysRevLett.97.087208 PMID:17026333

Chen, P., & Zhang, G. (2013). Carbon-based spintronics. *Sci. China Phys. Mech. Astron.*, *56*(1), 207–221. doi:10.1007/s11433-012-4970-8

Cornelissen, L. J., & van Wees, B. J. (2016). Magnetic field dependence of the magnon spin diffusion length in the magnetic insulator yttrium iron garnet. *Physical Review B: Condensed Matter and Materials Physics*, *93*(2), 020403. doi:10.1103/PhysRevB.93.020403

Crooker, S. A., Furis, M., Lou, X., Crowell, P. A., Smith, D. L., Adelmann, C., & Palmstrøm, C. J. (2007). Optical and electrical spin injection and spin transport in hybrid Fe/GaAs devices. *Journal of Applied Physics*, *101*(8), 081716. doi:10.1063/1.2722785

Daniels, C., Bullard, Z., Girao, E. C., & Meunier, V. (2014). Emergent magnetism in irradiated graphene nanostructures. *Carbon*, *78*, 196–203. doi:10.1016/j.carbon.2014.06.072

Das Sarma, S. (2001). Spintronics. *American Scientist*, *89*(6), 516–523. doi:10.1511/2001.6.516

Das Sarma, S., Fabian, J., Hu, X., & Zutic, I. (2000). Theoretical perspectives on spintronics and spin-polarized transport. *IEEE Transactions on Magnetics*, *36*(5), 2821–2826. doi:10.1109/20.908600

Datta, S., & Das, B. (1990). Electronic analog of the electro-optic modulator. *Applied Physics Letters*, *56*(7), 665–667. doi:10.1063/1.102730

Dietl, T. (2006). *Theory of Spin Transport and Spin Transfer in Modulated Structures of (Ga, Mn)As*. The 1st RIEC Int.l Workshop on Spintronics, Tohku, Japan.

Dietl, T. (2013). *Information devices: history and new concepts*. The Academia Europaea 25th conference, Wrocław, Poland.

Dietl, T. (2015). Origin of ferromagnetism in functional topological insulators and semimetals. *15th Int. Symposium Magnetism Optics Research MORIS2015*, Penang, Malaysia.

Dirac, P. (1928). Quantum theory of the electron. *Proceedings of the Royal Society of London. Series A, Containing Papers of a Mathematical and Physical Character*, *117*(778), 619–624. doi:10.1098/rspa.1928.0023

Dresselhaus, G. (1955). Spin-orbit coupling effects in zinc blende structures. *Physical Review*, *100*(2), 580–586. doi:10.1103/PhysRev.100.580

Dutta, S., & Pati, S. K. (2010). Novel properties of graphene nanoribbons: A review. *Journal of Materials Chemistry*, *20*(38), 8207–8223. doi:10.1039/c0jm00261e

Dyakonov, M. I. (2008). *Spin Physics in Semiconductors*. Berlin: Springer-Verlag. doi:10.1007/978-3-540-78820-1

Dyakonov, M. I., & Kachorovskii, V. Y. (1986). Spin relaxation of two-dimensional electrons in non-centrosymmertic semiconductors. *Sov. Phys. Semicond.*, *20*, 110–112.

Dyakonov, M. I., & Perel, V. I. (1971). Spin orientation of electronics associated with the interband absorption of light in semiconductors. *Soviet Physics, JETP*, *33*, 1053–1059.

Dyakonov, M. I., & Perel, V. I. (1972). Spin relaxation of conduction electrons in non-centrosymmetric semiconductors. *Sov. Phys. Solid State*, *13*, 3023–3026.

Dyson, F. J. (1956). General Theory of Spin-Wave Interactions. *Physical Review*, *102*(5), 1217–1230. doi:10.1103/PhysRev.102.1217

Ehrenfreund, E., & Vardeny, Z. V. (2012). Effects of magnetic field on conductance electroluminescence in organic devices. *Israel Journal of Chemistry*, *52*(6), 552–562. doi:10.1002/ijch.201100120

El Hajj, R. (2008). *Etude mathematique et numerique de modeles de transport: application a la Spintronique* (PhD thesis). Universite Paul Sabatier-Toulouse III, France.

El Hajj, R. (2014). Diffusion models for spin transport derived from the spinor Boltzmann equation. *Commun. Math. Sci.*, *12*(3), 565–592. doi:10.4310/CMS.2014.v12.n3.a9

Evgeny, Y. T., & Igor, Z. (2011). *Handbook of Spin Transport and Magnetism*. Chapman and Hall/CRC.

Fabian, J., Matos-Abiague, A., Ertler, C., Stano, P., & Zutic, I. (2007). Semiconductor Spintronics. *Acta Physica Slovaca*, *57*, 565.

Fabian, J., & Zutic, I. (2004). Spin-polarized current amplification and spin injection in magnetic bipolar transistors. *Physical Review B: Condensed Matter and Materials Physics*, *69*(11), 115314. doi:10.1103/PhysRevB.69.115314

Fabian, J., & Zutic, I. (2005). The Ebers-Moll model for magnetic bipolar transistors. *Applied Physics Letters*, *86*(13), 133506. doi:10.1063/1.1886251

Fert, A. (2008). Nobel lecture: Origin, development and future of Spintronics. *Reviews of Modern Physics*, *80*(4), 1517–1530. doi:10.1103/RevModPhys.80.1517 PMID:18626879

Fert, A. (2008). Nobel lecture: Origin, development, and future of spintronics. *Reviews of Modern Physics*, *80*(4), 1517–1530. doi:10.1103/RevModPhys.80.1517 PMID:18626879

Fert, A., & Campbell, I. A. (1976). Two-current conduction in nickel. *Physical Review Letters*, *21*(16), 1190–1192. doi:10.1103/PhysRevLett.21.1190

Gao, L. (2009). *Spin Polarized Current Phenomena in Magnetic Tunnel Junctions* (PhD Thesis). Stanford University

Garcia-Cervera, C. J., & Wang, X.-P. (2007). Spin-polarized currents in ferro-magnetic multilayers. *Journal of Computational Physics*, *224*(2), 699–711. doi:10.1016/j.jcp.2006.10.029

Geim, A. K. (2009). Graphene: Status and Prospects. *Science*, *324*(5934), 1530–1534. doi:10.1126/science.1158877 PMID:19541989

Geim, A. K., & Novoselov, K. S. (2007). The rise of graphene. *Nature Materials*, *6*(3), 183–191. doi:10.1038/nmat1849 PMID:17330084

Geng, R., & Tyler, T., Do, K., Luong, H.M., & Nguyen, T.D. (2016). A review on organic spintronic materials and devices: I. Magnetic field effect on organic light emitting diodes. J. *Science: Advanced Mat & Devices*, *1*, 128–140.

Giriat, W., & Furdyna, J. K. (1988). *Diluted Magnetic Semiconductors in Semi-Conductors and Semimetals* (J. K. Furdyna & J. Kossnt, Eds.). Boston: Academic Press.

Goudsmith, S., & Uhlenbeck, G. E. (1926). Article.. *Physica*, *6*, 273.

Hai, P. N., Ohya, S., & Tanaka, M. (2010). Long spin-relaxation time in a single metal nanoparticle. *Nature Nanotechnology*, *5*(8), 593–596. doi:10.1038/nnano.2010.130 PMID:20601943

Hirose, K., & Kobayashi, N. (2014). *Quantum Transport Calculations for Nano-systems*. Pan Stanford.

Hirsch, J. E. (1999). Spin Hall Effect. *Physical Review Letters*, *83*(9), 1834–1837. doi:10.1103/PhysRevLett.83.1834

Holstein, T., & Primakoff, H. (1940). Field Dependence of the Intrinsic Domain Magnetization of a Ferromagnet. *Physical Review*, *58*(12), 1098–1113. doi:10.1103/PhysRev.58.1098

Hsieh, D., Xia, Y., Qian, D., Wray, L., Dil, J. H., Meier, F., & Hasan, M. Z. et al. (2009). A tunable topological insulator in the spin helical Dirac transport regime. *Nature*, *460*(7259), 1101–1106. doi:10.1038/nature08234 PMID:19620959

Huang, B. (2008). *Vertical transport silicon spintronic devices* (PhD thesis). University of Delaware.

Jansen, R. (2012). Silicon spintronics. *Nature Materials*, *11*(5), 400–408. doi:10.1038/nmat3293 PMID:22522640

Johnson, M., & Silsbee, R. H. (1985). Interfacial charge-spin coupling: Injection and detection of spin magnetization in metals. *Physical Review Letters*, *55*(17), 1790–1793. doi:10.1103/PhysRevLett.55.1790 PMID:10031924

Jungel, A., Shpartko, P., & Zamponi, N. (2016). *Energy-Transport Models For Spin Transport In Ferromagnetic Semiconductors*. arXiv:1604.05480v

Jungwirth, T., Sinova, J., Mašek, J., Kučera, J., & MacDonald, A. H. (2006). Theory of ferromagnetic (III,Mn)V semiconductors. *Reviews of Modern Physics*, *78*(3), 809–864. doi:10.1103/RevModPhys.78.809

Katine, J. A., & Fullerton, E. E. (2008). Device implications of spin-transfer torques. J. Magnetics. *Magn. Matter.*, *320*(7), 1217–1226. doi:10.1016/j.jmmm.2007.12.013

Kikkawa, J. M., & Awschalom, D. D. (1998). Resonant Spin Amplification in n -type GaAs. *Physical Review Letters*, *80*(19), 4313–4316. doi:10.1103/PhysRevLett.80.4313

Kikkawa, J. M., & Awschalom, D. D. (1999). Lateral drag of spin coherence in gallium arsenide. *Nature*, *397*(6715), 139–141. doi:10.1038/16420

Kittel, C. (1995). *Introduction to Solid State Physics* (7th ed.). Wiley.

Klaer, P., Hoffmann, F., Woltersdorf, G., Arbelo Jorge, E., Jourdan, M., Back, C. H., & Elmers, H. J. (2011). Element-specific ferromagnetic resonance in epitaxial Heusler spin valve systems. *Journal of Physics. D, Applied Physics*, *44*(42), 425004. doi:10.1088/0022-3727/44/42/425004

Ko, V., Teo, K. L., Liew, T., Chong, T. C., MacKenzie, M., Mac-Laren, I., & Chapman, J. N. (2008). Origins of ferromagnetism in transition-metal doped Si. *Journal of Applied Physics*, *104*(3), 033912. doi:10.1063/1.2963485

Kohn, W., & Sham, L. J. (1965). Self-Consistent Equations Including Exchange and Correlation Effects. *Physical Review*, *140*(4A), A1133–A1138. doi:10.1103/PhysRev.140.A1133

Kuemmeth, F., Churchill, H. O. H., Herring, P. K., & Marcus, C. M. (2010). Carbon nanotubes for coherent Spintronics. *Materials Today*, *13*(3), 18–26. doi:10.1016/S1369-7021(10)70030-4

Landau, L., & Lifshitz, E. (1935). On the theory of the dispersion of magnetic permeability in ferromagnetic bodies. *Phys. Z. Sowjet.*, *8*, 153–169.

Landau, L. D., & Lifshitz, E. M. (1997). *Quantum mechanics*. Oxford, UK: Butterworth-Heinemann.

Lewis, R. A. (2007). Physical Phenomena in Electronic Materials in the Terahertz Region. *Proceedings of the IEEE*, *95*(8), 1641–1645. doi:10.1109/JPROC.2007.898902

Liu, C., Patel, S. J., Peterson, T. A., Geppert, C. C., Christie, K. D., Palmstrøm, C. J., & Crowell, P. A. (2016). Dynamic detection of electron spin accumulation in ferromagnet-semiconductor devices by ferromagnetic resonance. *Nature Com.*, *7*, 10296. doi:10.1038/ncomms10296 PMID:26777243

Mecklenburg, M., & Regan, B. C. (2011). Spin and the Honeycomb Lattice: Lessons from Graphene. *Physical Review Letters, 106*(11), 116803. doi:10.1103/PhysRevLett.106.116803 PMID:21469887

Mott, N. F. (1935). A discussion of the transition metals on the basis of quantum mechanics. *Proceedings of the Physical Society, 47*(4), 571–588. doi:10.1088/0959-5309/47/4/305

Ohno, Y., Young, D. K., Beschoten, B., Matsukura, F., Ohno, H., & Awschalom, D. D. (1999). Article. *Nature, 402*, 790. doi:10.1038/45509

Özer, M. (2010). *Handbook of Spintronic Semiconductors* (W. M. Chen & I. A. Buyanova, Eds.). Singapore: Pan Stanford Publishing.

Parkin, S. (2004). Spintronic Materials and Devices: Past, present and future! *Proceedings of IEEE Electron Device Modelling (IEDM) Conf. 2004*, 903.

Pauli, W. (1927). *Zur Quantenmechanik des Magnetischen Elektrons Zeitschrift für Physik*. Academic Press.

Pershin, Y. V., & Di Ventra, M. (2008). Voltage probe of the Spin Hall Effect. *Journal of Physics Condensed Matter, 20*(2), 025204. doi:10.1088/0953-8984/20/02/025204

Pershin, Y. V., Saikin, S., & Privman, V. (2005). Semiclassical Transport Models for Semiconductor Spintronics. *IEE Proc. Circuits, Devices & Systems, 152*, 366.

Possanner, S. (2012). *Modeling and Simulation of Spin-Polarized Transport at the Kinetic and Diffusive Level* (PhD Dissertation). Paul Sabatier-Toulouse III University and Graz University of Technology.

Pramanik, S., Bandyopadhyay, S., & Cahay, M. (2005). Spin Relaxation in the Channel of a Spin Field-Effect Transistor. *IEEE Transactions on Nanotechnology, 4*(1), 2–7. doi:10.1109/TNANO.2004.840140

Rashba, E. I. (2003). Spin currents in thermodynamic equilibrium: The challenge of discerning transport currents. *Physical Review B: Condensed Matter and Materials Physics, 68*(24), 241315. doi:10.1103/PhysRevB.68.241315

Rashba, E.I. (2005). Spin Dynamics and Spin Transport. *Journal of Superconductivity: Incorporating Novel Magnetism*, 137-144.

Sabri, S. S., Lévesque, P. L., Aguirre, C. M., Guillemette, J., Martel, R., & Szkopek, T. (2009). Graphene field effect transistors with parylene gate dielectric. *Applied Physics Letters, 95*(24), 242104. doi:10.1063/1.3273396

Saikin, S. (2004). Drift-diffusion model for spin-polarized transport in a non-degenerate 2DEG controlled by spin-orbit interaction. *Journal of Physics Condensed Matter, 16*, 5071–5081. doi:10.1088/0953-8984/16/28/025

Saikin, S., Shen, M., Cheng, M., & Privman, V. (2003). Semiclassical Monte Carlo model for in-plane transport of spin polarized electrons in III–V heterostructures. *Journal of Applied Physics, 94*(3), 1769–1775. doi:10.1063/1.1589581

Schmidt, G. (2005). Concepts for Spin Injection into SemiconductorsA Review. *Journal of Physics. D, Applied Physics, 38*(7), 107–122. doi:10.1088/0022-3727/38/7/R01

Schmidt, G., Ferrand, D., Molenkamp, L. W., Filip, A. T., & van Wees, B. J. (2000). Fundamental obstacle for electrical spin injection from a ferromagnetic metal into a diffusive semiconductor. *Physical Review B: Condensed Matter and Materials Physics*, *62*(8), R4790–R4793. doi:10.1103/PhysRevB.62.R4790

Shafir, E., Shen, M. & Saikin, S. (2004). Modulation of spin dynamics in a channel of a non-ballistic spin-field effect transistor. *Phys. Rev. B, Condens. Matter Mater. Phys.*, *70*, 241302(R) 1–4.

Slachter, A., Bakker, F. L., Adam, J.-P., & van Wees, B. J. (2010). Thermally driven spin injection from a ferromagnet into a non-magnetic metal. *Nature Physics*, *6*(11), 879–882. doi:10.1038/nphys1767

Song, Y., & Dery, H. (2010). Spin transport theory in ferromagnet /semi-conductor systems with non-collinear magnetization configurations. *Physical Review B: Condensed Matter and Materials Physics*, *81*(4), 045321–045328. doi:10.1103/PhysRevB.81.045321

Song, Y., & Dery, H. (2012). Analysis of phonon-induced spin relaxation processes in silicon. *Physical Review B: Condensed Matter and Materials Physics*, *86*(8), 085201. doi:10.1103/PhysRevB.86.085201

Stern, O., & Gerlach, W. (1922). Der Experimentelle Nachweis der Richtung Squantelung im Magnetfeld. *Zeitschrift fur Physik*, *9*(1), 349–352. doi:10.1007/BF01326983

Sun, Q. F., Xie, X. C., & Wang, J. (2008). Persistent spin current in nanodevices and definition of the spin current. *Physical Review B: Condensed Matter and Materials Physics*, *77*(3), 035327. doi:10.1103/PhysRevB.77.035327

Tanaka, M., & Higo, Y. (2001). Large Tunneling Magnetoresistance in GaMnAs /AlAs/GaMnAs Ferromagnetic Semiconductor Tunnel Junctions. *Physical Review Letters*, *87*(2), 026602–026605. doi:10.1103/PhysRevLett.87.026602

Thomas, L. H. (1926). The motion of the spinning electron. *Nature*, *117*(2945), 514. doi:10.1038/117514a0

Thompson, S. M. (2007). The discovery, development and future of GMR: The nobel prize 2007. *Journal of Physics. D, Applied Physics*, *41*(9), 093001. doi:10.1088/0022-3727/41/9/093001

Tombros, N. (2008). *Electron spin transport in graphene and carbon nanotubes* (PhD Thesis). Zernike Institute for Advanced Materials.

Tombros, N., Jozsa, C., Popinciuc, M., Jonkman, H. T., & van Wees, B. J. (2007). Electronic spin transport and spin precession in single graphene layers at room temperature. *Nature*, *448*(7153), 571–574. doi:10.1038/nature06037 PMID:17632544

Tse, W.-K., & Sarma, S. D. (2007). Coulomb Drag and Spin Coulomb Drag in the presence of Spin-orbit Coupling. *Physical Review B: Condensed Matter and Materials Physics*, *75*(4), 045333. doi:10.1103/PhysRevB.75.045333

Valenzuela, S., & Tinkham, M. (2006). Direct electronic measurement of the spin Hall Effect. *Nature*, *442*(7099), 176–179. doi:10.1038/nature04937 PMID:16838016

Wagner, R. (2009). *G-Factor, Effective Mass and Spin Susceptibility of a 2-Dimensional Electron Gas* (Master Thesis). University of Basel, Basel.

Weber, E. (1983). Transition Metals in Silicon. *Applied Physics. A, Materials Science & Processing*, *30*(1), 1–22. doi:10.1007/BF00617708

Weng, M. Q., & Wu, M. W. (2003). Kinetic theory of spin transport in n-type semiconductor quantum wells. *Journal of Applied Physics*, *93*(1), 410–420. doi:10.1063/1.1526936

Wolf, S. A. et al.. (2001). Spintronics: A spin-based electronics vision for the future. *Science*, *294*(5546), 1488–1495. doi:10.1126/science.1065389 PMID:11711666

Wu, Y., Hu, B., Howe, J., Li, A. P., & Shen, J. (2007). Spin injection from ferromagnetic Co nanoclusters into organic semiconducting polymers. *Physical Review B: Condensed Matter and Materials Physics*, *75*(7), 075413. doi:10.1103/PhysRevB.75.075413

Xu, J. (2008). *Modeling and simulation of spin transport and precession in silicon* (PhD Thesis). University of Delaware.

Yazyev, O. V. (2010). Emergence of magnetism in graphenematerials and nanostructures. *Reports on Progress in Physics*, *73*(5), 056501. doi:10.1088/0034-4885/73/5/056501

Zamponi, N. (2014). Analysis of a drift-diffusion model with velocity saturation for spin-polarized transport in semiconductors. *Journal of Mathematical Analysis and Applications*, *420*(2), 1167–1181. doi:10.1016/j.jmaa.2014.06.065

Zhou, S., & Schmidt, H. (2010). Mn-doped Ge and Si: A Review of the Experimental Status. *Materials (Basel)*, *3*(12), 5054–5082. doi:10.3390/ma3125054

Zhou, Y., Han, W., Chang, L.-T., Xiu, F., Wang, M., Oehme, M., & Wang, K. L. et al. (2011). Electrical spin injection and transport in germanium. *Physical Review B: Condensed Matter*, *84*(12), 125323. doi:10.1103/PhysRevB.84.125323

Zutic, I., Fabian, J., & Das Sarma, S. (2004). Spintronics: Fundamentals and applications. *Reviews of Modern Physics*, *76*(2), 323–410. doi:10.1103/RevModPhys.76.323

Zutiic, I., Fabian, J., & Das Sarma, S. (2002). Spintronics: Fundamentals and applications. *Reviews of Modern Physics*, *76*(2), 323–410. doi:10.1103/RevModPhys.76.323

ENDNOTES

1 The IBM's hard disk drive division was acquired by Hitachi in 2003 and renamed Hitachi Global Storage Technologies. In 2012 Hitachi sold the division to Western Digital who rebranded it as HGST

2 The Dirac equation combines relativity and quantum mechanics to explain the half-integer spin particles. This equation shows that relativistic quantum mechanics implies spin. However, there is no explanation of how electrons can have an angular (rotational) momentum, or why electron spin should have two values. In fact, spin cannot be a rotational property, since this requires that electron would need to spin (rotate) faster than the speed of light to give the observed magnetic moment (Mecklenburg & Regan, 2011).

3 The electron g-factor $g = 2(1+\alpha/2\pi +..) = 2.0023$. The value of 2 comes from the Dirac equation, while the rest is called the anomalous contribution, proportional to the fine structure constant, $\alpha = e^2/4\pi\hbar\varepsilon_0 c^4$ arising from quantum corrections

4 According to Johnson-Silsbee (1985): EMF appears in the proximity of a ferromagnetic metal and spin- polarized nonmagnetic metal (inverse of spin injection).

5 The state of the electron spin can be geometrically represented by any point on the Bloch sphere. The electron spin can have only two opposing projections on a fixed axis, and hence only two measurable states. These two states are spin up$|\uparrow\rangle$ and spin down $|\downarrow\rangle$

6 This model is sometimes referred to as "two-component drift-diffusion model".

7 Some references refer to to Baibich, as the discoverer of GMR. See Baibich et al, (1988).

8 We have already defined the *singlet*, *doublet* and *triplet* (spin) states in Chapter 8 (see Note 2 from Chapter 8).

Chapter 10
Plasmons, Polarons, and Polaritons Transport

1. OVERVIEW AND CHAPTER OBJECTIVES

In the previous Chapters, we presented and discussed the transport models of charge carriers, phonons, photons and spin. In this Chapter we turn our attention to other composite quasi particles, such as: polarons, excitons and polaritons. Table 1 summarizes the quantum quasit particles, which are used to describe interactions in solids.

We have already defined the physical meaning of simple (or elementary) quantum particles and quasi particles, in previous Chapters. As for the composite quasi particles, we summarize their definitions as follows.

- **Polaron:** Is a quasiparticle which represents the interactions between electrons and atoms vibration in solids. The polaron concept was first proposed by Lev Landau in 1933 to describe an electron moving in a dielectric crystal where the atoms vibrate around their equilibrium positions to effectively screen the electron charge, by a phonon cloud. This lowers the electron mobility and increases the electron's effective mass. Therefore, electron mobility in semiconductors can be greatly decreased by the formation of polarons. In particular, polarons are important in organic semiconductors. Polaronic effects appear in many guises and forms and can have a dramatic influence on the transport properties information carriers.
- **Exciton:** Is a bound state of an electron and an electron hole which are attracted to each other by the electrostatic Coulomb force. As we have seen in Chapter 8 of this Book, an exciton can form when a photon is absorbed by a semiconductor.
- **Magnon:** Is a quasiparticle, a collective excitation of the electrons' spin structure in a crystal lattice. The concept of a magnon was introduced in 1930 by Felix Bloch in order to explain the reduction of the spontaneous magnetization in a ferromagnet. As we have seen in Chapter 9 of

DOI: 10.4018/978-1-5225-2312-3.ch010

Table 1. Quasi particles in semiconductors and nanodevices

Quasi Particle	Description	Notes
Simple (or Collectivity of Same Type) Quasi Particles		
Holes	Missing electrons in valence band	In semiconductors
Phonons	Quanta of crystal lattice vibrations	large momentum little energy
Photons	Light quanta	
Magnons	Spins (Spin wave quanta)	Crystal magnetic moments
Plasmons	Electron collective	In nanocrystals
Composite Quasi Particles		
Cooper pairs	2 electrons coupled by phonons	Responsible for superconductivity
Polarons	e^- + phonon	Important in organic semiconductors
Excitons	Bound electron-hole pair	Produce light in LEDs
Polaritons	Photon + phonon	(Phononic Polariton)
	Photon + exciton	(Excitonic Polariton)
	Photon + electron(s)	(Plasmon Polariton)

this Book, magnons carry a fixed amount of energy and lattice momentum, as well as a spin of -1, indicating they have boson behavior.

- **Plasmon:** Is a quantum of plasma oscillation. As the light consists of photons, the plasma oscillation consists of plasmons. The plasmon can be considered a quasiparticle since it arises from the quantization of plasma oscillations, just like phonons are quantizations of mechanical vibrations. Thus, plasmons are collective (a discrete number) oscillations of the free electron gas density, for example, at optical frequencies. Plasmons can couple with a photon to create another quasiparticle called a plasma polariton

The above quasiparticles are particularly important in metals, dielectrics and semiconductors nanostructures. In fact, the electromagnetic properties of metal/dielectric interfaces undergo a steadily increasing interest in science and technology. In order to understand the dynamics of the above quasiparticles and their transport properties in these materials, we start studying the dielectric function of a semiconductor, since they are more or less coupled to their dielectric properties. Actually, the electrical permittivity is a measure of how an electric field interacts with a dielectric medium.

Upon completion of this Chapter the student will be able to understand the following interactions and the transport of their associated quasi particles

- Dielectric function,
- Electron-electron interactions (Screening & Plasma Oscillations),
- Electron plasma oscillation – (Plasmons),
- Electron-lattice interaction (Polarons),
- Photon-phonon interaction (Phonon-polaritons),
- Photon-exciton interactions (Excitonic polaritons),
- Insulators & Mott metal-insulator transition.

2. DIELECTRIC FUNCTION

When a dielectric is subjected to an alternating electric field, a displacement current will flow across the material. For a lossy material there will be a component of this current in the direction of applied field (resistive component) such that the permittivity ε becomes complex as follows:

$$\varepsilon = \varepsilon_1 - j\,\varepsilon_2 \tag{1}$$

The dielectric function of a material describes the electrical and optical properties versus frequency, wavelength, or energy. It describes the polarization and absorption properties of the material. The dielectric function is actually the electric permittivity, ε, which consists of two parts[1] ($\varepsilon = \varepsilon_1 - j\,\varepsilon_2$). The dielectric function is also related to the refractive index of the material by the equation: $\varepsilon/\varepsilon_0 = \varepsilon_r = n^2$, where ε_0 is the absolute value of permittivity in free space and ε_r is the relative permittivity (or dielectric constant). As the quantities ε_1 and ε_2 vary with the frequency of applied electric field, they are usually written as functional form: $\varepsilon_1(\omega)$ and $\varepsilon_2(\omega)$. Actually, the real part ε_1 represents how much a material becomes polarized when an electric field is applied due to creation of electric dipoles. When the applied field is oscillating, the sign of ε_1 may be positive or negative according to whether the induced dipoles are oscillating in phase or out of phase with the applied field. When the oscillations of induced dipoles in a material reach a certain frequency, it is possible for the material to start absorbing energy from the applied field. When absorption happens the term ε_2 becomes significant, whereas it is zero when the material is transparent. Thus, the imaginary part ε_2 represents absorption (or losses due to absorption) in a material.

We now present some particular models for the frequency dependence of the dielectric function of a solid, such as the Drudé model. Though very simple, this model is widely used to provide easy explanation of the physical origin behind dielectric and optical properties of matter. We also discuss the relaxation models of the dielectric function.

2.1 Drudé Model

One major contribution to the dielectric function is through the free carriers. Such free carrier contributions are very important in semiconductors and metals, and can be understood in terms of the simple classical conductivity model, due to Drudé. The Drudé model is based on the classical equation of motion of free charge carriers (gas of free electrons), under the effect of an external force $\boldsymbol{F}$ (electric or magnetic fields or both):

$$m^*(dv/dt) + m^*v/\tau = \boldsymbol{F} \tag{2a}$$

Here, the external force $\boldsymbol{F} = -e\boldsymbol{\zeta}$ is assumed sinusoidal or $\boldsymbol{\zeta} = \zeta_0 exp(-j\omega t)$. Also, the middle term (m^*v/τ) represents the friction or damping force. Assuming a constant damping[2] factor $(1/\tau)$, the solution of the above equation is: $v(t) = v_0.exp(-j\omega t)$. Substituting this solution into (2a) yields:

$$v(1 - j\,\omega\tau) = -\,e\tau.\zeta/m \tag{2b}$$

The charge carrier current density is given, by definition, as follows:

$$J = - e\, nv = e^2 n \tau .\zeta / [m(1- j\omega\tau)] \text{ or } J = \sigma.\zeta \tag{2c}$$

where the electrical conductivity $\sigma = \sigma_o/(1-j\omega\tau)$ with $\sigma_o = e^2 n (\tau / m)$. The dielectric function of the material is then expressed using Maxwell's equation:

$$\varepsilon_r (\omega) = D/\varepsilon_o \zeta = 1 + P/ \varepsilon_o \zeta \tag{2d}$$

where the polarization density $P = -enx = en.v/\omega = - e\tau.\zeta /[m\omega (1- j\omega\tau)]$

$$\varepsilon = \varepsilon_o - j \sigma_o / [\omega (1- j \omega \tau)] \tag{2e}$$

$$\varepsilon_r = \varepsilon /\varepsilon_o = 1 - j \sigma_o /[\varepsilon_{o.}\omega (1- j \omega \tau)] \tag{2f}$$

A more elegant derivation of the Drudé expression can be obtained starting from the Boltzmann transport equation (BTE), under the effect of electromagnetic field, as we did in Chapter 2. The Drudé model is accurate for the dielectric function of most metals in the optical and IR spectrum.

2.2 Lorentz Model

The Lorentz model of the dielectric function is based on the damped harmonic oscillator approximation in quantum mechanics. It describes the electronic interband transitions by assuming that electrons are bound to the positive ions of the material and oscillate about them. The Drudé-Lorentz model is suitable for modeling dispersive materials with multiple resonant frequencies. In this case, the dielectric function is given by:

$$\varepsilon(\omega) = \Sigma_k \left(\frac{g_k \omega_p^k}{(\omega_k^2 - \omega^2) + j\omega\gamma_k} \right) \tag{3}$$

where the subscript k changes from 1 to the total number of oscillators with frequency ω_k and strength g_k. Also, γ_k is the damping constant of the kth harmonic oscillator and ω_p is called the plasma frequency.

2.3 Drudé-Lorentz Model

The Drudé-Lorentz model of the dielectric function combines the features of Drudé and Lorentz models. It takes into account both free electrons and bound electrons to the positive ions of the material. According to this model, the dielectric function due to a single resonator is given by:

$$m(dv/dt) + m (v/\tau) + C.x = - e\zeta(t) \tag{4a}$$

where m is charge mass x is the deviation from its equilibrium position and the driving force $-e\zeta(t)$ oscillates with a characteristic frequency ω. Note that the harmonic oscillator force term (spring force

$-Cx$) is not present in the original ***Drudé*** model for free carriers. Taking the Fourier transform of this equation yields:

$$m[-\omega^2 - j(\omega/\tau) + C/m].\, x(\omega) = -e.\zeta(\omega) \tag{4b}$$

The solution is then:

$$x(\omega) = -(e.\zeta/m)/[-\omega^2 - j(\gamma\omega) + \Omega^2] \tag{4c}$$

where $\gamma = 1/\tau$ is the damping factor and $\Omega = (C/m)^{1/2}$ is the harmonic oscillator frequency. The dielectric function of the material is then expressed as follows:

$$\varepsilon(\omega) = D/\zeta(\omega) = \varepsilon_o + P/\zeta(\omega) \tag{4d}$$

$$\varepsilon(\omega) = \varepsilon_o + (e^2 n/m).[-\omega^2 - j\omega\gamma + \Omega^2] \tag{4e}$$

Such that the relative permittivity ε_r is given by:

$$\varepsilon_r(\omega) = \varepsilon(\omega)/\varepsilon_o = 1 + (e^2 n/\varepsilon_o m).[-\omega^2 - j\omega\gamma + \Omega^2] \tag{4f}$$

where $\boldsymbol{P} = -e.n.x$ is the polarization density and $\boldsymbol{D} = \varepsilon\,\zeta$ is the electric displacement. Expressing $\varepsilon = \varepsilon_1 - j\,\varepsilon_2$, we can calculate $\varepsilon_1(\omega)$ and $\varepsilon_2(\omega)$.

3. DIELECTRIC RELAXATION MODELS

Physically speaking, a material cannot polarize instantaneously in response to an applied field. Dielectric relaxation is the delay in the dielectric response of a material to a changing electric field. This relaxation is often described in terms of permittivity as a function of frequency.

3.1 Debye Relaxation

Debye relaxation is the dielectric relaxation response of an ideal, non-interacting population of dipoles to an alternating external electric field. According to Peter Debye, the variation of the out-of–phase component (ε_1) and the in-phase component (ε_2) of the complex dielectric constant with frequency, is given by (Debye, 1913):

$$\varepsilon_1(\omega) = \varepsilon_\infty + \left(\frac{\varepsilon_s - \varepsilon_\infty}{1 + \omega^2\tau^2}\right) \tag{5a}$$

$$\varepsilon_2(\omega) = \varepsilon_\infty + \left(\frac{\varepsilon_s - \varepsilon_\infty}{1 + \omega^2\tau^2}\right) \tag{5b}$$

where τ is called the dielectric relaxation time and ε_s and ε_∞ are the static (low) and infinite (high) frequency values of the dielectric permittivity. The above equations are sometimes called Debye equations.

Figure 1 depicts the frequency response of a dielectric, as described by (5). As shown in figure, the real part of ε relaxes at high frequency to a smaller value ε_{oo}. Also imaginary part ε_2 has a peak value at $\omega\tau = 1$.

As we pointed out above, equations (5) describe the relaxation of the real and imaginary parts of the complex permittivity of medium as a function of the field frequency ω. However, the Debye model is usually expressed in the complex permittivity of a dielectric material, as follows:

$$\varepsilon(\omega) = \varepsilon_\infty + (\varepsilon_s - \varepsilon_\infty) / (1 + j\omega\tau) \tag{6}$$

This is exactly equivalent to equations (5).

3.2 Havriliak–Negami Relaxation Model

The Havriliak–Negami model is an empirical modification of the Debye relaxation model, accounting for the asymmetry and broadness of the dielectric dispersion relation. The model was first used to describe the dielectric relaxation of some polymers, by adding two exponent parameters to the Debye equation, such that:

$$\varepsilon(\omega) = \varepsilon_\infty + (\varepsilon_s - \varepsilon_\infty) / [(1 + (j\omega\tau)^\alpha]^\beta \tag{7}$$

Figure 1. The dielectric relaxation of a polarization process near resonance

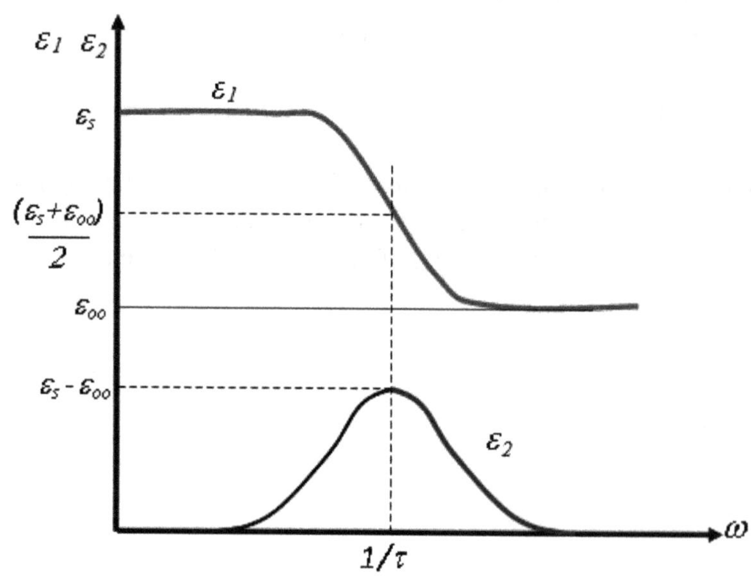

where the exponents α and β describe the asymmetry and broadness of the corresponding spectra. For β=1 the Havriliak-Negami equation reduces to the Cole-Cole equation. When α=1 we get the Cole-Davidson equation.

3.3 Cole-Cole Equation

The Cole-Cole equation is a dielectric relaxation model that constitutes a special case of Havriliak-Negami relaxation when the symmetry parameter β is equal to 1 - that is, when the relaxation peaks are symmetric:

$$\varepsilon(\omega) = \varepsilon_{\infty} + (\varepsilon_s - \varepsilon_{\infty}) / [(1+(j\omega\,\tau)^{1-\alpha}] \quad (8)$$

The parameter α, which takes a value between 0 and 1, is an experimentally determined correction factor. When α = 0, the Cole-Cole model reduces to the Debye model. Most polymers show dielectric relaxation patterns that can be accurately modeled by this equation.

3.4 General Frequency Response of Matter

Actually there is no a single resonant frequency at which the dielectric losses are maximum ($\omega\tau = 1$) and the frequency response shows many peaks. This is because most dielectric materials polarize in different ways in different ranges of frequency. For instance, at very high frequency, the ionic and electronic polarization may appear together. Figure 2 shows a part of the real permittivity characteristics $\varepsilon_1(\omega)$ over a wide scale of frequencies.

Figure 2. Characteristics of the real part of ε(ω) over a wide scale of frequencies

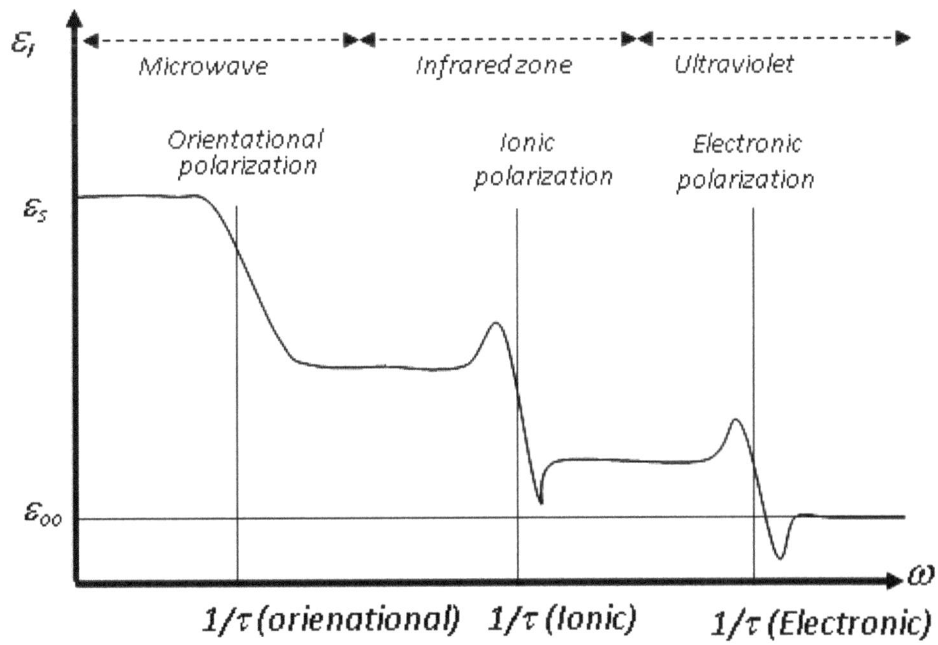

3.5 Kramers-Kronig Relations

According to Kramers and Kronig, the real and imaginary parts of the relative permittivity of a dielectric material are directly correlated to each other by the following equations (Froehlich, 1958):

$$\varepsilon_1(\omega) = 1 + \frac{2}{\pi} P \int_0^{\infty} \frac{\varpi.\varepsilon_2(\omega)}{\varpi^2 - \omega^2} .d\varpi \tag{9a}$$

$$\varepsilon_2(\omega) = -\frac{2}{\pi} P \int_0^{\infty} \frac{\varpi.\varepsilon_1(\varpi)}{\varpi^2 - \omega^2} .d\varpi \tag{9b}$$

where ***P*** denotes the principal value of the integral. To avoid explicit use of the principal value of a function, we can subtract out the singularity at a specific frequency *($\omega_o = 1/\tau$)*, as follows:

$$\varepsilon_1(\omega_o) = 1 + \frac{2}{\pi} \int_0^{\infty} \frac{\varpi.\varepsilon_2(\varpi) - \omega_o.\varepsilon_2(\omega_o)}{\varpi^2 - \omega_o^2} .d\varpi \tag{10a}$$

$$\varepsilon_2(\omega_o) = -\frac{2}{\pi} \int_0^{\infty} \frac{\omega_o.\varepsilon_1(\varpi) - \omega_o.\varepsilon_1(\omega_o)}{\varpi^2 - \omega_o^2} .d\varpi \tag{10b}$$

These relations show the *mathematical* causality and *linearity* of the dielectric function. *Causality* means that there is no output before an input; *linearity* means that twice the input produces twice the output.

4. PLASMA FREQUENCY

It can be shown from the above models that at very low frequencies the optical properties of semiconductors exhibit a metal-like behavior, while at very high frequencies their optical properties are like those of insulators. The frequency at which the material changes from a metallic to a dielectric response is called the plasma frequency ω_p, This is the frequency at which the real part of the dielectric function vanishes such that $\varepsilon_1(\omega_p) = 0$.

In order to relate the plasma frequency to other material properties, let's start from the Drudé model:

$$\varepsilon = \varepsilon_1 - j\varepsilon_2 = \varepsilon_0 - \frac{jne^2r}{mw(1 - jwt)} \tag{11}$$

We can then write the real and imaginary parts as:

$$\varepsilon_1(w) = \varepsilon_0 - \frac{ne^2r^2}{m(1+w^2r^2)}, \ \varepsilon_2(w) = \frac{1}{m}\frac{ne^2r}{m(1+w^2r^2)} \tag{12}$$

Since the damping factor ($1/\tau$) is usually small compared with ω_p, we can neglect this term[3] and identify the plasma frequency as follows:

$$w_p = \sqrt{\frac{ne^2}{\varepsilon_0 m}} \tag{13}$$

Note that the plasma frequency is a direct measure of the square root of the density of valence electrons in a solid. In fact, the contribution of holes in a semiconductor has the same sign as electrons. Therefore, we can write the contribution from electrons and holes as

$$\varepsilon = \varepsilon_0 - \frac{j}{w}\left[\frac{n_e e^2 r_h}{m_e(1-jw\tau_e)} + \frac{n_h e^2 \tau_h}{m_h(1-jw\tau_h)}\right] \tag{14}$$

It should be also noted that the dielectric function is negative below the plasma frequency (when $\omega < \omega_p$) and the field cannot transmit through the material. Alternatively, the light with frequency below the plasma frequency is entirely reflected.

5. SCREENING POTENTIAL AND PLASMA OSCILLATIONS

We have seen so far that the electron-electron interactions in a solid may be categorized into two basic types, namely:

1. Short range binary interactions (between individual electrons),
2. Long range interaction between collectivity of electrons (plasmons).

These interaction mechanisms were shortly illustrated, in the context of electron-electron scattering in Chapter 2 of this book. We review here their physical origins and impacts in semiconductors and nanostructures.

5.1 Screening Potential and Short-Range Electron-Electron Interaction

The binary electron-electron interactions appear in short range, for distances in the order of Debye length (λ_D). According to the screening potential theory (Thomas-Fermi, 1927), the Debye length (or screening length) due to the Coulomb potential, is given by:

$$\lambda_D = \sqrt{\frac{\varepsilon_s k_B T_L}{e^2 n_0}} \text{ For } T_L >> 0,\ \lambda_D = \sqrt{\frac{2\varepsilon_s E_F}{3e^2 n_0}} \text{ For } T_L \to 0 \quad (15)$$

where n_o is the equilibrium electron density, ε_s is the dielectric constant of the material and E_F is the Fermi energy level.

$$\varepsilon_F(r) \approx \frac{\hbar^2}{2m}\left(3\pi^2 n(r)\right)^{2/3} \quad (16a)$$

The Fermi energy level (E_F) is related to the electrochemical potential (μ):

$$\mu(r) = \varepsilon_F(r) - e\phi(r) = \frac{\hbar^2}{2m}\left(3\pi^2 n_0(r)\right)^{2/3} \quad (16b)$$

The screening potential ϕ(x) follows the following distribution:

$$\phi(r) = n\left(\frac{e^2}{4\pi\varepsilon_s r}\right)\exp\left[\frac{-r}{\lambda_D}\right] \quad (16c)$$

Therefore, the density of electrons can be enhanced due to screening potential, as shown in Figure 4. In fact, the experiments showed that the electron mean free path (the distance an electron can move before being scattered) in pure metals is about 10000 A at 300K and reaches 10 cm at 1 K! This is why e-e interaction can usually be ignored in metals, where average e-e separation is about 2A. In semiconductors, e-e interaction can be also neglected except for highly-doped materials.

5.2 Long-Range Electron-Electron and Interaction Plasmons

The other type of electron-electron interactions occurs at long range with collectivity of electrons or plasma oscillation. Plasma oscillations of free electron density with respect to the fixed positive ions can

Figure 3. The dielectric relaxation process

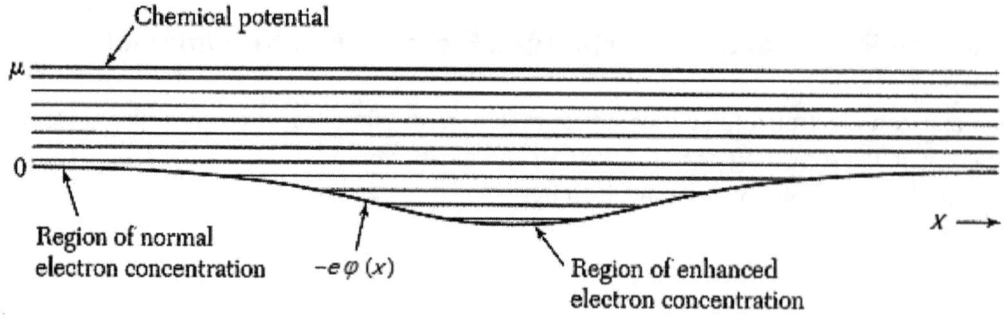

Figure 4. The dielectric relaxation process

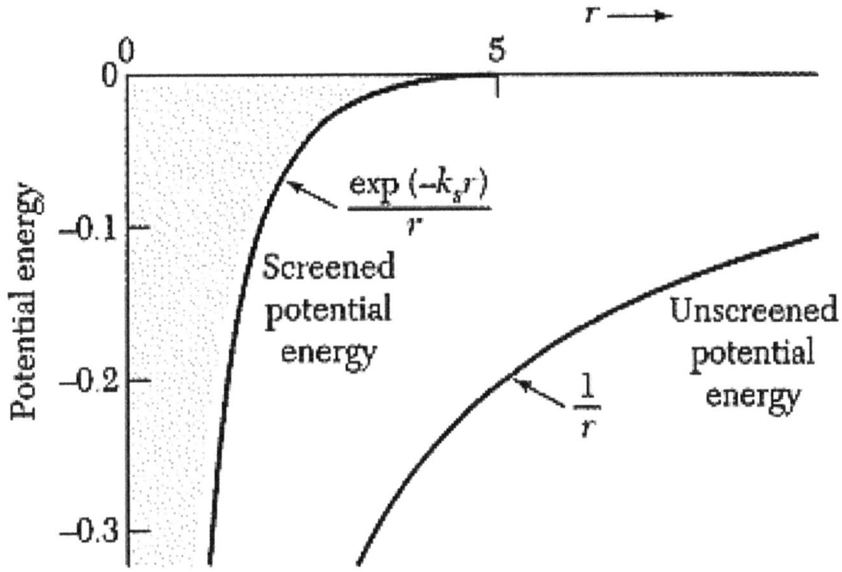

happen in metals and semiconductors. The quantum of plasma oscillation is called plasmon. Plasmons play a large role in the optical properties of metals.

Light of frequencies below the plasma frequency is reflected, because the electrons in the metal screen the electric field of the light. Light of frequencies above the plasma frequency is transmitted, because the electrons cannot respond fast enough to screen it. In most metals, the plasma frequency is in the ultraviolet, making them shiny (reflective) in the visible range. Some metals, such as copper and gold, have electronic interband transitions in the visible range, whereby specific light energies (colors) are absorbed, yielding their distinct color.

In semiconductors, the valence electron plasma frequency is usually in the deep ultraviolet, which is why they are reflective. Plasmons can be measured with EELS (Electron Energy Loss Spectroscopy). In EELS, the material is ecposed to a beam of electrons with known kinetic energy. Some of incident electrons are scattered inelastic, and lose energy. The amount of energy loss is measured by a spectrometer and interpreted in terms of what caused the energy loss.

Surface plasmons are confined to surfaces and interact strongly with light resulting in a polariton. They occur at the interface of a vacuum and material with a small positive imaginary and large negative real dielectric constant (usually a metal or doped dielectric). Recently, metallic carbon nanotubes and graphene have also shown to accommodate surface plasmons (Moradi, 2010). These plasmons are observed by near field infrared optical microscopy techniques and infrared spectroscopy. Potential applications of graphene plasmonics mainly addressed the terahertz to mid-infrared frequencies, such as optical modulators, photodetectors, biosensors. Plasmons can also couple with a photon to create another quasiparticle called a plasma polariton.

Figure 5. Plasma oscillations in Al

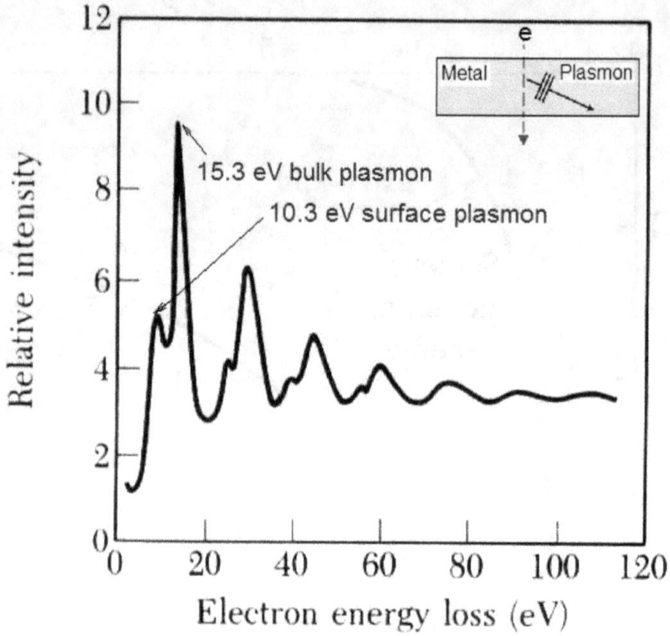

Figure 6. Schematic illustration of surface plasma oscillations in Al

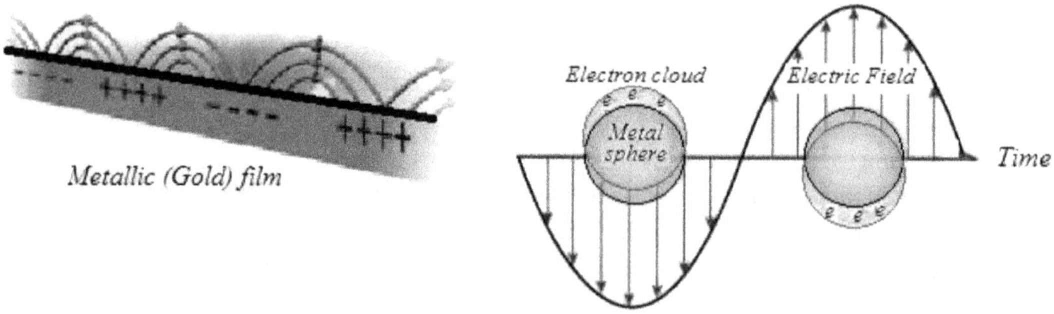

16. POLARONS AND ELECTRON-LATTICE INTERACTION

The composite object of an electron in deformed lattice (phonon cloud) is called polaron. Here, the vibrating ions drag and slow down moving electrons. For instance, a conduction electron in an ionic crystal or a polar semiconductor forms a polaron. Polarons are Fermions and have larger effect in polar crystals (like *NaCl*), and smaller effect in covalent crystals (like *GaAs*).

The basic theory of polarons was initiated by Landau (1933, 1948) and Pekar (1951). The polaron model for semiconductors was developed by Yamashita et al.(1958). Later, it was extended to organic semiconductors.by Holstein (1959).

Figure 7. Schematic view of a polaron showing how an electron couple with lattice vibrations (phonons) field

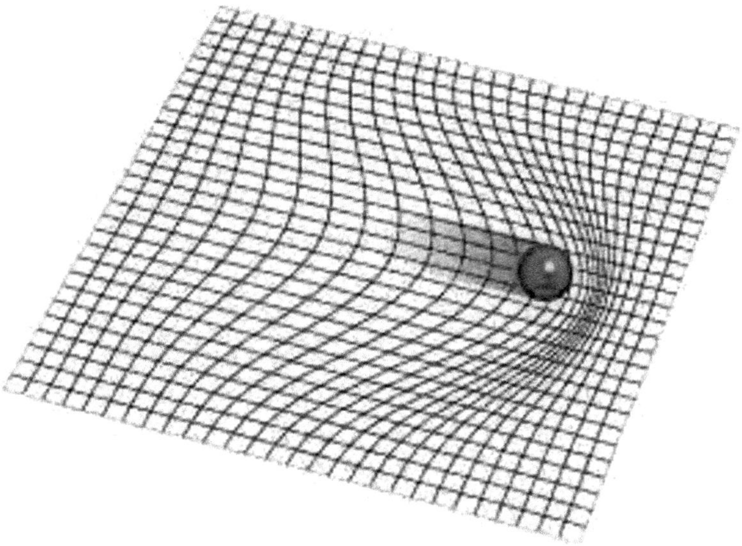

6.1 Polaron Self-Energy and Effective Mass

A polaron is characterized by its *self-energy* ΔE, and *effective mass* m_{pol} as well as its characteristic *response* to external electric and magnetic fields. When the electron-phonon cloud coupling is weak, the self-energy of the polaron can be approximated as:

$$\frac{\Delta E}{\hbar\omega} \approx -\alpha - 0.015919622\alpha^2, \tag{17}$$

where α is the coupling constant. In this case, the polaron mass m_{pol}, which can be measured by cyclotron resonance, is larger than the band mass m^* of the charge carrier without self-induced polarization:

Polaron effective mass $m_{pol} \approx m^* (1 + \alpha/6 + 0.0236\,\alpha^2)$ (18)

When the coupling is strong (α large), the Landau and Pekar theory (1963) shows that the self-energy is proportional to α^2 and the polaron mass scales as α^4

For the 2D structures, polaron system consists of an electron confined to a plane, interacting via the Fröhlich interaction with the LO phonons of a 3D surrounding medium. The self-energy and the mass of such a 2D polaron are no longer described by the 3D expressions (13). Rather, the self-energy for weak coupling can be approximated as follows in 2D:

$$\frac{\Delta E}{\hbar\omega} \approx -\frac{\pi}{2}\alpha - 0.06397\alpha^2 \tag{19}$$

$$\frac{m^*}{m} \approx 1 + \frac{\pi}{8}\alpha + 0.1272348\alpha^2 \tag{20}$$

Polaronic effects appear in many guises and forms and can have a dramatic influence on the physical properties of materials (Jørgensen et al, 2016). Generally speaking, when a particle (electron, hole, exciton, etc.) interacts with bosons from its environment (phonons, magnons, electron-hole pairs, etc) it becomes surrounded by a cloud of such excitations; the resulting composite object may be called a polaron. In other words, the term of polaron describes a quantum particle interacting with a bosonic environment. Extensions of the polaron concept include: acoustic polaron, piezoelectric polaron, spin polaron, Jahn-Teller polaron and bipolarons. The so-called bipolaron is a bound state of two polarons, mediated by exchange of bosons between their excitation clouds.

6.2 Polaron Transport and Modelling Approaches

The polaron properties are dependent on both the particle-environment coupling strength and polaron momentum. The Hamiltonian of a lattice polaron (electron+lattice distortion) is given by the following Holstein model (1959):

$$H = -t\sum_{\langle i,i\rangle \cdot \sigma}\left(c_{i\sigma}^{+}c_{j\sigma} + c_{j\sigma}^{+}c_{i\sigma}\right) + \Omega\sum_{i} b_i^{+}b_i + g\sum_{i} n_i b_i^{+}b_i \tag{21a}$$

where t is the hopping operator, $c^{\dagger}_{i\sigma}$ ($c_{i\sigma}$) is the creation (annihilation) operator of an electron at the ith site, and $b^{\dagger}_{i}$ (b_i) is phonon creation (annihilation) operator. The first term is the electron tight-binding kinetic energy. The second term describes the lattice degrees of freedom considered as a set of independent oscillators at each site, with frequency ω_0. The third term describes how the electrons couple through the density $n_{i\sigma} = (c^{\dagger}_{i\sigma} c_{i\sigma})$ to the local lattice displacement $x_i \sim (b^{\dagger}_I + b_i)$ with strength g. The above Hamiltonian can be also written in the momentum space as follows:

Table 2. Fröhlich coupling constants

Material	α	Material	α
InSb	0.023	*KI*	2.5
InAs	0.052	*TlBr*	2.55
GaAs	0.068	*KBr*	3.05
GaP	0.20	*RbI*	3.16
CdTe	0.29	$Bi_{12}SiO_{20}$	3.18
ZnSe	0.43	CdF_2	3.2
CdS	0.53	*KCl*	3.44
AgCl	1.84	$SrTiO_3$	3.77
α-Al_2O_3	2.40	*RbCl*	3.81

$$H = \sum_k \varepsilon(k) c_k^\dagger c_k + \omega_0 \sum_q b_q^\dagger b_q + \frac{g}{\sqrt{N}} \sum_{k,q} c_{k+q}^\dagger c_k \left(c_{-q}^\dagger c_q \right) \tag{21b}$$

The Holstein model is characterized by two relative parameters, the dimensionless coupling constant $\alpha = g^2/zt\omega_0$ and the adiabaticity ratio ω_0/zt. where z is the dimensionality of the problem

The eigenvalue problem of a polaron can be then written as follows:

$$H\left|1,k,\alpha\right\rangle = E_{1,k,\alpha}\left|1,k,\alpha\right\rangle \tag{22}$$

The subscript 1 means one particle, with momentum k, and a denotes any other quantum numbers.

6.3 Green's Functions Approach

The Green's function technique is appropriate for describing interactions with continuum states, such as the polaron problem. The Green's function or propagator of lattice polaron is defined as follows:

$$G(k,\omega) = \left\langle 0 \left| c_k \frac{1}{\omega - H + i\eta} c_k^+ \right| 0 \right\rangle = \sum_\alpha \frac{Z_{1,k,\alpha}}{\omega - E_{1,k,\alpha} + i\eta} \tag{23a}$$

where Z is the quasiparticle weight which measures how similar is the true wavefunction to a non-interacting (free electron, no bosons) wavefunction.

$$Z_{i,k,\alpha} = \left| \left\langle i,k,\alpha \left| c_k^+ \right| 0 \right\rangle \right|^2 \tag{23b}$$

Also, the spectral weight may be defined as:

$$A(k,\omega) = \frac{1}{\pi} \operatorname{Im} G(k,\omega) = \sum_\alpha Z_{1,k,\alpha} \delta\left(\omega - E_{1,k,\alpha}\right) \tag{23c}$$

Figure 8. One-dimensional representation of a phonon polaron in ionic crystals

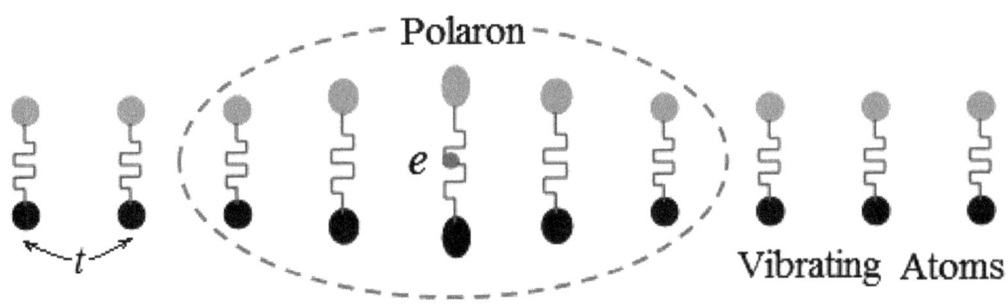

6.4 Monte Carlo Simulation of Polarons

Polaron transport can be simulated by quantum Monte Carlo (MC) method (Tempere et al, 2009). In fact, Feynman introduced a variational principle for path integrals to study the polaron (Feynmann, 1955). The analysis of 1D-polaron models has been carried out by Monte Carlo simulation and showed the accuracy of Feynman's path-integral approach. According to Mishchenko et al (2000), we can use the MC algorithm, with the aid of Feynmann diagrams, to calculate Green's function as follows:

$$G(k,\tau) = \left\langle 0 \left| c_k e^{-\tau H} c_k^{\dagger} \right| 0 \right\rangle = \sum_{\alpha} \left| \left\langle 1, k, \alpha \left| c_k^{\dagger} \right| 0 \right\rangle \right|^2 \xrightarrow[\tau \to \infty]{} Z_k e^{-\tau E_k} \tag{24}$$

In order to determine the polaron energy we need to calculate the behavior of the Green's function at large imaginary time, therefore we need to know $G(k,\tau)$ in many τ points. The figure 9 shows the quantum Monte Carlo (QMC) simulation of the Green's functions of a polaron. Here, $P(k,\tau)$ denotes the sum of all diagrams (Feynmann paths) with length τ.

Note that $P(k,\tau) \rightarrow exp(-E\tau)$ when $\tau \rightarrow \infty$, where E is the polaron energy. The plotted two curves represent $P(k,\tau).exp(\mu\tau)$ and $G(k,\tau).exp(\mu\tau)$, where μ is close to the polaron energy, versus imaginary time. The extrapolation of $G(k,\tau)$ at zero gives the quasiparticle weight $Z_0(k)$.

Figure 10 depicts the polaron dispersion relation, according to the Rayleigh-Schrödinger perturbation theory (Donsker & Varadhan1,983) and QMC (Temper et al 2009). Note that when the polaron momentum is small, the Rayleigh-Schrödinger theory works well. The QMC results show that the polaron energy continues to increase towards the threshold value $E(0)+\omega_0$, corresponding to the phonon continuum.

Figure 9. Monte Carlo simulation of polaron dynamics
After Temper et al (2009).

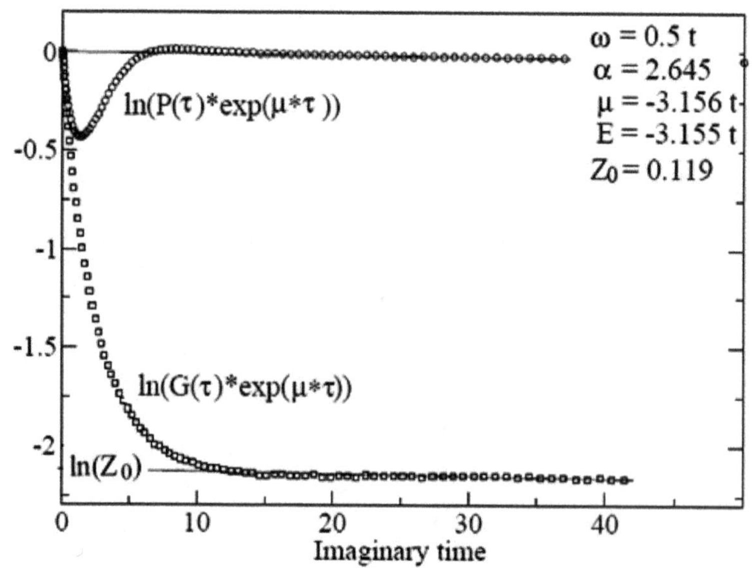

Figure 10. Monte Carlo simulation of polaron dynamics
After Svistunov et al (2002).

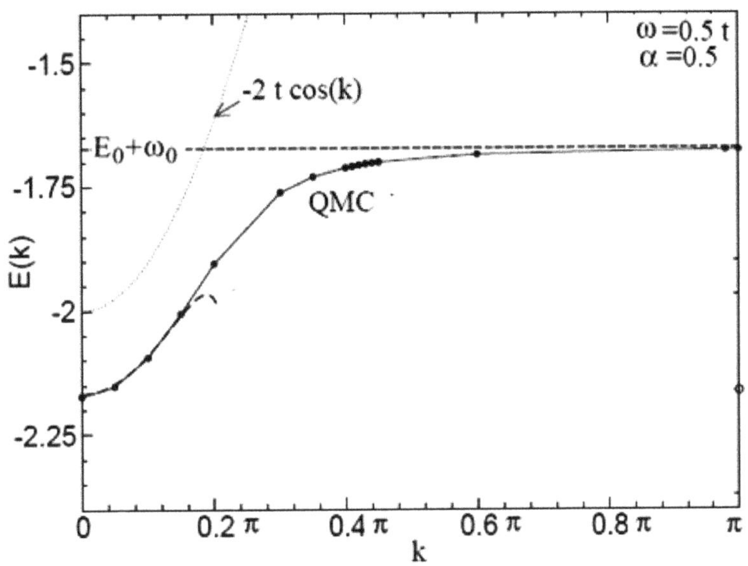

7. POLARITONS

Polaritons are Bosonic quasiparticles resulting from strong coupling of electromagnetic waves (photons) with an electric or magnetic dipole-carrying excitation. Therefore, the composite object of photon+ phonon is called polariton. At the point where the two dispersion relationships of light and excitation are crossing each other they have the same energy and therefore coupling occurs.

Figure 11. Phonon Polariton coupling point

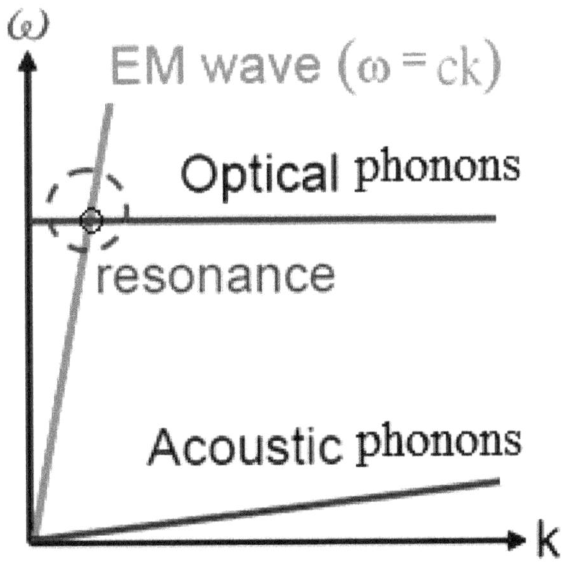

The most common types of polaritons are:

- **Phonon-Polaritons:** Where, EM waves (photons) are coupled with the optical phonons of a polar medium
- **Exciton-Polaritons:** Where, EM waves (photons) are coupled with excitons (bound electron-hole pairs) in a semiconductor
- **Plasmon-Polaritons:** Where, EM waves (photons) are coupled with plasma waves of a conducting medium

7.1 Phonon-Polaritons (Photon-Phonon Interaction)

Polariton can be formed with the coupling of photons to polar phonon in ionic solids and semiconductors. The composite object of photon+TO phonon is called polariton. In fact, resonance can only happen between photons and TO phonons. This is the most famous type of polaritons. Note that LO phonons do not couple with transverse EM waves, and do not show any dispersion, as illustrated in figure 12. Let's start with the dielectric function of a non-conducting polar material:

$$\varepsilon(\omega) = \varepsilon_o \left[(\omega^2 - \omega_{LO}^2)/(\omega^2 - \omega_{TO}^2) \right] \quad (25)$$

The dispersion of transverse polaritons can be derived from the dielectric function of the medium, with transverse EM wave, and given by:

$$\omega^2(\varepsilon(\omega)/\varepsilon_o) = q^2/c^2. \quad (26)$$

Figure 12. The dielectric function of a polar solid

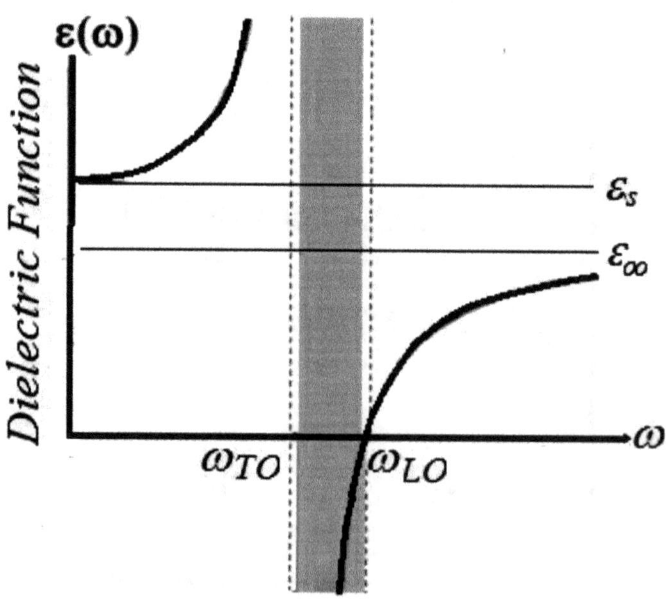

Substituting this equation into the dielectric function of a polar material (polar insulator or polar semiconductor) gives the following dispersion relation of transverse phonon-polaritons:

$$\omega^4 - \omega^2 \,[\omega_{LO}^{\;2} + q^2.c^2(\varepsilon_0/\varepsilon_\infty)] + q^2.c^2\omega_{TO}^{\;2}(\varepsilon_0/\varepsilon_\infty) = 0 \tag{27}$$

This dispersion relation is plotted in Figure 13

Phonons polaritons in insulating materials can be well measured with Raman spectrometry. In fact, Raman spectroscopy can be used to study vibrational, rotational, and other low-frequency modes in several materials. It relies on inelastic scattering (Raman scattering) of monochromatic light. Typically, a sample is illuminated with a laser beam. The light interacts with molecular vibrations or excitation modes in the sample. This results in energy shift of the light (up: Anti-Stokes-shift or down: Stokes-shift). The shift in energy can be measured and gives information about the material.

Figure 13. The phono-polariton dispersion relation

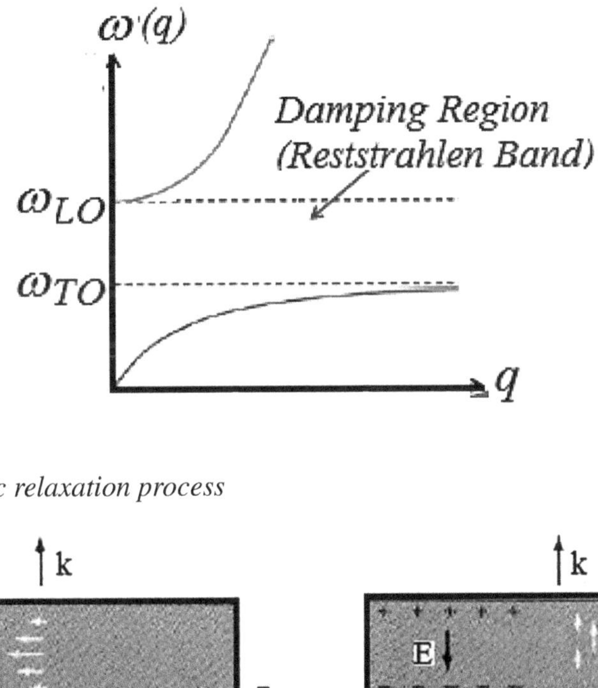

Figure 14. The dielectric relaxation process

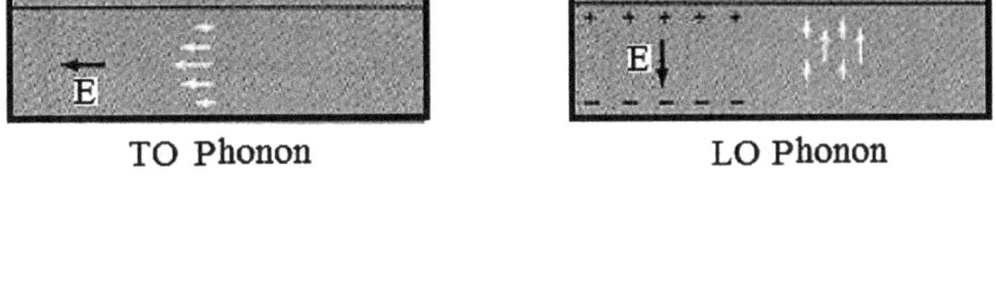

7.2 Excitonic Polaritons (Photon-Exciton Interaction)

Excitonic polariton constitutes an electric dipole which can efficiently couple to electromagnetic wave, forming the new quasi-particle. Being Bosons, exciton-polaritons can form condensates, at low temperature. The strong coupling of exciton-polaritons is particularly significant in semiconductor microcavities (e.g., of a laser) and photonic waveguides. There are several scattering mechanisms which can take away energy from the exciton-polaritons; especially via the phonon- exciton coupling. Figure 15 depicts the dispersion relation of excitonic polaritons.

The microcavity exciton-polaritons have reminiscent excitonic features. Figure 16 illustrates the microcavity structure, where a quantum well is embedded inside the cavity photon field. The strong light-matter coupling in the microcavity system exhibits anti-crossing behavior as a split to two polariton branches, namely: upper polaritons branch (UPB) and lower polaritons branch (LPB), as shown in figure 15. The energy difference between two branches is named as vacuum-Rabi splitting energy (Ω). The quantity Ω represents a collective dipole coupling strength depending on the exciton oscillator strength and the penetration depth of the Bragg mirror.

The so-called polariton laser has been realized in semiconductor micro-cavities (Kavokin, 2012). As shown in Figure 17, it consists of multilayer crystal structures in which light confined between two parallel mirrors and strongly interacts with excitons in the crystal.

This is actually a vertical-cavity surface-emitting terahertz laser, which is based on the laser polariton interaction. The unique properties of polariton laser devices make them suitable for spin switches and terahertz generators.

7.3 Plasmon Polaritons (Photon-Plasmon Interaction)

Plasmon Polariton is a coupled state between a plasmon and a photon. Plasmons may be excited in the bulk (volume) or at the surface of a metal. The so-called surface plasmon polaritons (SPPs) are electro-magnetic modes propagating along metal–dielectric (or metal-air) interfaces. The term surface

Figure 15. The dispersion relation of exciton-polaritons

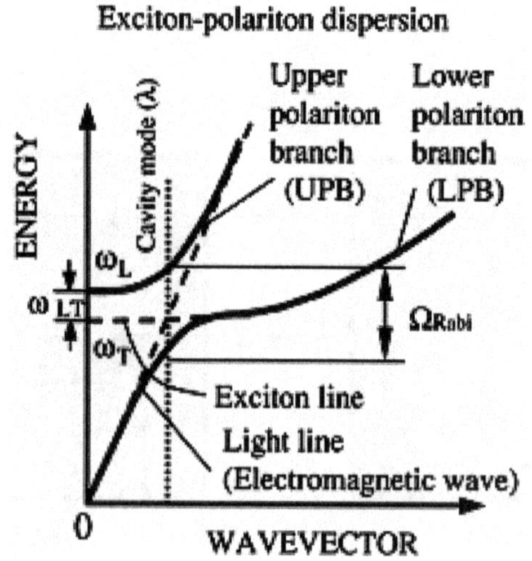

Figure 16. (a) Schematic of a microcavity quantum well structure; (b) Schematic of an electrically pumped polariton laser based on a GaN microcavity with embedded InGaN/GaN quantum wells
After Kavokin, (2012).

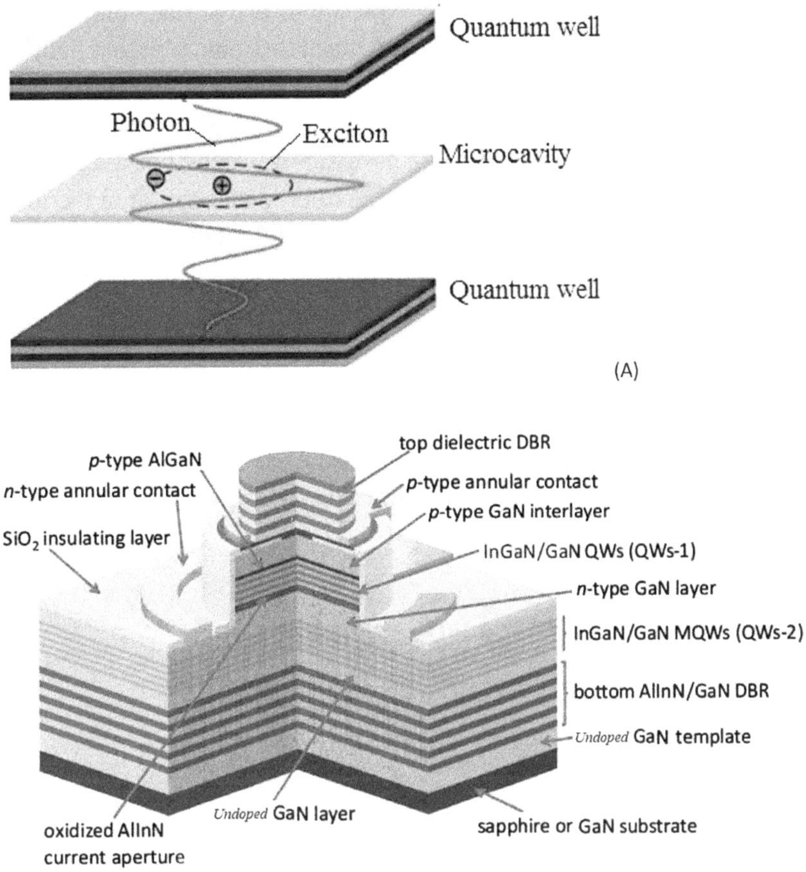

Figure 17. Volume and surface plasmon polaritons

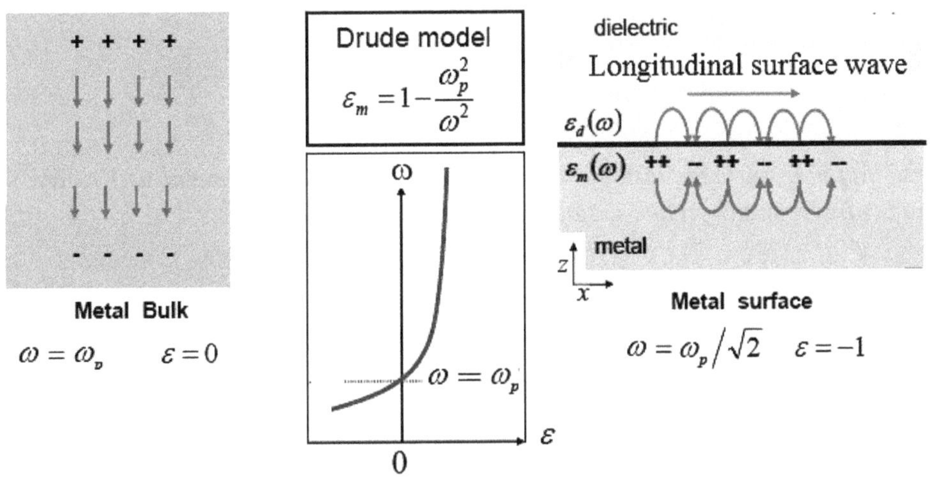

plasmon polariton explains that the wave involves both charge motion in the metal (surface plasmon) and electro-magnetic waves in the air or dielectric (polariton). Figure 17 illustrates the dielectric function of a metal (ε_m) and the frequency at which the volume and surface plasmons may be excited. The next figure depicts the dispersion relation of both volume plasmons and surface plasmon polaritons. SPPs are a type of surface wave, guided along the interface in much the same way that light can be guided by an optical fiber. SPPs are shorter in wavelength than the incident light (photons). Hence, SPPs can have tighter spatial confinement and higher local field intensity.

As SPPs are perpendicular to the interface, they have sub wavelength-scale confinement. An SPP will propagate along the interface until its energy is lost either to absorption in the metal or scattering into other directions (such as into free space). Therefore, the field modes on both sides are evanescent. Application of SPP enables sub-wavelength optics in microscopy and lithography (in IC processing) beyond the conventional diffraction limit.

SPPs can be excited by electrons or photons. Excitation by electrons is created by injecting electrons into the bulk of a metal. As the electrons scatter, energy is transferred to the electron plasma in the bulk. The component of the scattering vector parallel to the surface results in the formation of a surface plasmon polariton. For a photon to excite an SPP, both must have the same frequency and momentum. This coupling can be achieved using a coupling medium such as a prism or grating to match the photon and SPP wave vector, as shown in Figure 19.

Figure 18. Dispersion relations of volume and surface plasmon polaritons

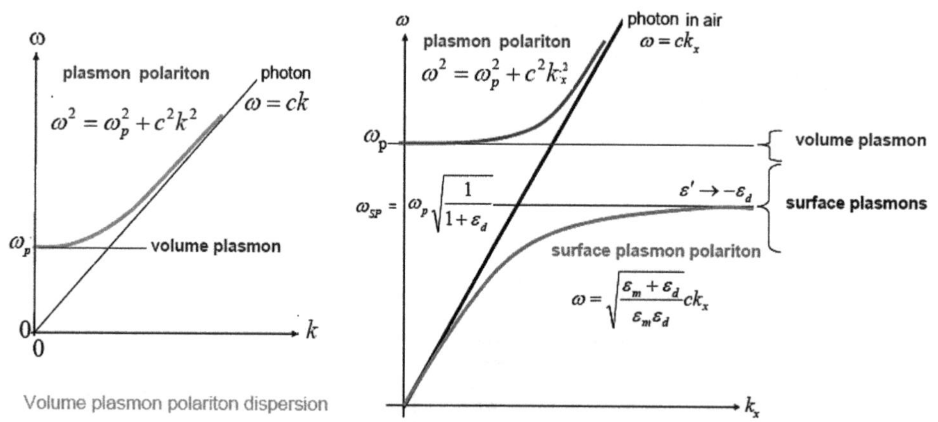

Figure 19. Excitation of surface plasma oscillations on the surface of a metal with different methods: Kretschmann configuration (left) and surface grating method (right)

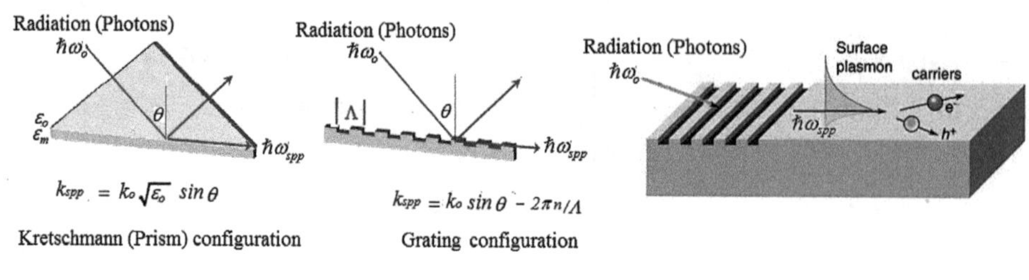

7.4 Polaritons in Photonic Waveguides

It is strongly believed that the collective behavior of excitons and polaritons at the interface of a metal/semiconductor and a dielectric, may bridge the gap between electronic and photonic devices. Some researchers observed the strong coupling between quantum well excitons and the guided mode of a planar film waveguide. Compared to using distributed Bragg reflectors (DBRs), optical confinement by total internal reflection (TIR) in principle provides for smaller losses and also larger Rabi splitting through better spatial overlap of optical field and quantum well(s). Waveguides naturally operate at large in-plane wavevectors and group velocities so that polaritons should propagate large distances within their lifetime.

In order to couple light in and out of the waveguide, we can employ a grating coupler, which consists of a periodic modulation of the cladding layer interface. Figure 20 depicts an indium tin oxide (ITO) photonic waveguide, with ITO periodic modulation corrugated structure. Here, an ultrafast laser beam is coupled into an ITO waveguide by diffraction. The guided photonic wave travels underneath a metallic corrugation, where it is channeled into a surface plasmon polariton (SPP). When the dispersion of the electromagnetic mode crosses that of the excitons and the coupling strength is large enough, there is an avoided crossing of the modes, which is the signature of the strong coupling regime.

The figure below shows another SPP-based photonic waveguide, which is composed of a quantum well and a top grating structure. The right figure shows the angle-resolved photoluminescence (PL) spectrum for excitation and detection at the patterned grating area. We observe the anti-crossing and the Rabi splitting, which comes from the single quantum well. Here, the polaritons propagate with a group velocity of 26μm/ps and a lifetime of 11.4 ps in resonance. The exciton binding energy provides a quantitative measure of Coulomb electron–hole interactions. This quantity is equal to the difference between the electrical and the optical gaps of the material

8. CASE STUDY 16: SPPD (SURFACE PLASMON POLARITON DIODE)

Figure 22 depicts the basic structure of Surface Plasmon Polariton Diode (SPPD), which was proposed by Vinnakota & Genov (2014). This device can be operated as a high-speed optical switch (up to 1THz

Figure 20. Schematic of ITO photonic waveguide, with a periodic corrugated structure

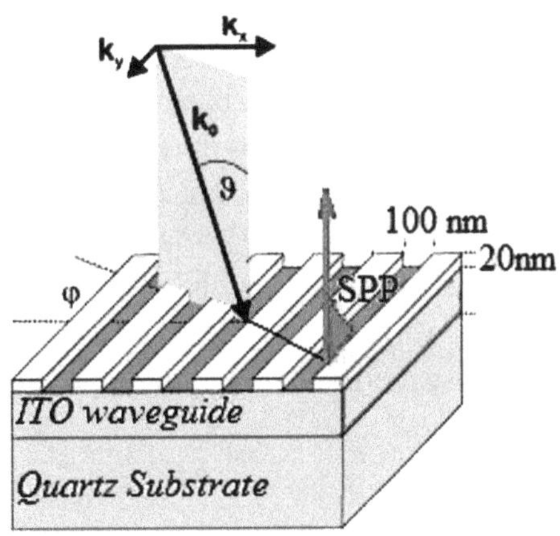

Figure 21. The dielectric relaxation process
After Walker et al, (2013).

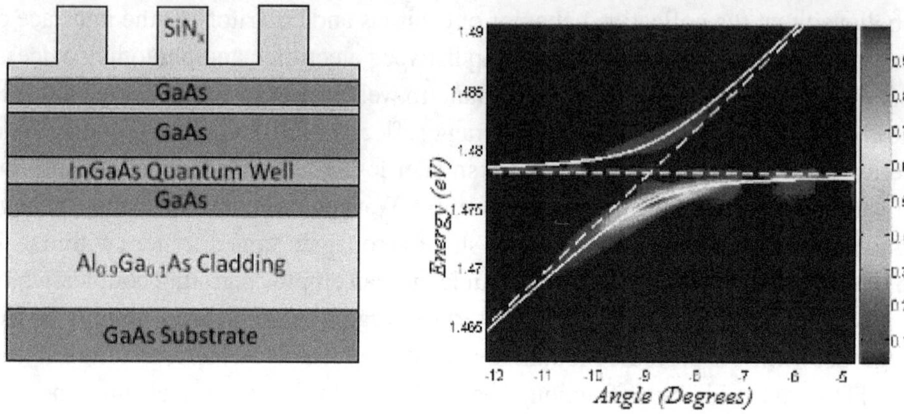

Figure 22. A schematic of an SPPD
After Vinnakota & Genov (2014).

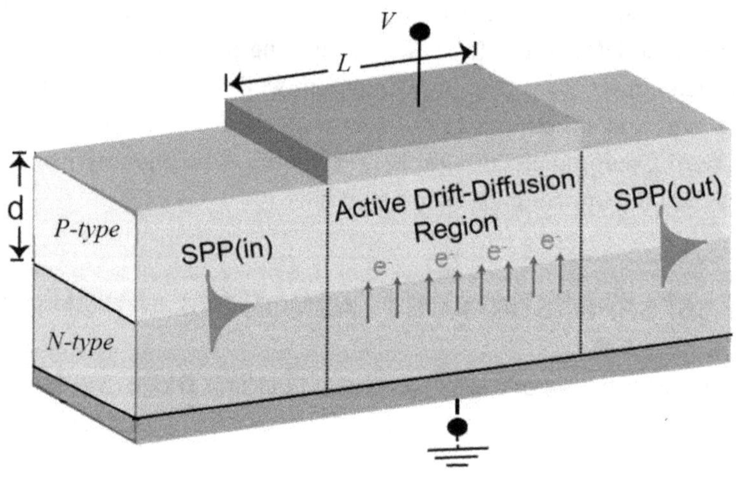

modulation ratio). As shown in figure 22, the SPPD is composed of a highly-doped P-N junction, which exploits surface plasmon polaritons (SPP's) at the junction boundaries. The device may be implemented in Si or better in *GaAs*, which has higher mobility and plasma frequency.

Figure 23 depicts the SPP dispersion curve $\omega(kspp)$ in *GaAs* for two doping concentrations. At low frequencies the SPP wave vector (k_{spp}) coincides with that in air. However, at frequencies close to the surface plasmon frequency $\omega_{sp}=\omega_p/(2\varepsilon_b)^{1/2}$, where ε_b=12.9 is the bound electrons contribution to the p-region permittivity, the wave experiences a resonance with $k_{ssp} >> 2\pi/\lambda$, where λ is the wavelength in free space. Consequently, the effective mode wavelength is dramatically reduced below the diffraction limit in the semiconductor. The displayed dispersion relation and output characteristics refer to two doping

Figure 23. The SPPD dispersion curves and transfer characteristics in air/p-type/n-type GaAs with two doping configurations; $N_A = 10^{19}$, $N_D = 5\times10^{19}\ cm^{-3}$ *(red line), and* $N_A = 10^{19}$, $N_D = 3\times10^{19}\ cm^{-3}$ *(blue line)*
After Vinnakota & Genov (2014).

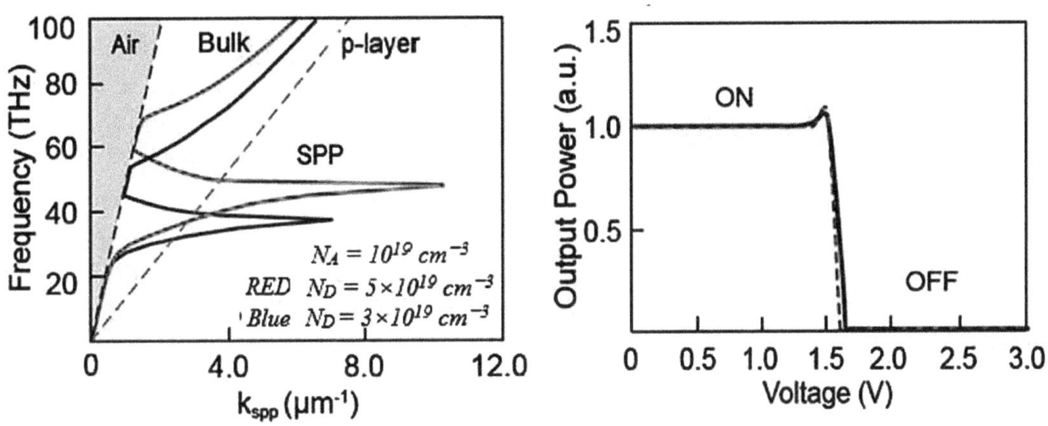

configurations: $N_A = 10^{19}$, $N_D = 5\times10^{19}\ cm^{-3}$ (red line), and $N_A = 10^{19}$, $N_D = 3\times10^{19}$ cm^{-3} (blue line). In this simulated device, the p-type layer thickness is d=0.5μm and the active region width is w=2μm. Note that for high-doping levels ($N_D > 10^{19}\ cm^{-3}$) the depletion region has a few nanometers width, and the excitation of weakly bound photonic modes is inhibited. The simulation of the SPPD is carried out by the *COMSOL* Multiphysics solver (2010) in conjunction with device simulator *SENTAURUS*, from Synopsys Inc.

The SPPD operation characteristics are studied here under steady state conditions. At zero applied bias ($V=0$), the electron concentration in the p-type layer is very low, causing this layer to behave as a dielectric. Therefore, the SPP propagation establish the ON state of the device. If a forward bias is applied ($V>0$), the electrons drift from the n-type layer into the p-type layer, and the minority concentration near the metallurgic junction increases exponentially with bias. At sufficient forward bias, greater than a critical value ($V > V_c$), the electron concentration in the active region will surpass a critical value at which the permittivity of the p-doped layer becomes negative such that $\varepsilon_p(\omega) <0$. Therefore, the p-layer layer behaves like a metal[4] and the SPP could no-longer propagate through the active region, and the OFF state of the device is established. For a given operation frequency, the critical voltage (V_c) can be obtained from the transparency condition: $\omega= \omega_p/(\varepsilon_b)^{1/2}$ or $\varepsilon_p(\omega, Vc) = 0$, and the carrier continuity equations, giving

$$V_c = (k_BT/e).ln\ [\ \varepsilon_b\, \omega^2 / \omega_{po}^{\ 2}] \tag{28}$$

where $\omega_{po} = (n_0e^2/\varepsilon_o m_n)^{1/2}$ is the plasma frequency of the minority carriers under thermal equilibrium and zero external bias, m_n is the electron effective mass in the p-type layer, and e is the electron charge. For $N_A = 5\times10^{19}$ cm^{-3} and $f = 35$ THz, the critical voltage becomes $V_c = 1.57$V. Once the applied bias is removed ($V\rightarrow0$), the excess electrons in the p-region diffuse out and their concentration falls below the critical value in about 5ps. At this point the SPPs can propagate through the device re-establishing the ON state.

The switching speed can be roughly calculated as

$$f = 1/(t_{ON} + t_{OFF}), \quad (29)$$

where

$$t_{ON} = l_{spp}^{\ 2}/16D_n,\ t_{OFF} = l_{spp}/v_d\ . \quad (30)$$

Here, the SPP field penetration length $l_{spp} = 1/k^z_{spp}$, with $k^z_{spp} = \sqrt{(k_{spp}^{\ 2} - \varepsilon_p . k_o^{\ 2})}$ and the electron diffusion constant in the p-type region $D_n = (k_B T/e)\mu_n$. Also the drift velocity may be estimated as $v_d = \mu_n (V-V_{bi})/x_p$, where x_p is the thickness of space-charge-region extension in the P-side and the built-in voltage $V_{bi} = V_T ln\ (N_A/n_i)$. Figure 24 shows the switched output power when the SPPD is switched with a square input voltage.

9. SUMMARY

In this Chapter we study the properties of polarons, plamons and polaritons quasiparticles, along with their application to quantum nano-photonic and nanoelectronic devices.

In order to understand the dynamics of the above quasiparticles and their transport properties in these materials, we reviewed the dielectric function of a semiconductor, since they are more or less coupled to their dielectric properties. Consequently, we study the following interactions and the transport of their associated quasi particles

- Electron-electron interactions & Plasma Oscillations,
- Electron-lattice interaction (Polarons),
- Photon-phonon interaction (Phonon-polaritons),
- Photon-exciton interactions (Excitonic polaritons).

Figure 24. The SPPD switching characteristics (output power), upon exposure to a square input voltage After Vinnakota & Genov (2014).

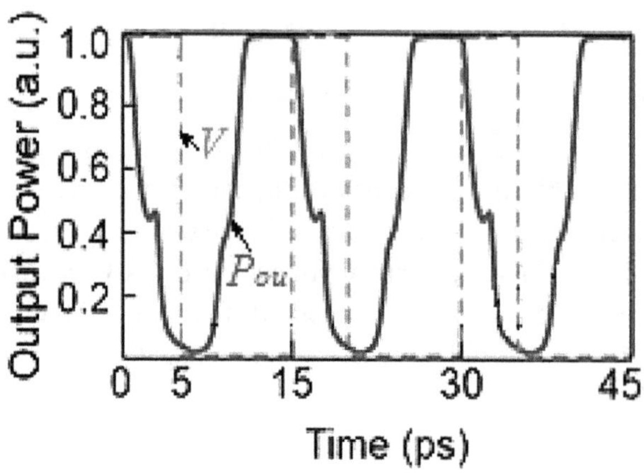

REFERENCES

Abdeldayem, H., Frazier, D. O., Paley, M. S., & Witherow, W. K. (2002). *Recent Advances in Photonic Devices for Optical Computing*. Retrieved from science.nasa.gov

Bond, W. L., Cohen, B. G., Leite, R. C., & Yariv, A. (1963). Observation of the dielectric-waveguide mode of light propagation in p-n junctions. *Applied Physics Letters*, *2*(3), 57–59. doi:10.1063/1.1753773

Buot, F. A. (2009). *Nonequilibrium Quantum Transport Physics in Nanosystems: Foundation of Computational Nonequilibrium Physics in Nanoscience and Nanotechnology*. World Scientific. doi:10.1142/6042

Cao, Y. H., Xie, J., Liu, Y. M., & Liu, Z. Y. (2014). Modeling and optimization of photonic crystal devices based on transformation optics method. *Optics Express*, *22*(3), 2725–2734. doi:10.1364/OE.22.002725 PMID:24663564

Colclaser, R., & Diehl-Nagle, S. (1985). *Materials and Devices for Electrical Engineers and Physicists*. New York: McGraw-Hill.

Das Sarma, S., & Mason, B. A. (1985). Optical phonon interaction effects in layered semiconductor structures. *Annals of Physics*, *163*(1), 78–119. doi:10.1016/0003-4916(85)90351-3

Datta, S. (1990). A simple kinetic equation for steady-state quantum transport. *Journal of Physics Condensed Matter*, *2*(40), 8023–8052. doi:10.1088/0953-8984/2/40/004

Debye, P. (1913). Article. Ver. Deut. Phys. Gesell., 15, 777.

Deng, H., Haug, H., & Yamamoto, Y. (2010). Exciton-polariton Bose-Einstein condensation. *Reviews of Modern Physics*, *82*(2), 1489–1537. doi:10.1103/RevModPhys.82.1489

Devreese, J. T. (2005). Polarons. In R. G. Lerner (Ed.), *Encyclopedia of Physics* (pp. 2004–2027). Weinheim: Wiley-VCH.

Devreese, J. T. L., & Evrard, R. (1964). On the excited states of a symmetrical polaron model. *Physics Letters*, *11*(4), 278–279. doi:10.1016/0031-9163(64)90324-5

Dolocan, V., Dolocan, A., & Dolocan, V. O. (2010). Some aspects of polaritons and plasmons in materials. *Communications in Nonlinear Science and Numerical Simulation*, *15*(1), 629–636. doi:10.1016/j.cnsns.2009.05.018

Donsker, M., & Varadhan, R. (1983). Asymptotic for the Polaron. *Communications on Pure and Applied Mathematics*, *36*(4), 505–528. doi:10.1002/cpa.3160360408

Feynman, R. P. (1955). Slow Electrons in a Polar Crystal. *Physical Review*, *97*(3), 660–665. doi:10.1103/PhysRev.97.660

Froelich, H. (1958). *The Theory of Dielectrics*. Oxford Press.

Fröhlich, H. (1954). Electrons in lattice fields. *Advances in Physics*, *3*(11), 325–361. doi:10.1080/00018735400101213

Fröhlich, H., Pelzer, H., & Zienau, S. (1950). Properties of slow electrons in polar materials. *Philosophical Magazine, 41*(314), 221–242. doi:10.1080/14786445008521794

Harrop, P. J. (1972). *Dielectrics*. Butterworths.

Heavens, O. S. (1991). *Optical Properties of Thin Solid Films*. New York: Dover Publications.

Heidel, T.D., Mapel, J.K., Celebi, K., Singh, M., & Baldo, M.A. (2007). Analysis of surface plasmon polariton mediated energy transfer in organic photovoltaic devices. *Proc. of SPIE, 6656*, 66560I1-8.

Hess, K. (2000). *Advanced Theory of Semiconductor Devices*. Piscataway, NJ: IEEE Press.

Hill, A., Mikhailov, S. A., & Ziegler, K. (2009). Dielectric function and plasmons in graphene. *Euro. Phys. Letters, 87*(2), 27005. doi:10.1209/0295-5075/87/27005

Hirose, K., & Kobayashi, N. (2014). *Quantum Transport Calculations for Nanosystems*. Pan Stanford.

Hopfield, J. (1958). Theory of contribution of excitons to the complex dielectric constant of crystals. *Physical Review, 112*(5), 1555–1567. doi:10.1103/PhysRev.112.1555

Huang, D., Easter, M., Gumbs, G., Maradudin, A. A., Lin, S.-Y., Cardimona, D. A., & Zhang, X. (2014). Controlling quantum-dot light absorption and emission by a surface-plasmon field. *Optics Express, 22*(22), 27576–27605. doi:10.1364/OE.22.027576 PMID:25401904

Jørgensen, N. B., Wacker, L., Skalmstang, K. T., & Parish, M. M. (2016). Observation of Attractive and Repulsive Polarons in a Bose-Einstein Condensate. *Phys. Rev. Lett., 117*(5), 055302. arXiv: 1604.07883

Kadanoff, L., & Baym, G. (1962). *Quantum Statistical Mechanics: Green's Function Methods in Equilibrium and Non Equilibrium Problems*. New York: W. A. Benjamin.

Kavokin, A. V., Shelykh, I. A., Taylor, T., & Glazov, M. M. (2012). Vertical cavity surface emitting terahertz laser. *Physical Review Letters, 108*(19), 197401. doi:10.1103/PhysRevLett.108.197401 PMID:23003086

Keldysh, L. V. (1964). Diagram technique for nonequilibrium processes. *Zh. Eksp. Teor. Fiz., 47*, 1515.

Kira, M., & Koch, S. W. (2006). Many-body correlations and excitonic effects in semiconductor spectroscopy. *Progress in Quantum Electronics, 30*(5), 155–296. doi:10.1016/j.pquantelec.2006.12.002

Kittel, C. (1996). *Introduction to solid state physics*. New-York.

Knox, R. S. (1963). Theory of excitons. In Solid state physics. Academic.

Koswatta, S. O., Hasan, S., Lundstrom, M. S., Anantram, M. P., & Nikonov, D. E. (2007). Nonequilibrium Greens Function Treatment of Phonon Scattering in Carbon-Nanotube Transistors. *IEEE Transactions on Electron Devices, 54*(9), 2339–2351. doi:10.1109/TED.2007.902900

Landau, L. D. (1933). Über die Bewegung der Elektronen in Kristalgitter. *Phys. Z. Sowjetunion, 3*, 644–645.

Landau, L.D., & Pekar, S.I. (1948). Effective mass of a polaron. *Zh. Eksp. Teor. Fiz., 18*, 419–423. (in Russian)

Lieb, E. H., & Thomas, L. E. (1997). Exact Ground State Energy of the Strong Coupling Polaron. *Communications in Mathematical Physics, 183*(3), 511–519. cond-mat/9512112 doi:10.1007/s002200050040

Low, T., Chaves, A., Caldwell, J. D., Kumar, A., Fang, N. X., Avouris, P., . . . Koppens, F. (2016). *Polaritons in layered 2D materials*. Available at: https://www.researchgate.net/publication/309191944

Mahan, G. D. (1990). *Many Particle Physics*. New York: Plenum. doi:10.1007/978-1-4613-1469-1

Maier, S. A. (2007). *Plasmons: Fundamentals and Applications*. Springer.

Maier, S. A. (2010). *Plasmonics: Fundamentals and Applications*. Springer.

Mishchenko, A., Nagaosa, N., Prokofev, N., Sakamoto, A., & Svistunov, B. (2002). Self-trapping of polarons in the Rashba-Pekar model. *Physical Review B: Condensed Matter and Materials Physics*, *66*(2), 020301. doi:10.1103/PhysRevB.66.020301

Mishchenko, A. S., Prokofev, N. V., Sakamoto, A., & Svistunov, B. V. (2000). Diagrammatic quantum Monte Carlo study of the Fröhlich polaron. *Physical Review B: Condensed Matter and Materials Physics*, *62*(10), 6317–6336. doi:10.1103/PhysRevB.62.6317

Miura, N., & Imanaka, Y. (2003). Polaron cyclotron resonance in II–VI compounds at high magnetic fields. *Physica Status Solidi*, *237*(1), 237–243. doi:10.1002/pssb.200301781

Moradi, A. (2010). Guided dispersion characteristics of metallic single-walled carbon nanotubes in the presence of dielectric media. *Optics Communications*, *283*(1), 160–163. doi:10.1016/j.optcom.2009.09.038

Moradi, A., & Khosravi, H. (2007). Plasmon dispersion in metallic carbon nanotubes in the presence of of low-frequency electromagnetic radiation. *Physics Letters. [Part A]*, *371*(1-2), 1–6. doi:10.1016/j.physleta.2007.05.110

Mott, N. F., & Gurney, R. W. (1940). *Electronic processes in ionic crystals*. London: Oxford University press.

Nikolaenko, A. E., Papasimakis, N., Chipouline, A., De Angelis, F., Di Fabrizio, E., & Zheludev, N. I. (2012). THz bandwidth optical switching with carbon nanotube metamaterial. *Optics Express*, *20*(6), 6068–6079. doi:10.1364/OE.20.006068 PMID:22418486

Peeters, F. M., Wu, X. G., & Devreese, J. T. L. (1988). Exact and approximate results for the mass of a two-dimensional polaron. *Physical Review B: Condensed Matter and Materials Physics*, *37*(2), 933–936. doi:10.1103/PhysRevB.37.933 PMID:9944589

Pekar, S.I. (1951). *Issledovanija po Elektronnoj Teorii Kristallo*. Gostekhizdat.

Raether, H. (1988). *Surface Plasmons on Smooth and Rough Surfaces and on Gratings Springer. Tracts in Modern Physics* (Vol. 111). Springer Berlin.

Sze, S. M., & Ng, K. K. (2006). *Physics of Semiconductor Devices*. Wiley-Interscience. doi:10.1002/0470068329

Tempere, J., Casteels, W., Oberthaler, M., Knoop, S., Timmermans, E., & Devreese, J. T. (2009). Feynman path-integral treatment of the BEC-impurity polaron. *Physical Review B: Condensed Matter and Materials Physics*, *80*(18), 184504. 0906.4455 doi:10.1103/PhysRevB.80.184504

Tikhodeev, S. G., Gippius, N. A., Christ, A., Zentgraf, T., Kuhl, J., & Giessen, H. (2005). Waveguide-plasmon polaritons in photonic crystal slabs with metal nanowires. *Solid State Physics*, *2*(2), 795–800.

Titantah, J. T., Pierleoni, C., & Ciuchi, S. (2001). Free Energy of the Fröhlich Polaron in Two and Three Dimensions. *Physical Review Letters*, *87*(20), 206406. cond-mat/0010386 doi:10.1103/PhysRevLett.87.206406 PMID:11690499

Vinnakota, R. K., & Genova, D. A. (2014). Terahertz Optoelectronics with Surface Plasmon Polariton Diode. *Scientific Reports*, *4*, 4899. doi:10.1038/srep04899 PMID:24811083

von Hippel, A. (Ed.). (1961). *Dielectric Materials and Applications*. Cambridge, MA: MIT Press.

Walker, P. M., Tinkler, L., Durska, M., Whittaker, D. M., Luxmoore, I. J., Royall, B., & Ritchie, D. A. et al. (2013). Exciton polaritons in semiconductor waveguides. *Applied Physics Letters*, *102*(1), 012109. doi:10.1063/1.4773590

Wannier, G. H. (1937). The structure of electronic excitation levels in insulating crystals. *Physical Review*, *52*(3), 191–197. doi:10.1103/PhysRev.52.191

Wannier, G. H. (1959). *Elements of Solid State Theory*. New York: Cambridge University Press.

Yu & Cardona. (1996). *Fundamentals of Semiconductors*. Springer Verlag.

Zahid, F., Paulsson, M., & Datta. S. (2002). Advanced Semiconductors and Organic Nano-Techniques. In *Electrical Conduction through Molecules*. Academic Press.

Zhanghua, H., & Bozhevolnyi, S. I. (2012). Radiation guiding with surface plasmon polaritons. IOP Publishing Ltd. *Reports on Progress in Physics*, *76*(1).

ENDNOTES

1. Note that ε_2 is sometimes considered as a negative quantity, and hence we write $\varepsilon = \varepsilon_1 + j\,\varepsilon_2$
2. Generally speaking, damping and friction forces are due to carrier scattering events. When the carrier scattering by impurities is dominant, the damping factor will be frequency dependent.
3. This is correct when there is no collisions (no damping), in an electron gas or plasma
4. Like metals, the optical response of highly doped *semiconductors* may be described by the Drudé model for dielectric function: $\varepsilon = \varepsilon_b - \omega_p^2/(\omega^2 + j\omega/\tau_n)$, where ε_b is the bound electrons contribution ($\varepsilon_b = 12.9$ in *GaAs*). The electronic relaxation time may be taken as the mean free time such that $\tau_n = m_n\mu_n/e$, where μ_n is the electron drift mobility (Bond et, 1963).

Chapter 11
Carrier Transport in Organic Semiconductors and Insulators

1. OVERVIEW AND CHAPTER OBJECTIVES

In this chapter we focus our attention on the electronic transport Models of organic semiconductors and insulator materials, which is an active research topic nowadays. Indeed, it may seem strange to talk about charge transport in insulators, and organic semiconductors, which are also inherently insulators. However, if one generates charge carriers in such materials, for example by optical excitation, they may move with a mobility that is comparable to that of conventional semiconductors or more. Obviously, an insulator can be converted into a semiconductor if free charge carriers are generated by either injection, or doping, or by optical excitation.

As we pointed out so far, some organic semiconductor crystals, such as polyacenes (anthracene, tetracene, and pentacene), and *fullerenes* (bucky balls), can conduct electricity and have attractive characteristics for electronic and optoelectronic devices. Such organic semiconductors are already used in photocopiers. They are also employed in biosensors, light emitting diodes (OLEDs) in flat panel displays, and about to enter the solar cell market. In fact, almost all main players in the electronics industry work on organic displays, nowadays. Technically speaking, organic semiconductors are relatively low cost materials, their fabrication techniques are quite simple (no need for clean-room or high temperature processing), and they can be deposited on various types of substrates.

Organic semiconductors (OSCs) can be classified into two main categories: small molecules or *oligomers* and *polymers*. A polymer is a macromolecule that is composed of many repeating units, the monomer units. The properties of OSCs differ from conventional inorganic crystalline semi-conductors in many aspects. A central point is the mechanisms related to charge transport. In order to understand charge transport in OSCs, we need to elaborate on the electronic structure of organic solids. As shown in Figure 1, the organic solids such as polymers, are made of molecular subunits.

Conductive OSCs are hydro-carbon molecules with a backbone of unsaturated carbon atoms (like --CH-CH-CH--, each atom is threefold coordinated). These chains may contain other elements, including

DOI: 10.4018/978-1-5225-2312-3.ch011

Figure 1. Schematic illustration of the two basic families of organic materials

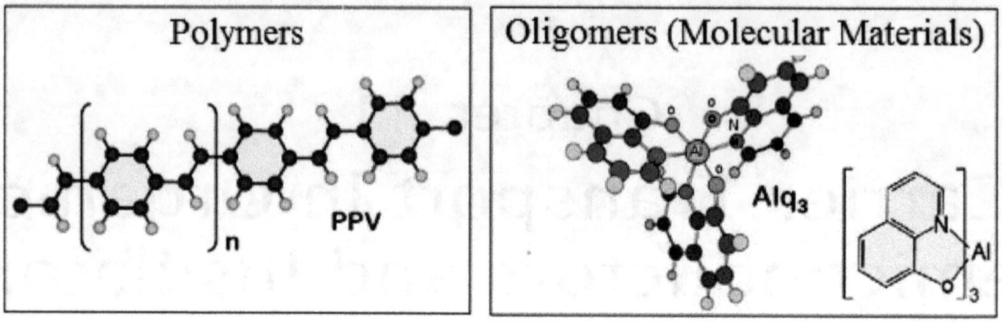

H, N, and O. Table 1 shows the chemical formulas of conducting polymers commonly used in electronic applications. Note that a polymer is a large molecule (macromolecule) composed of repeating structural units. The polymer subunits are usually connected by covalent bonds.

In order to understand the conduction mechanisms in organic semiconductors, it is necessary to review the concept of energy bands and hybridization of orbitals. The bonds that form the molecular backbone arise from sp^2 hybridized atomic orbitals of adjacent carbon atoms that overlap yielding a bonding and antibonding molecular σ and σ^* orbitals. The remaining atomic pz orbitals overlap to a lesser degree, so that the resulting molecular π and π^* orbitals are less binding or anti-binding, thus forming the frontier orbitals of the molecule. Therefore, the molecular orbitals are called σ-orbitals when the spatial probability density of the electrons is centered on the axis joining two atoms like a cylinder, and π-orbitals, when the electrons are most likely to be found above and below the line connecting two atoms. According to Hückel theory (1931), the calculations of band structure become much simpler, yet give still reasonably

Table 1. Molecular structures of typical conductive polymers

Polymer	Structure	Polymer	Structure
Polyacetylene	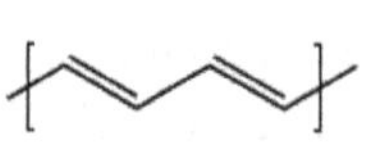	Polyaniline (PANI)	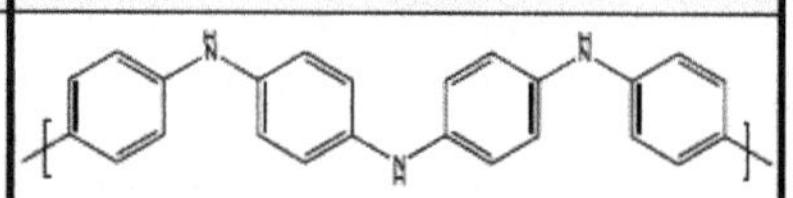
Polyethylenedioxythiophene (PEDOT)	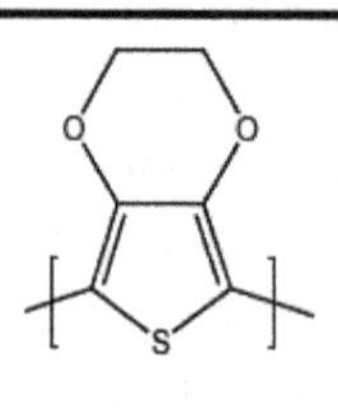	Polypyrrole (PPy)	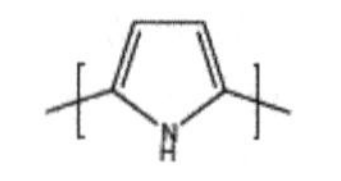
Polyalkyl thiophenes	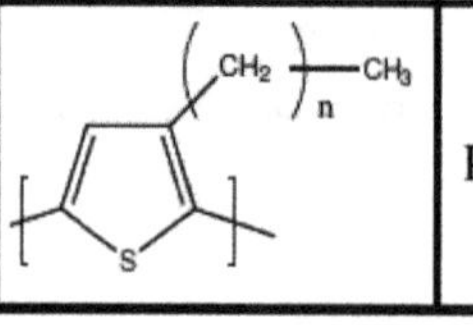	Poly(2-7carbazoles)	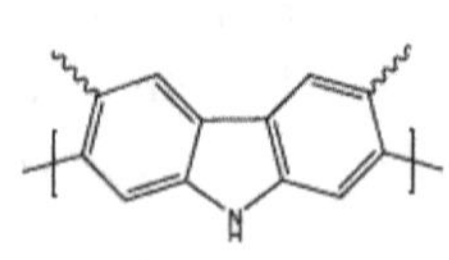

accurate results, when one treats the π-orbitals separately from the σ-orbitals. This Hückel-approximation works for many hydrocarbons with valence electrons in π-orbitals.

In the ground state of the molecule, all bonding orbitals up to the highest occupied molecular orbital (HOMO) are filled with two electrons of antiparallel spin while the antibonding orbitals, from the lowest unoccupied molecular orbital (LUMO) onwards, are empty. The band gap that is opened by this symmetry lowering is in the order of 2.5eV, which is in the range of semiconductors. Figure 2 depicts the orbital hybridization and band structure of organic molecules and how it is different from bulk semiconductor materials

With a band gap of about 2.5eV (between π -valence band and π*-conduction band) the polymer cannot conduct electrical current at all. For charge transport in organic solids to take place, there must be a charge on the molecular unit. This may either be an additional electron that is donated to an antibonding orbital (in π*-conduction band), or one that is removed from a bonding orbital (in π-conduction band). The molecule is then no longer in the ground state but rather in a charged (excited) state. Excited states can be formed in polymer molecules by variety of methods, like light absorption, where an electron is promoted from HOMO to LUMO.

In disordered organic semiconductors charge transport occurs mainly by *hopping* between nearby localized states which are induced by disorder. Disorders may be static, due to impurities and lattice imperfections or dynamic due to temperature fluctuations.

Note 1: Van der Waal Forces in Organic Molecular Structures

The concept of the van-der-Waals interaction between two molecules is based on the fact that the molecules have no static dipole moment, but that they have a charge distribution that is not entirely rigid. A

Figure 2. Schematic illustrations of the orbital hybridization (above) and the band structure (below) of organic molecules and polymers

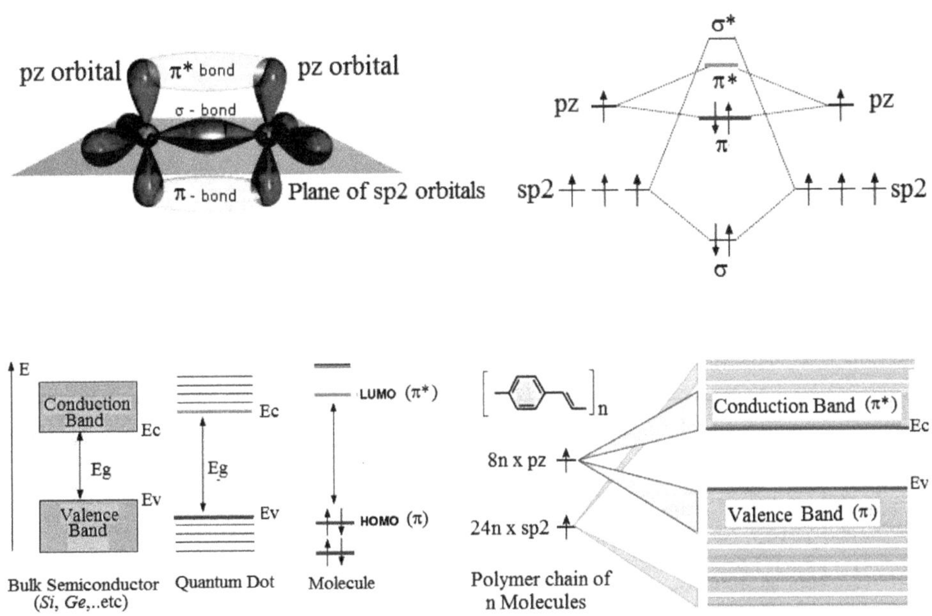

temporal fluctuation in the charge distribution in a molecule implies an associated temporary fluctuating dipole moment. This will induce a corresponding fluctuating dipole in the second molecule. The electrostatic interaction between the correlated fluctuating dipoles in the two molecules results in an attractive force, the van-der-Waals-attraction. This force depends strongly on the distance *r* between the molecules, and on the ability to induce dipole moments in a charge distribution, that is, the polarizability α of a molecule. Quantitatively, the potential energy associated with a van-der-Waals interaction is given by $VV_{dW} \propto \alpha 2/r6$, that is, the force is proportional to $r{-}7$·

This has two consequences. First, the dependence on polarizability implies that molecular crystals are preferentially formed by molecules that possess a filled outer orbital that is large and delocalized, so that many electrons can easily be moved over some distance on the molecule. This is the case for flat molecules with π-orbitals, like polyacenes (e.g., benzene, naphthalene).

In this chapter we present the transport mechanisms and models of both insulators and organic semiconductors.

As shown in Figure 3, the transport in insulators and organic semiconductors has been studied on the basis of the semiclassical and quantum approaches which are based on the Schrödinger equation, such as the *NEGF*. The combination of NEGF with band structure calculation methods, such as the *DFT*, has been also used for studying charge transport in molecular devices (e.g., by Stokpro, 2008). However, some approaches are macroscopic (such as GDM, EMA and MRT) and others are microscopic or molecular (such as Marcus theory and NEGF-DFT).

In order to appreciate how tiny is molecular devices and how much important is the accuracy of adopted transport model, look at the simple molecular wire device in the next figure. As shown, the device consists of a molecule of dithiol phenylene–ethynylene oligomer (popularly called a tour wire) coupled with two metal gold electrodes.

We start this Chapter by reviewing the conventional transport mechanisms in insulating materials, and thereupon, we continue with the recent transport models, which are dedicated for organic semiconductors. This arrangement is logic because, as we already illustrated, organic semi-conductors are insulators and become semiconducting by doping or when charges are generated by photoexcitation or injected from external electrodes. Most of the material of this Chapter is based on the review articles of Coropceanu et al (2007) and Bassler and Kohler (2012). We summarize the most important models and show their

Figure 3. Transport mechanisms and models in insulators and organic semiconductors

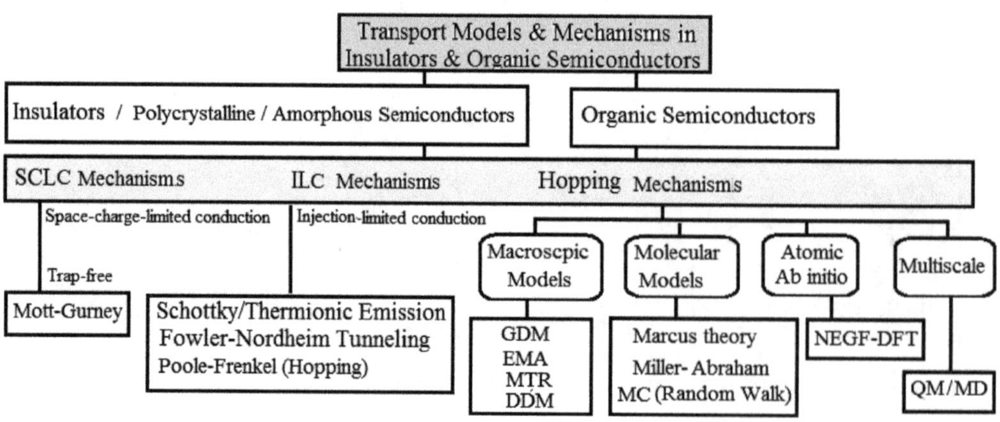

Figure 4. Schematic of a molecular wire, between two gold electrodes

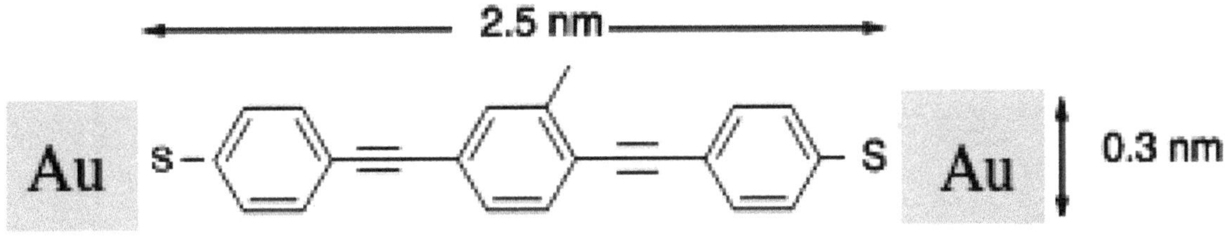

field of application and finish with practical case studies, showing how these models are applied to real organic semiconductor devices.

Upon completion of this Chapter, the reader will be able to

- Understand the different transport mechanisms of insulators and organic semiconductors
- Know how and when to apply the relevant transport models across insulators and organic semiconductors.
- Understand the mechanisms of charge carrier transport in organic semiconductors and nanodevices.

2. TRANSPORT MODELS IN INSULATORS AND DISORDERED MATERIALS

According to the energy band theory, conventional insulators and semiconductors have no free electrons at 0K. In additions to their classic use as isolators, insulators have other interesting features. The most important of these features is the ability of insulators to be polarized under the effect of electric field and to create electric dipoles.

The transport of charge carriers in insulators and disordered materials is different from conventional semiconductors. In general, it has a stronger field-dependence and may involve the motion of massive ions, by hopping, across vacancies and interstitial sites. This is true especially in view of the wide energy gap of insulator materials. In thin insulators, electrons may tunnel across the material as well. Hopping is important in disordered materials, such as polycrystalline and amorphous semiconductors as well as organic semiconductors. In the following subsections we present the physical models of these transport mechanisms

2.1 Classification of Insulators

Before discussing the transport and tunneling of charge carriers across insulating materials, it is wise to review the important types of insulators and their characteristics. There exist so many categories of insulating materials, among which one can distinguish the following types:

- Band Insulator, due to e-lattice interaction
- Mott Insulator, due to insulator-metal quantum transition
- Anderson insulator, due to disorder and localized states
- Topological insulators (TI).

Figure 5. Schematic illustration of the hopping conduction mechanism in insulators

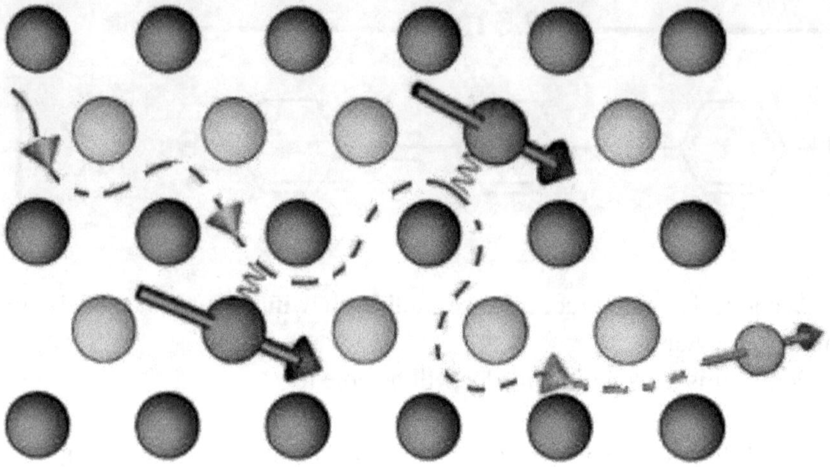

Figure 6 depicts the chronical evolution of discovery of different types of insulators. We present a squeezed definition of each type in the following subsections.

Band (Bloch-Wilson) Insulator s

Band insulators are the conventional insulators which have large energy gap, due to electron-periodic lattice field interactions (Wilson, 1931). Band insulators are simple insulators because the band theory of solids successfully accounts for their properties.

Mott Insulators

Mott insulators are a special type of insulators, due to insulator-metal quantum transition because of electron-electron interactions (Mott, 1937). These many-body interactions lead to an energy gap in the excitation spectrum of the materials. The vanadium oxide (V_2O_3) is a clear example of Mott insulators. As we outlined so far, in Chapter 4, the Hubbard model (which an extension to the TB model) is able to describe the interaction-driven transition from a metal to an insulator, commonly known as the *Mott transition.*

Figure 6. Basic types of insulators

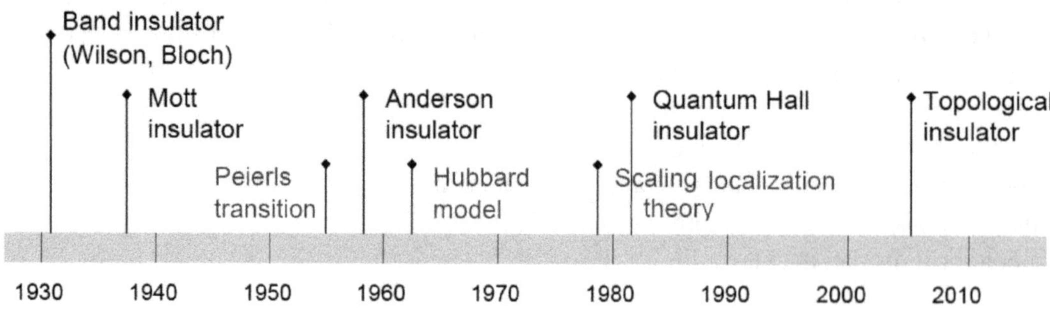

Figure 7. Band insulators vs. conductors

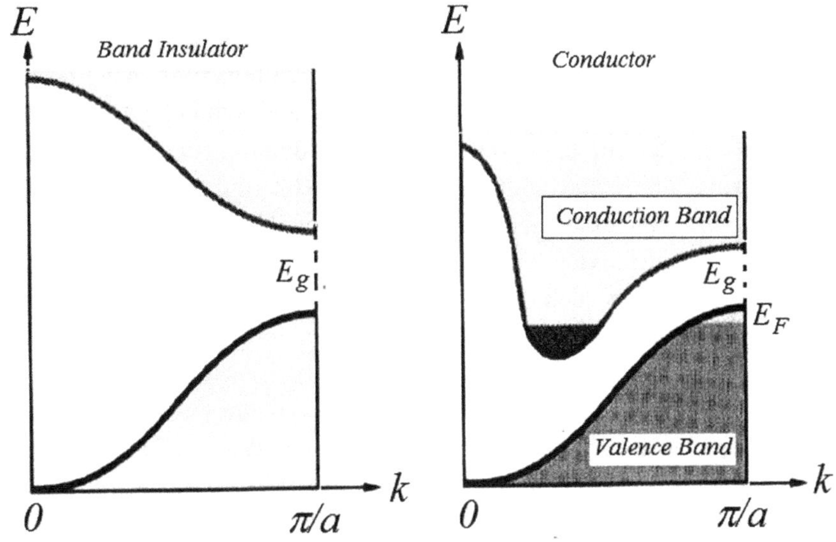

Anderson Insulators

Anderson insulators are due to localized states and disorders, such as impurities and lattice imperfections that present in certain materials. The phenomenon of a spatially localized wavefunctions in a disordered material is known as the Anderson localization (Anderson, 1958). Figure 9 depicts the Anderson insulator transition & Peierls instability of metals (1955).

Figure 8. Mott insulator transition due to e-e interaction at low temperature (left). Vanadium oxide structure as an example (right)

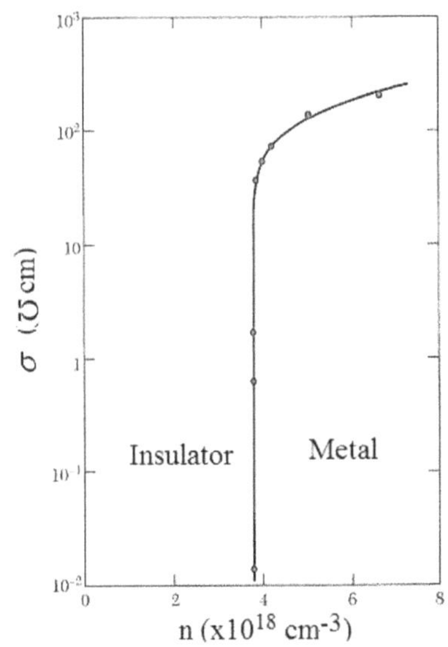

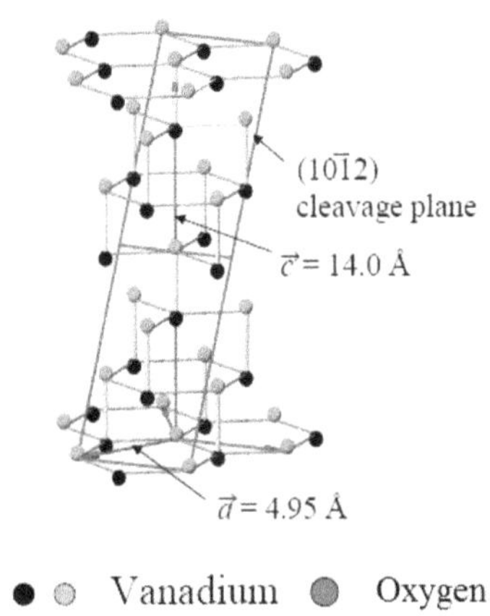

Topological Insulators

The topological insulator (TI) is a state of matter with unusual properties that appears on the surface of insulators or at the interface of narrow band gap materials with gapless (metallic) surface/interface states. The surface/interface states are excited by massless fermions (called *Kramers* partners). The spin of such fermions is locked at a right angle to their momenta due to spin-orbit coupling, making them protected against scattering. This means the larger the spin-orbit coupling, the better is the protection. Topological insulators are a subclass of narrow-gap semi-conductors and heterostructures where the band gaps are smaller than the relativistic corrections to the band structure. In fact, the heterostructures of some compound semiconductors such as $Bi_{1-x}\,Sb_x$ and $Pb_{1-x}Sn_xTe$ exhibit a strong topological phase. For instance, in $Bi_{1-x}Sb_x$, with increasing the molar fraction x, the gap at the L point decreases and the L^+ and L^- bands meet at x = 0:04. With further increase in the mole fraction x, the gap reopens with inverted band. Some 2D topological insulators, such as *HgTe* quantum wells, exhibit the quantum spin Hall Effect (SQHE). In the interface of the quantum well of such 2D TI's, the spin changes sign, thereby creating a pair of gapless helical edge states inside the energy gap.

2.2 Transport Across Thin Insulator Materials

Now, we review the principal methods used to formulate charge carrier transport across insulators. In fact, the transport in nanoscale devices cannot neglect the transport and tunneling of charge carriers across thin insulators and dielectric materials that essentially coexist with semiconductor active regions. Actually, charge carriers may move or tunnel across insulating materials by different mechanisms. In thin insulators, the most frequent transport mechanisms of charge carriers are *Fowler-Nordheim* and *Poole-Frenkel* mechanisms. In the following subsections we present the physical models of these transport mechanisms. An emphasis is made on the transport across thin oxide insulator, in metal-oxide-semiconductor (MOS) field effect transistors.

Figure 9. Anderson insulator transition and Peierls instability of metals

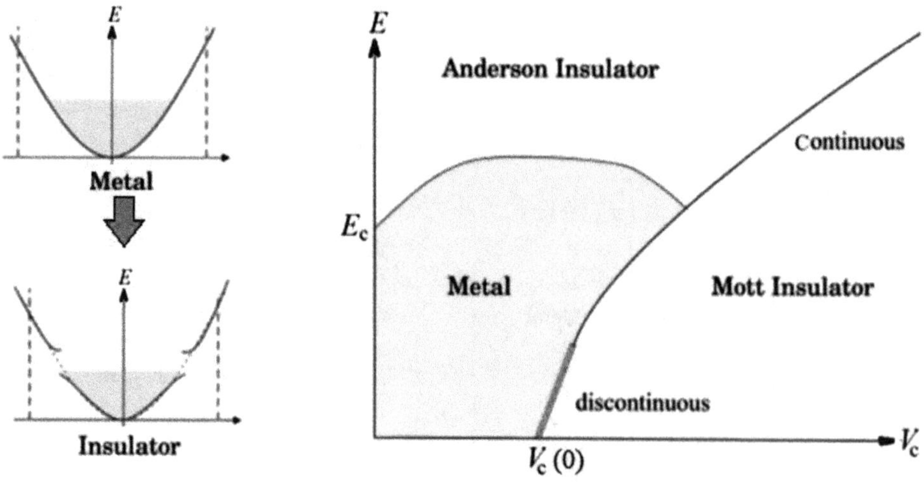

Figure 10. Band inversion between two end members in $Pb_{1-x}Sn_xTe$., and energy spectrum of the inverted contact

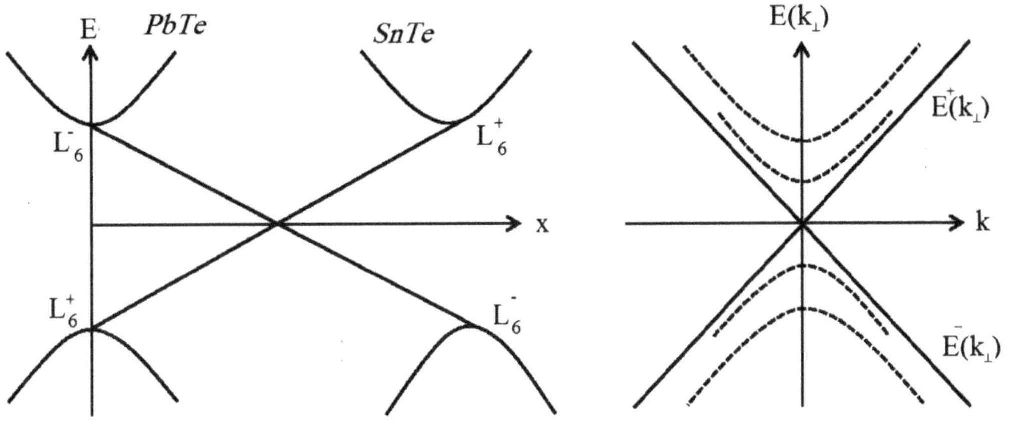

Poole-Frenkel Mechanism

The Poole-Frenkel model was introduced in 1938 by Frenkel (1938) to explain the increase of conductivity in insulators when high fields are applied. This mechanism is concerned with the motion of heavy impurity *ions* by *hopping* across vacancies and interstitial sites in the insulator material. For instance, in deposited insulators, impurity ions and defects are abundant and are drifted by this mechanism. In this case, the current density is proportional to the electric field inside the insulator material ($J = \sigma\, \zeta$). The electrical conductivity in this case is field-dependent according to the following relation:

$$\sigma = \sigma_{o.}\, exp\, [-\sqrt{(e.\zeta / \pi\, \varepsilon_r)} / k_B T\,] \tag{1a}$$

where ε_r is the relative permittivity of the insulator (ε_r =7 for Si_3N_4) and σ_o is a constant dependent on the temperature. The Poole-Frenkel current density is then given by:

$$J_{PF} = \sigma_0 \zeta \exp\left[-\frac{e(\Phi_t - \sqrt{e\zeta / \pi\varepsilon}}{k_B T} \right] \tag{1b}$$

Fowler-Nordheim Tunneling Mechanism

The Fowler-Nordheim (FN) tunneling is a dominant transport mechanism in metal-oxide-semiconductor (MOS) structures, especially for relatively-thick insulating oxides. In fact, the quantum mechanical tunneling from a conductor into an adjacent insulator defines the current through the structure. When the electrons tunnel into the insulator they are assumed free to move within the conduction band of the insulator. The following equation, relates the tunneling current density, J_{FN}, with the tunneling probability across the insulator:

$$J_{FN} = \int T(E_n)\, g_n(E_n)\, u_q^{\perp} f_n(E_n)\, dE_n \tag{2}$$

where, f_n is the electron distribution function, $g_n(E_n)$ is the density of states, $u_q^{\perp}$ is the electron normal velocity to the semiconductor/insulator interface (where $u_q^{\perp} \approx 1/4\, u_q$) and. Also, the tunneling probability $T(E_n)$ can be found by solving the Schrödinger equation in the insulator barrier, for instance, using the well-known Wentzel-Kramers-Brillouin (WKB) method or by the Gundlach technique (Gundlach et al, 1997).

The tunneling probability may be expressed by the WKB approximation in terms of the barrier height and shape. For a *trapezoidal barrier* of upper height ϕ_B and lower height ϕ_o, the tunneling probability $T(E_n)$ is given by:

$$T(E_n) = \exp\left[-\frac{4\sqrt{2m_{ox}}}{3e\hbar\zeta_{ox}}\left(e\phi_B - E_n\right)^{3/2}\right] \text{ for } \phi_B > E_n > \phi_o \tag{3}$$

and

$$T(E_n) = \exp\left[-\frac{4\sqrt{2m_{ox}}}{3e\hbar\zeta_{ox}}.\left(\left(e\phi_B - E_n\right)^{3/2} - \left(e\phi_o - E_n\right)^{3/2}\right)\right] \text{ for } E_n < \phi_o \tag{4}$$

Here, m_{ox} is the effective mass of electrons inside the insulator barrier (for SiO_2, $m_{ox} = 0.42m_0$).

Assuming the electron energy distribution function is Maxwellian and the energy bands are parabolic, we can derive the following Fowler-Nordheim formula for tunneling current:

$$J_{FN} = J_0 . \zeta^2 \exp\left[-\zeta_0 / \zeta\right] \tag{5a}$$

where J_0 and ζ_0 are constants, related to the energy barrier height ϕ_B between the insulator and the injecting conductor as follows:

$$\zeta_o = \frac{4\sqrt{2m_{ox}}}{3e\hbar}.\left(e.\phi_B\right)^{3/2} \tag{5b}$$

$$J_o = \frac{e^2}{16\pi^2 (m_{ox} / m_o).\hbar.\varphi_B}. \tag{5c}$$

The energy barrier height ϕ_B can be determined experimentally from the structure *J-V* characteristics by plotting $ln(J_{FN}/\zeta^2)$ versus ζ^{-1}. This is shown in Figure 12. For thin layers we may take $\zeta = V_G/d$, where V_G is the applied voltage across the MOS structure and d is the insulator thickness. Note that an important condition for *Fowler-Nordheim* tunneling is that the electric field inside the insulator should be

Figure 11. Illustration of the tunneling across oxide insulator barrier in a MOS structure

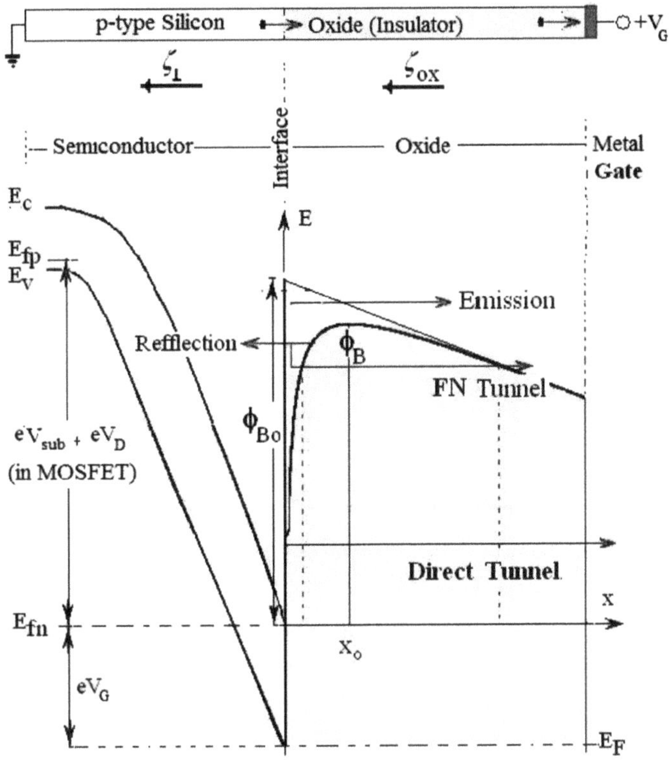

high enough such that $\zeta.d>\phi_{B'}$. Another condition is the possibility for electrons to move freely (without frequent collisions or recombination) inside the insulator. This is the case in thermally-grown silicon dioxide (SiO_2). However, in other insulators, like Si_3N_4, there exist huge amount of impurities and defects, which have deep energy levels within the insulator energy gap and act as trapping centers for electrons.

Other Tunneling Mechanisms

There exist a variety of tunneling processes that can be identified in silicon-dielectric-silicon structures in general and MOS devices in particular. When we consider the shape of the energy barrier alone, we can distinguish the Fowler-Nordheim (FN) tunneling and the direct tunneling. Other classification differentiates between electrons from the conduction band (ECB), electrons from the valence band (EVB), holes from the valence band (HVB), and trap-assisted tunneling (TAT) mechanisms, as shown in Figure 13.

The EVB process is due to electron tunneling from the valence band to the conduction band of two semiconductors, across a dielectric material. It thus creates free carriers at both sides of the dielectric, which, for MOSFET devices, gives rise to increased substrate current. The TAT process can either be elastic, which means that the energy of the carrier is conserved, or inelastic, where the carrier loses energy due to the emission of phonons. Furthermore, in dielectrics with a very high defect density, hopping conduction via multiple defects may occur. The tunneling current in such structures can be modeled by several methods, such as the *Tsu-Esaki* model. The Tsu-Esaki tunneling formula finds the

Figure 12. Oxide tunneling current to Si, as a function of oxide voltage

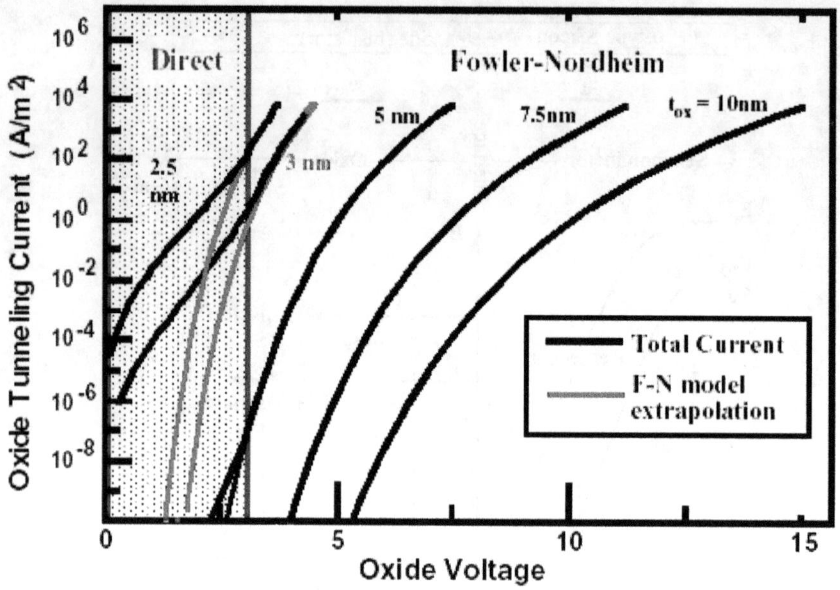

Figure 13. Schematic illustration of different tunneling mechanisms across a semiconductor-dielectric-semiconductor structure

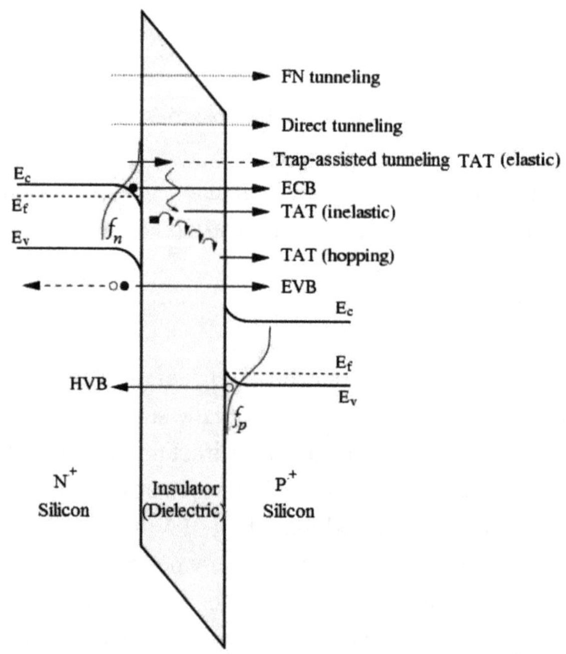

tunneling current density across MOS structures by integration in the energy domain. In the channel of an inverted MOS, we have:

$$J = 4\pi\, (e\, m_0 / h^3) \int T(E).N(E)\, dE \tag{6}$$

The *Tsu-Esaki* model allows to distinguish between the supply function, *N(E)*, which describes the supply of carriers for tunneling, and the transmission propability, *T(E)*, which characterizes the considered energy barrier.

$$N(E) = \int (f_1(E) - f_2(E))\, dE \tag{7}$$

The supply function depends on the energy distribution function of carriers in the semiconductor, *f(E)*. Actually, it is not easy to calculate the energy distribution function of carrier at non-equilibrium (in presence of bias voltage). However, one can model the shape of distribution function by a first-order approximation, such as the Fermi-Dirac distribution to get:

$$J = \frac{em^* k_B T}{2\pi^2 \hbar^3} \int_0^\infty T(E).\ln\left[\frac{1+\exp[(E_F - E)/k_B T]}{1+\exp[(E_F - E - eV)/k_B T]}\right] dE \tag{8}$$

In the above formula, we assume the Fermi-Dirac distribution function at equilibrium, which is a very crude approximation. We also assume a parabolic band structure for the semiconductor. In order to obtain more realistic results of tunneling currents, some authors assumed a non-Maxwellian distribution function and considered the a non-parabolicity of energy bands. As we pointed out in the above section, the transmission coefficient, *T(E)*, can be found by the solution of Schrödinger's equation around the considered region. The Wentzel-Kramers-Brillouin (WKB) and the Gundlach methods can be used as approximate solutions of the Schrödinger equation. If we adopt the analytical WKB approximation, for a triangular barrier we get:

$$T(E) = 4\eta / (4+\eta)^2 \tag{9a}$$

$$\eta = \exp\left[-2\int_{x1}^{x2} \left(\frac{2m^*}{\hbar^2}(V(x) - E)\right)^{1/2} dx\right] \tag{9b}$$

Here, x_1, x_2 are the two classical turning points for the potential barrier. If we take the classical limit of Planck's constant, as $\hbar \rightarrow 0$, we see that the transmission coefficient correctly goes to zero. In the case of metal-oxide-semiconductor structures, the barrier takes the triangular shape, as shown in Figure 13. Hence, the tunneling coefficient is given by (-9). Figure 14 depicts the tunneling coefficient across SiO_2 dielectric in an MOS structure as a function of electron energy. However, for the accurate simulation of transport across arbitrary barriers, there exist other advanced models, such as the transfer-matrix method (TMM) and the quantum transmitting boundary method (QTBM).

The TMM provide an important tool for investigating bound and scattering states in quantum structures (Grun et al, 1987). They are mainly used to solve the one-dimensional Schrödinger or effective mass equation, e.g., to obtain the quantized energies in hetero-structures and metal–oxide–semiconductor (MOS) structures or the transmission coefficient of potential barriers. We know that analytical expressions for the transfer matrices are only available in certain cases, as for constant or linear potentials and potential steps. An arbitrary potential can be treated by approximating it for example in terms of piecewise steps or linear segments, for which analytical transfer matrices exist. For constant potential steps, the matrices contain complex exponentials, while the linear potential approximation yields Airy functions.

3. CONDUCTION IN ORGANIC SEMICONDUCTORS

Organic semiconducting molecules are distinguished from insulating molecules by their large spatial extent of the frontier orbitals. In the case of hole transport, the orbital of relevance is the highest occupied molecular orbital (HOMO), whereas, for electron transport, it is the lowest unoccupied molecular orbital (LUMO). In insulating molecules, these orbitals are small, localized on just a few atoms. As a result, the spatial overlap between frontier orbitals on neighboring molecules is also small and so the probability of charge transfer between them is almost null. On the other hand, the frontier orbitals of semiconducting molecules are spatially delocalized, covering much of the molecule. This delocalization arises from the *conjugation* (strong interaction) of partially filled orbitals on neighboring atoms (often

Figure 14. Tunneling coefficient across gate dielectric barrier (of thickness t_{ox}) in a metal-oxide-semiconductor (MOS) structure as a function of electron energy

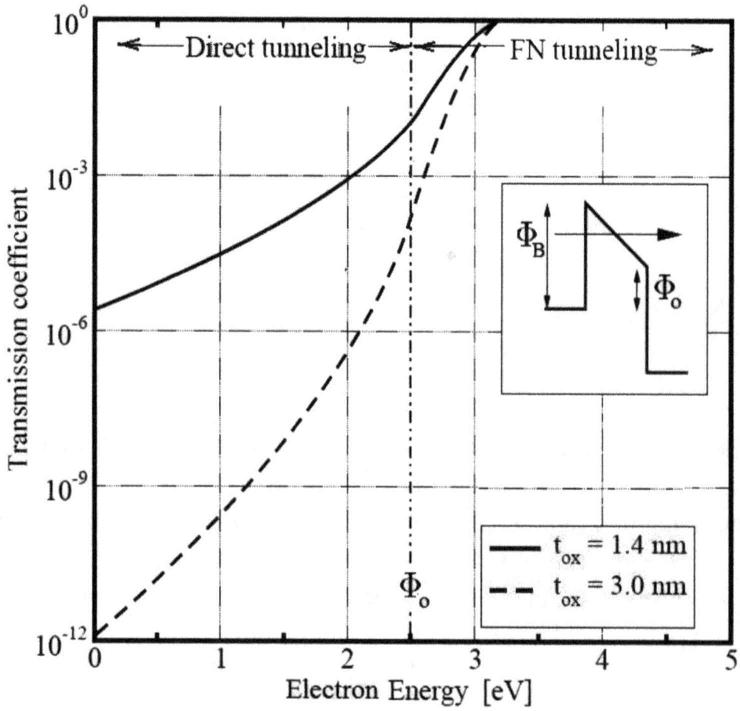

p-orbitals of *C, O, S,* or *N*). Because of delocalization, the neighboring orbitals on molecules overlap, and the probability of charge transfer is appreciable.

Now, let's demonstrate some of the limiting cases of charge carrier transport in organic semiconductors. When applying an electric field to an organic semiconducting device, any injected charge is drifted to the opposite electrode, giving a typical V^2 behavior of the conduction current. The resulting current is called a space-charge limited (SCL) current, because the current is impeded by the presence of the space charge in the device. The SCL current regime occurs when the equilibrium carrier concentration (before charge injection) is negligible compared to the injected carrier concentration. By solving the Poisson equation with the charge drift equation in a trap-free solid, the resulting current relation is called the *Mott–Gurney* law (Mott and Gurney, 1940):

$$J = (9/8)\, \varepsilon\mu\, (V^2 / L^3) \tag{10a}$$

where ϵ is the electric permittivity of the material, μ is the charge-carrier mobility (μ_n for electrons or μ_p for holes) and, L is the sample thickness. For bipolar devices, the injection of opposite charge carriers (electrons and holes) reduces net charge and allows for a much larger current. Moreover, electrons and holes can recombine, which reduces the current. For Ohmic contacts the bipolar current is given by (Parmenter and Ruppel, 1959):

$$J = (9/8)\, \varepsilon\, [2\pi\, \mu_n\, \mu_p(\mu_n + \mu_p)/\mu_r\,]^{1/2}\, (V^2 / L^3) \tag{10b}$$

Therefore, in a double-carrier device, the typical V^2 dependence of the current is also found. However, many other conduction mechanisms can occur in organic devices, and we discuss them in the following sections.

3.1 Charge Carrier Mobility in Organic Semiconductors

The most important parameter of carrier transport in organic devices is carrier mobility. The charge mobility is defined as the ratio of average carrier velocity v and the applied electric field ζ:

$$\mu = v/\zeta \tag{11}$$

The disordered nature of organic semiconductor films makes it difficult to accurately calculate μ with theoretical approaches. Instead, it is necessary to simulate the motion of charges through a large number of molecules in real chains. Fortunately, this is possible when intermolecular charge transfer occurs in the hopping regime, because the motion of a charge is simply a sequence of independent hops between neighboring molecules.

The molecular conductance can be measured by the mechanical controllable break junction (MCBJ) technique, usually between gold electrodes (Reed et al. 1997). The figure 14 depicts the evolution of mobility histograms, as obtained by the MCBJ method with Au point contacts (Yang et al, 2011).

The evolution of organic materials over the years is shown in the graph 15 for polycrystalline and crystalline organic FET (OFET). As shown in figure, the mobility of organic semiconductors has improved by five orders of magnitude over the past 20 years. The horizontal lines indicate the comparison guides to the main competitors – amorphous (*a-Si*) and polycrystalline silicon

Figure 15. Schematic of the mechanical controllable break junction (MCBJ) technique for conductance measurement of a molecular system

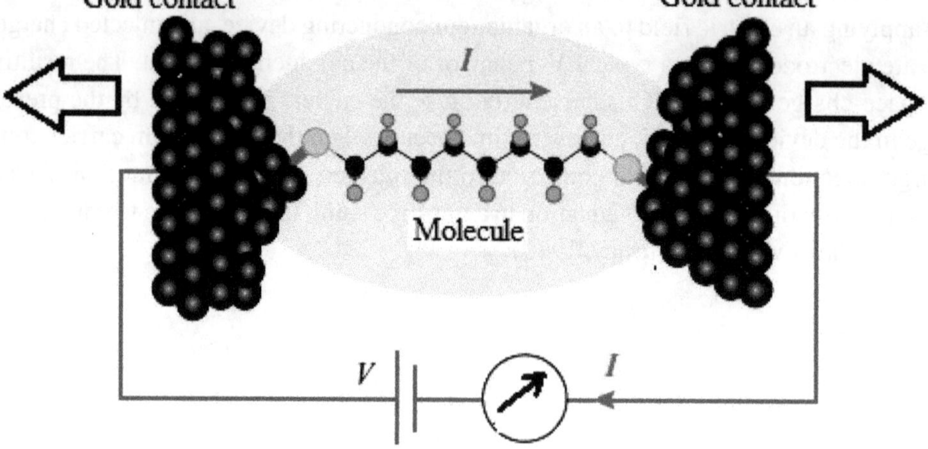

Figure 16. (a) Carrier mobility of promising organic semiconductors at 300K; (b) comparison of reported motilities of crystalline, polycrystalline and amorphous silicon, with respect to organic materials
After Yang et al (2011).

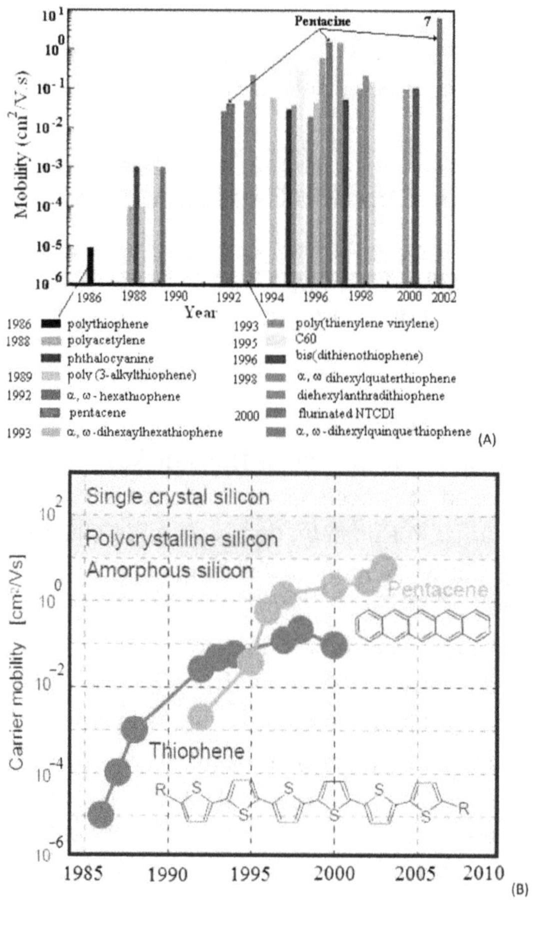

The graph reveals that the mobility in polycrystalline OFETs is comparable to that of a-Si whereas mobility in rubrene-based OFETs (20–40 cm^2/Vs) approaches that of best poly-silicon devices. These materials have very poor mobility at room temperature. However, the mobility of these materials is improved at low temperature and may even show superconductivity effects at very low temperatures. For instance, the pentacene has electron drift mobility of 100,000 cm^2/Vs at low temperature (Scott et al, 2000). The advantage of these materials, over metallic and ceramic superconductors, is their simple fabrication method at low cost. The performance of organic semiconducting devices has been improving greatly since the start of the 1990's. In 1997, pentacene was used instead of thiophene as organic material. In fact, pentacene has the smallest bandgap among all linear polyacenes and the highest mobility in OFETs, with μ=0.3-1.5 cm^2/Vs. This value is comparable to that of amorphous Si devices, as shown in Figure 15(c). Here, doping is performed by replacing a fraction of the TMTPD with their salt TMTPD:SbF6.

3.2 Conductivity of Doped Organic Semiconductors

The conventional way to increase the conductivity of a semiconductor is to introduce dopants that can act as electron donors or/and acceptors. In fact, significant progress with inorganic semiconductors could only be obtained once carrier transport is controlled by doping. Like other semiconductors, organic semiconductors may be either n-type or p-type, according to their majority carriers are electrons or holes. However, the underlying mechanisms of molecular doping in organic semiconductors are only little understood compared with their inorganic counterparts. Anyway, the most widely used organic semiconductors are p-type. For instance, the π-conjugated polymers are generally p-type materials.

Figure 17. Conductivity of TMTPD-PC, as a function of doping
After Scott et al (2000).

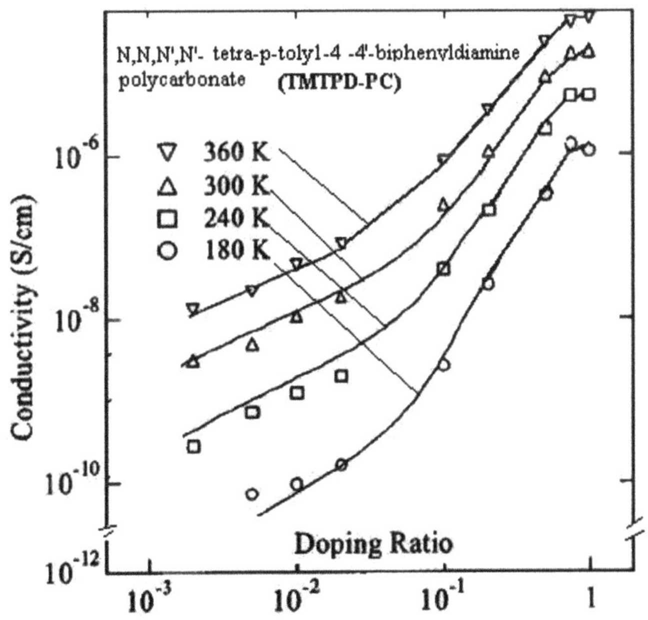

Figure 17

Nevertheless, some polymers with high electron affinity can be used as n-type semiconductors. The reason behind this is mainly because of the stability of p-type organic semiconductors in air. In fact, the controlled and doping stability can be accomplished in organic semiconductors but it imposes some problems. The problem is related to the level of transport states. P-type doping requires the transfer of an electron from the filled HOMO of the host to the LUMO of the dopant at no or only little energy expense. Correspondingly, the HOMO of the dopant has to be close to the LUMO of the host in order to promote n-type doping. This makes a serious constraint on the mutual energy levels. In most organic semiconductors the HOMO is around 5-6 eV below the vacuum level. Assuming an energy gap of 2.5eV, a p-type dopant therefore has to act as a very strong electron acceptor. On the other hand, the n-type dopants should have a HOMO level near 2.5-3.5eV, which is difficult to achieve. To resolve these difficulties, molecular doping emerged as a viable alternative, where larger organic molecular acceptors or donors were developed as p- or n-dopants (Salzmann et al, 2016). Figure 18 depicts the charge carrier mobility in P3HT as a function of the doping concentration. The experimental mobility is calculated from the steady state current at a given doping level

4. TRANSPORT MODELING IN ORGANIC SEMICONDUCTORS

The study of charge carrier transport in organic semiconductors has been started long time ago by Ely (1948), Akamatsu & Inokuchi (1952), Marcus (1956), Kallmann & Pope (1960) and LeBlanc (1961) and so many other scientists. A review of the cornerstones in this field along 60 years of research and development can be found in Hush & Ann (2003) and Nawla (2008).

Figure 18. Hierarchy of transport models for organic semiconductors From Rule et a (2011), with behavior

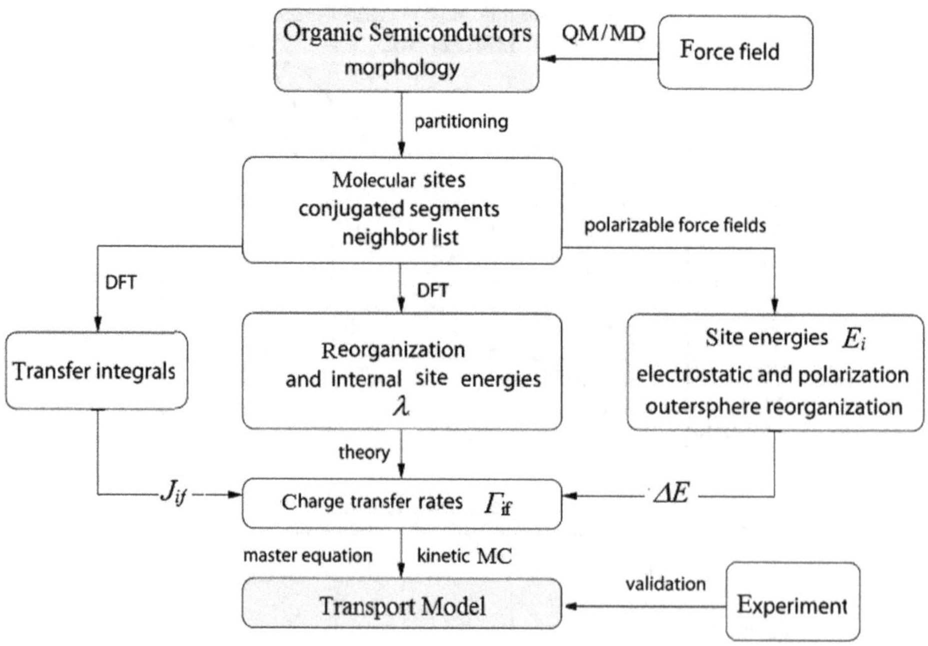

The electrical and thermal transport of a large class of organic semiconductors has been studied on the basis of the semiclassical Boltzmann transport equation (BTE) for wide-gap semiconductors. Alternatively, other studies are relying on the quantum approaches which are based on the Schrödinger equation, such as the nonequilibrium Green's functions (NEGF). In order to predict the molecular energy band structure (or molecular orbitals) of organic semiconductors, there are many approaches. Among the famous ones one can site: the Valence-bond (VB) method, density functional theory (DFT), the LCAO and Hückel-theory. In particular, the Hückel approach (1931) is much simpler, yet gives reasonably results for many hydrocarbons, when one can treats the π-orbitals separately from the σ-orbitals. The combination of NEGF with band structure calculation methods, such as the DFT, is also a promising framework for studying transport in molecular devices. For instance, the NEFG-DFT method has been applies to predict the I-V characteristics of a single molecule (Stokbro, 2008).

The so-called non-adiabatic molecular dynamic simulations have been applied to study the carrier transport of organic materials (Kubar & Marcus, 2013). In non-adiabatic simulations the electronic structure and nuclear motions are treated separately. This allows the nuclei to react with the instantaneous position of moving charge and vice versa.

On the device scale, one can distinguish between two basic categories of charge transport in disordered organic semiconductors, namely:

1. Macroscopic models, and
2. Molecular models.

The Macroscopic (such as Gaussian disorder model GDM, Polaronic model and Effective medium approach EMA) are making use of fitting parameters, which can be adjusted with experimental results. The molecular models, on the other hand, treats with the electronic interactions between small clusters of molecules. The molecular approach can be used to calculate the rate of charge transfer between two molecules. In fact, in almost all approaches of molecular charge transport, we need an expression of the charge transfer rate. This rate may be expressed by the Marcus theory or the Miller-Abraham model. The conventional tool to describe the electron transfer in complex molecular systems has been the Marcus theory (Marcus, 1956, 1964, 1985) and its extensions (Small, Matyushov & Voth, 2003). This theory is presented in the next section.

4.1 Charge Transfer Rate of Molecules in Organic Semiconductors

In order to understand the Marcus theory, consider the following charge transfer reactions, for electrons, and holes:

$$M_1^- + M_2 = M_1 + M_2^- \tag{12a}$$

$$M_1^+ + M_2 = M_1 + M_2^+ \tag{12b}$$

where M_1 and M_2 are two semiconducting molecules. If the coupling between M_1 and M_2 is strong, this reaction may occur so fast in both directions so that the charge is essentially delocalized across both molecules. It should be noted that charge transfer reactions that involve both holes and electrons are also important, for example in charge generation and recombination in OLEDs. Figure 19 depicts the initial, transition, and final states of an electron transfer reaction between identical molecules M_1 and M_2, where the nuclear coordinates of both molecules are described by Q_i,Q_t,, Q_f respectively.

In the initial state, the electron is in the lower unoccupied molecular orbital (LUMO) of M_1 and consequently, these molecules have longer bond lengths, since the LUMO is an anti-bonding orbital. In the final state, the electron has transferred to the LUMO of M_2 and the bond lengths have been adjusted accordingly. Both molecules have the same bond length in the transition state, and the charge can be considered to be on either molecule.

In order to describe the charge transfer process, it is useful to consider the simple case of two non-interacting (adiabatic) wavefunctions ϕ_i and ϕ_f that describe the distribution of electrons before and after charge transfer. In each case, the wavefunction describes the distribution of all electrons on both molecules for a certain configuration of their nuclei. Both are eigenstates of the electronic Hamiltonian $H_e^{(0)}$, in which charge transfer is not possible because molecules M_1 and M_2 aren't coupled:

$$H_e^{(0)} \phi_i = E_i \tag{13a}$$

$$H_e^{(0)} \phi_f = E_f \tag{13b}$$

where E_i and E_f are the energies of states ϕ_i and ϕ_f respectively. In order to consider the charge transfer, a more complete and coupled Hamiltonian is necessary, in which ϕ_i and ϕ_f are coupled by a transfer integral J_{if}

$$H_e = E_1[|\,\phi_1\rangle\langle\phi_1\,|] + E_f[|\,\phi_f\rangle\langle\phi_f\,|] + J_{if}[|\,\phi_i\rangle\langle\phi_f\,| + |\,\phi_f\rangle\langle\phi_i\,|] \tag{14}$$

where H_e is the electronic Hamiltonian and the transfer integral J_{if} is given by:

Figure 19. A schematic of electron transfer between molecules M_1 and M_2 showing initial, transition, and final states with nuclear coordinates $\mathbf{Q}_i$, $\mathbf{Q}_t$ and $\mathbf{Q}_f$ respectively. Charge transfer occurs at the transition state

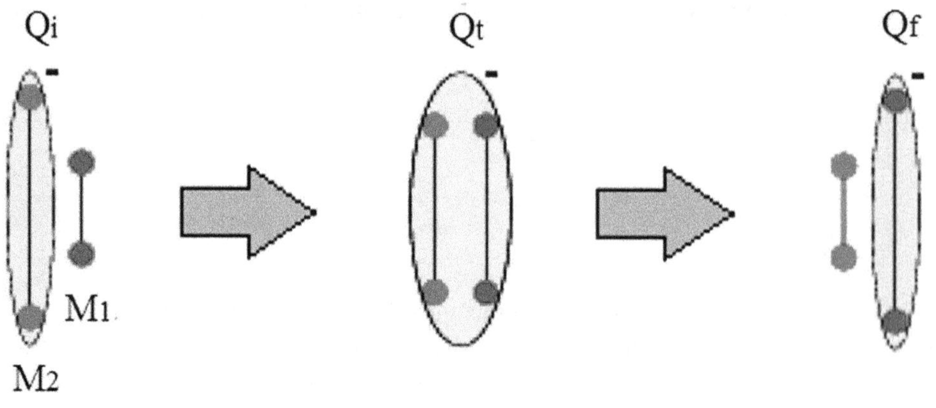

$$J_{if} = < \phi_i \,|H_e\,| \,\phi_i > \quad (15)$$

Figure 20 depicts the potential energy surfaces of the initial and final states as a function of the nuclear coordinates (Q). The transition state (at Q_t) is defined by the point where the two curves meet, and the potential energy required to reach this point from the optimum configuration of the initial state is E^*. One can define E^* in terms of the energy offset $\Delta E = E_f - E_i$ and the *reorganisation energy* $\lambda = ½K(Q_f + Q_i)^2$, where K is the force constant.

$$E^* = (\Delta E + \lambda)^2 / 4\lambda \quad (16)$$

Both ΔE, J_{if} as well as λ, can be calculated by quantum self-consistent field (SCF) methods, such as the Hartree-Fock or DFT. According to the degree of electron–phonon interaction (the ratio between the intermolecular electron coupling J_{if}, which is also called electron transfer integral, and the molecular re-organization energy λ), one can distinguish different charge transport models:

1. When $J_{if} >> \lambda$ for molecules in a crystal, the conventional band theory (with delocalized states) can be used to describe the charge carrier mobility. In this regime, the mobility is limited by phonon scattering. Some polyacenes single crystals indicate band-like charge transport (Machida et al, 2010).
2. When $J_{if} \approx \lambda$, the charge transfer can be still described in the framework of the adiabatic approximation.

Figure 20. Potential energy surfaces of the initial and final states as a function of position along the charge reaction pathway Q

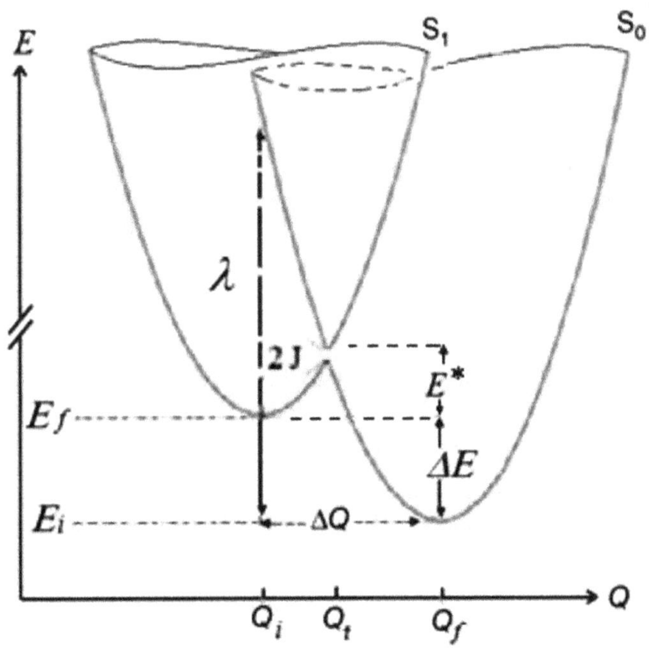

3. When $J_{if} << \lambda$, the electron interacts strongly with intramolecular vibrations which eventually lead to self-localization: This is the case of disordered semiconductors where the band description fails and hopping model applies.

In fact, it became clear since the early 1990s, that coulomb and exchange interactions between charges are so significant in π-conjugated polymers so that they cannot be neglected. Instead, coulomb and exchange interactions are central for a correct description of the electronic structure and carrier transport of organic semiconductors.

4.2 Semiclassical Marcus Theory

The Marcus theory (1956) is a seminal contribution to the field of charge transfer, and worthy of the 1992 Nobel Prize in Chemistry. In various forms, Marcus theory has been applied to charge transfer in many chemical and biological systems. In the semiclassical formalism, Marcus theory describes the rate of charge transfer (Γ_{if}) between an initial state S_o (with wavefunction ϕ_i) and final state S_1 (with wavefunction ϕ_f) as follows:

$$\Gamma_{if} = \left(\frac{2\pi}{\hbar}\right) | J_{if} |^2 (4\pi\lambda k_B T)^{-1/2} \exp\left[-\frac{\Delta E + \lambda)^2}{4\pi\lambda k_B T}\right] \tag{17a}$$

where λ is the re-organization energy and ΔE is the energy difference between the initial state ϕ_i and final state ϕ_f:

$$\Delta E = < \phi_f | H_e | \phi_f > - < \phi_i | H_e | \phi_i > \tag{17b}$$

When the difference ΔE between E_i and E_f is large, the mixing between ϕ_i and ϕ_f is small. In this case, ϕ is predominantly composed of the state with the lowest energy, and the charge is localized on a single molecule. However, the charge transfer reaction must proceed through the intersection between curves in order to conserve energy and satisfy the Franck-Condon principle[1] (1926). The Marcus theory has got great success due to its prediction that the transfer rate Γ_{if} doesn't increase indefinitely as the reaction becomes more exergonic, and is maximum when $\Delta E = -\lambda$, as shown in figure 19. This result was confirmed experimentally in 1984 (Karabunarliev et al., 2000).

4.3 Quantum Transport Modeling in Organic Semiconductors

In order to develop a microscopic quantum description of charge transport in Organic Semiconductors, it is important to evaluate the extended charge carrier wavefunction. This can be obtained by solving the time-independent Schrödinger equation (SE). Using the Born-Oppenheimer (adiabatic) approximation, the many-body SE can be reduced to one-electron form with effective (Hartee-Fock) potential, as explained in Chapter 4. We can then write the one-electron Hamiltonian for organic semiconductors, taking into account the specific effects of disorders and molecular transfer rate. In this Hamiltonian, we assume low carrier density, such that electron correlations and Coulomb interactions can be neglected (Atkins, 1983):

$$H = H_o + H_1 + H_2 + H_3 + H_4 \tag{18}$$

where, H_o is the electronic and vibrational (phonon) term,

$$H_o = \Sigma_n E_n .(a_\lambda^\dagger a_\lambda) + \Sigma_\lambda \hbar\omega_\lambda .(b_\lambda^\dagger b_\lambda) + ½) \tag{19}$$

Here, E_n is the electron energy in a perfectly ordered lattice, $a_n(a_n^\dagger)$ is the creation (destruction) operator of excited electron at the molecular n site, $b_\lambda(b_\lambda^\dagger)$ is the creation (destruction) operator of a phonon mode of energy $\hbar\omega_\lambda$,

The electron transfer term H_1 is given by:

$$H_1 = \Sigma_{n,m} J_{n,m} .(a_n^\dagger a_m) \tag{20}$$

Here, J_{nm} is the electronic coupling (transfer integral) between site m and n $(n \neq m)$ in perfectly ordered lattice. The dynamic diagonal disorder term H_2 is given by

$$H_2 = \Sigma_\lambda \Sigma_n g^2_{n,\lambda} \hbar\omega_\lambda .(a_n^\dagger a_n) .(b_\lambda^\dagger + b_\lambda) \tag{21}$$

The dynamic off-diagonal disorder term H_3 is given by:

$$H_3 = \Sigma_n \Sigma_\lambda f^2_{nm\lambda} \cdot \hbar\omega_\lambda (a_n^\dagger a_m) .(b_\lambda^\dagger + b_\lambda) \tag{22}$$

The static diagonal & off-diagonal disorder terms are lumped in H_4 which is given by:

$$H_4 = \Sigma_n \delta E_n .(a_n^\dagger a_n) + \Sigma_n \delta J_{n\neq m} (a_n^\dagger a_m) \tag{23}$$

where δE_n and δJ_{nm} are the variations of E_n and δJ_{nm} due to static disorder, and $g_{n\lambda}$ and $f_{nm\lambda}$ are coupling constants for the electron–phonon coupling.

Note that disorders may be static (due to fixed defects or impurities) or dynamic (due to temperature fluctuations, which affect the weak van der Waal bonds). Mathematically speaking, one can also classify disorder into two categories, namely: *diagonal disorder*, which reflects the fluctuations in site energies (i.e., the energies of the HOMO or LUMO levels of individual molecules); and *off-diagonal disorder*, which is related to fluctuations in the strength of interactions between adjacent molecules, i.e., to modifications of their relative positions and orientations (Coropceanu et al, 2007).

When the Hamiltonian term H_1 is small compared to H_2 and H_3, and H_4 is neglected, the transport is dominated by the coupling of the electronic excitation to molecular vibrations (phonons). We already know that a charge carrier coupled to lattice vibrations (phonons) is called polaron. Therefore, the interaction term H_2 causes a reduction of the site energy by the polaron binding energy. This energy should be overcome by thermal activation to enable charge transport. In this case, the charge transfer takes place by uncorrelated, phonon-assisted hopping process, which is determined by H_3.

4.4 Disorder-Based Transport Models

On the basis of the early work of Anderson on carrier transport in disordered materials (1958), we can conclude that charge carriers in organic semi-conductors (OSCs) are localized rather than extended throughout the material. Actually, OSCs are either statically disordered (due to crystal imperfections & impurities) or dynamically disordered (due to molecular vibrations) near room temperature. When the fluctuations in the intermolecular distances and their orientations give rise to a large variation in the site energy compared to other terms in the Hamiltonian, the static disorder dominates the charge transport. Therefore, charge carriers move by hopping. Of course, a thermal activation is required to overcome the energy differences between different sites.

A distinguished problem in disordered materials is their low charge carrier mobility, compared to crystalline semiconductors. The theory of electronic wavefunction localization in a disordered material was presented by Anderson in 1958. Anderson argued that scattering plays no role on carrier transport when the degree of disorder in a solid is greater than a critical point. The low mobility is attributed to the hopping distance, which is shorter than the Coulomb radius $r_c = e^2/4\pi\varepsilon k_B T$, where e is the electron charge, and ε is the dielectric permittivity. The charge carrier mobility in a disordered organic solid is usually modelled by the Gaussian disorder model (GDM). In the GDM, the system is considered as an array of structureless point-like hopping sites whose energies feature a Gaussian-type density of uncorrelated energy states (Bassler, 2008). Also, the model for the charge hopping rate follows that of Miller and Abrahams (1960):

$$\Gamma_{ij} = \Gamma_0 \, exp(-\gamma \, r_{ij}).exp\,[-(E_j - E_i)/k_B T] \; for \; E_j > E_i \tag{24a}$$

$$\Gamma_{ij} = \Gamma_0 \, exp(-\gamma \, r_{ij}) \; for \; E_j \leq E_i \tag{24b}$$

where $r_{ij} = a$ is the relative jump distance between site i and site j, a is the lattice constant, γ is the inverse localization radius related to the electronic coupling matrix element between adjacent sites, and Γ_0 is a frequency pre-factor. In an extended version of the hopping model, the positional (off-diagonal) disorder is employed, in addition to energetic (diagonal) disorder._

Figure 21. Schematic illustration of extended wave function in a crystalline material with negligible disorder and wavefunction localized by disorder

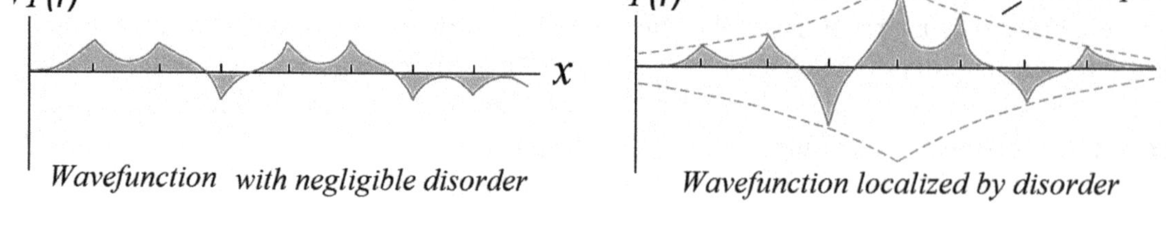

The different modes of transport have different temperature dependence of the charge carrier mobility. There are many approaches to predict the dependence of the charge carrier mobility on temperature and electric field. For instance, the effective medium approach (Fishchuk & Kadashchuk), stochastic hopping theory and Monte Carlo simulations can be used to fit the following model:_

$$\mu(\sigma,\zeta) = \mu_0 \exp(-4/9s^2).\exp[C(s^2 - \Sigma^2)\zeta^{1/2}] \text{ for } \Sigma \geq 1.5 \quad (25a)$$

$$\mu(\sigma,\zeta) = \mu_0 \exp(-4/9s^2).\exp[C(s^2 - 2.25)\zeta^{1/2}] \text{ for } \Sigma \leq 1.5 \quad (25b)$$

where the energetic disorder parameter$\sigma = k_B T$, ζ is the electric field, Σ is a positional disorder parameter and C is a constant that depends on the site separation. If the lattice constant a = 0.6nm, then C =2.9x10^4 cmV$^{½.}$ This equation resembles the Poole-Frenkel field dependence, which is obeyed experimentally within a large range of electric fields. The reason is that the energies of the hopping sites are essentially determined by the van der Waals interaction between a charged site and its polarizable neighbor sites which may carry an additional static dipole moment.

4.5 Polaronic Transport in Organic Semiconductors

Charge transport in organic semiconductors can be described by means of polarons. We have so far introduced the concept of polaron in semi-conductors and dielectric materials (Yamashita et al., 1958). A polaron is a quasiparticle composed of an electron plus phonon field. Therefore, a polaron is an electron dressed with phonons. The polaron model was extended to molecular crystals and conjugated polymers by Holstein (1959) and by Fesser et al. (1983). In organic polymers, polarons result from the deformation of the conjugated chain under the action of the charge. Holstein proposed a model to determine the mobility μ of polarons in the semiconductor, as a function of the lattice constant a, the electron transfer energy J, the reduced mass of the molecular site M, the frequency of the harmonic oscillators associated to the molecules ω_0, the polaron binding energy E_b=$A^2/(2M\omega_0^{\,2})$ and temperature T:

$$\mu = \sqrt{\frac{\pi}{2E_b}}\left(\frac{ea^2J^2}{\hbar}\right).\left(k_B T\right)^{-3/2} \cdot \exp\left(-E_b / 2k_B T\right) \quad (26)$$

This equation is valid for $T>\Theta$, where $k_B\Theta = \hbar\omega_0$ is the Debye energy. However, upon ionizing a molecule or a polymer chain, and hence adding an extra electron, there is a readjustment of bond lengths because the electron distribution changes. In optical transitions this effect is revealed by the coupling of the excitation to molecular vibrations. It determines the relaxation energy between the initially generated vertical Franck Condon transition and relaxed electronic state.

The lack of quantitative knowledge gave rise to a discussion in the literature on whether or not disorder effects or polaron effects control the temperature dependence of the charge carrier mobility. Actually, the analysis of the temperature and field dependence of the mobility solely in terms of polaronic effects requires fitting with unphysical parameters (notably unrealistic electronic overlap). In addition, polaron effects cannot explain the observation of dispersive transport at low temperature. Using the effective medium approach (EMA), Fishchuk et al. (2013), considered the effect of disorder and polaron effects. In this analytical theory, the charge transfer process is included using the symmetric Marcus rates instead of Miller–Abrahams rates. The predicted temperature and field dependence of the mobility is then:

$$\mu = \mu_{0.} \exp[-(E_a / k_B T) - (s / 8\eta^2].[(s^{1.5} / \eta^2)\sqrt{(ea\zeta / 8\sigma)}].\exp[ea\zeta / 4\eta^2 k_B T] \quad (27)$$

where E_a is half of the polaron binding energy, $\eta = (1-\sigma^2/ 8eak_BT_p)^{1/2}$ and $\sigma = 8ea(k_BT_p)$. This equation agrees qualitatively with the empirical expression derived from numerical simulations:_

$$\mu = \mu_{0.} \exp[-(E_a / k_B T) - 0.31s^2.\exp[0.78(s^{1.5} - 1.75)\sqrt{(ea\zeta / \sigma)}] \quad (28)$$

4.6 Strong Coupling and Tunneling Transport

Now, let's consider the strong electron–phonon interaction limit where the charge is regarded as localized in a single molecule. Then, the charge transport consists of successive hopping from molecule to molecule, overcoming the trapping caused by electron scatterings with intramolecular vibrations. Concerning, the transfer rate, the Marcus theory distinguishes between two limiting cases of weak coupling and strong coupling as shown in Figure 22 The weak coupling limit is the case when the potential energy surfaces of initial and final state are only weakly displaced so that the surfaces only intersect far away from the minimum of the upper state surface. This condition is fulfilled in aromatic hydrocarbons. Strong coupling applies when the minima of potential energy curves of initial and final state are strongly displaced so that the potential curves intersect near to the potential minimum of the high energy state.

Figure 22. Illustration of the charge carrier coupling in disordered materials

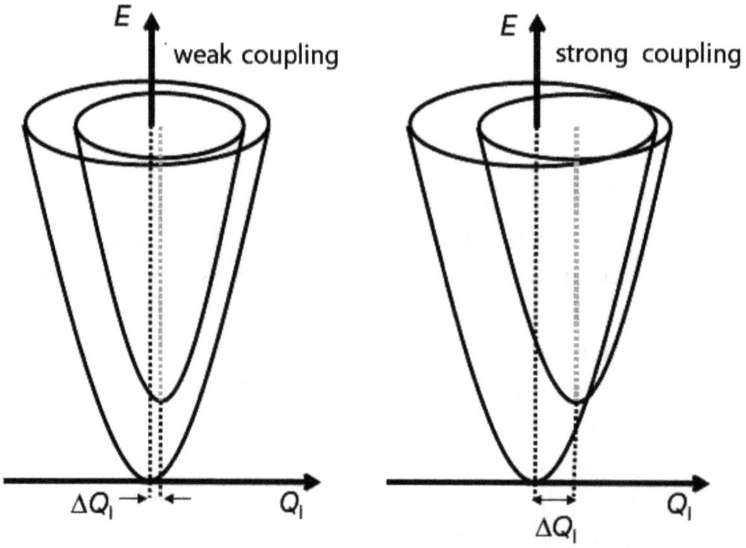

In order to evaluate the charge mobility within the hopping picture, there are two important rate processes at different spatial scales, namely, the electron transfer within molecular dimers and the electron diffusion in the organic material. The first process is characterized by the intermolecular electron transfer rate at the atomistic level, while the latter can be simulated at the molecular level by the random walk and Monte Carlo technique. We dully note that when the intermolecular electronic coupling (J) is much smaller than the reorganization energy (λ) of the electron transfer, the electron transfer rate falls into the hopping regime. For organic materials, we know that the intermolecular interaction is of van der Waals type. Therefore, J is smaller than λ, and the Marcus rate is usually adopted to evaluate the room-temperature mobility. However, beyond the semiclassical Marcus theory, the quantum nuclear tunneling effect should be also considered in the charge transfer process. According to Shuai et al (2014), this quantum effect is essential for interpreting the paradoxical experimental results between optical and electrical measurements of charge transport in organic semiconductors.

4.7 Charge Carrier Injection in Organic Semiconductors

We have stated earlier that the current conduction, in insulators may be space-charge limited (SCL) or injection limited (IL). Injection from the electrodes is the process by which charge carriers are generated in many organic devices, such as OLEDs and OFETs. Usually it is limited by an energy barrier between the Fermi-level of the electrode and the transport levels of the dielectric.

In conventional inorganic semiconductors, the relevant injection processes are either Richardson thermionic emission over Schottky barrier or Fowler–Nordheim (F-N) tunneling. The Richardson–Schottky (S-R) implies that a thermally excited electron from the Fermi level travels over the barrier of the electrostatic potential modified by the coulomb potential of the image charge and the applied electric

Figure 23. Illustration of the quantum nuclear tunneling effect in the charge transfer

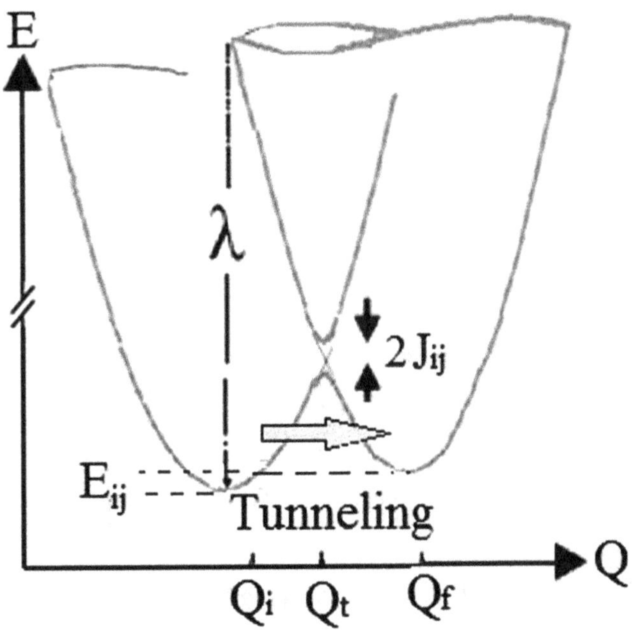

field without being scattered. It gives rise to an Arrhenius-type[2] of temperature dependence ($J_s = J_{so}$ $exp\ [-\ e\Phi_B / k_BT]$), and a Poole–Frenkel-type of field dependence ($\mu = \mu_{o.}\ exp\ [-\gamma \sqrt{\zeta}]$), where Φ_B is the effective barrier height, ζ is the applied field and γ is a constant.

In the classic Fowler–Nordheim model ($J_{FN} = J_0 .\zeta^2 \exp [-\zeta_0 / \zeta]$), we ignore the image potential and assume that an electron at the Fermi level of the metal tunnels through a triangular potential barrier set by the interface energy barrier and the applied potential. For both mechanisms, the crucial condition is that there is strong electronic coupling among the lattice elements that leads to wide valence and conduction bands. This implies that the scattering length of charge carriers is much larger than the interatomic separation. In organic solids this condition does not hold valid, because electronic coupling between molecules is weak (van der Waals bond). Accordingly, transport is collision-limited and of the hopping type. Thus, the Schottky theory condition of collision-free charge injection across the Schottky barrier potential does not hold in organic semiconductors.

There are many experimental evidences of the failure of the classic injection models in organic semiconductors. For instance, the temperature dependence of the injection current is weaker than expected using the estimated energy barriers. This excludes a classic Richardson–Schottky emission process, and also eliminates the Fowler–Nordheim tunneling mechanism since these observations extend to low fields were tunneling is negligible. In addition, the field dependence of the injection efficiency follows a Poole–Frenkel-type law and has a sub-linear temperature dependence with a mean activation energy that is significantly lower than one might suspect.

It comes from the above theoretical reasoning that the classical models for charge injection in organic semiconductors do not exactly fit the experimental observations. Therefore, more accurate models are needed. In order to cope with these problems, Burin and Ratner (2000) presented an injection model based on the idea that injection and transport occur through 1D straight paths. Also, the treatment of Arkhipov et al. (1999) gives sufficient results at room temperature. Later on, van der Holst et al. (2009) developed a 3D quantum model (master equation) that includes the disorder transport. Actually, all these models consider that the initial injection event is controlled by the existence of tail states of the energy of states (DOS) that is defined by energy versus hopping. The Gaussian distribution function is usually chosen to describe DOS in disordered organic semiconductors. Unlike the parabolic DOS found in conventional inorganic semiconductors, the Gaussian DOS does not have a simple band edge, as shown in Figure 25. The Figure 26 depicts the dependence of current on the barrier height. Therefore, one of the

Figure 24. Different injection and transport mechanisms across meta;-insulator barriers

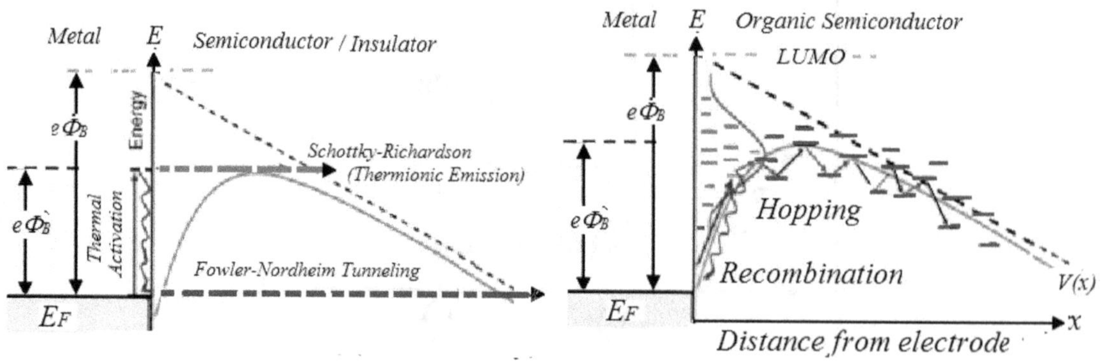

most determinant factors for carrier transport in organic semiconductor is the DOS within the bandgap of the organic material. Usually, the DOS energy dependence is assumed exponential or Gaussian.

Many traditional hopping transport models were developed when organic semiconductors were primarily amorphous, and the assumption of strong localization fit the experimental data. However, the assumption of strong localization is no longer valid in recently-developed organic semiconductors with almost perfect crystalline structure. Nevertheless, one can argue the nature of carrier localization in crystalline organic semiconductors as follows. The thermal energy at room temperature (about 25 meV) is comparable to the weak van der Waals bond energy, and therefore significant dynamic disorder can exist in an organic semiconductor, regardless of its crystalline structure.

The Figure 27depicts the results of different models for current injection from a metal electrode into a hopping system featuring a Gaussian DOS as a function of the electric field at different temperatures. The simulation parameters are the sample length L, the inter-site separation a and the injection barrier

Figure 25. Density of states and its tail at the metal-organic semiconductor interface

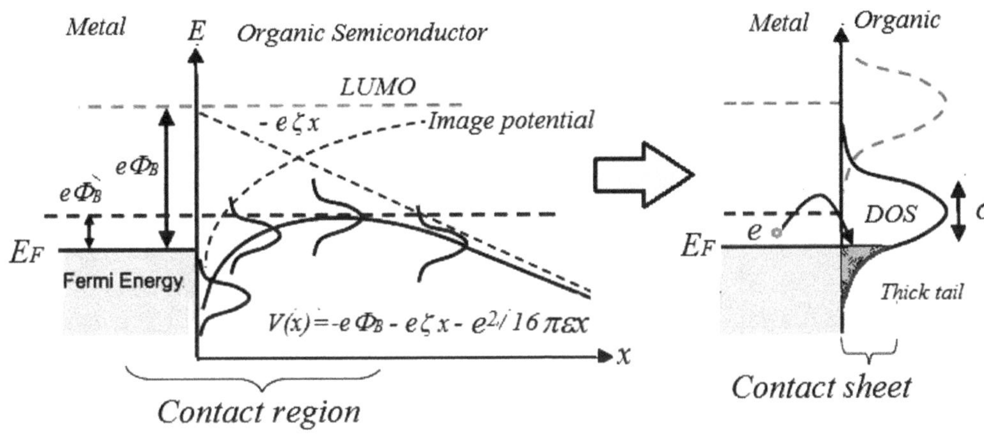

Figure 26. Simulation results of current-field characteristics in organic semiconductor

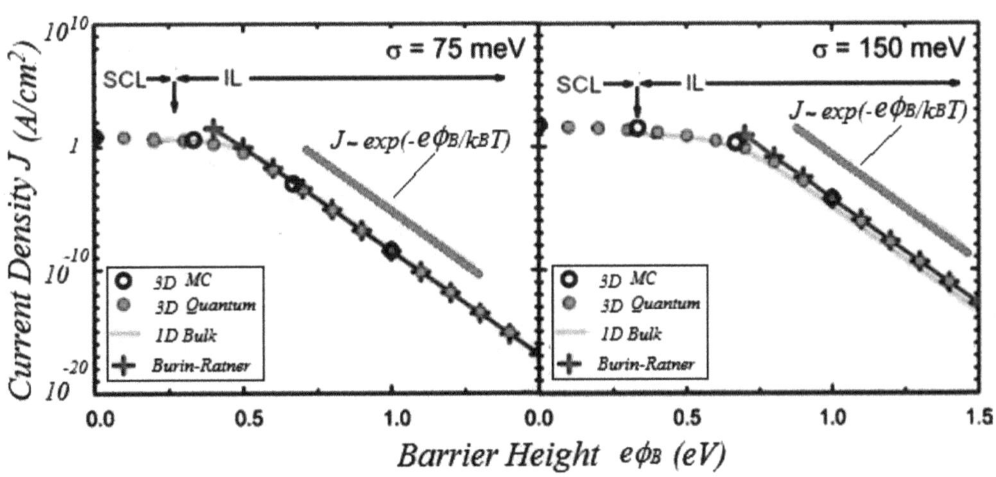

Figure 27. The comparison of the predictions of various models for current injection from a metal electrode into a hopping system featuring a Gaussian DOS of variance σ=75 meV as a function of electric field at different temperatures
After (Bassler et al., 2009).

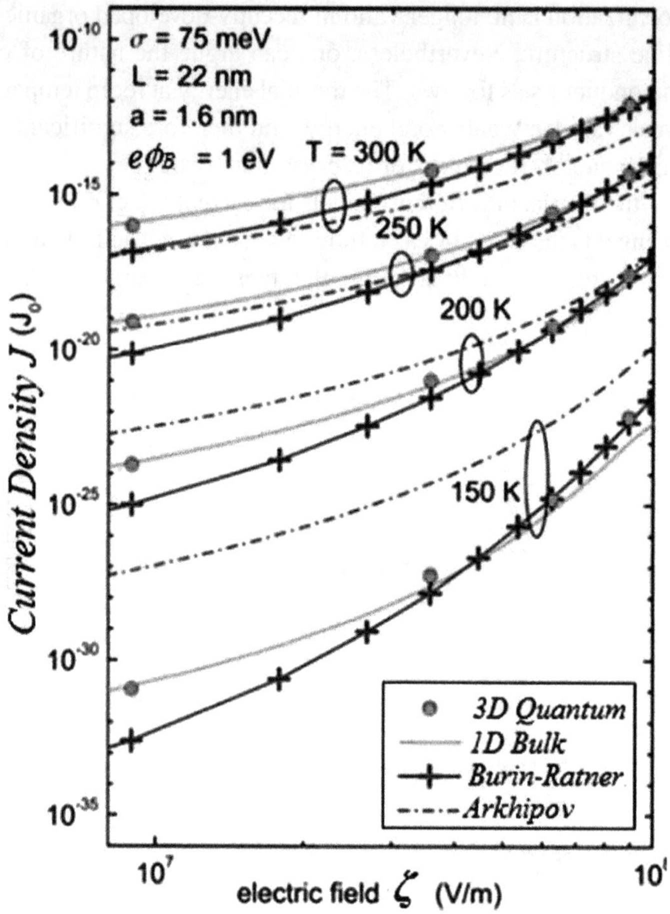

Φ_B. The simulation results labeled and the 3-D quantum (master equation) have been developed by van der Holst et al. (2009). The calculations are based on the models of Burin-Ratner (2000) and Arkhipov et al. (2005).

4.8 Kinetic Monte Carlo Method for Organic Semiconductors

Figure 28 depicts the KMC flowchar. It shows a comparison of the field dependent mobility between our method (symbols) and numerically solving the corresponding master equation (lines), for different values of σ and a relative charge carrier concentration of 10^{-3}. For increasing disorder and decreasing field, the charge transport characteristics deteriorate fast because of increasing energy differences between hop sites. A good agreement is found between the two different methods.

Figure 28. (a) Flowchart of the Kinetic Monte Carlo (KMC) method (b) Mobility as obtained by KMC method

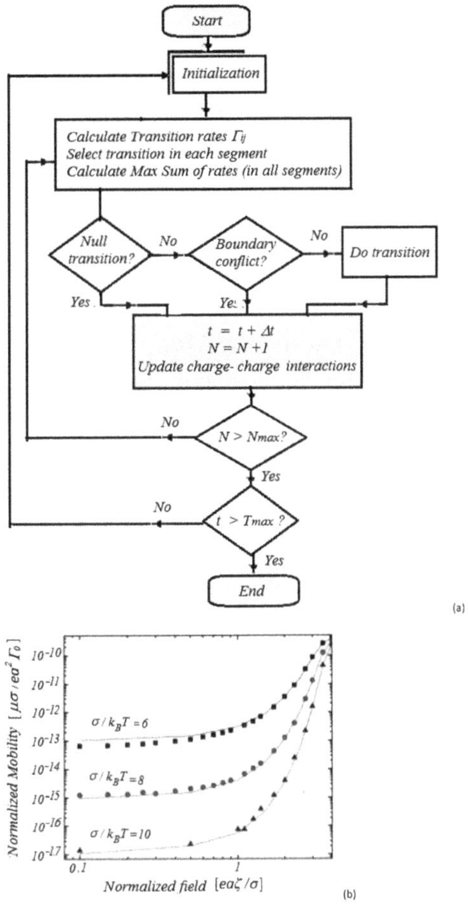

4.9 Drift-Diffusion Model for Organic Semiconductors

We have shown in previous sections that both the Richardson-Schottky (**R-S**) model for thermionic emission and the Fowler-Nordheim (**F-N**) tunneling current are not sufficient to simulate the carrier transport in disordered organic semiconductors. For instance, the F-N current has no temperature dependence. The drift–diffusion model (DDM) can be used to simulate and optimize organic semiconducting devices, if we could account for dispersion of carriers in such disordered structures. This may be done through the deliberated consideration of field-dependent carrier mobility and diffusion coefficient. The consideration of traps in the DOS of the device and their dynamic behavior is also important.

The DDM consists of the carrier continuity equations (for electrons and holes), the carrier current constitutive relations and the Poisson equation. This system was already presented in Chapters 1, 2 of this book. As we stated above, the drift mobility may be obtained on the basis of the GDM or its variants. These models take into account the distribution of localized states and the discrete hopping mechanism.

As for the diffusion coefficient, we can calculate it from the generalized Einstein relation: $D_n/\mu_n = e.n/[dn(E_F)/dE_F]$. Substituting the carrier density $n(E_F)$ results in the relation between the carrier mobility and diffusion coefficient. In the bulk of non-degenerate semiconductors, we can substitute $n=N_c \cdot exp[-(E-E_F)/k_BT]$ to get the conventional Eintein relation[3] $D_n/\mu_n= k_BT/e$. However, in 2D OSC layers, we rather consider the 2D electron density.

Alternatively, the electric field, temperature and carrier density dependence of mobility can be obtained from kinetic Monte Carlo (KMC) simulation in the bulk (Bolognesi et al, 2003) and may be also tabulated in look-ahead mobility tables and invoked upon request during simulation (Kordt, 2012).

4.10 Multiscale Simulations of Non-Adiabatic Molecular Dynamics

In non-adiabatic simulation, the electronic structure and nuclear motions are considered simultaneously. This allows the nuclei to react to the instantaneous position of the charge and allows the charge carrier to respond to the nuclear dynamics. The electronic motion is propagated with the time-dependent Schrödinger equation which allows delocalization of the charge carrier wave function, while the nuclear degrees of freedom are considered with classic (Newton-like) equations in a mean field. In fact, the nuclear vibrations which are important for modulating the transfer integrals have low-frequency modes. Therefore, it is quite straightforward to introduce molecular dynamics (MD) to describe their almost classical dynamics. As a first approximation, the quantum mechanical (QM) treatment is restricted to regions which are directly involved in the charge transport and describing the remainder of the system classically with molecular dynamics (MD) methods leading to a QM/MD multi-scale approach (Stafstrom, S. 2010).

5. PHOTONIC TRANSPORT IN ORGANIC SEMICONDUCTORS

Organic semiconductors may be well-suited to photonic applications, thanks to their ability to transmit, modulate and detect light in low cost, and flexible, structures. The dielectric constant of inorganic semiconductors is large (12 in Si) so that coulomb effects between electrons and holes are unimportant due to dielectric screening, and light absorption at room temperature creates free electrons and holes. In contrast, photogenerated electron–hole pairs in organic semiconductors are bound by coulomb attraction in the form of excitons. This is due to the large bandgap (about 2.5eV) and small dielectric constant (about 3.5) of organic materials. Therefore, the absorption and emission take place mostly in the range of 2–3 eV. Consequently, the electron–hole pair, which is created by optical excitation is bound by a coulomb energy of about 0.5–1eV (Sariciftci, 1997).

In order to design and characterize organic semiconductors for optoelectronic applications we first measure the absorption and photoluminescence their spectra. The Figure 28(a) shows the absorption spectra of radical cations of oligophenylene-vinylenes OPV of different chain lengths in CH_2C_{12} solution. Photonic parameters such as the fluorescence and phosphorescence lifetimes and quantum yields, as well as their spectra are usually measured in a film or in solution. Typical organic solvents for liquid solution include toluene, xylene chloroform (CHC13) and many others.

5.1 Excited States in Organic Semiconductors

Figure 30 depicts the energy band diagram of an organic semiconductor, showing charges, singlet excitons and triplet excitons. The vertical arrows are optical transitions, whereas wavy arrows are non-radiative transitions (vibrational relaxation in black, inter-system crossing in blue, charge generation in red, and

Figure 29. Absorption spectra of radical cations of oligo-phenylenevinylenes OPV of different chain lengths in CH_2C_{12} solution
After (Deussen,1993).

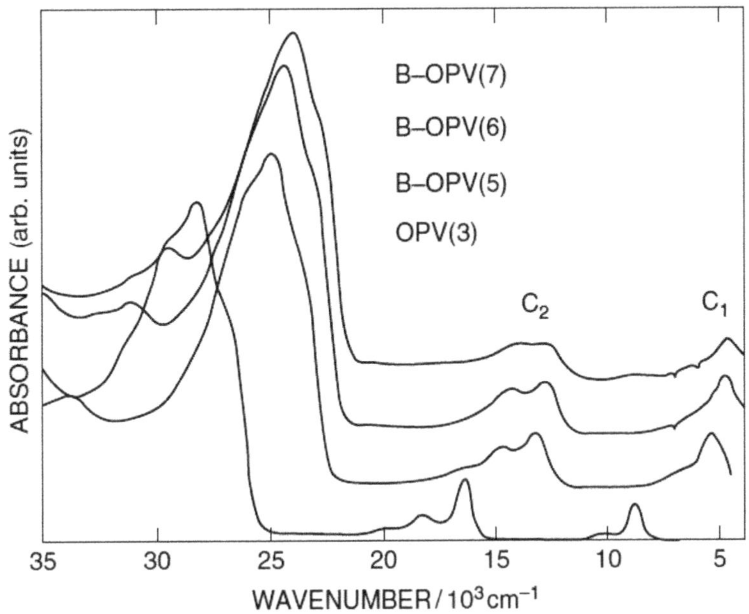

charge recombination in green). The absorption of light from the ground state S_0 to the first optically accessible excited state S_1 creates bound electron–hole pair (exciton). Adding more energy to the system by excitation to a higher state S_n can lead to exciton dissociation, resulting in the formation of free charges. These long-lived charges can recombine to reform excitons. In addition, triplet excitons can be formed through intersystem, as shown in Figure 28(b). Note that charge absorption overlaps with stimulated emission. The vertical arrows in figure are radiative transitions, and the wavy arrows are non-radiative transitions, due to vibrational relaxation (black), intersystem crossing (blue), charge generation (red) or recombination (green).

As shown in Figure 28(c), we note also that a singlet (or triplet) state are antiparallel (parallel), when the spin of electron in the π* orbital and that of the remaining electron in the π-orbital. Therefore, they add up to a total system spin of zero or one ($0, \hbar$). Then we number excited states in order, that is, S_1, S_2, or T_1, T_2, and so on, for the lowest energy or second lowest singlet or triplet excited states.

5.2 Optical Gain and Lasing

Stimulated emission occurs from the S_1 level, and lasing condition occurs when amplification is greater than loss. Optical amplification depends on the population inversion density and stimulated emission cross-section. A reduction in the population inversion density (depopulation of singlets) leads to reduction in amplification. This results mainly from annihilation processes and exciton dissociation. Losses include ground-state self-absorption, scattering in the cavity and absorption from triplet and charge excited states, whose absorption overlaps with the stimulated emission and whose long lifetimes allow significant population build-up. When optical gain is achieved in conjugated polymers, it can be

Figure 30. (a) Energy level diagram of organic semiconductors, showing charges, singlet excitons and triplet excitons; (b) idealized spectra of a conjugated polymer
After Clark & Lanzani (2010).

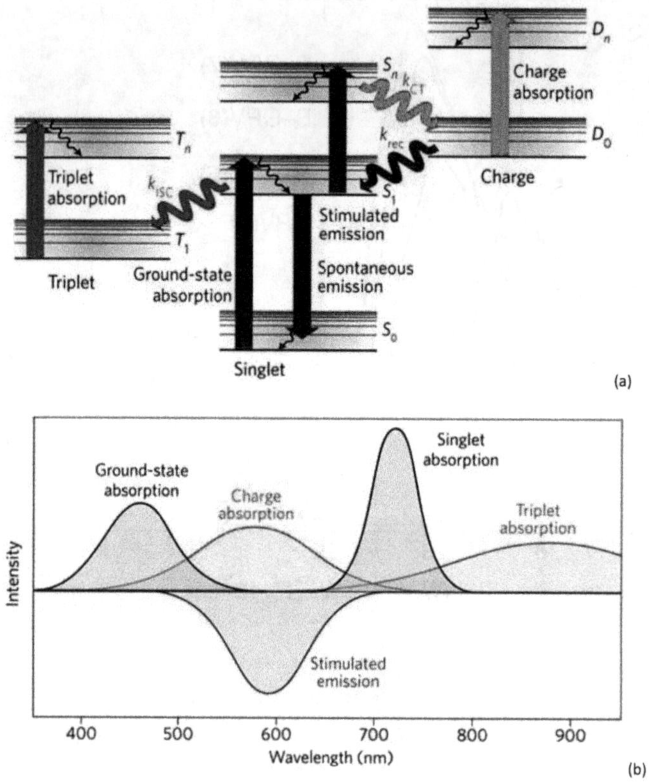

optically controlled by re-exciting the material to the S_n level (typically using infrared or near-infrared light), causing charge generation. These charges quench the stimulated emission and gain because their absorption spectrum overlaps with the stimulated emission band. Recovery of the emission depends on the lifetime of the charges, which can be sub-picosecond owing to quantum confinement. This paves the way to all-optical logic in plastic media

5.3 Nonlinear Optical Effects

Organic molecules can have large third-order optical nonlinearities, due to their large π-electron delocalization in organic semiconductors. In fact, the third-order component of the electric susceptibility of conjugated polymers is about 10^{-8} esu at resonance. This nonlinearity is useful for all-photonic devices (like switches and multiplexers) and can be also exploited in the implementation of electro-optic devices (modulators) via the Kerr effect.

6. THERMOELECTRIC TRANSPORT IN ORGANIC SEMICONDUCTORS

The interest in thermoelectricity has increased in the last few years since new applications in energy conversion had been envisaged. Carrier thermoelectric-transport theory has recently become of growing interest and numerous thermoelectric-transport models have been proposed for organic semiconductors. It is well known that organic materials have poor conduction of electricity and heat. However, it was recently reported that the thermoelectric figure of merit (zT) of PEDOT.Tos (poly-ethylene-dioxy-thiophene-tosylate) can be enhanced when the percentage oxidation is altered. This is attributed to change from lattice scattering to ionized impurity scattering in PEDOT.Tos.

The purpose of research is to provide a theoretical description of the thermoelectric Seebeck effect in organic semiconductors. Special attention is devoted to the effect of carrier concentration, temperature, and polarons on the Seebeck effect and its relationship to hopping transport theory. Table 2 summarizes the main thermal properties of some organic semiconductors, according to *Lu Li Liu (2016)*

7. MEASUREMENT OF MOBILITY OF ORGANIC SEMICONDUCTORS

The charge carrier mobility provides the figure of merit for organic Semiconductor devices. The mobility of charge-carriers in organic semiconductors can be measured by a number of techniques. In such measurements, the behavior of charge carrier mobility with temperature and applies field, provide a metric to quantify the carrier activation energy and extent of disorder in these materials.

Figure 31. Thermal transport and thermoelectric power in organic semiconductors. After Kim & Pop (2012)

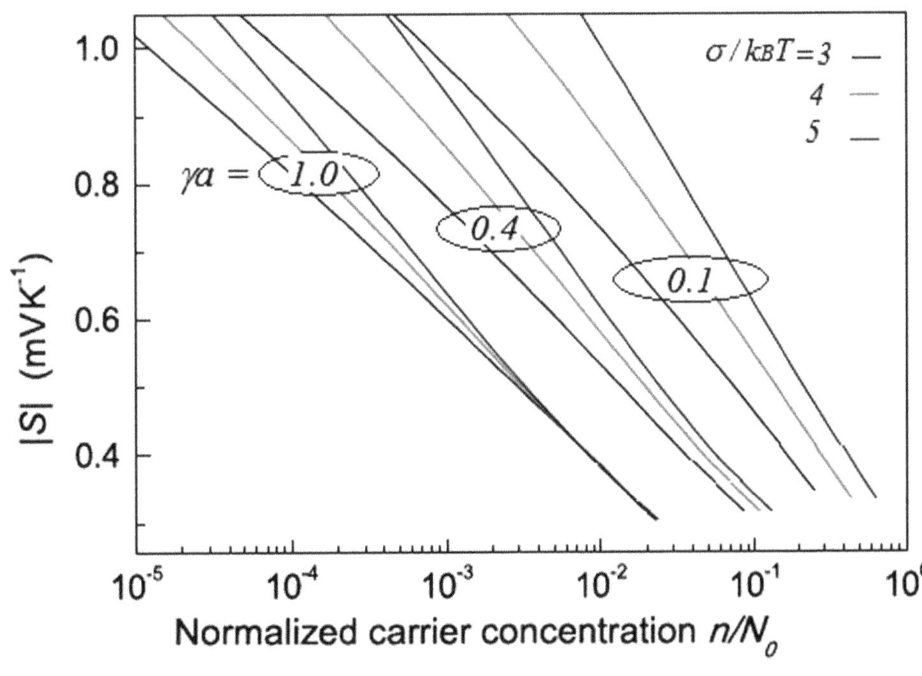

Figure 32. Thermoelectric power (Seebec coefficient) in organic semiconductors, as a function of carrier concentration
After Lu Li Liu (2016).

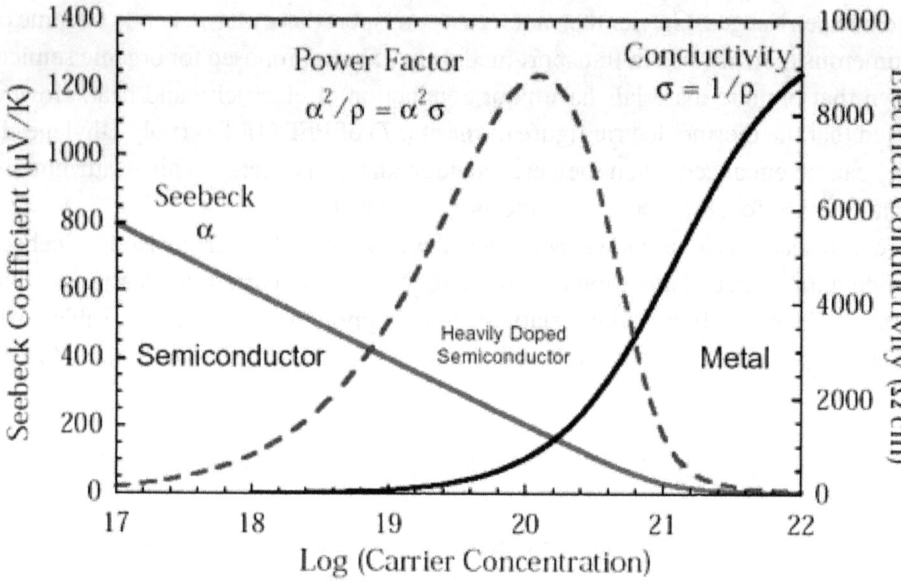

Table 2. Thermal properties of some organic semiconductors

Organic Semiconductor System	σ (Ʊ/cm)	S (μV/K)	k_{th} (W/mK)	ZT
PEDOT: Tos/ PEO-PPO-PEO electrochem. reduct.	~1200	~100		~1.02
PEDOT: PSS + EG treat.	~980	~70	0.23	0.4
PEDOS- C6 electrochem. reduct.	~200	~110		
PEDOT: PSS + EG treat. + hydrazine reduct.	~1300	~49	0.3	0.3
PEDOT: Tos + TDEA reduct.	~80	~290	0.37	0.25
PEDOT: BTFMSI + hydrazine reduct.	~1080	~37	0.19	0.22
PEDOT: PSS electrochem. reduct.	~25	~90	0.17	0.041
PEDOT: PSS + DMSO 5%	298	12.65		~0.001
PANI/CSA doping in m-cresol	220	~20		
PANI doped with H_3PO_4	40	~7		
Polyselenophene and its copolymers with 3-methyl-thiophene	0.1–54	20–98		0.034
Phenylenevinylene block copolymers & their blends with MEH-PPV	6×10^{-6}–14.4	7–531		
PEDOT: PSS + Polythiophenes Bilayered nanofilms	125–200	11–17		

In addition to the mobility and activation energy, the thermoelectric power (Seebeck coefficient) in molecular junctions can be used to provide complementary information distinguish organic materials in which disorders play a significant role. In fact, the structural disorders decrease the charge carrier mobility and makes the charge carrier mobility frequency dependent. However, many other parameters can influence the charge carrier mobility as well, including temperature, pressure, presence of impurities and size. However, we present here the most famous techniques to measure the charge carrier mobility in OSCs, and their main features.

The classic experiment to measure the mobility of charge carriers in organic semiconductors is called the time of flight (ToF) technique (Kepler, Leblanc, 1960). In the ToF method, the carrier drift velocity is determined by measuring the transit time of carriers through a semiconductor (here an organic semiconductor layer) of uniform electric field that extends across a sample of known width. Figure 28 shows a typical test structure that uses a thin film of organic semiconductor. The sample is sandwiched between two contacts (Schottky-Ohmic). The structure is then connected to a high reverse bias, such that the electric field is almost homogenous across the sample. The test sample is exposed to a short laser pulse of high-energy. The photogenerated electron-hole pairs drift under the effect of electric field. The generated holes are collected at the anode contact (Schottky-like contact[4]) while the electrons drift to the cathode (Ohmic-like contact). The electron-hole pairs may be rather generated in the ToF experiment by a high energy electron irradiation (1-10 keV).

The drift of electrons induces a current in the external circuit that continues to flow until carriers are collected at the contacts. Consequently, the duration of the current pulse is essentially equal to transit time of the electrons through the semiconductor region. The current is observed directly on a sampling oscilloscope from which the transit time is determined. The output waveform is shown in Figure 30. The field dependent transit time is obtained as the half amplitude width of the curve shown. Therefore, the electron drift velocity and mobility are given by:

$$V_d(\zeta) = d/\tau,\ \mu(\zeta) = V_d(\zeta)/\zeta = d^2/(\tau V_R) \tag{29}$$

where d is the is the distance between the electrodes, ζ is the electric field, τ is the averaged transient time, and V_R is the applied voltage.

Typically, one observes an initial spike followed by a plateau that falls off with a more or less pronounced kink. The initial spike reflects charge motion prior to the energetic relaxation in the DOS. The generated charges have a high hopping rate to neighbor sites since almost all neighbor sites are at lower energy. This high hopping rate translates in a high current. Near thermal equilibrium, the hopping rate is slower, and the current is moderate and constant. The ToF signals require that the dielectric relaxation time is large compared to the transit time t_{tr}, the thickness of the spread of the charge carriers is small compared to the film thickness, the concentration of charges is low enough that the charges do not interact, and there is no deep trapping,.

Another method to measure the charge carrier mobility in OSCs involves the injection of a space-charge-limited current from an Ohmic contact. In the absence of deep trapping, the current is given by Mott–Gurney law ($J=KV^2$). In presence of traps, the J-V curves become more complex. The carrier mobility can be also extracted from the electrical characteristics of an organic field-effect transistor (OFET), as reviewed by Horowitz (1998). In this technique the charge mobility is extracted from the I-V characteristics of the OFET. However, the mobility may be gate-voltage dependent, due to the presence of traps and/or impurities as well as carrier.

Figure 33. Measurement of charge carrier mobility by time of flight (ToF) method

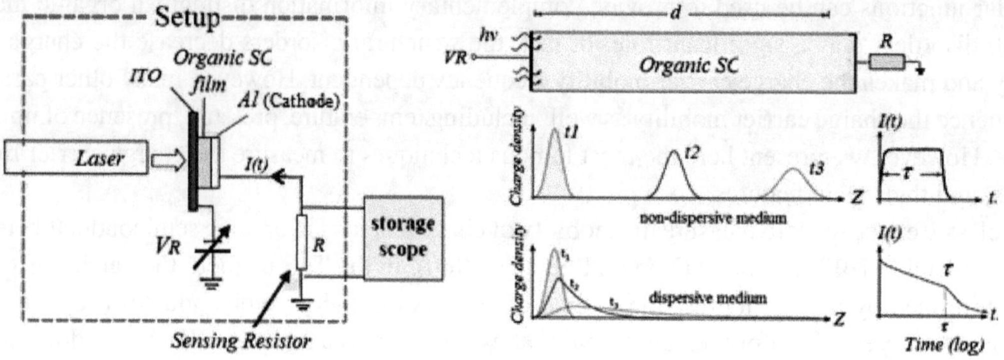

In conjugated polymers, the calculations predict carrier mobility, which is six to nine orders of magnitude larger than the measured values by ToF experiment. This discrepancy is related to the spatial scale over which transport is measured. It is reasonable to assume that transport is fast as long as it is not affected by scattering and localization. Figure 34 depicts the hole mobility in a film of poly-spiro-bifluorene-co-benzothiazole (PSF-BT) by ToF, after carrier generation by 130fs laser pulse at different applied voltages. The figure shows the mobility as a function of the averaged distance travelled by a carrier after a given time. The inset depicts the structure of PSF-BT.

We've seen so far that the observed temperature dependence of charge carrier mobility in single crystals semiconductors follows the law: $ln(\mu)\ \alpha\ T^n$, where n is a constant depending on the dominant scattering mechanism. In organic semiconductors, the situation is different. The Figure 32 depicts the effect

Figure 34. The hole mobility in a film of poly-spiro-bifluorene-co-benzothiazole (PSF-BT) as function of drift distance after carrier generation by laser
After Basler (2009).

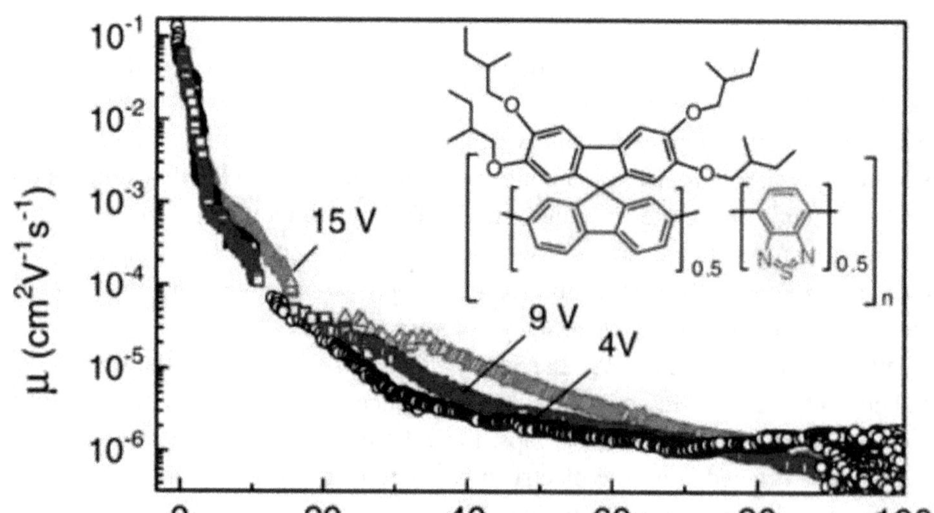

of ambient temperature on the charge carrier mobility of both electrons and holes in pure naphthalene. Transport measurements on molecular single crystals indicate that the charge mobility can be as high as a few hundred $cm^2/V.s$ at low temperature (up to300 $cm^2/V.s$ for holes in naphthalene at 10K). In addition, the mobility of electron and hole may be equally large in organic semiconductors (Coropceanu et, 2007).

8. TRANSPORT IN POLYCETSTALLINE AND AMORPHOUS MATERIALS

The fabrication technologies of organic semiconductor materials, such as vacuum deposition of molecules or solution processing of polymers, often lead to disordered amorphous or polycrystalline structures. However, in spite of their lower performance, in terms of electronic conduction, organic semiconductors are still less expensive than amorphous Si. Figure 36 illustrates the structure of crystalline and amorphous materials and their lattice potential (in one dimension)

In pure semiconductors, electrons are delocalized and their wavefunctions are extended over large distances. In this case, transport occurs in the bands, and this regime of transport is called band transport. The band transport of charge carriers has been discussed in the previous chapters of this Book. On the other hand, the transport of charge carriers in disordered materials is mainly by hopping between nearby localized pockets. Hopping requires activation to surmount the barriers between localized states. Figure

Figure 35. Temperature dependence of charge carrier mobility in naphthalene
After Warta, & Karl (1985).

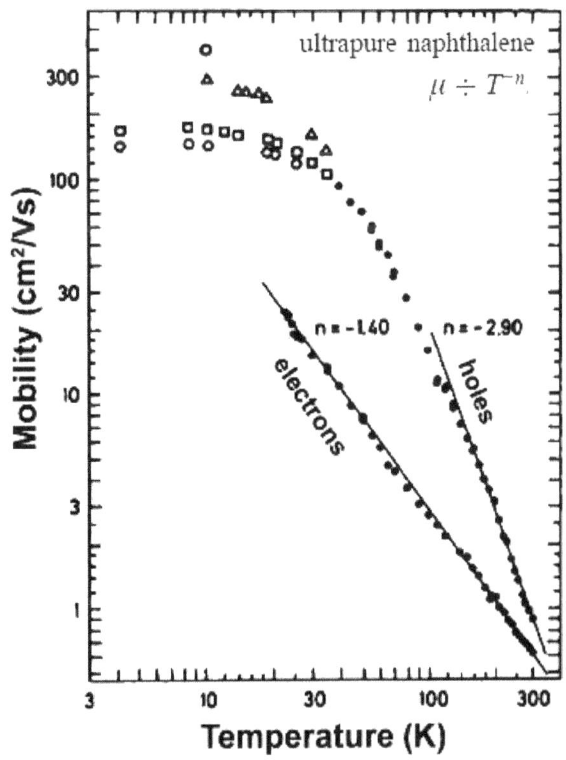

36 illustrates the nature of transport in crystalline (low-disorder) materials and polycrystalline materials and insulators with high degree of disorder.

In spite of the great progress of amorphous and polycrystalline thin film devices, their physical limits are not well established yet, since their intrinsic properties are masked by the complicated effects of disorder and grain boundaries. However, the transport of charge carriers across amorphous semiconductors may be treated in much the same way as insulators. Actually, the DC conductivity of several amorphous semiconductors follows the general form:

$$\sigma = \sigma_o \, exp\,(-\Delta E/k_B T) \quad (30)$$

where ΔE is the activation energy and the pre-exponential factor σ_o is the conductivity at high temperature, due to localized states. The activation energy ΔE is generally between 0.5-1.0 eV and σ_o is in the order of 1000 mho/cm.

Figure 36. Schematic illustration of the structure of crystalline and amorphous materials and their lattice potential (in one dimension)

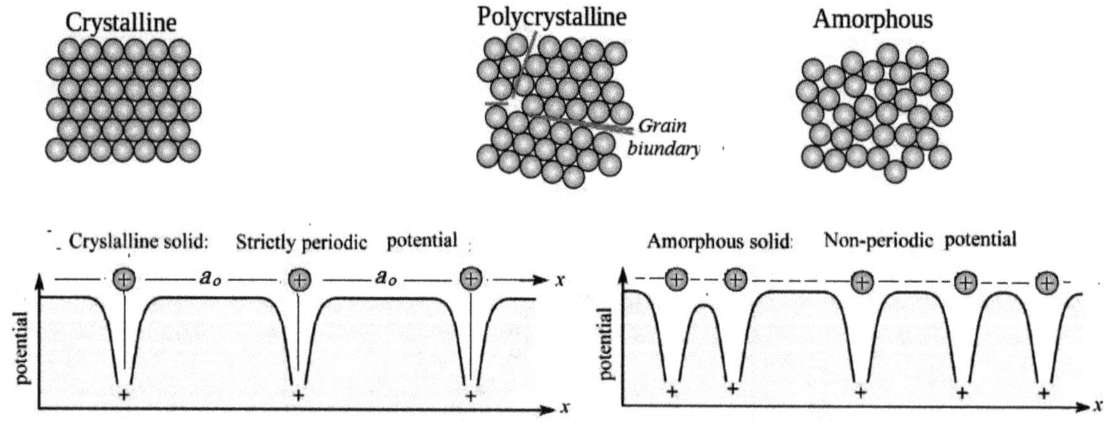

Figure 37. Schematic illustration of band transport and hopping transport regimes

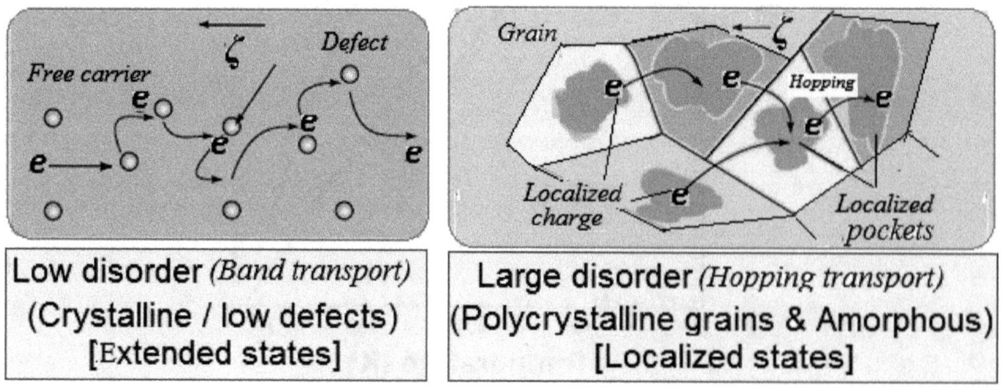

8.1 Kubo Formula

The electrical conductivity, due to localized states inside amorphous materials, may be modeled using the following Kubo formula[5]:

$$\sigma = \frac{2\pi e^2}{3m^2\omega\Omega}\sum_{i,k}(f_i - f_k)\left|M_{i,k}\right|^2 \delta(E_k - E_i - \hbar\omega) \quad (31)$$

where $M_{i,k} = <\psi_i|p|\psi_k>$ is the matrix element of momentum operator between eigen-states ψ_i and ψ_k and Ω is the cell volume, and E_i is the energy of the ith state, f_i is the Fermi-Dirac distribution function evaluated at energy E_i. By inspection of the Kubo formula, it is clear that the localization of the states, and their overlap and proximity to each other in energy and to the Fermi level, determine the DC conductivity. Figure 38 depicts the DC conductivity in amorphous semiconductors, with localized states.

8.2 Multiple Trapping and Release Model (MTR)

The multiple trapping and release (MTR) model is a phenomenological model that describes the drift mobility in amorphous semiconductors, under the effect of electric field. The MTR model was originally proposed by Shur and Hack (1984) to describe the drift mobility in hydrogenated amorphous silicon (H:α-Si), and was later extended to organic semiconductors by Horowitz et al. (2000). The MTR model assumes that charge transport occurs in extended states, but that most of the carriers injected in the semiconductor are trapped in states localized in the forbidden gap. These traps can be deep (trap energy level near the middle of the energy gap), or shallow (trap energy level near the conduction or valence band). The model shows a dependence of the mobility of the carriers on temperature, the energy level of traps, as well as on the carrier density (and therefore on the applied field). For a single trap, the charge drift mobility is given by:

Figure 38. The simulated DC conductivity of amorphous Si, as a function of energy computed with the Kubo formula

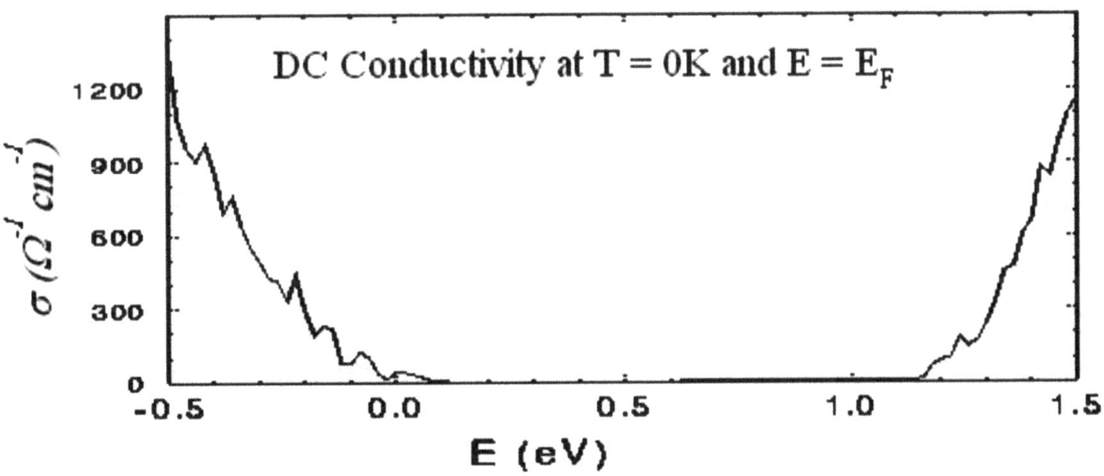

$$\mu = \mu_o \exp(-E_{tr} / k_B T) \tag{32}$$

Using this model, the polycrystalline semiconductor may be described as trap-free grains separated by boundaries with high trap density. When the grain size is lower than the Debye length, the distribution of the traps can be considered uniform. However, if the grains are much larger than the Debye length, charges move through the grain boundaries. At high temperature, this occurs via thermionic emission, which makes the drift mobility depend on the temperature. At low temperatures, the charges can tunnel through the grain boundaries, and the mobility becomes temperature independent. At intermediate temperatures, charge transport is determined by thermally activated tunneling. As we'll see in the next case study, traps can distort the transfer curves of organic semiconductor devices (linear $I_{ds} \sim V_g$ become power-law, $I_{ds} \sim V_g^{n}$) and make the mobility temperature and bias dependent.

9. CASE STUDY 16: ORGANIC SEMICONDUCTOR TRANSISTORS

Nowadays, the technological interest in molecular crystals arises mostly from the field of organic field effect transistors (OFETs). OFETs have seen rapid development in many applications. There are many advantages to OFETs, such as the flexibility of the plastic substrate and the low fabrication cost by ink-jet

Figure 39. (a) Multiple trapping and release in organic semiconductors (b) the distribution of trap states in the energy band

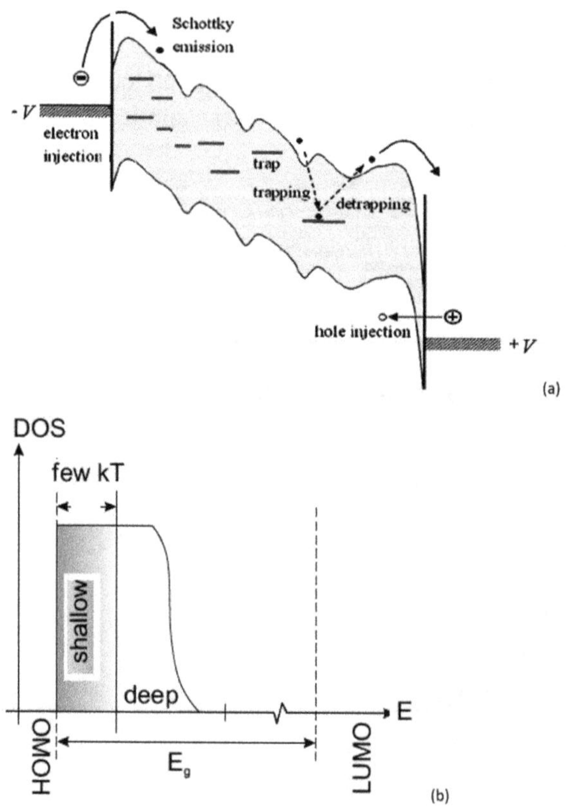

printing. One of the most widely studied organic semiconductor materials used for OFETs is Pentacene. Pentacene-based OFETs have a typical field effect mobility of around $1cm^2/Vs$. This is of comparable value to hydrogenated amorphous silicon. Therefore, the organic thin-film transistors (OTFTs) are expected to replace hydrogenated amorphous silicon TFT in flat panel displays. In fact OTFTs do not require a glass substrate as amorphous silicon does and can be made on a piece of plastic

9.1 Device Structure and Models

The numerical simulation is helpful in understanding OTFT's basic device operation and the optimization of device structures. In order to simulate I-V characteristics of OTFTs, it is important to consider how carriers transport in organic semiconductors. In many cases, the space-charge limited current (SCLC) model is successful in explaining the conduction current of organic semiconductors. This approach is acceptable in organic light-emitting diodes (OLEDs) and OTFTs. In fact, the disorder effect on charge carrier motion is diminished in OTFTs because current flow is confined in a thin layer. Therefore, the tail states of the DOS distribution are filled and the activation energy for charge carrier hopping is trivial. In addition, the SCLC model is suited for use in conjunction with the conventional drift-diffusion model (DDM) in device simulators.

The experimental I_D-V_D curve of a Pentacene OTFT is numerically simulated by *ATLAS* using the SCLC model and a field-dependent mobility model. In the SCLC model, the carriers are self-trapped. In addition, one of the most determinant factors for carrier transport characteristics is the energy distributions of density of states (DOS) within the bandgap. In this example, a Pentacene TFT reported by Lin, et al. is simulated with *ATLAS* simulator and then compared to their experimental I_D-V_D curves. The device structure is shown in Figure 40. A staggered inverted structure is adopted. The thickness of the grate dielectric layer is 400nm and the Pentacene active layer is 50nm. The channel length and width are 20 um and 220 um, respectively. The charge accumulation in the idealized device structure is shown in Figure 41.

9.2 Effects of Traps on the Electrical Characteristics of TFTs

Traps are localized electronic states that can capture free carriers. For instance, a trap can capture a hole in a reaction and the resulting positive charge becomes immobile and unavailable for conduction. It is obvious that traps have a significant effect on the performance of organic devices, be it an OTFT

Figure 40. Cross section of an organic thin-film transistor (OTFT)

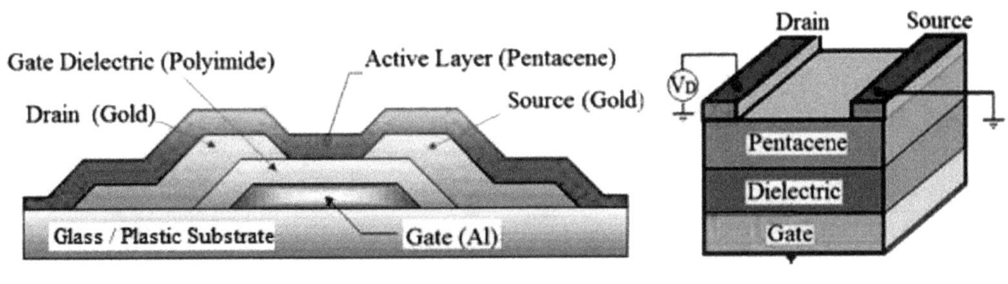

Figure 41. Charge accumulation in an idealized (top-gated) OTFT

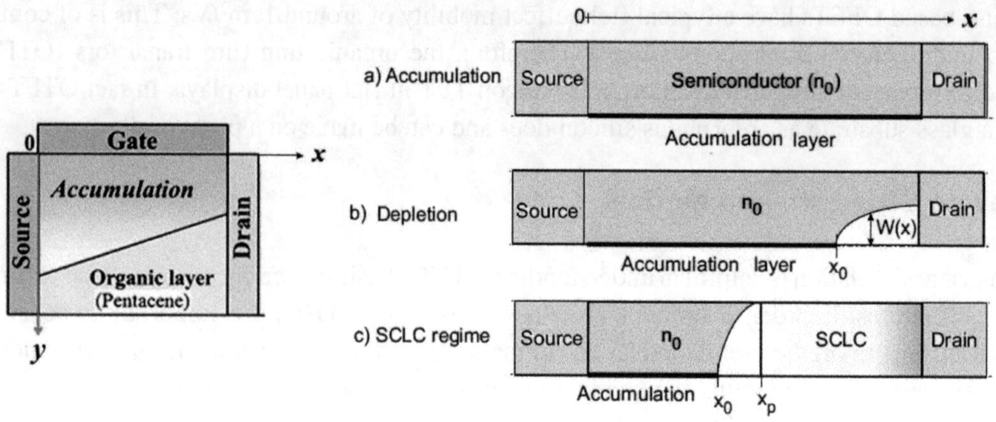

or an OLED and their consideration is very important. The effects of traps on the electronic properties of TFTs can be summarized as follows:

- Reduced mobility (Poole-Frenkel)
- Temperature dependence of mobility
- Non-linearity in I-V curves
- Non linearity in transfer curves; (power-law, $I_{ds} \sim V_g^{\,n}$)
- activation energy of mobility depends on bias (Meyer-Neldel rule)
- Transients; I_{ds} depends on time, often following multi-exponentials, resulting in so-called stretched exponential, $I_{ds} \sim exp[-(t/\tau)^{\beta}]$

9.3 Simulation Results

In simulation process, the energy band gap of Pentacene is taken from optical bandgap data as 2.8eV. Also, the extracted mobility is 0.62 $cm^2/V.s$. The density-of-states in the gap of pentacene can be calculated from the electrical characteristics of thin-film transistors measured at different temperatures (Puigdollersa et al., 2010). In the illustrated case, the acceptor tail DOS is expressed by an exponential function of energy, as shown in Figure 42.

Figures 43 shows the simulation results (green lines) compared with the experimental data (red lines). The gate bias-dependent saturation drain currents appear to agree with the experiment, but the currents in transition regions, from linear to saturate, are still different.

Though a constant mobility model is used in the illustrated case, most organic semiconductor materials have electric field-dependent carrier drift mobility. The famous mobility model of Pool-Frenkel model has a square root dependence of electric field and can be expressed as follows:

$$\mu = \mu_{o.}\, exp[-E_a/k_B T]\; exp[\sqrt{\zeta}\,(\beta/T-\gamma)] \quad (33)$$

Figure 42. Illustration of the density of states in an OTFT

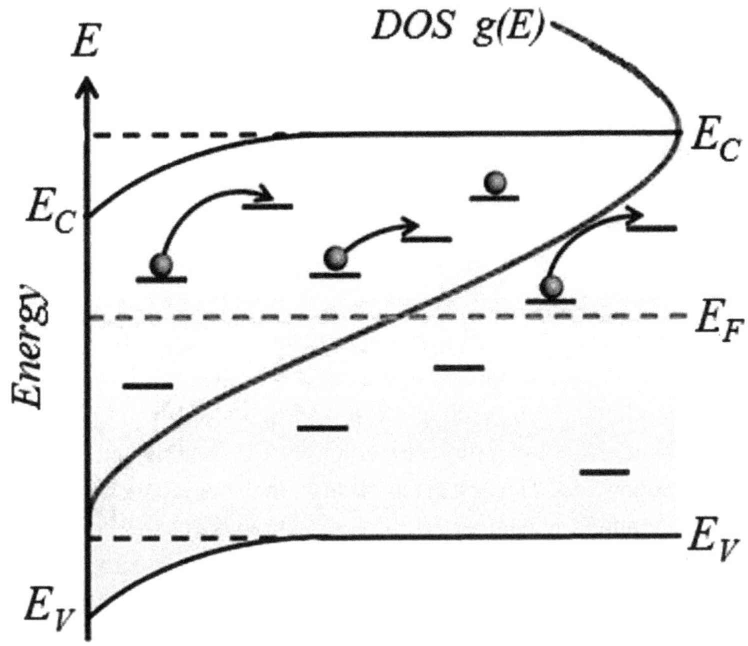

Figure 43. Simulated (green) I_D-V_D curves using constant mobility model and the experimental (red) curve
After Kymissis, Dimitrakopoulos & Purushothaman (2001).

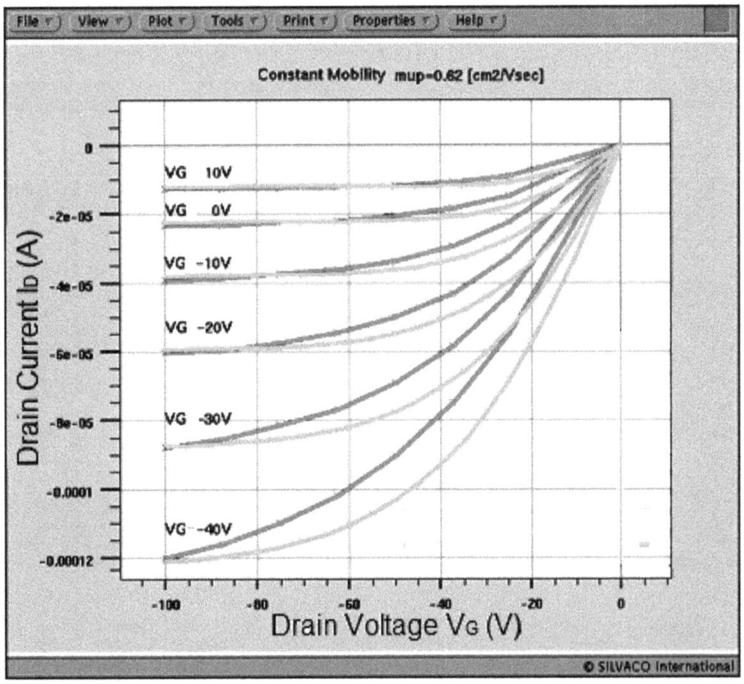

where β and γ are fitting parameters. The results in Figure 44 are obtained with this mobility model and agree with experimental data. Figure 45 shows the hole density and current flow lines at V_D =-100V and V_G =- 40V

10. CASE STUDY 17: ELECTRON TRANSPORT THROUGH MOLECULAR WIRES

If a long chain of carbon atoms is formed, the van der Waal π-bonds become delocalized along the chain and form a one-dimensional molecular system. The electron transport through such molecular wires takes place via tunneling and thermally-activated hopping.

Different levels of transport models, starting from semiclassical methods up to the ab initio methods, have been already introduced and can be applied in the context of nanowires (e,g, see Hasnip et al, 2014). For instance, the first principle methods, such as DFT, and many-electron GW calculations have been applied for simulating the band structure of molecular nanowires. The different levels of theory belong to two major computational frameworks. The first calculation framework is called the complex band structure (CBS) and can estimate the tunneling parameters. The second makes use of the Landauer–Buttiker (L-B) transport approach to computes the transmission spectra and ballistic current. The so-called contact block reduction (CBR) is a ballistic variant of the NEGF quantum approach that considers the effects of contact geometry (Ryu & Klimec, 2008). As an example of the CBS approaches, we show here the application of the DFT-TB method to study transport through the molecular device shown in Figure 46

Figure 44. Simulated I_D-V_D (green) curves using field dependent mobility and experimental (red) curves
After Kymissis, Dimitrakopoulos & Purushothaman (2001).

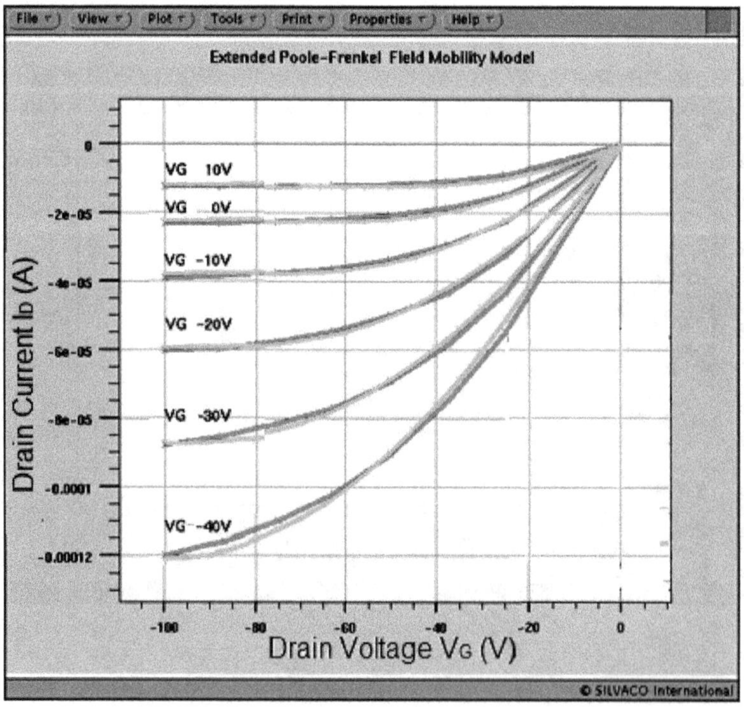

Figure 45. Hole concentration distribution and Current flow lines in the OFET
After Kymissis, Dimitrakopoulos & Purushothaman (2001).

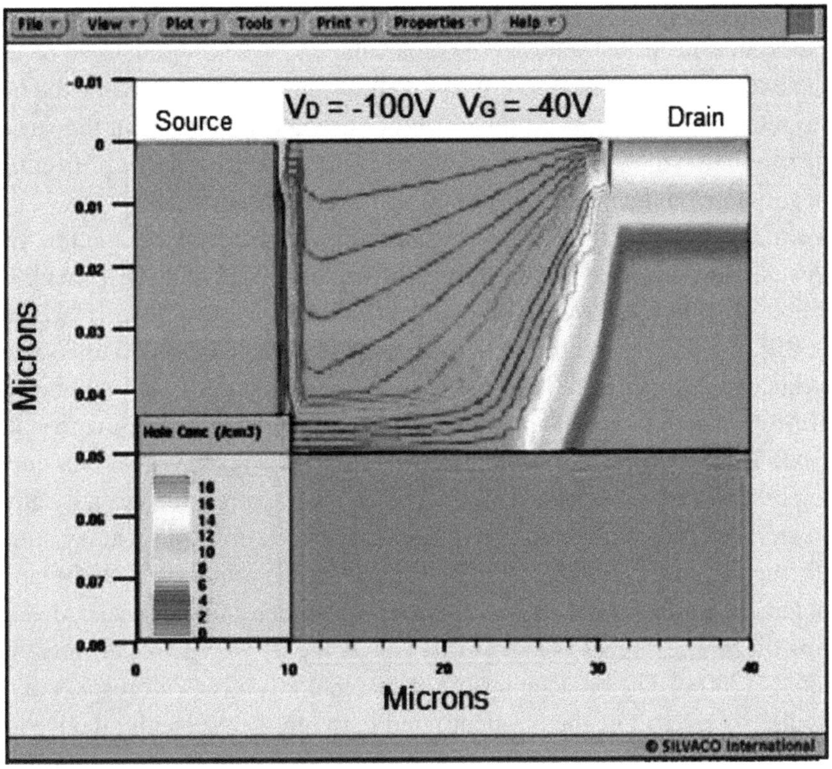

Figure 46. A molecular device of 5 OPV rings, between 2 gold contacts via 2 sulfur atoms

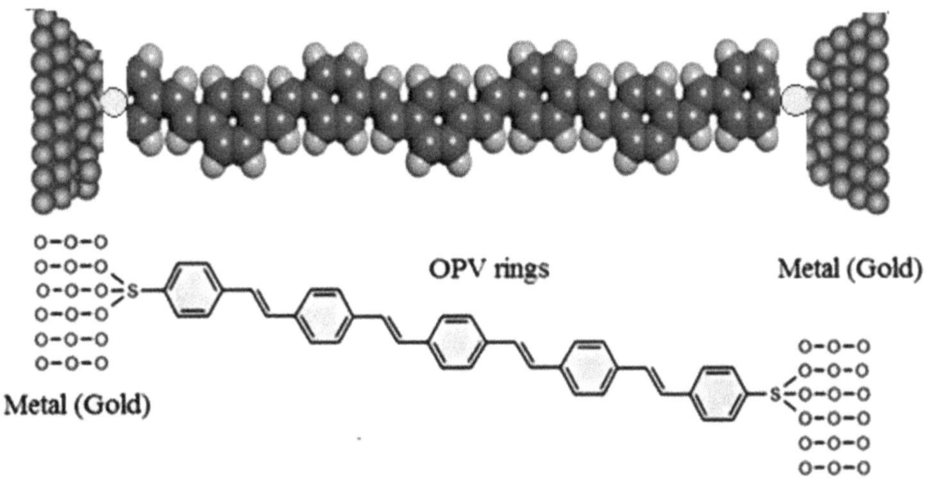

The device is composed of a five oligopolyvinylene (OPV) chains anchored to metallic (gold) contacts. The contact between the molecule and the Au electrodes is established by covalent S-Au bonds, which is often done experimentally. The electronic transport characteristics of the OPV molecules are governed by the available electronic states in the molecular chain. These states can be described by the local density of states (LDOS), which represents the equilibrium density of states in the molecular device broken down into the contributions of single atoms or ensembles of atoms in the structure. Using an appropriate many simulation tool, we can calculate the LDOS of each atom by projecting the density of states of the total structure on the molecular orbitals of this atom.

There are so many simulation tools that perform the band structure calculation and conductance modeling of molecular devices. Among these tools, one can cite: TURBOMOLE (Ahlrichs et al, 1989), SIESTA (Ordejón et al, 1996), NEMO (Klimeck et al, 2008) and NextNano (Hackenbuchner, 2002; Andlauer, 2009). SIESTA is based on the DFT method and has been extended to compute the molecular conductance via Green's functions in another package called: TranSIESTA (Brandbyge et al, 2002). The so-called NEMO simulator can calculate the atomic structure from semi-empirical TB model and makes use of the NEGF to calculate the terminal current. Also, Nextnano is a famous quantum simulator that analyzes the physical properties of semiconductor nanostructures, for virtually any geometry and combination of materials. The combination DFT-NEGF has been utilized by many authors to simulate the behavior of molecular wires, (Stokbro, 2008). The advantage of the DFT-NEGF approach is that it can include the interaction effects between electrons with phonons and photons, which are important in phononic and photonic devices. Figure 47 shows the LDOS for OPVs wires of different lengths, according to Schuster et al., (2008). The broadening of the molecular levels (with respect to the sharp peaks of isolated molecules) is a result of the covalent bond between the molecular device and the contacts or the adjacent sulfur atoms. The charge density and terminal current can be calculated from the local density of states (LDOS), by a variety of methods, such as the CBR (Briner, 2011).

Figure 47. Local DOS of four unfunctionalized OPV molecules. The Fermi level (Ef) of the gold contacts at equilibrium is shown
After Schuster et al., (2008).

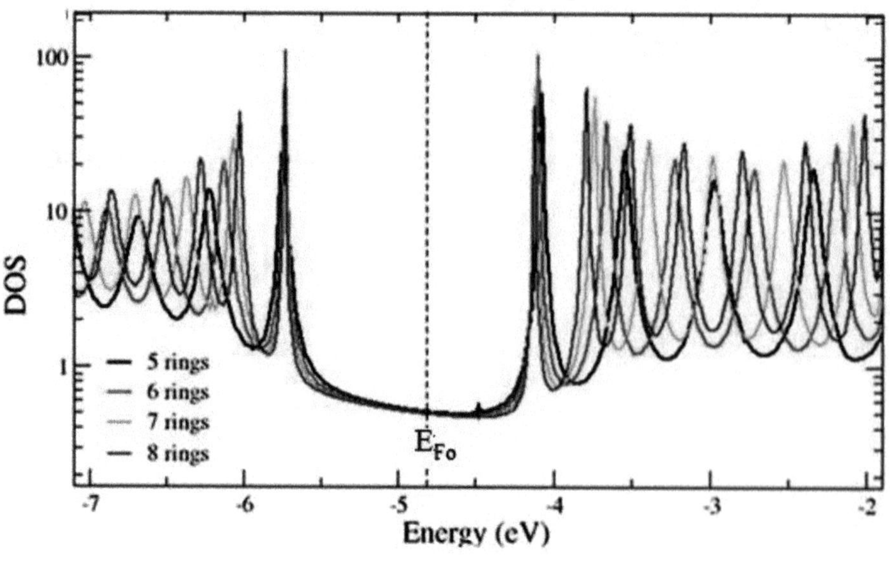

The Figure 48 shows the *I–V* characteristics of OPV chains with different lengths (OPV5–OPV8). All chains display higher currents in the resonance regime as opposed to the tunneling mechanism. The threshold voltage does not change (about 1.7V) as the transmission gap of all wires is almost constant. However, the current decreases with wire length. The observed shift of the transmission peaks is also due linear potential drop on the chain

11. SUMMARY

The term *organic semiconductors* is used to describe organic materials which possess the ability of conducting electric current. In fact, it was shown that certain organic materials (plastics) can conduct electricity. Organic materials are well-suited to certain electronics applications due to their low cost, low weight, and flexibility, but are less desirable for other applications due to poor conduction of electricity and heat. In order for organic semiconductors to be widely spread in electronic devices, the mobility of charge carriers should be improved. Compared to inorganic semiconductors (like Si), the carrier mobility in organic semiconductors is very low, the highest value on the order of 1 $cm^2/V{\cdot}s$. The improvement of the conduction properties requires more advanced models of charge transport in disordered organic semiconductors.

Organic semiconductors can be broadly classified into two groups on the basis of their molecular weight: conjugated polycyclic compounds of molecular weight less than 1000, and heterocyclic polymers with molecular weight greater than 1000. Polymers are useful materials for semiconductors because of the ease with which they form thin films with large surface area

In organic semiconductors the molecules are held together by weak van der Waals bonds (π-bonding). The properties of organic semiconductors have much in common with amorphous materials. Some organic semiconductor crystals, such as polyacenes and fullerenes, have attractive characteristics for

Figure 48. I-V characteristics of the OPV molecular device
After Schuster et al., (2008).

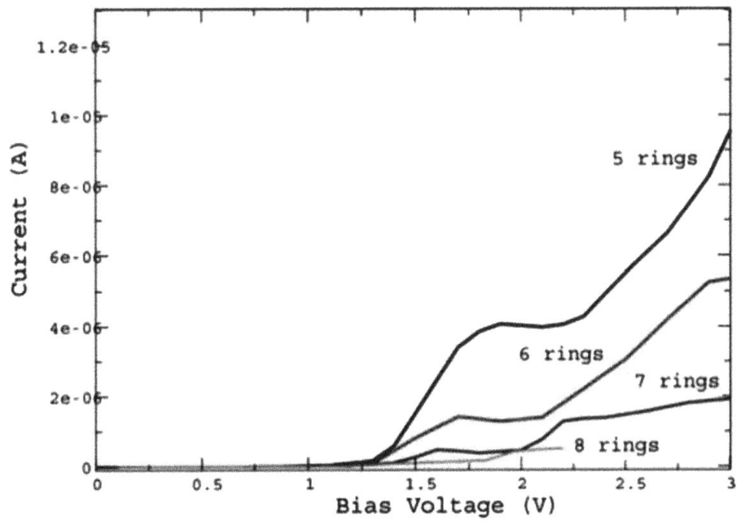

electronic and optoelectronic devices. For instance, pentacene has the smallest bandgap among the linear polyacenes, and the highest effective mobility in polycrystalline organic thin film FETs, with μ= 0.3-1.5 $cm^2/V.s$. This value is comparable to that of amorphous Si. The organic semiconductors have, therefore, very poor mobility at room temperature. However, the mobility of such materials is improved at low temperature and may even show superconductivity.

Charge transport in organic semiconductors occurs through hopping. Charges hop between localized sites that are disordered in position and energy. The charges are called polarons and can interact and form e–h pairs, excitons and bipolarons. The current in a typical organic device is limited by space charge, resulting in typical *J(V)* characteristics.

The transport modeling of charge carriers across organic semiconductors is sometimes treated in much the same way as insulators. The DC conductivity of organic semiconductors has the form:

$$\sigma = \sigma_o \exp(-\Delta E/k_B T)$$

where ΔE is the activation energy and σ_o is the conductivity at high temperature, due to localized states.

The role of disorder, that is inevitably present in organic semiconductors, is not completely understood. In fact, charge carriers in organic semiconductors are spatially localized due to several reasons. Any mechanism that can destroy the periodicity of lattice field $V(\mathbf{r})$ to the level of critical randomness over a certain interval can cause Anderson localization. For example, a large concentration of crystal imperfections (e.g., impurities, grain boundaries, dangling bonds) destroys the periodicity of a crystal. For a localized carrier, external energy is needed to escape the localized state. A theoretical description for the charge carrier tunneling process is hopping transport.

Many traditional hopping transport models have been developed when organic semiconductors being studied were primarily amorphous, and the assumption of strong localization fit the experimental data. However, this assumption of strong localization has failed recently in several organic semiconductors with high crystallinity, and the nature of carrier localization in these crystalline organic semiconductors is now under academic debate. Nevertheless, one can argue the nature of carrier localization in crystalline organic semiconductors as follows. The thermal energy at room temperature is comparable to the weak van der Waals bonding energy, and therefore a dynamic disorder can exist in an organic semiconductor, regardless of its crystallinity, which causes Anderson localization.

In conclusion, our understanding of the charge transport mechanism of organic materials is at a rudimentary phase, very similar to the phase of "*cat's whisker diode*" in semiconductor research. Apparently we have some understanding of the macroscopic transport process of charge carriers in disordered organic solids. However, our understanding on the microscopic scale, in particular on very short distance and time, is still incomplete.

REFERENCES

Ahlrichs, R., Bar, M., Haser, M., Horn, H., & Kolmel, C. (1989). Electronic structure calculations on workstation computers: The program system turbomole. *Chemical Physics Letters*, *162*(3), 165–169. doi:10.1016/0009-2614(89)85118-8

Akamatsu, H., & Inokuchi, H. (1952). Article. *The Journal of Chemical Physics*, *20*, 1481. doi:10.1063/1.1700784

Albuquerque, R. Q., Hofmann, C. C., Köhler, J., & Köhler, A. (2011). Diffusion limited energy transfer in blends of oligofluorenes with an anthracene derivative. *The Journal of Physical Chemistry B*, *115*(25), 8063–8070. doi:10.1021/jp202333w PMID:21634399

Anderson, P. W. (1958). Absence of Diffusion in Certain Random Lattices. *Physical Review*, *109*(5), 1492–150. doi:10.1103/PhysRev.109.1492

Arkhipov, V. I., Emelianova, E. V., & Adriaenssens, G. J. (2001). Effective transport energy versus the energy of most probable jumps in disordered hopping systems. *Physical Review B: Condensed Matter and Materials Physics*, *6412*(12), 125125. doi:10.1103/PhysRevB.64.125125

Arkhipov, V. I., Emelianova, E. V., Heremans, P., & Bassler, H. (2005). Analytic model of carrier mobility in doped disordered organic semiconductors. *Physical Review B: Condensed Matter and Materials Physics*, *72*(23), 235202. doi:10.1103/PhysRevB.72.235202

Arkhipov, V. I., Wolf, U., & Bassler, H. (1999). Current injection from metal to disordered hopping system. II. Comparison between analytic theory and simulation. *Physical Review B: Condensed Matter and Materials Physics*, *59*(11), 7514–7520. doi:10.1103/PhysRevB.59.7514

Atkins, P. W. (1983). *Molecular Quantum Mechanics*. Oxford University Press.

Bagmut, A. G. (2012). Classification of the Amorphous Film Crystallization Types with Respect to Structure and Morphology Features. *Technical Physics Letters*, *38*(5), 488–491. doi:10.1134/S1063785012050197

Bassler, H., & Kohler, A. (2012). *Charge Transport in Organic Semiconductors. Top Curr Chem.*, *312*, 1–66. PMID:21972021

Birner, S. (2011). *Modeling of semiconductor nanostructures and Semiconductor–electrolyte interfaces* (Ph.D. Dissertation). Technischen Universität München.

Bolognesi, A., Di Carlo, A., Lugli, P., & Conte, G. (2003). Large drift-diffusion and Monte Carlo modeling of organic semiconductor devices. *Synthetic Metals*, *138*(1-2), 95–100. doi:10.1016/S0379-6779(02)01318-8

Brawand, N., Vörös, M., Govoni, M., & Galli, G. (2016). Generalization of dielectric dependent hybrid functionals to finite systems. *Physical Review X, 6*(4), 041002. doi:10.1103/PhysRevX.6.041002

Buot, F. A. (2009). *Nonequilibrium Quantum Transport Physics in Nanosystems: Foundation of Computational Nonequilibrium Physics in Nanoscience and Nanotechnology*. World Scientific. doi:10.1142/6042

Clark, J., & Lanzani, G. (2010). Organic photonics for communications. *Nature Photonics, 4*(7), 438–446. doi:10.1038/nphoton.2010.160

Coehoorn, R., & Bobber, P. A. (2012). Physics of Organic Semiconductors. *Phys. Status Solidi A, 209*(12), 2354–2377. Available at: https://www.researchgate.net/publication/258692334_Physics_of_Organic_Semiconductors

Condon, E. (1926). A theory of intensity distribution in band systems. *Physical Review, 27*, 640.

Coropceanu, V., Cornil, J., Filho, D., Olivier, Y., Silbey, R., & Bredas, J. (2007). Charge Transport in Organic Semiconductors. *Chemical Reviews, 107*(4), 926–952. doi:10.1021/cr050140x PMID:17378615

Debdeep, J. (2004). *Charge Transport in Semiconductors*. Univ. of Notre Dame.

Deussen, M., & Bassler, H. (1993). Anion and cation absorption-spectra of conjugated oligomers and polymer. *Synthetic Metals, 54*(1-3), 49–55. doi:10.1016/0379-6779(93)91044-3

Eley, D. D. (1948). Phthalocyanines as Semiconductors. *Nature, 162*(4125), 819. doi:10.1038/162819a0 PMID:18103107

Emin, D. (2008). Generalized adiabatic polaron hopping: Meyer-Neldel compensation and Poole-Frenkel behavior. *Physical Review Letters, 100*(16), 166602. doi:10.1103/PhysRevLett.100.166602 PMID:18518230

Fishchuk, I., & Kadashchuk, A. (2013). Effective Medium Approximation Theory. Description of Charge-Carrier Transport in Organic Field-Effect Transistors. In Small Organic Molecules on Surfaces, Springer Series in Materials Science, (Vol. 173, pp. 171-201). Springer-Verlag Berlin.

Fishchuk, I., Kadashchuk, A., Hoffmann, S. T., Athanasopoulos, S., Genoe, J., Bässler, H., & Köhler, A. (2013). Unified description for hopping transport in organic semiconductors including both energetic disorder and polaronic contributions. *Physical Review B: Condensed Matter and Materials Physics, 88*(12), 125202. doi:10.1103/PhysRevB.88.125202

Forrest, S. R. (2004). The path to ubiquitous and low-cost organic electronic appliances on plastic. *Nature, 428*(6986), 911–918. doi:10.1038/nature02498 PMID:15118718

Fowler, R. H., & Nordheim, L. (1928). Electron emission in intense electric fields. *Proceedings of the Royal Society of London A: Mathematical, Physical and Engineering Sciences, 119*(781), 173-181 doi:10.1098/rspa.1928.0091

Franck, J., & Dymond, E. G. (1926). Elementary processes of photochemical reactions. *Transactions of the Faraday Society, 21*(February), 536–542. doi:10.1039/tf9262100536

Frenkel, J. (1938). On pre-breakdown phenomena in insulator and electronic semiconductors. *Physical Review, 54*(8), 647–648. doi:10.1103/PhysRev.54.647

Friend, R. H., & Ruppel, W. (1959). Two-Carrier Space-Charge-Limited Current in a Trap-Free Insulator. *Journal of Applied Physics*, *30*(10), 1548–1558. doi:10.1063/1.1734999

Gonzalez, G. (2014). Quantum theory of space charge limited current in solids. *Journal of Applied Physics*, *117*(8), 084306–084310. 1411.4659v1 doi:10.1063/1.4913512

Grün, N., Muhlhans, A., & Scheid, W. (1982). Numerical treatment of the time-dependent Schrödinger equation in rotating coordinates. *Journal of Physics. B, Atomic and Molecular Physics*, *15*(22), 4042. doi:10.1088/0022-3700/15/22/007

Gundlach, D. J., Lin, Y. Y., Jackson, T. N., Nelson, S. F., & Schlom, D. G. (1997). Pentacene organic thin-film transistors-molecular ordering and mobility. *IEEE Electron Device Letters*, *18*(3), 87–89. doi:10.1109/55.556089

Hasnip, P. J., Refson, K., Probert, M. I. J., Yates, J. R., Clark, S. J., & Pickard, C. J. (2014). Density functional theory in the solid state. *Phil. Trans. R. Soc. A*, *372*(2011), 20130270. doi:10.1098/rsta.2013.0270 PMID:24516184

Heeger, A. J. (2001). Semiconducting and metallic polymers: The fourth generation of polymeric materials (Nobel lecture). *Angewandte Chemie International Edition*, *40*(14), 2591–2611. doi:10.1002/1521-3773(20010716)40:14<2591::AID-ANIE2591>3.0.CO;2-0 PMID:11458348

Hirao, A., Nishizawa, H., & Sugiuchi, M. (1995). Diffusion and drift of charge carriers in molecularly doped polymers. *Physical Review Letters*, *75*(9), 1787–1790. doi:10.1103/PhysRevLett.75.1787 PMID:10060391

Hirose, K., & Kobayashi, N. (2014). *Quantum Transport Calculations for Nano systems*. Pan Stanford.

Holstein, T. (1959). Studies of polaron motions - part i: The molecular-crystal model. *Annals of Physics*, *8*(3), 325–342. doi:10.1016/0003-4916(59)90002-8

Horowitz, G. (1998). Organic field-effect transistors. *Advanced Materials*, *10*(5), 365–377. doi:10.1002/(SICI)1521-4095(199803)10:5<365::AID-ADMA365>3.0.CO;2-U

Horowitz, G. (2015). Validity of the concept of band edge in organic semiconductors. *Journal of Applied Physics*, (118): 11.

Horowitz, G., Hajlaoui, M. E., & Hajlaoui, R. (2000). Temperature and gate voltage dependence of hole mobility in polycrystalline oligothiophene thin film transistors. *Journal of Applied Physics*, *87*(9), 4456–4463. doi:10.1063/1.373091

Hückel, E. (1931). Quantentheoretische Beiträge zum Benzolproblem I. Die Elektronenkonfiguration des Benzols und verwandter Verbindungen. *Zeitschrift fur Physik*, *70*(3-4), 204–286.

Hückel, E. (1934). Theory of free radicals of organic chemistry. *Transactions of the Faraday Society*, *30*(0), 40–52. doi:10.1039/TF9343000040

Hush, N. S., & Ann, N. Y. (2003). An Overview of the First Half-Century of Molecular Electronics by Acad. *Sci.*, *1006*, 1–20. PMID:14976006

Ilatikhemeneh, H., Salazar, R. B., Klimeck, G., Rahman, R., & Appenzeller, J. (2015). *From Fowler-Nordheim to Non-Equilibrium Green's Function Modeling of Tunneling*. arXiv:1509-08170v1

Inoue, Y. (2005). Organic Thin-file transistors with high electron mobility based on perfluoropentacene. Japan Society of Applied Physics, 44(6A), 3663-3668.

Ishii, H., Hayashi, N., Ito, E., Washizu, Y., Sugi, K., Kimura, Y., . . . Seki, K. (2004). *Kelvin probe study of band bending at organic semiconductor / metal interfaces: examination of Fermi level alignment*. Available from: https://www.researchgate.net/publication/227998798

Kallmann, H. P., & Pope, M. (1960). Positive Hole Injection into Organic Crystals. *The Journal of Chemical Physics*, *32*(1), 300–301. doi:10.1063/1.1700925

Kane, E. O. (1961). Theory of Tunneling. *Journal of Applied Physics*, *32*(1), 83–91. doi:10.1063/1.1735965

Kepler, R. G. (1960). Charge Carrier Production and Mobility in Anthracene Crystals. *Physical Review*, *119*(4), 1226–1229. doi:10.1103/PhysRev.119.1226

Khoo, K., Chen, Y., Liacd, S., & Quek, S. Y. (2015). Length dependence of electron transport through molecular wires- a first principles perspective. *Phys. Chem.*, *17*, 77–96. PMID:25407785

Klimeck, G., Ahmed, S. S., Bae, H., Kharche, N., Clark, S., Haley, B., & Boykin, T. B. et al. (2007). Atomistic simulation of realistically sized nanodevices using NEMO 3-D—Part I: Models and benchmarks. *IEEE Transactions on Electron Devices*, *54*(9), 2079–2089. doi:10.1109/TED.2007.902879

Kohler, A., & Beljonne, D. (2004). The singlet-triplet exchange energy in conjugated polymers. *Advanced Functional Materials*, *14*(1), 11–18. doi:10.1002/adfm.200305032

Kordt, P. (2016). *Charge Dynamics in Organic Semiconductors: From Chemical Structures to Devices*. Berlin: Walter de Gruyeter. doi:10.1515/9783110473636

Kordt, P., Stodtmann, S., Badinski, A., Al Helwi, M., Lennartz, C., & Andrienko, D. (2015). Parameter-free continuous drift–diffusion models of amorphous organic semi-conductors. *Physical Chemistry Chemical Physics*, *17*(35), 22778–22783. doi:10.1039/C5CP03605D PMID:26267617

Koswatta, S. O., Hasan, S., Lundstrom, M. S., Anantram, M. P., & Nikonov, D. E. (2007). Nonequilibrium Greens Function Treatment of Phonon Scattering in Carbon-Nanotube Transistors. *IEEE Transactions on Electron Devices*, *54*(9), 2339–2351. doi:10.1109/TED.2007.902900

Kubar, T., & Elstner, M. (2013). Efficient algorithms for the simulation of non-adiabatic electron transfer in complex molecular systems: Application to DNA. *Physical Chemistry Chemical Physics*, *15*(16), 5794–5813. doi:10.1039/c3cp44619k PMID:23493847

Kwiatkowski, J. J. (2008). *From molecules to mobilities: Modelling charge transport in organic semiconductors* (PhD Dissertation). Imperial College London, UK.

Kymissis, I, Dimitrakopoulos, C.D. & Purushothaman, S. (2001). Two-Dimensional ATLAS Device Simulation of Pentacene Organic Thin-film Transistors. *IEEE -TED, 48*(6), 1060.

Leblanc, O. H. Jr. (1960). Hole and Electron Drift Mobilities in Anthracene. *The Journal of Chemical Physics*, *33*(2), 626. doi:10.1063/1.1731216

LeBlanc, O. H. Jr. (1961). Band Structure and Transport of Holes and Electrons in Anthracene. *The Journal of Chemical Physics*, *35*(4), 1275–1280. doi:10.1063/1.1732038

Li, L., Meller, G., & Kosina, H. (2007). Temperature and field-dependence of hopping conduction in organic semiconductors. *Microelectronics Journal*, *38*(1), 47–51. doi:10.1016/j.mejo.2006.09.022

Li, V. J., Nardes, A. M., Liang, Z., Shaheen, S. E., Gregg, B. A., & Levi, D. H. (2011). Simultaneous measurement of carrier density and mobility of organic semiconductors using capacitance techniques. *Organic Electronics*, *12*(11), 1879–1885. doi:10.1016/j.orgel.2011.08.002

Lu, N., Li, L., & Liu, M. (2016). A review of carrier thermoelectric-transport theory in organic semiconductors. *Physical Chemistry Chemical Physics*, *18*(29), 19503–19525. doi:10.1039/C6CP02830F PMID:27386952

Machida, S.-i., Nakayama, Y., Duhm, S., Xin, Q. A., Funakoshi, A., Ogawa, N., & Ishii, H. et al. (2010). Highest-Occupied-Molecular-Orbital Band Dispersion of Rubrene Single Crystals as Observed by Angle-Resolved Ultraviolet Photoelectron Spectroscopy. *Physical Review Letters*, *104*(15), 156401. doi:10.1103/PhysRevLett.104.156401 PMID:20482000

Marcus, R. A. (1956). On the Theory of Oxidation-Reduction Reactions Involving Electron Transfer. 1. *The Journal of Chemical Physics*, *24*(5), 966–978. doi:10.1063/1.1742723

Marcus, R. A. (1964). Chemical + Electrochemical Electron-Transfer Theory. *Annual Review of Physical Chemistry*, *15*(1), 155–196. doi:10.1146/annurev.pc.15.100164.001103

Marcus, R. A., & Sutin, N. (1985). Electron Transfers in Chemistry and Biology. *Biochimica et Biophysica Acta*, *811*(3), 265–322. doi:10.1016/0304-4173(85)90014-X PMID:4074748

Miller, A., & Abrahams, E. (1960). Impurity Conduction at Low Concentrations. *Physical Review B: Condensed Matter and Materials Physics*, *120*(3), 745–755. doi:10.1103/PhysRev.120.745

Mott, N. F. (1956). The transition to metallic conduction in semiconductors. *Canadian Journal of Physics*, *34*(12A), 1356–1368. doi:10.1139/p56-151

Mott, N. F. (1969). Conduction in non-crystalline materials. Localized states in a pseudo-gap and near extremities of conduction and valence bands. *Philosophical Magazine*, *19*(160), 835–852. doi:10.1080/14786436908216338

Mott, N. F., & Gurney, R. W. (1940). *Electronic processes in ionic crystals*. Oxford University Press.

Nalwa, H. S. (2008). *Handbook of Organic Electronics and Photonics (3-Volumes)*. American Scientific Publishers.

Oelerich, J. O., Jansson, F., Nenashev, A. V., Gebhard, F., & Baranovskii, S. D. (2014). Energy position of the transport path in disordered organic semiconductors. *J. Physics: Cond. Mat*, *26*(25). PMID:24888582

Okamato, Y., & Brenner, W. (1964). *Organic Semiconductors*. New York: Reinhold Publishing Corporation.

Parkinson, P., Lloyd-Hughes, J., Johnston, M. B., & Herz, L. M. (2008). Efficient generation of charges via below-gap photoexcitation of polymer-fullerene blend films investigated by terahertz spectroscopy. *Physical Review B: Condensed Matter and Materials Physics, 78*(11), 115321. doi:10.1103/PhysRevB.78.115321

Pecchia, A., & Di Carlo, A. (2004). Atomistic theory of transport in organic and inorganic nanostructures. *IOP Publishing Ltd, 67*, 8.

Pecchia, A., Penazzi, G., Salvucci, L., & Di Carlo, A. (2008). Non-equilibrium Green's functions in density functional tight binding: method and applications. New Journal of Physics, 10.

Prasongkit, J., Grigoriev, A., Pathak, B., Ahuja, R., & Scheicher, R. H. (2011). Transverse conductance of DNA nucleotides in a graphene nano gap from first principles. *Nano Letters, 81*(5), 1941–1945. doi:10.1021/nl200147x PMID:21495701

Prasongkit, J., Grigoriev, A., Wendin, G., & Ahuja, R. (2010). Cumulene molecular wire conductance from first principles. *J. Phys. Rev. B, 81*(11), 115404. doi:10.1103/PhysRevB.81.115404

Prasongkit, J., Grigoriev, A., Wendin, G., & Ahuja, R. (2011). Interference effects in phtalocyanine controlled by H-H tautomerization: Potential two-terminal unimolecular electronic switch. *Physical Review B: Condensed Matter and Materials Physics, 84*(16), 165437. doi:10.1103/PhysRevB.84.165437

Puigdollersa, J., Marsala, A., Cheylanb, S., Voza, C., & Alcubilla, R. (2010). Density-of-states in pentacene from the electrical characteristics of thin-film transistors. *Organic Electronics, 11*(8), 1333–1337. doi:10.1016/j.orgel.2010.05.007

Reed, M. A., Zhou, C., Muller, C. J., Burgin, T. P., & Tour, J. M. (1997). Conductance of a Molecular Junction. *Science, 278*(5336), 252–254. doi:10.1126/science.278.5336.252

Roichman, Y. Preezant, Y. & Tessler, N. (2004). Analysis and modeling of organic devices. *Physica Status Solidi a-Applied Research, 201*, 1246-1262.

Roy, A., Geng, R., Zhao, W., Subedi, R. C., Li, X., Locklin, J., & Nguyen, T. D. (2016). Engineering of Spin Injection and Spin Transport in Organic Spin Valves Using π-Conjugated Polymer Brushes. *Advanced Functional Materials, 26*(22), 3999–4006. doi:10.1002/adfm.201504201

Ryu, H., & Klimeck, G. (2008). Contact Block Reduction Method for Ballistic Quantum Transport with Semi-empirical sp3d5 Tight Binding Band Models. *Proceedings of 9th Int. Conf. on Solid-State and Integrated- Circuit Technology (ICSICT 2008)*, 349-352.

Salzmann, I., Heimel, G., Oehzelt, M., Winkler, S., & Koch, N. (2016). Molecular Electrical Doping of Organic Semiconductors: Fundamental Mechanisms and Emerging Dopant Design Rules. *Accounts of Chemical Research, 49*(3), 370–378. doi:10.1021/acs.accounts.5b00438 PMID:26854611

Sariciftci, E. N. S. (1997). *Primary Photoexcitations in Conjugated Polymers: Molecular Exciton Versus Semiconductor Band Model*. Singapore: World Scientific.

Schein, L. B. (2012). Organic electronics—Silicon device design without semiconductor band theory. *Russian Journal of Electrochemistry, 48*(3), 281–290.

Schön, J. H., Kloc, Ch., Laudise, R. A., & Batlogg, B. (1998). Electrical Properties of Single Crystals of Rigid Rod like Conjugated Molecules. *Physical Review B: Condensed Matter and Materials Physics, 58*(19), 12952–12957. doi:10.1103/PhysRevB.58.12952

Schuster, S., Scarpa, G., Latessa, L., & Lugli, P. (2008). Charge transport in oligophenylenvinylene molecules. *Physica Status Solidi, C5*(1), 390–393. doi:10.1002/pssc.200776567

Scott, J. C., Brock, P. J., Salem, J. R., Ramos, S., Malliaras, G., Carter, S. A., & Bozano, L. (2000). Charge transport processes in organic light-emitting devices. *Synthetic Metals, 111–112*, 289–293. doi:10.1016/S0379-6779(99)00449-X

Scott, J. C., & Malliaras, G. G. (1999). Charge injection and recombination at the metal-organic interface. *Chemical Physics Letters, 299*(2), 115–119. doi:10.1016/S0009-2614(98)01277-9

Shuai, Z., Geng, H., Wei, X., Liaoc, Y., & Andréa, J.-M. (2014). From charge transport parameters to charge mobility in organic semiconductors through multiscale simulation. *Chemical Society Reviews, 43*(8), 2662–2679. doi:10.1039/c3cs60319a PMID:24394992

Shur, M., & Hack, M. (1984). Phyisics of amorphous silicon based alloy field-effect transistors. *Journal of Applied Physics, 55*(10), 3831–3842. doi:10.1063/1.332893

Small, D. W., Matyushov, D. V., & Voth, G. A. (2003). The theory of electron transfer reactions: What may be missing? *Journal of the American Chemical Society, 125*(24), 7470–7478. doi:10.1021/ja029595j PMID:12797822

Stafstrom, S. (2010). Electron localization and the transition from adiabatic to non-adiabatic charge transport in organic conductors. *Chemical Society Reviews, 39*(7), 2484–2499. doi:10.1039/b909058b PMID:20520911

Stallinga, P. (2001). *Theory of electrical characterization of organic semiconductors*. Academic Press.

Stokbro, K. (2008). First-principles modeling of electron transport. *Journal of Physics Condensed Matter, 20*(6), 064216–064222. doi:10.1088/0953-8984/20/6/064216 PMID:21693878

Stokbro, K., & Smidstrup, S. (2013). Electron transport across a metal-organic interface: Simulations using nonequilibrium Greens function and density functional theory. *Physical Review B: Condensed Matter and Materials Physics, 88*(7), 075317–075327. doi:10.1103/PhysRevB.88.075317

Sun, H., Ryno, S., Zhong, C., Ravva, M. K., Sun, Z., Körzdörfer, T., & Bredas, J.-L. (2016). Ionization Energies, Electron Affinities, and Polarization Energies of Organic Molecular Crystals: Quantitative Estimations from a Polarizable Continuum Model (PCM)–Tuned Range-Separated Density Functional Approach. *Chem. Theory Comp.*, *12*(6), 2906–2916. doi:10.1021/acs.jctc.6b00225 PMID:27183355

Tiwari, S., & Greenham, N. C. (2009). Charge mobility measurement techniques in organic semiconductors. *Optical and Quantum Electronics*, *2*(41), 69–89. doi:10.1007/s11082-009-9323-0

Torricelli, F., Zappa, D., & Colalongo, L. (2010). Space-charge-limited current in organic light emitting diodes. *Applied Physics Letters*, *96*(11), 113304. doi:10.1063/1.3358147

Troisi, A. (2011). Charge transport in high mobility molecular semiconductors: Classical models and new theories. *Chemical Society Reviews*, *40*(5), 2347–2358. doi:10.1039/c0cs00198h PMID:21409232

Van der Holst, J. J. M., van Oost, F. W. A., Coehoorn, R., & Bobbert, P. A. (2009). Electron-hole recombination in disordered organic semiconductors: Validity of the Langevin formula. *Physical Review B: Condensed Matter and Materials Physics*, *80*(23), 235202. doi:10.1103/PhysRevB.80.235202

Venkateshvaran, D., Broch, K., Warwick, C. N., & Sirringhaus, H. (2016). Thermoelectric transport properties of high mobility organic semiconductors, Proc. SPIE 9943. *Organic Field-Effect Transistors*, *XV*, 99430U.

Wang, L., & Beljonne, D. (2013). Charge transport in organic semiconductors: Assessment of the mean field theory in the hopping regime. *The Journal of Chemical Physics*, *139*(6), 064316. doi:10.1063/1.4817856 PMID:23947864

Wannier, G. H. (1937). The structure of electronic excitation levels in insulating crystals. *Physical Review*, *52*(3), 191–197. doi:10.1103/PhysRev.52.191

Warta, W., & Karl, N. (1985). Hot holes in naphthalene: High, electric-field-dependent mobilities. *Physical Review B: Condensed Matter and Materials Physics*, *32*(2), 1172–1182. doi:10.1103/PhysRevB.32.1172

Wu, Y., Hu, B., Howe, J., Li, A. P., & Shen, J. (2007). Spin injection from ferro-magnetic Co nanoclusters into organic semiconducting polymers. *Physical Review B: Condensed Matter and Materials Physics*, *75*(7), 075413. doi:10.1103/PhysRevB.75.075413

Yang, Y., Liu, J.-Y., Chen, Z.-B., Tian, J.-H., Jin, X., Liu, B., & Tian, Z.-Q. et al. (2011). Conductance Histogram Evolution of an EC–MCBJ Fabricated Au Atomic Point Contact. *Nanotechnology*, *22*(27), 275313. doi:10.1088/0957-4484/22/27/275313 PMID:21613733

Zahid, F., Paulsson, M., & Datta, S. (2002). *Advanced Semi-conductors and Organic Nano-Techniques, Chapter Electrical Conduction through Molecules. Academic Press.*

ENDNOTES

1 The Franck–Condon principle explains the intensity of vibronic transitions, which are simultaneous changes in electronic and vibrational energy levels of a molecule due to the absorption or emission of a photon. The Franck–Condon principle states that during electronic transitions, a change from one vibrational energy level to another is more probable if the two vibrational wavefunctions overlap more significantly.

2 The Arrhenius equation is a formula for the temperature dependence of reaction rates. $k = A\ exp(-Ea/k_B T)$

3 This relation, which is also known as Einstein–Sutherland-Smoluchowski relation, is actually an early example of a fluctuation-dissipation relation.

4 According to Ishi et al, (2004), the band bending used in metal-inorganic semiconductor interfaces is still valid for doped organic semiconductors (OSC) although much thicker films are often necessary to achieve bulk Fermi level alignment However, for thin OSC layers the situation is different.

5 *1*– This formula was originally developed for AC conductivity, where ω is the applied field frequency.

Index

N

O

P

Q

S

T

W

www.ingramcontent.com/pod-product-compliance
Lightning Source LLC
LaVergne TN
LVHW080617300926
849162LV00005B/1189
* 9 7 8 1 5 2 2 5 2 3 1 2 3 *